The comprehensive sourcebook for locating and identifying
chemical tradename product lines
in the international marketplace

Chemical Tradename Dictionary

Compiled by Michael and Irene Ash

Contains over 14,000 entries
for chemical tradename product lines
currently sold throughout the world

VCH

Michael Ash
Irene Ash
Synapse Information Resources, Inc.
1247 Taft Ave.
Endicott, NY 13760

This book is printed on acid-free paper. ∞

Library of Congress Cataloging-in-Publication Data

Ash, Michael
    Chemical Tradename Dictionary/Compiled by Michael and Irene Ash
        p.   cm.
      1. Chemicals—Dictionaries 2. Chemicals—Trademarks.
    I. Ash, Irene.   II. Title.
TP9.A73   1992   92-35154
660´.03—dc20   CIP

Printed in the United States of America.

ISBN 1-56081-625-2 VCH Publishers

Printing history
10 9 8 7 6 5 4 3

Published by:

VCH Publishers, Inc.	VCH Verlagsgesellschaft mbH	VCH Publishers (U.K.) Ltd.
220 East 23rd St.	P.O. Box 10 11 61	8 Wellington Court
New York, New York 10010	D-6490 Weinheim	Cambridge CB1 1HZ
	Federal Republic of Germany	United Kingdom

Preface

This key reference serves as the most comprehensive source for identifying product lines in the international chemical marketplace. The chemical industry is rapidly expanding in product innovation and specialty manufacturing. Any professional involved in the purchasing of tradename products has experienced problems in identifying and locating these chemicals and is usually forced to spend long and tedious hours consulting a multitude of sources. This moderately priced sourcebook, which is regularly updated, provides brief, accurate descriptions of the product lines including the chemical classification, function, and/or application. Proprietary chemicals are identified by function alone.

The scope of this reference includes tradename chemicals from the entire spectrum of chemical materials used in manufacturing. Some of the areas covered are:

> cosmetic additives
> catalysts
> adhesives and sealants
> paint additives
> detergent materials
> wetting agents
> emulsifiers
> cutting oils
> agricultural chemicals
> colors and pigments
> fillers, modifiers, and reinforcing materials
> films
> plastic compounds, resins, and additives
> natural and synthetic elastomers, and additives
> textile specialty chemicals

The second part of this book contains a detailed Manufacturers Directory including all the necessary contact information needed by the user to obtain technical and material handling data sheets on individual products directly from the manufacturers.

The information provided in this book is the culmination of many years of research and direct contact with over 2300 chemical manufacturers. We are especially grateful to Roberta Dakan for her skill and dedication in the development and maintenance of the tradename database that generated this reference work. Her talent and dedication have been instrumental in the success of this project.

M. & I. Ash

NOTE

The information contained in this series is accurate to the best of our knowledge; however, no liability will be assumed by the publisher for the correctness or comprehensiveness of such information. The determination of the suitability of any of the products for prospective use is the responsibility of the user. It is herewith recommended that those who plan to use any of the products referenced seek the manufacturer's instructions for the handling of that particular chemical.

ABBREVIATIONS

ABS	acrylonitrile-butadiene-styrene
absorp	absorption
ACN	acrylonitrile
agric.	agricultural
AMP	2-amino -2-methyl-1- propanol
anhyd.	anhydrous
applic(s)	application(s)
aq.	aqueous
ASA	acrylic-styrene-acrylonitrile
ATH	alumina trihydrate
aux.	auxiliary
BMC	bulk molding compound
BP	British Pharmacopeia
BR	butadiene rubbers, polybutadienes
B/S	butadiene/styrene
C	degrees Centigrade
CFC	chlorofluorocarbon
char.	characteristic
compd.	compound
conc.	concentrated, concentration
coeff.	coefficient
compr.	compression
conduct.	conductive
CP	Canadian Pharmacopeia
CPE	chlorinated polyethylene
CPVC	chlorinated polyvinyl chloride
CR	chloroprene rubber, polychloroprene
CTFA	Cosmetic, Toiletry, and Fragrance Association
DEA	diethanolamide, diethanolamine
deriv.	derivative(s)
dielec.	dielectric
DMC	4,4´-dichloro(methylbenzhydrol)
DMDM	dimethylol dimethyl
DNA	deoxyribonucleic acid
DOP	dioctyl phthalate
DTPA	diethylene triamine pentaacetic acid
DVB	divinylbenzene
EDTA	ethylene diamine tetraacetic acid
elec.	electrical
EP	extreme pressure
E/MA	ethylene-methyl acrylate
EMC	electromagnetic conductive
EMI	electromagnetic interference
EO	ethylene oxide
EPDM	ethylene-propylene-diene rubber
EPM	ethylene-propylene rubber
EPR	ethylene-propylene rubber
equip.	equipment
ESD	electrostatic discharge
esp.	especially
ETFE	ethylene tetrafluoroethylene
EVA	ethylene vinyl acetate
exc	excellent

FCC .... Food Chemicals Codex
FD&C .... Foods, Drugs, and Cosmetics
FEP .... fluorinated ethylene propylene
FRP .... fiberglass-reinforced plastics
GRP .... glass-reinforced plastics
HAF .... high abrasion furnace carbon black
HCl .... hydrochloric acid
HEDTA .... hydroxyethylenediamine triacetic acid
HDPE .... high-density polyethylene
HIPS .... high-impact polystyrene
HPLC .... high performance liquid chromatography
HT .... heat transfer
IC .... integrated circuit
IIR .... isobutylene-isoprene rubber
incl .... including
ingred .... ingredient(s)
inj. .... injection
inorg. .... inorganic
IPA .... isopropyl alcohol
IR .... isoprene rubber (synthetic)
IV .... intravenous
LDPE .... low-density polyethylene
liq. .... liquid
LLDPE .... linear low-density polyethylene
lt. .... light
MA .... methacrylic acid
MCPA .... (4-chloro-2-methylphenoxy) acetic acid
MDI .... methylene diphenylene isocyanate
MDM .... monomethylol dimethyl
MDPE .... medium density polyethylene
MEA .... monoethanolamine, monoethanolamide
mech. .... mechanial
mfg. .... manufacture
mixt .... mixture(s)
m.w. .... molecular weight
nat .... natural
NBR .... nitrile-butadiene rubber
NC .... nitrocellulose
NF .... National Formulary
N/F .... nonflammatory
NR .... isoprene rubber (natural)
NTA .... nitrilotriacetic acid
OTC .... over-the-counter
o/w .... oil-in-water
PA .... polyamide
PABA .... p-aminobenzoic acid
PAN .... polyacrylonitrile
PBT .... polybutylene terephthalate
PC .... polycarbonate
PCA .... 2-pyrrolidone-5-carboxylic acid
PCTFE .... polychlorotrifluoroethylene
PE .... polyethylene
PEEK .... polyetheretherketone

PEG ....polyethylene glycol
PEK ....polyetherketone
PEI ....polyetherimide
PES ....polyether sulfone
PET ....polyethylene terephthalate
petrol. ....petroleum
PFA ....perfluoroalkoxy
pH ....hydrogen-ion concentration
pkg ....packaging
PMMA ....polymethyl methacrylate
POE ....polyoxyethylene, polyoxyethylated
POM ....polyoxymethylene
POP ....polyoxypropylene, polyoxypropylated
powd ....powder
PP ....polypropylene
PPE ....polyphenylene ether
PPG ....polypropylene glycol
PPO ....polyphenylene oxide
PPS ....polyphenylene sulfide
prep ....preparation(s)
prod ....product(s), production
PS ....polystyrene
pt. ....point
PTFE ....polytetrafluoroethylene
PU ....polyurethane
PVAc ....polyvinyl acetate
PVAL ....polyvinyl alcohol
PVB ....polyvinyl butyral
PVC ....polyvinyl chloride
PVDC ....polyvinylidene chloride
PVDF ....polyvinylidene fluoride
PVP ....polyvinylpyrrolidone
quat. ....quaternary
resist. ....resistance
RFI ....radio frequency interference
RIM ....reaction injection molded/molding
RNA ....ribonucleic acid
RT ....room temperature
RTV ....room temperature vulcanizing
SAN ....styrene-acrylonitrile
S/B ....styrene/butadiene
SBR ....styrene/butadiene rubber
SBS ....styrene-butadiene-styrene
SDA ....specially denatured alcohol
SE ....self-emulsifying
SEBS ....styrene-ethylene/butylene-styrene
sec ....secondary
SMA ....styrene maleic anhydride
SMC ....sheet molding compound
sol. ....soluble, solubility
sol'n. ....solution
solv(s). ....solvent(s)
SPF ....sun protection factor

STPP .... sodium tripolyphosphate
syn .... synthetic
tech .... technical
temp .... temperature
TBHQ .... tert-butyl hydroquinone
TDI .... toluene diisocyanate
TEA .... triethanolamine, triethanolamide
tens. .... tensile
tert. .... tertiary
TFE .... tetrafluoroethylene
TMC .... transfer molding compound
TPO .... thermoplastic polyolefin
TPR .... thermoplastic rubber
UHF .... ultra high frequency
UHMW .... ultra high molecular weight
UHMWPE .... ultra high molecular weight polyethylene
unsat. .... unsaturated
USP .... Unites States Pharmacopeia
uv .... ultraviolet
VA .... vinyl acetate
VAE .... vinyl acetate ethylene
VC .... vinyl chloride
VHF .... very high frequency
visc .... viscous, viscosity
VLDPE .... very low density polyethylene
VTR .... video tape recorder
w/o .... water-in-oil
XLPE .... crosslinked polyethylene

CONTENTS

A

A- . [La Roche Chem.] Activated aluminas; selective adsorbent.

A-1. [Monsanto] N,N´-Diphenylthiourea; primary accelerator for latex and repair stocks, CR, CR latex, and EPDM sponge compds.

A-17. [Phillips] Butane; hydrocarbon propellant.

A-20FG-0100, A-30FG-0100. [Thermofil] Glass-reinforced polystyrene.

A-0020 Series. [Goldsmith & Eggleton] SBR/carbon black masterbatches.

A-31. [Phillips] Isobutane; hydrocarbon propellant.

A-108. [Phillips] Propane; hydrocarbon propellant.

A-625/641. [ICI Am.] Sorbitol; nutrient and dietary supplement, food additive; bodying agent for paper, textile, liq. pharmaceuticals; in mfg. of sorbose, ascorbic acid, propylene glycol, synthetic plasticizers, resins; as humectant, sequestrant.

A25656, 26903, 27529. [Polycom Huntsman] Brominated organic dispersions; flame retardants for plastics.

AA. [CasChem] Castor oil; emollient for industrial, cosmetic, pharmaceutical applics.

AA. [PMC Specialties] Anthranilic acid; antioxidant for fats, greases, lube oils, and polyamides; sludge preventative in furnace and lube oils; chelating agent and sequestrant; corrosion inhibitor; stabilizer of can lacquers, oils, and lubricants.

Aatex. [Evode Speciality Adhesives Ltd.] Latex.

AA White Factice. [Am. Cyanamid] Sulfur chlorinated vegetable oil; extender, plasticizer, softener for NR, SR; facilitates extrusion; retards cure.

AB. [Cuyahoga Plastics] Thermoset polyester, some glass or mineral-reinforced; composite molding materials.

AB®. [Angus] 2-Amino-1-butanol; pigment dispersant, neutralizing amine, corrosion inhibitor, acid-salt catalyst, pH buffer, chemical and pharmaceutical intermediate, solubilizer.

Abalyn®. [Hercules] Methyl rosinate; resin with compatibility, surf.-wetting properties, visc., and tack used in rubbers, lacquers, inks, paper coatings, varnishes, adhesives, sealing compds., plastics, wood preservatives, and perfumes; plasticizer, softener, tackifier.

Abate®. [Am. Cyanamid/Ag] Temephos; gran. and emulsifiable conc. herbicide for control of mosquito larvae in standing water, ponds, etc.

Abco Binder. [Abco Industries] Nonyellowing, self-crosslinking acrylic emulsion pigment printing binders for textile printing.

Abco Blockout. [Abco Industries] Clay-based opacifiers for pigment printing of textiles.

Abco Conc. [Abco Industries] Pre-neutralized synthetic thickening agents for printing of pigments, acid dyes, disperse dyes, and fiber reactive dyes.

Abcohesive. [Abco Industries] Polymers; printing blanket adhesives for textile printing.

Abco Pak. [Abco Industries] Granular warp sizing agents for spun yarns.

ABC-Trieb®. [BASF AG] Ammonium bicarbonate; baking raising agent.

Abcure S-40-25. [Abco Industries] Benzoyl peroxide; catalyst for unsat. polyester resins.

Abex. [Rhone-Poulenc Surf.] Anionic surfactants; emulsifier for polymerization; detergent, rewetting agent.

Abex VA 50. [Chem-Y GmbH] Octoxynol-33, sodium laureth sulfate.

ABG. [BASF] Plasticizers for plastics

industry.

Abil®. [Goldschmidt; Goldschmidt AG] Silicone compds.; surfactants used in personal care prods.; as foam formers, lubricants, glossers, refatting agents, conditioner, emollient, emulsifier.

Abil®-Quat 3270, 3272. [Goldschmidt] Quaternium-80; conditioner, antistat for shampoos and hair rinses; refatting agent for skin cleansers.

Abil®-Wax. [Goldschmidt; Goldschmidt AG] Silicone waxes; spreading, penetrating, and emollient properties for skin care prods.; pigment solubilizer; water barrier; thickener.

Abiol. [3-V; Sigma Prodotti Chimici] Imidazolidinyl urea NF; preservative.

Abitol®. [Hercules] Dihydroabietyl alcohol; resinous plasticizer and tackifier in plastics, rubbers, lacquers, inks, and adhesives; chemical intermediate.

A-Black. [Polymer Valley] Carbon black; filler, reinforcer for rubber liners, extruded goods.

Ablaphene. [Rhone-Poulenc Plastiques] Phenolic resins.

Abluhide. [Taiwan Surf.] Blends; emulsifier, degreaser, rewetting agent, fatliquoring agent, dyeing auxiliary for leathers.

Ablumide. [Taiwan Surf.] Fatty acid alkanolamide; foam stabilizer, thickener for cosmetics, liq. detergents.

Ablumine. [Taiwan Surf.] Alkyl dimethyl benzyl ammonium chloride; quats. as algicide for industrial cooling towers, swimming pools; retarder for acrylic fiber; antistat and hair conditioner.

Ablumox. [Taiwan Surf.] Amine oxides or ethoxylates; foamers, wetting agent, foam stabilizer, antistat, emollient; leveling agent for dyes.

Ablumul. [Taiwan Surf.] Nonionic/anionic blend; dispersant for wettable powd. agric. pesticides.

Ablunol. [Taiwan Surf.] Ethoxylated ethers, esters, sorbitan esters, or castor oil or glycerol esters; emulsifier, dispersant, detergent, wetting agent, lubricant, antistat, leveling agent for textile processing, cosmetics, metalworking compds., agric. preps., industrial cleaners, paints, pharmaceuticals, foods, plastics.

Ablunol NP. [Taiwan Surf.] Nonoxynol series; detergent, dispersant, emulsifier, wetting agent for petrol. oils, textile, paper, leather industries, metal processing; intermediate; coupling agent.

Abluphat. [Taiwan Surf.] Phosphate esters; antistat for syn. fibers.

Ablupol HT. [Taiwan Surf.] Organo compd.; water-dilutable antifoamer for high-temp. jet dyeing.

Ablupol SAE. [Taiwan Surf.] Silicone compd.; antifoam.

Ablusoft. [Taiwan Surf.] Fatty polyamide; softeners, antistats for textiles.

Ablusol. [Taiwan Surf.] Sulfonates or sulfosuccinates; wetting agent, emulsifier, dispersant for textile, emulsion polymerization.

Abluter. [Taiwan Surf.] Betaines or glycines; antistat, softener, germicide, spreading-wetting agent, thickener for shampoos, liq. detergents.

Abluton. [Taiwan Surf.] Dye carriers for textile industry.

Abluwax EBS. [Taiwan Surf.] Ethylene bis-stearamide; lubricant for ABS, PS, PVC; defoamer and mold release agent.

Abocast. [Abatron] Casting compounds and adhesives.

Abocoat. [Abatron] Coatings.

Abocure. [Abatron] Catalysts, curing agents.

Abojet. [Abatron] Structural crack-injection resins.

ABS. [Mitsui Petrochemicals] ABS copolymers; molding, extrusion grades for automotive, appliance, consumer goods, pipe, electroplating, etc.

A.B.S. 87%. [Triantaphyllou] Alkylbenzene sulfonic acid; detegent intermediate.

Absafil. [Akzo Engineering Plastics] ABS.

Abselex. [Royalite Plastics Ltd.] ABS.

ABS-G1FG-2. [Washington Penn Plastics] Glass-filled ABS; for automotive components.

Absize. [Abco Industries] Warp sizing agents for spun yarn.

Absol. [Surpass] Nonionic and anionic surfactants; detergents, wetting agents.

Absynt. [ABS Tech. Ltd.] Nylon block copolymers.

ABT-2500®. [Pfizer] Talc; antiblocking agent for polyolefin films.

AC-. [AluChem] Alumina trihydrate; flame retardants for plastics.

Acardite 2. [Lowi] N-Methyl-N´,N´-diphenylurea; stabilizer improving storage stability of powders and propellants; plasticizer for celluloid.

Accelerate Harvest Aid. [Atochem N. Am.] Harvest aid for cotton.

Accelerator. [Akrochem] Dithiocarbamate and disulfide accelerators; for EPDM, NR and SR latexes; stabilizer and antioxidant in uncured rubber.

Accelerator. [Akzo] Cobalt octoate mixtures; accelerators for cure of unsat. polyester resins at R.T. and elevated temps.

Accelerator. [Harwick] Accelerator for elastomers.

Accelerator. [Polymerics] Accelerator for elastomers.

Accelerator 399. [Texaco] Epoxy curing promoter for use with amine hardeners.

Accelerator 808. [Elastochem] Accelerator for elastomers.

Accelerator 3711. [Manufacturers Chems.] Proprietary blend of acids; pH control and acid generator for use as a textile dyeing auxiliary in continuous and batch operations.

Accelerator D. [Anchor UK] Cyclo amine blend; accelerator for shoe soling industry.

Accelerator E, G, R. [Scott Bader] Cobalt sol'ns. in styrene; accelerators.

Accelerit®. [Finetex] Dye carriers for polyester and blends.

Accobetaine. [Karlshamns] Coco betaine; detergent, wetting agent, emulsifier, high foaming agent, solubilizer for household and cosmetic uses.

Accobond. [Am. Cyanamid] Cellulosic film resins.

Accomeen. [Karlshamns] Ethoxylated fatty amines; emulsifier, antistat, surfactant, dispersant.

Accomid. [Karlshamns] Fatty acid DEA; nonionic surfactants; detergent, viscosity improver, foam booster/stabilizer, emulsifier for shampoos, liq. soaps, dish detergents, bubble bath prods.

Acconon. [Karlshamns] Nonionic surfactants; used as emulsifier, lubricant, dispersant, solubilizer, visc. control agent, foaming and wetting agent, chemical intermediate for cosmetics, pharmaceuticals, and industrial applics.

A-C® Copolymer. [Allied-Signal] EVA and ethylene-acrylic acid copolymers; plastics lubricant and processing aid, pigment dispersant.

Accoquat. [Karlshamns] Dicocodimonium chloride; cationic emulsifier, coupling agent; used for car spray waxes, dust control oil, spot removal.

Accosize. [Am. Cyanamid] Synthetic sizes.

Accosoft. [Apollo] Cationic softeners for pad or exhaust on all types of fibers.

Accosoft. [Stepan] Diamidoamine quaternaries; cationic fabric softeners for textile processing, household and industrial applics.

Accosperse. [Karlshamns] Ethoxylated sorbitan fatty acid esters; emulsifier, solubilizer.

Accostrength. [Am. Cyanamid] Paper resins.

Accrox. [Am. Chrome & Chem.] Chromium oxide.

Accudri SF6. [Allied-Signal] Sulfur hexafluoride.

Accufluor CFX. [Allied-Signal] Fluorinated carbon.

Acculose. [Accurate Chem. & Scientific] Cellulose ion exchange resins for research laboratory.

Accusand. [Unimin Specialty Minerals] Silica sands.

Accusil. [Accurate Chem. & Scientific] Silica gel; drying agent, for chromatography, in research laboratories.

Accusorb. [Accurate Chem. & Scientific] Adsorbents for separation and chromatography in research laboratories.

Accuthane® UR-1. [H.B. Fuller] One-component urethane; adhesive for difficult-to-bond substrates.

Accu-Way. [Harwick] Pre-weighed chemicals.

ACDI. [Aakash Chem. & Dye-Stuffs] Textile dyes and pigments.

Aceko. [John Campbell] Textile dyes and pigments.

Acelan. [Fabriquimica] Lanolin derivs.

Acetadeps. [Westbrook Lanolin] Acetylated lanolin; emollient in baby oils, skin oils sunscreen oils, bath oils; superfatting agent in soaps.

Acetamin. [Kao Corp. SA] Fatty acid amine acetate; surface coating agent for pigments, anticaking agent for fertilizer; emulsifier, dispersant, and softening agent for textiles; min. flotation reagent.

Aceto. [Aceto] 3,5-Di-t-butyl p-hydroxy benzoic acid; uv stabilizer for plastics.

Acetol® 1706. [Henkel/Emery/Cospha] Cetyl acetate, acetylated lanolin alcohol; water repellent, emollient, penetrant, and cosolv. use in suntan preparations and baby prods.

Acetoquat. [Aceto] Quaternary ammonium salts; germicide, sanitizing agent.

Acetorb A. [Aceto] 2-Hydroxy-4-methoxy benzophenone; UV stabilizer for plastics.

Acetulan®. [Amerchol; Amerchol Europe] Cetyl acetate, acetylated lanolin alcohol; binder for pressed powds.; emollient, plasticizer, cosolv. for personal care prods.; lubricant for clay, talc, and starch; stabilizer for lanolin; solubilizer in aerosols; penetrant and spreading agent.

Aciculite. [Kaopolite] Acicular aluminum silicate; for plastics, molding compds., brake linings, asphalt coatings, abrasives; replaces asbestos in sealants.

Acid Aid. [Crown Tech.] Acid extenders and inhibitors.

Acidan. [Grindsted Prods.] Monoglyceride citric acid ester; emulsifier, surfactant for food industry.

Aciderm. [Miles/Organic Prods.] Textile dyes and pigments.

Acid Felt Scour. [Hart Chem. Ltd.] Pulp and paper felt cleaner.

Acid Foamer. [Exxon/Tomah] Quaternary ammonium chloride; cationic foaming agent for acids; used in aluminum trailer cleaner, brightener, acid inhibitors and cleaners, chrome plating baths.

Acid Leveler H. [Leatex] Acid dye leveler for nylon hosiery; allows rapid dye penetration.

Acidol®. [BASF; BASF AG] Acid dyes or metal complex dyes for textile dyeing and printing.

Acid Thickener. [Exxon/Tomah] Surfactant, visc. builder, corrosion inhibitor for acid-based cleaners.

Acihib. [ICI Australia] Amine deriv.; corrosion control in HCl, sulfuric, phosphoric, and other acids; acid restrainer.

Acintene. [Arizona] Terpene prods.

Acintol®. [Arizona] Tall oil acids and rosins; softener and tackifier for IIR, SBR, CR, acetal, polypropylene.

ACL. [Monsanto] Chloroisocyanurates; bleaching compd., sanitizer, disinfectant, oxidizer, detergent

Aclan. [Acla-Werke] Polyurethane homogeneous.

Aclarat. [Sandoz] Fluorescent whiteners for synthetics and wool for laundry and cleaning prods.

Aclar Films. [Allied-Signal] Fluorocarbon thermoplastic; high moisture barrier flexible film for pkg. and container liners for pharmaceutical and cosmetic pkg., clean room, DOD, electronic applics.

Aclathan. [Acla-Werke] Polyurethane homogeneous.

Aclon. [Acton Tech.] Wettable, bondable TFE fluorocarbon fiber.

AClon. [Allied-Signal] PCTFE; for extrusion and inj. molding.

AClyn®. [Allied-Signal] Ethylene/acrylate copolymers; additives for adhesives and coatings.

ACM-. [AluChem] Magnesium hydroxide; flame retardants for plastics.

Acme-Bond. [Acme Resin] Urethane no-bake binders.

Acme-Flow. [Acme Resin] Urethane cold box binders.

Acofor. [Reichhold] Tall oil fatty acids; latex stabilizer, softener, plasticizer; mold lubricant; dispersant (as soap) for pigments and fillers.

Aconew. [Reichhold] Tall oil fatty acids; plasticizer, softener, mold lubricant for NR, SR, latexes.

Aconol. [Hart Chem. Ltd.] Ethoxylated tallate; emulsifier, low-foaming surfactant for built detegent systems.

Acosix. [Reichhold] Distilled tall oil; plasticizer, softener, tackifier for NR, SR, latexes.

Acpol. [Cook Composites & Polymers] Thermosetting polymers.

A-C® Polyethylene. [Allied-Signal] Polyethylene and oxidized polyethylene; wax for polishes, finishes and emulsions; processing lubricant, mold release agent, plastics lubricant.

Acra-500. [Exxon/Tomah] Fatty amine complex; asphalt wetting and antistripping agent.

Acraconc® BN. [Miles/Organic Prods.] Synthetic thickener for textile printing.

Acrafix® MA. [Miles/Organic Prods.] Crosslinker for Acramin binders.

Acralane. [PPG Industries/Adhesives] One-component structural adhesive for plastics.

Acralen. [Bayer] Self-crosslinking binder for textile industry.

Acramin®. [Miles/Organic Prods.] Leveling agents, binders for pigment dyeing and padding, defoamers, fixatives, softeners, thickeners for textile use.

Acrawax®. [Lonza] Amide wax; lubricants for waxes, plastics, textiles; processing aid; resin plasticizer; pigment dispersant.

Acriflex. [Rohm GmbH] Acrylate adhesive.

Acrilan. [Monsanto] Acrylic textile fiber.

Acrilev. [Finetex] Phosphate ester salts; detergent, wetting agent, dye leveler for textiles.

Acrisint. [3-V; Sigma Prodotti Chimici] Carbomer. emulsifier, thickener, stabilizer, suspending agent for cosmetics.

Acritamer. [RITA] Carbomer. suspending and viscosity agent.

Acronal®. [BASF; BASF AG] Acrylic ester copolymer dispersions; binders for paper and board coating, binders and coating agents for prod. of materials based on leather fibers.

Acrosol®. [BASF AG] Acrylic acid ester copolymers; cobinders for paper coating.

Acrycal®. [Continental Polymers] Acrylic resins, pellets, sheet.

Acryclean G. [Evode-Tanner Industries] Antisticking agent for acrylic and vinyl acrylics.

Acrylafil. [Akzo Engineering Plastics] Styrene-acrylonitrile, glass-reinforced.

Acrylic Resin AS. [ICI Surf. UK] Acrylic copolymer aq. dispersion; finishing agent for fibers.

Acrylite®. [Cyro Industries] Cast acrylic sheet; used in industrial plants and building prods., e.g., window glazing, skylights, tub enclosures.

Acrylocoat. [Rohm & Haas] Pesticide binder/sticker.

Acryloft. [Rhone-Poulenc] Quaternary ammonium compd.; textile softener

Acryloid®. [Rohm & Haas] Acrylic and methacrylate resins; thermosetting and thermoplastic resins for finishes and coatings applics.

Acrylon. [Arakawa] Sizing agent for textiles.

Acrylron. [Synthron] Dispersing and leveling agents.

Acrylux. [Sullivan Chem. Coatings] Thermoset acrylic resin; for coatings.

Acrymul. [Protex] Acrylates copolymer.

Acryrene. [J. Gaillon SA] ABS.

Acrysol®. [Rohm & Haas; Rohm & Haas France] Water-sol. acrylic compds. thickener, binder, pigment suspending agent, stabilizer; for paints, inks, other coatings, latexes, adhesives.

ACS 60. [Witco SA] Ammonium cumene sulfonate; hydrotrope, solubilizer, coupling agent for detergent formulations; cloud pt. depressant; antiblocking agent.

Actabs. [Stewart Hall] Slime preventives for air conditioner drain pans.

Actafoam®. [Uniroyal] Fatty acid salts dispersion; activator-stabilizer for vinyl foams, sponge rubber; gas release accelerator.

Actan. [Troy] Tannic acid conc.; corrosion inhibitor; base for applic. of coatings systems.

Actana. [Aetna] Boiler water treatment.

Actene. [Aetna] Fuel oil additive; dispersant, stabilizer; emission control.

ACter. [Allied-Signal] Ethylene/acrylic acid/vinyl acetate copolymer.

Actibon®. [OxyChem] Activated carbon.

Acticarbone. [Atochem N. Am.] Activated carbons; for mineral industry.

Acticide. [Thor] Biocides.

Actiflo®. [Central Soya] Lecithin; emulsifier, wetting agent, dispersant; for food industry applics.

Actigen. [Active Organics] Collagen and elastin products.

Actiglow. [Active Organics] Hydrolyzed mucopolysaccharides.

Actilane. [Harcros Chem. Scandia ApS] Oligomers and monomers; for radiation curing.

Actimer FR. [Dead Sea Bromine] Flame retardants, crosslinking agents.

Actimet. [Atochem N. Am.] Acid salts.

Actimex FR. [Eurobrom BV] Reactivated bromine monomers.

Actimoist. [Active Organics] Sodium hyaluronate.

Actipol®. [Amoco] Activated polybutene; used in adhesives, sealants, coatings, unsat. polyesters, elec. compds., foams, and other applics.

Actiron®. [Synthron] Epoxy accelerators; peroxide stabilizers for bleaching operations.

Actiron NX. [Protex] 2,4,6-Tri-dimethylaminomethylphenol; accelerator for solvent based and powder coating (epoxy, polyurethane).

Actirox. [Colores Hispania SA] Noncorrosive nontoxic pigments.

Activ-8. [R.T. Vanderbilt] 1,10-Phenanthroline sol'ns.; drier accelerator/stabilizer for coating systems which cure by oxidative polymerization.

Activate Plus. [Terra Int'l.] Nonionic surfactants.

Activator 1102. [Anchor UK] Dibutyl ammonium oleate; accelerator/activator for natural and synthetic rubbers.

Activator STAG. [Akrochem] Complex sec. amine; activator for thiazole-type accelerators; accelerator for natural rubber and SBR.

Active. [Blew Chem.] Fatty acid DEA.

Activex. [Manufacturers Chems.] Heavy duty solvent scour for removing oil, grease, graphite, tint and oxidation stains; for cleaning print rolls.

Activol. [Harry Miller] Pickling acid inhibitor.

Acto. [Exxon] Sodium alkylaryl sulfonate; detergent, wetting agent, emulsifier, rust preventative.

ACtone. [Allied-Signal; NV Allied Corp. Int'l. SA] Proprietary ionomer; pigment wetting additive, dispersion aid to enhance colors.

ACtone P. [Allied-Signal] Low m.w. branched polyethylene.

Actox. [Zinc Corp. of Am.] Zinc oxide; activator for NR, SR, latexes; reinforcer; inorganic color for resins and rubber.

Actrabase. [Climax Performance] Soap sulfonate and petroleum sulfonate; emulsifier, rust inhibitor for oil systems, metalworking fluids.

Actracor. [Climax Performance] Corrosion inhibitor for metalworking, hydraulic fluids.

Actrafoam. [Climax Performance] Blend of glycols, fatty acids, and nonionic surfactants in a hydrocarbon base; defoamer for general purpose and water sewage applics.

Actrafos. [Climax Performance] Phosphate esters; lubricant, emulsifier, coupling agent, hydrotrope, solubilizer for cutting oils, cleaner formulations

Actralube. [Climax Performance] Esters; lubricant, corrosion inhibitor, emulsifier for metalworking fluids.

Actramide. [Climax Performance] Alkanolamides; emulsifier for sol. oils, metalworking fluids and emulsion cleaners; corrosion inhibitor; foam booster/stabilizer, thickener for shampoos and cleaners.

Actran. [Climax Performance] Lard oils.

Actrasol. [Climax Performance] Sulfated or sulfonated anionic surfactants; lubricant, emulsifier in metalworking fluids, cleaners, textiles, paper processing; fat liquor for leather; mold release; rust preventative.

Actrol. [Climax Performance] Ethoxylated nonionic esters; emulsifier and lubricity additive for metalworking.

ACuflow AF-1. [Allied-Signal; NV Allied Corp. Int'l. SA] Ethylene copolymer and aluminum stearate mixt.; processing additive.

Aculyn 22. [Rohm & Haas] Acrylates/steareth-20 methacrylate copolymer; thickener for cosmetics and toiletries, hair care prods., hand creams, lotions, waterless hand cleaners.

Acumer. [Rohm & Haas] Polyacrylic acids or their sodium salts; dispersant, thickener, anti-scale deposition agent for NR, SR latexes, for water treatment; calcium phosphate stabilizer.

ACumist. [Allied-Signal; NV Allied Corp. Int'l. SA] Polyethylene and oxidized polyethylene waxes.

Acusol®. [Rohm & Haas] Acrylic acid and salts; thickener, emulsion stabilizer for NR, SR latexes; suspending agent for pigments; detergent polymers.

ACX-0011 Oxidized Polyethylene. [Allied-Signal] Polyethylene/calcium silicate blend; mixing aid, processing aid, release agent, lubricant for fluoroelastomers.

Acylan. [Croda Inc.; Croda Chem. Ltd.] Acetylated lanolin; lipid emollient for personal care and pharmaceutical prods.

Acylglutamate. [Ajinomoto] Glutamates; detergent, emollient for personal care prods.

Adaphax. [Am. Cyanamid] Sulfurless vulcanized vegetable oils; extender for millable urethane and CR formulations; aids processing, calendering.

Adco. [Adco] Liquid soaps.

Addaroma. [Merix] Antistats, stabilizers, fragrances for plastics and rubber.

Additin 30. [Miles/Polysar Rubber] Phenyl-α-naphthylamine; staining antioxidant for rubber tech. goods and heavily stressed goods; antiflexcracking agent for NR and IR; storage stabilizer for petrol. prods.

Additive-A. [LignoTech] Lignosulfonates; clay conditioners for prod. of bricks and tiles; plasticizers, lubricants, binders, antiscumming agents, defloculants.

Additive GP. [Manufacturers Chems.] Blend of inorganic compds. and potassium permanganate; abrasive and high light bleaching agent to achieve acid wash look on denim.

Additol. [Hoechst Celanese] Additives for paints, printing inks, rubber industries.

Adeka. [British Traders & Shippers] Epoxy resins and hardeners.

Adeka GH-200. [Asahi Denka Kogyo] PPG-24-glycereth-24.

Adell. [Adell Plastics] Nylon 6, 6/6, and 6/12, polypropylene resins; inj. molding resins.

ADF Oleile. [Vevy] PPG-25-laureth-25.

Adilen. [Transformaciones Quimico] Polymeric plasticizers.

Adimoll. [Bayer] Adipates; plasticizer for PVC articles

Adinol. [Croda Chem. Ltd.] Sodium methyl taurates; detergent, emulsifier for cosmetics and pharmaceuticals.

Adiprene®. [Uniroyal; Uniroyal Chem. Ltd.] Urethane elastomers; for cast films and coatings, footwear, cable jackets, belts, hose, coated fabrics, rolls, molded goods.

Adjunct B. [Drew Ind. Div.] Disodium phosphate; deposit inhibitor for boiler water treatment.

Adjust 4. [United Catalysts] Liq. thixotrope, antisettling and antisag agent for paints.

Adjuzyme. [Enzyme Development] Enzyme for brewing.

ADM. [Climax Performance] Ammonium dimolybdate; corrosion inhibitor for vapor phase inhibitor applics.

Adma®. [Ethyl] Alkyl dimethyl amines; intermediates for quaternary ammonium compds., amine oxides, and betaines.

Admerol. [Reichhold] Modified oils.

Admex®. [Hüls Am.] Polymeric plasticizers.

Admiral®. [Aqualon] Fluidized polymer suspensions.

Admox® 1214. [Ethyl] Alkyl dimethylamine oxide; foaming agent, visc. modifier, emollient for anionic surfactants.

Admul. [Quest Int'l.] Glycerides, lactylates, esters; emulsifier for food prods.

Adogen®. [Sherex] Quaternary ammonium salts; specialty quat.; emulsifier, dispersant, fabric softener, antistat, flocculant; corrosion inhibitor formulations

Adogen® MA, S. [Sherex] Dimethyl amines; neutralizer, conditioner, co-emulsifier.

Adol®. [Sherex] Fatty alcohols; coemulsifier, lubricant, foam control agent, cosolvent, plasticizer, stabilizer, emollient, intermediate; for metal lubricants, inks, textiles, emulsions, paper, cosmetics, mineral processing, oil field chemicals, detergents, fabric softeners.

Adox. [Int'l. Dioxcide] Sodium chlorite; antimicrobial for water and waste water treatment.

Adpro. [Genesis Polymers] Polypropylene homopolymer and copolymers; for inj. molded automotive parts, exterior equip. components, tool boxes, appliances, furniture, consumer goods, medical components, and extreme impact items; impact modifier.

Adrub RTV Molding Rubber. [Adhesive Prods.] Two-component urethane; used for molds, gaskets, sealants, caulking, embossing rollers.

Adsee®. [Witco/Organics] Ethoxylated ethers; agricultural surfactant, penetrant.

Adsorbit. [Barnebey & Sutcliffe] Activated carbon; odor remover.

Advaco. [Advance Coatings] Unsat. polyester resins.

Advantage. [Hercules] Water-based, oil-based, or water-extended foam and entrained air control agents; maximizes drainage and prod. in brownstock washing systems, screen room systems, bleach plants.

Advantage. [ISP] Vinyl acetate/butyl maleate/isobornyl acrylate copolymer, ethanol SDA-40B.

Advantage. [Milliken] Colorants for opaque polyolefin applics., inj. molded prods., lids and closures.

Advantage®. [Drew Ind. Div.] Deposit and corrosion inhibitor, boiler water treatment.

Advapak. [Morton Int'l.] Blend; lubricant, stabilizers for plastics.

Advastab®. [Morton Int'l.; Morton Int'l. NV SA] Organotin compd.; lubricating stabilizer for PVC pipe, rigid sheeting, moldings, extrusions, bottles, profiles, PVC-PVAc systems; catalyst for rigid and flexible systems.

Advawax®. [Morton Int'l.] Ethylene bis-amides; synthetic waxes used as plastics processing lubricant, release agent, antistat, melting point modifier; coupling agent; pigment dispersant; used in adhesive tapes, coatings, food pkg. materials.

Advex. [BFGoodrich/Geon Vinyl] Rigid PVC cube profile extrusion compd.; interior grade.

Advitagel. [Quest Int'l.] Blends of distilled monoglyceride and alpha-tending emulsifiers; for aeration of sponge cakes.

Advitrol. [United Catalysts] Thixotrope.

AE-. [Proctor & Gamble] Ethoxylated alcohols; detergent, emulsifier, lubricant, wetting agent for textiles; intermediate in mfg. of surfactants.

AEPD®. [Angus] 2-Amino-2-ethyl-1,3-propanediol; pigment dispersant, neutralizing amine, corrosion inhibitor, acid-salt catalyst, pH buffer, chemical and pharmaceutical intermediate, solubilizer.

Aerex®. [BASF AG] De-icing and anti-icing fluid for airplanes.

Aero. [Am. Cyanamid] Xanthates, melamine, thiocarbanilide, or cyanides; for flotation, gold and silver recovery.

Aero Calcium Carbide. [Cyanamid Canada] Calcium carbide.

Aero Calcium Cyanamide. [Cyanamid Canada] Calcium cyanamide.

Aero Cyanamide 50. [Cyanamid Canada] Hydrogen cyanamide.

Aero Desulfurizing Reagents. [Cyanamid Canada] Desulfurizing agents.

Aerodet. [Aerochem] Scouring agents and stabilizers for textile dyeing and bleaching.

Aero Dicyandiamide. [Cyanamid Canada] Dicyandiamide.

Aerodri. [Am. Cyanamid] Dewatering agents.

Aero-Duster®. [Miller-Stephenson] Air spray cleaner.

Aerodye. [Aerochem] Scouring and leveling agents, dyeing assistants for textiles.

Aerofix N. [Aerochem] Aromatic condensate; dye fixing agent for acid dyes on nylon.

Aerofloat. [Am. Cyanamid] Promoters for flotation.

Aerofloc. [Am. Cyanamid] Reagents.

Aerofroth. [Am. Cyanamid] Frothing agents.

Aeroguard. [Am. Cyanamid] Flame retardants.

Aerolam® F Board. [Ciba-Geigy Plastics UK] Rigid, stable, lightweight honeycomb panels for construction of tool framework.

Aerolene Oil C. [Aerochem] Anionic blend; wetting agent and dyeing assistant for direct dyes.

Aeromine. [Am. Cyanamid] Promoters for flotation.

Aeronox. [Am. Cyanamid] Antioxidants.

Aerosil®. [Degussa; Degussa AG] Silica and fumed silica; anticaking and free-flow agent with high absorption capacity; reinforcer, thickener for silicone rubber industry; antiblock for PET, PP and PE films and tapes.

Aerosize. [Am. Cyanamid] Sizes for paper, building board.

Aerosoft. [Aerochem] Fatty amide; softeners for textiles.

Aerosol®. [Am. Cyanamid; Cyanamid BV] Sulfonates, sulfosuccinamates and sulfosuccinates; emulsifier, dispersant, foamer, detergent, wetting agent, solubilizer for soaps and surfactants; cleaner formulations, emulsion polymerization, agriculture, electroplating, textile, rubber, petrol., paper, metal, paint, plastics.

Aerospray. [Am. Cyanamid] Chemical binders.

Aerotex. [Am. Cyanamid] Buffers, resins, softeners, water repellent.

Aerothene. [Dow] Solvent and vapor pressure depressant.

Aerotru. [Am. Cyanamid] Sizing emulsion.

Aerowhite. [Aerochem] Optical brightener for textiles.

Aerozine A-50. [Olin] Rocket propellant.

Aethoxal® B. [Henkel/Emery/Cospha; Henkel KGaA] PPG-5-laureth-5; superfatting agent, emollient for bath oils, skin and personal care prods.

AF. [GE Silicones] Silicone antifoams; for food applics., industrial systems incl. textiles, soap mfg., resin prod., adhesives, inks, paint, petroleum processing.

AF. [Harcros] Silicone or nonsilicone antifoams; for agric., cutting oils, drilling muds, effluent treatment, solvs., adhesives, latex paints, inks, food applics., detergents, textiles.

AF-. [Harwick] Silicone emulsions or compds.; antifoams for natural and synthetic latexes, adhesives, paper, textile, leather applics.

A-Fax®. [Hercules] Amorphous polypropylene; thermoplastic used in wire and cable saturants, caulks and sealants, hot-melt adhesives, carpet backing, modification of polyolefin, wax, and asphalt.

Afflair® Lustre Pigments. [EM Industries] Mica platelets coated with titanium dioxide and/or iron oxide; pigments.

AF Foam Reducer. [Adhesive Prods.] Antifoaming agent for NR, SR latexes.

Afilan. [Hoechst Celanese/Colorants & Surf.; Hoechst AG] Esters; textile auxiliaries.

Aflas. [3M] Tetrafluoroethylene/propylene copolymer; used for o-rings, shaft seals, gaskets, packings, diaphragms, elec. connectors, wire and cable insulation, flexible joints, fabric-reinforced parts, hose and tubings, profiles, pipe, moldings.

Aflux. [Rhein Chemie] Esters; dispersant and lubricant for rubber molded and extruded goods.

AFP 2000. [Solvay Enzymes] Protease; enzyme.

Afpol. [Cal Polymers] SBR copolymer, hot emulsion polymerized; for adhesives mfg.

Afranil®. [BASF AG] Alcohol or fatty acid derivs.; grease and foam inhibitors, pulp deaerators for papermaking.

Afron 22. [Vevy] MEA-lauryl sulfate, potassium phosphate, magnesium aspartate, PEG-8.

AG-12-28. [Neville] Modified hydrocarbon resins; for printing ink vehicles, flushing and grinding applics.

Ageflex. [CPS] Acrylates, methacrylates, and glycidyl ethers; used in peroxide cure of elastomers.

Agefloc. [CPS] Quaternary ammonium salt; flocculant and coagulant, dewatering aids in centrifugation, filtration, and flotation of both industrial and municipal waste sludges, potable water treatment.

Agenap. [CPS] Naphthenic acid.

Agent 2A-2S. [Norman, Fox] Modified amide; air-entraining admixture.

Agent 765. [Witco SA] Fatty amine salt; corrosion inhibitor, pigment grinding and dispersion aid; antistripping agent for paints.

Agent #5032. [Polymer Research Corp. of Am.] Antitarnishing agent for pigment roller printing; solvent, lubricant, antistat, conditioner for textile fibers.

Agequat. [CPS] Polyquaternium; cationic polymers for personal care formulations, as drainage and retention aids, sludge dewatering.

Agerite®. [R.T. Vanderbilt] Antioxidants, polymer stabilizers for rubbers, latex, automotive, appliance prods.

Agesperse. [CPS] Polyacrylic acids and salts; dispersant, emulsifier, stabilizer for pigments in paints, coatings, carpet backcoating, paper, rubber, mining, textiles, ceramic slip, detergents, boiler and cooling water compds., adhesives.

Agestat. [CPS] Polymers; retention aid, pigment dispersant, drainage aid, stabilizer, raw and waste water clarifier for paper industry.

Agrico. [Agrico] Fertilizers.

Agrilan®. [Harcros UK] Emulsifier, wetter, dispersant, stabilizer for agricultural toxicants.

Agrimer. [ISP] Polyvinylpyrrolidone and copolymers.

Agrimul. [Henkel/Emery; Henkel-Nopco] Agricultural emulsifiers.

Agrisol®. [Harcros UK] Aromatic alkoxylates; cosolvents for agrochemical toxicants.

Agrisynth BLO. [ISP] γ-Butyrolactone; solvent for PAN, PS, fluorinated hydrocarbons, cellulose triacetate, shellac; used in paint removers, petrol. processing, specialty inks; intermediate for aliphatic and cyclic compds.; reaction and diluent solvent for pesticides.

Agriwet. [Henkel/Emery] Phosphate ester; wetting agent, detergent, emulsifier, lubricant, coupling agent, resuspension aid for pesticides.

Agrohyd. [Alexander Scheiner AG] Water absorbent for soil.

Agron. [Pfanstiehl Labs] Agric. chelates.

Agrosil®. [BASF AG] Colloidal silicate; agricultural aid; encourages intensive root development, improves irrigation efficiency, improves soils.

Agrox. [ICI Am.] Seed dressings.

Agsol. [ISP] Pyrrolidone.

AgsolEx. [ISP] Pyrrolidone derivs.; solvent.

Agsorb. [Oil Dri Corp. of Am.] Carriers for pesticides.

AH. [Huntsman] Polystyrene; for thermal insulation board.

Ahco. [ICI Am.] Emulsifier, surfactant, dispersant, detergent, wetting agent, solubilizer, lubricant, antistat; for textile use.

Ahcovel Base. [ICI Am.] Textile softener, lubricant.

Ahcowet. [ICI Am.] Sulfated fatty acid ester; wetting agent, dispersant, detergent, dye leveling agent for textiles.

AHP. [Henkel] Sodium azacycloheptane diphosphonate.

AH Salt. [BASF AG] Hexamethylenediamine adipate; monomer for prod. of nylon 6/6.

Aidall. [Disco] Internal lubricant for processing elastomeric compds.

Airbond. [Air Prods.] Acrylic and vinyl-acrylic copolymers.

Airflex®. [Air Prods.] VAE and ethylene-vinyl chloride polymers and copolymers; binder and saturant for paper, adhesives; vehicle base for paints, caulks, mastics, masonry coatings in the building industry.

Airlift®. [Air Prods.] Water-based release agents.

Aironil. [Manufacturers Chems.] Defoamer for textile use, brown stock paper defoamer for sewage treatment plants.

Airout. [Furane Prods.] Silicone reacted with tallate; air release agent in rubbers and resins; used in water or solvent-based systems.

Airpad. [Aero Consultants (UK) Ltd.] Nonsilicone rubber.

Air-Plas®. [Westvaco] Amine tallate; air entraining agent for masonry cement applics.

Airrol CT-1. [Toho Chem. Industry] Dialkyl sulfosuccinate; wetting agent, dyeing assistant, penetrant for agric. pesticides.

Air-Scent. [Air-Scent Int'l.] Fragrances and odor counteractants.

Air-Sweet. [Uncle Sam Chem.] Perfumed deodorizing blocks.

Airthane®. [Air Prods.] Polyurethane prepolymers.

Air-Trol. [Aqua Process] Odor neutralizer for industrial use.

Airvol®. [Air Prods.] Polyvinyl alcohol; offers high tens. str. and ease of film formation, exc. adhesive chars; super hydrolyzed grades for max. water and humidity resistance.

Airx. [Bullen Chem.] Odor counteractants.

Ajicoat. [Ajinomoto] Sodium polyglutamate; surface modifier, coemulsifier, codispersant.

Ajicure®. [Ajinomoto] Proprietary; accelerator for latent epoxy resin systems.

Ajidew. [Ajinomoto] PCA or sodium salt; humectant used in cosmetics, soaps, dentifrices, medicinal supplies, tobacco, cellulose film, paper prods., fiber prods., paints; additive to dyeing agent, softening agent, finishing agent, and antistatic agent; intermediate for synthesis.

Akachrome. [Aakash Chem. & Dye-Stuffs] Textile dyes and pigments.

Akacid. [Aakash Chem. & Dye-Stuffs] Textile dyes and pigments.

Akasol. [Aakash Chem. & Dye-Stuffs] Textile dyes and pigments.

Akasperse. [Aakash Chem. & Dye-Stuffs] Textile dyes and pigments.

Akaustan® A. [BASF AG] Flameproofing agent for vegetable and animal fibers.

Akavat. [Aakash Chem. & Dye-Stuffs] Textile dyes and pigments.

Akemi. [Hastings Plastics] Polyester filler.

Akofect. [Karlshamns] Partially hydrogenated fats from soybean or cottonseed oil; cocoa butter replacements.

Akoleno. [Karlshamns] Partially hydrogenated fats from soybean and/or cottonseed oils; bakeable filling fats, candy centers, caramel.

Akolidine. [Lonza] Alkyl pyridine mixture; corrosion inhibitor for metals, acid pickling, industrial acid cleaning, oilfield processes and treatment of oil refinery equip.

Akopol. [Karlshamns] Fractionated fats from soybean or cottonseed oil; cocoa butter replacements.

Akorex. [Karlshamns] Partially hydrogenated soybean and/or cottonseed oils; used as candy centers, color/flavor carriers, spray coatings, antidusting agents.

Akorine. [Karlshamns] Fractionated lauric-based fats from coconut and/or plam kernel oils; cocoa butter substitutes.

Akorol®. [BASF AG] Brake fluid components.

Akoyl. [Aakash Chem. & Dye-Stuffs] Textile dyes and pigments.

Akreact. [Aakash Chem. & Dye-Stuffs] Textile dyes and pigments.

Akrocal. [Akrochem] Calcium oxide; dessicant to control trace moisture in rubber processing.

Akrochem® Accelerators and Activators. [Akrochem] Accelerators and activators for rubber and latexes.

Akrochem® Antifoam. [Akrochem] Food-grade emulsifiers and silicone compds.; high-strength aq. system for systems requiring persistent antifoam properties.

Akrochem® Antioxidant. [Akrochem] Antioxidant, stabilizer, and antiozonant in polymers, incl. nat. and syn. rubber and latex, plastics, waxes.

Akrochem® Antiozonant. [Akrochem] Antiozonant protecting rubber polymers against heat, oxidation, and flexcracking; copper, manganese inhibitor and SBR stabilizer.

Akrochem® Calcium Stearate. [Akrochem] Calcium stearate; mold release agent.

Akrochem® Carnauba Wax. [Akrochem] Used in rubber compding., as stiffener for cured stocks, to improve surface of molded goods, as mold release and polish.

Akrochem® Ceresin Wax. [Akrochem] Used in polishes, insulating compds., waterproofing, to improve processing and surface finsih of molded rubber goods.

Akrochem® EW, HC, SC. [Akrochem] Kaolin clays; reinforcers, fillers for NR, SBR, CR, NBR, latexes.

Akrochem® Hydrated Alumina. [Akrochem] Alumina trihydrate; inert filler and flame-retardant smoke suppressant for latex foam, rubber, carpet backing, epoxies, reinforced polyesters, phenolics, and urethane foam.

Akrochem® PEG. [Akrochem] PEG; activator for rubber industry.

Akrochem® Peptizer. [Akrochem] Peptizing agents for rubber; processing aid for chloroprene.

Akrochem® Pigments. [Akrochem] Organic and inorganic pigments, fluorescent pigments.

Akrochem® Plasticizer. [Akrochem] Naphthenic rubber process oil; plasticizer.

Akrochem® Proaid. [Akrochem] Aromatic resins; homogenizing agents for elastomers; peptizers.

Akrochem® P Series. [Akrochem] Phenolic resins; tackifier, plasticizer, reinforcing agent for rubber compds. used in tire construction, mechanical goods, adhesives.

Akrochem® Retarder. [Akrochem] Retarders for rubber processing.

Akrochem® Rubbermakers Sulfur. [Akrochem] Sulfur and treated sulfur; vulcanizing agents for rubber industry.

Akrochem® RubberSil. [Akrochem] Precipitated amorphous silica; reinforcement for rubber mech. goods, tires, footwear.

Akrochem® Silicone Emulsions & Fluids. [Akrochem] Silicone emulsions and fluids; lubricants, release agents for tires and mech. goods, wire and cable, plastics.

Akrochem® SWS-201. [Akrochem] Dimethicone compd.; defoamer.

Akrochem® Zinc Oxide. [Akrochem] Accelerator-activator for rubber goods, adhesives.

Akrochem® Zinc Stearate. [Akrochem] Zinc stearate; mold release agent.

Akrochex. [Akrochem] Polymeric rubber color masterbatches.

Akrochlor. [Akrochem] Chlorinated paraffins; flame retardant for natural and synthetic rubbers.

Akrodip. [Akrochem] Slab dips.

Akrodye. [Akrochem] Dyes.

Akrofax. [Akrochem] Vulcanized vegetable oils; extender, processing aid, and softener in natural and synthetic rubbers.

Akroflock. [Akrochem] Cellulose, cotton, nylon, polyester, rayon, sisal, wood flours and blends; inert filler for natural and synthetic resins, rubbers.

Akroform®. [Akrochem] Accelerators, vulcanizers for elastomers.

Akro-Gel®. [Akrochem] Sodium salt of a fatty acid ester; mold release agent, lubricant.

Akrolease®. [Akrochem] Silicone polymers; release agent for plastics; corrosion inhibitor.

Akro-Mag® Bar. [Akrochem] Dispersed magnesium oxide; accelerator, vulcanizer, acid acceptor for polychloroprene, Hypalon,

Akron Chemical Beeswax. [Akrochem] Refined beeswax; used in rubber compding. to improve surface finish of molded goods, as mold lubricant.

Akroplast®. [Akrochem] Thermoplastic color concs.

Akrosorb. [Akrochem] Dry liq. dispersions incl. plasticizers, polymers, and resins.

Akrosperse®. [Akrochem] Pigment pastes, accelerator, vulcanizer in masterbatch form for synthetic elastomers.

Akrotak. [Akrochem] Pentaerythritol ester; tackifier for synthetic and natural rubber applics.

Akrowax. [Akrochem] Petroleum waxes; antiozonant, antisunchecking waxes for rubber vulcanizates.

Akro-Zinc® Bar. [Akrochem] Zinc oxide dispersion; activator, vulcanizer for rubber industry.

Aktiplast. [Rhein Chemie] Blend of zinc salts; peptizing and dispersing agents for tire compds., molded, extruded, and hard rubber; vulcanization accelerator; reclaiming agent for scrap rubber; processing promoter.

Aktisil®. [Hoffmann Min.] Silane compds.; filler for thermosets and thermoplastics.

Akucel. [Akzo] Sodium carboxymethyl cellulose; in drilling muds; in detergents as soil-suspending agent; in emulsion paints, adhesives, inks, textile sizes; as protective colloid; food stabilizer, binder, thickener; cosmetics; in pharmaceuticals as suspending agent, excipient, viscosity modifier

Akulon. [Akzo Engineering Plastics BV] Nylon 6 and 66 polymers.

Akuloy. [Akzo Engineering Plastics] Polyamide alloy; reinforced and unreinforced engineering plastics.

Akupol. [Alpe SRL] Polyamide 6 and 66, modified and flame retardant.

Akuvil. [Alpe SRL] PVC compds.; rigid and flexible compds. for engineering.

Akwilox. [Am. Chem. Services] Brominated soybean oil; food additive in soft drinks for visc. adjustment.

Akypo®. [Chem-Y GmbH] Carboxylates; surfactants for cosmetics, industrial cleaning; detergent, emulsifier, thickener.

Akypogene. [Chem-Y GmbH] Surfactant blends; base for rug shampoo and upholstery cleaner with antistatic and anticorrosive properties; disinfectant.

Akypomine. [Chem-Y GmbH] Mineral flotation surfactant.

Akypo®-Muls 400. [Chem-Y GmbH] Stearamide ether carboxylic acid; emulsifier for cosmetics.

Akypopress. [Chem-Y GmbH] Synthetic polymer; depressant for metal ions.

Akypoquat. [Chem-Y GmbH] Quaternary ammonium chloride; raw material for mfg. of laundry softeners, cosmetic hair care prods.; softener, antistat.

Akyporox. [Chem-Y GmbH] Ethoxylates; solubilizer, emulsifier, wetting agent for cosmetics, textiles.

Akyposal. [Chem-Y GmbH] Sulfates; emulsifier, detergent, shampoo base for cosmetics, dishwashing, bath prods.,textiles, emulsion polymerization.

Akyposept B. [Chem-Y GmbH] Benzylhemiformal.

Akypo® Soft. [Chem-Y GmbH] Lauryl polyglycol ether carboxylic acid salts; emulsifier, wetting agent, detergent for cosmetics.

Akypostat MA 35. [Chem-Y GmbH] Myreth-5 carboxylate; antistatic and antifogging prods.

Akyver. [Irpen SA] Cellular polycarbonate.

Akzo. [Chemach NV] Accelerators for epoxy.

Al-. [M&T Harshaw] Alumina; catalyst support, drying agent; dehydration reactions; fluid bed operations.

AL 2070. [ICI Am.] POE/POP block polymer; dispersant for agric. formulations.

Alacen. [New Zealand Milk Prods.] Whey protein.

Alacid. [New Zealand Milk Prods.] Casein.

Alacsan®. [Rhone-Poulenc Surf.] Imidazoline quats.; surfactants.

Alamgear. [Lubrication Engineers] Extreme pressure gear lubricants.

Alamine. [Henkel] Fatty amines.

Alanap. [Uniroyal] Herbicide.

Alanate. [New Zealand Milk Prods.] Sodium and potassium caseinates.

Alaren. [New Zealand Milk Prods.] Casein.

Alarsol AL. [Auschem SpA] Dodecylbenzene sulfonic acid; liq. detergent.

Alatal. [New Zealand Milk Prods.] Whey protein.

Alathon®. [OxyChem] HDPE homopolymer and copolymers; injection molding, blow molding, and film resins.

Alazate. [New Zealand Milk Prods.] Hydrolyzed casein.

Alba. [Alba Int'l.] Glycerin.

Alba. [Witco/Sonneborn] Petrolatum.

Albacar. [Pfizer] Precipitated calcium carbonate; reinforcing filler for SR, resins, NR, plastisols; for footwear, hose, belting, extruded and molded goods, sundries.

Albagel. [Whittaker, Clark & Daniels] Bentonite.

Albagen. [Whittaker, Clark & Daniels] Bentonite.

Albaglos. [Pfizer] Precipitated calcium carbonate; reinforcing filler for SBR, NR, CR, IIR, vinyl plastisols, resins; for extrusions, moldings, plastisols, belting, footware, sundries.

Albalan. [Westbrook Lanolin] Lanolin wax; emollient, emulsifier.

Albaryt. [Sachtleben Chemie GmbH] Micronized white barytes.

Albatex®. [Ciba-Geigy/Dyestuffs] Leveling agent, penetrant for textile dyes.

Albegal®. [Ciba-Geigy/Dyestuffs] Dyeing assistant, penetrant, leveling agent.

Alberger Salt. [Akzo Salt] Sodium chloride; for household and personal care prods.

Albigen® A. [BASF AG] Dye-affinitive stripping, brightening and leveling assistant for cellulose fiber dyeing and printing.

Albion Clay. [Albion Kaolin] Kaolin clay; filler, reinforcer, extender, mold release, lubricant for rubbers.

Albon. [Finetex] Defoamers for aq. systems; used in textile dyeing and finishing operations.

Albone. [DuPont] Hydrogen peroxide.

Albrite®. [Albright & Wilson Am.] Organic phosphites and phosphates; antioxidants, flame retardants, stabilizers for plastics, agric.

Alcalase®. [Novo Nordisk] Protease; enzyme for laundry detergents, prespotting; food grade for hydrolysis of protein.

Alcamine. [Allied Colloids] Softening agents.

Alcamizer. [Kyowa Chem. Industry] Magnesium aluminum carbonate; heat stabilizer for PVC.

Alcan AA. [Alcan] Activated alumina; in selective absorption processes; as starting material for catalyst; as dessicant for drying of gases.

Alcan Alumina Hydrate. [Alcan] Alumina hydrate; intermediate for mfg. of low-iron aluminum sulfate, sodium aluminate, hydrated aluminum chloride, other aluminum chemicals.

Alcan Aluminum Fluoride. [Alcan] Aluminum fluoride; as electrolyte in reduction of alumina to aluminum metal; as flux in remelting and refining of aluminum and its alloys; opacifier aid in production of ceramic enamels, glass, and glazes.

Alcan Aluminum Sulphate. [Alcan] Alumium sulfate; in pulp and paper mills, water purification plants, leather, textile, wallboard gypsum treatment, component in fire retardants.

Alcan C Series. [Alcan] Calcined alumina; raw material for mfg. of fused alumina abrasives, high alumina refractories, ceramic fiber insulation, hot topping compds., frits, glazes; in abrasives, polishes; as catalyst support.

Alcan FRF. [Alcan] Alumina hydrate.

Alcan H. [Alcan] Alumina trihydrate; filler and extender in plastics, resins, rubber, latex foams.

Alcan Recovered Cryolite. [Alcan] Sodium fluoroaluminate (87-90%), aluminum oxide (2%), sodium sulfate (4%), sodium carbonate (1-1.5%); source of fluorine; as ceramic flux and opacifier aid in the prod. of vitreous enamels, glass, glazes, abrasives; metallurgical flux in refining of aluminum and its alloys.

Alcan Superfine. [Alcan] Alumina hydrate.

Alcan Ultrafine. [Alcan] Alumina hydrate.

Alchemix. [Alchemie Ltd.] Silicone rubber, polyester resins, epoxy resins, polyurethane.

Alchlordrate. [UPI] Aluminum chlorohydrate; antiperspirant and deodorant; water purification; treatment of sewage and plant effluent.

Alcodet®. [Rhone-Poulenc Surf.] POE thioethers; emulsifier, detergent for carbon soil and grease cleaning, metal cleaning, textile scouring, insecticides.

Alcofix®. [Allied Colloids] Fixing agent for textile dyeing.

Alcogum. [Alco] Polyacrylates; thickener for latexes, adhesives, paints, paper coatings; flocculant; dye assistant.

Alcojet®. [Alconox] Sodium metasilicate, sodium carbonate, POE ester of mixed fatty and resin acids; detergent for mechanical washers.

Alcolan®. [Amerchol] Lanolin/petrolatum blends; emulsifier, emollient base for cosmetics and pharmaceuticals.

Alcolec®. [Am. Lecithin] Lecithin; wetting agent, emulsifier, stabilizer, release agent, diet supplement; for paints, cosmetics, pharmaceuticals, food applics.

Alconate®. [Rhone-Poulenc Surf.] Sulfosuccinates; surfactants, wetting agent, dispersant, penetrant, thickener, conditioner for personal care prods., rug shampoos, industrial and mining applics.; emulsifier in emulsion polymerization; antimicrobial for dandruff shampoos.

Alconox®. [Alconox] Blend of alkylaryl sulfonates, lauryl alcohol sulfates, phosphates, carbonates; detergent, wetting agent, sequestrant; for labware cleaners.

Alcophor. [Henkel; Henkel KGaA] Tannin compds.; corrosion inhibitor used in petroleum-based paint systems.

Alcopol. [Allied Colloids] Wetting agents.

Alcoprint®. [Allied Colloids] Binder, pigment dispersant, softening agent, thickener for pigment printing on textiles.

Alcoramnosan. [Vevy] Hydroxyethylcellulose; thickener, suspending agent; stabilizer for vinyl polymerization; binder in ceramic glazes; used in paper and textile sizing.

Alcoset® B2. [Allied Colloids] Self-crosslinking formaldehyde-free acrylic resin; stiffener, fabric stabilizer.

Alcosperse. [Alco] Polyacrylates; stabilizer, dispersant for pigments, high solids slurries, paper coatings, paint, latex, textile, mining, ceramic applics.

Alcosperse®. [Allied Colloids] Dispersant, leveling agent for textile dyeing.

Alcotabs®. [Alconox] Blend of alkylaryl sulfonates, lauryl alcohol sulfates, phosphates, carbonates; detergent with wetting, sequestering and synergistic agents.

Alcotex. [Harlow Chem. Co. Ltd.] Polyvinyl alcohol.

Alcryn®. [DuPont; DuPont UK] Halogenated ethylene interpolymer alloy; melt processable thermoplastic elastomer for seals, gaskets, weatherstripping, coated fabrics, sheet goods, belting, mech. goods, tubing, wire and cable.

Alcudia. [Petroplas Ltd.] HDPE, LDPE, PP, or EVA.

Aldehyd. [Henkel/Emery/Cospha] Aldehydic aroma chemicals for fragrance compounding in cosmetic, personal care, detergent industries.

Aldo®. [Lonza] Glyceryl or propylene glycol esters; coupling agent, stabilizer, emollient, consistency builder, emulsifier for food, cosmetics, industrial use.

Aldolyte. [Aldoa] Electroplating brighteners.

Aldomax. [UOP] Immobilized amyloglucosidase; enzyme.

Aldor. [Alpine Aromatics] Natural and synthetic fragrance materials; covers undesirable odors during rubber processing.

Aldor. [Vikon] Masking odorants to cover formaldehyde and amine odors on resin finished fabrics.

Aldosperse®. [Lonza] Glyceryl stearate and/or ethoxylated glyceryl stearates; emulsifier, solubilizer, suspending and dispersing agent used in personal care prods., food industry, textiles.

ALE-56. [Union Carbide] Amino bispropyl dimethicone.

Alexis Antibloom. [Albright & Wilson Am.] Proprietary metal finishing additive; for hot water sealing of anodized aluminum to prevent sealing bloom.

Alf. [Alfordshire Ltd.] PVC compds.

Alfedox. [Shieldalloy Metallurgical] Ferro aluminum; deoxidant for steel industry.

Alfol®. [Vista] Fatty alcohols; emulsifier, wetting agent, surfactant intermediate; lubricant for metal rolling oils; emollient for cosmetics; defoamer.

Alfonic®. [Vista] Ethoxylated linear alcohols or their sulfate salts; nonionic; detergent, wetting agent, emulsifier, foaming agent; surfactant intermediate.

Alfrimal. [Alpha Calcit Fullstoff GmbH] Aluminum hydroxide; flame retardant.

Alftalat. [Hoechst Celanese/Fine Chem.] Polyester and alkyd resins; for coat-

ings and inks.

Algaesil. [Mason] Colloidal silver; algicide.

Algaetrol 76. [Harcros] Copper TEA compd.; algicide for swimming pool use.

Algard. [Witco] Stamping lubricants, die films, die lubricants, drawing compds.

Algene. [Bond] Algicide for cooling towers.

Algepon. [Sandoz] Fatty quaternary ammonium compd.; detergent stripping compd.

Algisium. [Exsymol] Sodium mannuronate methylsilanol; Lipolytic, cutaneous hydration agent for cosmetic and health prods.

Algoflon HC. [Ausimont; Montefluos] PTFE; for molding, extrusion applics.

Algogen. [Grant] Water, algae extract, hydrolyzed actin.

Algon. [Auschem SpA] Ethoxylated fatty esters; emulsifier, thickener for cosmetics, pharmaceuticals.

Algonina AC. [Auschem SpA] Cyclic and amino amide compds.; laundry softener for natural and synthetic fibers.

Alguard® NS. [Allied Colloids] Sulfonated aromatic condensate; stain repellent for nylon carpet.

Alicep®. [BASF AG] Chloridazon, chlorbufam; for pre- and post-emergence weed control in crops.

Aliette. [Rhone-Poulenc/Ag] Fungicide.

Alipal®. [Rhone-Poulenc Surf.] Ethoxylated alcohol sulfate salt; detergent, emulsifier, stabilizer, lime-soap dispersant, wetting agent, foamer; dishwashing formulations, scrub soaps, car washes, rug and hair shampoos, emulsion polymerization, concrete, petrol. waxes, textile wet processing, cosmetics, pesticides.

Aliquat. [Henkel] Fatty quaternary ammonium chloride; phase transfer agent for catalyzing reactions.

Alkadet. [Rhone-Poulenc Surf.] Formulated drycleaning charge soap base.

Alkafilm. [W.R. Grace/Dearborn] Corrosion inhibitor.

Alkaflo®. [Sybron] Patented phosphate salt mixture; alkalizer for reactive dyes.

Alkafoam. [Rhone-Poulenc/Textile & Rubber] Ethoxylated alcohol; foaming agent for foam finishing applics. for textiles.

Alkali Surfactant NM. [Exxon/Tomah] Caustic surfactant hydrotrope used in alkaline formulations for hard surface cleaning; solubilizer.

Alkameen. [W.R. Grace/Dearborn] Corrosion inhibitor.

Alkamerce. [Hart Prods. Corp.] Wetting agents.

Alkamide®. [Rhone-Poulenc Surf.] Alkanolamides or alkyl amidopropyl dimethylamines; surfactant, detergent, foam booster/stabilizer, superfatting agent, thickener, emulsifier for personal care and laundry prods., cleaners; lubricant for cutting oils.

Alkaminox®. [Rhone-Poulenc Surf.] Ethoxylated fatty acid amine; textile scouring, dyeing assistant, softener, antistatic agent; used as corrosion inhibitor in steam generating and circulating systems; coemulsifier.

Alkamox®. [Rhone-Poulenc Surf.] Amine oxides; detergent, wetting agent, foaming agent and foam stabilizer for rug shampoos, laundry detergents, dishwashing, shampoos, cleaners, antistatic softeners; foam stabilizer in foam rubber, electroplating, paper coatings.

Alkamuls®. [Rhone-Poulenc Surf.] Ethoxylated, sorbitan, or glycol esters; w/o emulsifier, dispersant, defoamer, coemulsifier, coupling agent, solubilizer, thickener, lubricant, opacifier, pearlescent, wetting agent, antistat in metalworking and textile oils, pesticides, cosmetics; as paper softener.

Alkanol®. [DuPont] Wetting agent, emulsifier, detergent, penetrant, antiprecipitant, dyeing assistant, dye solubilizer, dispersant.

Alkanox 240. [Enichem Synthesis SpA] Hydrolytically stable phosphite; antioxidant.

Alkaphos®. [Rhone-Poulenc Surf.] Phosphate esters; surfactants.

Alkapol. [Rhone-Poulenc Surf.] PEG; intermediate for surfactants; binder/lubricant in pharmaceuticals; plasticizer; paper softener; humectant; solvent; antistat; for cosmetics, textile, plastics processing, dyes and inks.

Alkaquat®. [Rhone-Poulenc Surf.; Rhone-Poulenc France] Quaternary compds.; substantive antistat, softener, wetting agent, emulsifier, biocide, conditioner for laundering, beverage, dairy and food industries, water treatment, paper industry, pest control, preservatives.

Alkaquest EDTA. [Rhone-Poulenc Surf.] Tetrasodium EDTA; chelating agent.

Alkaril. [Rhone-Poulenc/Textile & Rubber] Stabilizers for peroxide bleaching, lubricants for textile dyeing, soil release finishes.

Alkasil®. [Rhone-Poulenc Surf.] Silicone polyalkoxylate block copolymer; intermediate for prod. of rigid polyurethane foams; surfactants used in cosmetics, toiletries, textiles, coatings, and as release agent.

Alkasperse®. [Rhone-Poulenc Surf.] Polyacrylates; polymeric dispersant, antiscalant for water treatment, household and industrial detergents, paints, drilling muds.

Alkasurf®. [Rhone-Poulenc Surf.] Anionic surfactants; detergent, emulsifier, wetting agent for toiletries, carpet shampoos, textiles, resin treatments, emulsion polymerization, household and industrial cleaners.

Alkaterge®. [Angus] Oxazoline deriv.; detergent, emulsifier, wetting agent, antifoamer, antioxidant, dispersant, corrosion inhibitor; used in salt, soap, paper, textiles, and metal cleaners; emulsion stabilizer; acid scavenger; pigment grinding and dispersion.

Alkateric®. [Rhone-Poulenc Surf.] Amphoterics; foaming agent, wetting agent, emulsifier for industrial and detergent formulations, personal care prods.; conditioners, antistats, emollients.

Alkathene Polythene. [Mersey Plastics Ltd.] HDPE, LDPE, MDPE. antioxidant.

Alkatronic PGP. [Rhone-Poulenc/Textile & Rubber] Block/graph copolymer surfactant; scouring and defoaming agents for textiles.

Alkawet®. [Lonza] Amphoterics; emulsifier, surfactant intermediate.

Alkazid® Lye. [BASF AG] For removal of hydrogen sulfide and carbon dioxide from synthesis gas and cracked gas for prod. of ethylene.

Alkazine®. [Rhone-Poulenc Surf.] Imidazolines; emulsifier, antistat, corrosion inhibitor, softener for textiles, plastics, asphalt, water repellent treatment, cutting oils, slime control additive in paper processing, wetting and flocculating agent, lubricant; pigment dispersant.

Alken Disperse Formulas. [Alken-Murray] Boiler and cooling water dispersants.

Alken K-P Formulas. [Alken-Murray] Chelant/polymer boiler scale inhibitors.

Alken P700 Formulas. [Alken-Murray] Wastewater treatment polymers.

Alkenyl Succinic Anhydrides. [Humphrey] Intermediates for prod. of amide and imide rust inhibitor lube oil additives; detergent and dispersant additives in gasoline.

Alkon. [Apollo] Penetrant/scouring agent, mercerizing agent for textile processing.

Alkorcell. [Alkor Plastics UK] Copolymer.

Alkorfol ABS. [Alkor Plastics UK] ABS.

Alkorphan. [Alkor Plastics UK] PVC.

Alkorplan. [Alkor Plastics UK] PVC; for roofing applics.

Alkox. [Tropag GmbH] High m.w. polyethylene oxide.

Alkozell. [Alfelder Kunststoffwerke] Foamed polyethylene.

Alkuloy. [Akzo Engineering Plastics] Polyamide alloy.

Alkylate. [Monsanto] Linear alkylbenzenes; detergent, intermediate.

Alkylene. [Hart Prods. Corp.] Phosphated esters; anionic wetting agents and emulsifiers.

Alkylox. [Seppic] Alkyl polyethoxy ether; textile detergent.

Allabond. [Bacon] Epoxy adhesive.

Allantoina. [Sigma Prodotti Chimici] Allantoin (2,5-dioxo-4-imidazolidinyl urea); emollient for skin care.

All Aqueous Clear Conc. 7711 Series. [CNC Int'l.] Clear concs. for roller and screen printing.

All Aqueous Liquid Clear. [CNC Int'l.] Pourable liq. for roller and screen printing.

All-Chem. [Procedyne] Antistat for plastics.

Allguard. [Dow Corning] Elastomeric waterproof coating; for masonry, construction, stucco, stone.

Allied Whiting. [Akrochem] Calcium carbonate; filler for NR, SR.

Alloprene. [ICI Resins] Chlorinated rubber.

Allo-Scour. [Scholler] Ethylene oxide adducts; nonionic surfactants for low foaming scouring operations; scour/dye assistants for all fibers.

All-Soft®. [Dycho] Blends of cationic and modified cationic fatty acid; softeners-lubricants for cotton and synthetic fibers.

All Wet. [W.A. Cleary] Ethoxylated nonylphenyl ether; detergent, wetting agent, emulsifier for pesticides.

Almaflex. [Alma SA] Polypropylene.

Almagard. [Lubrication Engineers] Multipurpose lubricant.

Almaklar. [Alma SA] Clear polystyrene.

Almapack. [Alma SA] ABS.

Almaplex. [Lubrication Engineers] Industrial lubricants.

Almasol. [Lubrication Engineers] Specialty lubricants.

Almatek. [Lubrication Engineers] General purpose lubricants.

Almatex. [Anderson Development] Acrylic resins.

AlNel. [Advanced Refractory Tech.] Aluminum nitride; filler for elastomeric, epoxy, polymeric, and other filled systems.

Alnovol. [Hoechst Celanese/Fine Chem.] Phenolic resins; for coatings, inks, rubber industry.

Aloe-Con. [Florida Food Prods.] Aloe vera gel; moisturizer for first aid preps., acne and facial preps., creams, shaving prods., depilatories, shampoos, conditioners, suncare prods.

Aloe Gel Stabilized. [Terry Labs] Aloe vera gel; moisturizer for first aid preps., acne and facial preps., creams, shaving prods., depilatories, shampoos, conditioners, suncare prods.

Aloe HS. [Alban Muller] Propylene glycol, aloe extract.

Aloe-Moist. [Terry Labs] Aloe vera compd.; moisturizer for first aid preps., acne and facial preps., creams, shaving prods., depilatories, shampoos, conditioners, suncare prods.

Aloe Phytogel 1:199 Powd. [Lipo] Aloe vera gel; moisturizer for first aid preps., acne and facial preps., creams, shaving prods., depilatories, shampoos, conditioners, suncare prods.

Aloe Powd. 1-200. [Chemetics Labs] Aloe; skin moisturizer and palliative; film-forming.

Aloe Vera Conc. 40 Fold. [Meer] Aloe vera gel; moisturizer for first aid preps., acne and facial preps., creams, shaving prods., depilatories, shampoos, conditioners, suncare prods.

Aloe Vera Conc. I-40. [Chemetics Labs] Aloe vera gel; moisturizer for first aid preps., acne and facial preps., creams, shaving prods., depilatories, shampoos, conditioners, suncare prods.

Aloe Vera Powd. 200XXX Extract-Microfine. [Tri-K Industries] Aloe vera gel; for cosmetic, health, and pharmaceutical indutries.

Alox®. [Alox] Film-forming corrosion preventative; lubricant in metalworking oils; mold release agent for concrete; penetrant.

Alo-X-11. [Warren] Aloe vera gel; moisturizer for first aid preps., acne and facial preps., creams, shaving prods., depilatories, shampoos, conditioners, suncare prods.

Aloxime. [Henkel] Oil-sol. metal chelating compds.

Alpco. [Am. Lignite Prods.] Montan wax.

Alperox. [Atochem N. Am.] Lauroyl peroxide; initiator for polymerization, curing of polyester and acrylic resins.

Alpex. [Hoechst Celanese/Fine Chem.] Cyclized rubber resins; for coatings and inks.

Alpha. [Alpha Chem. & Plastics] Vinyl compds.; for inj. molding, extrusion, medical, wire and cable, automotive trim.

Alpha. [Katalistiks Int'l.] Fluid cracking catalysts.

Alpha 500. [Paniplus] Monoglyceride, polysorbate 60; emulsifier.

Alphadim®. [Am. Ingredients] Monoglyceride; food additive, processing aid, stabilizer.

Alphaflex®. [Alphaflex] PTFE and molybdenum disulfide; reinforcing agent, friction-reducing additive for rubber.

Alphamod. [Am. Ingredients] Glyceryl stearate; nonionic sec. o/w emulsifier for creams and lotions; visc. booster for emulsions.

Alpha-Rez. [Lawter Int'l.] Hydrocarbon resin.

Alpha-Step®. [Stepan] Alpha sulfo methyl esters; surfactants for dishwashing liqs., fine fabric washes, hard surface cleaners, bubble baths.

Alphenate. [Henkel-Nopco] Alkyl polyglycol ethers and esters; detergent for textiles dyeing and printing; dye leveling agent; wetting agent, emulsifier for emulsion polymerization.

Alphoxat. [Zschimmer & Schwarz] Ethoxylated fatty acid ester; emulsifier, dispersant, defoamer for textile, ceramic, and chemical technical industries.

Alpol PA. [Lion] Pearlescent/anionic surfactant blend; for pearl-toned shampoos, dishwashing detergents.

Alresat. [Hoechst Celanese] Modified maleinic resin; wetting agent for lustering emulsions, printing inks.

Alrodyne. [Ciba-Geigy] Polyoxyalkylene fatty ester polymer; emulsifier.

Alrosol. [Ciba-Geigy] Fatty acid alkanolamide; foam booster/stabilizer, wetting agent, dispersant, emulsifier used in liq. soaps and detergents, rubber and latexes.

Alrosperse. [Ciba-Geigy] Fatty acid amide; emulsifier, lubricant used in hair rinses, hand modifiers for textiles, spreading agent in paste waxes and polishes; antistat for rubbers and polyolefins.

Alrowet® D-65. [Ciba-Geigy] Dioctyl sodium sulfosuccinate; wetting and rewetting agent for textiles; dispersant, antistat, wetting agent for NR and SR latexes.

Alscoap. [Toho Chem. Industry] Alkyl sulfate salts; foaming agent, base material for hair shampoos, toothpaste.

Alsibronz. [Franklin Mineral Prods.] Wet ground muscovite mica.

Alsilan. [Alpha Calcit Fullstoff GmbH] Mica.

Alsimica. [Franklin Mineral Prods.] Wet ground muscovite mica; release agent, mold lubricant for rubbers.

Altafin. [Astor Wax] Paraffin waxes; processing aid, antiozonant for NR and SR.

Altalc. [Cyprus Industrial Min.] Talc; filler and pigment in rubber, paints, soaps, ceramics, cosmetics, pharmaceuticals; dusting agent, lubricant, electrical insulation.

Altax®. [R.T. Vanderbilt] Benzothiazyl disulfide; accelerator for natural and synthetic rubbers.

Altcofast. [Crompton & Knowles] Textile dyes and pigments.

Altowhite. [Dry Branch Kaolin] Calcined aluminum silicate; extender pigment for paints.

Altuglas. [Atochem UK] Polymethacrylate.

Alubond. [Universal Scientific] Bonded aluminas.

Alubrasoft®. [PPG/Specialty Chem.] Cationic fatty polyamide; softener, lubricant, antistat, conditioner, fiber finish for textiles.

Alubrasol®. [PPG/Specialty Chem.] Fiber lubricant and processing aid for natural and synthetic fibers.

Alubraspin® 100-P. [PPG/Specialty Chem.] Lubricant/antistat for natural and producer dyed nylon and polypropylene filament and polyester staple.

Aludone®. [UCIB] Aluminum PCA; for deodorants, shower gels, hair comb-out balm.

Alumagel. [Witco] Gelling agent for organic fluids.

Aluminum Oxide C. [Degussa] Alumina; free-flow and anticaking agent.

Aluminum Silicate. [Degussa] Precipitated aluminum silicate; clay used in reinforced plastics, dental cements, glass industry, paint filler, mfg. of precious stones, enamels.

Aluminum Stearate EA. [Witco] Aluminum distearate; binder, emulsifier, anticaking agent for food applics.

AluMystique. [U.S. Cosmetics] Metal soap; surface treatment for cosmetics; binder for pressed powds.

Alutrat. [Vevy] Aluminum citrate.

Alveocel. [Alveo AG] Noncrosslinked polyethylene foam.

Alveolen. [Alveo AG] Continuously extruded crosslinked closed-cell polyethylene foam.

Alveolit. [Alveo AG] Continuously extruded crosslinked closed-cell polyethylene foam.

Alveolux. [Alveo AG] Batch molded chemically crosslinked closed cell polyethylene foam.

Alvinox. [3-V; Sigma Prodotti Chimici] Hindered phenolics; stabilizers, antioxidants for thermoplastics, elastomers, adhesives.

Alviscal. [Alpha Calcit Fullstoff GmbH] Very fine calcium carbonate.

Ama. [Vinings Industries] Slimicide.

Amaplast. [Color-Chem Int'l.] Textile dyes and pigments.

Amasil®. [BASF AG] Formic acid; for ensiling and feed preservation.

Amax®. [R.T. Vanderbilt] Oxydiethylene benzothiazole sulfenamide products; accelerator for SBR, NR, IR, BR.

Amazin®. [Aceto] Maneb with zinc; fungicide.

Amberex. [Am. Cyanamid] Vulcanized

vegetable oils; processing aid for NR, SBR, CR, NBR.

Ambergard. [Rohm & Haas] Absorbent, ion exchange resin.

Ambergum®. [Aqualon] Cellulose deriv.; replaces gum arabic in lithographic printing processes.

Amberlac. [Reichhold] Modified alkyd resin sol'n.

Amberlig. [LignoTech] Calcium lignosulfonate; fly ash retarder, handling and processing aid.

Amberlite®. [Rohm & Haas] Ion exchange resin; for industrial water treatment, pharmaceutical, chemical, and food processing industries.

Amberlyst. [Rohm & Haas] Macro-reticular ion exchange resin; for nonaq. fluid treatment.

Ambersep. [Rohm & Haas] Inert polymer bead; ion exchange resin.

Ambersil. [Ambersil Ltd.] Mold release agents.

Ambersorb. [Rohm & Haas] Carbonaceous adsorbents.

Ambidex. [Morton Int'l.] Cationic; wetting agent, emulsifier.

Ambiflo. [Dow] Lubricants.

Ambiteric D40. [Rhone-Poulenc UK] Alkyl dimethyl betaine; amphoteric foam booster/stabilizer.

Ambitrol. [Dow] Coolants.

Ambrocide. [Chapman] Insecticidal wood preservative.

Ambroxan. [Henkel/Cospha] 8-Alpha-12-oxido-13,14,15,16 tetra-norlabdane; fragrance raw material.

AMC-2. [Aerojet Propulsion] Curing agent for epoxy resins.

Amcar. [Am. Emulsions] Carriers for textile dyes.

Amcat. [Activated Metals & Chem.] Amine coated catalyst.

Amdro®. [Am. Cyanamid/Ag] Hydramethylnon; insecticide for control of fire ants in pastures, lawn, turf.

Amdye PH-12. [Am. Emulsions] Inorganic salts; replacement for trisodium phosphate in liq. form; for textile applics.

AME 4000®. [Ashland] High performance marine resin.

Ameenex. [Chemron] Tall oil amido-amine; industrial surfactant, corrosion inhibitor, wetting agent, emulsifier, antistripping agent for mineral flotation, asphalts, boiler corrosion control.

Amercell Polymer HM-1500. [Amerchol] Nonoxynol hydroxyethylcellulose; thickener for hair and skin care prods.

Amerchol®. [Amerchol] Lanolin and/or sterol blends; aux. emulsifier, emulsion stabilizer for cosmetics, pharmaceuticals; conditioner, emollient, lubricant.

Amercor®. [Drew Ind. Div.] Blend of volatile amines; corrosion inhibitor for boiler water treatment.

Amerel. [Drew Ind. Div.] Defoamer.

Ameret. [Drew Ind. Div.] Evaporator scale inhibitor.

Amerfeed. [Drew Ind. Div.] Automatic feed system.

Amerfloc®. [Drew Ind. Div.] Polymers; flocculant and coagulant aid for sludge conditioning, dewatering industrial slurries, water clarification.

Amergel®. [Drew Ind. Div.] Surfactant blends; antifoam for aq. processing operations, esp. pulp and paper, nonwovens, effluent systems.

Amergy®. [Drew Ind. Div.] Organometallic blend; deposit and corrosion inhibitor for boilers, fuel additive.

AmeriBond 2000. [Borregaard LignoTech] Lignosulfonate; animal feed binder, processing aid.

Ameripol. [Ameripol Synpol] Styrene-butadiene rubber.

Ameripol Synpol. [Ameripol Synpol] Styrene-butadiene rubber.

Amerlate®. [Amerchol; Amerchol Europe] Lanolin acids and esters; emulsifier, stabilizer, emollient, conditioner, moisturizer, lubricant, opacifier, pigment dispersant, wetting agent for cosmetics, pharmaceuticals, wax systems, household prods.

Ameroxol®. [Amerchol; Amerchol Europe] Ethoxylated fatty alcohol ethers; emulsifier, solubilizer, stabilizer, dispersant, cosolvent for cosmetics and toiletries.

Ameroyal. [Drew Ind. Div.] Evaporator

scale inhibitor.

Amerpac. [Drew Ind. Div.] Recyclable semibulk containers consisting of HDPE on expanded polystyrene foam cushioning and protected by galvanized steel outer shell.

Amerplex®. [Drew Ind. Div.] Deposit and corrosion inhibitor for boiler water treatment.

Amerscent. [Drew Ind. Div.] Neutralizing/masking agent for odor control.

Amerscol USP. [Amerchol] Octyl dimethyl PABA; sunscreen agent.

Amerscreen. [Amerchol] UV absorbers.

Amersep®. [Drew Ind. Div.] Sodium dimethyldithiocarbamates blend; metals precipitant for plating and metal finishing operations.

Amershield. [Ameron Protective Coatings] Self-priming, high-build polyurethane.

Amersil. [Amerchol] Silicone derivs.; emulsifier for prep. of water-in-silicone oils for personal care prods.

Amersite®. [Drew Ind. Div.] Corrosion inhibitor for boiler water treatment.

Amersperse. [Drew Ind. Div.] Evaporator/vacuum pan scale and deposit inhibitor.

Amerstat®. [Drew Ind. Div.] Combination of isothiazolin compds.; antimicrobial, preservative in industrial water systems, latex, adhesives, oil recovery systems, metalworking fluids; sugar and paper mill slimicide.

Amerthane. [Ameron Protective Coatings] Elastomeric polyurethane.

Amertrol®. [Drew Ind. Div.] Deposit and corrosion inhibitor.

Amerzine®. [Drew Ind. Div.] Hydrazine sol'n.; corrosion inhibitor for feed and boiler waters.

Amfix FRL. [Am. Emulsions] Formaldehyde-free fixing agent; improves wetfastness of cellulosics dyed or printed with direct or reactive dyes.

Amfotex. [Pulcra SA] Detergent, foaming and wetting agent, solubilizer, thickener, conditioner, corrosion inhibitor for personal care products, metal cleaning, textile softeners, paint emulsifiers.

AMG 200 L. [Novo Nordisk] Amyloglucosidase derived from Aspergillus niger; enzyme used in starch industry.

Amgard®. [Albright & Wilson Am.] Tributoxyethyl phosphate; flame retardant.

Ami Bioprotector. [Alban Muller] Oak-apple extract; soothing prod.

Amical®. [Angus] Diiodomethyl-p-tolyl sulfone; preservative, mildewcide, fungicide for adhesives and sealants, and in lumber, construction, home improvement, textile, leather, and automotive industries.

Amical®. [Franklin Industrial Minerals] Calcium carbonate; used in latex rubber systems.

Amicon®. [Emerson & Cuming] Conductive and nonconductive epoxy, polyimide, or silicone; adhesive for IC assembly; die coating; encapsulant.

Amicure®. [Air Prods.] Amine-based; catalyst for coatings, adhesives, sealants; crosslinker for polyurethane coatings; accelerator.

Amidan. [Grindsted Prods. Denmark] Monoglycerides; emulsifier, dough conditioner, starch complexing agent for food industry.

Amide CD 2:1. [Hysan] Cocamide DEA, diethanolamine.

Amidex. [Chemron] Fatty acid alkanolamide; antistat, humectant, conditioner, thickener, visc. builder, foam stabilizer, wetting agent for shampoos, cosmetics, skin prods., cleaners, bubble baths, industrial oils and lubricants, degreasers, cleaners.

Amidine. [John Campbell] Textile dyes and pigments.

Amido Betaine C. [Zohar Detergent Factory] Coconut amido betaine; industrial foamer, component of personal care prods.

Amidox®. [Stepan] Ethoxylated alkylolamides; emulsifier, detergent for dishwashing, shampoos, emulsions.

Amiema. [Nihon Emulsion] Myristoyl methyl alanates; oil-phase materials.

Amiet. [Kao Corp. SA] Ethoxylated fatty acid amines; antistat, textile dyeing assistant, softening and dispersing agent.

Amifat. [Ajinomoto] Pyrrolidone carboxylic acid esters; surfactant used as emulsifier in personal care prods.

Amiflex. [Takeda USA] Flavor enhancer.

Amigase®. [Int'l. Bio-Synthetics] Amyloglucosidase; enzyme for hydrolyzing dextrins to dextrose.

Amigel. [Alban Muller] Sclerotium gum; natural gellant.

Amihope® LL-11. [Ajinomoto] Lauroyl lysine; surface modifier, coemulsifier, codispersant; in cosmetics, medical, painting and other fields; filler for ink and paint; chelating agent.

Amiladin. [Dai-ichi Kogyo Seiyaku] PEG alkyl amine ether; dispersant, washing agent, dye leveling agent.

Amilon. [Ikeda] Silica, lauroyl lysine.

Amine. [Berol Nobel] Fatty amines; emulsifier, corrosion inhibitor; surfactant intermediates.

Amine. [Ciba-Geigy] Imidazolines; emulsifier, dispersant, detergent; used in acid cleaners; antistat for textiles, plastics; paints; corrosion inhibitor; cutting and rust preventive oils.

Amine CS-1135®. [Angus] Oxazolidine; corrosion inhibitor, alkaline prod. for metalworking fluids and aq. systems.

Amine CS-1246. [Angus] Oxazolidine; formaldehyde substitute, crosslinking agent, corrosion inhibitor, synthesis applics.

Amine D®. [Hercules] Dehydroabietylamine; asphalt additive; flotation reagent; for prod. of algicides, preservatives, corrosion inhibitors.

Amine D® Acetate. [Hercules] Dehydroabietylamine acetate; asphalt additive; flotation reagent; for prod. of algicides, preservatives, corrosion inhibitors.

Amine Guard. [UOP] Corrosion inhibitor.

Amine PMT®. [Angus] Aliphatic ditert. amine; alkaline material, corrosion inhibitor for aq. systems.

Amino Acid Gelatinization Agent.
[Ajinomoto] N-Acyl glutamic acid diamide; gelatinization agent for oil for solidifying almost all oils ranging from petrol. to veg. oils.

Amino-Collagen. [Maybrook] Collagen amino acids; substantivity agent, moisturizer, penetrant for hair and skin care prods.

Aminodermin CLR. [Henkel/Cospha] Amino acid conc.; conditioner for damaged hair.

Aminofoam C. [Croda Inc.; Croda Chem. Ltd.] TEA-lauroyl animal collagen or keratin amino acids; detergent, conditioner, foaming agent for skin and hair care cleansing systems.

Amino Gluten MG. [Croda Inc.; Croda Chem. Ltd.] Maize gluten amino acids, sodium chloride; conditioner for skin and hair creams and lotions; humectant for cosmetics, pharmaceuticals.

Aminol. [Chem-Y GmbH] Alkyl polyglycol ether carboxylate; thickening agent.

Aminol. [Finetex] Fatty acid alkanolamides; softening agent, emulsifier, soap additive, foam booster/stabilizer, thickener; cosmetic emulsions, household cleaners, industrial applics.

Amino-Silk SF. [Maybrook] Silk amino acids; substantivity agent, moisturizer for elegant skin and hair care preps.

Aminox®. [Uniroyal] Diphenylamineacetone reaction prod.; antioxidant for CR, NBR, NR, SBR, latexes, insulated wire.

Aminoxid. [Goldschmidt; Goldschmidt AG] Amine oxides; detergent, emulsifier, wetting agent, softener, foam stabilizer for detergents, cosmetics, pharmaceuticals.

Amiox. [Alban Muller] Natural antioxidizing agent.

Amipal. [Takemoto Oil & Fat] Nonionic softener.

Amisoft. [Ajinomoto] Glutamates; emulsifier, emollient for cosmetics.

Amisol. [Amico] Lecithin and blends.

Amiter. [Ajinomoto; Nihon Emulsion] Glutamates; emulsifier, emulsion stabilizer for cosmetics.

Amizyme. [PMP Fermentation Prods.] Alpha-amylase; enzyme with starch liquefying and heat stability properties; for paper coatings and sizes.

Amjet A-4. [Am. Emulsions] Leveling carrier for pressure dyeing of stock yarn and piece goods.

Amlev. [Am. Emulsions] Leveling agents, retarding agents for textiles.

Amlight M-2. [Am. Emulsions] Inorganic blend; peroxide bleaching stabilizer for cellulosic fibers and blends.

Amlube. [Am. Emulsions] Textile lubricant.

Ammo. [FMC/Ag] Insecticide.

Ammonium Cumene [Sulfonate 60. [Hüls Am.] Ammonium cumene sulfonate; hydrotrope, solubilizer, coupler for liq. detergents.

Ammonium Stearate 33%. [Hart Chem. Ltd.] Ammonium stearate; foam stabilizer, hand modifier for frothed latex systems.

Ammonyx® 4, CA, CETAC, KP, LKP. [Stepan] Quaternary ammonium chlorides; emulsifier, conditioner, softener, emollient for cosmetics.

Ammonyx® CDO, CO, DMCD, LO, MCO, MO, OAO, SO. [Stepan] Amine oxides; wetting agent, foamer, foam stabilizer, conditioner, visc. builder for personal care prods., household and janitorial prods.

Amoco®. [Amoco] Petroleum additives.

Amoco® Ethylene/Propylene Copolymers. [Amoco] Ethylene/propylene copolymers; profile extrusion, blow molding, film applics.

Amoco® Polybutene. [Amoco] Polybutene; tackifier, strengthener, extender in adhesives, as plasticizer for rubber, as vehicle and binder for coatings, as cling additive for LLDPE stretch wrap films, as reactive intermediate for specialty chemicals, in caulks, sealants, and glazing compds.

Amoco® Polypropylene. [Amoco] Polypropylene resin; injection molding, extrusion, fiber/film, powder grades.

Amoco® Polystyrene. [Amoco] Polystyrene resin; crystal, high-impact grades. for food containers, automotive parts, inj. molding.

Amoco® Resin 18. [Amoco] Poly-α-methylstyrene; extrusion and molding process aid in ABS, PVC, CPVC, and semirigid vinyl, thermoplastic urethanes, molded rubbers, and thermoplastic elastomers; modifier and reinforcer in adhesives, thermoplastic powd. coatings, hot-melt coatings.

Amodel®. [Amoco] Polyphthalamide; semicrystalline thermoplastic resin; for automotive under-the-hood parts, gears, bearings, and industrial parts.

Amojell Petrolatum. [Amoco Lubricants] Petrolatum.

Amolit. [Alpha Calcit Fullstoff GmbH] Very pure fine chalk.

Amollan®. [BASF AG] Wetting agents, emulsifiers, lubricants for leathers, furs.

Amonyl. [Seppic] Betaines; amphoteric detergents for shampoos; quats. for use as algicide, fungicide, germicide.

Amorphous Microcrystalline Wax. [Int'l. Wax Refining] Antioxidant for rubbers.

AMP. [Angus] 2-Amino-2-methyl-1-propanol; emulsifier, catalyst; dispersant for pigments and latex paints; corrosion inhibitor; stabilizer; resin solubilizer.

Ampacet. [Ampacet] Antistats, flame retardants, stabilizers for plastics.

Ampal®. [Ciba-Geigy GmbH] Unsat. polyester; molding compds.

AMPD. [Angus] 2-Amino-2-methyl-1,3-propanediol; pigment dispersant, neutralizing amine, corrosion inhibitor, acid-salt catalyst, pH buffer, chemical and pharmaceutical intermediate, solubilizer.

Amphionic. [Rhone-Poulenc Ltd.] Ampholytic biocide; chelating agent, solubilizer for nonionic surfactants.

Amphisol. [Bernel] Cetyl phosphate salts; anionic emulsifiers.

Ampho. [Karlshamns] Amphoteric surfactants; detergent, wetting and foaming agents, solubilizer.

Amphocerin®. [Henkel/Emery/Cospha; Henkel KGaA] Fatty alcohols and oil blends; cream base for creams and ointments; emulsifier.

Ampholak. [Berol Nobel] Amphoteric surfactants; foam booster, visc. modifier, thickener, wetting agent for industrial and household cleaners, toiletries.

Ampholak 7TX. [Amphoterics Int'l.] Sodium carboxymethyl tallow polypropylamine.

Ampholan®. [Harcros] Betaines; amphoteric detergent, foaming and stabilizing agent for personal care prods., cleaners, cement, industrial applics.

Ampholyt. [Hüls AG] Glycinates or betaines; amphoteric surfactant for cosmetics, shampoos, detergents.

Ampholyte KKE. [Berol Nobel] N-Cocoalkylamiopropionic acid; used in detergents and toiletries.

Amphomer®. [Nat'l. Starch & Chem.] Octylacrylamide/acrylates/butylaminoethyl methacrylate copolymer; hair fixative resin for setting lotions, conditioners.

Amphonyl. [Chemag] Sodium amphopropionates.

Amphoram. [Ceca SA] Amphoteric coco derivs.; detergent, bactericide, emulsifier, foaming agent; for cosmetics.

Amphoset. [Mitsubishi Int'l.] Methacryloyl ethyl betaine/methacrylates copolymer.

Amphosol. [Stepan; Stepan Europe] Betaines; visc. builder, foam booster, base for cosmetics, liq. detergents; lime soap dispersant.

Amphosperse. [Grant Industries] Dispersant, wetting agent, penetrant, emulsifier, detergent for synthetics and cellulosics and dyes.

Amphoteen 24. [Berol Nobel] C12-14 alkyl dimethyl betaine; amphoteric for low-irritation shampoos, cleansers, hard surface cleaners, vehicle cleaners.

Amphotensid. [Zschimmer & Schwarz] Amphoterics; detergent, wetting agent for personal care prods., liq. cleaners.

Amphoterge®. [Lonza] Amphoteric surfactants; wetting agent and detergent for shampoos, cleansers, dishwashing; textile softener.

Amphoteric. [Exxon/Tomah] Amphoterics; detergent, foam booster/stabilizer, wetting agent, coupler for liq. detergents; defoamer for paints; corrosion inhibitor in metalworking lubricants; leather lubricant.

Ampitol. [Dexter] Softeners for textile and paper industries.

AmpliWax PCR Gems. [Perkin-Elmer] Specially formulated wax beads; enhances PCR specificity and replaces mineral oil.

Ampro. [W.R. Grace] Amines; solvent.

Amquest. [Am. Emulsions] Sequesterants for textile dyeing; dye solubilizer for carpet dyeing.

AMR. [Petrokem] Insecticides.

Amres®. [Georgia-Pacific] Polyamide wet strength resin; for paper/paperboard industry.

Amresco. [Amresco] Reagent chemicals, ultrapure chemicals, specialty chemicals, clinical reagents.

AMS-33. [Hefti Ltd.] Glycol stearate; nonionic emulsifier for cosmetics; opacifier, thickener for shampoos, foaming baths, lotions.

Amsco. [Unocal] Hydrocarbon solvents; solvents for NR, SR processing.

Amsco-Res. [Unocal] Emulsion polymers.

Amset. [Georgia-Pacific] Ink fixative.

Amsil N. [Nutex] Long chain quaternary/amino-functional silicone emulsion; softener for textiles.

Amsoft. [Am. Emulsions] Softener for synthetic and natural fibers.

Amsoft. [Nutex] Quaternary compds.; softener for textiles.

Amsol GMS. [Am. Emulsions] Nonionic surfactant blend; antigelling agent for conc. dye sol'ns.

Amsperse. [Amspec] Antimony oxide dispersions; flame retardants for plastics.

Amsperse 109. [Am. Emulsions] Dispersing assistant for disperse dyes.

Amsperse F-R 21. [Glo-Tex] Decabromodiphenyl oxide and antimony trioxide dispersion; durable flame retardant for natural, synthetic, and blended fabrics.

Amstar. [Amspec] Antimony trioxide; synergist in flame retardant rubber and polymer compds.

Amsul. [Harcros] Amine dodecylbenzene sulfonate; emulsifer for industrial and agric. formulations.

Amterge TC. [Am. Emulsions] Organic solvents, detergents, and emulsifiers blend; scouring and cleaning aid for textile processing.

Amtex. [Prochem] Cleaner, degreaser for textile processing equipment.

Amwet. [Am. Emulsions] Wetting agent, detergent, penetrant for textile processing.

Amyl Zimate®. [R.T. Vanderbilt] Zinc diamyldithiocarbamate; rubber accelerator.

Amyx. [Clough] Quats. and amine oxide surfactants; mild foam booster/stabilizer for personal care, household, and janitorial prods.

AN-. [Compounding Tech.] Amorphous nylon, glass and carbon reinforced; composites with low moisture absorp., good dimensional stability.

Anadoucissant. [Ceca SA] Fiber softeners.

Anar. [Anar] Anionic surfactant; detergent.

Anatol. [Lanaetex Prods.] Lanolin alcohol; w/o emulsifier, stabilizer, softener, emollient, gelling agent, thickener, plasticizer, moisturizer for absorption bases, cosmetics, cleansing preps.

Ancadride®. [Pacific Anchor] Anhydrides; curing agent; intermediate for alkyds, plasticizers, insect repellents, rust inhibitors, hardener in epoxy resins.

Ancamide®. [Air Prods.; Pacific Anchor; Anchor UK] Amido amine and polyamides; curing agents for epoxy resins.

Ancamine®. [Air Prods.; Pacific Anchor; Anchor UK] Aliphatic, cycloaliphatic and aromatic polyamines; curing agents for epoxy and urethanes, antioxidants, stabilizers, catalysts, intermediates, corrosion inhibitor, epoxy hardening agent.

Ancaref®. [Air Prods.] Unidirectional fibers for composites.

Ancarez®/ [Pacific Anchor] Epoxy resin and flexibilizer.

Anchoid. [Anchor UK] Sodium naphthalene formaldehyde sulfonate; dispersant for latex, cement, concrete, ceramics.

Anchor. [Pacific Anchor] Epoxy curing agents, accelerators, antioxidants for rubber, plastics, latex, cements, adhesives, carpet backings, footwear, cable insulation, mechanical goods; peptizer.

Ancomer®. [Air Prods.] Acrylic monomers and oligomers.

Ancor®. [Air Prods.] Amine/acetylenic alcohol; corrosion inhibitor for ferrous and ferric environments; steel pickling; electroplating brightener.

Anderol. [Hüls Am.] Synthetic lubricants.

Andrez. [Anderson Development] High styrene butadiene resin.

Andur. [Anderson Development] Polyurethane prepolymers.

Anew X. [Alex C. Fergusson] Aluminum brightener.

Anfomul. [Croda Chem. Ltd.] Emulsifier for emulsion explosives.

Angamol 75. [Lubrizol] Zinc dialkyl dithiophosphate; antiwear agent for hydraulic fluids; lubricant.

Anglamol. [Lubrizol] Gear oil additives.

Anhydrol. [Union Carbide] Industrial solvents.

Anhydrous Lanolin. [Westbrook Lanolin] Refined wool fat; for cosmetics, pharmaceuticals.

Anhydrous Lanolin HP-2050. [Henkel/ Emery/Cospha] Lanolin; ointments, leather finishing, soaps, face creams, facial tissues, hair set and sun-tan preps.

Aniline. [Uniroyal] Intermediate.

Anionyx® 12S. [Stepan] Disodium oleamido PEG-2 sulfosuccinate; detergent for personal care prods.

Annonyx SO. [Stepan] Stearamine oxide; conditioner, emulsifier, visc. modifier with wetting, foaming, foam stabilization properties.

Anox 20, PP18. [Enichem Synthesis SpA] High m.w. phenolic antioxidant.

Anox TB Blends. [Enichem Synthesis SpA] Antioxidant blends.

Anoxsyn 442. [Atochem N. Am.] Bis alkyl sulfide; antioxidant for ABS, polyester, PE, PP, PS, EVA.

Anquamine®. [Pacific Anchor] Modified liq. polyamide; for water-dispersible coatings.

Ansar. [ISK Biotech] Postemergence herbicide.

Anscor. [Steetley Magnesia Prods. Ltd.] Magnesium oxide.

Ansilex. [Engelhard] Anhydrous kaolin; extender.

Anstac. [CDC Int'l.] Antistat for plastics.

Anstex. [Toho Chem. Industry] Phosphate; antistat for synthetic fibers, plastics.

Antara®. [Rhone-Poulenc Surf.] Phosphate acid esters; lubricant, extreme pressure additive, rust inhibitor, wetting agent, emulsifier; for cutting fluids, hydraulic fluids, rolling oils.

Antarol. [Aquatec Quimica SA] Mineral oil/organic blends; antifoamer for metalworking fluids, synthetic adhesives, latex systems, agric. formulations, mining, polymerization, paints, fermentation industries.

Antaron. [ISP] Surfactants and PVP copolymers.

Antarox®. [Rhone-Poulenc Surf.] Linear aliphatic and aromatic polyethers; detergent, wetting agent for metal cleaning, rinse aids, textiles, household and industrial cleaners.

Antex. [Lanaetex Prods.] Glycerin, propylene glycol blends.

Anthiol®. [Pacific Anchor] Epoxy resin and flexibilizer.

Anthium Dioxcide®. [Int'l. Dioxcide] Chlorine dioxide; broad spectrum biocide, preservative.

Anthosin®. [BASF AG] Based on acid dyes; for paper coloring in papermaking.

Anthoxan. [Henkel/Emery/Cospha] 4-Isopropyl-5, 5-dimethyl-1, 3-dioxane; raw material for herbal fragrances.

Antiblaze®. [Albright & Wilson Am.] Halogenated organic phosphates and phosphonate esters; flame retardants.

Anticake 17. [Exxon/Tomah] Amine; anticaking agent for fertilizers.

Antichlor O. [Manufacturers Chems.] Antichlor for last rinse following chlorine bleach.

Anticurl PVT. [Dooley] Anticurl agent for processing knit fabrics.

Antiflana. [Croxton & Garry Ltd.] Fire retardant masterbatches.

AntiFoam. [Yorkshire Pat-Chem] Silicone and nonsilicone defoamers; for textile wet processing, dyeing, printing, finishing.

Antifoam. [Am. Emulsions] Defoamers and antifoams for textile processing.

Antifoam. [Chem-Tex Labs] Silicone or nonsilicone defoamers for textile processing or effluent control.

Antifoam. [CNC Int'l.] Silicone and nonsilicone emulsions; antifoam agents, foam suppressants.

Antifoam. [Glo-Tex] Defoamers for textile bleaching, dyeing, and finishing operations.

Antifoam. [Grant Industries] Defoaming agents for textile processing systems for all types of fabric.

Antifoam. [Harcros] Silicone blend; defoamer for industrial processing, carpet cleaning, wood pulping, paper coating, textile mfg.

Antifoam. [Manufacturers Chems.] Antifoams for textile processing.

Antifoam. [Reliance Chem. Prods.] Defoaming agents for scouring, dyeing, and finishing baths.

Antifoam. [Soluol] Silicone defoamers for textile applics.

Anti-Foam. [Stewart Hall] Defoamers.

Antifoam 396, 488, 2382. [Reilly-Whiteman] Silicone and nonsilicone defoamers for textile applics.

Antifoam 906, M-90. [Gresco Mfg.] Silicone emulsion; eliminates foam externally and internally; for textile applics.

Antifoam BTB. [Ivax Industries] Silicone emulsion; foam control agent for bleach and scour baths, becks, jets, package dyeing and finishing, free rinsing.

Antifoam E-20. [Kao] Emulsified denatured silicone; defoamer for tech. applics.

Anti-Foam TP. [BASF AG] Antifoam for pigment printing of textiles.

Antifog/Antistatic MCG. [Merix] Antistat for plastics.

Antifume. [CNC Int'l.] Finishing agent for cellulose acetate to protect dyed fabric from fading.

Antifume T. [Sybron] Self-emulsifiable aromatic amino compd.; gas fading inhibitor for textile processing.

Antil® 141. [Goldschmidt; Goldschmidt AG] PEG-55 propylene glycol oleate; thickener for surfactant systems; solubilizer for essential oils, fragrances.

Antilux. [Rhein Chemie] Paraffins and micro waxes; sun-checking, anti-weathering and ozone protective wax for tech. molded and extruded rubber goods, cellular rubber, tires, belting; lubricant.

Antimigrant. [Chemonic Industries] Synthetic and/or alginate gums; antimigrant yielding even dyeing of soluble or insoluble dyes in continuous or pad procedures.

Antimigrant 2L. [Eastern Color & Chem.] Propoxylated organics sol'n.; antimigrant for pigments during drying of pigment padded fabric.

Antimigrant 80. [Chem-Tex Labs] Dye migration inhibitor for use in pigment dyeing.

Antimigrant 100. [Auralux] Antimigrant for pigments during padding and drying operations; promotes level dyeing.

Antimigrant 157, MC-10. [Catawba-Charlab] Antimigrant for continuous dyeing of textiles.

Antimigrant 1020. [Applied Textile Tech.] Synthetic antimigrant for washing directs.

Antimigrant A. [CNC Int'l.] Padding vehicle to prevent migration of dyes during textile processing.

Antimigrant C-45, MA, SA. [Yorkshire Pat-Chem] Antimigrants for dyeing.

Antimussol WLN Liq. [Sandoz] Non-

ionic silicone-free defoamer for all textile wet processing.

Antiox. [Campine SA] Antimony oxide.

Antioxidant. [Sovereign] Antioxidants for rubbers.

Antioxidant. [Uniroyal] Antioxidant for plastics.

Antioxidant 235. [Akrochem] 2,2'-Methylenebis (4 methyl-6-t- butylphenol); antioxidant for elastomers for medical equip., rubber thread, latex.

Antioxidant 425. [Am. Cyanamid] 2,2'-Methylenebis (4 ethyl-6-t- butylphenol); antioxidant for molded prods., NR, SBR and CR; stabilizer for acrylics and ABS.

Antioxidant 431, 449N, 445. [Uniroyal] Nonstaining antioxidant for latex, foam, molded and mechanical goods, sundries, polymer mfg.

Antioxidant G-2. [Provital] Min. oil, tocopheryl acetate, ascorbyl palmitate, lecithin.

Anti-Oxidant MBP-5P. [Aceto] 2,2'-Methylenebis (4-methyl-6-t-butyl phenol); antioxidants for natural and synthetic rubbers.

Antioxydant NV 3. [BASF AG] Alkylphenol; for tire industry and tech. goods.

Antiprex. [Allied Colloids] Antiscalant.

Anti-Scale S-MS. [Stewart Hall] Corrosion and scale inhibitor for cooling towers.

Antisettle. [Ketro A/S] Paint additive.

Antislip 239. [Piedmont Chem. Industries] Colloidal silica; stabilizer for yarns and fabric preventing fiber shifting and seam slippage.

Antislip ASE. [Yorkshire Pat-Chem] Improves seam slippage and pile retention on textiles.

Antistat. [Am. Emulsions] Antistat for textiles.

Antistat. [CNC Int'l.] Antistatic agents for textile use.

Antistat. [Cooper] Phosphate esters or various glycols; antistats enhancing spinning performance.

Antistat. [Hunt] Blended phosphate esters; antistat for synthetic yarns.

Antistat. [Polycom Huntsman] Internal

antistat for polyolefins.

Antistat. [Specialty Chem.] Antistatic agents for yarn and fabric processing.

Antistat BH-201. [Fibertint & Chems.] Phosphated multi-chain alcohol salt; cohesive type antistat for synthetic and natural fibers and fabrics.

Antistat BT, GW. [Yorkshire Pat-Chem] Antistats for textiles.

Antistat Conc. [Ferro] Antistats for plastics.

Antistat Conc. [HiTech Polymers] Antistats for plastics.

Antistat Conc. [Santech] Antistats for plastics.

Antistat JS. [Reilly-Whiteman] Cationic long chain fatty amide; antistatic agent for textiles.

Antistat PC. [Chem-Tex Labs] Antistatic agent for electrical charge build-up on yarns and fibers of natural or synthetic substrates.

Antistat RD2-2351. [Marlowe-Van Loan] Quaternized fatty amine; antistat for textile fibers.

Antistat S. [Aerochem] Amine condensate; liq. antistatic agent for synthetic fibers.

Antistatic Agent 575. [E.F. Houghton] External antistat for plastics.

Antistatic Coating 1412. [Coating Systems] External antistat for plastics.

Anti-Static Con #79. [Merix] Antistats.

Antistatic KN. [3-V] Antistat for plastics.

Antistatic Spray. [Price-Driscoll] External antistat for plastics.

Antistatico KN. [Sigma Prodotti Chimici] Stearamidopropyl dimethyl-β-hydroxyethyl ammonium nitrate; external antistat for PVC, polystyrenes.

Antistaticum. [Rhein Chemie] Fatty alkyl ether of polyethylene glycol; antistatic plasticizer for NR and syn. rubber, PVC plastics.

Antistick 1. [High Point] Complex wax dispersion; additive to reduce finish build-up on rolls, dry cans, etc.

Antistick 12. [Yorkshire Pat-Chem] Lubricant and release emulsion to prevent finish buildup on textile processing equip.

Antisun®. [Frank B. Ross] Antiozonant, anticracking and sunchecking wax for rubber industry.

Anti Terra. [BYK-Chemie USA] Wetting/dispersing additives.

Antitrack. [Ditta Salt SpA] Melamine powds.

Anti UVA. [Aceto] 2-Hydroxy-4-methoxy benzophenone; sunscreen agent.

Anti UVK. [Aceto] t-Butyl phenyl salicylate; UV stabilizer for plastics.

Antiwick 87. [Morton Int'l.] Stops wicking on prefinished polyester/cotton pigment prints.

Antiwick OP, WP. [Yorkshire Pat-Chem] Antiwick, thickener for pigment print pastes.

Antozite®. [R.T. Vanderbilt] Phenylene diamines; antiozonant for rubber industry.

Antracol®. [Bayer] Propineb; fungicide for crops.

AO. [Exxon/Tomah] Amine oxide; foam booster and stabilizer for industrial and household detergents.

AO-. [Ciba-Geigy/Additives] Antioxidants.

AOB. [Allied-Signal] Methylethyl ketoximes.

AOM. [Climax Performance] Calcium molybdate; flame retardants for plastics.

Aosyn. [Hüls Am.] Specialty lubricants.

Apachi®. [Air Prods.] Fuel gas.

APCI PVC. [Air Prods.] PVC; for inj. molding, pipe, film, sheet, wire and cable, and medical applics.

Apco DA. [Apollo] Surfactant/deaerator blend; wetting agent, penetrant for textile processing.

Apcoclear SP. [Apollo] Scouring agent, afterclear for removing excess dye.

Apcolev. [Apollo] Leveling agent for textile dyeing.

Apcon 255. [Apollo] Nonionic low foaming surfactant; scouring agent for preparation and desizing operations.

Apcoquest. [Apollo] Chelating agent for scouring, kier boiling, dyeing and bleaching operations.

Apcosist A-1. [Apollo] Water softener

for textile processing.

Apcosolve AS. [Apollo] Low foaming solvent scouring agent for synthetic and cellulosic fibers.

Apcosperse. [Apollo] Dispersant, scouring agent for textiles.

Apec®. [Bayer; Miles] Polycarbonate; thermoplastic resin for automotive elec. components, lamp housings, lamp reflectors, domestic appliance components.

Apescour 1609. [Apex] Scouring agent for cotton and polyester/cotton blends.

Apex. [Evans Clay] Hydrated aluminum silicate; reinforcing filler in NR, SR, latexes, resins, plastics; color and processing aid; used in coated materials, footwear, flooring, V-belts, rolls, belts, matting, molded and extruded goods, foam goods, o-rings, seals, sundries.

Apex Backote FR 316. [Apex] Organic/ inorganic blend; durable flame retardant backcoating for synthetic fabrics and blends.

Apex Defoamer 421. [Apex] Nonsilicone hydrocarbon/wax emulsion; defoamer for jet dyeing.

Apex Flameproof. [Apex] Flame retardants for textiles.

Apex Mordantine. [Apex] Fixative for acid dyes on nylon.

Apex Pentrapex 1923. [Apex] Penetrant for use with Apex salt type fire retardants to improve penetration.

APG® Series. [Henkel/Emery] Decyl and lauryl polyglucose; degreaser, emulsifier, dispersant, wetting agent, visc. modifier, detergent active; for general purpose and hard surface cleaners.

Aphrogene 5001. [ICI Surf. UK] Silicone emulsion; antifoam for textile dyeing and finishing.

Aphrogene Jet. [ICI Surf. UK] Polydimethylsiloxane emulsion; surfactant for dyeing in jet, beam, and package machines.

Apicell. [API Applicazion Plastiche Industriali] Polyol for polyurethane.

Apicerol 2/014081. [Dragoco] Petrolatum, mineral oil, beeswax, polyglyceryl-2 sesquioleate, lanolin, isopropyl palmitate.

Apicolor. [API Applicazion Plastiche Industriali] Masterbatches.

Apifac. [Gattefosse; Gattefosse SA] Polyglyceryl-2 isostearate; nonionic base for w/o creams.

Apifil®. [Gattefosse; Gattefosse SA] PEG-8 beeswax; self-emulsifying nonionic base for o/w emulsions in cosmetics and pharmaceuticals.

Apiflex. [API Applicazion Plastiche Industriali] PVC.

Apilon 52. [API Applicazion Plastiche Industriali] Linear polyurethane.

Apirex. [API Applicazion Plastiche Industriali] Isocyanates.

Apizero. [API Applicazion Plastiche Industriali] Crosslinkable polyolefins.

Apocleaner EC. [Apollo] Nonsolvent heavy-duty cleaner for high temp. textile dyeing equip.

Apollo Anti-Spot. [Apollo] All-purpose solvent cleaner for fabrics.

Apollo Chlor 200. [Apollo] General purpose water softener and chlorine neutralizer.

Apollo Extender 119. [Apollo] Water repellent fluorocarbon extender for synthetic and natural fibers.

Apolloscour 951. [Apollo] Liq. after-soap for cleaning loose reactive dyes from cotton and cotton/polyester blends.

Apollo Stabilizer ES-1. [Apollo] Organic peroxide stabilizer for non-silicate systems.

Apparelock. [Bostik Div./Emhart] Thermoplastic adhesives.

Appretan Grades. [Hoechst AG] Polymer dispersions; wash-fast finishing agents.

Appryl. [Atochem Deutschland GmbH]; Atochem UK] Polypropylene homopolymers and copolymers.

Apscom. [Akzo Engineering Plastics BV] Specialty compds. for conductivity, lubrication, flame retardance, reinforcement for polymers.

Aptex Product. [Applied Textile Tech.] Cleaners, lubricants, dispersants, leveling agents, wetting agents, afterscours for textile processing.

Aptex Wetter SP-6. [Applied Textile Tech.] Nonionic wetting agent for garment and beck dyeing.

Apyral. [Wilfrid Smith Ltd.] Aluminum trihydroxide; fire retardant.

Aquabase. [Westbrook Lanolin] Fatty alcohols and PEG ester; base for o/w emulsions.

Aquabead. [Micro Powders] Water beading/repellent wax.

Aquabrom. [Great Lakes] Swimming pool sanitizer.

Aquacat. [Ultra Additives] Ink, paint, and varnish driers.

Aquadag. [Acheson Colloids] Colloidal graphite in water; lubricant additive.

Aquafilm. [CNC Int'l.] Fluorocarbon; water and greaseproofing for specialty paper and nonwovens.

Aquafloc. [W.R. Grace/Dearborn] Flocculant.

Aquafoam. [Aquaness] Ethoxy ether sulfates; foaming agent for foam drilling.

Aquagard. [Soluol] Silicone emulsions; curing water repellent emulsion for fibers and fabrics.

Aqua Hue®. [Blackman Uhler] Textile auxiliaries, binders, pretreats, softeners.

Aqualease. [George Mann] Silicone emulsions; release agents for casting molds, cast urethane elastomer and integral skin urethane foams, epoxies, filled polyester, vinyl ester resins.

Aqualine. [Dexter/Frekote] Water-based release agent; semipermanent mold release interface for natural and synthetic rubber compds., thermosets and thermoplastics.

Aqualipid. [Central Soya] Lecithin; for marine animal feeding.

Aqualizer EJ. [Kolmar Labs] Poly-amino sugar condensate, urea.

Aqualon®. [Aqualon] Carboxymethyl cellulose prods.; binder, thickener, stabilizer, suspending agent, film-former, rheology control aid, fabric size, water-retention aid used for adhesives, ceramics, coatings, detergents, lithography, paper, textiles, and tobacco.

Aqualose. [Westbrook Lanolin] Ethoxy-lated lanolins and lanolin derivs.; emollient, emulsifier, plasticizer, solubilizer for personal care prods., perfumes, germicides; conditioner, superfatting agent.

Aqualox®. [Alox] Surfactant, corrosion inhibitor, lubricant, antiwear additive for metalworking formulations.

Aquametal. [Industeel] Iron complex and alloy; creates chelation sol'n. for organic plants when added to water.

Aquamid. [Aquafil SpA] Polyamide 6; for inj. molding and extrusion applics.

Aquamix. [Harwick] Halogenated hydrocarbon dispersions, chlorinated paraffins emulsions; flame retardants, stabilizers, accelerators, activators, vulcanizing agents, antioxidants, tackifiers for plastics.

Aquamolin Brands. [Hoechst UK] Tetrasodium EDTA; sequestering agent.

Aquamul. [Aquaness] Modified fatty acids; oil-sol. emulsifiers for drilling fluids.

Aquanel®. [Schenectady] Water-borne polyesters; insulating varnishes.

Aquanox. [Baker Perf. Chem.] Demulsifiers.

Aqua Pac. [Aqualon] Polyanionic cellulose.

Aquapel®. [Hercules] Reactive sizes; for paper industry.

Aquaperle D34. [ICI Surf. UK] Paraffin wax/aluminum and zirconium salt emulsion; waterproofing agent for natural and synthetic fibers with resistance to weathering.

Aquaphil K. [Westbrook Lanolin] Lanolin, lanolin alcohol; emollient, emulsifier.

Aquaplus. [Miles/Organic Prods.] Textile dyes and pigments.

Aquaplus. [Rohm & Haas] Polymer for floor care.

Aquapoly. [Micro Powders] Water-compatible polyethylene.

Aquasan. [Reilly-Whiteman] Water repellents and fluorochemical extenders for textiles.

Aquasol. [Chemetics Labs] Aloe derivs.

Aquasol®. [Am. Cyanamid/Textiles] Sulfonated castor oil; wetting, penetrating, leveling agent for dyeing textiles; emulsifier for self-scouring fiber and yarn lubricants.

Aquasorb®. [Aqualon] Carboxymethylcellulose; absorbent for urine, blood, body fluids for feminine hygiene prods., medical disposables.

Aquasperse. [Hüls Am.] Color dispersants.

Aquastab. [Eastman] Stabilizers protecting materials from oxidative and uv degradation, static buildup, acidic catalyst residues.

Aquasurf. [Baker Perf. Chem.] Surfactants.

Aquatac®. [Arizona] Glycerol ester of rosin aq. dispersion; water-sol. adhesive tackifier.

Aquatec®. [Arizona] Water-based resin; freeze/thaw stable resin for adhesive applics., labels.

Aqua-Tein C. [Maybrook] Animal collagen amino acids and acetamide MEA; moisturizer, substantivity agent for hair and skin care prods.

Aquathane. [CNC Int'l.] Fluorochemical polyurethane copolymers; textile finishes.

Aquathol. [Atochem N. Am.] Aquatic herbicide.

Aquatreat® AR. [Alco] Acrylic acid derivs.; dispersant, sludge conditioner, antiscalant for water treatment.

Aquatreat® DNM, KM, SDM. [Alco] Dithiocarbamate salts; biocide, fungicide, algicide for water treatment, paper and sugar mills, petrol. applics.

Aquazym®. [Novo Nordisk] Bacterial diastase; enzyme for starch desizing in textile industry.

Aqucar Microbiocides. [Union Carbide] Glutaraldehyde; biocide for water treatment.

Aquene 212. [Leatex] Organic amino deriv.; chelating agent for iron and other polyvalent metal ions; prevents rust stains or iron spots in peroxide bleach systems with optical brighteners.

Arakote®. [Ciba-Geigy/Plastics] Polyester resins.

Arakyd. [Arakawa] Alkyd resins.

Araldite®. [Ciba-Geigy/Plastics; Ciba-Geigy Plastics UK; Ciba-Geigy GmbH] Solid and liq. epoxy resins and adhesives.

Aramide. [Aquaness] Alkanolamine condensate; visc. builder for all-purpose cleaners, emulsifiers, and lubricants.

Aramide. [Baker Perf. Chem.] Surfactants.

Aranox®. [Uniroyal] Amines; stabilizers, antioxidants for polypropylene, polyethylene, EVA, polyamides; used in fabric proofing, clothing, wire insulation, latex, foam.

Arasan. [DuPont] Seed protectants.

Arasorb. [Arakawa] Super-absorbent polymer.

Aratex. [Van Den Bergh Foods] Hydrogenated vegetable oil; icing stabilizers; for dry bakery mixes, syrups, donut glazes.

Aratronic®. [Ciba-Geigy/Plastics] Epoxy and amine resins; electronic grades.

Aravite®. [Ciba-Geigy Plastics UK] Cyanoacrylate; adhesive.

Arazate®. [Uniroyal] Zinc dibenzyl dithiocarbamate; accelerator for NR, SBR, latexes, cements, foam.

Arbocaulk. [Adshead Ratcliffe] Acrylic/emulsion sealant.

Arbocrylic. [Adshead Ratcliffe] Acrylic sealant.

Arboflex. [Adshead Ratcliffe] Butyl glazing compd.

Arbokol. [Adshead Ratcliffe] Polysulfide; sealants.

Arbosil. [Adshead Ratcliffe] Silicone; sealants.

Arbreak. [Aquaness] Polyol, resin ester oxyalkylates; emulsion breakers.

Arbyl. [Grünau] Surfactants for textiles.

Arbylen. [Grünau] Surfactant for textiles.

Arc Ease. [Am. Resin] Release agent.

Arcel. [Arco] Polyethylene copolymers; resins for molded resilient foam pkg.

Archer. [Witco] General agric. lubricating oils and greases.

Arc Kleer. [Am. Resin] Adhesives, bonding resin.

Arconate® Propylene Carbonate. [Arco] Solvent for pigments and dyes; extracting agent; intermediate for organic syntheses.

Arcopure. [Arco]

Arcosolv®. [Arco] Propylene glycol ether or ether acetate solvents; for coatings, cleaners, inks, agric. prods., electronic, cosmetics, chemical intermediate applics.

Arcton 13. [ICI Am.] Chlorotrifluoromethane; refrigerant.

Ardel®. [Amoco] Polyarylate resin; thermoplastic engineering polymer for inj., extrusion, and blow molding at high temps.; for electronics/elec., mech., lighting/optical, medical applics., household elec. goods.

Ardet. [Baker Perf. Chem.] Surfactants.

Ardril. [Aquaness] Suspending agents, filtration control agents for invert mud systems.

Ardril. [Baker Perf. Chem.] Drilling additives.

Areca. [Polychim] Masterbatches and compds.; for thermoplastic elastomers, polyolefins.

Aremsan C40. [Ronsheim & Moore] Cocamidopropyl dimethyl benzyl ammonium chloride; bactericide for prep. of hair conditioners.

Aremsol. [Ronsheim & Moore] Lauryl sulfate salts; shampoo materials.

Aremul. [Arol Chem. Prods.] Surfactant blends; detergent, emulsifier.

Arflow 168. [Aquaness] Sodium alkyl sulfate; paraffin wash or remover.

Arfoam. [Aquaness] Blend; foaming agents to remove water and drilled solids during drilling operations.

Arfoam. [Baker Perf. Chem.] Foamers.

Argidone®. [UCIB] Arginine PCA. moisturizing adjuvant; activates cell metabolism; stabilizes water; for nutritive and regenerative creams and lotions for skin care.

Argobase. [Westbrook Lanolin] Lanolin extracts; emollient, emulsifier, stabilizer, absorption base for personal care prods.

Argo Brand. [Corn Prods.] Corn starch; source of glucose; filler in baking pow-

der; thickening agent in food prods.; adhesives, coatings; additive in plastics.

Argon. [BASF AG] Flushing gas for degassing of molten metal; blanketing gas for welding, casting, melting, forging, and plating of metals.

Argonol. [Westbrook Lanolin] Lanolin oil derivs.; emulsifier, emollient, lubricant for personal care prods.

Argowax. [Westbrook Lanolin] Lanolin alcohols and acids; gelling agent, emulsifier.

Argus. [Witco/Argus] Thiodipropionates; Antioxidants for plastics and rubber, additive for high-pressure lubricants and greases, plasticizer and softening agent, antioxidant for edible fats and oils.

Ari Cel. [Astro Industries] Catalysts for thermoplastic or thermosetting resins; for textile industry.

Aridry. [CNC Int'l.] Durable and nondurable water repellents for textiles.

Arikrome S. [CNC Int'l.] Liq. chrome complex; water repellent for paper industry.

Aristoflex. [Hoechst Celanese] Vinyl acetate/crotonic acid copolymers; hair fixative for aerosols, hair sprays, setting lotions, conditioners.

Aristol. [Pilot] Substituted C20-24 benzene; lubricant, lube oil additive; chemical feedstock for sulfonation to produce emulsifiers and corrosion preventatives.

Aristonate. [Pilot] Oil-sol. petroleum sulfonates; for formulating drycleaning soaps, cutting oils, textile oils, leather oils, rust preventive, fuel oil compositions; ore flotation; emulsifier for agric. sprays.

Arite. [Consos] Sodium hexametaphosphate; chelating agent and water softener for textile use; prevents precipitation of lime soaps.

Arizona. [Arizona] Disproportionated rosin derivs.; emulsifier, detergent, wetting agent; plasticizer, softener, tackifier for rubber industry.

Arklone. [ICI Am.] 1,1,2-Trichoro-1,2,2,-trifluoroethane and derivs. refrigerant; dry-cleaning solvent; fire extinguishers; mfg. of chlorotrifluoroethylene; blowing agent; polymer intermediate; solvent drying.

Arkofil. [Hoechst AG] Synthetic polymers; sizing agents for spun yarns.

Arkofix. [Hoechst AG] Reactant self-crosslinking resins; for resin finishing.

Arkomon. [Hoechst AG] Basic materials for detergents; corrosion inhibitor for metalworking fluids.

Arkon. [Arakawa] Hydrog. hydrocarbon resin.

Arkopal N. [Hoechst Celanese; Hoechst AG] Nonoxynol series; detergent, wetting agent, emulsifier for industrial use.

Arkophob NCS. [Hoechst AG] Polysiloxane emulsion; water-repellent finishing agent.

Arkopon Brands. [Hoechst AG] Fatty acid condensation prods. basic material for detergent mfg.; auxiliary for metal and rubber industries.

Arkopon T. [Hoechst AG] Sodium oleoyl methyltaurides; surfactant, wetting agent.

Arlacel®. [ICI Spec. Chem.; ICI Surf. Belgium] Esters; surfactants, emulsifiers for personal care prods., pharmaceuticals.

Arlacide. [ICI Spec. Chem.; ICI Surf. Belgium] Chlorhexidine salts; preservative, bactericide.

Arlagard E. [ICI Surf. Belgium] Polyhexamethylene biguanide hydrochloride; cationic cosmetic preservative and bactericide.

Arlamol®. [ICI Spec. Chem.; ICI Surf. Belgium] Emollient, refatting agents, solvents for personal care prods.

Arlasolve®. [ICI Spec. Chem.; ICI Surf. Belgium] Surfactant, emulsifier, solubilizer, solvent for cosmetics and pharmaceuticals.

Arlatone®. [ICI Spec. Chem.; ICI Surf. Belgium] Ethoxylated fatty acid esters; surfactant, coupling agent, solubilizer, antistat, lubricant for cosmetics, perfumes, textiles.

Arlex. [ICI Am.] Sorbitol prods.

Arloy®. [Arco] Blends of Dylark resins and polycarbonate; engineering thermoplastic resins for automotive instrument panels, seat belt retractors, speaker grills, surgical appliances, camera components, power tool housings.

Armac®. [Akzo] Aliphatic amine acetates; wetting agent, emulsifier, corrosion inhibitor, flotation reagent, lubricant for metal processing, mineral flotation, flocculation

Armeen®. [Akzo] Primary and secondary amines; emulsifier, flotation agent, corrosion inhibitor; lubricant for metal treatment, rubber processing and mold release.

Armeen® DM. [Akzo] Dimethyl (tert.) amines; surfactant intermediates.

Armid®. [Akzo] Aliphatic amides; lubricant, slip agent, release agent for plastics, rubber, coatings, films.

Armix. [Witco/Organics] Wetting, sticking, spreading agent, penetrant for agric. formulations, high caustic industrial cleaners; foaming agent.

Armoblen®. [Akzo] Long chain fatty amine derivs.; pigment wetting agent for paints and coatings.

Armocure®. [Akzo] Aliphatic polyamine; curing agent for epoxy resins; for industrial applics., adhesives, laminating resins.

Armofilm. [Akzo BV] Fatty amines; Filming amine in water, boiler systems; corrosion inhibitor.

Armoflo®. [Akzo; Akzo Chem. BV] Anticaking, antidusting, and conditioning agents for hygroscopic materials, in fertilizers; corrosion inhibitor for iron and steel equip.

Armoflote. [Akzo BV] Mineral flotation agents.

Armogard. [Akzo BV] Corrosion inhibitors for oil industry.

Armohib®. [Akzo] Fatty amines; acid corrosion inhibitor used in metal cleaning and pickling; antifouling agents.

Armomist. [Akzo] Foaming agents for petrol. industry.

Armorgloss. [Southern Coatings] High-build urethane.

Armoslip®. [Akzo] Fatty acid amides;

internal lubricants, mold release agents, antiblock and slip agents for plastics.

Armosoft L. [Akzo] Ditallow dimonium chloride; lubricant, fabric softener for household and commercial use.

Armostat®. [Akzo; Akzo Chem. BV] Fatty acid amines; antistats for plastics.

Armotan®. [Akzo BV] Sorbitan esters and ethoxylated sorbitan esters; wetting agent, emulsifier, solubilizer for cosmetics, pharmaceuticals, cutting oils.

Armoteric LB. [Akzo BV] Lauryl betaine; amphoteric for shampoo formulations.

Armul. [Witco/Organics] Emulsifiers for industrial and agric. applics., S/B latexes; degreaser, dye penetrant for leathers; detergent, solubilizer for drycleaning mixtures; lime soap dispersants; cosmetics, textiles.

Arneel®. [Akzo] Fatty acid nitrile compds.; detergent, wetting agent, rust inhibitor for metal processing; plasticizer for rubbers and plastics.

Arnica Oil CLR. [Henkel/Cospha] Arnica extract, soybean oil, tocopherol; emollient, conditioner; protective skin and hair care prods.

Arnite. [Akzo Engineering Plastics BV] PET or PBT polyester.

Arnitel®. [Akzo Engineering Plastics; Akzo Engineering Plastics BV] Thermoplastic copolyether-ester elastomer; for automotive, elec./electronic, industrial, consumer durables, and sporting goods.

Arnox. [Witco/Organics] Surfactants; detergent, emulsifier, defoamer intermediate, wetting agent, corrosion inhibitor; for personal care prods., household, paint, oilfield, refinery, agric., textile, leather, plastics applics.

AroCy®. [Rhone-Poulenc] Dicyanate monomers and prepolymers; co-reacts with and cures epoxy resins; reactive diluent for adhesives, prepregs; laminating resin.

Arodet. [Arol Chem. Prods.] Surfactant blends; detergent, wetting agent, emul-

sifier, coupler, dispersant, dye assistant, penetrant; for dishwashing, personal care prods., textiles, foods, cleaners.

Arofene®. [Ashland] Phenolic resin; for impregnating/specialty adhesives.

Aroflat. [Reichhold] Alkyd resin.

Aroflint. [Reichhold] Polyester-epoxy reactive systems.

Arofos. [Arol Chem. Prods.] Phosphate surfactant blends; detergent, wetting agent, emulsifier, penetrant, dye leveling agent, dispersant.

Arogrip. [Ashland] Adhesives.

Arol Defoamer. [Arol Chem. Prods.] Complex alkyl ethoxylate blend; defoamer.

Arolite TD Conc. [Arol Chem. Prods.] Thiourea dioxide; reducing agent for textiles.

Arolon. [Reichhold] Alkyd resins.

Arolterge. [Arol Chem. Prods.] Fatty acid alkanolamides/phosphate ester blend; detergent, emulsifier, wetting agent.

Aromatic 100, 150, 200. [Exxon] Aromatic hydrocarbon solvents.

Aromatic Oil 745. [Neville] Aromatic plasticizer; used in adhesives, rubber (cements, mech. and molded goods, tires), caulks.

Aromatic Solvent. [Texaco] Aromatic solvents; for paint, protective coatings, herbicide and pesticide carrier, synthetic resin mfg., degreasing, fuel additive treatments, cleaning compds., oilfield applics.

Aromox®. [Akzo; Akzo Chem. BV] Amine oxides; wetting agent, emulsifier, stabilizer, antistat, foaming agent for detergents, shampoos, cosmetics, textiles, metal plating, petrol. additives, paper, plastics, rubber.

Aron. [Simlak Ltd.] PVC compd.

Aroplaz. [Reichhold] Alkyd and polyester resins.

Aropol. [Ashland] Thermoset polyester resins; for gel coats, filled molding compds., filament winding, SMC and BMC applics., elec. laminates, pipe, ducts and housing, marine and transportation equipment, appliance, construction materials.

Aroset®. [Ashland] Acrylic resin; for pressure-sensitive adhesives.

Arostit® LF Powd. [Sandoz] For bleaching of wool in presence of dyes.

Arosurf®. [Sherex] Alkoxylated fatty alcohols and quats.; emulsifier, detergent, emollient, coupling agent, emulsion stabilizer; for personal care prods.; fabric softener for laundry and textile processing.

Arotap®. [Ashland] Resins for textile and paper.

Arothix. [Reichhold] Alkyd resins.

Arotran. [Ashland] Thermoset polyester resin; for resin transfer molding, automotive body panels, inner structures.

Arowet. [Arol Chem. Prods.] Sulfonated ester blend; wetting agent.

Arpak. [Arco] Expanded polyethylene beads; for closed-cell resilient foam.

Arphos. [Baker Perf. Chem.] Scale inhibitors.

Arpro. [Arco] Expanded polypropylene beads; for closed-cell foam, dynamic cushioning applics.

Arquad®. [Akzo; Akzo Chem. BV] Quaternary ammonium salts; emulsifier, foaming, wetting, dispersing agents, corrosion inhibitor, softener, dyeing aid, antistat for textiles, paper, cosmetics; industrial, agriculture, plastics, rubber, petrol. industry, acid pickling baths; bactericide, algicide.

Arsan. [Arco] Molded resilient resin; for closed-cell foam.

Arsenal®. [Am. Cyanamid/Ag] Nonselective broad-spectrum herbicide for weed control on roads, railroad right-of-ways.

Arsonax. [Climax Performance; Humphrey] Organic-inorganic additive; flame retardants for plastics, rubbers, paper/paperboard.

Arstim RRC. [Aquaness] Sulfonate and surfactant; for cleaning emulsion blocks, foreign solids, water blocks, drilling mud.

Arsul. [Aquaness] Dodecylbenzene sulfonic acid; intermediate; derivs. as emulsifiers, dispersants, cleaning compds., emulsion breakers, corrosion inhibitors.

Arsurf. [Aquaness] Surfactants and proprietary blends; wetting agents, emulsifiers, detergents, dispersants, penetrants, solubilizers.

Artilene. [Sandoz] Textile dyes and pigments.

Artisil®. [Sandoz] Pigments for coloring oils, waxes, and solvents.

Artodan. [Grindsted Prods.] Sodium stearyl lactylates; emulsifier for foods, cosmetics, cutting oils.

Artoflex. [Arto Chem. Ltd.] Polychloroprene.

Artoflex PC. [Arto Chem. Ltd.] Polycarbonate.

Artomag. [Arto Chem. Ltd.] Magnesium oxide.

Artonyl. [Arto Chem. Ltd.] Nylon 6 or 6/6.

Artrads. [Aquaness] Dimer trimer polybasic acids; intermediate for reaction with diols, polyols, and polyamines to produce corrosion inhibitors.

Arylan®. [Harcros; Harcros UK] Surfactants; detergent, wetting agent; emulsifier for emulsion polymerization, pesticides; intermediate.

Arylene. [Hart Chem. Ltd.] Sulfosuccinates; wetting agent, filtration aid, dewatering surfactant.

Arylon®. [DuPont] Polyarylate resin; high-performance thermoplastic engineering polymer; for electronics/elec., mech., lighting/optical, medical applications, household elec. goods.

Arylpon® F. [Henkel/Emery; Henkel KGaA] Ethoxylated fatty alcohol; thickener with solvent character for shower baths, shampoos, bubble baths.

AS-. [Compounding Tech.] ABS, glass, carbon, nickel or stainless steel reinforced; thermoplastic for business machine housings, cabinetry, tool housings, appliances, automotive, construction materials.

ASA. [Ethyl] Alkenyl succinic anhydride; intermediate for defoamers, demulsifiers, emulsifiers, foam boosters, wetting agents, detergents, dispersants; sizing agent for paper.

Asana. [DuPont/Ag] Insecticide.

Asarco. [Asarco] Cadmium, silver, selenium, antimony, bismuth and their compds.

Asbury. [Asbury Graphite Mills] Graphites; fillers, lubricants, extrusion aids for rubbers.

Ascend. [Huntington Labs] Germicidal detergent.

Ascorbosilane C. [Exsymol] Ascorbyl methylsilanol pectinate; increases cell membrane resistance; for skin maintenance and tissue regeneration in anti-aging, after sun bath, and burn treatment formulations.

Ascort. [Vita Cortex Ltd.] Antistatic urethane foam.

Ascote. [Kao Corp. SA] Fatty amines and polyamines; antistripping agents for road applics.

Asfier. [Kao Corp. SA] Fatty amines and diamines; emulsifier for asphalt emulsions.

Ashlene®. [Ashley Polymers] Nylon 6 or 6/6 resins, some glass- or mineral-reinforced, lubricated; inj. molding and extrusion resins, economy grades.

Aska. [Akzo] Antiskinning agents.

Askofene/Arotap. [Ashland-Südchemie-Kernfest GmbH] Phenol and alkylphenol-based resins. `

Askopur/Askopox. [Ashland-Südchemie-Kernfest GmbH] Tooling resins.

Asol. [Lucas Meyer] Lecithin fraction; release agent, emulsifier for food, cosmetics, pharmaceuticals.

Asp. [Lawrence Industries PLC] Aluminum silicate; pigments.

ASP®. [Engelhard] Hydrous aluminum silicates; extender pigment in paints, inks; filler providing opacity in plastics; in adhesives for corrugated board, laminated fiber board, paper bag seams; in rubber industry, in mastics, putties, caulks, textile and chemical conditioning.

Aspect®. [Phillips; Phillips Petrol. Chem. SA/NV] Thermoplastic polyester compds.; for automotive applics., elec. connectors, underhood, dash, housings.

Asperzyme. [Enzyme Development] Enzyme for starch conversion.

Aspun. [Dow] Fiber grade resins.

Assault. [West Chem. Prods.] Weed killer.

Assert. [Am. Cyanamid/Ag] Herbicide.

Assure. [DuPont/Ag] Herbicide.

AST-1001. [Merix] Light stabilizer, antistat for plastics.

Astab. [Croxton & Garry Ltd.] PVC stabilizer systems.

Astacin®. [BASF AG] Finishes, softeners for leather and fur industry.

Asterite. [ICI Chem. & Polymers Ltd./ Acrylics] Acrylic dispersions.

Aston. [Rhone-Poulenc/Textile & Rubber] Thermosetting polyamine; antistat, softener for plastics, textile and industrial applics.

Astor. [Astor Wax] Petrol. wax; used for rubber processing, graphic arts.

Astra. [Miles/Organic Prods.] Textile dyes and pigments.

Astra Bar. [ECC Int'l.] Barium sulfate; rubber additive; imparts high density, chemical inertness, sound deadening, chemical resistance.

Astra-Brite. [ECC Int'l.] Kaolin; paper coatings.

Astra-Cote. [ECC Int'l.] Kaolin; paper coatings.

Astra-Fil. [ECC Int'l.] Kaolin; delaminated filling clay.

Astragal®. [Miles/Organic Prods.] Migrating agent, leveling agent, retarders for textile dyeing.

Astra-Glaze. [ECC Int'l.] Kaolin; paper coatings.

Astramyl. [Henkel-Nopco] Complex fatty amidoamines; corrosion inhibitor, asphalt additive.

Astrawax. [Astor Wax] Bisstearamide-based wax; lubricant for inks, paints, rubber, and plastics industries.

Astrazon. [Miles/Organic Prods.] Textile dyes and pigments.

Astro Add Wgt. [Astro Industries] Ethoxylated alcohol; weighter for textile applics.

Astro Back. [Astro Industries] Compounded acrylic or other polymers; backcoating for fabrics, mattress ticking.

Astroclean 26-A. [Glo-Tex] Aqueous

copolymer sol'n.; soil release agent for polyester/cotton blends.

Astroclear 200. [Glo-Tex] Detergents, alkali and reducing agents; clearing, stripping and cleaning agent for textile processing.

Astrodye. [Glo-Tex] Carrier for textile dyes.

Astro Fix. [Astro Industries] Fixative for direct dyes.

Astro Floctite. [Astro Industries] Compounded acrylic or other polymers; flocking adhesives.

Astro Foam S-1. [Astro Industries] Ammonium stearate; liq. foaming or frothing agent.

Astro Foamkill ST. [Astro Industries] Solvent/emulsifier blend; defoamer for oil and water repellent textile finishes.

Astro Fume ET. [Astro Industries] Antifuming agent for acetate fabrics.

Astrol. [Tribol] High-performance lubricants.

Astro Leveller ND-4. [Astro Industries] Leveling agent for dyeing of nylon fabrics.

Astro Lube. [Astro Industries] Textile lubricant and softener.

Astro Mel. [Astro Industries] Crosslinker for polymer systems, oil phase printing systems.

Astromid. [Alco] Alkyl sulfosuccinamates; emulsifier for polymerization; foaming, frothing agent for elastomeric compds.

Astro Nap. [Astro Industries] Napping and sanding softener-lubricant.

Astro Pel NS. [Astro Industries] Water repellent and fluorocarbon extender.

Astro Pon KNB-1. [Astro Industries] Scouring and desizing agent for PVA-starch sizes.

Astro Pyro-Pruf. [Astro Industries] Textile flame retardants.

Astro React. [Astro Industries] Dimethyloldihydroxyethyleneurea; reactants for textile processing.

Astro Scour. [Astro Industries] Ethoxylated phosphated alcohol; continuous scour and bleach assistant.

Astro Set. [Astro Industries] Hand modifier, stabilizer for viscose, cotton, blends.

Astro Sof. [Astro Industries] Lubricant, softener, antistat for textiles.

Astro Stat PO. [Astro Industries] Antistat for synthetic fibers.

Astrotherm. [Glo-Tex] Antimigrant for continuous dyeing with disperse, vat, fiber reactive, sulfur, direct and pigment dyes.

Astrowet. [Alco] Sulfosuccinates; polymerization emulsifier, wetting agent.

Astro Wet. [Astro Industries] Wetting agent for textile processing.

Astrozine OC-70. [Glo-Tex] Anionic surfactant; leveler/retarder for dyeing nylon with neutral premetallized dyes.

Astryn®. [Himont] Polypropylene homopolymer and copolymers, some glass and mineral-reinforced; for automotive parts, housewares, lawn and garden tools/housings, appliance housings, ABS replacement.

Asulox. [Rhone-Poulenc/Ag] Agric. and industrial herbicide.

AT-. [Compounding Tech.] Acetal, PTFE and silicone lubricated or glass filled; offers lubricity and chemical and hot water resistance for automotive, hardware, plumbing applics.

AT-. [Procter & Gamble] Dimethyl fatty acid amines; cationic detergent; corrosion inhibitor; acid-stable emulsifier; chemical intermediate, raw material for surfactants.

AT-1 Anatase. [Sovereign] Titanium dioxide; additive for paper, rubber, ceramics, and paints.

ATBC. [Croda Surf. Ltd.] Acetyl tributyl citrate; plasticizer for food contact applics.

ATC-3. [Aerojet Propulsion] Curing agent for epoxy resins.

Atcowet. [Yorkshire Pat-Chem] Surfactant; detergent, wetting agent for textile applics.

Atebin®. [Boehme Filatex] Polyethylene; softener for textile processing.

Ateprint® VP 972. [Boehme Filatex] Aminoplast resin mixture; pigment pad binder.

Atflow. [ICI Am.] Surfactants.

ATH Nycoat®. [Nyco Minerals] Chemically surface-modified alumina tri-hydrate; flame retardants for plastics.

Atlac®. [Reichhold/Reactive Polymers] Thermoset polyester resins; for fiber-glass-reinforced structures, glass-reinforced coatings, mortars in pulp and paper, caustic-chlorine, metal treatment, filament winding, centrifugal casting of large tanks and pipe.

Atlas EM. [Atlas Refinery] Glycol ester; fiber lubricant, emulsifier.

Atlas EMJ. [Atlas Refinery] Ethoxylated nonyl phenyl ether; detergent, wetting agent, emulsifier.

Atlas G. [ICI Am.; ICI Surf. Belgium] Surfactants, emulsifiers, thickeners for foods, cosmetics, textile, agric. use.

Atlas L-801-LF. [Atlas Refinery] Sulfated and ethoxylated oils; lubricant, dye leveler, scouring agent.

Atlas M 130. [Degussa] Methacrylate-based resin filled with aluminum; mass cast tooling resin.

Atlas Solvent Scour. [Astro Industries] Self-emulsifying solvent detergent for scouring and stain removal on cellulosic and synthetic fibers.

Atlas Sul. [Atlas Refinery] Sulfated neatsfoot or fish oils; lubricant, emulsifier.

Atlas WA-100. [Atlas Refinery] Dioctyl sulfosuccinate; wetting agent.

Atlasol. [Atlas Refinery] Sulfated alcohols or oils; emulsifier, fiber lubricant.

Atlasol. [Crompton & Knowles] Textile dyes and pigments.

Atlastan. [Atlas Refinery] Acrylic polymer.

Atlosol. [Aquaness] Drilling mud emulsifier concs.

Atlox. [ICI Am.; ICI Surf. Belgium] PEG sorbitan and glyceride esters or sulfonates; emulsifier, wetting agent, coupling agent, suspending agent, dispersant for agric. formulations, textiles.

Atmer®. [ICI Polymer Additives; ICI Surf. Belgium] Antifog agent, antistat, lubricant, cling agent for plastics, films; nucleating agent, plasticizer.

Atmos. [ICI Am.; ICI Surf. Belgium] Mono or diglycerides; emulsifiers for food applications.

Atmos. [Witco/Humko] Surfactants, food emulsifiers, plastics antistats.

Atmul. [Witco/Humko] Mono or diglycerides; food emulsifiers.

Atolene. [Dexter] Oleic acid sulfated ester; lubricant, wetting agent, dye leveler used in textile processing.

Atolex. [Standard Chem. UK] Dyebath auxiliary, detergent, wetting agent, leveling agent, dispersant for textiles.

Atomite®. [ECC Int'l.] Calcium carbonate; used in paints, rubber, plastics, floor coverings.

Atomsolve. [Atomergic Chemetals] HPLC solvents.

Atpet. [ICI Am.; ICI Surf. Belgium] Sorbitan esters; surfactants, rust inhibitor, petroleum additives.

Atplus. [ICI Am.] Surfactant, foaming agent, coupling agent, stabilizer, wetting agent for agric. formulations.

ATP Nucleotides. [Croda Inc.] Propylene glycol, collagen amino acids, adenosine triphosphate.

Atranonic Polymer. [Atramax] Ethoxylated prods.; dispersant for textile applics.

Atra Polymer 10. [Atramax] Ammonium carboxylate, styrene copolymer; film-forming dispersant for pigments; vehicle for flexo ink.

Atrasein 115. [Atramax] Ammonium caseinate; dispersant, leveling agent.

Atrinon. [Henkel/Emery/Cospha] Woody aroma chemical.

Atryl. [Owens/Corning Fiberglas] Low profile resin for achieving class A surface in compression-molded automotive parts.

Atsolyn. [Atsaun] Surfactants; emulsifier, wetting agent, dispersant, pigment grinding aid, antistat, corrosion inhibitor, solubilizer for fatliquors; for polymerization, textiles, metal treatment.

Atsowet. [Atsaun] Sulfosuccinate; wetting and rewetting agent.

Atsurf. [ICI Am.] Polyol or sorbitan esters; surfactant, emulsifier, for paper coatings, polishes, fiberboard, industrial use.

Attaclay. [Engelhard] Pesticide carrier

and diluent.

Attacote. [Engelhard] Attapulgite clay; anticaking, anti-agglomerating agent for conditioning ammonium nitrate and sulfate crystals, urea, gran. fertilizers.

Attaflow. [Engelhard] Attapulgite clay; replacement for dry gelling clay.

Attagel. [Engelhard] Attapulgite clay or colloidal attapulgites; thickener, gellant, stabilizer, thixotrope, suspending agent; used for drilling muds, paints, adhesives, sealants, and mastics.

Attane. [Dow Plastics] Ultra low density linear polyethylene; for film extrusion.

Attapulgus. [Engelhard] Attapulgite clay.

Attasorb. [Engelhard] Conditioner, anticaking agent.

Attllev TRA. [Applied Textile Tech.] Nonionic polyester leveling agent.

Attlpoly Lev. [Applied Textile Tech.] Polyester carrier.

Attlscour. [Applied Textile Tech.] Scouring and wetting agents for textiles.

Attlsperse. [Applied Textile Tech.] Dispersant for disperse dyes for polyester, acetate, or their blends.

Atwet. [ICI Am.] Surfactants.

Auracote. [Hilton Davis] Plasticizer flushed colorants for vinyl.

Auralube. [Auralux] Softener, lubricant, sewing assistant.

Auramel. [Auralux] Melamine-formaldehyde resin; stabilizing resin for textiles.

Aurantiol®. [BASF AG] Hydroxycitronellal/methylanthranilate blend; sweet floral fragrance.

Aurapel. [Auralux] Water repellent, softener, lubricant for textiles.

Aurarez. [Auralux] Reactant for textile processing.

Aurascour LN. [Auralux] Detergent for removal of graphite in lace industry.

Auraset. [Auralux] Anticurling agent, leveling agent, soft hand binder for textiles.

Aurasoft®. [Auralux] Softener for textiles.

Aurawet. [Auralux] Nonrewetting penetrant, detergent, scouring agent, stabilizer, foaming agent for textile operations.

Aurum. [Advanced Web Prods.] Thermoplastic polyimide resin.

Austin Black. [Coal Fillers; Harwick] Bituminous fine black; filler, extender, reinforcer for rubbers, fluoroelastomers, reclaims, PVC.

Autan. [Acla-Werke] Cellular polyurethane.

Auto-Dri. [Engelhard] Oil and grease absorbent.

Autofroth. [Olin] Rigid foam systems (froth-in-place) for insulation, flotation, molding.

Automate. [Morton Int'l.] Textile dyes and pigments.

Autopak. [Olin] Rigid/flexible foam system (pour-in-place) for pre-mold pkg.

Autopoon. [Zschimmer & Schwarz] Cationic surfactants, solvents, and solubilizers; water-repellent conc. for preps. for lacquered surfaces.

Autopour. [Olin] Rigid foam system (pour-in-place) for insulation, molding.

Autopur. [Zschimmer & Schwarz] Nonionic/cationic surfactants; basic material for car shampoos.

Auxiliary BCY. [Gresco Mfg.] Self-emulsifiable oil; softener, penetrant, and color smoothness for screen and roller print.

Auxiliary PR. [Catawba-Charlab] Antiwicking agent, wetting agent, thickener, rheology modifier for textile pigment printing systems.

Avadex®. [BASF AG] Triallate emulsion conc.; herbicide.

Avadyne. [Pierce & Stevens] Water or alcohol dilutable two-part urethane prepolymer emulsion adhesive; for lamination of film-to-film and film-to-metallized film structures.

Avamid 150. [Mona Industries] Avocadamide DEA, avocado oil; self-emulsifying foam stabilizer, visc. builder, lubricant for conditioning shampoos, creams and lotions.

Avanel®. [PPG/Specialty Chem.] Sodium C12-15 pareth sulfonate; anionic surfactants, emulsifiers for personal care, household, and institutional prods., industrial processes.

Avasol. [Alframine] Surfactants.

Avatech. [Avatar] White mineral oil, tech.

Avenge®. [Am. Cyanamid/Ag] Difenzoquat methyl sulfate; wild oat herbicide for barley and wheat crops.

Avicel. [FMC] Microcrystalline cellulose; binder, disintegrant, flow aid, and filler for pharmaceuticals and animal health prods.; absorbent; peptizing agent; anticaking agent for oils.

Avicol. [Takemoto Oil & Fat] Nonionic softener.

Avilene. [Nutex] Fatty condensation prod.; highly conc. liq. softener.

Avirol®. [Henkel] Sulfates or esters; wetting agent, stabilizer, foaming agent, suspending agent, surfactant base, emulsifier, dispersant; for dyes, cleaners, cosmetics, polymerization, plastics, rubber, adhesives, paints, inks, agric. industries.

Aviscour. [Albright & Wilson Australia] Sulfonates or surfactant blends; detergent conc. for hard surface, laundry, household, car, upholstery cleaning.

Aviso®. [BASF AG] Metiram, cymoxanil; fungicide for potatoes, vines, and other crops.

Avistin®. [Hüls Am.; Hüls AG] Fatty acid polyamine condensates; bases for textile auxiliary agents.

Avitex. [DuPont] Complex higher alkylamine; softener, antistat for textiles.

Avitone®. [DuPont] Sodium alkyl sulfonates; softeners for textiles, paper, leather, elastomers.

Avivan®. [Ciba-Geigy/Dyestuffs] Fatty acid amide condensate; softener for cellulosic and synthetic fibers.

Avocado Oil CLR. [Henkel/Cospha] Avocado fatty oil; emollient, conditioner for skin preps.

Avoid. [Petrokem] Insecticides.

Avolan®. [Miles/Organic Prods.] Dispersant, leveling agent for textile dyeing.

Avron. [VT Plastics Ltd.] Acrylic composites.

Avtel. [Phillips Petrol. Chem. SA/NV] High-performance composites.

Axel. [Aquatec Quimica SA] Fatty alcohol and fatty esters; antifoamer for paper machines.

Axol®. [Goldschmidt AG] Glyceryl mono/di esters; food emulsifiers; lubricants, solvents, plasticizers, coating materials for foodstuffs, cosmetics, plastics, and rubber.

Axxis PC. [Erta Cestidur Industries] Polycarbonate.

Ayrcryl. [Rohm & Haas] Latex.

Az-Cup. [Hercules] Azidosilane sol'n.; coupling agent for bonding most siliceous reinforcing fillers to most polymers.

Azdel®. [Azdel] Fiberglass/polypropylene composites; thermoplastic sheet for automotive applics.

AZDN. [Aceto] Azobisisobutyronitrile; catalyst.

Azo. [Asarco] Zinc oxide; antimicrobial, fungistat for paints; uv stabilizer, accelerator-activator and reinforcing pigment in plastics, rubbers, latex, wire and cable compds., ceramics.

Azoanthrene. [Crompton & Knowles] Textile dyes and pigments.

Azocel. [Fairmount] Blowing agents for plastics, rubber.

Azo D. [Dong Jin] Azodicarbonamide; foaming agent for plastics.

Azodox. [Asarco] Zinc oxide; activator.

Azofoam. [Biddle Sawyer] Azodicarbonamide; foaming agent for plastics.

Aztec®. [Catalyst Resources] Peroxide derivs.; crosslinking agent, catalyst in polyethylene crosslinking, styrene polymerization, polyester resins; vulcanizing agent for SBR, NBR, EPDM, EPM, silicones; initiator in polymerization.

B

B- . [Wacker Silicones] Silicone elastomers.

B-. [Midwest Rubber Reclaiming] Reclaimed rubber; used in carcass, sidewall, and undertread of passenger, lt. truck, and off-road tires, and general purpose mech. goods.

B-. [Thermofil] Glass-reinforced styreneacrylonitrile.

B-40. [Van Den Bergh Foods] Partially hydrog. soybean oil with mono and diglycerides; kosher.

B-80. [Thiele Kaolin] Hydrous aluminum silicate-kaolin; used in inks, plastics, paint, adhesives, and pharmaceutical applics.

B-400, -410. [Grindsted Prods.] Food industry additives.

B2500 Series. [Hüls Am.] Silane compds.; blocking agent; silylating reagent.

B-8880-50%. [Ethox] High m.w. polyoxyalkylene polymer; dyeing assistant for acrylics; stabilizer for aq. emulsions.

BA-. [Polyvel] Azodicarbonamide concs.; foaming agent for plastics.

Babinar. [Marubishi Oil Chem.] Fatty acid polyethylene polyamine condensate; softener, antistat for textiles, esp. polyacrylonitrile fiber.

Baco. [BA Chem. Ltd.] Alumina trihydrate; prod. of aluminum, abrasives, refractories, ceramics, elec. insulators, catalysts and catalyst supports, paper, spark plugs, crucibles and lab ware, adsorbent for gases/water vapors, chromatographic analysis, heat-resist. fibers, food additives.

Bactiram. [Ceca SA] Bactericide for water treatment.

Bactistep. [Stepan Europe] Dialkyl dimethyl ammonium methoxy sulfate; sanitizer.

Bactosol. [Sandoz] Enzyme for desizing and deweighting operations.

Bad Air Sponge. [Mateson] Odor/fume adsorbent.

Baerostab®. [Bärlocher GmbH; R.T. Vanderbilt] Barium zinc complexes; stabilizer for PVC.

Bafixan®. [BASF; BASF AG] Disperse dyes for transfer paper printing in the textile industry.

Bahydrol. [Miles] Polyurethane dispersion.

Bakelite. [Union Carbide] Polyethylene; for cable jacketing, blow molding, inj. molding and extrusion.

Balab. [Witco; Witco SA] Silicone-free organics; defoamers for paints, adhesives, food processing.

Ballkyd. [Ranbar Tech.] Resin.

Ballrez. [Ranbar Tech.] Resin.

Baltane. [Atochem Deutschland GmbH] 1,1,1-Trichlorethane.

BAN. [Novo Nordisk] Alpha-amylase; enzyme used for partial breakdown of gelatinized starch into dextrins.

Banox ES. [UOP] BHT, lecithin, soybean oil, mineral oil; food grade antioxidant.

Bapolan®. [Bamberger Polymers] PS, ABS, SAN, and PC resins.

Bapolene®. [Bamberger Polymers] Polyethylene or polypropylene resins.

Bapolon®. [Bamberger Polymers] Nylon resins.

Baquacil. [ICI Am.] Swimming pool sanitizer and algicide.

Baragel®. [Rheox] Clays; rheological additive for greases; PVC stabilizer; antiplating agent.

Barbalube 366. [Apollo] Nonionic dyebath lubricant for natural and synthetic fibers to prevent crack marks, chafing and rope marks in all types of wet processing equip.

Barbe. [Barbe Am.] Liq. antitack agents; diluted sol'ns. providing release and antitack properties to unvulcanized rubber.

Barchlor. [Lonza] Alkyl chlorides.

Barco B. [Barium & Chems.] Barium sulfate; inert filler, coloring ingredient for rubbers.

Bardac®. [Lonza] Quaternary ammonium compds.; disinfectant, sanitizer, bacteriostat.

Barden R. [J.M. Huber] Hydrated aluminum silicate; filler and reinforcing agents for rubber.

Bar-Dust®. [Petrolite/Polymers] Polymeric anticaking aid for urea and NPK fertilizers.

Bareco®. [Petrolite/Polymers] Plastic microcrystalline waxes; for laminating adhesives, hot-melt coatings in contact with oily or greasy foods, elec. insulation, candles, cosmetics, rubber compounding, casting wax for metals, leather treatment, rust-preventive coatings, plasticizers.

Barex®. [BP Chem. Inc.] Acrylonitrile-methacrylate copolymer, rubber modified; high barrier resins in inj. molding and extrusion grades for film, sheet, blow and inj. molding applics.

Barfire. [Apollo] Flame retardant and carrier for synthetic and natural fibers.

Barfoamkil. [Apollo] Defoamer for textile dyeing, printing, and finishing processes.

Barimite. [Cyprus Industrial Min.] Fine ground, unbleached barite.

Barisol®. [Dexter] Potassium alcohol phosphate; textile wetting agent, dye leveling agent, dispersant; pectin removal from cottons.

Barium Lithol. [BASF] Textile dyes and pigments.

Barium Petronate. [Witco/Sonneborn] Barium salts; emulsifier, lubricant additive, rust preventative for industrial oils and fuels.

Barlene®. [Lonza] Alkyl dimethyl amines; cationic detergent; corrosion inhibitor; emulsifier; chemical intermediate, raw material for surfactants.

Barlox®. [Lonza] Amine oxides; detergent, visc. builder, emollient.

Bärodur. [Bärlocher GmbH] Impact modifiers.

Bärolub. [Bärlocher GmbH] Lubricants, wax.

Bäropan. [Bärlocher GmbH] Stabilizer, lubricant.

Bärorapid. [Bärlocher GmbH] Processing aids.

Bärostab. [Bärlocher GmbH] Stabilizers, epoxy plasticizers.

Bärostat. [Bärlocher GmbH] Visc. modifiers, antistatic agents.

Bar-Ox. [Devoe Coatings] Alkyd coating.

Barpel. [Apollo] Fluorochemical oil and water repellent for use on synthetic and natural fibers.

Barquat®. [Lonza] Quaternary ammonium salts; algicide, germicide, disinfectant, sanitizer for swimming pools, water treatment; antistat for plastics.

Barre® Common Degras. [RITA] Wool grease deriv.; emollient, leather softener, corrosion preventive.

Barrierta® Greases. [Kluber Lubrication N. Am.] Extreme high temp. perfluorinated greases for the severest environments.

Bar-Rust. [Devoe Coatings] Surface-tolerant epoxy.

Barstat PA2. [Apollo] Cationic antistat for natural and synthetic fibers, yarns, and fabrics.

Bartex®. [Hitox] Barium sulfates; extender pigment for coatings, plastics, rubber goods, ceramics.

Bartyl® F-2. [Givaudan-Roure] Anti-skinning agents for oil-based systems.

Basacid®. [BASF AG] Anionic dyes for prod. of inks, pigmentation of cleaners, wood preservatives in paint and varnish industry.

Basacryl®. [BASF; BASF AG] Cationic dyes for dyeing and printing polyacrylonitrile fibers and anionically modified polyester fibers; leveling agents and retarders.

Basagran®. [BASF AG] Bentazon and blends; for postemergence control of broadleaf weeds in cereal crops.

Basamid®. [BASF AG] Dazomet; soil

fumigant for control of nematodes, soilborne fungi, soil insects and germinating weeds.

Basammon® extra 25. [BASF AG] Ammonium sulfate nitrate, dicyandiamide (stabilizer); for improved nitrogen utilization and sustained action for agric. crops.

Basantol®. [BASF AG] Anionic dyes for paint and varnish industry.

Basazol®. [BASF AG] Basic dyes for papermaking, leather.

Bascal®. [BASF AG] Aliphatic dicarboxylic acids, deliming agent, pickling acid mixture; for processing furs.

Base. [CNC Int'l.] Fabric softener bases and concs.

Base. [Ferro/Keil] Petroleum sulfonate/emulsifier blends or methyl esters; emulsifier, extreme pressure agents for metalworking and lubricating oils.

Base 3059 E. [Henkel-Nopco] Ethoxylated alkylamidoamine; corrosion inhibitor, asphalt additive, antistripping agent.

Base CR. [Auschem SpA] Nonionic blend; pearlescent for shampoos, bubble baths.

Basensol®. [BASF AG] Functional block polymers based on propylene oxide, ethylene oxide, or polyalkylene glycol ether; sensitizing agents for polymer dispersions.

Base Wax. [Eastern Color & Chem.] Substantive amide wax conc., softener for textiles.

BASF Alkali Blue®. [BASF AG] Conc. pigment pastes with triphenylmethane pigments; for inks, typewriter ribbon inks, carbon paper.

Basfapon®. [BASF AG] Dalapon-sodium; postemergence systemic herbicide for control of grasses in annual and perennial crops.

BASF Catalyst. [BASF AG] Catalysts for various industrial uses.

BASF Lutexal® TX-401. [BASF] Thickener, leveling agent, emulsifier and curing agent for textile pigment printing.

Basfoliar®. [BASF AG] Liq. foliar fertilizers.

BASF Reactive Resist Liquid. [BASF] Auxiliary for resist prints with Basilen P dyes under Primazin dyes.

BASF Wax. [BASF; BASF AG] Polyethylene, montan ester, carnauba waxes; used for plastics and rubber processing, printing inks, hot melts, polishes, paints, cleaners, as lubricant.

Basicop. [Griffin] Basic copper sulfate; wettable powd. agric. fungicide for crops.

Basic Silicate White Lead. [Akrochem] General stabilizer for vinyl chloride plastics; paint pigment.

Basilen®. [BASF; BASF AG] Reactive dyes for cellulosic fibers.

Basin®. [BASF AG] Radiator antifreeze containing propylene glycol.

Baso®. [BASF AG] Dye bases; for brightening gravure inks for illustrations.

Basocoll® CM. [BASF AG] Anionic melamine-formaldehyde condensate-based; sizing assistant for papermaking.

Basoflex®. [BASF] Aq. pigment dispersions; for aq. printing inks for pkg. materials and wallpaper.

Basoform®. [BASF AG] Formaldehyde bonding agent; formaldehyde catcher for cured urea-formaldehyde foams.

Basogal® C. [BASF] Nonfoaming cationic leveling agent for vat dyes.

Basojet®. [BASF] Dispersant, leveling agent for disperse dyes on polyester fibers.

Basokol®. [BASF] Sequestrant, dispersant, protective colloid for textile dyeing and printing.

Basol® WS. [BASF] Condensation prod. of an aromatic sulfonic acid; anionic dye dispersant for textile processing.

Basolan®. [BASF; BASF AG] Assistants for nonfelting finishing of wool.

Basomol®. [BASF AG] Phosphoric acid-based synthetic wetting agent; for mfg. of synthetic resin foam in conjunction with Basopor®.

Basonat®. [BASF AG] Isocyanates; hardener component for polyols, prep. of prepolymers, polyurethane adhesives and binders, paints and varnishes.

Basonyl®. [BASF AG] Cationic dyes; for prod. of inks, carbon paper coatings, for dyeing natural fibers, paints and varnishes, prod. of fluorescent pigments and gloss paints.

Basopal®. [BASF AG] Anionic detergents for textile industry washing and cleaning processes.

Basophen®. [BASF/Fibers] Wetting agent, detergent for textile applics., pkg. dyeing.

Basophob®. [BASF AG] Aq. paraffin or polyethylene wax dispersions, sol'ns. of fatty acid derivs.; hydrophobic agents for internal or surface treatment of paints, mortars, concrete, and paper.

Basophor. [BASF/Fibers] Fatty acid ester; emulsifier.

Basoplast®. [BASF AG] Anionic and cationic copolymers or alkylketene diamide; internal and surface sizes for papermaking.

Basopon®. [BASF/Fibers] Alkylaryl polyglycol ether; detergent, dispersant for textiles.

Basopor®. [BASF AG] Urea-formaldehyde resin precondensate; for prod. of synthetic urea-formaldehyde resin foam, thermal insulation in construction, protection against explosions in mining.

Basoset®. [BASF AG] Epoxy resin (epichlorohydrin/aliphatic polyol); casting resin used with or without filler for elec. potting and encapsulating, for bonding glass-reinforced polyester parts to each other, for binding expanded polystyrene, for gluing foam or metal parts, glass, ceramics, brick.

Basosoft®. [BASF AG] Brighteners and softeners for textile spinning, dyeing.

Basotect®. [BASF AG] Open-cell resilient foam plastic based on melamine resin; used for soundproofing, thermal insulation with fireproofing characteristics.

Basotol®. [BASF] Auxiliary for protecting dyes against reduction; oxidizing agent for vat dyes.

Basotronic®. [BASF] For electroplating and electronics industry.

Basotrope® W. [BASF] Auxiliary for

producing white discharges on dyes grounds and for stripping dyed shades.

Basovit®. [BASF AG] Anionic dyes for foodstuffs and cosmetics industry.

Bastamol®. [BASF AG] Dyeing auxiliaries; for leather and fur industry.

Basyntan®. [BASF AG] Synthetic tanning agents.

Batan BTZ 100. [Hoechst AG] Bacterial protease; bating agent.

Bath Oil Dispersant CB0684. [Croda Chem. Ltd.] Nonionic blend; dispersant for bath oils.

Bavistin®. [BASF AG] Carbendazim; systemic fungicide for control of fungus in vines, fruits, vegetables, ornamentals.

Bayblend®. [Bayer; Miles; Albis UK Ltd.] PC/ABS blends, some glass reinforced; thermoplastic resins for business machine and automotive markets.

Baybond. [Miles] Polyurethane aq. dispersions.

Baycoll. [Bayer; Miles] Linear hydroxyl polyesters and polyethers; polyols for adhesive formulation.

Baycor®. [Bayer] Bitertanol; broad spectrum fungicide.

Baydur®. [Bayer; Miles] Polyurethane structural foam (RIM).

Bayer 5072. [Bayer] p-Dimethylaminobenzenediazo sodium sulfonate; fungicide.

Bayer SBR Latex. [Bayer] Styrene-butadiene latex; for blended latex goods, paper impregnation, carpet backing, needle felt reinforcement; binder.

Bayferox. [Bayer; Miles] Synthetic iron oxide; pigments.

Bayfidan®. [Bayer] Triadimenol; systemic fungicide.

Bayfill®. [Bayer; Miles] Polyurethane, semirigid foam (RIM).

Bayfit®. [Bayer; Miles] Flexible polyurethane foam systems.

Bayflex. [Bayer; Miles] Integral skin or elastomeric polyurethane foam (RIM).

Bayfol®. [Bayer] PC blend film and sheet.

Baygal®. [Bayer] Polyether or polyester polyols; for prod. of polyurethane casting compds. for electronics and elec.

engineering; as flexibilizer in polyurethane prod.

Baygard®. [Miles/Organic Prods.] For antisoiling and pile stabilizing treatment of textiles.

Baygenal. [Miles/Organic Prods.] Textile dyes and pigments.

Baygon®. [Bayer] Propoxur; broad spectrum insecticide for control of household and hygiene pests.

Bayhibit®. [Miles/Organic Prods.] Corrosion and scale inhibitor in cooling systems; multifunctional textile auxiliary.

Bayhydrol. [Miles] Polyurethane dispersion; adhesion promoters in metal/plastic composites, textile and leather coatings, primers for rigid surface caotings.

Baylan NT. [Miles/Organic Prods.] Assistant for low temp. dyeing of wool.

Bayleton®. [Bayer] Triadimefon; systemic fungicide for control of diseases on field, fruit, vegetable crops and turf.

Baylith. [Miles] Synthetic teolites.

Bayluscide®. [Bayer] Clonitralid; molluscicide for control of water snails.

Baymer®. [Bayer] Rigid polyurethane foams.

Baymidur®. [Bayer] Diisocyanates; for prod. of polyurethane cast compds. for electronics and elec. engineering; binder for foundry sands.

Baymod. [Bayer] ABS, SAN, ASA, EVA, and thermoplastic polyurethanes.

Baynat®. [Bayer] Rigid polyurethane foams.

Baypac. [IGI Baychem] Polyolefin; hot melt adhesive for multiwall bag and pkg. applics.

Baypren. [Bayer; Miles] Polychloroprene and latexes; used for moldings and extrudates, reinforced hoses, belting, cable insulation, sponge rubber, sheeting, fabric proofings, footwear, food-contact goods, adhesives; latex for dipped goods, coatings, lamination.

Bayreduct® PRN. [Miles/Organic Prods.] Process regulator to remove peroxide after bleaching prior to fiber reactive dyeing.

Bayrusil®. [Bayer] Diethchinalphion; insecticide for the control of biting and sucking pests.

Bayscript. [Miles/Organic Prods.] Textile dyes and pigments.

Baysilone. [Miles] Silicone raw materials to produce hot and cold room temperature vulcanizing rubbers.

Baysport®. [Bayer] Polyurethane elastomers.

Baystabil®. [Miles/Organic Prods.] Stabilizer for peroxide bleaching.

Baytac. [IGI Baychem] Polyolefin; hot melt adhesives for flexible web lamination.

Baytan®. [Bayer] Triadimenol; systemic fungicidal seed dressing for cereals.

Baytec®. [Bayer; Miles] Polyurethane cast elastomer.

Baytex®. [Bayer] Fenthion; insecticide for control of hygiene pests.

Baytherm®. [Bayer; Miles] Urethane rigid foam system.

Baythion®. [Bayer] Phoxim; insecticide for the control of stored-product pests.

Baythroid®. [Bayer; Miles/Ag] Cyfluthrin; synthetic pyrethroid with insecticidal activity; for cotton crops.

BB. [Compounding Tech.] Glass-reinforced Bayblend resin.

BBH44. [Atochem N. Am.] Brominated organic; flame retardants for plastics.

BBS. [Karlshamns] Partially hydrog. vegetable oil.

BBTS. [Akrochem] N-t-butyl-2-benzothioazole sulfenamide; delayed-action accelerator for natural and synthetic rubbers.

BCA. [Pierce Chem.] Protein assay reagent.

BCI Nylon. [Belding] Nylon 66 resins; used in textile, wood, metals, rubber industries for protective coating and bonding, sewing thread, fabric finishes, barriers in commercial and military aircraft fuel tanks.

BDMC. [Henley] Bismuth dimethyl dithiocarbamate; ultra accelerator for SBR, butyl and natural rubbers.

Beacon. [Witco] Degreasers, heat treating salts, tempering oils, blackening oils and salts, heat treating salts, quenching oils.

Bead-Gel. [W.R. Grace/Davison] Spherical silica gel.

Beam Deaerator 124. [Reilly-Whiteman] Deaerating and penetrating agent for dyebaths and alkaline bleaching liquors.

Bearflex® LAO. [Witco/Golden Bear] Aromatic hydrocarbon oil; extender oil for elastomer compding.

Beaucoup. [Huntington Labs] Germicidal detergent.

Beaulight. [Sanyo Chem. Industries] Sulfosuccinates or carboxylates; bases for shampoos.

Beaverwhite. [Cyprus Industrial Min.] Magnesium silicate (talc); dusting agent for SR bales and crumbs; filler for SBR and NR foams, plastic resins.

Be Be Bond. [Bostik Div./Emhart] Thermoplastic adhesive.

Be Be Tex. [Bostik Div./Emhart] Latex adhesives.

Beckacite. [Arizona] Phenolic modified rosin ester resins; for lithographic printing inks.

Beckamine. [Reichhold; Reichhold Chemie AG] Urea-formaldehyde resins.

Becklube 3811. [Dooley] Nonionic lubricant for scouring and dyeing operations.

Beck Lubricant 1183. [Piedmont Chem. Industries] Nonionic lubricant for scour and dyebaths.

Beckolin. [Reichhold] Synthetic oil.

Beckosol. [Reichhold; Reichhold Chemie AG] Alkyds and oil-free polyesters.

Beefeater. [Champlain Industries] Protein hydrolysates.

Beesyn White Wax. [Int'l. Wax Refining] Antioxidant for rubbers.

Beetafil. [BIP Chem. Ltd.] Urea-formaldehyde resins.

Beetafin. [BIP Chem. Ltd.] Polyurethane resins.

Beetle. [Am. Cyanamid] Urea-formaldehyde; molding compd.; crosslinking agent.

Beetle. [BIP Chem. Ltd.] Urea-formaldehyde, nylon 6 and 66 compds., PET, PBT, unsat. polyester, or polycarbonate; molding compds.

Befmate. [Champlain Industries] Protein hydrolysates.

Befsate. [Champlain Industries] Protein hydrolysates.

Beftone. [Champlain Industries] Protein hydrolysates.

Belclene®. [Ciba-Geigy] Algicides for water treatment; scale and deposit inhibitor.

Belcor 575. [Ciba-Geigy] Hydroxy phosphinocarboxylic acid; corrosion inhibitor for ferrous metals.

Belfasin®. [Henkel/Textile] Cationic softeners for textiles.

Belgard. [Ciba-Geigy] Scale and deposit control additive.

Belite. [Ciba-Geigy] Antifoam.

Bellacide. [Ciba-Geigy] Algicide.

Bellasol. [Ciba-Geigy] Scale inhibitor.

Belros. [Ciba-Geigy] Scale and deposit control additive.

Belsil. [Wacker Chemie GmbH; Wacker Silicones] Silicone compds.; conditioner for hair and skin care prods., decorative cosmetics.

Belsperse. [Ciba-Geigy] Phosphinocarboxylic acid; dispersant for industrial water systems.

Belzak. [Belzak] Glucoheptonates.

Benathix. [Rheox] Montomorillonite clay; rheological additive for unsat. polyester.

Benecel®. [Aqualon] Cellulose prods.; thickener, stabilizer, rheology control agent, film-former, suspending agent, water-retention aid, binder for food, pharmaceutical, and cosmetic industries.

Benlate. [DuPont] Fungicides.

Benol. [Witco] White mineral oil NF; lubricant for food, drug, and cosmetics industries.

Benox. [Norac] Benzoyl peroxide; initiators.

Bentolite. [Southern Clay Prods.] White montmorillonite.

Bentone®. [Rheox] Gellant, thixotrope, rheological additive, thickener, suspending agent, antisettling agent for inks, coatings, cosmetics, antiperspirants.

Benvic. [Solvay & Cie] PVC compds.

Benzafix. [Shyamac Int'l.] Textile dyes

and pigments.

Benzamate. [Harwick] Zinc dibenzyl dithiocarbamate; ultra accelerator for NR, SBR latexes and cements.

Benzasol. [Shyamac Int'l.] Textile dyes and pigments.

Benzate-Z. [Novachem] Zinc dibenzyl dithiocarbamate.

Benzo. [Miles/Organic Prods.] Textile dyes and pigments.

Benzoflex®. [Velsicol] Benzoate derivs.; plasticizer for plastisols, PVC, cast urethane, adhesives, caulks, paints; process aid, modifier, extender.

Berbond. [Bercen] Latex saturants.

Bercen Catalyst. [Bercen] Buffered magnesium chloride sol'n.; catalysts for use with glyoxal-based reactants.

Berchem. [Bercen] Softeners, lubricants.

Bergum [Bercen] Modified natural gum; high solids gum for control of dye and pigment migration.

Bernachrome. [Berncolors-Poughkeepsie] Textile dyes and pigments.

Bernacid. [Berncolors-Poughkeepsie] Textile dyes and pigments.

Bernacron. [Berncolors-Poughkeepsie] Textile dyes and pigments.

Bernacryl. [Berncolors-Poughkeepsie] Textile dyes and pigments.

Bernactive. [Berncolors-Poughkeepsie] Textile dyes and pigments.

Bernalan. [Berncolors-Poughkeepsie] Textile dyes and pigments.

Bernalizarine. [Berncolors-Poughkeepsie] Textile dyes and pigments.

Bernamine. [Berncolors-Poughkeepsie] Textile dyes and pigments.

Bernatan. [Berncolors-Poughkeepsie] Textile dyes and pigments.

Bernazine. [Berncolors-Poughkeepsie] Textile dyes and pigments.

Bernel. [Bernel] Esters; emollients, film formers, solubilizers, dispersants for cosmetics.

Bernoil. [Berncolors-Poughkeepsie] Textile dyes and pigments.

Bernylon. [Berncolors-Poughkeepsie] Textile dyes and pigments.

Berol. [Berol Nobel] Ethoxylated fatty ethers or amines, sulfates, phosphate esters, etc.; detergent, antistat, corrosion inhibitor, emulsifier, wetting agent, hydrotrope; intermediate for agric., leather, textiles, metalworking, and plastic industries.

Berol Finetex 572. [Berol Nobel] Quaternary ammonium compd.; textile softener.

Berol PEG. [Berol Nobel] Polyethylene glycols; stabilizer for wood.

Berscour®. [Bercen] Detergent for textile applics.

Berset. [Bercen] Resins.

Bersil®. [Bercen] Silicone emulsions; imparts luxurious hand on rayon and blends; durable water repellent and fabric softener.

Bersize. [Bercen] Paper sizes and textile water repellents.

Bersoft®. [Bercen] Polyethylene or fatty ester or amide derivs.; softeners for textiles.

Berwax 4201. [Bercen] Wax blend; softener for cellulosic and blended fabrics; for napping and high sheen calendering.

Berwet®. [Bercen] Wetting agent for textile finishing baths.

Be Square®. [Petrolite/Polymers] Microcryst. wax; barrier properties; for adhesives, coatings, inks, antisunchecking agents for rubber, leather treating agents, water repellents for textiles, cosmetic ingreds., as plasticizer, binder.

Betacote. [ECC Int'l.] Kaolin; for paper coatings.

Betacure. [Acme Resin] Catalyst for phenolic ester cold box binder.

Betadyr. [Lunds of Bingley] Polyethylene.

Betagloss. [ECC Int'l.] Kaolin; paper coating clay.

Beta Lite®. [Arizona] Hydrocarbon resin; thermoplastic for printing inks.

Beta Plus. [Van Den Bergh Foods] Hydrog. soybean oil, sodium stearoyl lactylate, ethoxylated mono and diglycerides, butylhydroquinone; fluid shortening for breads.

Betaprene®. [Arizona] Olefinic resin; tackifier for SR, natural rubber; for elec. compding., inks, coatings, mastics, construction adhesives.

Beta Res. [Arizona] Hydrocarbon resin; thermoplastic for inks.

Beta Series. [Katalistiks Int'l.] Fluid cracking catalysts.

Betaset. [Acme Resin] Catalyst for phenolic ester cold box binder.

Beta-Tac®. [Arizona] Polyolefinic hydrocarbon resin; used for inks, mastics, construction adhesives.

Betathane. [Essex Specialty Prods.] Castable elastomers.

Bethamin® GFL. [Boehme Filatex] Quaternary fatty acid amide deriv.; cationic textile softener for cottons and blends.

Betricing. [Van Den Bergh Foods] Partially hydrog. veg. oil (soybean, cottonseed), mono and diglycerides, < 0.9% polysorbate 60; emulsified shortening for icings, fillings, yeast-raised prods.

Betrkake. [Van Den Bergh Foods] Partially hydrog. veg. oil (soybean, cottonseed), mono and diglycerides; emulsified shortening for cakes, icings.

Beutene®. [Uniroyal] Butyraldehyde-aniline reaction prod.; accelerator for CMNR, SBR, molded and mechanical goods; intermediate.

Bexfilm MP. [Cadillac Plastic Ltd.] Heat-stabilized polyester.

Bexloy®. [DuPont; DuPont UK] Thermoplastic polyester elastomer; for automotive trim.

BExM. [Exxon/Tomah] Cationic modifier for use in colloidal bituminous emulsions, esp. coal tar and asphalt emulsions.

Beycopen EC. [Ceca SA] Sodium/TEA alkylaryl sulfonate; antistat for polystyrene, ABS, SAN.

Beycopon. [Ceca SA] Alkylaryl sulfonate salts; emulsifier, degreaser, release agent.

Beycostat. [Ceca SA] Phosphate esters or ethoxylates, or sulfates; dispersant, antistat, degreaser, wetting agent, release agent, emulsifier for polymerization.

BFG. [BFGoodrich/Spec. Polymers] Vinyl compds.; for inj. molding, extrusion, medical, wire and cable, automotive trim.

BFP. [Am. Ingredients] Mono- and diglycerides; crumb softener, aeration aid, emulsifier for food applics.

Biafol. [Tiszai Vegyi Kombinát] Oriented PP film.

BIBBS. [Akrochem] N,N-diisopropyl benzothiazole-2-sulfenamide; delayed-action accelerator.

Bicor®. [Mitsui Petrochemicals] Oriented polypropylene coated film; for laminations, overwrap, barrier applics.

Bidrin. [DuPont/Ag] Insecticide.

Big Horn. [Wyo-Ben] Wyoming bentonite.

Biju. [Mearl] Bismuth oxychloride; colorant and pearlescent for frosted cosmetics, nail enamels.

BIK®. [Uniroyal] Surface-coated urea; activator for CR, IIR, NBR, NR, SBR; activator for nitrogen-type blowing agents, thiazole, thiuram and dithiocarbamate accelerators.

Bikorit. [Hüls Am.] White aluminum oxide.

Bi-Lite®. [Van Dyk] Bismuth oxychloride/mica; pearlescent pigments for cosmetic makeup.

Bi-Loft. [Monsanto] Acrylic fibers.

Bilt-Cote. [R.T. Vanderbilt] Kaolin clay; agric. conditioner, anticaking agent, suspending agent for paper coatings.

Bilt-Plates. [R.T. Vanderbilt] Suspending agent for paper coatings.

Bilt-Rex. [R.T. Vanderbilt] Paper resins.

Bina QAT. [Ciba-Geigy] Quaternary ammonium chloride; conditioner and emulsifier for personal care prods.

Binder. [Glo-Tex] Pigment binders.

Binder 813 Series. [Gresco Mfg.] Synthetic resin emulsion with thickeners; water phase metallic binders.

Binder QF. [Carib Int'l.] Self-crosslinking polymer; durable binder for printing; finishing agent for synthetics and cellulosics producing a soft hand.

Binder SH. [Yorkshire Pat-Chem] Soft hand pigment print binder for textiles.

Bioban®. [Angus] Biocide, preservative for metalworking fluids, latex paints, caulks, adhesives, oilfield water systems, drilling muds, textiles.

Bio-Beads. [Bio-Rad Labs] Styrene-divinylbenzene copolymers.

Biobor JF. [U.S. Borax & Chem.] Microbicide.

Biobrom. [Dead Sea Bromine] 2,2-Dibromo-3-nitrilopropionamide; biocide for industrial water systems, pulp and paper effluents, oil-recovery systems, metal-cutting coolants, air conditioning systems.

BioCare®. [Amerchol; Amerchol Europe] Hyaluronic acid derivs.; emollient, humectant, conditioner, softener, moisturizer, lubricant for hair and skin care prods.

Biocheck. [Calgon] Industrial preservative.

Bio-Chelated. [Bio-Botanica] Botanical extracts for pharmaceuticals, cosmetics, food ingredients.

Biodac. [Golden Cat] Biodegradable pesticide carriers and diluents.

Bio-Dac. [Lonza] Quaternary ammonium chlorides; antimicrobial, disinfectant, sanitizer.

Biodet. [Auschem SpA] Polyglycol ether citrates and tartrates; biodegradable raw materials for detergents.

Biodynes® TRF. [Brooks Industries] LIve yeast cell deriv.; moisturizer for skin care cosmetics; promotes wound healing; anti-inflammatory effects.

Biolase Brands. [Hoechst Celanese; Hoechst AG] Bacterial amylases; enzymatic desizing agent.

BioMeT. [Atochem N. Am.] Antimicrobial, antifoulant.

Biomet. [Ketro A/S] Antifoulants.

Biomin®. [Brooks Industries] Mineral protein derivs.; moisturizer for hair and skin care cosmetics.

Biopal®. [Rhone-Poulenc Surf.] Iodine complexes; germicide, for formulating cleaners, sanitizers.

Biopen. [Aquatec Quimica SA] Microbicide for adhesives, latexes, paints, metalworking fluids, leather preservation.

Biophos. [Brooks Industries] Soluble glycophosphoproteins; for skin and hair care cosmetics.

Bioplas. [A. Schulman] Biocide additive.

Bioplex RNA. [Brooks Industries] Propylene glycol, hydrolyzed RNA, hydrolyzed DNA; for skin and hair care cosmetics.

Biopol®. [Brooks Industries] Dermal tissue extract; for skin and hair care cosmetics.

Bio-Pol® OE. [Brooks Industries] Sodium C8-16 isoalkylsuccinyl lactoglobulin sulfonate; oil absorbing polymer, film former, pigment dispersant, color enhancer for skin prods.

Bio-Pruf®. [Morton Int'l.] Antimicrobials for plastic prods.

Biopure 100. [Nipa Labs] Imidazolidinyl urea; antimicrobial preservative for cosmetics.

Bio-Rex. [Bio-Rad Labs] Resins.

Bio-Sil. [Bio-Rad Labs] Silicic acid.

Bio-Sil. [Sil-Med] Peroxide-catalyzed silicone rubber.

Bio-Soft®. [Stepan; Stepan Canada; Stepan Europe] Alkylate sulfonic acids, alcohol ethoxylates, or olefin sulfonates; detergent, foamer, wetting agent, emulsifier; detergent base.

Biosperse®. [Drew Ind. Div.] Antimicrobials for water treatment, oil field aq. systems; preservative in aq. metalworking fluids.

Biosulphur. [Henkel/Cospha] Hydro alcohol solubilized sulfur; conditioner for skin prods.

Bio-Surf. [Lonza] Iodophor conc.; emulsifier, dispersant, germicide, detergent, sanitizer, disinfectant, wetting agent, emollient.

Bio-Terge®. [Stepan; Stepan Europe] Olefin sulfonates; detergent, foaming agent for personal care, commercial and industrial formulations.

Biothane Systems. [CasChem] Polyurethane systems; for biomedical applics.

Biovinil. [Bioplast SRL] Plasticized PVC compds.

Bioweld. [Ambersil Ltd.] Weld spatter release agent.

Biozan. [Hercules] Xanthan gum; thickener, suspending agent, emulsifier for slurry explosives, foundry coatings, cleaning compds., cosmetics, pharmaceuticals, oil field chemicals, aq. systems.

BIP. [CPS Kemi Aps] Nylon 6 or 66, PET or PBT polyester.

Bipax®. [Tra-Con] Two-part packaging for epoxy adhesives.

BIP Phthalimide. [Novachem] Alkyl phthalimides; carrier and leveling agent for cotton polyester blends.

Bisate. [Novachem] Bismuth dimethyl dithiocarbamate; accelerator for natural rubber and SBR.

Bismate®. [R.T. Vanderbilt] Bismuth dimethyldithiocarbamate; ultra accelerator for rubbers.

Bismet. [Akrochem] Bismuth dimethyldithiocarbamate; accelerator for rubbers.

Bismica. [Presperse] Mica/bismuth oxychloride blends; pearlescent pigments.

Bisoflex. [BP Chem. Ltd.] Plasticizers.

Bisulf 21. [Reilly-Whiteman] Bisulfited fish oil deriv.; for soft leather treatment.

Bitrex. [Henley] Denatonium benzoate; aversive (bitter) agent used to minimize danger of prod. ingestion; denaturant for ethanol.

BL 3. [Releasomers] Semipermanent release agents for flexible and metal molds.

Black 103. [Presperse] Iron oxides, bismuth oxychloride.

Black Out®. [R.T. Vanderbilt] Toluene/pigment blends; decorative and protective finishing agent for rubber; coating materials.

Black Pearls®. [Cabot] Pelleted carbon black; for plastics, coatings.

Bladafum®. [Bayer] Sulfotepp; insecticide; fumigant for control of greenhouse pests.

Bladex. [DuPont/Ag] Herbicide.

Blanc fixe. [Sachtleben Chemie GmbH] Barium sulfate; inert filler for plastics.

Blancol®. [Rhone-Poulenc Surf.] Sodium naphthalene-formaldehyde condensates; thinning, dispersant, peptizing agent, dye-leveling agent used in paper industry for slime control, improved retention of fillers or fines, improved sizing.

Blancosoft Conc. [Specialty Chem.] Softener for cotton knit goods, shirtings, sheetings, and yarns.

Blandol. [Witco/Sonneborn] White mineral oil NF; lubricant for food, drugs, and cosmetics industries.

Blanket Adhesive. [Carib Int'l.] Water-soluble polymers; adhesive for holding fabric to print blanket and automatic screen printing machines.

Blanket Adhesive H-98. [Catawba-Charlab] Blanket adhesive for flat bed and rotary screen print machines.

Blankit®. [BASF AG] Sodium dithionite; stabilized bleaches for textile, paper and pulp industries.

Blankophor. [Miles/Organic Prods.] Textile dyes and pigments.

Blaxon LT. [Eastern Color & Chem.] Acid phosphating conc.; for coating of metals with a corrosion resistant finish.

Blazer®. [BASF AG] Acifluorfen; for post-emergence control of broad-leaved weeds and annual grasses in soybean crops.

BLE®. [Uniroyal] Diphenylamine-acetone reaction prod.; antioxidant for CR, NBR, NR, SBR, latexes.

Bleachaid 930, 930 SP. [Evode-Tanner Industries] Chelating agents and stabilizers for hydrogen peroxide.

Bleachassist. [Reilly-Whiteman] Peroxide stabilizers for batch and continuous bleaching of textiles.

Bleach Assist 29. [Reilly-Whiteman] Chelating agents and stabilizers blend; chelating agent and peroxide stabilizer at high caustic concs. and temps.; for batch and continuous bleaching systems.

Bleach Assist 1260. [Chem-Tex Labs] Additive to peroxide/silicate bleaching sol'ns. to stabilize pH, ion conc., and chelate harmful metallic ions.

Bleach Guard. [General Chem.] Ferric nitrate sol'n.

Bleachit®. [BASF] Reductive bleaching agent; auxiliary for bleaching and optical whitening of wool and polyamide.

Bleachite XX Plus. [Leatex] Inorganic peroxygen compd.; bleaching assistant for cellulosics and blends.

Bleach SNO. [Stockhausen] Sodium

percarbonate compd.; oxidizing and bleaching agent.

B Lead Free Zinc Oxide. [Smith Chem.] Zinc oxide; activator for NR, SR.

Blendex®. [GE Specialty] ABS and blends; modifier resin for PVC, epoxies, polyurethane, PC, polyesters.

Blendmax. [Central Soya] Lyso lecithin. antioxidant, emulsifier.

Blendur®. [Bayer] Polyether polyol.

Blensil. [GE Silicones] Silicone rubber compds.; used for general purpose applics. incl. rubber rolls, diaphragms, extruded parts, o-rings.

BLO®. [ISP] gamma-Butyrolactone; solv. used in paint removers, petrol. processing, hectograph process, specialty inks; intermediate for aliphatic and cyclic compds.; reaction and diluent solv. for pesticides.

Block-Out A-SF. [Catawba-Charlab] Smoke-free auxiliary in pigment printing to achieve a discharge effect on dyed grounds.

Blo-Foam. [Rit-Chem] Blowing agents.

Blue EVA. [Cabot Plastics Ltd.] EVA copolymer; static dissipative compd.

Blue-J®. [Atochem N. Am./Textiles] Fixatives, scouring agents, optical brighteners, enzymes, wetting agents, bleaches, softeners for textiles.

Blue Shield. [Cuproquim] Copper hydroxide; agric. fungicide.

BMC. [BMC] Thermoset polyester; bulk molding compds.

Bochek. [IGI Boler] Antichecking rubber waxes.

Bocolene E. [Boliden Compd.] Polyethylene compd.

Bocolene P. [Boliden Compd.] Polypropylene compd.

BOE®. [General Chem.] Ammonium fluoride or hydrofluoric acid blends; buffered oxide etchants.

Boehringer Ingelheim. [Andertech Plastteknik A/S] Blowing and foaming agent.

Boflex. [IGI Boler] Micro waxes; for lamination.

Bohrmittel Hoechst. [Hoechst Celanese/ Colorants & Surf.] Sodium alkylsulfamido carboxylate; o/w corrosion inhibitor, lubricant, and emulsifier for metalworking fluids.

Boilerguard. [Calgon] Boiler water treatment.

Boisambrene. [Henkel/Emery/Cospha] Formaldehyde cyclododecylacetals; raw material for fragrances.

Boll Popper. [Western Nutrients] Defoliant additive.

Bolstar. [Miles/Ag] Insecticide for control of bollworm and other pests on cotton.

Boltaron. [GenCorp Polymer Prods.] PVC; for film and sheet applics.

Boltaron 5535. [GenCorp Polymer Prods.] Flame-retarded polypropylene sheet.

Bonarox. [Henkel/Emery/Cospha] Fruity aroma chemical.

Bonda Clear Casting. [Bondaglass-Voss Ltd.] Polyester resin.

Bonda Epovoss. [Bondaglass-Voss Ltd.] Epoxy resin.

Bonda Foam-Buoyancy. [Bondaglass-Voss Ltd.] Polyurethane foam.

Bonda G4 Damp & Floor Sealant. [Bondaglass-Voss Ltd.] Polyurethane resin.

Bondaprene. [Bondabelt] Neoprene rubbers.

Bonda Resin. [Bondaglass-Voss Ltd.] Polyester resin.

Bondatex. [Bondabelt] Solid rubbers.

Bondathane. [Bondabelt] Polyurethanes.

Bondedgegum. [Apollo] Nonyellowing edgegums for knit goods.

Bondesil. [Analytichem Int'l.] Bonded phase silica gels.

Bondica. [Mearl] Treated mica powd.; for plastics.

Bonding Agent. [Uniroyal] Adhesive.

Bondogen. [King Industries; Struktol] Sulfonic acid and mineral oil blend; rubber plasticizer and peptizing agent.

Bondseal. [H.B. Fuller] Sealants, protective coatings.

Bond-Tite. [U.S. Gypsum] Calcium sulfate hemihydrate; pellet binder.

Bondwave. [Flexible Reinforcements Ltd.] Nylon-reinforced PVC.

Booster Kut. [Castrol Industrial East] Extreme pressure lubricant bases.

Boot Hill. [LiphaTech] Rodenticide.

Borax. [U.S. Borax & Chem.] Sodium tetraborate; dispersant, wetting agent for NR, SR latexes; mold lubricant for general dry rubber molding.

Borester. [U.S. Borax & Chem.] Boric acid esters.

Borg-Warner. [Barkley Plastics Ltd.] ABS.

Bor-Nitrophoska®. [BASF AG] Complex fertilizer with 13% nitrogen, 13% phosphate, 21% potash, 0.1% boron; for agric. crops requiring boron and horticultural crops not sensitive to chloride.

Borogard. [U.S. Borax & Chem.] Corrosion inhibitor.

Borol. [Morton Int'l.] Bleaching agent for mech. pulps.

Boron #10. [In-Cide Tech.] Borate-based; fire retardant and preservative for wood.

Borrebond. [Borregaard LignoTech] Calcium lignosulfonate; binder, dispersant.

Borrechel. [Borregaard LignoTech] Lignosulfonate with trace elements; micronutrients.

Borresperse. [Borregaard LignoTech] Calcium or sodium lignosulfonate; dispersant, filler for pesticide formulations, concrete admixtures.

Borrewell. [Borregaard LignoTech] Chrome, ferrochrome, or iron lignosulfonates; conditioners in water-based oilwell drilling mud systems.

Bortek. [Covan Ltd.] Sodium borohydride; source of H_2 and other borohydrides; bleaching wood pulp; blowing agent for plastics; decolorizer for plasticizers.

Boscodur. [Bostik Div./Emhart] Curing agents.

Bostik. [Bostik Div./Emhart] Adhesives, coatings, polymers.

Bourbonal. [Haarmann & Reimer] Perfume and flavor.

Bovinol 30. [RITA] Serum albumin; whole protein skin, hair, and nail conditioner.

Bowax. [IGI Boler] Microcrystalline waxes.

Bower Brand. [Nat'l. Ammonia; Bower Ammonia & Chem.] Anhydrous ammonia.

Bowflon. [WJP Engineering Plastics Ltd.] Filled PTFE.

Bozemine. [Hoechst AG] Polyethylene or silicone derivs.; softener for cellulosic and synthetic fibers; improves abrasion resistance and tear strength in resin-finished fabrics.

BQDD. [Prochimie] 1,4-Benzoquinone dioxime; accelerator for rubbers.

BR®. [Am. Cyanamid] Epoxies and phenolics; potting compds.

Bradford Enzymes. [Original Bradford Soap Works] Bacterial alpha-amylase; enzymes for starch removal.

Bradford Soaps. [Original Bradford Soap Works] Fatty acid sodium or potassium salts; detergents, lubricants in flaked or liquid forms.

Bradsyn. [Original Bradford Soap Works] Scouring and wetting agents, hand modifier, lubricant for textile industry.; wet processing chemicals for textile and paper applics.

Bravo. [ISK Biotech] Chlorothalonil; broad-spectrum agric. fungicide.

Braze®. [R.T. Vanderbilt] Xylene, dimethylbenzene, xylol; rubber-to-metal bonding agent.

Breaker. [Rhone-Poulenc/Perf. Resins & Coatings] Enzyme prod. for hydrolyzing soluble polysaccharides (derivatized guar, cellulose ethers).

Breakerase. [Int'l. Bio-Synthetics] Mannan depolymerase; enzyme for controlled hydrolysis of guar gum and related polymers.

Breaxit. [Exxon] Demulsifiers.

Breon. [Aceto] NBR/conductive carbon materbatch; used for oil and chemical resistant electrically conductive elastomeric prods.

Breox. [BP Chem. Ltd.] Polyalkylene glycols; brake fluids, lubricants.

Bretol®. [Zeeland] Cetyldimethylethylammonium bromide.

Brewer's Fermex. [Enzyme Development] Enzymes for brewing.

Brewer's Mylase. [Enzyme Development] Enzymes for brewing.

Brichox. [Prince Mfg.] Manganese dioxide.

Bright BJ-5. [Dooley] Hydrogen peroxide bleach bath stabilizer for batch or continuous bleaching; wetting agent.

Brij®. [ICI Spec. Chem.] Ethoxylated alcohol ethers; surfactant, emulsifier for cosmetics, pharmaceuticals; solubilizer for fragrances.

Briquest®. [Albright & Wilson Am.] Phosphonic acids and salts; scale/corrosion inhibitors; for water treatment, oil-drilling muds, powd. detergents, photographic applics.; for peroxide stabilization in pulp bleaching and de-inking; in liq. detergents and oil-field chemicals.

Britesil®. [PQ Corp.] Sodium silicates; alkaline detergent builder.

Britesorb®. [PQ Corp.] Silica hydrogel; preservative, stabilizer for beer.

Britex. [Auschem SpA] Ethoxylated ethers; emulsifiers for cosmetics and pharmaceuticals.

Britol®. [Witco/Sonneborn] Mineral oil USP; binder, carrier, conditioner, defoamer, dispersant, extender, heat transfer agent, lubricant, moisture barrier, plasticizer, protective agent, and/or softener in adhesives, agric., cleaning, cosmetics, food, pkg., plastics, textiles.

Britomya. [Croxton & Garry Ltd.] Chalk whiting; fillers.

BRJ-473. [Schenectady] Phenol-formaldehyde one-step resin sol'n.; bonding agent for use in NBR structural adhesives.

Brom 55. [Great Lakes] Biocide.

Bromat®. [Zeeland] Cetrimonium bromide; cationic surface active agent, detergent; antiseptic; laboratory reagent; germicide.

Bromicide. [Great Lakes] Antibacterial agent for cooling towers.

Brominex. [Great Lakes] Flame retardant.

Bromobutyl. [Exxon] Brominated isobutylene-isoprene elastomer. for tire inner liners and pharmaceutical stoppers; fast curing.

Brom-O-Gas. [Great Lakes] Space, soil fumigant.

Bromoklor. [Ferro/Keil] Halogenated aliphatic liqs.; flame retardants for plasticized PVC.

Brom-O-Sol. [Great Lakes] Soil fumigant.

Bronate. [Rhone-Poulenc/Ag] Herbicide.

Bronidox® L. [Henkel/Cospha; Henkel KGaA] 5-Bromo-5-nitro-1,3-dioxane in propylene glycol; preservative for shampoos, foam baths, surfactant preps.

Bronopol. [Angus] 2-Bromo-2-nitropropane-1,3-diol; broad-spectrum antimicrobial for cosmetics and personal care prods.

Brookosome®. [Brooks Industries] Phospholipids blends; moisturizer for cosmetics.

Brookswax. [Brooks Industries] Emulsifying waxes; for hair care cosmetics.

Brophos. [Brooks Industries] Phosphate esters; cosmetics ingredients.

Brosco®. [Scholler] Textile auxiliaries.

Brosco-Nol®. [Scholler] Blends of inorganic salts and sequestrants or blends of nonionic and high polymeric compounds; processing aids, activators, water softeners, sequestrants, compatibilizers, corrosion inhibitors for textile industry.

Brown. [Presperse] Iron oxides, bismuth oxychloride.

Brucimag-S. [Premier Services] Magnesium hydroxide powd.; flame retardant and smoke suppressant for rubbers.

Brudet. [Bruce Chem.] Dispersant, lubricant, wetting agent, penetrant, emulsifier, dyeing assistants for textile processing.

BSP. [Parkland Engineering Ltd.] Polypropylene.

BSWL. [Eagle-Picher] Basic silicate white lead; pigment, heat stabilizer for plastics; rust-inhibitive pigment in the automobile industry; used in industrial or maintenance paints.

BTC®. [Stepan] Quaternary compds.; antimicrobials, disinfectants, sanitizers.

Bubble Breaker®. [Witco/Organics] Defoamers, corrosion inhibitor, detergent; for coatings, textile processing, agric. chemicals, inks, adhesives, effluent water, petroleum processing.

Bubblekup®. [Malvern Minerals] Surface-modified glass microspheres.

Bubond. [Buckman Labs] Cationic polymers.

Bubreak. [Buckman Labs] Defoamers and emulsion breakers.

Buca®. [Engelhard] Hydrous aluminum silicate; reinforcing extender for rubber and polymer systems.

BUC Benzoate. [Novachem] Ethylene glycol monobutyl ether benzoate; carrier.

Bucotive. [Blackman Uhler] Textile dyes and pigments.

Buctril. [Rhone-Poulenc/Ag] Herbicide.

Budene®. [Goodyear] Polybutadiene; used in tire treads, belting, hose covering, footwear, sponge rubber, mech. goods.

Bueno 6. [ISK Biotech] Postemergence herbicide.

Buffalo. [Corn Prods.] Corn starches.

Buffer L/Buffer H-2. [Mfg's Chem. & Supply] Buffering system effective at pH 2.5-10.0.

Buffer MSP. [Piedmont Chem. Industries] Phosphate-free monosodium phosphate replacement for buffering disperse or acid dyebaths for polyester.

Buffer NP-61. [Crompton & Knowles] Nonphosphated acid donor for atmospheric exhaust applic.

Bufloc. [Buckman Labs] Flocculating aid.

Bug-A-Bye II. [West Chem. Prods.] Aerosol insecticide.

Bugcheck®. [Angus] Simplified bacterial and fungal viable counts.

Bug Remover Conc. [Sherex] Surfactant blend; bug and tar remover conc. for automotive industry.

Bulab. [Buckman Labs] Water treatment chemical.

Bulab Flambloc. [Buckman Labs] Barium metaborate; flame retardants for plastics.

Bulk-Aid 30. [Grefco] Perlite, fused sodium potassium, aluminum silicate; inert filler, processing aid, dusting agent for NR, SR, latexes, resins.

Bullite. [Buller Plastics Ltd.] Blown low-density PVC.

Bumyr. [Amerchol] Butyl myristate; emollient; cosmetic ingredient.

Buna AP. [Hüls Am.] EPDM rubber; for molded articles and hoses; blend component; modifier for polyolefins.

Buna BL. [Bayer; Miles] Styrene-butadiene rubber, lithium.

Buna CB. [Bayer; Miles] Butadiene rubber compds.; used in tires, belting, footwear soles, transmission belting; for blending with NR in buffers, roll covers, seals, and inj. molded goods.; in rubber blends.

Bunatak. [Morton Int'l.] Polymeric resins; tackifier and plasticizer for neoprene and SBR.

Bunaweld. [Morton Int'l.] Polymeric resin; tackifier for Hypalon, neoprene, SBR, nitrile rubbers.

Bunnell. [Artilabo SA] PFA, FEP.

Bur-A-Loy®. [Mach-1 Compounding] NBR/PVC blends; can be vulcanized, used as a thermoplastic or as a hardness enhancer; used for extruded or molded prods.

Burcane. [Burn Tubes Ltd.] ABS, EVA, HDPE, LDPE.

Burco. [Burlington Chem.] Surfactants; emulsifier for metal degreasing, textile scouring, hard surface cleaning, personal care prods., household cleaners, dishwashes, food plant cleaning.

Burco Acrylic DC&L. [Burlington Chem.] Self-crosslinking acrylic emulsion polymer; for textile coating industry.

Burco BSH-400. [Burlington Chem.] Sodium glucoheptonate sol'n.; chelating agent.

Burco Color-Loc. [Burlington Chem.] Prevents migration of direct dyes, hydrolyzes reactive dyes prior to aftertreatment or during drying operations.

Burcocryl. [Burlington Chem.] Acrylic emulsions; binders, hand modifiers, finishing aid, water repellent aid; latex ingredient in flocking and laminating adhesives.

Burco Direct Leveler GRA. [Burlington Chem.] Leveling agent for direct and reactive dyeing of cotton and rayon.

Burco Duc/TX-199. [Burlington Chem.]

Low foaming detergent for textile processing.

Burcofac. [Burlington Chem.] Phosphate ester; wetting agent, detergent, emulsifier, hydrotrope for industrial and household detergent formulations.

Burco Finish. [Burlington Chem.] Softener, lubricant for surface treatment of textiles.

Burcofix. [Burlington Chem.] Fixative reducing bleeding and crocking of direct dyes; improves wetfastness.

Burcofluor. [Burlington Chem.] Produces brilliant white effects on textiles.

Burco Glyd. [Burlington Chem.] Softener for hand improvement of nylon, cotton, and blends.

Burcolev. [Burlington Chem.] Leveling agents, dyeing assistants for textiles.

Burco Lubricant G-LF/BLA. [Burlington Chem.] Lubricant for textile processing.

Burcolye. [Burlington Chem.] Liq. replacement for sodium carbonate in reactive dyeing.

Burconap. [Burlington Chem.] Fabric softener for cellulosic blends and other fabrics; napping aid.

Burco Resin. [Burlington Chem.] Pigment printing binders.

Burcorez. [Burlington Chem.] Acrylic copolymer; textile finish for durability to washing and drycleaning.

Burco Scour. [Burlington Chem.] Detergents, wetting agents, textile scouring agents.

Burcosoft. [Burlington Chem.] Softeners, lubricants for cotton and blends.

Burcosol ADS-40. [Burlington Chem.] Multifunctional component; chelating agent, antiredeposition agent, dispersant.

Burcosolv TM. [Burlington Chem.] Water-sol. solvent mixt.; solvent, emulsifier for grease cutting formulations.

Burcosperse. [Burlington Chem.] Sodium polyacrylate; chelating agent, antiredeposition agent for detergent formulations.

Burcotase. [Burlington Chem.] Proprietary detergent enzymes.

Burcoterge. [Burlington Chem.] Detergent conc.; detergent, emulsifier for laundry and hard surface cleaners.

Burcotex. [Burlington Chem.] Deaerator and penetrant for textile wet processing.

Burcotreat. [Burlington Chem.] Polyacrylic acid; for detergent formulations.

Burcovel. [Burlington Chem.] Softener and lubricant for cotton and polyester/cotton blends; replacement for silicone softeners.

Burcowet. [Burlington Chem.] Alkoxylated alcohols; wetting agent, detergent.

Burcowhite. [Burlington Chem.] Textile auxiliary for pad and exhaust applic. on cellulosics and their blends.

Burcowite. [Burlington Chem.] For bleaching applics.

Burgess. [Burgess Pigment] Anhydrous and hydrous aluminum silicates, some treated; reinforcing fillers for rubbers, resins.

Burmol®. [BASF AG] For stripping of dyeings and for removing discolorations and stains from textiles; for commercial laundries.

Burst®. [Hydrolabs] Silicone or nonsilicone defoamers for textile wet processing.

Busan. [Buckman Labs] Microbicides; UV stabilizers for plastics.

Buspense. [Buckman Labs] Fatty acid amides; lubricants for plastics.

Busperse. [Buckman Labs] Dispersants.

Butac®. [Whitney & Oettler] Resin acids-amine resin soaps blend; tackifier, molding aid for SBR rubbers, nitrile compds.; activates cure slightly.

Butacite®. [DuPont] Polyvinyl butyral resin; sheeting with uv absorbent; for safety glass laminates for automotive windshields and architectural applics.

Butaclor®. [A. Schulman] Polychloroprene.

Butaplex. [Sullivan Chem. Coatings] Cellulose acetate butyrate coatings.

Butasan. [Harwick] Zinc dibutyl dithiocarbamate; sec. accelerator in NR, SBR, EPDM, latexes; antioxidant for noncuring applics. and adhesives.

Butazate®. [Uniroyal] Zinc dibutyl dithiocarbamate; accelerator for EPDM;

stabilizer; for automotive extruded sponge, latex, foam.

Butazin. [Atochem N. Am.] Zinc dibutyldithiocarbamate; accelerator for EPDM, latexes; activator for thiazole accelerators.

Butex. [Midwest Rubber Reclaiming] Reclaimed rubber from butyl inner tubes; used in tire inner liners, inner tubes, butyl tapes and sealants.

Butisan®. [BASF AG] Metazachlor; systemic herbicide controlling annual grasses, broadleaf weeds.

Butofan®. [BASF AG] Polymer dispersions based on butadiene; binders and coating agents for papers, fibers, textile coating, adhesives, leather-based materials.

Butonal®. [BASF AG] Butadiene/styrene polymer dispersions; binders for prod. of adhesives, treatment of asphalt.

Butoxyne. [ISP] Butynediol hydroxyethyl ethers; corrosion inhibitor, pickling inhibitor, nickel brightener; for electroplating applics.

Butrol. [Buckman Labs] Corrosion and scale inhibitor.

Butvar. [Monsanto] Polyvinyl butyral.

Butyl. [Exxon Chem. Mediterranea SpA] Isobutylene elastomers.

Butyl Carbitol®. [Union Carbide] Butoxydiglycol; solvent.

Butyl Carbitol® Acetate. [Union Carbide] Diethyene glycol butyl ether acetate; solvent.

Butyl Cellosolve®. [Union Carbide] Butoxyethanol; solvent.

Butyl Cellosolve® Acetate. [Union Carbide] Butoxyethanol acetate; solvent.

Butyl Diglyme. [Ferro/Grant] Diethylene glycol dibutyl ether; solvent used in electrochemistry, polymer and boron chemistry, gas absorption, extraction, stabilization; in fuels, lubricants, textiles, pharmaceuticals, pesticides.

Butyl Dioxitol. [Shell] Butoxydiglycol; solvent for use in lacquers, inks; coupling solvent for cleaning sol'ns. and cutting oils; coalescing agent in latex paints; component of brake fluids in automotive industry.

Butyl Eight®. [R.T. Vanderbilt] Activated dithiocarbamate; ultra accelerator for NR, SBR.

Butyl Oleate C-914. [C.P. Hall] Butyl oleate; plasticizer.

Butyl Oxitol. [Shell] Butoxyethanol; solvent used in surf. coating formulations such as lacquers and enamels; coupling agent in cleaners and cutting oils.

Butyl Propasol. [Union Carbide] Propylene glycol butyl ether; solvent for coatings, cleaners, electronic, and ink applics., water-reducible polyester and alkyd resin prod.

Butyl Stearate C-895. [C.P. Hall] Butyl stearate; solvent, spreading and softening agent in plastics, textiles, cosmetics, rubbers.

Butyl Tuads®. [R.T. Vanderbilt] Tetrabutylthiuram disulfide; accelerator for natural and polyisoprene rubbers.

Butyl Zimate®. [R.T. Vanderbilt] Zinc dibutyldithiocarbamate; rubber accelerator.

Butyrac. [Rhone-Poulenc/Ag] Herbicide.

B-W. [PQ Corp.] Sodium silicate.

BWT. [Stewart Hall] Boiler compd. for steamline corrosion prevention, oxygen scavenging.

BXA®. [Uniroyal] Diarylamine-ketone-aldehyde reaction prod.; protects against heat and oxygen in CR, NBR, SBR, in tires, inner tubes, insulated wire, soles, heels, mech. goods.

BY-59-18. [Neville] Hydrocarbon resin. used as replacement for rosin derivs., and in printing ink vehicles, flushing and grinding applics.

Byco. [Croda Inc.] Gelatin NF; sizing, textile and paper adhesives, cements, capsules for medications, matches; clarifying agent; protective colloid in ice cream; stabilizer, thickener, texturizer in food.

BYK. [BYK-Chemie USA] Wetting agent, dispersant, flow, leveling, antimar and slip agent, defoamer, catalyst additive.

Bykanol. [BYK-Chemie USA] Viscosity reducing additive.

Byketol. [BYK-Chemie USA] Leveling additive.

Byktone. [BYK-Chemie USA] Wetting/

dispersing additive.

Bykumen. [BYK-Chemie USA] Wetting/dispersing additive.

Bynel®. [DuPont; DuPont UK] Co-extrudable adhesive resins.

Byrasol. [Acrol Ltd.] PVC plastisols.

BZW-70. [Witco/Argus] Benzoyl peroxide; initiator for polymerization of vinyl monomers, unsaturated polyester resins, suspension and emulsion polymerization.

C

C-. [Bacon] Epoxy resin; potting compd. for potting and casting applics.

C-. [Procter & Gamble] Fatty acids and alcohols; intermediates for mfg. of soaps, amides, esters, surfactant and nonsurfactant applics.

C-. [Wacker Silicones] Silicone elastomers.

C-20. [Feldspar] 20-mesh soda feldspar.

C-90, -95. [Hoffmann-La Roche] Ascorbic acid.

C-560 Dispersion. [R.T. Vanderbilt] Zinc oxide blend.

CA-. [Eastman] Cellulose acetate.

CA0397 Series. [Hüls Am.] Silane derivs.; coupling agent, chemical intermediate, blocking agent, release agent, lubricant, primer, reducing agent.

CAB. [Eastman] Cellulose acetate butyrate. used in lacquers, hot melts, powd. coatings, printing inks, cloth coatings for airplanes, wire, leather, and plastics.

Cabelec®. [Cabot Plastics Ltd.] Thermoplastic blends with carbon black; semiconductive compd. for shielding of power cables, pkg. and prod. handling applics., inj. molding and sheet extrusion.

Cab-O-Sil®. [Cabot] Fumed silica; for mfg. of glass, water glass, refractories, abrasives, ceramics, enamels, petrol. prods.; filler in cosmetics; rubber reinforcing agent; as anticaking and defoaming agent; abrasive; thickener, dispersant.

Cab-O-Sperse®. [Cabot] Fumed silica aq. dispersion; thickener, rheological control agent for aq. systems.

Cabot®. [Cabot Plastics Ltd.] LDPE masterbatches with processing aids.

Cacahlot®. [M. Michel] Fatty alcohols; emollient, conditioner, lubricant for cosmetics; corrosion inhibitor for lube oils; mold release/processing aid for plastics and rubber.

CA DBS 50 SA. [Witco SA] Calcium dodecylbenzene sulfonate in aromatic solvent; emulsifier for pesticides.

Caddy. [W.A. Cleary] Liq. cadmium turf fungicide.

Cadet®. [Akzo] Benzoyl peroxide; initiator for curing unsat. polyester resins.

Cadflon. [Cadauta sas di R Fornasero & C] Polytetrafluoroethylene.

Cadfluorene. [Cadauta sas di R Fornasero & C] Polyvinylidene fluoride.

Cadmate®. [R.T. Vanderbilt] Cadmium dithiocarbamates; accelerator for natural and polyisoprene rubbers.

Cadmolith. [SCM] Cadmium lithopone; pigments.

Cadon®. [Monsanto] Styrene-maleic anhydride terpolymers; engineering thermoplastic for inj. moldings.

Cadoussant. [Ceca SA] Fiber softeners.

Cadoussant AS. [Ceca SA] Softening detergent for wool and cotton.

Cadox®. [Akzo] Organic peroxide compds.; initiator for curing polyester resins; crosslinking agent for curing silicone rubbers.

Cadplen. [Cadauta sas di R Fornasero & C] Polypropylene.

Cadra EXP. [Leatex] Complex opthalic acid deriv., organic ester; odorless atmospheric carrier for polyester and polyester blends.

CAE. [Ajinomoto] PCA ethyl cocoyl arginate; cationic surfactant, foamer, antistat, preservative, antiseptic, germicide, disinfectant in cosmetics, detergents, dentifrices, medical supplies.

Cairox. [Carus] Potassium permanganate.

Cake Mix 96. [Van Den Bergh Foods] Partially hydrog. soybean oil, propylene glycol mono and diesters of fats and fatty acids, mono and diglycerides, optional lecithin; shortenings for cake mixes.

Calamide®. [Pilot] Fatty acid alkanol-amides; nonionic emollient, lubricant, foam stabilizer, visc. builder, detergent, solubilizer for personal care prods., industrial applics.

Calbex. [Calbar] Epoxy coatings, adhesives, sealants.

Calcicat. [Ultra Additives] Ink, paint, and varnish driers.

Calcidar. [Omya GmbH] Calcium carbonate.

Calciplast. [Alpha Calcit Fullstoff GmbH] Fine, high purity calcium carbonate.

Calcium Petronate. [Witco/Sonneborn] Calcium petroleum sulfonate; anionic detergent, rust inhibitor, emulsifier for lube oil additives.

Calcium Stearate Regular. [Witco] Calcium stearate; lubricant for ceramics, foundry sands, plastics; plasticizer/lubricant for paper coatings; aux. emulsifier.

Calcium Sulfonate. [Witco/Sonneborn] Calcium sulfonates; extreme pressure additive, rust preventive, industrial detergent, lube oil additive.

Calco. [BASF] Textile dyes and pigments.

Calcocid. [BASF] Textile dyes and pigments.

Calcoda. [Calbar] Silicone; masonry water repellent.

Calcofluor. [BASF] Textile dyes and pigments.

Calcophyl. [BASF] Textile dyes and pigments.

Calcotone. [BASF] Textile dyes and pigments.

Calcozine. [BASF] Textile dyes and pigments.

Calendula Oil CLR. [Henkel/Cospha] Soybean oil, calendula extract, tocopherol; emollient, conditioner for skin care preps.

Calester. [Pilot] Alpha sulfo methyl laurate; surfactant for toilet soaps, laundry detergents, automotive cleaners, foamers, emulsifiers.

Calfax®. [Pilot] Sodium alkyl diphenyl oxide disulfonate; surfactant, solubilizer, dispersant for dye bath leveling, pigment dispersion, cleaners, latex emulsification.

Cal-Flor-Dry®. [Floridin] Fuller's earth; industrial absorbent.

Calfoam. [Pilot] Alcohol and ethoxy sulfates; detergent, foam booster/stabilizer, wetter, emulsifier for detergent systems, personal care prods., emulsion polymerization.

Calgon®. [Calgon Carbon] Activated carbon; filter medium.

Cal-Grid I, II. [Quigley] Calcium-aluminum additive.

Caliban. [Alpha Calcit Fullstoff GmbH] Fine, high purity calcium carbonate.

Calibre. [Dow Plastics] Polycarbonate resin; engineering thermoplastic for food contact, medical, transportation, appliance, housewares, business machines, recreation, and service industries.

Califlux®. [Witco/Golden Bear] Aromatic processing oil; extender oil for elastomer compounding.

Calight RPO. [Calumet] Naphthenic process oil.

Calimulse PRS. [Pilot] Isopropylamine dodecylebenzene sulfonate; anionic emulsifier, solubilizer, pigment dispersant for dry cleaning, degreasers, agric. sprays; latex emulsifier.

Calixin®. [BASF AG] Tridemorph compds.; systemic fungicide.

Callaway. [Callaway] Dye fixatives for direct, reactive, sulfur and pigment dyes; lubricants in textile finishing; hand modifiers.

Cal-Max. [Ash Grove Cement] Pulverized quicklime.

Calnox. [Baker Perf. Chem.] Sodium polyacrylate; scale inhibitors.

Calomel. [Thor] Mercurons chloride.

Calon. [Lion] Polystyrene sulfonic acid, sodium salt; antistatic agent for paper.

Calona. [Purflo DTL SA] Polyester.

Caloria. [Exxon] Heat-transfer fluid.

Caloxol. [SA Sturge] Dehydrating agents.

Caloxylate N-9. [Pilot] Nonoxynol-9; detergent, emulsifier for hard surface cleaners, liq. laundry prods., general purpose cleaning compds.; biodeg.

Calquest. [Mfg's Chem. & Supply] Sequestrant blend.

Calsan. [PPG Industries] Calcium stearate dispersions.

Cal-Seal. [Calbar] Butyl, acrylic, polysulfide sealants.

Calsoft. [Pilot] Alkylaryl sulfonic acid and salts; surfactant, detergent, emulsion stabilizer, wetting and foaming agent for household and industrial detergents, agric. formulations, emulsion polymerization, cosmetics, textile scouring.

Calsol. [Calumet] Naphthenic process oil; for rubber industry, resin extending, PVC, textiles, caulking compds.

Calsolene. [ICI Am.] Sulfated fatty acid ester; anionic surfactant, emulsifier, wetting agent, penetrant, lubricant, leveling agent.

Calstar. [FMC] Calcium phosphate dibasic.

Calsuds. [Pilot] Surfactant blends; foamer, wetting agent, visc. modifier, emulsifier for liq. detergents, shampoos, textile scours, agric. sprays.

Calthane. [Cal Polymers] Two-component urethane elastomer; for flexible and rigid castings, prosthetics, underwater sports equipment, optical lenses, art objects, mech. devices, flexible windshields, industrial tubing or sheeting, toys, adhesives.

Cal'Therm. [Ouest Isol] Silicate.

Cal/Tint. [Hüls Am.] Universal tinting colors.

Caltreat®. [General Chem.] Liq. calcium chloride.

Calwet No. 435. [Saramco] Castor oil ethoxylate; surfactant.

Calwhite®. [Georgia Marble] Calcium carbonate; extender filler for liq. polyester systgems.

Camacryl. [John Campbell] Textile dyes and pigments.

Camacyl. [John Campbell] Textile dyes and pigments.

Camel-CAL®. [Genstar Stone Prods.] Calcium carbonate; filler for water-based coatings and inks, paper, and PVC pipe.

Camel-CARB®. [Genstar Stone Prods.] Calcium carbonate; filler/extender used in interior flat paint and exterior house paints, rubber compds., putty and caulk, ceramics, adhesives, linoleum, floor tile, and textile coatings.

Camel-FIL. [Genstar Stone Prods.] Calcium carbonate; filler for high loadings in glass-reinforced polyester, in PVC, PP, rubber automotive goods and floor tiles, caulks, sealants, and adhesives.

Camelon. [John Campbell] Textile dyes and pigments.

Camel-TEX®. [Genstar Stone Prods.] Calcium carbonate; filler for paints, primers, sealers, polyester-fiberglass premixes, preforms, and hand lay-up gel coats, rubber automotive prods., household prods., tubing, medical prods., closures, putty, caulk, adhesives.

Camel-WITE®. [Genstar Stone Prods.] Calcium carbonate; filler for paint, paper, paper coating, PVC, rubber (automotive goods, footwear, medical supplies), thermoplastics, thermosets, and in caulks, glazing compds., ceramics, adhesives, food processing.

Camester. [John Campbell] Textile dyes and pigments.

Camie. [Camie-Campbell] Aerosol and bulk adhesives; for textile screen printing, pkg., floor covering, construction, apparel mfg., maintenance, automotive, upholstery, and furniture applics.

Campamio. [Union Camp] Amidoamines/imidazolines.

Cam Soft. [Riechem/Mt. Vernon Mills] Polyethylene emulsion; softener for resin finishing.

Canamulse. [Canada Packers/Edible Oils] Propylene glycol mono fatty acid esters or mono- and diglycerides; food emulsifiers.

Canasperse. [Canada Packers/Food Ingreds.] Lecithin; food emulsifiers, dispersants, wetting agents.

Can Bond SPA. [Astro Industries] Modified acrylic polymer; slipproofing agent for all fibers.

Candex®. [Mendell] Dextrose blend;

used in chewable tablets.

Canfelzo. [Pigment & Chem.] Zinc oxide; pigment.

Canfelzo Photozinc. [Pigment & Chem.] Zinc oxide; electrophotographic grade.

Canguard®. [Angus] Preservative for latex paints, emulsions, caulks, adhesives, textiles.

Canopy. [DuPont/Ag] Herbicide.

Cantrece. [DuPont] Nylon.

CAO®-. [PMC Specialties] Antioxidants for rubber elastomers, polymerics; additive for oils, fats, greases; stabilizer for petrol. prods., waxes, insecticides.

CAP-. [Eastman] Cellulose acetate propionate; used in printing inks, paper coatings, cloth coatings, grease barrier applics.

Capa. [Interox Chem. Ltd.] Caprolactone monomer and polymers.

Capcure® Emulsifier. [Henkel/Functional Prods.] Ethoxylates; emulsifier for epoxy resins.

C-A-P Enteric Coating Polymer. [Eastman] Cellulose acetate phthalate.

Capital. [Karlshamns] Vegetable oil prods. and fatty acids;

Caplube. [Karlshamns] Esters or ethoxylates; emulsifier, dispersant, lubricant, antistat, coupler for textiles, food, industrial applics.

Capmul®. [Karlshamns] Glyceryl esters, sorbitan esters, propylene glycol esters and their ethoxylates; food emulsifier, pigment dispersant, defoamer, stabilizer, internal lubricant for cosmetics.

Capow®. [Kenrich Petrochemicals] Powdered coupling agents.

Capran®. [Allied-Signal] Nylon 6 film; thermoplastic film.

Capran® Unidraw®. [Allied-Signal] Oriented nylon 6; carrier web for molding fiberglass-reinforced panels.

Capri Line LF Series. [Astro Industries] Blend of natural and acrylic polymers; hand modifier and slipproofing agent for textiles.

Caprolactam. [Bayer] Fiber raw material; anionic polymerization.

Caprolan®. [Allied-Signal/Fibers] Nylon 6; tire yarn.

Caprol®. [Karlshamns] Polyglyceryl esters; nonionic emulsifiers, solubilizer, humectant, lubricant, dispersant, thickener, stabilizer for foods, textiles, cosmetics.

Capron®. [Allied-Signal; Croxton & Garry Ltd.] Nylon 6 homopolymers, copolymers, glass and mineral-reinforced grades; for inj. molding and extrusion.

Caps®. [Kenrich Petrochemicals] Pelleted coupling agents.

Cap-Sil. [Ronald T. Dodge] Microencapsulated silicone oil.

Captan. [Natural Gas Odorizing] Gas odorant prods.

Captax®. [R.T. Vanderbilt] 2-Mercaptobenzothiazole; accelerator for natural and synthetic rubbers.

Captax®-Tuads Blend. [R.T. Vanderbilt] 2-Mercaptobenzothiazole/tetramethylthiuram disulfide blend; accelerator for butyl rubbers.

Captex®. [Karlshamns] Propylene glycol esters, fatty acid triglycerides; carrier, coupler, solvent, emollient, moisturizer, fixative, lubricant, vehicle, extender for cosmetics, pharmaceutical, nutritional prods.

Capture. [FMC/Ag] Insecticide/miticide.

Caracet. [Carolina Color & Chem.] Textile dyes and pigments.

Caracid. [Carolina Color & Chem.] Textile dyes and pigments.

Caracryl. [Carolina Color & Chem.] Textile dyes and pigments.

Carafloc. [Carus] Polymers.

Caralan. [Carolina Color & Chem.] Textile dyes and pigments.

Caranthrene. [Carolina Color & Chem.] Textile dyes and pigments.

Caraster. [Carolina Color & Chem.] Textile dyes and pigments.

Car-A-Van. [Morton Int'l.] EP lubricant additives.

Carbaicar. [Costa Floros O.E.] Urea-formaldehyde molding material.

Carbam. [Vinings Industries] Ferbam; accelerators.

Carbamate. [FMC/Ag] Fungicide.

Carbapon®. [Hoechst Celanese/Colorants & Surf.] Dispersant, sequesterant, oxidative desizing agent for textiles.

Carbaryl 90 DF. [Terra Int'l.] Dry flowable insecticide.

Carbavert. [Henkel/Emery/Cospha] Fruity, banana-like aroma chemical.

Carbaway. [Oakite Prods.] Solvent cleaner.

Car-Bel-Rez. [Harwick] Hydrocarbon/ sulfur reaction prod.; processing aid in elastomers.

Carbital®. [ECC Int'l.] Calcium carbonate; paper coating and filling applics.

Carbitol®. [Union Carbide] Ethoxydiglycol; industrial solvent.

Carbobond. [BFGoodrich/Spec. Polymers] PSA emulsion; adhesive polymers.

Carbochem. [Chemtrade Inc.] Activated carbon.

Carbocite. [Shamokin Filler] Industrial carbon.

Carbofil. [Shamokin Filler] Carbon filler.

Carb-O-Fil. [Shamokin Filler] Pulverized anthracite; filler for NR, SR, adhesives, mech. goods, thermosetting resins.

Carboflex. [Ashland-Südchemie-Kernfest GmbH] Carbon fibers.

Carboflow. [BFGoodrich/Spec. Polymers] Additives for inks and coatings.

Carboglas. [Carboline] Polyester lining.

Carbokup. [Malvern Minerals] Surface-modified calcium carbonate.

Carbolon. [Exolon-Esk] Silicon carbide.

Carbomastic. [Carboline] Epoxy coal tar coating.

Carbomix®. [Copolymer Rubber] SBR/ carbon black masterbatches; for extruded and molded mechanical goods, tires.

Carbon Detergent K. [Henkel/Emery/ Cospha] Fatty nitrogen compd.; detergent for removing carbon from metal surfaces.

Carbonic. [Carbonic Industries] Carbon dioxide.

Carbo-Pentro 133. [Organic Dyestuffs] Fatty alcohol sulfate; detergent with wetting and emulsifying properties for carbonizing.

Carbopol®. [BFGoodrich/Spec. Polymers] Polyacrylic acids; emulsifier,

stabilizer, moisturizer, thickener, suspending agent, gellant for cosmetics, pharmaceuticals, detergents, drilling muds, water treatment, textiles.

Carbore. [Omya GmbH] Calcium carbonate.

Carbose. [Carbose] Carboxymethyl cellulose; tech. and detergent grades.

Carboset®. [BFGoodrich/Spec. Polymers] Acrylic resin sol'ns.; thermoplastic and thermoset film-forming resin used in protective metal coatings, paints, ceramics, adhesives, textiles, paper, leather, cosmetics, floor polishes, chemical specialties.

Carbostat 2203. [Hoechst Celanese/ Colorants & Surf.] Cationic quaternary amine; textile and fiber chemicals.

Carbotac. [BFGoodrich/Spec. Polymers] PSA emulsion; adhesive polymer.

Carbotam. [Shieldalloy Metallurgical] Boron additives; for steelmaking.

Carbotex S. [Manufacturers Chems.] Soaps and phsphates; fulling detergent, soaping-off agent for naphthol dyes.

Carbowax®. [Union Carbide] PEG; binder, lubricant, intermediate, antistat, release agent, plasticizer; for ceramics, powd. metallurgy, toilet bowl cleaners, adhesives, inks, mining, soaps, detergents, creams and lotions, dentifrices.

Carbowax® Sentry. [Union Carbide] PEG; emollient base for creams, lotions, pharmaceuticals and makeup; carrier for medicinals in cosmetics and pharmaceuticals.

Carbo Weld. [Carboline] Zinc primer.

Carboxide. [Union Carbide] Fumigants.

Carbo Zinc. [Carboline] Zinc coating.

Carding Oil A-4 Super. [Takemoto Oil & Fat] Nonionic surfactant, neutral oil; carding oil.

Cardipol®. [Petrolite/Polymers] Oxidized polyethylene; for formulating hot-melt adhesives.

Cardis®. [Petrolite/Polymers] Oxidized microcrystalline wax; used in formulation of emulsions, polishes, coatings; modifier for solvent systems.

Cardolite®. [Petrolite/Polymers] Raw material for surfactants, antioxidants, anticorrosives; lubricant additive;

cosolvent for insecticides; coupling agent; resin modifier; curing agent; reactive flexibilizer; extender/diluent for epoxy mfg.

Cardox. [Cardox] Carbon dioxide.

Caredon. [Carey Industries] Textile dyes and pigments.

Carefoam DX. [Dunlopillo UK] Fire-retardant foam.

Carefree SS. [Sybron] Textile softener.

Caresine® 2000. [BASF AG] Bentazon, isoproturon, dichlorprop; binders for prod. of adhesives, treatment of asphalt.

Carester. [Carolina Color & Chem.] Textile dyes and pigments.

Cargill High Purity PHG Evaporated Salt. [Cargill/Salt] High purity evaporated salt; carrier or diluent for dyes; conditioning agent for fabrics, water softener regeneration.

Cargill Hi-Tex® Evaporated Salt. [Cargill/Salt] High purity evaporated salt; carrier or diluent for dyes; conditioning agent for fabrics, water softener regeneration.

Carib-Antimigrant CK. [Carib Int'l.] Antimigrant producing uniform dyeings; stabilizers dye bath.

Carib-Binders. [Carib Int'l.] Resins for printing of natural and synthetic fibers.

Carib-Extender Concs. [Carib Int'l.] Concs. to prepare printing pastes for roller, rotary screen, and flat bed screen printing.

Carib-Fast Colors. [Carib Int'l.] Resin-bonded pigment system for roller, rotary screen, and flat bed screen printing of natural and synthetic fibers.

Carib-Fire Retardants. [Carib Int'l.] Durable and nondurable fire retardants for textiles.

Carib-Hide Colors. [Carib Int'l.] Ready-to-print colors for problem printing areas.

Carib-Pad Colors. [Carib Int'l.] Pad dyeing colors for dyeing polyester, cotton, glass fiber, rayon, acrylic, modacrylic and synthetic/natural fiber blends.

Carib-Pad Emulsions. [Carib Int'l.] Resins for pad dyeing of natural and synthetic fibers.

Carib-Sperse Colors. [Carib Int'l.] Nonionic dispersions for printing and pad dyeing of natural and synthetic fibers.

Carib-Tank and Machine Washes. [Carib Int'l.] Biodegradable water-based cleaners for printing and pad dyeing operations.

Cariflex BR. [H. Muehlstein] Butadiene rubber.

Cariflex Sol'n. SBR. [H. Muehlstein] Sol'n. SBR.

Carmine. [Presperse] Carmine and blends; natural colors.

Carmisol-50. [Presperse] Carmine.

Carnation. [Witco/Sonneborn] White mineral oil NF; emollient and lubricant for cosmetics, baby oil.

Carnauba Spray 200. [Sherex] Quaternary blend; car rinse.

Carnebon 200. [Int'l. Dioxcide] Stabilized chlorine dioxide.

Carocet. [Carolina Color & Chem.] Textile dyes and pigments.

Carochrome. [Carolina Color & Chem.] Textile dyes and pigments.

Carodirect. [Carolina Color & Chem.] Textile dyes and pigments.

Carofix. [Carolina Color & Chem.] Textile dyes and pigments.

Carolid®. [Sybron] Textile dye carriers.

Carolube. [Chemol] Fatty acid esters; softener, fiber lubricant for textiles.

Caronic. [Chemonic Industries] Emulsified solvents; dye carriers with leveling and wetting properties for atmospheric or pressure equip.

Carosol. [Carolina Color & Chem.] Textile dyes and pigments.

Carpenter. [Carpenter Tech.] Steel and other alloys.

Carriant. [Toho Chem. Industry] Dyeing assistants, carriers for polyester fibers.

Carrier. [Nikko Chem. Co. Ltd.] Carriers for polyester/wool blended fabrics and yarn.

Carrier 35. [A. Harrison] Biphenol type; carrier for acrylics.

Carrier 300, 326, LO. [Gresco Mfg.] Modified biphenyl and solvent blend; self-emulsifiable liq. carrier for polyesters and blends.

Carrier DPO, LO4, NOC, R54, RF25. [Piedmont Chem. Industries] Carriers for textile dyeing.

Carrier LO, SG. [Mfg's Chem. & Supply] Biphenyl carrier for dyeing polyester and blends.

Carrier SSL. [Reilly-Whiteman] Modified cyclic organic compd.; liq. carriers for use with polyester in becks, pkg. machines, skein dyeing.

Carroll 40% Coconut Hand Soap. [Carroll] Potassium coconut oil soap; for hand soaps, shampoos.

Carrot Oil CLR. [Henkel/Cospha] Soybean oil, carrot oil, carrot extract, beta-carotene, tocopherol; emollient, conditioner, superfatting agent; emulsified and oily preparations for care of skin and hair;

Carrybon. [Sanyo Chem. Industries] Polycarboxylic acid type surfactants; dispersant for pigments, paints.

Carryol. [Sanyo Chem. Industries] Fatty acid derivs.; cold flow improver and dispersant for fuel oils.

Carsamide®. [Lonza] Fatty acid alkanolamides; emulsifier, thickener, detergent, dispersant, wetting agent, foam booster/stabilizer for textile scouring, industrial, cosmetic, and household cleaners.

Carsamine®. [Lonza] Amines; chemical intermediates, acid detergents, metal and textile processing aids, personal care additives.

Carsofoam®. [Lonza] Surfactants; detergent conc. for shampoos, all-purpose cleaners, paint strippers, metal cleaners.

Carsol. [Carey Industries] Textile dyes and pigments.

Carsonol®. [Lonza] Alcohol sulfates or alcohol ether sulfates; detergent, wetting agent, foaming agent, emulsifier for cosmetics, dishwashing, soaps, chemical specialties.

Carsonon®. [Lonza] Alcohol alkoxylates; wetting agent, detergent, emulsifier, stabilizer, dispersant used for maintenance and institutional cleaners, in textile, paper, and paint industries.

Carsoquat®. [Lonza] Quaternary ammonium compds. and blends; corrosion inhibitor, surfactant, conditioner, antistat; for cosmetics, pharmaceuticals, textiles.

Carsosoft®. [Lonza] Quaternaries; fabric softener bases.

Carsosulf. [Lonza] Sulfonic acids and salts; coupler, solubilizer for liq. detergent systems.

Carsperse. [Carey Industries] Disperse dyes.

Carspray. [Sherex] Cationic rinse aids for automatic car washes.

Carstab®. [Morton Int'l.] Benzophenones and thiodipropionates; uv light stabilizer, antioxidant for PP, HDPE, LDPE, EVA, PVC, PC, acetals, epoxies, some cellulosics, in cosmetics, protective coatings, food use.

Carta. [Sandoz] Textile dyes and pigments.

Cartaretin F-4. [Sandoz] Adipic acid/ dimethylaminohydroxypropyl diethylene triamine copolymer; substantive polymer for hair care prods., shampoo systems.

Cartasol. [Sandoz] Textile dyes and pigments.

Carudan. [Grindsted Prods.] Locust bean gum; controls syneresis, improves texture in processed cheese.

Carulite. [Carus] Catalyst.

Carum. [Exxon] Chemical-resistant grease.

Carusorb. [Carus] Catalyst.

Carvat. [Carey Industries] Vat dyes.

Casacure M. [Chemtrade Int'l. BV] Polyurethane curing agent.

Casamid®. [Air Prods.; Pacific Anchor] Water-dispersible epoxy curing agents.

Casarez. [Chemtrade Int'l. BV] Epoxide resins.

Casathane. [Chemtrade Int'l. BV] Polyurethane prepolymers.

Cascade. [Union Carbide] Industrial gases.

Cascarol. [Dr. Madis Labs] Cascara glycosides.

Casoron. [Uniroyal] Herbicide.

Cassapret SRH. [Hoechst AG] Polyester copolymer; hydrophilizing agent.

Cassastat. [Hoechst AG] Quaternary

ammonium salt; antistat for synthetic fibers.

Cassofix®. [Hoechst Celanese/Colorants & Surf.] Fixing agent for improving waterfastness and washfastness of direct and reactive dyeings.

Cassulfon® Laking Agent PF. [Hoechst Celanese/Colorants & Surf.] Amine deriv.; alkylating agent for aftertreatment of sulfur dyes to give improved wetfastness properties.

Cassurit®. [Hoechst Celanese/Colorants & Surf.; Hoechst AG] Reactant resins; for resin finishing of cellulosic fibers.

Castethane. [Dow] Urethane elastomer systems.

Castorwax®. [CasChem] Hydrogenated castor oil; wax for formulating antiperspirant sticks; release agent, suspending aid.

Castrol. [Castrol Industrial East] Metalworking fluids.

Castrol Forge. [Castrol Industrial East] Hot forging lubricants.

Castrol Kleen. [Castrol Industrial East] Industrial and chemical cleaners.

Castrol Quench. [Castrol Industrial East] Quenching oils.

Castung. [CasChem] Dehydrated castor oil; binders for paints, caulks, sealants, inks.

Casul® 70 HF. [Harcros] Calcium dodecylbenzene sulfonate; coemulsifier for o/w and w/o formulations.

Cata-Chek 820. [Ferro] Dibutyltin dilaurate; heat and light stabilizer for flexible vinyl formulations.

Catadiene®. [Air Prods.] Catalysts.

Cataflot. [Ceca SA] Flotation reagent for mineral industry.

Catafor. [Rhone-Poulenc Ltd.] Quaternary ammonium ethosulfates; antistat.

Catafor®. [Aceto] Conductivity additive for the plastics industry, coatings, inks, adhesives.

Catalpo®. [Engelhard] Hydrous aluminum silicate; reinforcing extender for rubber and polymer systems.

Catalyst. [Auralux] Catalysts for silicone water repellents, thermosets.

Catalyst. [CNC Int'l.] Catalysts for thermosets used in textile processing.

Catalyst. [Eastern Color & Chem.] Catalysts for thermosetting resin finishing, for water repellent silicone finishing, textile resins.

Catalyst. [Hickson Danchem] Magnesium chlorides; catalyst for durable press resins.

Catalyst. [Ivax Industries] Catalyst for reactants and resin systems.

Catalyst 3P, N. [Sidney Springer] Buffered salt sol'n.; resin catalyst.

Catalyst 9. [BASF] Fast cure catalyst for resin finishes.

Catalyst 42C, CT. [Sybron] Catalysts for reactants or resins.

Catalyst 110. [Riechem/Mt. Vernon Mills] Modified magnesium salt sol'n.; activated catalyst for resins and reactants.

Catalyst 531, KR. [Sequa] Catalysts for resin/reactant systems.

Catalyst NKS. [Hoechst AG] Metal salts/organic acids blend; acid donor for crosslinking of reactant resins.

Catalyst PAT. [ICI Surf. UK] Alkanolamine hydrochloride; for use with acid-curing (thermosetting or reactant type) finishing resins.

Catamine. [Reilly-Whiteman] Cationic softener for textiles.

Catamine 101. [Exxon/Tomah] Fatty amine complex; asphalt emulsifier.

Catapal®. [Vista] Alumina monohydrate; catalyst binders and supports.

Catavat. [Catawba-Charlab] Vat dyestuffs.

C-A-T Enteric Coating Polymer. [Eastman] Cellulose acetate trimellitate.

Cat Floc. [Calgon] Cationic polymer.

Catigene®. [Stepan Europe] Quaternary ammonium chlorides; germicidal, algicidal, fungicidal, deodorizing, and antistatic agents.

Catimine. [Yoshimura Oil Chem.] Cationic surfactant; antistat for synthetic fabrics.

Catimuls. [ScanRoad] Emulsifiers for asphalt and oil.

Catinal. [Toho Chem. Industry] Quaternary ammonium salts; disinfectant, germicide, antistat; dyeing assistant; base material for hair rinses.

Catinex. [Pulcra SA] Ethoxylated nonyl phenyl and octyl phenyl ethers; detergent intermediate, emulsifier, dispersant, wetting agent, stabilizer, drycleaning agent; for insecticides, household and industrial cleaners, textile processing, petroleum prods., paints, metal pickling; solubilizer for perfumes.

Cation. [Sanyo Chem. Industries] Quaternary compds. hair rinse base, fabric softener, lubricant, conditioner.

Cationic Collagen Polypeptides. [Maybrook] Cationic collagen polypeptides; substantivity agent, film former, protective colloid for hair and skin care prods.

Cationic Guar C-261. [Henkel/Emery/ Cospha; Henkel Canada] Guar hydroxypropyl trimonium chloride; thickener, emulsion stabilizer, additive for shampoos.

Cationico SCL. [Auschem SpA] Benzalkonium chloride; antimicrobial.

Catisol AOC. [Stepan Europe] Oleamine acetate; emulsifier, wetting agent, antistat, corrosion inhibitor, lubricant for mineral and synthetic fibers.

Catofin. [Air Prods.] Catalysts.

Catomer. [Sybron] Polyvinyl acetate emulsion; textile stiffener.

Caytur. [DuPont] Rubber curing agents.

CB-4-34. [Neville] Aromatic plasticizer; used in adhesives, rubber (cements, mech. and molded goods, tires), caulks.

CB-97. [Dooley] Nonionic wetting agent for nylon carpet.

CB2100 Series. [Hüls Am.] Silane derivs.; coupling agent, chem. intermediate, blocking agent, release agent, lubricant, primer, reducing agent.

CBA. [Hernon Mfg.] Chip bonding adhesives; for bonding surface mounted devices to printed circuit board assemblies.

CBTS. [Akrochem] N-cyclohexyl-2-benzothiazole sulfenamide; delayed-action accelerator for natural, reclaim, and syn. rubbers.

CC-. [ECC Int'l.] Calcium carbonate. source of lime; neutralizing agent; opacifying agent in paper; fortification of bread; putty; tooth powds.; antacid; whitewash; portland cement; paint; rubber; plastics; insecticides; in chemical analysis.

CC-603. [Stepan/PVO] Blend of carrageenan, guar gum and dextrose; food grade stabilizer.

CC3005 Series. [Hüls Am.] Silane derivs.; coupling agent, chem. intermediate, blocking agent, release agent, lubricant, primer, reducing agent.

CCA Type C Wood Preservative. [CSI] Chromate copper arsenate; wood preservative.

CCH. [Olin] Calcium hypochlorite; dry chlorinator.

CD480, 492. [Sartomer] Acrylates; monomers for electronic applics. (dry film resists, solder masks), wood, metal, and vinyl flooring coatings, printing inks, overprint varnishes.

CD3770 Series. [Hüls Am.] Silane derivs.; coupling agent, chem. intermediate, blocking agent, release agent, lubricant, primer, reducing agent.

CDA 19-. [Quantum/USI] Completely denatured ethanol.

CDB. [Olin] Chloroisocyanuric acids and salts; chlorinating agent.

CDF. [Northern Industrial Plastics Ltd.] Polystyrene.

CDP. [Betz Industrial] Coagulant aids; flocculating agents.

CE-. [Procter & Gamble] Methyl esters.

CE-2000. [Siltech] Nonsilicone oil phase ingredient for personal care prods.

CE6250 Series [Hüls Am.] Silane derivs.; coupling agent, chem. intermediate, blocking agent, release agent, lubricant, primer, reducing agent.

Cecarbon. [Atochem N. Am.] Activated carbons; for mineral industry.

Cecasorb. [Atochem N. Am.] Activated carbons; for mineral industry.

Cedar-Aire. [Petrokem] Deodorants.

Cedar Plus. [Petrokem] Moth control prods.

Cedemide AX. [Stepan Canada] Lauramide DEA; foam stabilizer and visc. modifier for liq. detergents and shampoos.

Cedepal. [Stepan; Stepan Canada] Sulfates; detergent, dispersant, wetting agent for shampoos, dishwashing.

Cedepal CA. [Stepan Canada] Ethoxylated octyl phenyl ethers; emulsifier, detergent, dispersant, surfactant.

Cedepal CO. [Stepan Canada] Ethoxylated nonyl phenyl ethers; emulsifier, detergent, dispersant, surfactant, stabilizer, intermediate.

Cedepal® TD. [Rhone-Poulenc Surf.] Sulfates; detergent, wetting agent.

Cedephos®. [Stepan; Stepan Canada] Phosphate ester; detergent, emulsifier, wetting agent, corrosion inhibitor, coupling agent for industrial cleaners, metal cleaners, janitorial prods., textiles.

Cedepon. [Stepan Canada] Sulfonic acid and salts; emulsifier, detergent, dispersant for shampoos and detergents.

Cedepon® LS-30PM. [Rhone-Poulenc Surf.] Sodium lauryl sulfate; emulsifier, detergent, dispersant for shampoos and detergents.

Cederan®. [BASF AG] Phosphate fertilizers.

Cefkanat®. [BASF AG] Mineral single-feed for poultry.

Cefkaphos®. [BASF AG] Calcium phosphate; feed phosphate for the mixed feed industry.

Cegemett®. [Grünau] Mono- and diglycerides, partially esterified with lactic and/or citric acid; prevention of gelation and fat separation in sausage mfg.

Cegepal®. [Grünau] Fat powder, spray-dried fat-carrier compds.; fatty components in premixes, cake ready mixes, soups, and yeast-raised baked goods.

Cegeprot®. [Grünau] Protein concs.; emulsifier, stabilizer, protein enrichment for sausage and meat mfg.

Cegeskin®. [Grünau] Acetic acid esters of mono- and diglycerides of edible fatty acids; coating for raw and cooked sausages.

Cegesoft®. [Henkel KGaA] Esters; emollient, solubilizer.

Cegesol®. [Grünau] Medium chain triglycerides; prevention of dust for spice mixtures and other powder blends in sausage and meat mfg.

Cegesterin®. [Grünau] Monoglyceride hydrate dispersions; antistaling effect, formation of starch complexes for food industry.

Ceilcote. [Master Builders] Corrosion control prods.

Ceilthane. [Flexible Prods.] One- and two-component polyurethane coatings; for elastomers, encapsulation, consolidation, structural enhancement of media.

Celanese® Nylon. [Hoechst Celanese/ Engineering Plastics; Hoechst UK] Nylon 6/6 resins, some glass-reinforced, heat-stabilized, lubricated; for inj. molding and extrusion of mech. parts, gears, bearings, hardware, automotive parts, monofilaments.

Celanex®. [Hoechst Celanese/Engineering Plastics; Hoechst UK] PBT polyesters, some glass and/or mineral filled; for automotive body components, under-the-hood applics., computers, business machines, power tools, marine motor housings, connectors, switches, terminal boards, bobbins, relays, heavy-duty conveyance equip.

Celcon®. [Hoechst Celanese/Engineering Plastics; Hoechst UK] Acetal copolymers, some glass, mineral, or carbon-filled, lubricated; thermoplastic engineering resin used for automotive and industrial applics., antistatic applics.

Celero. [Vyse Gelatin] Gelatin.

Celeste. [Ubbink Nederland BV] Polycarbonate/acrylic.

Celite®. [Celite] Diatomaceous silica; filter aids, catalyst carriers, filler, extender pigment, grinding aid, conditioner, processing aid, flatting agent for paints, agric. chemicals, paper, rubber, polishes, cleaners.

Celkate T-21. [Celite] Hydrous magnesium silicate; functional filler.

Cell-Aire. [Plastec Iberica SA] LDPE foam.

Cellamine. [Eastern Color & Chem.] Gas fading inhibitors for textile dyeing.

Cellasto®. [BASF AG] Cellular polyurethane elastomers; casting elastomer for sealants, clutch linings, industrial springs.

Cellidor. [Albis UK Ltd.] Cellulose acetate butyrates and propionates; thermoplastic for automotive, tool, building, furniture industries, domestic appliances, lighting, elec., radio and TV industries, optical, photographic, stationery, toy, toiletries, and pkg. applics.

Cellitazol®. [BASF AG] Disperse dye for dyeing and printing acetate, triacetate, synthetic fibers.

Celliton®. [BASF; BASF AG] Disperse dye for dyeing and printing acetate, triacetate, polyester fibers and furs.

Cellobond®. [BP Chem. Inc.] Phenolic resin; for use in hand lay-up, spray-up, filament winding, RTM, and pultrusion.

Cellokyd. [Reichhold] High solids alkyds.

Cellolyn®. [Hercules] Esters; thermoplastic resin used in lacquers, ink vehicles, varnishes, adhesives.

Cellosize. [Union Carbide] Hydroxyethyl cellulose.

Cellosolve®. [Union Carbide] Ethoxy ethanol; industrial solvent.

Cellosolve® Acetate. [Union Carbide] Ethoxy ethanol acetate; solvent.

Cellovar. [Reichhold] Varnishes, modified oils.

Celluclast. [Novo Nordisk] Cellulase; enzyme for breakdown of cellulosic material for prod. of fermentable sugar and for municipal cellulosic waste treatment.

Celluferm. [Finnsugar Bioprods.] Cellulase; enzyme for fruit and vegetable processing.

Celluflow. [Presperse] Cellulose and derivs.; microporous spherical beads for use as cosmetic ingredients; oil absorbents, moisture retention aids, lubricant.

Cellusoft. [Novo Nordisk] Cellulase; for fabric treatment.

Celluzyme®. [Novo Nordisk] Cellulase; enzyme for laundry detergents and additives.

Celmar. [Amari Plastics] Polypropylene.

Celogen® AZ. [Uniroyal; Uniroyal Chem. Ltd.] Azodicarbonamide; chemical blowing agent for thermoset

and thermoplastic polymers, for inj.-molding structural foam, extrusion of profiles, sheet, pipe, and wire coatings, and vinyl plastisol, coating, and calendering.

Celogen® OT, RA, TSH. [Uniroyal; Uniroyal Chem. Ltd.] Sulfonyl hydrazides;

Celquat®. [Nat'l. Starch & Chem.] Polyquaterniums; cationic polymers for cosmetics and toiletries.

Cel-Soft. [CNC Int'l.] Cationic/nonionic softener/lubricant for paper industry.

Celstran®. [Polymer Composites; Hoechst UK] Acetal copolymer, nylon 6/6, nylon 6, PBT, PET, PC, PP, PPS, PU, SMA; glass-reinforced composites for automotive parts, elec. components, plumbing fittings, mech. handling, hydraulic parts, appliance parts and tools.

Celynol. [Rhone-Poulenc France] Fatty sucroglycerides or blends; emulsifier, stabilizer, dispersant for food applics.

Cem-All. [Mooney Chems] Synthetic driers for paints, inks.

Cenblo. [Central Chem. Co. Ltd.] Blowing agent.

Cenegen®. [Crompton & Knowles] Retarding and leveling agent, dyeing assistant for textiles.

Cenekol®. [Crompton & Knowles] Sulfonated phenolic condensate; acid dye fixative for nylon.

Cenpro 70. [Central Soya] Edible soy protein conc.

Cenpurge. [Central Chem. Co. Ltd.] Purging agent.

Centex. [Central Soya] Textured soy protein.

Centracote. [Central Soya] Refined edible vegetable oil.

Centracreme. [Central Soya] Refined edible vegetable oil.

Centrafry. [Central Soya] Refined edible vegetable oil.

Central. [Central Soya] Soybean meal.

Centralite. [Lowi] Diphenylureas; stabilizer for powders and propellants; plasticizer for celluloid.

Centrasoy. [Central Soya] Vegetable oil and shortening.

Centrex®. [Monsanto; Monsanto Europe] High gloss weatherable polymers; for inj. molding, extrusion, marine market, recreational vehicles, camper tops, automotive interior and exterior parts, spas, hot tubs, lawn and garden equip.

Centrocap®. [Central Soya] Lecithin; natural emulsifier.

Centrol®. [Central Soya] Lecithin; natural general purpose emulsifiers, dispersants.

Centrolene®. [Central Soya] Hydroxylated lecithin; o/w emulsifiers.

Centrolex®. [Central Soya] Lecithin granules, deoiled; emulsifiers, stabilizers, suspending agents for foods.

Centromix®. [Central Soya] Lecithin, water-dispersible.

Centrophase®. [Central Soya] Lecithin; multifunctional ingredient; food substance; lubricant and release agent for heated surfaces; wetting agent for magnetic media industry.

Centrophil®. [Central Soya] Lecithin; low flavor, sprayable grades.

Centura. [Vyse Gelatin] Gelatin.

Century. [Union Camp] Fatty acids.

Cenwax®. [Union Camp] Castor oil derivs.; lubricant, wax modifier; used for coatings.

Cerac. [Cerac] Specialty inorganic chemicals, evaporation materials.

Cerafil® DMK. [Boehme Filatex] Ethoxylated deriv.; leveling agent for dyeing of wool, polyamide fibers and their blends with acid and metal complex dyes.

Ceraflux. [Hammond Lead Prods.] Lead bisilicate.

Ceralan®. [Amerchol] Lanolin alcohols; emollient, w/o emulsifier.

Ceraloid. [Rhone-Poulenc] Paraffin; wax for yarn and thread lubricant.

Ceralox®. [Condea Chemie GmbH] Ultra-pure aluminas; used for translucent ceramics, synthetic sapphire glass, in the laser industry, for heavy-duty cutting tools, for bioceramics.

Ceramcote®. [Ashland/Foundry Prods.] Core and mold coatings.

Ceramer®. [Petrolite/Polymers] Polyolefin wax adducts; emulsifiable waxes used in the mfg. of emulsion coatings, release agents, floor polishes.

Ceramitalc. [R.T. Vanderbilt] Talc; used in ceramic wall tile and artware.

Ceramol. [Aceto] Partially sulfonated fatty alcohol from hydrog. vegetable oil; self-emulsifying wax, emulsifier for cosmetics.

Ceranine®. [Sandoz; Sandoz Prods. Ltd.] Softener, antistat for synthetic fibers; emulsifier, conditioner for skin care prods.

Ceraphyl. [Van Dyk] Esters and quats.; emollient, binder for pressed powds., solubilizer for makeup; wetting agent, cleansing agent.

Cera Size DL-30. [Chem. Processing] Textile sizing compd.

Cerasynt. [Van Dyk] Emulsifier, dispersant, wetting agent, thickener, opacifier, pearlescent for cosmetics, pharmaceuticals.

Cerax. [Frank B. Ross] Waxes for mfg. of solvent, liquid and paste polishes, leather finishes, lubricating sticks.

Cercron. [Pfizer] Talc; for ceramics.

Cereflo®. [Novo Nordisk] Bacterial betaglucanase; enzyme for brewing, malt and barley prods.

Cerelene. [Stevenson Bros.] Microcrystalline wax blends.

Cerelon. [C & R Landsman] Plastics material and components.

Cerelose. [Corn Prods.] Dextrose, anhydrous dextrose, liq. dextrose.

Ceres. [Miles/Organic Prods.] Textile dyes and pigments.

Cerevase. [Pfizer] Chillproofing enzyme.

Cerex. [Auschem SpA] Ethoxylated oils or esters; solubilizer for active ingredients, essential oils, vitamins; emulsifier for cosmetics and pharmaceuticals.

Cerfak. [E.F. Houghton] Alkyl POE ether; detergent, wetting agent for textiles.

Cerfax N-100. [E.F. Houghton] Fatty acid polyethanolamine condensate; detergent, wetting agent for textile scouring and dyeing.

Cerita. [M. Argueso & Co.] Industrial waxes.

Ceroflex. [Frank B. Ross] High m.p. wax

with excellent solvent retention in solvent paste waxes.

Cerol. [Auschem SpA] Wax emulsions or other blends; anticaking and antidusting agent for fertilizers, water repellent for textiles and nonwovens.

Cerone. [Rhone-Poulenc/Ag] Plant growth regulator.

Cerox. [Thermal Ceramics] High alumina brick.

Ceroxin. [Henkel] Hydrogenated castor oil; thickener used in petroleum solvent paint systems.

Certincoat. [Atochem N. Am.] Coating system consisting of pump, hood, and coating material for glass bottle mfg.

Cestidur. [Cestidur Industries; Erta Cestidur Industries] Ultra-high m.w. polyethylene.

Cestilite AST. [Cestidur Industries; Erta Cestidur Industries] Ultra-high m.w. polyethylene, antistat.

Cetal. [Amerchol] Cetyl alcohol; emollient for emulsions, oils, makeup.; auxiliary emulsifier.

Cetalox. [Witco UK; Witco SA] Ethoxylated fatty alcohol; nonfoaming detergent.

Cetamoll®. [BASF AG] Plasticizers for polyamide-based surface coating resins and for polyamides.

Cetats. [Zeeland] Cetrimonium p-toluene sulfonate; quaternary.

Cetax. [Aquatec Quimica SA] Fatty alcohols; emollient, consistency agent for creams and lotions; superfatting agent for hair prods.

Cetina. [Robeco] Cetyl esters and stearamide DEA; emulsifier, lubricant, emollient for cosmetics.

Cetiol®. [Henkel/Emery/Cospha; Henkel KGaA] Esters; emollient, superfatting agent, vehicle, lubricant for cosmetics, pharmaceuticals, makeup, bath oils.

Cetodan. [Grindsted Prods.; Grindsted Prods. Denmark] Acetylated monoglycerides; food emulsifier, aerating agent; plasticizer for chewing gum base; also for resins, thermoplastics.

Cetol®. [Zeeland] Cetyl quaternary salts.

Cetomacrogol 1000 BP. [Croda Inc.; Croda Chem. Ltd.] Ceteareth-20; emulsifier for pharmaceuticals.

Cetostearyl Alcohol. [Croda Inc.] Cetearyl alcohol; for cosmetics, pharmaceuticals.

CF. [Custom Fibers] Cellulose fibers; reinforcing filler and visc. control agent; for epoxy grouts, gasket material, mastics; as asbestos replacement.

C-Flakes. [Karlshamns] Hydrog. cottonseed oil.

C-Flex®. [Concept Polymer] Thermoplastic elastomers incl. SEBS, SBS; for medical, laboratory, food, and industrial fields.

CH7250 Series. [Hüls Am.] Silane derivs.; coupling agent, chem. intermediate, blocking agent, release agent, lubricant, primer, reducing agent.

Champagard. [Chem-Pak] Rust preventatives.

Champlex. [Champlain Industries] Protein hydrolysates.

Chardol. [Cook Composites & Polymers] Saturated polyesters.

Chardot. [Ceca SA] Leveling agent for dispersed dyes on polyamide.

Chargemaster®. [Grain Processing] Corn deriv.; paper furnish chemical.

Chargepac®. [Drew Ind. Div.] Coagulants for water and wastewater clarification.

Charguard 329. [Great Lakes] Intumescent fire retardant.

Charlab Antimigrants. [Catawba-Charlab] Antimigrants for continuous dyeing operations.

Charlab Binders. [Catawba-Charlab] Acrylic latex binders for printing with aq. or solvent pigment systems.

Charlab Blanket Adhesives. [Catawba-Charlab] Blanket adhesives for rotary and flat bed screen printing machines.

Charlab CHT. [Catawba-Charlab] Stabilizers, gliding agents, wetting agents, dispersants, chelating agents, detergents, lubricants, thickeners, fixatives, and softeners for textile processing.

Charlab Clear Concs. [Catawba-Charlab] Synthetic thickeners for printing all-aq. colors.

Charlab Defoamers. [Catawba-Charlab] Defoaming agents for dyeing or pigment printing systems.

Charlab Dyefixes. [Catawba-Charlab] Amino-aldehyde condensate; resinous fixing agents for direct dyes.

Charlab Roller Cleaner OS. [Catawba-Charlab] Emulsifiable solvent cleaner.

Charlab Scours. [Catawba-Charlab] Aq. and solvent scours for textile equip. and applics.

Charlab Thickeners. [Catawba-Charlab] Synthetic thickeners for aq. dye or pigment printing systems.

Charlab Winding Lubricants. [Catawba-Charlab] Special winding lubricant free of oils and waxes.

Checkmate. [Drew Ind. Div.] Photometer.

Cheelox®. [Rhone-Poulenc Surf.] Sequestrants for calcium, magnesium, textiles; clarifier for industrial cleaner sol'ns., herbicide formulations.

Chek-Mate. [Ferro] Cadmium pigments for plastic.

Chek-Rust. [Flame Control Coatings] Automotive and structural steel rustproofing.

Chel®. [Ciba-Geigy/Dyestuffs] Chelating agent used in bar soaps, photographic developer baths, cosmetics, textiles, and min. separations.

Chelate Series. [Chem-Tex Labs] DPTA, HEDTA, EDTA, or NTA; chelating agents.

Chelex. [Bio-Rad Labs] Ion exchange resin.

Chelfac. [Dyetech] Blended chelates; multifunctional chelate with antichlor and dispersants; for textile use.

Chelon. [Rhone-Poulenc Basic] Chelating agent for heavy metal and alkaline earth ions; for soaps, cosmetics, germicides, agric., metal finishing.

Chemag 108. [Manufacturers Chems.] Proprietary salt sol'n.; antimigrant for direct dyes, acid dyes, and hydrolyzed reactive dyes prior to aftertreatment or drying.

Chemal. [Chemax] Ethoxylated fatty ethers, esters, sulfosuccinates; wetting agent, penetrant, emulsifier, lubricant, detergent, dispersant, solubilizer for textile processing, cosmetics, polishes,

cutting oils, polymerization.

Chemax. [Chemax] Ethoxylated fatty acids, castor oil, or alkyl phenols; emulsifiers, lubricants, dispersants, solubilizers, wetting agents, detergents, dye assistants, softeners, antistats, coupling agents, stabilizers.

Chemazine. [Chemax] Imidazolines; softener, antistat, emulsifier, corrosion inhibitor, filming agents.

Chembee. [Frank B. Ross] Beeswax substitute; for candle mfg.

Chembetaine. [Chemron] Betaines; mild surfactants, foam boosters, visc. builders for baby personal care prods., other cosmetics, industrial drilling fluids.

Chem-Calk® 500. [Bostik Div./Emhart] Two-component polyurethane; high performance elastomeric joint sealant.

Chemcarb. [Engelhard] Ground calcium carbonate.

Chemclean. [Chem-Tex Labs] Scouring agents for dyehouse and latex equip.

Chemclear 320, LFND. [Reilly-Whiteman] Clearing agents for dyestuff clearing on textiles.

Chemcogen®. [Rhone-Poulenc/Textile & Rubber] Nonionic surfactant; antiprecipitant, retarding and leveling agent for dyes.

Chemcoloft®. [Rhone-Poulenc/Textile & Rubber] Nonionic polyethylene emulsion; for cotton softening and finishing.

Chem Copp. [Am. Chemet] Cuprous oxides.

Chemcor. [Chem. Corp. of Am.] Specialty polyethylene and wax emulsions.

Chemcor Zac. [Chem. Corp. of Am.] Zinc ammonium carbonate.

Chemeen. [Chemax] Ethoxylated fatty amines; emulsifier, antstat, dye leveler, wetting agent, dispersant in textiles, metal buffing, rubber compds.; lubricant for fiberglass.

Chemester. [Atlas Minerals & Chem.] Vinyl ester; floor topping and mortar.

Chemetics. [Evans Chemetics/W.R. Grace] Organic compds. or divalent sulfur compds.; for cosmetics.

Chemfac. [Chemax] Phosphate esters; detergent, emulsifier, wetting agent, lubricant, antistat, hydrotrope; emulsion polymerization; defoamer for oilfield applics.

Chemfax. [Chemfax] Hydrocarbon resins; thermoplastic resins for adhesives, coatings, tackifiers, etc.

Chem-Firm. [Ivax Industries] Polyvinyl acetate emulsion; nonreactive hand builder, bodying agent.

Chem Fish. [Tifa Ltd.] Rotenone; pisciscide.

Chemfix. [Chem-Tex Labs] Dye fixing agent for nylon and wool fibers.

Chemfroth. [Chem-Tex Labs] Frothing auxiliary and stabilizer for foam finishing, dyeing and latex back coating.

Chemglaze®. [Lord] Polyurethane coatings; increases environmental resistance of elastomer prods. for molded automobile parts, extruded tubes and hoses, EPDM weatherstripping.

Chemical 39 Base. [Sandoz] Stearamidoethyl ethanolamine; base, emulsifier, conditioner, lubricant for toiletries, hair and skin prods.

Chemical Base 6532. [Sandoz] Stearamidoethyl diethylamine; emulsifier, emollient, conditioner for skin and hair care prods.

Chemictive. [Shyamac Int'l.] Textile dyes and pigments.

Chemidex. [Chemron] Alkyl amidopropyl dimethylamine; emulsifier substantive to proteins and cellulosics.

Chemidize. [Witco] Chemical for pretreating and forming protective oxide coating on aluminum and aluminum alloy surfaces.

Chemie Rohstoffe. [Varobel BVBA] Additives for raw materials.

Chemiflex. [Sanyo Chem. Industries] Blocked isocyanate; adhesion promoter for PVC plastisols.

Chemigum®. [Goodyear; Goodyear Europe] Acrylonitrile-butadiene rubbers and latexes.

Chemilene. [Shyamac Int'l.] Textile dyes and pigments.

Chemistat. [Sanyo Chem. Industries] Quaternary ammonium polymers or surfactants; antistat, electroconductive agent; for facsimile paper, printing ink.

Chemistat Latex. [Goodyear] Saturated nitrile latex; for gasketing, latex dipping, adhesives, as binder.

Chemlease. [Chemlease] Silicone, fluoropolymer, or other blends; release agents and lubricants.

Chemlease SP. [Chemlease] Semipermanent release systems for plastic and rubber release.

Chemlev. [Chem-Tex Labs] Leveling agent, migrator, penetrant for dyes on nylon, acetate, polyester.

Chemlink®. [Sartomer] Acrylated compds.; curing agents.

Chemlock. [Chemonic Industries] Resinated silica; aids in stabilizing shrinkage, slippage and pilling of textiles.

Chemlok®. [Lord] Bonding agent, adhesive for rubber, urethanes, polyolefins, thermoplastic elastomers.

Chemlube. [Chem-Tex Labs] Napping and shearing lubricant for fibers.

Chemlube. [Chemonic Industries] Ethoxylated sulfated blends; dye assistant, wetting/nonrewetting agent, and detergent for use on polyester, cotton, nylon or blends.

Chem-Master. [Syn. Prods.] Polymeric dispersions of rubber chemicals.

Chemocarrier®. [Sybron] Carrier for cationic dyes on polyester.

Chemol. [Chemol] Glycerol esters of saturated fatty acids; fiber lubricant, sizing wax.

Chemos Buffer 102. [Reilly-Whiteman] Buffering agents for textiles.

Chemoxide. [Chemron] Alkyl amine oxides; foam booster, visc. builder, wetting agent for mild cosmetics, hair colorants, gels, permanent waves.

Chem-Oxy. [Ivax Industries] Oxidizing agent for sulfur and vat dyes.

Chempakool. [Chem-Pak] Soluble cutting oil.

Chempakut. [Chem-Pak] Metal cutting oil.

Chempasyn. [Chem-Pak] Synthetic metalworking coolant.

Chemphonate. [Chemron] Phosphonic and phosphoric acids and salts; scale inhibitors for oilfield prod.

Chemphos. [Chemron] Ethoxylated ether phosphates and salts; emulsifier, solubilizer, lubricant, detergent for hair care prods., drilling fluids.

Chemplex. [Sullivan Chem. Coatings] Chemical-resistant coatings.

Chempol. [Cook Composites & Polymers] Alkyd and polyurethane compounds; used in industrial finishes, machinery enamels, primers, appliance finishes, coatings.

Chemprene. [Chemfax] Thermoplastic isoprenoidal polymer; used in compounding of syn., natural, and reclaim rubber; softener, tackifier, and reinforcing agent; used in calendering and extruding, hard rubber, molded goods, tubes, tire stocks, rubber cements, wire and cable insulation, caulks.

Chempruf. [Atlas Minerals & Chem.] Epoxy, polyester, vinyl ester, furan linings.

Chemquat. [Chemax] Ethoxylated quaternary ammonium chlorides; corrosion inhibitor, antistat, sensitizer for latex foam prods.; visc. depressant for softener formulations.

Chemquest. [Chemonic Industries] EDTA or salt blends; water softening agent.

Chemrez. [Chemonic Industries] Binders, flame retardant finishes for textiles.

Chem-Rez®. [Ashland/Foundry Prods.] Resin foundry core binders.

Chem Rice. [Tifa Ltd.] Propanil; rice herbicide.

Chemsalan. [Chemsal] Sodium lauryl sulfates or ether sulfates; surfactants.

Chemscour. [Chem-Tex Labs] Stripper, reducing agent, scouring agent, detergent, emulsifier for textile applics.

Chem/Serv. [Chem/Serv] Additives for food, cosmetic, and paper mfg.

Chemset Albidur. [R.F. Bright Enterprises Ltd.] Silicone rubber modified thermosetting resins.

Chemset. [R.F. Bright Enterprises Ltd.] Thermosetting resins.

Chem-Silate®. [Chem. Prods.] Sodium silicate; Adhesive, bleaching assistant, corrosion inhibitor, detergent for textile processing.

Chem Sof. [Glo-Tex] Softener for textiles.

Chemsoft. [Chem-Tex Labs] Softeners and lubricants for yarn, fabric, and carpeting.

Chemsperse. [Chemron] Glycol or glyceryl esters; emulsifier, stabilizer, emollient, opacifier, pearlescent, thickener for personal care prods.

Chemstat®. [Chemax] Sulfonates, ethoxylated amines, amides; antistats for polyolefins, ABS, HIPS.

Chemstrip. [Chemonic Industries] Stripping auxiliary for removal of direct, acid and developed colors for wool, cotton, silk, rayon, acetate and other fabrics.

Chemsulf. [Chemax] Sulfates; emulsifier, detergent, foaming agent.

Chemtec. [Southern Coatings] Zinc-rich primer.

Chemterg. [Chemonic Industries] Surfactant blend; multi-purpose detergents and scouring agents for textiles.

Chemteric. [Chemron] Coco amphoterics; surfactants.

Chemtex. [Chem-Tex Labs] Surfactants, penetrants, scouring agents, binders, antifoams, leverlers, retarders, softeners, lubricants, oxidizing agents, emulsifiers for textile applics.

Chemtherm AM. [Chem-Tex Labs] Polymeric antimigrator for use in continuous dyeing of cotton and polyester with vats, sulfur, direct, and disperse dyes.

Chemtox-S. [Chapman] Sodium pentachlorophenate, tech.

Chemulose 10. [Chem-Tex Labs] Thickener, film-former, foam stabilizer, antisoil redeposition agent, warp or backsize agent and binder for textile applics.

Chemwax. [Chem. Processing] Textile wax sizing.

Chem-Weld. [Ivax Industries] Finishing aid to eliminate shifting of pile fabrics and to control curling of knits.

Chemwet. [Chem-Tex Labs] Penetrants, wetting agents, dispersants for textile applics.

Chemwet. [Chemonic Industries] Ethoxylated surfactant blends; wetting agents for textile processing.

Chemzoline. [Chemron] Imidazolines; industrial surfactants, intermediate, corrosion inhibitors, dispersant, emulsifier.

Chem-Zyme. [Ivax Industries] Bacterial alpha amylase sol'ns.; enzymes for desizing.

Chestnut. [Climax Performance] Tanning extract.

Ches® 500. [CasChem] Nonfat drymilk, xanthan gum, propylene glycol, alginate, glyceryl stearate, sodium glyceryl oleate phosphate; food grade stabilizer; emulsifier for cosmetics and pharmaceuticals.

Chevron Refined Waxes. [Chevron] Paraffin waxes.

Chevron Saturating Waxes. [Chevron] Refined paraffin waxes and additives; for increasing strength of corrugated board.

Chil By. [Childers Prods.] Sealant.

Chil Glas. [Childers Prods.] Reinforcing membrane.

Chil Joint. [Childers Prods.] Sealant.

Chil-Lag. [Childers Prods.] Lagging adhesive.

Chill-Garde. [Crosfield] Silica gel.

Chillsa FE®. [Arco] Inhibited propylene glycol;

Chillsafe. [Arco] Heat transfer fluid.

Chil-Perm. [Childers Prods.] Vapor barrier coatings.

Chil-Seal. [Childers Prods.] Lagging adhesive.

Chil Spray. [Childers Prods.] Insulation adhesive.

Chil Stix. [Childers Prods.] Insulation adhesive.

Chimassorb®. [Ciba-Geigy/Additives] Polymeric hindered amine; light and heat stabilizer for polyolefins.

Chimin. [Auschem SpA] Betaines, sulfosuccinates, esters, phosphates; conditioner, emulsifier, detergent, dispersant, wetting agent, antistat for cosmetics, leather processing, textiles, paper.

Chimipal. [Auschem SpA] Dispersant, wetting agent, detergent, thickener, foam stabilizer, conditioner, superfatting agent for cosmetics, detergents, emulsion polymerization.

Chimipon. [Auschem SpA] Sulfonates; emulsifier, detergent base, frothing agent for detergents, pesticides, emulsion polymerization.

Chipco 26019. [Rhone-Poulenc/Ag] Turf and ornamental fungicide.

Chipco Ronstar G. [Rhone-Poulenc/Ag] Turf and ornamental herbicide.

Chloracel®. [Reheis] Sodium aluminum chlorhydroxy lactate complex; deodorant.

Chlorasol. [Union Carbide] Fumigant.

Chlorasorb. [Huntington Labs] Stabilized chlorine; absorbent and deodorizer.

Chlorez®. [Dover] Resinous chlorinated paraffin; flame retardant for paints, printing inks, plastics, foams, adhesives, paper and fabric coatings.

Chlorhydrol®. [Reheis] Aluminum chlorohydrate; antisperspirant.

Chlorobutyl. [Exxon] Chlorinated isobutylene-isoprene rubber; used for tire inner liners, sidewalls, pharmaceutical stoppers.

Chloro-Chem. [Steelcote Mfg.] Chlorinated rubber prods.

Chloroflo®. [Dover] Chlorinated paraffin; lubricant additive.

Chlor-O-Pic. [Great Lakes] Space, soil fumigant.

Chloropren-Faktis. [Rhein Chemie] Sulfur chloride factice; processing aid for molded and extruded CR, NBR goods.

Chlorothene®. [Dow; Dow Europe] Solvents for degreasing, adhesives, paints.

Chlorovis. [Dover] Chlorinated paraffin.

Chlor-Tergent. [Oakite Prods.] Chlorinated detergent.

Choclin. [Van Den Bergh Foods] Cocoa butter equivalent; coating fat, center fat for confectionery, cosmetics, pharmaceuticals.

Chocorama. [David Michael] Chocolate flavors and cocoa prod. enhancer.

Chocothin. [Lucas Meyer] Lecithin; visc. reducing agent for milk chocolates.

Chocotop. [Lucas Meyer] Fractionated soya lecithins; natural visc. reducing agents esp. for chocolate mfg.

Cholestatin. [Traco Labs] Phytosterols.

Cholesterol NF. [Croda Inc.; Croda Chem. Ltd.] Cholesterol NF; moisturizer, emollient, emulsifier.

Cholfeed-S. [Heterochem] 50% dry choline chloride.

Chop-Pak. [Celite] Fiber glass reinforcement.

Chopper. [Am. Cyanamid/Ag] Herbicide.

CHP. [ISP] N-Cyclohexyl-2-pyrrolidone; solvent.

CHP-158. [Witco/Argus] Cumene hydroperoxide; initiator for vinyls, unsat. polyester resins.

CHPTA. [Chem-Y GmbH] 3-Chloro-2-hydroxypropyltrimethyl ammonium chloride; cationic starch modifier for textiles.

Chroma. [Cote Color] Textile dyes and pigments.

Chroma-Chem. [Hüls Am.] Industrial colorants.

Chromaflo. [Plasticolors] Low visc. colorants for pumping and pouring.

Chroma-Kote. [Crompton & Knowles] FD&C lake dispersions.

Chroma-Lite. [Van Dyk] Pearlescent pigment.

Chromaset®. [Henkel/Textile] Fixative for dyes on nylon, wool, cellulosics.

Chromasist®. [Henkel/Textile] Dispersant, leveling agent for disperse dyes.

Chromastral. [ICI Am.] Textile dyes and pigments.

Chromative. [Cote Color] Textile dyes and pigments.

Chromazol. [Cote Color] Textile dyes and pigments.

Chrome Green 106. [Presperse] Chromium hydroxide, bismuth oxychloride.

Chromesaver. [Rohm & Haas] Synthetic tanning agent.

Chrometan. [Am. Chrome & Chem.] Basic chrome sulfate.

Chromicoat. [Oakite Prods.] Coating.

Chromitan®. [BASF AG] Basic chromium salts; for tanning and retanning leather and furs.

Chromoprotulines. [Exsymol] Hydrolyzed vegetable proteins; moisturizers for skin care prods., oily cosmetics, regeneration and skin treatments, anti-aging and anti-wrinkle creams.

Chromosol. [Nippon Senka] Silicon complex; fixing agent for acid dyes on wool.

Chromosorb. [Celite] Chromatographic supports.

CHT Activator NB. [Catawba-Charlab] Hydrogen peroxide; activator for low pH bleaching.

CHT Antifoam MI. [Catawba-Charlab] Mineral oil-based defoaming agent.

CHT Biavin. [Catawba-Charlab] Gliding, anticreasing, and leveling agents for cotton.

CHT Carrier GR-A. [Catawba-Charlab] Carrier for dyeing of polyester and PES blends.

CHT Contavan. [Catawba-Charlab] Surfactant-free peroxide stabilizer for textile bleaching.

CHT Cotoblanc. [Catawba-Charlab] Scouring auxiliary, chelating agent, dispersant for textiles.

CHT Defoamer SC. [Catawba-Charlab] Deaerating agent for transfer printing inks for textiles.

CHT Egasol SP. [Catawba-Charlab] Leveling and penetrating agent for rapid dyeing of polyester.

CHT Felosan. [Catawba-Charlab] Stain remover, detergent, solvent scour for textiles.

CHT Heptol. [Catawba-Charlab] Sequestrant, iron dispersant; for textile washing.

CHT Intensol. [Catawba-Charlab] Dyestuff solvent and machine cleaner.

CHT Lavotan DS. [Catawba-Charlab] Wetting, washing and cleaning surfactant for textile industry.

CHT Lustraffin BA. [Catawba-Charlab] In-bath dye lubricant for knitting and weaving.

CHT Meropan. [Catawba-Charlab] Peroxide neutralizer, protective colloid, sequestrant for textiles.

CHT Prisulon. [Catawba-Charlab] Thickener for transfer paper printing.

CHT Rapidoprint. [Catawba-Charlab] Leveling agent, binder for transfer paper printing.

CHT Retinol M. [Catawba-Charlab]

Stripping and leveling agent with dyestuff affinity.

CHT Rewin MRT. [Catawba-Charlab] Cationic dye fixative.

CHT Sarabid. [Catawba-Charlab] Dyeing auxiliary for leveling and dispersing.

CHT Subitol. [Catawba-Charlab] Surfactant, wetting agent, detergent for continuous bleaching.

CHT Thickener. [Catawba-Charlab] Thickener for pigment prints.

CHT Tubingal. [Catawba-Charlab] Textile softeners.

CHT Tubiprint. [Catawba-Charlab] Pigment printing pastes for pearlescent or softening effects.

CHT Viscavin DMS. [Catawba-Charlab] Anticreasing agent, leveling agent, dispersant for polyester.

Chupol. [Takemoto Oil & Fat] Water reducing agent for concrete.

Chy-Max. [Pfizer/Brewery & Dairy Prods.] Chymosin milk clotting enzyme.

CI-. [Bacon] Two-part epoxy compds.; coil impregnant for elec. components; for casting and coating applics.

CI7810 Series. [Hüls Am.] Silane derivs.; coupling agent, chem. intermediate, blocking agent, release agent, lubricant, primer, reducing agent.

Cibacet. [Ciba-Geigy/Dyestuffs] Textile dyes and pigments.

Cibacron. [Ciba-Geigy/Dyestuffs] Textile dyes and pigments.

Cibafast®. [Ciba-Geigy/Dyestuffs] Photostabilizer for polyamide fibers; uv absorber.

Cibanone. [Ciba-Geigy/Dyestuffs] Textile dyes and pigments.

Cibaphasol®. [Ciba-Geigy/Dyestuffs] Sulfuric acid ester; leveling and penetrating agent for continuous dyeing and printing of nylon.

Cida-Stat. [Huntington Labs] Antimicrobial sol'n.

CIL. [Drew Ind. Div.] Silica; corrosion inhibitor for cooling water treatment.

Cilbond. [Compounding Ingredients Ltd.] Bonding agents for rubbers to metal, etc.

Cilon BV/DET. [Auschem SpA] Sodium acrylate; sequestering agent.

Cilrelease. [Compounding Ingredients Ltd.] Mold release agents for rubbers and plastics.

Cil Release 1000. [Synair] Mold release.

Cimflx. [Cyprus Industrial Min.] Talc; reinforcement for plastics.

Cimpact. [Cyprus Industrial Min.] Talc; impact modifier for polyolefins.

Cinclear T. [Stockhausen] Thiourea dioxide; afterclearing and stripping agent.

Cindet 558. [Stockhausen] Hydrogen peroxide scavenger proprietary blend of acids and bases.

Cindye. [Stockhausen] Carrier for dyeing polyester fiber and blends.

Cinlevel. [Stockhausen] Leveling agent for acrylics and blends.

Cinquasia. [Ciba-Geigy Pigments UK] Organic quinacridone pigments.

Cinsoft. [Stockhausen] Softener and lubricant for textiles.

Cintex. [Stockhausen] Sequestrant, pH buffering agent, softener for textiles.

Ciptane. [PPG Industries] Precipitated silica.

Cirami. [Alban Muller] Beeswax, candelilla wax, shea butter; natural thickener.

Cirol. [Van Den Bergh Foods] Hydrogenated vegetable oil; shortening for food industry.

Cirrasol®. [ICI Am.; ICI Surf. Belgium] Fatty acid amides or ethoxylated esters; fiber lubricant; emulsifier, antistat; for fiberglass.

Cisdene®. [Am. Syn. Rubber] Stereospecific polybutadiene; used in mixtures with SBR or NR to enhance properties.

Citation. [Avatar] White mineral oil USP and NF.

Cithrol. [Croda Surf. Ltd.] Ethoxylated esters; surfactants, antistat, emulsifier, dispersant, lubricant, wetting agent, cosolvent for textiles, paper, cutting oils, polishes, metal cleaners, cosmetics.

Citmol. [Bernel; Heterene] Citrates; noncomedogenic emollient, pigment wetter, castor oil replacement for lipsticks, film former.

Citowett®. [BASF AG] Alkylaryl polyglycol ether; wetting and sticking agent.

Citranox®. [Alconox] Blend of organic acids, anionic and nonionic surfactants, alkanolamines; liq. detergent.

Citrikleen. [Penetone] Nonpetroleum solvent degreaser.

Citroflex. [Morflex] Citrates; plasticizers for cellulosics, vinyls for medical and other applics.

Citrosynth. [Florasynth] Imitation citrus oils.

Citrus Belt. [Florasynth] Citrus oils.

CG6710 Series. [Hüls Am.] Silane derivs.; coupling agent, chem. intermediate, blocking agent, release agent, lubricant, primer, reducing agent.

Clairex. [Day-Glo Color] Clear aq. coatings.

Clairolite L. [Kaopolite] Calcium bentonite; plasticizing and binding agent, moisture absorbent, anticaking agent for white firing ceramics.

Claradex. [Syn. Rubber Tech.; Shin-A Chem. Mfg.; Tolson Holland BV] ABS resins.

Clarase®. [Solvay Enzymes] Alpha-amylase; enzyme for hydrolysis of starch and dextrins.

Clarate. [Clark] Sequesterants for textile processing, dyeing, bleaching, and scouring.

Clarcal. [Whittaker, Clark & Daniels] Whiting-calcium carbonate.

Clarene. [Solvay & Cie] Ethylene vinyl alcohol copolymers.

Claret Kut. [Castrol Industrial East] Cutting oils.

Clarex. [Clark] Low foam wetting agent, detergent, penetrant, emulsifier for textile scouring, desizing.

Clarex®. [Solvay Enzymes] Pectinase; enzyme for depectinization of fruit juices, grape processing, wine prod.

Clarifex. [ICI Polymer Additives] Nucleating agent.

Clarifloc. [Polypure] Polymers, defoamers, emulsion breakers, frothers, drying aids.

Clarion. [Am. Colloid] Bleaching earth.

Clar+ion®. [General Chem.] Enhanced coagulants.

Clarol NA-5. [Clark] Solvent scouring agent for use in fabric prep.

Clar (S+) Ion. [General Chem.] Liq. cationic coagulants.

Class 10®. [General Chem.] Low particle electronic grade chemicals.

Classic. [DuPont/Ag] Herbicide.

Classic Colors. [Spectra Polymer] Universal color concs. in pellet form.

Clavanol. [Dexter] Surfactants.

Clavodene®. [Dexter] Detergents, scouring agents for textile operations.

Clayphos. [FMC] Suspending agent.

Claytone. [Southern Clay Prods.] Gelling and suspending agents.

Clean All. [Specialty Chem.] All purpose solvent scour for removing oil, waxes and tar from fabrics, yarn and equip.

Clean-Bond. [Clean-Chem] Zinc phosphates; pre-paint treatment.

Clean-Coat. [Clean-Chem] Iron phosphates; cleaning and coating.

Cleaner. [Glo-Tex] Solvent cleaners for removal of resins, binders, dye and finishing agents from equip.

Cleaner NPR. [Piedmont Chem. Industries] Cleaning agent for removal of exhaust pigment dyes, silicone finishes, etc.

Cleaner NX. [Riechem/Mt. Vernon Mills] Dispersible d-limonene; scouring agent for polyester/cotton and pad rolls.

Clean-Floc. [Clean-Chem] Paint killers and flocculants.

Clean Front. [West Agro] Iodophor conc.

Clean-Grease. [Clean-Chem] Grease and wax removers.

Clean-Guard. [Clean-Chem] Corrosion prevention compds.

Clean-N-Dry. [Clean Room Prods.] Detergent cleaner and sanitizer for clean rooms.

Cleanol. [Lea Mfg.] Metal cleaners.

Clean-Phos. [Clean-Chem] Zinc phosphate; heavy rust preventative, lubricant.

Clean-Safe. [Clean-Chem] Demoisturizing compds. for elec. equip.

Clean-Shield. [Clean-Chem] Water-based rust preventives.

Clean-Strip. [Clean-Chem] Paint, varnish and wax stripper.

Clean-Wash. [Clean-Chem] Multipurpose cleaning conc.

Clean Wiz. [Axel Plastics Research Labs] Metal, plastic mold cleaners.

Cleapact. [Dainippon Ink & Chem.] Styrene-based copolymer; transparent high impact thermoplastic for inj. molding and extrusion of business machine parts, pkg. materials for electronics or food.

Clearate. [W.A. Cleary] Lecithin; dispersant, wetting agent, antiflocculant for pigments, paints, inks, foods.

Clearbreak. [Chemron] Polymeric amine; emulsion breaker for oilfield applics.

Clearcol. [Croda Inc.; Croda Chem. Ltd.] Soluble animal collagen; protein for moisturizing and conditioning of facial systems.

Clear Conc. 7174, GR-1. [Catawba-Charlab] Synthetic thickener for aq. pigment printing systems.

Clear Conc. P-90. [Gresco Mfg.] Blend of synthetic resins and thickeners; Aq. concs. for water phase pigment printing.

Clearlan®. [Henkel/Emery] Anhydrous lanolin; emollient for cosmetics.

Clearmol. [McLaughlin Gormley King] Insecticide conc.

Clearon®. [Aceto] Terpene tackifying resin.

Clear Ryne R227. [Amax Industrial Prods.] Water-soluble disinfectant and sanitizer.

Clear Sorb. [W.A. Cleary] Nontoxic spill absorbent.

Clearspray. [W.A. Cleary] Antitranspirant and foliage retainer for use on transplants, shrubs, evergreens, and turf to protect against wilt and winter kill.

Clear-Stat. [PolyColors] Amine-free antistatic conc. for use with polyolefins.

Clear-Thru. [Oil Dri Corp. of Am.] Filter aid.

ClearTint. [Milliken] Transparent polymeric colorants for polyolefins.

Cleartuf. [Goodyear; Goodyear Europe] PET polyester resin; thermoplastic providing clarity for bottles and cans.

Cleary's Waterless Hand Cleaner. [W.A. Cleary] Conc. heavy duty, gentle cleaning agent for removal of pesticide residue, grease, oil, tar, ink, paint, carbon, adhesives from hands and work clothes, car interiors.

C-Lex R587. [Amax Industrial Prods.] Selective broadleaf weed killer.

Click. [Petrokem] Insecticides.

Climalene. [Malco Prods.] Laundry detergent additive.

Climax. [Climax Molybdenum; Climax Molybdenum UK] Molybdenum-containing chemicals.

Clinco. [ADM Corn Processing] Corn starches.

Clincor. [ADM Corn Processing] Corn starches.

Clineo. [ADM Corn Processing] Corn starches.

Clink. [Yoshimura Oil Chem.] Surfactant blends; cleaning agent for metal surfaces, dye machines, heavy duty applics.

Clin-Link. [ADM Corn Processing] Corn starches.

Clintose. [ADM Corn Processing] Dextrose.

Clinvert. [ADM Corn Processing] Corn starches.

Cloisonne. [Mearl] Cosmetic pearl powds. in deep colors.

Clorafin®. [Hercules] Chlorinated paraffin; used to flameproof and waterproof textiles; plasticizer for vinyls, resins, film-formers; addtiive in cutting oils.

Closyl. [Clough] Sodium alkyl sarcosinates; wetting, foaming detergent for personal care and household prods.

CLSP. [Van Den Bergh Foods] Vegetable oils; coating fats, center fats for confectionery, peanut butter fillings.

Clysar®. [DuPont] Polyolefin film; for shrink film applics.

CM. [DuPont] Flame retardant.

CM. [Spartan Flame Retardants] Flame retardant.

CM-66. [Solem Industries] Alumina

trihydrate; filler for mfg. of cultured onyx; surface treatment.

CM-88®. [Georgia-Pacific] Amine lignosulfonate; modifier; emulsifier.

CM8450 Series. [Hüls Am.] Silane derivs.; coupling agent, chemical intermediate, blocking agent, release agent, lubricant, primer, reducing agent.

CMC-7 Series. [Hercules] Water-soluble polymers for surface treatment of paper.

CMC Warp Size. [Hercules] Sodium carboxymethylcellulose; warp sizing for cotton and synthetic yarns.

CMD. [Rhone-Poulenc] Epoxy, polyamide, or acrylic resin sol'ns. used for industrial coatings, flame-retardant epoxy compds.

CMI. [Alban Muller] Isolated plant fractions; anti uv agents for sunscreening prods.

C- Micro Mica. [KMG Minerals] Micronized mica.

CM Wetting Agent. [Solem Industries] Visc. suppressant, antisettling agent, processing aid.

CN. [Sartomer] Epoxy, polybutadiene, and urethane acrylates and blends; for paper clear coatings, wood top coatings, screen inks, litho inks, polyethylene coatings, metal decorative coatings, adhesive papers, wood fillers, solder masks and photoresists.

CNC Aquafilm®. [CNC Int'l.] Water, oil and soil release fluorochemical finishes for textiles.

CNC. [CNC Int'l.] Detergents, wetting agents, dispersants, antifoams, emulsifiers, dyeing assistants; for textile and paper processing.

CNComerse IMP. [CNC Int'l.] Alky phospho-sulfate; mercerizing agent, wetting agent, penetrant for textiles.

C.N.O. Series. [Stepan Europe] Refined coconut oil; emollient, superfatting agent, solvent.

CO-. [Procter & Gamble] Fatty alcohols; emollients, intermediates for mfg. of alkyl sulfates, ethoxylates, alkyl halides, esters.

CO9745 Series. [Hüls Am.] Silane derivs.; coupling agent, chem. interme-

diate, blocking agent, release agent, lubricant, primer, reducing agent.

Coa Butta. [Ruger Chem.] Synthetic suppository base.

Coad. [Norac] Metallic stearates.

Coagulant. [Bayer; Miles] Coagulant for dipped goods.

Coarse Zink. [Am. MicroTrace] Zinc sulfate monohydrate + 35.5% Zn; water-soluble nutrient for animal feeds.

Coasol B®. [Aceto] Diisobutyl esters of dicarboxylic acid blend; solvent.

Coateric. [Colorcon] Aq. enteric film coating.

Coatmaster®. [Grain Processing] Corn deriv.; binder.

Cobalume. [Atochem N. Am.] Plating process.

Cobee. [Stepan/PVO] Coconut oils; emollient, superfatting agent, clouding agent; for food, cosmetics, pharmaceuticals.

Coberine. [Van Den Bergh Foods] Cocoa butter equivalent; coating fat, center fat for confectionery, cosmetics, pharmaceuticals.

Cobox®. [BASF AG] Ammoniacal copper polyacrylate; contact fungicide for coffee, cotton, fruit crops.

Cobra®. [Miller-Stephenson] Non-CFC aerosol solvent spray brush or duster.

Cobratec®. [PMC Specialties] Benzotriazoles and tolyltriazoles; chelating agent and sequestrant; copper and steel corrosion inhibitor; metalworking fluids.

COC/COCDF. [Cuproquim] 50% copper oxychloride.

C O Cleaner. [Dycho] Built anionic detergent; scouring agent esp. for cotton and cotton blends.

Coconad. [Kao/Edible Fat & Oil] Caprylic and/or capric triglycerides; food emulsifier.

Coconut Oils®. [Stepan/PVO] Coconut oils.

Code 321. [Van Den Bergh Foods] Hydrogenated soybean oil; shortening for filler, prepared foods, spray oil.

Code 8059. [Hart Prods. Corp.] Fatty acid condensate; base for car water beading agent.

Co-Fiber. [Columbia Filter] Nonasbestos filter fiber.

Cogeviil. [Padanaplast SpA] PVC compds.; standard, modified, flame retardant, low smoke emission.

Cohedur. [Bayer] Methylene donor and resorcinol; bonding agent for fabrics and cord bonded to rubber.

Coherex. [Witco/Golden Bear] Dust control agent, soil conditioner and stabilizer; for quarries and mines.

Cohesive Finish No. 1. [Fibertint & Chems.] Built finish for nylon, polyester and polypropylene.

COHRlastic®. [CHR Industries/Furon] Silicone elastomers, some glass-reinforced, aluminum/rubber composites, silicone sponge rubber; dimensionally stable, flexible fabric and molded sheet for belting, vacuum blankets, thermal shielding applics.

Cojoba. [Costec] Lubricant specialty.

Colacryl. [Bonar Polymers Ltd.] Acrylic polymers; sol'n. and bead.

Cold Stripper-New. [Ivax Industries] Emulsifier/solvent blend; stripper for removing photosensitive lacquer from copper rolls and screens.

Collacral®. [BASF AG] Thickeners for polymer dispersions, aq. systems, paints, adhesives, sealants, prod. of nonwovens.

Collagen 15K. [Brooks Industries] Hydrolyzed mixed glycosaminoglycans, hydrolyzed collagen, hydrolyzed dermal proteins; for skin and hair care cosmetics.

Collagen CLR. [Henkel/Cospha] Carrier of native soluble collagen in weakly acid hydrophilic medium; prods. for aging skin, wrinkle, and after-sun treatments.

Collagen Hydrolyzate Cosmetic. [Maybrook] Hydrolyzed collagen; protective colloid, film former, moisturizer, substantivity agent for hair and skin care prods.

Collagen Native Extra 1%. [Maybrook] Soluble collagen; moisturizer, film former, protective barrier for skin care prods., anti-aging, anti-wrinkle formulations.

Collamino. [Brooks Industries] Collagen amino acids or blends; for skin and hair care cosmetics.

Colla-Moist. [Brooks Industries] Collagen derivs.; moisturizing protein complexes for cosmetics.

Collapur. [Henkel KGaA] Soluble collagen; humectant for increasing moisturizing effect on skin.

Collapuron. [Henkel KGaA] Soluble collagen; humectant for increasing moisturizing effect on skin.

Collasol. [Croda Inc.; Croda Chem. Ltd.] Soluble collagen; humectant, hygroscopic film former, conditioner for skin care prods.

Collex G. [Borregaard LignoTech] Fermented calcium lignosulfonate; concrete additive.

Colloid. [Rhone-Poulenc/Perf. Resins & Coatings] Polycarboxylate salts, acrylics; dispersant, visc. stabilizer for paints; defoamer; latex processing aid.

Collone. [Rhone-Poulenc Ltd.] Emulsifying waxes.

Colonial. [Colonial Chem.] Fatty acid sulfate salts or ether sulfate salts; surfactants.

Colorant. [Colorant GmbH] Color concs.

Colorflol. [Colorco] Liq. color dispersions for use in inj. molding, blow molding, and extrusion of plastic resins.

Colorific. [Monofil Tech. Ltd.] Polyamide 66 monofilament.

Colorin. [Sanyo Chem. Industries] Polyether polyol; antifoamer for acrylonitrile refining, fermentation processes.

Colorlok. [Marlowe-Van Loan] Inorganic salts (liq.); migrating inhibitor for direct dyes on cellulosics.

Colormatch®. [Plasticolors] High pigment-loading dispersions for coloring.

Color-Max. [W.J. Ruscoe] Color dispersions.

Colorol. [Lucas Meyer] Phosphoamino compds.; film-forming agent, wetting and dispersing agent, binder, antisettling agent, flatting agent for lacquers, varnishes, leather finishes, electroplating.

Colorol Aquasorb. [Lucas Meyer]

Phosphoamino compd.; improves anticorrosive properties and adhesion on slightly damp surfaces for coating films.

Colorol Rustbinder. [Lucas Meyer] Compd. based on modified oils and carboxylic ester; corrosion inhibitor.

Colorplast. [Color-Chem Int'l.] Textile dyes and pigments.

Colorpure. [Hall Chem.] Chemicals for pigments, ceramics.

Colorquid. [Colorco] Liquid color dispersions for use in injection molding, blow molding, and extrusion of plastic resins.

Colorsperse No. 1, 2, 4. [Fibertint & Chems.] Ethoxylated aryl phenols; nonionic surfactants for dispersing organic and inorganic pigments, acid, disperse and fiber reactive dyes.

Colour Diamond. [Chinghall Ltd.] Pigment dispersions.

Colsol. [Scher] Fatty amide; cationic cotton and wool softener and lubricant for textile industry; finishing agent.

Columbia Soap. [Reilly-Whiteman] Medium titer soap; detergent, scouring and fulling assistant, dispersant, dyeing assistant.

Colvinal® 226. [Allied Colloids] Acrylic copolymer; size with high adhesion to synthetic fibers and in conjunction with starch.

Combat®. [Carborundum] Boron nitride; powds. exhibiting high thermal conduct. and high dielec. str.; used as additives to silicone and epoxy resins, fluorinated hydrocarbons, silicone oils, etc.

Com-Bat. [Betz Industrial] Alkaline cooling water corrosion and scale inhibitors.

Comboloob. [Astor Wax] PVC lubricant.

Combustrol. [Calgon] Fuel conditioner.

Come Clean. [Castrol Industrial East] Industrial cleaners.

Comiel. [Chemach NV] Synthetic wax for print and plastic industries.

Comite. [Uniroyal] Miticide.

Command. [FMC/Ag] Herbicide.

Commence. [FMC/Ag] Herbicide.

Compactrol®. [Mendell] Calcium sulfate NF; binders, vehicles.

Compass. [Huntington Labs] Aerosol deodorant germicide.

Compel®. [Polymer Composites] Thermoplastic pellets; for compr. and transfer molding, structural, harsh environments, medical, recreation, aerospace, transportation, complex shapes.

Comperlan®. [Henkel/Emery/Cospha; Henkel KGaA; Pulcra SA] Alkanolamide and quaternary ammonium salts; foam booster/stabilizer, thickener, emulsifier, conditioner for cosmetics, pharmaceuticals; fungicide.

Compete. [Rohm & Haas] Herbicide.

Complementos. [Resinas y Complementos Castro] Catalysts.

Complemix®. [Am. Cyanamid; Cyanamid BV] Sulfosuccinates; emulsifier, wetting agent, clouding agent, dispersant, solubilizer for food and beverage industry.

Complexion. [Pulcra SA] Tetrasodium EDTA; chelating agent for calcium, magnesium and other ions.

Componere. [Poly Organix] Bismaleimide resins.

Compound 170. [Henkel/Emery/Cospha] Blended organic phosphates; emulsifier for metalworking fluids; imparts lubricity, anticorrosion, mild EP props.; grease additive.

Compound 535. [A. Harrison] Coconut acid and amine; wetting and fulling agent.

Compound 9264B. [Stepan/PVO] Ethoxylated glyceryl stearate; softener.

Compound MS-, SBC. [Rhone-Poulenc Surf.] Surfactant blends.

Compound PIN. [Nutex] Nonionic fatty condensate; softener, lubricant, dust control agent for textile industry.

Compound S.A. [A. Harrison] Cocamide DEA modified; wetting, fulling, and scouring agent for wool and worsted fabrics with soda ash.

Compriband. [Comprifalt] Molded flexible polyurethane foam.

Compritol 888. [Gattefosse; Gattefosse SA] Glyceryl di/tribehenate; food emulsifier and additive for tablet mfg.

Comsol. [Angus] Nitroparaffin; for solvent blends in inks and coatings.

Conacure®. [Conap] Curing agents for urethane and epoxy systems.

Conap®. [Conap] Polyester, acrylic, silicone, urethane resin systems; conformal coatings, casting resins.

Conapoxy®. [Conap] Epoxy systems; for dielectric, potting and encapsulation applics., as structural adhesives.

Conastic®. [Conap] Polyurethane adhesive; adhesive for bonding thermoplastics, dielectric applics.

Conathane®. [Conap] Polyurethane elastomers; conformal coating, potting and encapsulating system, casting system.

Con-BACN. [Tosoh] Bromoacenaphthylene; flame retardant for rubbers and resins, wire and cable.

Conco X-200. [Continental Chem.] Sodium alkylaryl polyether sulfonate; detergent, wetting agent, emulsifier, dye leveling agent, latex post-stabilizer; for pickling and plating baths.

Concresive. [Master Builders] Epoxy adhesives.

Condat Vicafil. [Etna Prods.] Wire drawing lubricants.

Condensate 6117, 7509, CC. [Manufacturers Chems.] Fatty alkanolamides; emulsifier, detergent, wetting agent, dye assistant.

Condensate PC. [Nat'l. Starch & Chem.] Cocamide DEA; wetting and rewetting agent, detergent, foaming agent, emulsifier, thickener, for textiles, dishwashing.

Condensol®. [BASF AG] Catalysts for textile finishing.

Conditioner CLC. [Piedmont Chem. Industries] Inorganic antichlor; for textile prep., dyeing, afterscouring and finishing.

Conditioner P6, P7. [Sigma Prodotti Chimici] Polyquaternium 6 or 7; hair conditioner.

Conductex. [Columbian Chem.] Carbon black.

Conductomer. [Syn. Rubber Tech.] Conductive thermoplastic resins; for electromagnetic shielding, electrical charge dissipation, control of static electricity.

Conductor. [Calgon] Cooling water treatment.

Confor. [E-A-R] Urethane slow-recovery foam; for shock isolation in electronic equip., padding and cushioning in medical devices.

Coning Oil C Special. [Hoechst AG] Surfactant/paraffin hydrocarbon mixtures; coning oil for imparting antistatic properties to textured PA and PES yarns.

Conlok. [Synair] Caulking compd.

Connector Plus®. [Miller-Stephenson] Non-CFC aerosol connector cleaners.

Conoptic®. [Conap] Polyurethane; casting system for prototype parts, windows, display devices.

Con-O-Shine. [Crain] Carnauba; floor polish.

Conox 2X. [Chemclean] Hydrochloric acid; fume suppressant.

Conox 95. [Chemclean] Sulfuric acid; pickling bath extender.

Conpol P. [Consolidated Polymers Ltd.] Polypropylene homopolymers and copolymers.

Conquor. [Calgon] Cooling water treatment.

Con-Sal. [Church & Dwight] Sodium carbonate decahydrate.

Consamine. [Consos] Detergent, emulsifier, intermediate, detergent base, thickener, fulling agent used in textiles.

Consil. [Handy & Harman] Silver alloys.

Consista. [A.E. Staley Mfg.] Modified corn starch.

Consofix #1. [Consos] Cationic resin; fixative for use improving washfastness and crockfastness of medium and heavy shades; antimigrant during drying.

Consoft. [Consos] Quaternaries; softener for acrylics and nylon, napping lubricant and leveling agent.

Consolevel. [Consos] Surfactant blends; leveling agent for textiles.

Consolube GP. [Consos] Sodium polyacrylate; low foaming biodegradable dyebath lubricant.

Consoluble 71. [Consos] Fatty deriv.; lubricant, dispersant, leveling agent for dyeing synthetic knit goods.

Consonyl. [Consos] Alkyl naphthalene sulfonate; fixative for nylon dyeing.

Consos Buffer N. [Consos] Salts of inorganic acid; buffer for dye systems.

Consos Castor Oil. [Consos] Sulfated castor oil; retarder and leveling agent for textile dyeing; dispersant, lubricant.

Consoscour. [Consos] Alkylaryl sulfonate; detergent, scouring, and leveling agent for dyeing applics.

Consos Defoamer JM. [Consos] Petroleum hydrocarbons/nonionic surfactants blend; defoamer for dyeing processes.

Consos Liq. Alkali. [Consos] Blend of inorganic salts; phosphate-free source of alkali for applic. of reactive dyes.

Consos Stabilizer B-22. [Consos] Stabilizer, chelating agent; aids in keeping calcium and magnesium salts from depositing on fabric and equip.

Consostat. [Consos] POE alkyl amine; antistat for wool and synthetic fibers.

Consoswhite CH. [Consos] Fluorescent whitening agent for nylon/spandex blends and cellulosic fibers.

Consotard. [Consos] Quaternary; cationic retarder for dyestuffs on acrylic fiber.

Construction 1200. [GE Silicones] Silicone sealant.

Contact Re-Nu®. [Miller-Stephenson] Non-CFC aerosol contact cleaners.

Contain. [Am. Cyanamid/Ag] Herbicide.

Continental® Clay. [R.T. Vanderbilt] Kaolin; filler for agric. applics.

Continental Polymers. [Lermont Plastics SA] Methacrylate sheet, polymethacrylate granules.

Continex. [Witco/Concarb] Furnace process carbon blacks.

Contractors 1000®, 2000®. [GE Silicones] Silicone construction sealants for glazing and general purpose sealing applics.

Contraspum. [Zschimmer & Schwarz] Fatty acids, paraffin oil, nonionic emulsifier; defoamer for yeast and alcohol prods.

Control. [Grain Processing] Corn starch prods.; for drilling mud systems.

Controx®. [Henkel/Emery/Cospha; Henkel KGaA; Grünau] Tocopherol blends; anitoxidant for oils and fats.

Cook-Cool. [Atochem N. Am.] Synthetic soluble oils.

Cook-Cut. [Atochem N. Am.] Cutting oils.

Cook-Draw. [Atochem N. Am.] Drawing oils.

Cook-Dri. [Atochem N. Am.] Vanishing oils.

Cookerzyme. [Enzyme Development] Enzyme for brewing.

Coolanol. [Monsanto] Dielectric heat transfer media.

Cool-Treet. [Schaefer Tech.] Cooling water treatment.

Coomassie. [ICI Am.] Textile dyes and pigments.

Copac® E. [BASF AG] Ammoniacal copper sulfate; for control of bacterial diseases in pears, vegetables, ornamentals.

Cope®. [Air Prods.] For sulfur recovery.

Copherol®. [Henkel/Emery/Cospha; Henkel KGaA] Tocopherol derivs.; natural source of vitamin E for skin care preps.

Coplast. [Mooney Plastics Ltd.; Industriplas Ltd.] PVC foam core.

Copo®. [Copolymer Rubber] Styrene-butadiene rubber; general purpose rubbers.

Copolymer 186. [CasChem] Dehydrated castor oil; vehicle for topcoats, primers; modifier for alkyds ; drying oil for concrete antispalling compds., binders, caulks, sealants.

Copolymer 845, 937, 958. [ISP] PVP/dimethylaminoethyl methacrylate copolymer; film-forming resin, conditioner for hair and skin care prods.

Coporex. [Toho Chem. Industry] Petroleum resins; binder, tackifier, base material for traffic paints, synthetic rubbers, adhesives.

Cop-O-Zinc 25-25. [Cuproquim] 25% basic copper sulfate.

Copperbrite. [London Chem.] Copper cleaning compds.

Coppercide. [Old Bridge Chem.] Copper hydroxide; fungicide.

Copper-Count-N. [Mineral Research & Development] Fungicide.

Coppershield. [Atochem N. Am.] Protective carrier coatings and antistaining agents.

Copperskin. [Atochem N. Am.] Metalworking compds.

Copper Wets. [Alpha Metals] Surfactants.

COPS®. [Rhone-Poulenc Surf.] Allyl ether sulfonate; copolymerizable surfactant.

Coptal WA OSN. [ICI Surf. UK] Sodium dioctyl sulfosuccinate in a high flash solvent; wetting agent, penetrant for continuous processes.

Corabond. [PPG Industries/Adhesives] Nonepoxy antiflutter and/or structural adhesive.

Coracoat. [PPG Industries/Adhesives] Water borne, air dry underbody coating.

Corafoam. [Corradini Gustavo & C SpA] Polyurethane prepolymer.

Coralon. [Hoechst AG] Aromatic condensate, neutral salt; leveling agent for dyeing, chrome leather.

Coravol. [Western Water Management] Steam system treatment.

Corax. [Henkel] Wax; rubber lubricant for o-rings.

Co-Rax. [Prentiss Drug & Chem.] Coated warfarin.

Corbel®. [BASF AG] Fenpropimorph compds.; systemic fungicide.

Corblok. [M. Michel] Corrosion inhibitors.

Cordex. [Finetex] Ethoxylated amine; antistat, leveling agent for synthetic fibers.

Cordinhib. [Croda Chem. Ltd.] Borate derivs.; corrosion inhibitor for metalworking fluids.

Cordon. [Finetex] Sulfates; lubricant, detergent, leveling agent, emulsifier, dispersant, penetrant for textiles, dyeing; plasticizer.

Coremat. [Tradex Colori SAS] Raw materials.

Coresize. [Hercules] Hydrocarbon emulsion; for gypsum board mfg.

Corexit. [Exxon] Degreaser, multipurpose cleaner, oil field chemicals.

Corfil®. [Am. Cyanamid] Microballoon-filled epoxy; potting compd.

Corflam. [British Cork Mills Ltd.] Isocyanurate foam.

Corfree. [DuPont] Dibasic acid mixtures; corrosion inhibitor raw materials.

Coriacide. [ICI Am.] Textile dyes and pigments.

Corial®. [BASF AG] Leather and fur finishes.

Corialbinder®. [BASF AG] Polymer dispersions; binders for leather finishes.

Corian®. [DuPont] Acrylic-based resin; premium solid surf. material for kitchen and bath counter tops, commercial work surfs., sinks and bowls, shower and tub surrounds, wainscoting and windowsills.

Coripol. [Stockhausen] Fatliquors.

Cornelowax. [Frank B. Ross] Waxes for mfg. of solvent, liq., and paste polishes, leather finishes, lubricating sticks.

Corn-Pro 35. [Brooks Industries] Hydrolyzed corn protein; for skin and hair care cosmetics.

Corofil. [Pre-Formed Components Ltd.] Glass-reinforced polyester.

Corogard. [3M] Coatings.

Corona. [Croda Chem. Ltd.] Anhydrous lanolin; conditioning emollient, moisturizer, cosolvent, plasticizer, w/o emulsifier, superfatting agent, wetting/dispersing agent; for cosmetics, pharmaceuticals.

Coronet. [Croda Chem. Ltd.] Lanolin; conditioning emollient, moisturizer, cosolvent, plasticizer, w/o emulsifier, superfatting agent, wetting/dispersing agent; for cosmetics, pharmaceuticals.

Corps Celeri. [Henkel/Emery/Cospha] Aromatic herbal aroma chemical.

Corroclean F. [Stewart Hall] Corrosion and deposit inhibitor for potable water systems.

Corrogen. [Betz Industrial] Catalyzed sodium sulfite.

Corrolite. [Reichhold] Vinyl ester resin.

Corromin. [Kao Corp. SA] Cationics; corrosion inhibitor for crude oil refining systems, acid cleaning; additives for fuel oils, hydraulic fluids.

Corr-Shield. [Betz Industrial] Cooling corrosion inhibitor.

Corsyrene. [Corstyrene] Insulating polystyrene.

Cortex. [Vita Cortex Ltd.] Polyurethane foams.

Corthane. [British Cork Mills Ltd.] Rigid urethane.

Cor-Trol. [Betz Industrial] Boiler corrosion inhibitor.

Cortx. [Vita Cortex (NI) Ltd.] Polyurethane.

Cortymol® LP. [BASF AG] Phenol-free organic fungicide; for preventing mold formation on leather.

Corvel. [Morton Int'l.] Powd. coatings.

Corvic. [European Vinyls Corp. GmbH] Vinyl polymers; used for records, flooring, battery separators, plastisols for coating and dipping applics., blow molding bottles, inj. molding of pipes, electrical/electronic parts, flexbile and rigid extrusions, cable, footwear, calendering.

Cosept. [Costec] Preservatives.

Cosmedia® Guar. [Henkel/Emery/ Cospha; Henkel KGaA] Guar gum derivs.; visc. builder, stabilizer, conditioner for personal care prods.

Cosmedia Polymer HSP 1180. [Henkel/ Cospha] Polyacrylamidomethylpropane sulfonic acid; agent for creams, lotions, antiperspirants, shaving creams, soaps, nail polish removers to give smooth feel.

Cosmetic Lanolin Anhydrous USP. [Croda Inc.] Lanolin.

Cosmetol®. [CasChem] Castor oil; emollient for cosmetics.

Cosmic Blacks. [Ebonex] Bone black pigments.

Cosmic® FL. [BASF AG] Maneb, tridemorph, carbendazim; broad spectrum cereal fungicide.

Cosmocil. [ICI Am.] Cosmetics biocide.

Cosmoline. [E.F. Houghton] Rust preventatives.

Cosmopon. [Auschem SpA] Sulfates and sulfosuccinates; detergent for shampoos, cleaners, bubble bath; emulsifier for emulsion polymerization; foaming agent for latex emulsions; antigelling and cleaning agents for paper mill felts.

Cosmowax. [Croda Inc.; Croda Chem. Ltd.] Fatty alcohol blends; emulsifying wax, stabilizer for personal care prods., pharmaceuticals.

Costausol. [Hercules/FFIG] Aromatic chemical.

Cottestren® C. [BASF AG] Mixed dyes for polyester/cellulose fiber blends.

Cotton-Pro®. [Griffin] Prometryn suspension; flowable herbicide for weed control in cotton and celery crops.

Coumadin. [DuPont] Crystalline warfarin sodium.

Counter.®. [Am. Cyanamid/Ag] Terbufos; systemic insecticide/nematicide for corn corps.

Coupler. [Vega Biotech.] Peptide synthesizers.

Coursemin. [Sanyo Chem. Industries] Anionic/nonionic surfactants; dyebath lubricant for dyeing, scouring, and bleaching.

Covbrite. [Covan Ltd.] Stabilized sodium borohydride sol'n.; for pulp bleaching.

Covera. [Costec] Emollient.

Cover-Up. [Mateson] Asbestos encapsulant.

Covi-ox. [Henkel/Cospha] Tocopherols; antioxidant and blocking agent for cosmetics, food industries.

Covipherol. [Henkel/Cospha] Tocopherols; natural source of vitamin E; antioxidant in food and cosmetic prods.

Covitol. [Henkel] Tocopherol and derivs.; antioxidant for pharmaceuticals; natural source of vitamin E.

Covtek. [Covan Ltd.] 12% sodium borohydride alkaline sol'n.; reducing agent.

CP-. [Eastman] Chlorinated polyolefin; used as primer for polyethylene, polypropylene, metals to promote adhesion.

CP-. [ICI Acrylics] Acrylic resin; for inj. molding and extrusion.

CP0110 Series. [Hüls Am.] Silane derivs.; coupling agent, chem. intermediate, blocking agent, release agent, lubricant, primer, reducing agent.

CPA [ECC Int'l.] Calcium carbonate; construction prods.

CPA-Alpha®. [Dixo] Anionic/nonionic

detergent blend; detergent, wetting agent, emulsifier for water and dry-cleaning solvents.

CPB. [Uniroyal] Accelerator for sundries.

CPF-. [Witco/Argus] Chlorinated paraffin; plasticizer for use in paints, coatings, caulks, sealants, rubber, and adhesives applics.

CP Filler. [ECC Int'l.] Calcium carbonate; construction prods.

CPH-. [C.P. Hall] Ethoxylated, glyceryl, and glycol esters; solubilizer, emulsifier, lube additive, rust preventive, wetting agent, dispersant.

CPR. [DuPont] Chlorinated polyethylene; elastomer for wire and cable industry.

C-Probond 9E. [Chemurgy Prods.] Carbohydrate ether; finishing agent, hand builder for textiles.

C-Prodye DP 150. [Chemurgy Prods.] Dyebath additive improving dye penetration, prevents agglomeration of dye in anionic/cationic systems.

C-Profix NF. [Chemurgy Prods.] Polyquaternary amine; fixing agent improving wetfastness of direct and reactive dyes on cellulosic fibers;

C-Proguard WSR. [Chemurgy Prods.] Modified polyester emulsion; treatment imparting soil release and comfort properties.

C-Prolube NNS. [Chemurgy Prods.] Blend of synthetic lubricants; fiber lubricant and knitting oil.

C-Proscour 66. [Chemurgy Prods.] Surfactants/solvent blend; high temp. scouring agent.

C-Prosoft. [Chemurgy Prods.] Nonionic or cationic softener bases.

C-Prostop P. [Chemurgy Prods.] Polyethoxylated emulsifier blend; heat activated antimigrant for pigment dyeing.

C-Protard AC. [Chemurgy Prods.] Cationic surfactant; leveler for cationic dyes on acrylic and most cationic dyeable polyesters.

CPS Series. [Hüls Am.] Silane derivs.; coupling agent.

Cr-. [M&T Harshaw] Chromium alumina; incineration catalyst for dehydrogenation of butane to butene.

CR-39. [PPG Industries] Thermoset resins.

Crafol. [Henkel KGaA; Pulcra SA] Phosphate esters; antistat for textiles; spinning oil for carpets; conditioner for cosmetics; corrosion inhibitor for cutting fluids.

Crain Aim-100. [Crain] Detergent surfactant.

Crainchem 75. [Crain] Surfactant for oil and gas industry.

Cralane. [Pulcra SA] Ethoxylated lanolin; emollient, wetting, cleaning, and superfatting agents, o/w emulsifier, conditioner, solubilizer, and plasticizer for personal care prods. and germicides.

Crapol. [Pulcra SA] Quaternaries; germicide, disinfectant, algicide for swimming pools, industrial waste water; retarding and leveling agent for dyeing; antistat for creams, hair conditioner.

Crastine®. [Ciba-Geigy GmbH] PBT polyester, some glass and/or mineral reinforced.

Craston®. [Ciba-Geigy GmbH] Polyphenylene sulfide; engineering thermoplastic.

Cravenette®. [Yorkshire Pat-Chem] Silicone water and oil repellents for textiles.

Craynor. [Sartomer] Epoxy acrylate compds.; for use in uv and eb curing compositions incl. paper and wood coatings, inks, metal decorating.

Crayvallac. [Cray Valley Prods.; Ketro A/S] Thixotropic additives for paints.

Creamoyl®. [Scholler] Finishes, textile softeners.

Creamtex. [Van Den Bergh Foods] Hydrogenated vegetable oil; shortening for baked goods.

Creek-O-Flow. [Golden Cat] Anticaking agent.

Creek-O-Nite Clay. [Golden Cat] Pesticide carrier and diluent.

Crelan. [Miles] Powd. coating raw materials.

Cremba. [Croda Inc.; Croda Chem. Ltd.] Mineral oil, petrolatum, lanolin alcohol, lanolin; emollient, moisturizer, and emulsifier for cosmetics and pharmaceuticals.

Cremodan. [Grindsted Prods. Denmark] Emulsifier/stabilizer blends; for ice cream and related prods.

Cremophor®. [BASF; BASF AG] Ethoxylated fatty ethers and esters; emulsifier, solubilizer, thickener, stabilizer, lubricant for cosmetics, pharmaceuticals.

Crepetrol®. [Hercules] Cationic polymer; creping and adhesion aid for tissue machines.

Crestalans. [Croda Chem. Ltd.] Lanolin derivs.; emollient, moisturizer, conditioner for personal care prods.; plasticizer for hair sprays and lacquers; superfatting agent.

Crestawhip. [Croda Food Prods. Ltd.] Fatty esters; improver for sponge cakes, bread, rolls.

Crester. [Croda Food Prods. Ltd.] Polyglyceryl blends; food emulsifier, visc. modifier.

Crestex. [Reilly-Whiteman] Additive for cleaning of pressure equip. in textile industry.

Crestodull M-3. [Reilly-Whiteman] Dulling agents for textiles.

Crestofix. [Reilly-Whiteman] Fixatives for cellulosics.

Crestoil. [Reilly-Whiteman] Emulsifier, fatliquoring agent for textile and leather industries.

Crestolan NF. [Reilly-Whiteman] Detergents for textile scouring.

Crestomer®. [Scott Bader] Unsaturated urethane acrylates; impact modifier; enhances performance of polyester resins, compds., and laminates.

Crestopal E. [Reilly-Whiteman] Dyeing assistant for polyester.

Crestopene 5X. [Reilly-Whiteman] Wetting agents for cellulosics and blends.

Crestosolve 630. [Reilly-Whiteman] Detergents for textile scouring.

Crest Retarder G. [Reilly-Whiteman] Dyeing assistant, retarding agents for acrylics.

CRF. [Witco/Golden Bear] Crack filler.

Cri-Line. [Cri-Tech] Fluoroelastomer compds.; for molding, calendering, extrusion, and precision applics.

Criliprint. [Sybron] Acrylic copolymers; pigment printing and pad dye binder.

Crilitex. [Sybron] Acrylic copolymers; emulsions for textile processing; pigment printing and pad dye binder-stiffener.

Crill. [Croda Inc.; Croda Surf. Ltd.] Sorbitan esters; emulsifier, dispersant, cosolvent, wetting agent, antifoam, visc. reducer, mold release, antiblock agent, corrosion inhibitor, lubricant, antistat; for cosmetics, foods, agric., leather, metalworking, paints, inks, pharmaceuticals, plastics.

Crillet. [Croda Inc.; Croda Chem. Ltd.] Ethoxylated sorbitan esters; solubilizer, emulsifier, dispersant, wetting agent for cosmetics, foods, household prods., agric., metalworking, paints, inks, pharmaceuticals, textiles.

Crillon. [Croda Surf. Ltd.] Fatty alkanolamides; detergent, foam stabilizer, emulsifier, antistat, anticorrosive, lubricant for personal care prods., cutting oils.

Cri-Loy. [Cri-Tech] Fluorosilicone or polyacrylate rubber alloys; for o-rings, diaphragms, hoses, stem seals, valve cover gaskets.

Cri-Spersion. [Cri-Tech] Calcium hydroxide/magnesium oxide/fluorocarbon elastomer blends; activator for fluoroelastomer compounding.

Cristyc. [Tradex Colori SAS] Polyester resins.

Croda Absorption Base. [Croda Chem. Ltd.] Cholesterol-rich base; emulsifier, conditioner, moisturizer, emollient; binder for cosmetic pressed powds.

Crodacel. [Croda Inc.; Croda Chem. Ltd.] Alkyl dimonium hydroxyethyl cellulose; conditioner improving foaming and imparting body for skin and hair care prods.

Crodacid B. [Croda Inc.] Behenic acid; gellant for stick formulations.

Crodacol. [Croda Inc.] Fatty alcohols; emulsifier, thickener, opacifier, emollient, lubricant, structural agent in anhyd. stick systems, hair and skin prods.

Croda Conditioner Base. [Croda Chem.

Ltd.] Proprietary prod.; hair conditioners.

Crodacreme. [Croda Food Prods. Ltd.] Emulsifier blend; ice cream emulsifier/ stabilizer.

Crodafos. [Croda Inc.; Croda Chem. Ltd.] Phosphates; emulsifier, antistat, stabilizer, conditioner, gellant for personal care prods.; corrosion inhibitor.

Crodalan. [Croda Inc.; Croda Chem. Ltd.] Acetylated lanolin alcohol blends; emollient, penetrant, wetting agent, conditioner, superfatting agent, solubilizer, conditioner, plasticizer for cosmetics, pharmaceuticals, detergent systems.

Crodamet. [Croda Chem. Ltd.] Ethoxylated primary amines; emulsifier, antistat for plastics and fibers.

Crodamide. [Croda Universal Ltd.] Fatty acid amides; slip and antiblock agents for polyolefins; release agent for molded thermoplastic polymers.

Crodamine. [Croda Universal Ltd.] Fatty amines and dimethyl amines; emulsifier for cosmetics, herbicides, ore flotation, pigment dispersion, textiles, leather, rubber, plastics, and metal industries.

Crodamol. [Croda Inc.; Croda Surf. Ltd.; Croda Food Prods. Ltd.] Fatty acid esters; thickener, opacifier, emollient, stabilizer, superfatting agent, coupling agent for personal care prods.

Crodapearl. [Croda Inc.; Croda Surf. Ltd.] Hydroxyethyl stearamide-MIPA blends; pearlescent for shampoos, bubble baths, detergent systems.

Crodapur. [Croda Chem. Ltd.] Lanolin; plasticizer for tape adhesives, mastics, putty.

Crodarom. [Croda Inc.] Botanicals for cosmetics, pharmaceuticals.

Crodascoop. [Croda Food Prods. Ltd.] Mono/diglycerides of fatty acids; emulsifier, stabilizer for soft ice cream.

Crodasinic. [Croda Chem. Ltd.] Sarcosinates; foaming and wetting agent, detergent, lubricant, antistat, corrosion inhhibitor, bacteriostat, penetrant for dental care preps., pharmaceuticals, personal care prods., household and industrial applics.

Crodasone. [Croda Inc.] Hydrolyzed wheat protein polysiloxane copolymer; film-forming and substantivity agent, lubricant, conditioner for skin and hair care prods.

Crodatem. [Croda Food Prods. Ltd.] Diacetyl tartaric acid ester; Antistaling agent, starch complexing agent, emulsifier for food industry.

Crodateric. [Croda Surf. Ltd.] Fatty acid derivs.; amphoteric surfactants.

Crodax. [Croda Chem. Ltd.] Blends; used in rust preventives, lubricants.

Crodazoline. [Croda Universal Ltd.] Fatty acid imidazolines; lubricating metal processing aids.

Croderol. [Croda Universal Ltd.] Glycerin; conditioner, humectant, moisturizing agent in cosmetics and pharmaceuticals, diluent and plasticizer for many polar materials.

Crodesta. [Croda Inc.; Croda Surf. Ltd.] Sucrose esters; dispersant, emulsifier, wetting agent, solubilizer, thickener, suspending agent, detergent in cosmetics, toiletries, pharmaceuticals.

Crodet. [Croda Chem. Ltd.] Ethoxylated fatty esters; o/w emulsifier for cosmetics and pharmaceutical creams, lotions and ointments, wetting agent, solubilizer for perfumes or aq. alcoholic preparations; dispersant; plasticizer for hair setting sprays.

Crodex. [Croda Chem. Ltd.] Cetearyl alcohol blends; emulsifying wax for pharmaceuticals and cosmetics; wetting agent, penetrant.

Crodinhib. [Croda Chem. Ltd.] Borate derivs.; corrosion inhibitor for metalworking fluids and oils.

Crodolene. [Croda Universal Ltd.] Oleic acid; dewatering protective.

Croduret. [Croda Chem. Ltd.] Ethoxylated hydrogenated castor oil; emulsifier, solubilizer, emollient, superfatting agent, detergent used for cosmetics, pharmaceuticals, textiles, metalworking fluids, emulsion polymerization, insecticides, herbicides, household detergents.

Crodyne BY19. [Croda Inc.; Croda

Chem. Ltd.] Gelatin NF; protective colloid, moisturizer, conditioner for skin and hair care prods.; humecant, thickener for pharmaceutical and food applics.

Crolactil. [Croda Surf. Ltd.] Acyl lactylates; o/w emulsifiers, ingredients in food prods.

Crolastin. [Croda Inc.; Croda Chem. Ltd.] Hydrolyzed elastin; conditioner, moisturizer for skin care prods. and cleansers.

Crolec. [Croda Chem. Ltd.] Modified lecithin; conditioner, superfatting agent, emulsifier for cosmetics and pharmaceuticals.

Cromeen. [Croda Chem. Ltd.] Substituted alkylamine deriv. lanolin acids; multifunctional surfactant, foamer, detergent, emulsifier for shampoos, detergents, cleansers, shaving foams.

Cromoist. [Croda Inc.; Croda Chem. Ltd.] Hydrolyzed collagen blends; moisturizer, conditioner for skin care prods.

Cromophthal. [Ciba-Geigy/Pigments; Ciba-Geigy Pigments UK] High-performance organic pigments.

Cromul. [Croda Chem. Ltd.] Ethoxylated fatty ethers; emulsifier, opacifier for cosmetics, pharmaceuticals.

Cronectin. [Croda Chem. Ltd.] Hydrolyzed fibronectin; conditioner for skin and hair care prods.

Croneton®. [Bayer] Ethiofencarb; insecticide for control of aphids.

Cronox. [Baker Perf. Chem.] Corrosion inhibitors.

Cropepsol. [Croda Chem. Ltd.] Hydrolyzed collagen; conditioner for skin and hair care prods.

Cropeptide W. [Croda Inc.] Hydrolyzed wheat protein, wheat oligosaccharides; film-former, conditioner for hair care prods.

Cropeptone. [Croda Chem. Ltd.] Hydrolyzed collagen; conditioner for skin and hair care prods.

Crop Mag. [Martin Marietta Magnesia Spec.] Magnesium oxide; fertilizer grade.

Crop Plus. [Western Nutrients] Neutralized complex fertilizer for foliar applics.

Cropol. [Croda Surf. Ltd.] Sulfosuccinate; dewatering protective, emulsification aid.

Cropotex®. [Bayer] Flubenzimine; acaricide.

Croquat. [Croda Inc.; Croda Chem. Ltd.] Quaternary deriv. of hydrolyzed keratin or collagen; cationic conditioner for hair and nail care prods.

Crosilk. [Croda Inc.; Croda Chem. Ltd.] Silk derivs.; protein for solid make-up; conditioner, humectant for skin and hair care prods.

Crosilkquat. [Croda Inc.; Croda Chem. Ltd.] Cocodimonium silk amino acids; conditioner for skin and hair care prods.

Crossential EPO. [Croda Inc.] Evening primrose oil; contains essential fatty acids for skin care.

Crosterene. [Croda Universal Ltd.] Stearic acid; pearlescent in cosmetic creams and lotions; binder; buffing compds., candles, lubricants.

Crosterol. [Croda Chem. Ltd.] Proprietary blend; superfatting and emollient agent for soap.

Crosultaine. [Croda Inc.] Alkyl amidopropyl hydroxysultaine; amphoterics.

Crotein. [Croda Inc.; Croda Chem. Ltd.] Animal and wheat protein derivs.; conditioner, film modifier, foam booster/stabilizer, dye leveling agent for hair care prods., nail enamels; moisturizer for skin care prods.

Crothix. [Croda Inc.] Polyol alkoxy ester; thickener for aq. systems.

Crovol. [Croda Inc.; Croda Chem. Ltd.] Ethoxylated vegetable glycerides; emulsifiers, dispersants, solubilizers for cosmetics.

Crown. [Penford Prods.] Corn starches.

Crown Clay. [Southeastern Clay] Hard rubber clay.

Cru Carrier. [Crucible] Textile dye carriers.

Cru Chem. [Crucible] Silicone; for cosmetics and cleaning applics.

Cru Cleaner. [Crucible] Cleaners.

Cru Ester. [Crucible] Specialty esters.

Cru Fac. [Crucible] Nonyl phenol and

fatty ethoxylated compds. and phosphates.

Cru Fluid. [Crucible] Silicone fluid.

Cru Lev. [Crucible] Leveling agent, retardants for textiles.

Cru Lube. [Crucible] Silicone lubricants.

Cru Moco. [Crucible] Textile printing assistants.

Cru Nyl. [Crucible] Nylon dye levelers and lubricants.

Cru Ox. [Crucible] Nonyl phenol and fatty ethoxylated compds. and phosphates.

Cru Pro. [Crucible] Flame retardant textile dye carriers.

Cru Release. [Crucible] Release agents.

Cru Scour. [Crucible] Textile scouring agents.

Cru Sil. [Crucible] Silicone surfactants.

Cru Soft. [Crucible] Textile softeners.

Cru Sol. [Crucible] Solvents and cleaners for textiles.

Cru Sperse. [Crucible] Textile wetting agents.

Cru Stat. [Crucible] Antistats.

Cru Thix. [Crucible] Polyacrylates; thickeners.

Cru Wet. [Crucible] Nonyl phenol and fatty ethoxylated compds. and phosphates.

Crylcon. [DuPont] Methacrylate bonding agent.

Cryocide. [Int'l. Dioxcide] Disinfectant.

CryOfine®. [Midwest Elastomers] Cryogenically ground fine particle rubber (butyl, EPDM, neoprene, nitrile, SBR, mixed and natural rubbers); filler for rubber compding.

Cryoflex®. [Sartomer] Dibutoxyethoxyethyl formal; plasticizer.

Cryosept. [Avatar] Antimicrobial preservative.

Cryovac®. [Cryovac] Polyolefin shrink film; for barrier applics.

Cryscoat. [Oakite Prods.] Coating.

Crystal®. [CasChem] Castor oil; specialty oils, suspending and wetting agents for cosmetics.

Crystalac. [Mantrose-Haeuser] Food-grade confectioners and pharmaceutical glaze.

Crystal Inhibitor #5. [Harcros] Blend; retards formation of crystals in 2,4-D amine formulations.

Crystaliron. [Synergy Prod. Group] Ferrous sulfate heptahydrate; crystalline form.

Crystalor. [Phillips] Polymethylpentene homopolymers and copolymers, some glass filled; for pkg., housewares, microware, appliances, transportation, elec./electronic, medical, films, and coatings.

Crystal White. [A.M. Todd] Oil of peppermint.

Crystamet®. [Crosfield; Rhone-Poulenc Basic] Sodium metasilicate pentahydrate; for formulating specialty detergents and contributing buffering capacity and corrosion inhibition of soft metals and ceramic glazes.

Crystex®. [Akzo] Sulfur and insoluble sulfur; prevents sulfur bloom on uncured rubber surfaces; detackifier; retards bin scorch; minimizes sulfur migration.

Crystic®. [Scott Bader] Thermoset polyester resins; for aircraft industry moldings, in elec. industry for switchgear, generator insulation, in chemical industry for chimneys, ducting, pipes, for building, land transport, boat hulls, GRP, filament winding, extrusion, gelcoat applics.

Crystic® Cleaner. [Scott Bader] Acetone with evaporation suppressant.

Crystic Copperclad. [Ferro] Antifouling coating.

Crystic® Fireguard. [Scott Bader] Thermoset polyester resin; surface coating resin giving fire protection to GRP laminates, timber, porous substrates.

Crystic® Kollerdur. [Scott Bader] One-pack polyurethane resins; for primers, sealers, coatings, or formulations such as aggregate-filled floors, pitch-extended roofing systems.

Crystic® Prefil. [Scott Bader] Antimony trioxide/polyester resin compd.; additive for mfg. of reduced fire hazard moldings.

Crystic® Pregel. [Scott Bader] Silica/polyester compd.; resin additive to confer thixotropic properties to general purpose polyester resins for laminating and gelcoat applics.

CS1590. [Hüls Am.] Silane derivs.; coupling agent, chem. intermediate, blocking agent, release agent, lubricant, primer, reducing agent.

CSGS. [Drew Ind. Div.] Corrosion inhibitor.

C2® Sodium Chlorite. [Olin] Sodium chlorite; bacterial slimicide for aq. paper mill systems.

CT-88 Aerosol. [Chem-Trend] Fluoropolymer/organic binder disp.; dry-film release agent and lubricant for moldings, casting, potting, hand layup, encapsulation, filament winding.

CT1750 Series. [Hüls Am.] Silane derivs.; coupling agent, chem. intermediate, blocking agent, release agent, lubricant, primer, reducing agent.

CTC-3300. [Compounding Tech.] Nylon 66, glass-reinforced.

CT-N. [Stewart Hall] Boiler compds., corrosion inhibitors.

CTX-. [Compounding Tech.] Glass-reinforced thermoplastics.

CTXC-. [Compounding Tech.] Reinforced thermoplastics; for ESD applics.

Cu-. [M&T Harshaw] Copper chromite; catalyst for dehydrogenation of alcohols, decomposition of methanol, oxygen scavenging, dehydrogenation of piperazines to pyrazines, methyl ester hydrogenolysis and dehydrogenation.

Cubond. [SCM Metal Prods.] Copper brazing paste.

Cudis. [Jonas Chem.] Controlled undistilled diisocyanate.

Culinox 999. [Morton Salt] High purity salt.

Culminal®. [Aqualon] Hydroxypropyl methyl cellulose and methyl cellulose.

Cumal. [Mitsubishi Gas] p-Isopropylbenzaldehyde; intermediate for pharmaceuticals, fragrances.

Cumar®. [Neville] Coumarone-indene resin; for adhesives, aluminum paints, varnishes, rotogravure inks, and rubber compds.

Cumate®. [R.T. Vanderbilt] Copper dimethyldithiocarbamate; ultra accelera-

tor for SBR, IIR.

Cumberland. [Tecindes-Technicas Industriales Espanolas] Additives and mold releases for plastics industry.

Cunilate®. [Morton Int'l.] Antimicrobial, mildewproofing agent for textile applics., pulp slurries, paperboard, concrete safes.

Cunimene. [Morton Int'l.] Mildewproofing agent.

Cuniphen®. [Morton Int'l.] Antimicrobial, mildewproofing agent for textile applics., in-can preservation.

CUPL PIC. [Bernel; Heterene] Dipropylene glycol isoceteth-20 acetate; fragrance solubilizer, emulsifier, stabilizer; imparts sheen to emulsions.

Cupramidine. [John Campbell] Textile dyes and pigments.

Cuprolignum. [Rudd] Marine coating and wood preservatives.

Cuprophenyl. [Ciba-Geigy/Dyestuffs] Textile dyes and pigments.

Cuprorite. [Rite Industries] Textile dyes and pigments.

Cuprose. [Nalco] Algicide.

Cuprostat. [Calgon] Copper corrosion inhibitor.

Cupsac. [Am. Cyanamid] EPDM accelerator.

Cupuran. [Giulini Adolfomer SA] Copper hydroxide.

Curasol. [Hoechst Celanese] Specialty polymer for soil erosion prevention.

Curaterr®. [Bayer] Carbofuran; insecticide, nematocide.

Curbeton 0550. [Borregaard LignoTech] Desugared hardwood calcium/magnesium lignosulfonate; water-reducing and strength-increasing concrete additive.

Cure. [Malco Prods.] Laundry detergent.

Curebond. [Poly Organix] Anhydride; curing agents.

Curene. [Anderson Development] Crosslinking agents, chain extenders; urethanes, epoxies.

Cure-Rite® 18. [Akrochem] Thiocarbamyl sulfenamide; nonstaining primary accelerator for EPDM, SBR, nitrile, natural and butyl rubbers.

Cure-rite Accelerator. [BFGoodrich

UK] Thiocarbamyl sulfenamide; accelerator for SBR, EPDM, other rubbers.

Curezol®. [Air Prods.; Pacific Anchor; Anchor UK] Imidazoles; epoxy curing agent for adhesives, composites, filament winding, solder-resistant inks, potting compds., insulating powds., transfer molding powds.; accelerator for dicyandiamide and anhydrides.

Curimid. [Poly Organix] Epoxy curing agents.

Curing Agent C. [Am. Bio-Synthetics] Epoxy curing agents.

Curite. [Nat'l. Starch & Chem.] Magnesium chlorides; catalyst for textile finishing operations.

Curithane®. [Air Prods./Polyurethanes] liq. catalysts; amine curing agent for polyurethane; intermediate in mfg. of polyamides, polyimides, coatings, plastics.

Cur-Rx. [Mooney Chems] Drier for high-solids paint resins.

Curtexil. [Borregaard LignoTech] Hardwood calcium/magnesium lignosulfonate; general binder and dispersant.

Custom Age 625 Plus®. [Carpenter Tech.] Nickel-base alloy; for deep sour gas wells, refineries, chemical process industry environments, high-temp. high-purity nuclear water.

Cutavit Richter. [Dr. Kurt Richter; Henkel/Cospha] Complex of vitamins A and E and essential fatty acids in lipophilic medium; vitamin-based prod. for dry skin and hair.

Cutina®. [Henkel/Emery/Cospha; Henkel KGaA] Esters, wax blends, ethoxylated esters; emollient, pearlescent, opacifier, visc. builder/stabilizer, lubricant, base for cosmetics; beeswax and spermaceti subsitutes; emulsifier.

Cutinox. [KMZ Chem. Ltd.] Biocides.

CV4720 Series. [Hüls Am.] Silane derivs.; coupling agent, chem. intermediate, blocking agent, release agent, lubricant, primer, reducing agent.

CW- Series. [Ferro/Keil] Chlorinated paraffins; corrosion inhibitor, extreme pressure additives for metalworking fluids.

CWAB. [R.T. Vanderbilt] Glass mold release.

CWT. [Drew Ind. Div.] Corrosion, scale inhibitor, slime preventives, and antifoams for cooling towers.

CXA. [DuPont] EVA terpolymer; coextrudable adhesive resin for flexible pkg.

Cyaf. [Am. Cyanamid] Antifoulants.

Cyaguard. [Cyanamid BV] Flame retardant.

Cyanacryl®. [Am. Cyanamid] Acrylic elastomer.

Cyanacure. [Am. Cyanamid] Curing agent for urethane elastomers.

Cyanalube® Softener TSI Spec. [Am. Cyanamid] Polyethylene emulsion; textile softener; improves tear strength of resin-treated cellulosic fibers.

Cyanamer. [Am. Cyanamid; Cyanamid BV] Polyacrylamide; dispersant, anti-precipitant, solubilizer for adhesive tapes, agriculture, cement, ceramics, coatings, detergents, dyes, inks, processing, textile, leather; binder, sizing, flocculating and suspending agent, lubricant, thickener, stabilizer.

Cyanaprene®. [Air Prods.] Polyurethane prepolymer; used for solid tires, coated materials, rolls, footwear, molded mech. goods, castings.

Cyanasize. [Am. Cyanamid] Sizing agent.

Cyanasol. [Am. Cyanamid] Surfactants for rubber latex.

Cyanatex®. [Am. Cyanamid] Dyeing assistants, antimigrants, fixatives, softeners.

Cyanatrol. [Am. Cyanamid] Mobility control polymer.

Cyanobrik. [DuPont] Sodium cyanide.

Cyanogran. [DuPont] Sodium cyanide.

Cyanox®. [Am. Cyanamid] Antioxidants/stabilizers for plastics, pipe, adhesives, household appliances, automotive, film, fibers, molded goods; polymerization inhibitor in chemical processes.

Cyasorb®. [Am. Cyanamid] Benzophenones; light stabilizer/uv absorber for films, coatings, elastomers, and plastics in automotive, greenhouse, construction, solar applics.

Cyastat®. [Am. Cyanamid] Antistatic agents for plastics, surface coatings, paper, glass, textiles, and other materials; emulsifier, settling, dispersing, and rewetting agent.

Cybond®. [Am. Cyanamid] Two-part epoxy structural adhesive; industrial metal adhesives.

Cycat. [Am. Cyanamid] Catalysts.

Cyclamber. [Henkel/Emery/Cospha] Woody aroma chemical.

Cyclanon® R. [BASF AG] After-treatment agent for polyester dyeings and prints.

Cyclochem®. [Rhone-Poulenc Surf.] Esters; emulsifier, emollient, thickener, opacifier, pearlescent, lubricant, detergent, dispersant for cosmetics, soaps.

Cyclochem 326A. [Baxenden] Wax blend; emulsifier, beeswax substitute for cosmetic cold creams, night creams, lipstick, mascara.

Cyclo-Flo. [Taiwan Surf.] Fuel oil additive; sludge preventive.

Cyclofor. [Baxenden] Surfactant blend; antisettling and anticaking agent for paint systems.

Cyclogen. [Witco] Asphalt pavement recycling agent.

Cyclol. [Witco SA] Fatty esters; spermaceti substitute for cosmetics and pharmaceuticals.

Cyclolube®. [Witco/Golden Bear] Naphthenic oils; process oils for general-purpose compounding of elastomers.

Cyclomatic Dur. [Ceca SA] Wetting agent for use in hydrosulfite sol'n.

Cyclomethylenecitronellol. [Firmenich] Fragrance ingredient.

Cyclomox®. [Rhone-Poulenc Surf.] Amine oxides; foaming agent, thickener, conditioner, emollient for dishwash and shampoo formulations; stabilizer for bleaching formulations.

Cyclonap. [Witco] Process and extender oils.

Cyclonox® BT-50. [Akzo] Cyclohexanone peroxide; initiator for ambient cures of polyester; for automotive body putty, hobby and automotive kits.

Cyclopal. [Akzo] Paper sizing agent.

Cyclophos®. [Rhone-Poulenc Surf.] Phosphate esters; surfactants.

Cycloryl. [Rhone-Poulenc Surf.] Surfactant blends; detergent, visc. builder, solubilizer, base for shampoos; thickener for agric. formulations.

Cyclosheen 202. [Rhone-Poulenc Surf.] Glycol stearate and emulsifiers; pearl conc., visc. builder, foam booster, conditioner.

Cyclo Sol. [Shell] Aromatic solvents.

Cycloteric®. [Rhone-Poulenc Surf.] Amphoterics for foaming, conditioning, and thickening for cosmetics, soaps, industrial applics.

Cycloton®. [Rhone-Poulenc Surf.] Quaternary ammonium salts; conditioner, emulsifier, base for hair conditoners, fabric softeners, antistatic treatments.

Cyclovertal. [Henkel/Emery/Cospha] 3, 6-Dimethyl-3-cyclohexene-1-carbaldehyde; general fragrance raw material; tart aroma.

Cycocel®. [BASF AG] Clormequat chloride, choline chloride; agric. chemical for oat, rye, and wheat crops.

Cyco-Gel. [Chattem] Aluminum acylate.

Cycolac®. [GE Plastics; GE Plastics Ltd.] ABS resins; for inj. molding, extrusion, thermoforming; for business machine, computer equip., elec. applics., housewares, luggage, automotive, appliances, decorative office and home furnishings, hardware, toys.

Cycolin®. [GE Plastics] Chemically resistant resins.

Cycoloy®. [GE Plastics; GE Plastics Ltd.] PC/ABS alloys; for automotive, appliance, and elec. applics.

Cycom®. [Am. Cyanamid] Thermoset epoxy, phenolic, polyester or polyimide resins; laminating materials for primary structural applics.

Cycor. [Am. Cyanamid] Corrosion inhibitors.

Cy-Fana. [Monsanto] Acrylic fibers.

Cyflo. [Am. Cyanamid] Rosin sizes for paper.

Cyfloc. [Am. Cyanamid] Flocculant.

Cyfluthrin. [Bayer] Cyfluthrin; synthetic pyrethroid for control of household, public health, and stored-product pests.

Cyfor. [Am. Cyanamid] Rosin sizes for paper.

Cyform®. [Am. Cyanamid] Composite tooling prepreg. synthetic pyrethroid for control of household, public health, and stored-product pests.

Cyglas®. [Am. Cyanamid] Thermoset glass-reinforced polyester composites; molding compds.

Cygon® 400. [Am. Cyanamid/Ag] Dimethoate; systemic insecticide-miticide for field, vegetable, and fruit crops.

Cykelin. [Reichhold] Linseed oil.

Cylco. [C. Lever] Textile dyes and pigments.

Cylcobright. [C. Lever] Basic dye.

Cylcofast. [C. Lever] Direct dye.

Cylcolite. [C. Lever] Solvent dyes.

Cylcosperse. [C. Lever] Textile dyes and pigments.

Cylcotone. [C. Lever] Acid dyes.

Cylink. [Am. Cyanamid] Monomer; crosslinking agent.

Cylinlock. [Hernon Mfg.] Retaining and mounting anaerobic adhesives and sealants.

Cylok. [Lord] Cyanoacrylate adhesives.

Cyme. [Am. Cyanamid] Crosslinking agent.

Cymel. [Am. Cyanamid] Melamine resins; crosslinking agent for cationic resins, emulsions, coating systems, electroplating, industrial finishes.

Cyncal® 80%. [Hilton Davis] Myristalkonium chloride; antimicrobial, antistat, disinfectant, sanitizer for use in food, beverage processing, for industrial and farm use, fabrics.

Cynoff. [FMC/Ag] Insecticide.

Cynol. [Am. Cyanamid] Defoamers, rewetting agents, softeners, resins.

Cypan. [Am. Cyanamid] Drilling mud additive.

Cypel. [Am. Cyanamid] Paper resins.

Cyplex. [Am. Cyanamid] Modified polyester coatings, resins.

Cypres. [Am. Cyanamid] Surface sizes for paper.

Cyprex® 65-W. [Am. Cyanamid/Ag] Dodine; fungicide for protection of fruits.

Cypronat. [Henkel/Emery/Cospha] Honey aroma chemical.

Cyquest. [Am. Cyanamid] Antiprecipitant.

Cyracure. [Union Carbide] Cycloaliphatic epoxy resins; diluents and flexibilizers for uv-cured coatings and adhesives.

Cyrez. [Cyanamid BV] Melamine-formaldehyde resins.

Cyroflex®. [Cyro Industries] Polycarbonate double-skinned sheet.

Cyrolite®. [Cyro Industries] Acrylic multipolymer; inj. molding, blow molding, vacuum forming, and extrusion compds.

Cyrolon®. [Cyro Industries] PC sheet; used for safety glazing, machine guards, panels for vending machines, signs, RV windscreens and windows, displays.

Cytame. [Am. Cyanamid] Formation aid.

Cythane. [Am. Cyanamid] Polyisocyanate resin.

Cythion®. [Am. Cyanamid/Ag] Malathion compds.; insecticide for crops and ornamentals.

Cytolase. [Genencor Int'l.] Fruit macerating enzymes.

Cytop. [Asahi Glass] Fluoropolymer; highly transparent polymer for optics, electronics, chemical industry, etc.

Cytox. [Am. Cyanamid] Industrial biocides.

Cyuram. [Am. Cyanamid] Tetramethyl thiuram.

D

D-. [Olin] PPG diol; lubricant in two- and four-cycle engines; used in the processing of rubber, to inhibit foam; used in brake fluids, cosmetics, oil and grease compds., pesticides, urethane foams, urethane coatings, adhesives, elastomers, and sealants/caulks.

D1-SEA 210 Silicone. [GE Silicones] Two-component silicone elastomeric adhesive.

D1SHP401. [GE Silicones] Primer used as adhesion promoter for weatherable abrasion-resistant silicone hardcoat on Lexan polycarbonate.

D12F. [Grain Processing] Oxidized corn starch; component for paper/paperboard.

D33-73F. [Cosmic Plastics] Diallyl ortho phthalate resin, glass and/or mineral-reinforced; for military and commercial applics.

D-70. [Georgia Marble] Ground calcium carbonate; noncolor-critical filler.

D99. [Tiodize] Cleaner/degreaser.

D-201. [La Roche Chem.] Activated alumina; low dust desiccant.

D3770, 3780, 6219.5. [Hüls Am.] Low m.w. siloxanes; cleaning, polishing, and damping media.

D4165 Series. [Hüls Am.] Silane derivs.; blocking agent, silylating reagent.

DA-. [Exxon/Tomah] Diamines; for cationic emulsification, corrosion inhibition, ore flotation, fuel additives.

Dabco®. [Air Prods.] Chain extender; catalyst for polyurethane, coatings, foam; crosslinking agent.

Dabco® DC. [Air Prods.] Silicone surfactants; additives for flexible molded polyurethane foams.

Dacamine 4D. [ISK Biotech] Post-emergence herbicide.

Daco. [Crompton & Knowles] Flavors.

Daconate 6. [ISK Biotech] Postemer-gence herbicide.

Daconil. [ISK Biotech] Chlorothalonil; broad-spectrum fungicide for use on golf courses.

Dacor. [Barnebey & Sutcliffe] Activated carbon; disposable odor remover.

Dacospin. [Henkel] Emulsifier, solubilizer, lubricant, antistat, wetting agent for textiles.

Dacron. [DuPont] Polyester fiber.

Dacthal. [ISK Biotech] Preemergence herbicide for turf, agronomic and vegetable crops.

Dag®. [Acheson Colloids] Colloidal graphite sol'ns.; lubricating specialty coatings.

Dai Cari. [Vikon] Butyl benzoate; carrier for disperse dyes on polyesters.

Daisurf. [Dai-ichi Kogyo Seiyaku] Nonionic detergent, wetting agent, penetrant for textiles.

Dalamar. [Cookson Pigments] Textile dyes and pigments.

Dalamar. [DuPont] Azo yellow.

Dalcon. [Dalau Ltd.] PTFE.

Dalon®. [Rhone-Poulenc/Textile & Rubber] Surfactants; leveling agents for acid or disperse dyes.

Dalpad. [Dow] Coalescing agents for latex.

Dama®. [Ethyl] Dialkyl methylamines; intermediates for quaternary ammonium compds.

Damox® 1010. [Ethyl] Didecyl dimethylamine oxide; nonionic wetting agent, emulsifier, dispersant, visc. modifier, hair conditioner.

Danamid. [Magyar Viscosagyár] Nylon 6.

Danar®. [Furon] Polyetherimide, FEP, PES films; high performance films.

Dan-Dee. [Vitamins, Inc.] Feed supplements.

Danoxyl SC. [Hickson Danchem] Alkalis/surfactants blend; nonsolvent cleaner for printing screens and machine surfaces.

Danscour VN. [Hickson Danchem] Phosphate ester salt; surfactant for wetting and scouring in alkaline baths.

Danset 384. [Hickson Danchem] Modified glyoxal resin; durable press resin for pre- or post-cured goods.

Dansperse N. [Hickson Danchem] Sodium lignosulfonate; dispersant for dyes on yarns or fabrics.

Dantobrom®. [Lonza] Halogenated hydantoin; biocides for water treatment, pools, spas.

Dantochlor®. [Lonza] Halogenated hydantoin; water treatment biocides.

Dantocol®. [Lonza] Dimethyl hydantoin glycols; intermediates for epoxies, urethane resins, and antistatic lubricants for the textile and plastic industries.

Dantoest®. [Lonza] Dimethyl hydantoin esters intermediates for epoxies, urethane resins, and antistatic lubricants for the textile industry.

Dantogard®. [Lonza] MDM hydantoin and DMDM hydantoin; industrial preservative.

Dantoin®. [Lonza] Hydantoin compds.; intermediate for custom chemical synthesis, cosmetics, textiles, laundry bleach formulations, and automatic dishwashing compds.; film former.

Dantosperse®. [Lonza] Ethoxylated hydantoin esters; intermediate for epoxies, urethane resins, and antistatic lubricants for textile industry.

DAP. [U.S. Gypsum] Adhesives, caulks, sealants, surface treatments, wood preservatives.

Dapex. [Rogers] Diallyl phthalate.

Daphne. [Etna Prods.] Lubricants.

Daplen. [PCD Polymere GmbH] Polypropylene and polyethylene.

Dapral®. [Akzo; Akzo Chem. BV] Polymers; detergent, wetting agent, emulsifier, thickener for making of cutting and sol. oils, for household, cosmetic, textile, and industrial uses.

Daran®. [W.R. Grace/Organics] PVDC emulsion; high barrier, high adhesion coatings for paper and films; laminat-

ing adhesive; binder for textile use.

Darasperse. [W.R. Grace/Dearborn] Dispersants.

Daratak®. [W.R. Grace/Organics] PVAc emulsions; adhesive emulsion for textiles, surface coatings, heat sealing, mastics, food pkg. , paper saturation, lamination.

Darathane®. [W.R. Grace] Urethane prepolymer; dispersant, emulsifier; latex polymerization.

Dar Chem. [Unichema] Fatty acids; emollient, lubricant for cosmetics, detergent systems, laundry prods., textile specialties, candles, buffing compds.

Darco. [Am. Norit] Activated carbon.

Darex®. [W.R. Grace/Organics] Nitrile, SAN, SBR latexes and resins; for bonding and impregnation of fibers, webs, and films, for paper and felt saturation, water treatment for scale and sludge inhibition, shoe soles, floor tiles, household goods, repair compds., surface coatings.

Dariloid®. [Kelco] Alginates; stabilizer, emulsifier, thickener for dairy industry.

Dari Lube. [Dryden Oil] Lubricants for the dairy industry.

Dark Green No. 2. [Witco] Petrolatum.

Darocur. [EM Industries] uv photoinitiator.

Darvan®. [R.T. Vanderbilt] Dispersant for latexes, agric., mineral slurries, electronics, ceramics, rubber; stabilizer, wetting agent, emulsifier; mold lubricant for elastomers; corrosion inhibitor.

DAS. [Sigma Prodotti Chimici] 4,4′-Diamino-2,2′-stilbene disulfonic acid; chemical intermediate for optical brighteners.

Dastar. [Croda Chem. Ltd.] Cholesterol; conditioner, emollient, emulsifier for skin and hair care prods.

Datagel. [Croda Chem. Ltd.] Diacetyl tartaric ester monoglyceride; food emulsifier.

Datamuls®. [Goldschmidt AG] Diacetyl tartaric ester monoglyceride; food emulsifier.

Davacast. [Davathane Ltd.] Castable polyurethanes.

Davaspray. [Davathane Ltd.] Sprayable polyurethanes.

Davathane. [Davathane Ltd.] Polyurethane elastomers.

Davis Colors. [Frank D. Davis] Cement and mortar colors.

Davisil. [W.R. Grace/Davison] Chromatographic silica gel.

Daxad®. [W.R. Grace/Organics] Naphthalene sulfonate salts, polymethacrylates, sodium polyisobutylene maleic anhydride copolymer; dispersant for agric., mastics, caulks, sealants, pigment slurries, inks, paper coatings, dyestuffs, paints, ore flotation, tanning, emulsion polymerization, cement, gypsum, water treatment.

Daxcel. [Dacar] Adhesive prods.

Dax Sealant. [Dacar] Sealants.

Dayflex. [Deutsche Exxon Chem. GmbH] Plasticizers.

Day-Glo. [Day-Glo Color] Daylight fluorescent pigments.

DB. [OxyChem] Butoxydiglycol.

DB Acetate. [OxyChem] Diethylene glycol butyl ether acetate.

DBE-. [DuPont] Dibasic ester mixture; solvent for coatings, cleaners, inks, textile lubricants, urethane prod.; plasticizer; polymer intermediate for polyester polyols for urethanes, wet-strength paper resins, polyester resins; specialty chemical intermediate.

DBM. [Textile Rubber & Chem.] Dibutyl maleate; plasticizer for vinyl resins, intermediate for mfg. of surfactants, in copolymerization reactions with PVC and vinyl acetates, latex paint formulations.

DB Oil. [CasChem] Castor oil; for sonar fluid, liquid dielectric, urethane reactions.

DBQDO®. [Lord] Dibenzoyl-p-quinone dioxime; vulcanizing agent for natural and synthetic elastomers.

DC 2900 HW. [Hardman] Two-part epoxy compound for elec./electronic applics.

D-Chlor Tablets. [Eltech Int'l.] Sodium sulfite tablets; used to remove chlorine from waste water or cooling tower blowdown.

DCI Plus. [Lubrication Engineers] Diesel fuel additive.

D-C-Tex 211. [Dooley] Flame retardant for cottons and blends.

DD-. [Dover] Brominated and chlorinated paraffins; flame retardant for plastics, rubber, textiles, coatings, carpet backing.

DDBS 100, 100 SP, Special. [Zohar Detergent Factory] Alkylbenzene sulfonic acid; detergent intermediate.

DDI. [Henkel/Functional Prods.] Liq. aliphatic diisocyanate.

DE. [OxyChem] Ethoxydiglycol; solvent.

DE Acetate. [OxyChem] Diethylene glycol ethyl ether acetate.

De-Airex. [E.F. Houghton] Defoamers.

Deam Deaerator 124. [Reilly-Whiteman] Dyeing assistant for nylon.

Dearcide. [W.R. Grace/Dearborn] Algicide.

Dearmeen. [W.R. Grace/Dearborn] Corrosion inhibitor.

Deartreat. [W.R. Grace/Dearborn] Waste and process water treatment.

Deatron®. [Nicca USA; Nikko Chem. Co. Ltd.] Alkyl phosphate; antistat for nylon fabrics.

Debonder. [Akzo] Quaternary; softener.

Decalin. [DuPont] Decahydronaphthalene; solvent and stabilizer for shoe creams, floor waxes, paints, lacquers, resins, rubbers, asphalt.

Decaltal®. [BASF AG] Nonswelling organic acids and their salts; deliming and masking agents for chrome tanning.

Decanox-F. [Atochem N. Am.] Decanoyl peroxide; initiator for polymerization, curing elastomers and polyester resins.

Deccoquin. [Atochem N. Am.] Fungicides.

Deccosalts. [Atochem N. Am.] Fungicides.

Deccoscalds. [Atochem N. Am.] Fungicides.

Deceresol® Surfactant. [Am. Cyanamid/Textiles] Detergent, emulsifier, wetting agent, scouring agent, penetrant for textile industry.

Dechlorane® Plus. [OxyChem] Chlorinated cycloaliphatic compd.; flame retardant for thermoplastic, thermoset, and elastomeric systems.

Declar®. [Sheffield Plastics] Thermoplastic sheet; for aircraft cabin interiors.

Decolorized Aloe Vera Gel. [Terry Labs] Aloe vera gel with color removed.

Deconyl. [Plastic Coatings Ltd.] Nylon powder coatings.

Decrolin®. [BASF AG] Zinc formaldehyde sulfoxylate; reducing agent after catalytic bleaching of fur skins; reducing and discharge agent for textile printing.

Dedevap®. [Bayer] Dichlorvos; insecticide and acaricide for control of biting and mining pests.

Dedico. [Unichema] Dehydrated castor oil fatty acids; for surface coating resins.

Dee Fo Series. [Ultra Additives] Defoamers.

Dee-Tac. [Olin] Detackifier for uncured rubber.

Def 6. [Miles/Ag] Defoliant for cotton crops.

Deflavit®. [BASF AG] Stripping agent for wool and nylon dyeings.

Defoam. [Chemonic Industries] Silicone or nonsilicone defoamers for dyeing, scouring or finishing operations.

Defoamer. [Chem-Y GmbH] Alkyl polyglycol ether carboxylic acid; defoamer for phosphoric acid prod.

Defoamer. [Crompton & Knowles] Ethoxylated fatty acid blends; defoamer for paper industry.

Defoamer. [Exxon] Petroleum-based; defoamer for activated sludge sewage treatment plants.

Defoamer. [Hart Chem. Ltd.] Proprietary blend; multipurpose defoamers for effluent, pulp and paper industry.

Defoamer. [Henkel/Textile] Defoamers for textiles, coating/latex applics.

Defoamer. [Hercules] Hydrocarbon oil; defoamer for chemical processing, aq. systems.

Defoamer. [Hoechst Celanese] Phosphoric acid esters; defoamer for waste water.

Defoamer. [Piedmont Chem. Industries] Antifoams and defoamers for textiles, dyeing, wastewater systems.

Defoamer. [Reilly-Whiteman] Silicone and nonsilicone defoamers for textile applics.

Defoamer 10. [Applied Textile Tech.] Nonsilicone antifoam for use with disperse dye and/or systems that will be treated with a fluorocarbon in finishing.

Defoamer 019-B, NOS. [Paradigm Labs] Silicone and nonsilicone defoamers for textile applics.

Defoamer 48, 72, 1100, FK-10. [Dooley] Silicone defoamers for textile processing.

Defoamer 167. [A. Harrison] Silicone emulsion; antifoaming agent for dyeing, finishing and printing.

Defoamer 544-C. [CNC Int'l.] Blend of silicone and silica; defoamer for jet and pressure dye machines.

Defoamer 2000, PA 510. [Leatex] Defoamers for dyeing processes.

Defoamer 7030, 9010, 9703. [Fibertint & Chems.] Emulsifiable silicone compounds; defoamers for general textile applics.

Defoamer A-69, D-10, K-20. [Yorkshire Pat-Chem] Silicone defoamers for textiles.

Defoamer G-280. [Stockhausen] Nonsilicone defoamer for jet dyeing.

Defoamer/Drainage Aid. [Crompton & Knowles] Fatty acids blends; defoamer and drainage aid for paper industry.

Defoamit. [Atochem N. Am.] Defoamer for wire mill processing.

Defomax. [Toho Chem. Industry] Nonionic surfactant/wax emulsion blend; antifoaming agent for rubbers and plastics.

Degacure®. [Degussa] Binder for uv-curable coatings and inks; photoinitiator for uv-curing of binder systems.

Degadur. [Degussa] Acrylic resin; industrial flooring systems.

Degadur®. [Degussa AG] Polymethyl methacrylates.

Degalan®. [Degussa; Degussa AG] Acrylic resin; molding resins.

Degament. [Degussa] Acrylic resin; used in concrete/polymer composites.

Degras. [Croda Inc.] Wool grease.

Degreez. [Alzo] Methyl ester blend; degreasing agent for grease removal from metal, textiles, etc.

Degressal®. [BASF AG] Defoamer for cleaners, chemical and sugar industries.

Degudent. [Degussa] Type III gold alloy; for dental applic.

Degulor. [Degussa] Type III gold alloy; for dental applic.

D.E.H. [Dow] Aliphatic polyamines; curing agent for epoxy resins; used for civil engineering, adhesives, grouts, casting and elec. encapsulation.

DEHA. [W.R. Grace/Dearborn] Diethylhydroxylamine; passivates metal surfaces, scavenges dissolved oxygen, volatilizes for complete boiler protection.

Dehesive Silicones. [Wacker Silicones] Paper release agents.

Dehpa®. [Albright & Wilson Am.] Metal extractants.

Dehscofix. [Albright & Wilson Am.; Albright & Wilson UK] Naphthalene sulfonic acid salts or condensates; dispersant for agric. formulations; superplasticizer for concrete.

Dehscotex. [Albright & Wilson Am.] Formulated auxiliaries for textile and leather prep., dyeing and finishing processes.

Dehscoxid. [Albright & Wilson Am.] Ethylene oxide condensates; for synthetic alcohol for agrochemical formulations and other industrial applics.

Dehydag® Wax. [Henkel; Henkel KGaA] Fatty alcohols or sulfates; emulsifier, consistency factor for cosmetics, pharmaceuticals.

Dehydat. [Henkel KGaA] Antistatic agents for plastics.

Dehydol®. [Henkel/Emery/Cospha; Henkel Canada; Henkel KGaA; Pulcra SA] Ethoxylated fatty alcohols; detergent, wetting agent, emulsifier, solubilizer, antifoam, dispersant for insecticides, household and industrial cleaners, cosmetics, metalworking fluids.

Dehydran. [Henkel; Henkel KGaA] Defoamer for industrial cleaners, dishwashing agents, waste water treatment, coatings.

Dehydril. [Henkel KGaA] Antiskinning agents, oximes.

Dehydrophen. [Henkel KGaA; Pulcra SA] Alkylaryl polyglycol ether; detergent, wetting agent, emulsifier, dispersant, stabilizer.

Dehygant. [Henkel KGaA] In-can preservative for aq. coatings.

Dehylan ET 40. [Henkel KGaA] Protein hydrolysate; skin conditioner for hand cleaners, dishwashing agents.

Dehymuls®. [Henkel/Emery/Cospha; Henkel Canada; Henkel KGaA] W/o emulsifier for cosmetics, pharmaceuticals.

Dehypon. [Henkel/Emery; Henkel Canada; Henkel KGaA] Ethoxylated, propoxylated fatty alcohols; detergent, wetting agent, emulsifier, dispersant for industrial cleaners, carpet cleaners.

Dehyquart®. [Henkel/Emery/Cospha; Henkel KGaA; Pulcra SA] Quaternaries; emulsifier for emulsion polymerization; softener, conditioner, bactericide, antistat, corrosion inhibitor, sequestrant for cleaning formulations, personal care prods.

Dehysan. [Henkel] Defoamer for cane and beet sugar prod.

Dehysol. [Henkel KGaA] Thickener, wetting agent, antisagging agent for alkyd paints.

Dehyton®. [Henkel Canada; Henkel KGaA; Pulcra SA] Amphoterics; detergent, foamer, conditioner, thickener, solubilizer for personal care prods., dishwashers, textiles, paints.

Dekol®. [BASF AG] Protective colloids for dyeing cellulosic fibers and blends, and furs.

Dekopan. [Dexter] Defoamers.

Delac®. [Uniroyal] Benzothiazole sulfenamides; delayed-action accelerator for natural and synthetic rubbers; used in tire treads, carcass, mechanicals, and wire jackets.

Delaflo. [J.W.S. Delavau] Calcium sulfate; directly compressible granulation; excipient.

Delatab. [J.W.S. Delavau] Precipitated calcium carbonate USP; directly compressible.

Delestat. [Seppic] Quaternary; resistivity control for painting.

Delfloc®. [Hercules] Cationic polymer sol'n.; retention aid and flocculant for the paper industry.

Delion. [Takemoto Oil & Fat] Phosphated and sulfonated surfactants; antistat, lubricant, wetting agent, detergent, finishing oil for textiles.

Delios®. [Grünau] Medium chain triglycerides; solvent for flavors and oil-soluble food additives, surface treating agent for dried fruits, confectionery, dietary foodstuffs.

Delrin®. [DuPont] Acetal resins; engineering thermoplastic used in machinery, agric. equipment, interior automotive door handles, clock mechanisms, ballcock valves, videocassettes, and other molding applics.

Delsette. [Hercules] Polyamide resin; hair fixative.

Deltagloss. [ECC Int'l.] Chemically structured kaolin.

Deltatex. [ECC Int'l.] Calcined clay-kaolin.

Delta-Therm®. [Flexible Prods.] Rigid polyurethane systems; for dispensing froth foams through pressurized equip.; for sandwich panels, coolers/freezers, trucks, tanks, pipeline, boats.

Deltyl® Extra. [Givaudan] Isopropyl myristate; emollient and auxiliary emulsifier for cosmetics.

Delvan. [Dexter] Gums for textile printing.

Delvanol. [Dexter] Detergents.

Delvet. [Henkel] Nonionic surfactant, flame retardant for chlorinated paraffin.

Delvocid. [Int'l. Bio-Synthetics] Natamycin; food grade mold and yeast inhibitor.

Demelan. [Pulcra SA] Sulfate and sulfonate blends and EO/PO block polymers; detergent, coupling agent for dishwashers, heavy-duty detergents, hand cleaners, textile scouring.

Deminal 90. [Mutchler] Demineralized whey powd.

Demix®. [Arizona] Disproportionated tall oil prods.; emulsifiers, as base stock for prod. of sodium and potassium soaps of fatty acid/rosin acid emulsifier systems in SBR mfg.

Demol. [Kao] Dispersant for dyes, pigments, clay, agric. chemicals.

Demulfer. [Toho Chem. Industry] Demulsifier for crude oil prod.; desalting agent for oil refinery.

Demulsifier 3837. [Hoechst Celanese/ Colorants & Surf.] Polymer; demulsifier for use at alkaline pH.

D.E.N. [Dow] Epoxy-novolac resin; used in high-performance adhesives, structural and elec. laminates, potting and molding compds., coatings and castings for elevated temp. service, and filament wound pipe.

Denflex. [Dennis Chem] Plastisols, organosols, epoxies, polyurethanes.

Denflex. [Diversified Compounders] Natural rubber latexes and compds.; used for adhesive, dipping, coating, foam, molded materials.

Denimax. [Novo Nordisk] Glucanase; enzyme for processing denim fabric.

Denlube. [Graden] Antiblocking agent, lubricant, processing aid, frothing aid, foam stabilizer for rubber and latex processing, coatings.

Denox. [Carus] Brightener.

De-NOx Catalyst. [Mitsubishi Kasei] Nitric acid plant exhaust gas catalyst.

Denphos. [Graden] Phosphate esters; emulsifier for emulsion polymerization; wetting agent, detergent, hydrotrope for household and industrial cleaners, pesticides.

Densodrin®. [BASF AG] Natural and synthetic oils; waterproof agents; also for tanned leather.

Densol. [Graden] Sulfates or block polymers; thickener, stabilizer, emulsifier, defoamer, dispersant, wetting agent for latexes, adhesives, paper, textiles.

Densotroptic. [Anderson Labs] Organic solvent formulation.

Denstone. [Norton Chem. Process Prods.] Inert catalyst bed support.

Densulf. [Graden] Sulfonated tallow;

softener, fatliquoring agent, plasticizer, defoamer for textiles, leather, glues.

Denwet. [Graden] Sulfonated surfactants; wetting agent, emulsifier, coupling agent, penetrant, solubilizer for emulsion polymerization, cleaners, agric. chemicals.

Denzox. [Eagle Zinc] Dense zinc oxide; for ceramic applic.

Deodorant. [Andrea Aromatics] Mixture of fragrance materials; deodorant for PVC processing and finished prods.

Deodorant Richter/K. [Henkel/Cospha] Tetrabromo-o-cresol; combats body odor due to sweat.

Deodorizer 78. [Acme Resin] Deodorizing agent for foundry resins.

Deoxo. [Engelhard] Gas purification systems.

Deoxy-Sol. [Fairmount] 35% hydrazine.

Depasol. [Pulcra SA] Sulfosuccinate; mild detergent for cosmetic formulations; dispersant for pigments; wetting and penetrating agent; used in drycleaning detergents and emulsion polymerization; rewetting agent for textile and paper industries.

Depco. [Gresco Mfg.] Leveling retarders.

Depcolevel 1252 Series. [Gresco Mfg.] Quaternary compd.; migrating agents for basic dyes.

Depcosoft NP. [Gresco Mfg.] Nonionic ester; napping softener, lubricant for synthetic fiber with scroopy hand.

Depcosol. [Gresco Mfg.] Softeners for cotton, rayon, wool, knit cotton/polyester blends.

Depcosperse LQD. [Gresco Mfg.] Nonionic fatty condensate; emulsifier, dispersant, lubricant for high pressure dyeing of polyester.

Deplastol. [Henkel KGaA; Pulcra SA] Ethoxylated lauric acid; visc. reducer for organosols.

Depsodye. [ICI Surf. UK] Stabilizer, dye assistant, carrier, leveling agent for dyes, textiles.

Depsolube. [ICI Surf. UK] Anticrease lubricant for textile processing, dyeing.

Depuma®. [Ciba-Geigy/Dyestuffs] Defoamers for atmospheric or pressure equipment.

Dequest®. [Monsanto] Phosphonic acids and salts; process aid for paper, textiles, metals, industrial cleaners; scale and corrosion inhibitor.

D.E.R. [Dow] Epoxy resins; for adhesive, civil engineering, casting, potting, encapsulation, filament winding, powd. coatings, and wet lay-up applics.

Derakane®. [Dow] Vinyl ester resin; thermoset used in chem. processing industry, pulp and paper mills, pipe, filament winding.

Deriphat®. [Henkel/Emery/Cospha; Henkel Canada; Henkel KGaA] Propionic acids or salts, betaines; amphoteric surfactant, wetting agent, emulsifier, detergent, corrosion inhibitor, solubilizer, stabilizer for hard surface cleaning, textiles, emulsion polymerization, petroleum processing.

Derma. [Sandoz] Textile dyes and pigments.

Dermacor. [Sandoz] Textile dyes and pigments.

Dermalcare®. [Rhone-Poulenc Surf.] Esters or blends; emulsifier, emulsion stabilizer, base, emollient, moisturizer, lubricant for cosmetics, cleaners, antiperspirants, conditioners.

Dermalight. [Sandoz] Textile dyes and pigments.

Dermasome®. [Microfluidics] Liposomal preps. containing cosmetic ingreds. entrapped within the lipid spheres; for skin care prods.

Dermatein GSL. [Hormel] Fluid matrix containing glycosphingolipids, phospholipids, cholesterol; skin lipid for barrier renewal and moisturization; for night creams, lip protectants.

Derminol. [Hoechst AG] Fatliquoring agents for leather, fur processing.

Dermoblock. [Alzo] Fatliquoring agents for leather, fur processing.

Dermol. [Alzo; Bernel] Esters; emollient, anti-irritant for cosmetics; dispersant, solubilizer.

Dermolan GLH. [Alzo] Glycereth-7.5 hydroxystearate; emollient, visc. builder, conditioner for cosmetic creams, lotions, shampoos.

Dervacid. [Union Derivan SA] Fatty acids.

Desadipol. [Seppic] Alkyl polyethoxy ether; detergent for skin prods.

Desamidocollagen. [Henkel/Emery/ Cospha; Henkel KGaA] Soluble collagen; moisturizer and film former for toiletry emulsions.

Deselex. [Guardian Labs] Detergent phosphate replacement.

Desical®. [Harwick] Calcium oxide dispersions; desiccants.

Des-I-Cate. [Atochem N. Am.] Vine killer and harvest aid.

Desiccite 25. [Engelhard] Alumina sulfur recovery catalysts.

Desi Pak. [United Desiccants-Gates] Clay-based; desiccant for adsorption of moisture.

Desisphere. [Universal Scientific] Desiccants.

Desludgit. [Stewart Hall] Fuel oil additives for sludge removal.

Desmocap®. [Bayer; Miles] Urethane polymer.

Desmocoll. [Bayer; Miles] Polyurethane resin; for use in adhesives for bonding plastics, rubber, leather, wood, textiles, paper, footwear, laminating, pkg.

Desmodur®. [Bayer; Miles] Polyurethane and isocyanates; adhesive for bonding textiles, leather, wood, metal, plastics; crosslinking agent for adhesives.

Desmoflex®. [Bayer] Polyurethane elastomers.

Desmolac®. [Bayer] Polyurethane coating raw materials.

Desmopan®. [Bayer; Miles; Albis UK Ltd.] Thermoplastic polyurethane; inj. molding and extrusion resins used for heavy-duty engineering, automotive, mech., and apparatus components.

Desmophen®. [Bayer; Miles] Polyester, polyether, or acrylic polyols.

Desmotherm®. [Bayer] Polyurethane coating raw materials.

Desmuldo. [Henkel-Nopco] Emulsion breaker for crude petroleum.

DeSomeen TA. [Witco/Organics] Ethoxylated tallow amines; emulsifier, dispersant, textile scouring, dyeing assistant, desizing assistant, softener, antistat.

DeSonate. [Witco/Organics] Sulfonic acid and sulfonates; detergent intermediate and component for shampoos, bubble baths, liq. soaps.; emulsifier.

DeSonic®. [Witco/Organics] Ethoxylated alcohol and alkyl phenyl ethers, sorbitan and ethoxylated sorbitan esters; defoamer, detergent, emulsifier, wetting agent, dispersant for textile, paper, agric., polishes, metal cleaning, detergents.

DeSophos. [Witco/Organics] Phosphate esters; surfactant, emulsifier for industrial cleaners, drycleaning, lubricants, pesticides, textiles, polymerization; softener, antistat, coupling agent, corrosion inhibitor.

DeSotan. [Witco/Organics] Sorbitan and ethoxylated sorbitan esters; emulsifier, wetting agent, antistat, lubricant, softener.

Desox. [Katalistiks Int'l.] SO_x reduction catalyst.

Desulco. [Superior Graphite] Desulfurized petroleum coke.

DET. [Swastik] Alkylaryl sulfonates and additives; household cleanser, laundry detergent.

Deterflo. [Ceca SA] Ethoxylated alcohol; degreaser.

Detergent 8®. [Alconox] Alkanolamine, glycol ether, alkoxylated fatty alcohol blend; liq. machine detergent.

Detergent #34. [Reilly-Whiteman] Detergents for textile scouring.

Detergent ADC. [ICI Surf. UK] Sodium dodecylbenzene sulfonate; general-purpose anionic detergent.

Detergent Conc. 840. [Mona Industries] Modified alkanolamide; used in car wash, bubble baths, all-purpose cleaners.

Detergent CR. [Arol Chem. Prods.] Fatty ethanolamine condensate; detergent, wetting agent, emulsifier, thickener, penetrant, leveling agent for textiles, personal care prods., food, and household prods.; pigment dispersant.

Detergent E. [CNC Int'l.] Anionic coconut amine condensate; surfactant, detergent, wetting and rewetting agent for textile industry.

Detergent Honey. [Dixo] Drycleaning detergent.

Deterpal. [Ceca SA] Alkyl sulfate/solvent blend; degreaser, emulsifier, dispersant, antisettling agent.

Detersol. [Finetex] Blends; solvent scour, detergent for textile, wax and size removal, machine cleaning; wetting and rewetting agent.

Deteryl. [ICI Surf. UK] Detergent for scouring textiles; antifrosting agent in printing and dyeing of carpets.

Det-Lub. [Reilly-Whiteman] Fatty acid esters; lubricants for use in scour, dye, and finish baths.

Det-O-Jet®. [Alconox] Highly alkaline detergent blend; low sudsing detergent for ultrasonic and mechanical washers.

DET-Washmatic. [Swastik] Sodium alkylaryl sulfonate; detergent for washing machines.

Deva. [Degussa] Gold/palladium alloy; for dental applic.

Devchlor. [Devoe Coatings] Chlorinated rubber coating.

Deventer. [Deventer Benelux BV] PVC.

Devflex. [Devoe Coatings] Acrylic latex coatings.

Deviscol VS. [Henkel KGaA] Viscosity stabilizer.

Devprep. [Devoe Coatings] Cleaners.

Devran. [Devoe Coatings] Epoxy coating.

Devtar. [Devoe Coatings] Epoxy coating.

Devthane. [Devoe Coatings] Polyurethane coating.

DEWT L. [Drew Ind. Div.] Chromate/organic blend; corrosion inhibitor for closed recirculating water systems.

Dexene®. [Dexter] Softeners, lubricants for textile processing.

Dex-Lo®. [Int'l. Bio-Synthetics] Alpha amylase; enzyme for liquefaction of starch.

Dexolene. [Dexter] Surfactants.

Dexopal®. [Dexter] Polyether; detergent, penetrant, wetting agent for dye applics.

Dexstar. [Dexter] Weatherable finishes.

Dextrajet. [Dexter] Scouring compds.

Dextralube 193. [Dexter] High m.w. polymer; lubricant for prevention of bruises, cracks, creases, and chafe marks in dyeing.

Dextranase Novo 25 L. [Novo Nordisk] Dextranase; enzyme used in sugar industry.

Dextraset. [Dexter] Finishing agents.

Dextrasol. [Dexter] Leveling agents.

Dextrasperse COM-100. [Dexter] Complex polyether mixture; dispersants, antiprecipitants, compatibilizers for dyeing operations.

Dextrol®. [Dexter] Phosphate esters; detergent, wetting agent, dispersant, stabilizer, penetrant, emulsifier for pesticides, emulsion polymerization, textiles; corrosion inhibitor.

Dextrozyme. [Novo Nordisk] Pullulanase/amyloglucosidase; enzyme for starch industry.

Dezyme. [Novo Nordisk] Starch desizing enzyme.

DF 1040. [GE Silicones] Reactive silicone fluid; forms water-repellent fluids for textile, particle treatment, etc.

D-Floc. [Baker Perf. Chem.] Dispersants.

DI-43. [Ferro/Keil] Polymeric lubricity additive.

Diabase Developer®. [Mitsubishi Kasei] Azoid bases used with Diathol Grounder prepares.

Diablack®. [Mitsubishi Kasei] Carbon black; for rubber industry.

Diable AS. [Kao Corp. SA] Amine-based; slurry seal and emulsifier.

Diacelliton. [Hoechst Celanese] Textile dyes and pigments.

Diacelliton Dye®. [Mitsubishi Kasei] Disperse dyes for acetate and nylon.

Diacid Dye®. [Mitsubishi Kasei] Acid dyes for wool, silk, and nylon.

Diaclear®. [Mitsubishi Kasei] Polyacrylamides or polymethacrylates; synthetic flocculants for industrial wastewater treatment, industrial processing (sedimentation, filtration, dewatering, centrifugal processes), enhanced oil recovery.

Diacotton Dye®. [Mitsubishi Kasei] Direct dyes for vegetable fiber and viscose rayon.

Diacron Dye. [Mitsubishi Kasei] Hottype reactive dyes for printing and continuous dyeing method.

Diacryl. [Akzo] Methacrylate monomers.

Diacryl Dye®. [Mitsubishi Kasei] Cationic colors for dyeing polyacrylonitrile fiber, polyester.

Diacupro Dye®. [Mitsubishi Kasei] Direct dyes.

Diadavin® BL. [Miles/Organic Prods.] Stabilizer for peroxide bleaching.

Diadol. [Mitsubishi Kasei] Higher alcohols; bases for oily preps., superfatting agents, plasticizers, surfactants for detergents and shampoos.

Diafil. [CR Minerals] Diatomaceous earth prods.

Diaformer. [Sandoz; Mitsubishi Petrochem.] Carboxylic betaine amphoteric acrylic polymers; fixative, conditioner, film-former, antistat for hair care prods.

Diagum/Diaprint. [Multi-Kem] Modified mannogalactan gums; print thickeners for acid, basic, cationic, disperse and vat colors.

Diahold. [Sandoz; Mitsubishi Petrochem.] Anionic methacrylate polymers; fixative, film-former for hair care prods.

Diahope®. [Mitsubishi Kasei] Activated carbon; for wastewater treatment, water, solvent, and chemical purification, sugar refining, food additives purification, gas purification, deodorization, solvent recovery.

Diaion®. [Mitsubishi Kasei] Ion exchange resins; for softening and demineralization of water; metal recovery and separation; refining of chemicals, sugar, dextrose, formalin, and amino acid; dehydration of organic solvent; catalyst; decolorization; dealkalization; prep. of antibiotic medicines.

Diak. [DuPont] Rubber curing agents.

Diak No. 4. [R.T. Vanderbilt] 4,4-Methylenebis (cyclohexylamine) carbamate.

Diakon. [ICI Chem. & Polymers Ltd. / Acrylics] Molding resins.

Dialead®. [Mitsubishi Kasei] Coal tar pitch carbon fiber; for thermoset plastics, cement, metal, rubber industries.

Dialen. [Mitsubishi Kasei] Alpha olefins; comonomers for polyolefin; chemical intermediate.

Dialuminous Dye®. [Mitsubishi Kasei] Direct dyes with good to excellent lightfastness.

Diamiet. [Kao Corp. SA] Ethoxylated propylene alkyl fatty diamine; emulsifier, dispersant, corrosion inhibitor, wetting agent.

Diamin. [Kao Corp. SA] Diamine derivs.; dispersant, asphalt emulsifier, corrosion inhibitor, antistripping agent.

Diamine. [Berol Nobel] Diamine derivs.; emulsifier, corrosion inhibitor.

Diamine® H extra. [BASF AG] 1,6-Hexanediamine; intermediate for prod. of AH salt.

Diamonada. [Atomergic Chemetals] Zircon crystals.

Diamond. [Miles/Organic Prods.] Textile dyes and pigments.

Diamond. [Van Den Bergh Foods] Hydrogenated vegetable oils and blends; emulsified shortening for cakes.

Diamond K. [Reichhold] Linseed and soybean oils.

Diamond Quality®. [CasChem] Castor oil; emollient for cosmetics.

Diamonine B. [ICI Surf. UK] Urea-formaldehyde resin; thermoset producing durable press finishes on cellulose fibers and stiff finishes on synthetic fibers.

Dianal. [British Traders & Shippers] Acrylic resins.

Dianix. [Hoechst Celanese] Textile dyes and pigments.

Dianix Dye®. [Mitsubishi Kasei] Disperse dyes for dyeing synthetic fiber, esp. polyesters.

Dianol. [Dai-ichi Kogyo Seiyaku] Fatty acid DEA; detergent, solubilizer, emulsifier, dispersant for pigments, waxes, solvents.

Dianol®. [Akzo] Ethoxylated bisphenol A diol; reactive modifiers for saturated and unsaturated polyesters, vinyl esters, and polyurethane resin formulations.

Diapol®. [Mitsubishi Kasei] SBR/carbon black masterbatch; for rubber industry.

Diaproof®. [Mitsubishi Kasei] Paraffin emulsion; water-repellent modifier for construction materials, paper, agric. and other applics.

Diaresin Dye®. [Mitsubishi Kasei] Solvent dyes for thermoplastic and thermosetting plastics.

Diatami. [Alban Muller] Peeling products.

Diathol Grounder®. [Mitsubishi Kasei] Azoic coupling components used on cotton and rayon prior to combining with Diabase Developer to form insoluble dyes; also combined with diazotized bases to produce organic pigment.

Diazital. [Henkel; Henkel-Nopco] Fatty alcohol polyglycol ethers; dyeing assistant.

Diazogas. [Michlin Diazo Prods.] Anhydrous ammonia.

Diazol. [ICI Am.] Textile dyes and pigments.

Diazon-7. [Molecular Rearrangement] Polymethylene p-diazo diphenylamine zinc chloride.

Diazopon®. [Rhone-Poulenc Surf.] Ethoxylated alkylphenol; surfactant; anticrock and soaping agent for naphthol dyes; dyeing assistant; solubilizer and stabilizer for fast color salts; accelerates diazotization of fast color bases.

Diazorb. [Michlin Diazo Prods.] Absorbers.

Diazyme®. [Solvay Enzymes] Glucoamylase; enzyme for hydrolysis of starch dextrins to glucose.

DIBA. [Croda Surf. Ltd.] Diisobutyl adipate; plasticizer for plastics and synthetic rubbers.

DIBM. [Textile Rubber & Chem.] Diisobutyl maleate; plasticizer for vinyl resins, intermediate for mfg. of surfactants, in copolymerization reactions with PVC and vinyl acetates, latex paint formulations.

DIBS. [Am. Cyanamid] Accelerators for rubber.

Dicalite. [Grefco; Steetley Minerals Ltd.] Diatomite or perlite; Filter aids, fillers.

Dicaperl. [Grefco; Steetley Minerals Ltd.] Lightweight fillers and extenders.

Dica-Sorb. [Grefco] Absorbents.

Dichan 100. [Olin] Dicyclohexylamine nitrite; corrosion inhibitor for ferrous metals; used for hot water heating systems, nuclear reactor heat-exchange units, gas recovery systems, jet aircraft engine compressors, internal combustion engines.

Dichevrol. [Shrieve Chem. Prods.] Dielectric insulating fluid.

Dicrylan® BSR. [Ciba-Geigy/Dyestuffs] Urethane emulsion; nonionic; durable finish with soft hand on cotton and polyester/cotton blends; binder for stain release finish on cotton fabrics.

Di-Cup. [Hercules; Hercules Europe] Dicumyl peroxides; vulcanizing agent, polymerization catalyst for rubber and plastics.

Dicyanex®. [Pacific Anchor] Dicyandiamide; epoxy curing agent for prepregs, composites, printed circuit board laminates.

Die Gard. [Reilly-Whiteman] Drawing compds.

Diene®. [Firestone Syn. Rubber] Polybutadiene rubber; used to produce tires, retreads, molded and extruded goods; impact modifier for ABS and PS.

Diesel Fuel Gard. [Gard] Diesel fuel additive.

Dieselmax. [Arol Chem. Prods.] Diesel fuel conditioner and pour pt. depressant.

Diesel Pep. [Spray Prods.] Diesel fuel conditioner.

Diglycolamine® Agent (DGA®). [Texaco] 2-(2-Aminoethoxy) ethanol; solvent for removal of CO_2 or H_2S from gases, for recovery of aromatics from refinery streams, for prep. of foam stabilizers, wetting agents, emulsifiers, and condensation polymers.

Diglyme. [Ferro/Grant] Diethylene glycol dimethyl ether; solvent used in electrochemistry, polymer, and boron chemistry.

Dihydroxy-acetone. [Int'l. Bio-Synthetics] Dihydroxyacetone; enzyme used as cosmetic stain and reagent in chemical synthesis.

Dikar. [Rohm & Haas] Fungicide and

miticide.

Di Kleer PE-2. [Leatex] Diester solvent; afterclearing agent for polyester dispersed dyes; for soaping off of reactive dyes and afterclearing of dispersed dyes at same time.

Dikssol. [Dai-ichi Kogyo Seiyaku] Anionic-nonionic activators; emulsifier for agric. formulations.

Dilasoft. [Sandoz Prods. Ltd.] Fatty acid deriv.; softener for textiles.

Dilev. [Ivax Industries] Leveling agent for textile applics.

Dilexo®. [Condea Chemie GmbH] Copolymer dispersions; base materials for paints, plasters, adhesives, and sealants.

Diluex®. [Floridin] Attapulgite clay; absorbent carrier for pesticides.

Dilyn. [Bruce Chem.] Leveling agents, optical brighteners for textiles.

Dimension. [Allied-Signal] Nylon 6 alloy; for inj. molding applics. incl. wheel covers, mirror housings, automotive body panels and fenders, large industrial housings and enclosures, pump housings, impellers

Dimetcote. [Ameron Protective Coatings] Inorganic zinc coatings.

Dimilin®. [Uniroyal] Diflubenzuron; insecticide for forestry ornamentals, fruit, field crops, horticulture.

Dimodan. [Grindsted Prods.; Grindsted Prods. Denmark] Vegetable oil monoglycerides; emulsifier for foods, adhesives, cosmetics, cutting oils, fabric softener, lubricants, pharmaceuticals, printing inks, resins.

Dimul. [Witco/Humko] Distilled monoglycerides.

Dinar. [Bruce Chem.] Wetting and rewetting agents, surfactants for textile dyeing and finishing.

Dinoram. [Ceca SA] Alkyl propylene diamines; chemical intermediate, corrosion inhibitor; emulsifier for bitumen; antistripping agent for road making.

Dinoramox. [Ceca SA] Ethoxylated alkyl propylene diamines; dispersant, wetting agent, emulsifier, corrosion inhibitor used in detergent, paint, agric., chemical, and textile industries.

Diofan®. [BASF; BASF AG] Polymer dispersions based on vinylidene chloride; for prod. of chemically resistant, heat sealable pkg. material; binders for bonding fiber webs, textile coatings, vapor barrier coatings for construction.

Diolpate. [MacPherson Polymers] Polymeric plasticizers.

Dioltech. [Drew Ind. Div.] Molybdate-based; corrosion and scale inhibitor for cooling water treatment.

DIOM. [Textile Rubber & Chem.] Diisooctyl maleate; plasticizer for vinyl resins, intermediate for mfg. of surfactants, in copolymerization reactions with PVC and vinyl acetates, latex paint formulations.

Dional. [Hoechst Celanese/Fine Chem.] Fluorinated hydrocarbon; solvent for dry cleaning.

Dion® Cor-Res. [Reichhold] Thermoset polyester resin; used in transportation, construction, elec., corrosion control applics., chemical process equip.

Dion® FR. [Reichhold] Thermoset polyester resin; flame retardant grade for corrosion applics. in filament winding and compression molding/SMC/BMC processes.

Dion® VER. [Reichhold/Reactive Polymers] Vinyl ester resin.

Dionil®. [Hüls Am.; Hüls AG] Fatty acid amide polyglycol ether and blends; detergent for lt. and heavy-duty detergents, dishwashing agents, cosmetic preps.; component in textile auxliaries; superfatting agent.

Dion-Iso®. [Reichhold] Isophthalic polyester resin; resin for transportation, marine, elec., and corrosion control applics.; for sprayup/layup, filament winding, moldings.

Diorez. [MacPherson Polymers] Polyester polyols.

Dioxitol. [Shell] Ethoxydiglycol; solvent for resins, dyes, fats used in automotive brake fluids, paints, textiles, cleaners; coupling agent; intermediate.

Diphacinone Conc. [Bell Labs] Conc. for formulating rodenticides.

Diphone. [Yorkshire Nachem] 4,4-Dihydroxydiphenyl sulfone.

Diprane. [MacPherson Polymers] Polyester-based polyurethane systems.

Dipropasol. [Union Carbide] Industrial solvents.

Diprosin. [Toho Chem. Industry] Dispro-portionated rosins; polymerization emulsifier for synthetic rubbers and plastics.

Dipsal. [Scher] PPG-2 salicylate; uv absorbent for suncreen bases, polymers, dyestuffs, toiletries, pharmaceuticals, hair prods.

Direx®. [Griffin] Diuron; flowable herbicide for control of many weeds and grasses in a variety of crops.

Dirubin. [Hüls Am.] Aluminum oxide pink.

Discharge Agent DP. [BASF] Stabilized sulfoxylate deriv.; reducing agent for discharge and discharge resist printing on polyester, acetate, triacetate, and their blends with polyamide.

Disco 727. [Callaway] Nonionic wax for use with cationic fixatives in slashing.

Discodye. [Callaway] Retarding agent, antimigrant for textile dyeing.

Discofix. [Callaway] Cationic fixative for garment desizing to reduce dye bleeding, for home laundry use.

Discol. [Callaway] Dye fixatives, defoamers, suspending agents, scouring aid, stabilizer.

Discoloc. [Callaway] Fixing agent and color control aid.

Discolube. [Callaway] Lubricant for bleaching, dyeing, and finishing operations.

Discopen. [Callaway] Penetrant for textile use.

Discosoft. [Callaway] Softeners, sanforizing lubricants, plasticizers for textile applics.

Discoterge. [Callaway] Surfactants for desizing, alkaline scouring, peroxide bleaching, wetting, dispersing.

Discozone. [Callaway] Cationic antiozonant softener.

Disflamoll. [Bayer; Miles/Polysar Rubber] Phosphates; flame retardant, gelatinizing plasticizer for PVC and cellulosic; used for extruded and inj. molded parts, mechanical foam, belts, imitation leather, coatings, cable sheathing.

Dis-Mist. [McGean-Rohco] Chromium fume suppressant.

Dismulgan Brands. [Hoechst AG] Polyamine containing condensation prod.; reverse emulsion breaker for o/w emulsions.

Dispal®. [Vista] Alumina monohydrate; for paper antiskid, surfactant viscosifier, sol gel ceramics.

Dispatex G. [Nikko Chem. Co. Ltd.] Aromatic condensate; dispersant for polyester dyeing.

Dispel. [Vinings Industries] Dispellant.

Disperal®. [Condea Chemie GmbH] Highly dispersible aluminas; base materials for abrasives and high temp.-resistant ceramics, as auxiliaries for separation processes.

Disperbyk. [BYK-Chemie USA] Wetting/dispersing additives.

Dispercoll. [Bayer; Miles] Polychlorobutadiene and polyurethane latexes; for formulation of water-based contact cements, laminating and mastic adhesives for the automotive, construction, furniture, footwear, and pkg. industries.

Disperplast. [BYK-Chemie USA] Wetting/dispersing additives.

Dispersal. [Aquatec Quimica SA] Chelating agent for complexing calcium, magnesium, and iron in alkaline sol'ns.; dispersant.

Dispersant. [Hart Chem. Ltd.] Dispersing agents for pulp and paper industry.

Dispersant. [Hoechst AG] Modified polymers; reduces visc. and gelling properties of water-based and gypsum saturated drilling fluids.

Dispersant 912. [Dooley] Aromatic sulfate; dispersant and antiagglomerant for disperse dyes.

Disperse Black 00-6607. [BASF AG] Pigment prep. for prod. of Indian inks.

Disperse White 00-2207. [BASF AG] Casein-free pigment for coloring polishes.

Dispersing Agent SS Dry. [Hoechst AG] Polymeric organic sulfonic acid; dispersant for biocidal wettable powds.

Dispersite. [Rite Industries] Textile dyes and pigments.

Disperso. [Witco] Metallic stearates;

water-wettable.

Dispersogen. [Hoechst Celanese/Colorants & Surf.; Hoechst AG] Dispersant, dyeing auxiliary for textiles, leather.

Dispersol. [Akzo] Blends for cleaning and pitch dispersing.

Dispersol. [ICI Am.] Textile dyes.

Dispersol XP-100. [Bruce Chem.] Agent to assist in the removal and prevention of scale buildup on bleaching equip.

Dispersols. [CNC Int'l.] Anionic liq. pitch dispersants for clay slurries, pigments, and coatings in the paper industry.

Dispex. [Allied Colloids] Dispersant.

Displasol DP. [Am. Emulsions] Quaternary ammonium compd. blend; displaces acid dyes on nylon; used for producing novel styling effects.

Dispolene. [Seppic] Anionic/nonionic surfactant blends; oil spill dispersant.

Disponil. [Henkel/Functional Prods.; Henkel KGaA; Pulcra SA] Emulsifier for emulsion polymerization; dispersant for paints; surfactant for shampoos.

Dispose. [Bell Labs] Odor control agent.

Disrol SH. [Nippon Nyukazai] Sodium bis (naphthalene) sulfonate; dispersant, suspending agent.

Dissolvan Brands. [Hoechst AG] EO/PO block polymers and/or oxyalkylated resins; emulsion breaker for w/o emulsions; dehydration and desalting agents for crude oil.

Dissolver GX Grades. [Hoechst AG] Alkylaryl polyglycol ethers; surfactants, hydrotropes for liq. cleaners.

Dissolvine®. [Akzo; Akzo Chem. BV] EDTA, NTA, HEDTA, or DTPA-based; chelating agents, corrosion and scale inhibitors.

Distal. [Procter & Gamble] Oleic acids.

Distilled Lipolan. [Lipo] Hydrogenated lanolin; emollient, lubricant, and conditioner used in cosmetics, toiletries, and topical pharmaceuticals.

Disyston®. [Bayer; Miles/Ag] Disulfoton; systemic herbicide and acaricide.

Di-Tab. [Rhone-Poulenc Basic] Unmilled dicalcium phosphate dihydrate.

Dithane. [Rohm & Haas] Fungicide.

Divergan®. [BASF AG] Stabilizing agent for drinks.

Divers. [European Master Batch] Dispersions of pigments in plasticizers; dispersion of foaming agent in PVC soft foam.

Diwatex. [Borregaard LignoTech] Kraft lignin derivs.; dispersant for dyestuffs, pesticides.

Dixie® Clay. [R.T. Vanderbilt] Kaolin, hydrated aluminum silicate; reinforcing mineral filler and extender for plastics.

Dixo. [Dixo] Dry cleaning chemicals.

Dixon Compd. [Furon] PTFE compds.

Dizad® 6. [Blackman Uhler] Polyoxyethylene ester of fatty acids; nonfoaming stabilizer for diazo bath in naphthol dyeing and printing.

Dizene. [PPG Industries] Emulsifiable orthodichlorobenzene; aromatic used in industrial plants, stockyard pens, restaurants, hotels, hospitals, camps, parks, land-fill areas, cesspools, septic tanks, grease traps, dry wells, and soil; solv. for oil and grease.

DK Ester. [Dai-ichi Kogyo Seiyaku; Grünau] Sucrose fatty acid ester; food additive for stabilizing emulsions; emulsifier, dispersant, solubilizer.

DL-. [Dover] Chlorinated esters, fatty acids, oils; lubricant additives.

DLC®. [Natrochem] Dry liq. concs.

DM. [OxyChem] Methoxydiglycol; solvent.

DMAMP-80®. [Angus] 2-Dimethylamino-2-methyl-1-propanol; solubilizer, emulsifier, corrosion inhibitor, urethane catalyst, synthesis applics.

DMP. [Rohm & Haas] Substituted amino-methyl phenol.

DMP. [Schering Berlin Polymers] 2,2-Dimethoxypropane; chemical intermediate for pharmaceuticals; dehydrating agent.

DMPA®. [Rhone-Poulenc] Dimethylolpropionic acid; in mfg. of alkyd resins, surfactants, chemical intermediates, syn. lubricants, plasticizers, pharmaceuticals, cosmetics.

DMS-33. [Hefti Ltd.] PEG-2 stearate; emulsifier for creams and ointments, shoe creams, paraffin wax, pigment suspensions.

DN 25L®. [Novo Nordisk] Dextranase; enzyme for reduction of dextran content in sugar juices.

DND Compd. [Anchor UK] Permanently plastic filler compd. for ships' cable; completely prevents ingress of moisture; also used as a caulking compd. for sealing junction boxes.

DO-. [Dover] Chlorinated olefin; lubricant additive.

DO-33-F. [Hefti Ltd.] Sorbitan dioleate; emulsifier for creams and ointments; pigment stabilizer for paint emulsions, anticorrosion oils.

DOA. [Monsanto] Dioctyl adipate; plasticizer for PVC and synthetic rubbers.

Dobanol. [Shell UK] Primary alcohols or primary alcohol ethoxylates; nonionic detergent, wetting agent, emulsifier, intermediate; for cosmetic, specialty, laundry, and dishwashing formulations.

Doccolin. [Degen] Alkyd resin.

Dodecenyl Succinic Anhydride. [Humphrey] Alkenyl succinic anhydrides; intermediate for producing amide and imide rust inhibitors, sludge dispersant for lube oils and greases, industrial cleaners; epoxy hardener, mercurial fungicide.

Dodicor. [Hoechst Celanese/Colorants & Surf.; Hoechst AG] Quaternary ammonium chloride; corrosion inhibitor for oil and gas industry.

Dodiflood Brands. [Hoechst AG] Ether sulfonates, ether carboxylates; anionic surfactants for enhanced oil recovery and microemulsion flooding.

Dodiflow Brands. [Hoechst AG] Ethylene copolymers and wetting agents; paraffin inhibitors for prod. of crude oil, flow improvers for middle distillates.

Dodifoam Brands. [Hoechst AG] Surfactant blends; foaming agents for drilling.

Dodigen Brands. [Hoechst AG] Nitrogen-containing compds. or quaternary ammonium salts; disinfectant, antimicrobial; corrosion inhibitor for oil and gas industry; conditioner for hair rinse prods.

Dodilube Brands. [Hoechst Celanese] Nitrogen-containing condensate; lubricant for water-based drilling muds.

Doittol. [Henkel-Nopco] Surfactant, detergent, wetting agent, leveling agent, dyeing assistant; antiblocking and mold release agent for rubber; textile antistat.

Dokipel. [Ina-Oki] Expandable polystyrene.

Doki Polystyrene. [Ina-Oki] General-purpose and impact polystyrene.

Dolchem. [Dolphin Paint & Chem.] Adhesives, coatings, dispersions, organosols, paints, primers, putty.

Dolflex. [John C. Dolph] Insulating compds.

Dolite. [Pfizer] Dolomitic limestone.

Dolocron. [Pfizer] Ground dolomite.

Dolofil. [Steetley Minerals Ltd.] Dolomite fillers.

Dolowhite. [Nat'l. Refractories & Minerals] Dolomite; high brightness grade.

Dolphon. [John C. Dolph] Epoxy and polyester resins.

DOM. [Textile Rubber & Chem.] Dioctyl maleate; plasticizer for vinyl resins, intermediate for mfg. of surfactants, in copolymerization reactions with PVC and vinyl acetates, latex paint formulations.

Dominol. [Dominion Prods.] Beverage clouding agent.

Doocopen 152-B. [Dooley] Fatty deriv.; dyeing assistant and surfactant.

Doocosist. [Dooley] Hand modifiers, dyeing assistants, antistat for textiles.

Doocosoft. [Dooley] Softener for textile applics.

Doocosolv. [Dooley] Oil, grease, and tar spot remover; aids dye solubility.

Doocovel. [Dooley] Amphoteric softener for long bath applics. on synthetics and blends.

Doolex. [Dooley] Sequestrants for textile wet processing.

Doolon. [Dooley] Substantive softeners for natural and synthetic fibers and blends.

Door-Ease. [Am. Grease Stick] Stainless lubricant.

Doresco. [Dock Resins] Synthetic resins.

Dorifex. [S. Dory Ltd.] PVC.

Doriflex. [S. Dory Ltd.] PVC.

Dorika. [S. Dory Ltd.] PVC.

Dorox®. [Condea Chemie GmbH] Aluminum alcoholates, alkenyl succinic anhydride, or derivs.; for mfg. of pharmaceuticals, agrochemicals, dyes, fuel and lubricant additives.

Dortan N. [Exxon] Metalworking lubricant.

DOSS-70. [Manufacturers Chems.] Sulfosuccinate; fast wetting and rewetting agent for textiles.

DOSS 70%. [Fibertint & Chems.] Dioctyl sulfosuccinate; wetting agent, component for fiber processing aids; for antisoil carpets.

Double/Bubble. [Hardman] Two-part epoxy pkgs.

Douglas. [Penford Prods.] Corn starches.

Dovanox. [Lion] Ethoxylated alcohols; detergent, penetrant, emulsifier, dispersant.

Doverchlor. [Dover] Chlorinated olefins; lubricant additives.

Doverguard®. [Dover] Chlorinated, brominated, or bromochlorinated paraffin; flame retardants.

Doverlub. [Dover] Heavy-duty extreme-pressure oil.

Doverphos. [Dover] Phosphites; heat and color stabilizers, antioxidants, chelating agents for plastics, rubber; lubricant additives; aids curing and hardening in epoxies.

Dowanol®. [Dow; Dow Europe] Glycol ethers; solvents.

Dow Antimicrobial 7287. [Dow Europe] Dibromonitrile propionamide; biocide.

Dowclene. [Dow] Solvent; for dry cleaning.

Dow Corning®. [Dow Corning; Dow Corning France SA] Silicone compds.; surfactants, release coatings, antifoams, lubricants; detackifier and plasticizer for hair fixative resins; for urethane foam, personal care prods., industrial processing.

Dow Corning® ACH, AZG. [Dow Corning] Aluminum or aluminum/zirconium chlorhydrates and glycines; active ingredient in antiperspirants.

Dow Corning® Z-. [Dow Corning] Silane compds.; coupling agent for thermoplastic and thermoset resins.

Dowetch. [Dow] Photoengraving chemicals.

Dowex. [Dow] Catalysts for prod. of motor fuel oxygenates; ion exchange resins.

Dowfax. [Dow; Dow Europe] Sulfonate salts; detergent, emulsifier, wetting agent, solubilizer for electroplating, dyeing, pesticides, emulsion polymerization.

Dowflake. [Dow] Calcium chloride.

Dowfrost. [Dow] Heat transfer medium, antifreeze.

Dowfroth. [Dow] Flotation aid for mining.

Dowfume. [Dow] Fumigants.

Dowgard. [Dow] Antifreeze, coolant.

Dowicide. [Dow] Antimicrobials.

Dowicil®. [Dow; Dow Europe] Quaternium-15 or blends; antimicrobial, preservative for adhesives, latex emulsions, paints, metal cutting fluids, drilling muds, detergents, paper coatings.

Dowlex. [Dow] PE or LLDPE resin; resin for blown and cast film extrusion, inj. molding.

Dow Medical Packaging Film 2000. [Dow] Thermoformable medical device pkg.

Downright. [Dow] Latex additives.

Dowper. [Dow] Dry cleaning solvent.

Dowtherm. [Dow] Heat transfer fluids.

Doxifor. [Int'l. Dioxcide] Stabilized chlorine dioxide powd.

DP-. [Dover] Phosphites.

DPG. [Uniroyal] Diphenylguanidine; sec. accelerator for rubber industry.

DPNR. [H.A. Astlett] Deproteinized NR; engineering grades for off-shore and underwater use, brake pads, vibration mounts, pharmaceutical prods.

D.P.P.G. [Gattefosse; Gattefosse SA] Propylene glycol dipelargonate; emollient and oily rancidless additive for cosmetics and pharmaceuticals.

DPR®. [Hardman] Polyisoprene; used for potting, sealants, caulk, adhesives, flexible rubber molds.

DPTT. [Akrochem] Dipentamethylene thiuram hexasulfide; accelerator for rubbers; vulcanizing agent for heat-resistant latex.

DRA-1500. [Toho Chem. Industry] POE disproportionated rosin ester; emulsifier, dispersant.

Dragnet. [FMC/Ag] Insecticide/termiticide.

Drainaid. [Hart Chem. Ltd.] Drainage aid for pulp and paper industry.

Drakeol. [Penreco] Mineral oil; plasticizer, lubricant, emollient for cellulosics, plastics, personal care prods., textile, paper applics.

Draketex. [Penreco] Light mineral oil; coning and finishing oil base for nylon and rayon prod.

Drapex®. [Witco/Argus] Epoxidized oils or esters; plasticizer for vinyl compds.

Draw Free. [Castrol Industrial East] Drawing lubricants.

Draw-Well. [G. Whitfield Richards] Metal drawing lubricants.

Drench. [Mackenzie Chem Works] All-purpose cleaner.

Dresinate®. [Hercules; Hercules BV] Rosin soaps; emulsifier, pigment wetting agent, dispersant, foaming agent, stabilizer; for adhesives, polymerization, metalworking, drilling muds, asphalt, industrial and household cleaners.

Dresinol. [Hercules] Rosin derivs.; modifier for polymer film-formers in adhesives, sizings.

Dress All. [Van Den Bergh Foods] Hydrogenated soybean oil, TBHQ blend; fluid shortening for dressing oil, vegetables, frying, prepared foods.

Dress'n. [A.E. Staley Mfg.] Modified corn starch.

Drewamine. [Drew Ind. Div.] Amine based; corrosion inhibitor for boiler water treatment.

Drewbrom. [Drew Ind. Div.] Bromide ion aq. sol'n.; precursor for oxidizing agents, algicides, bactericides, fungicides, slimicides used in disinfectants, sanitizers, cooling water and wastewater treatment systems.

Drewchlor®. [Drew Ind. Div.] Sodium chlorite sol'n.; algicide and precursor for generation of chlorine dioxide for industrial cooling water treatment.

Drewclean®. [Drew Ind. Div.] Alkaline cleaner; ion exchange resin, filter media cleaning and maintenance.

Drewcor®. [Drew Ind. Div.] Corrosion inhibitor for boiler water treatment.

Drewfax®. [Drew Ind. Div.] Sulfosuccinates or fatty acid derivs.; wetting agent, penetrant, surfactant, dispersant for textile, cosmetic, paper, metal, paint, rubber, plastics, petrol. and agric. industries.

Drewfloc®. [Drew Ind. Div.] Anionic polymers; flocculant for water clarification applics.

Drewgard®. [Drew Ind. Div.] Corrosion inhibitor for cooling water treatment.

Drewlate 30. [Stepan/PVO] Acetylated tartrated vegetable oil monoglyceride; food emulsifier.

Drewlene 10. [Stepan/PVO] Propylene glycol stearate; food emulsifier.

Drewmulse®. [Stepan/PVO; Stepan Europe] Polyglyceryl, glycol, or sorbitan esters or blends; emulsifier, solubilizer, dispersant, stabilizer, opacifier, and visc. builder in cosmetics and pharmaceuticals; food emulsifier.

Drewplast. [Stepan/PVO] Plastics additives.

Drewplex. [Drew Ind. Div.] Boiler water treatment.

Drewplus®. [Drew Ind. Div.] Defoamers for latex, rubber, textiles.

Drewpol®. [Stepan/PVO; Stepan Europe] Polyglyceryl esters; food emulsifier and additive; flavor and color solubilizer.

Drewpone®. [Stepan/PVO] Ethoxylated sorbitan esters; food emulsifier, foaming agent, dough conditioner.

Drewsoft 100. [Stepan/PVO] Glyceryl stearate; softener for fibers and blends.

Drewsorb®. [Stepan/PVO] Sorbitan esters; food emulsifier.

Drewsperse®. [Drew Ind. Div.] Sodium polyacrylate; pigment and paint dispersant; deposit control agent, wetting agent.

Drewspray Assembly. [Drew Ind. Div.] System for completely water rinsing the interior of 55-gal drums.

Drewtain. [Drew Ind. Div.] Retention

aid polymer.

Drewtrol®. [Drew Ind. Div.] Deposit and corrosion inhibitor for boiler water treatment.

Dridex. [Dexter Spec. Coatings] Powd. coatings.

Drierite. [W.A. Hammond Drierite] Desiccants and drying equip.

Drifilm. [GE Silicones] Silicone; water repellent.

Drifix DR. [Leatex] Organometallic compd.; antimigrant for direct and reactive dyes on cotton.

Drift Proof. [W.A. Cleary] Drift control agent, spreader-sticker, pesticide deposit builder.

Drikalite®. [ECC Int'l.] Calcium carbonate; extender pigment.

Drilev CB, JH. [ICI Am.] Organometallic compds.; antimigrants for direct and fiber reactive dyes during batch operations.

Drillers Choice. [Akzo Salt] Salt used for compounding well drilling slurries.

Drimarene®. [Sandoz] Specialty dyes for coloring aq. media, wood stains.

Drimax. [Allied Colloids] Filter aids.

Driocel. [Engelhard] Desiccant.

Dri-Out. [Petrokem] Air dehumidifer.

Driox. [Union Carbide] Industrial gases.

Dri-Pax. [W.R. Grace/Davison] Desiccants.

Dripstop. [Hernon Mfg.] Pipe thread sealants.

Dri-Quik. [Meridian Petroleum] Absorbents.

Dri-Rx. [Mooney Chems] Paint drier accelerator.

Dri-Seal. [Crompton & Knowles] Spray-dried flavors, perfumes, fragrances.

Drisoy. [Reichhold] Soybean oil.

Dri Spot. [Golden Cat] Oil and grease absorbents.

Dri-Tac. [Adhesive Prods.] Pressure-sensitive adhesives.

Driton. [Sybron] Colloidal silica; antislip agent for fibers.

Dri Up. [Petrokem] Absorbents.

Dri-White. [Meridian Petroleum] Absorbents.

Drizone. [Engelhard] Mineral; poultry litter material.

Druspin. [Stepan/PVO] Esters or blends; textile lubricant, antistat, emulsifier.

Dry. [Merix] Dehydrating agent.

Dry-Blend® NCG Fiber. [Am. Cyanamid] Nickel-coated graphite fiber in ABS/PC, ABS, or PC.

Dry-Crisp. [Merix] Dehydrating agent.

Drydene. [Dryden Oil] Lubricating oils.

Dryflex. [Drycolor AB] Masterbatches.

Dry Flo®. [Nat'l. Starch & Chem.] Aluminum starch octenyl succinate; for body powds., antiperspirants, feminine hygiene sprays, foot powds.

Drymaster. [BDJ (England) Ltd.] Dehumidifying dryers.

Drymet®. [Crosfield; Rhone-Poulenc Basic] Sodium metasilicate anhydrous; soap builder and detergent; for formulating industrial cleaners; aids buffering capacity and corrosion inhibition of soft metals and ceramic glazes.

Drynol. [Rhone-Poulenc Geronazzo SpA] Calcium dodecylbenzene sulfonate/ ethoxylated alkylphenol/stabilizer blend; emulsifier for pesticides.

Dryol®. [Synthron] Water repellent, waterproofing agent, fluorochemical extender.

Dryon® M. [Yorkshire Pat-Chem] Zirconium wax compd.; semidurable water repellent.

Dry Pexol®. [Hercules] Rosins; dry size for paper/paperboard mfg.

Dryphite. [Kano Labs] Dry graphite lubricant.

Dryplast. [Chrostiki SA] Dry colors.

Dry Size. [Hercules] Rosin/paraffin wax blend; dry size for paper/paperboard mfg.

Drysperse. [Witco Israel] Polyoxyalkylene glycol blends; dispersant for agric. formulations.

Drytech. [Dow] Super adsorbent polymer.

DS-90. [Continental Sulfur] 90% elemental sulfur and clay blend; granular material for agric. use.

D.S.H. C. [Exsymol] Dimethylsilanol hyaluronate; hydrating, regenerative agent for skin care and anti-aging formulations.

DSMA Liq. [ISK Biotech] Postemergence herbicide.

D Sodium Silicate. [PQ Corp.] Sodium silicate; corrosion inhibitor for water systems in industrial plants, textile mills, laundries, office bldgs., municipalities, oil refineries.

DSP-A. [Monsanto] Anhydrous disodium phosphate; buffer agent in dyebaths, textile processing.

DSX. [Henkel/Coating Chem.] Urethane associative thickener; rheology modifier, thickener for paints, adhesives, coatings.

D-Tox Flowable Charcoal. [W.A. Cleary] Flowable activated carbon; for spill decontamination, soil decontamination, plant and seed protection, spray equip. cleaning.

D-Trans Allethrin. [McLaughlin Gormley King] Synthetic pyrethroid.

Dualite. [Pierce & Stevens] Hollow composite microspheres with PVDC/acrylonitrile copolymer shell and calcium carbonate coating; low density filler for use in plastics, coatings, adhesives, BMC, SMC, rubber compding., paper mfg.

Dual Shrink. [Zeus Industrial Prods.] Heat-shrink TFE/FEP dual layers.

Duflex. [Duflex Ltd.] Flexible open-cell PVC foam.

Dulceta. [ICI Surf. UK] Pourable polyethylene emulsions; improves sewability and tear strength of resin-treated fabrics.

Dulectin. [Solvay Duphar BV] Lecithin; surfactant, emulsifier, dispersant, and stabilizer for pharmaceuticals and cosmetics.

Duller P. [Ivax Industries] Delustering, softening aid.

Dumoflex. [Dumo Plastics NV] Polyether foam.

Dumofoam. [Dumo Plastics NV] Polyether foam.

Dumoform. [Dumo Plastics NV] Polyether foam.

Dumolux. [Dumo Plastics NV] Polyether foam.

Duocrome. [Mearl] Iridescent colors.

Duo-Cure. [Henkel/Functional Prods.] Resin system.

Duofol. [Hart Chem. Ltd.] Sulfated ester; wetting agent, dispersant, penetrant, finishing agent for textiles.

Duo Kote. [Reilly-Whiteman] Phosphate coating and lubricant.

Duolite. [Rohm & Haas] Ion-exchange resins; acid adsorbent, org. scavenger, sugar deionization, decolorization, removal of acids in chemical processing, deionization of starch hydrolysates, etc.

Duomac®. [Akzo; Akzo Chem. BV] Alkyl propanediamine diacetates; emulsifier, corrosion inhibitor, flotation reagent, bactericide; for drilling fluids, pigment flushing, flocculation.

Duomeen®. [Akzo; Akzo Chem. BV] Alkyl propane diamines or salts; chemical intermediate, corrosion inhibitor, fuel oil additive, flotation agent; used in metals, textiles, plastics, herbicides, paints, bitumen emulsions; epoxy curing agent.

Duoquad®. [Akzo] Diquaternary ammonium salts; detergent, corrosion inhibitor, metal cleaner, emulsifier for sec. oil recovery.

Duoteric. [Rhone-Poulenc Ltd.] Nonionic/anionic blend; emulsifiers for agrochemical formulations.

Duothane. [Synair] Castable urethane systems.

Duplosan®. [BASF AG] Mecoprop-P or dichlorprop-P; herbicides for control of broadleaf weeds in cereals.

Duponol. [DuPont] Surfactants.

Duponol. [Witco/Organics] Sulfates; wetting agent, dispersant, penetrant, softening agent, emulsifier used for cosmetics, pharmaceuticals, textile, metal, emulsion polymerization, and leather processing;

DuPont 20 Series. [DuPont] Polyethylene resin; blow molding, extrusion, and inj. molding resins.

Dura Beau. [Scholler] Textile finishes.

Durabond. [Borregaard LignoTech] Calcium/magnesium lignosulfonate; pellet binder for animal feeds; contributes some nutritive value.

Durabond. [Eastern Color & Chem.] Stiffening agent, hand modifier for fabrics.

Durabond. [U.S. Gypsum] Adhesives.

DuraCap®. [BFGoodrich/Geon Vinyl] Exterior vinyl capstock compds.; for weathering protection in exterior building prods.

Duracarb. [PPG Industries] Polycarbonate polyols.

Duracryn®. [DuPont] Thermoplastic elastomer; for inj. molding, extrusion, and blow molding of reinforced hose, flexible and supported tubing, seals, gaskets, mech. goods, profile extrusions, wire and cable jacketing, automotive trim and under-the-hood parts.

Durad. [FMC] Lubricant additive.

Duradene®. [Firestone Syn. Rubber] Sol'n. SBR; used for mech. rubber goods, tire treads, as processing and extrusion aid for other elastomers.

Duraflex®. [Shell] Polybutylene resins; for hot-melt adhesives and sealants.

Dura Flex. [Scholler] Cyclic nitrogen compd.; cellulosic reactant for wash/wear, crease resistant and shrink control finishes.

Dura-Glaze. [Porter Int'l.] Epoxy topcoat; low odor, water-borne.

Dura-Jel. [A.E. Staley Mfg.] Modified corn starch.

Duralink HTS. [Monsanto] Disodium hexamethylene bisthiosulfate; post-vulcanization stabilizer for sulfur cures of NR, IR, SBR, and NBR; used in tire treads, sidewalls, belting and inj. molded goods.

Dura Lube. [Scholler] Blends of emulsified and synthetic waxes and lubricants; softener/lubricants for hosiery and synthetic knits.

Duramac. [ScanRoad] Asphalt sealant.

Duramite®. [ECC Int'l.] Calcium carbonate; carpet backing, roofing compounds, spackles, coatings.

Dura Nap. [Scholler] Napping and brushing agents for synthetic textiles.

Durane®. [Raffi & Swanson] Catalyzed finishes and polyurethane coatings.

Duraphos. [Albright & Wilson Am.] Phosphites; lubricant additives.

Duraplex. [Reichhold] Synthetic resin.

Duraplus®. [Rohm & Haas] Emulsion polymer; for floor polishes with high durability.

Dura-Pox Epoxy. [Southern Coatings] Polyamide epoxy.

Dura Seal. [Scholler] Crosslinking acrylic resin emulsions; dimension stabilizers, body builders, binders, washable and drycleanable finishes for textiles.

Durasoft. [Manufacturers Chems.] Softeners, lubricants, antistat for textiles.

Dura Soft. [Scholler] Softeners, lubricants for textiles.

Durastat. [PPG Industries] Antistatic concs.

Dura Stat. [Scholler] Nondurable antistats for control of plant and consumer static problems.

Durastrength 200. [Atochem N. Am.] Acrylic; impact modifier.

Dura Stretch. [Scholler] Acrylic resin; durable body building finishes for stretch fabrics, webs, and tapes.

Duratex. [Van Den Bergh Foods] Partially hydrogenated cottonseed oil; lubricant for tablets, release agent, candy dusting and coating for hydroscopic materials.

Durawax. [Frank B. Ross] Wax additives to increase hardness of other waxes in cosmetics.

Dura-Wear. [Uncle Sam Chem.] Polyurethane finish.

Durax®. [R.T. Vanderbilt] N-Cyclohexyl-2-benzothiazole sulfenamide; rubber accelerator.

Durazol. [ICI Am.] Textile dyes and pigments.

Durazone 37. [Uniroyal] 2,4,6-Tris-(N-1,4-dimethylpentyl-p-phenylenediamino)-1,3,5-triazine antiozonant/antioxidant for natural and synthetic rubbers, tires, hose, footwear, mech. goods, roofing, wire and cable.

Durazym. [Novo Nordisk] Protease; enzyme for laundry detergents, dishwashing powds.

Durel. [Monsanto] Acrylic fibers.

Durel®. [Hoechst Celanese/Engineering Plastics] Polyarylate resin; engineering thermoplastic for glazing and lighting applics., lenses, automotive panels, appliance parts, solar collection panels.

Dur-Em®. [Van Den Bergh Foods] Glyceryl esters or mono- and diglyceride blends; emulsifier, stabilizer for foods, cosmetics. and pharmaceuticals; lubricant for plastics, textiles, paints, insecticides.

Durethan®. [Bayer; Miles] Nylon 6 and 6/6 resins, some glass reinforced; engineering plastic for automotive, household appliances, building construction, furniture manufacture, and packaging applics.; for inj. molding and extrusion.

Durez®. [OxyChem/Durez] Phenolic, alkyd, or phthalate resins, some glass or mineral filled; thermosets for applics. incl. automotive transmission parts and braking systems, communication products; for compr., transfer, and inj. molding applics.

Durfax®. [Van Den Bergh Foods] Ethoxylated sorbitan esters; emulsifier for foods, personal care, and household prods.; antistat, lubricant, softener; dispersant for pesticides; antifog for plastics and polishes.

Duribbon. [Nat'l. Starch & Chem.] Industrial and architectural sealants.

Durkex. [Van Den Bergh Foods] Hydrogenated vegetable oil blends; food industry additives; carrier, lubricant, moisture barrier, antidusting agent; also for cosmetics and pharmaceuticals.

Durko. [Van Den Bergh Foods] Hydrogenated vegetable oil; emulsified shortening for yeast-raised sweet goods.

Durkote. [Van Den Bergh Foods] Encapsulated ingredients; for acidification in food industry.

Durlac®. [Van Den Bergh Foods] Glyceryl lacto esters; food emulsifier; starch gelling agent in industrial processes.

Dur-Lec®. [Van Den Bergh Foods] Lecithin (phospholipids); emulsifier for foods, chewing gum.

Durlite. [Van Den Bergh Foods] Hydrogenated vegetable oil blends; shortening systems for bakery items.

Dur-Lo®. [Van Den Bergh Foods] Mono- and diglycerides with BHA and citric acid; food emulsifier.

Durmont. [Astor Wax] Montan wax deriv.

Duro. [Lyondell Petrochemical] Hydraulic oils.

Dur-O-Bond. [Nat'l. Starch & Chem.] Emulsions for building prods.

Dur-O-Cote. [Nat'l. Starch & Chem.] Textile coating compds.

Dur-O-Cryl®. [Nat'l. Starch & Chem.] Acrylic copolymer emulsions; backcoating, pigment binders.

Durola. [Van Den Bergh Foods] Hydrogenated canola oil; shortening for baked goods; filler fat for cream centers, whipped toppings.

Duro-Lok. [Nat'l. Starch & Chem.] Adhesives.

Duroloy®. [Rogers] Self-lubricating ball bearings with thermoplastic polyimide retainers.

Duromaize. [Van Den Bergh Foods] Hydrogenated corn oil blend.

Duromel. [Van Den Bergh Foods] Hydrogenated cottonseed oil; coating fat for vegetable dairy, confectionery, bakery.

Dur-O-Set®. [Nat'l. Starch & Chem.] Polyvinyl acetate emulsions or ethylene vinyl acetate copolymers; finishing, sizing, stiffening agents for fabrics; pigment binder.

Durosil. [Degussa] Precipitated silica; reinforcing agent for rubber compds.

Durostabe. [Harcros Chem. Scandia ApS] Stabilizers.

Durotex®. [Morton Int'l.; Morton Int'l. NV SA] Antimicrobial, mildewproofing agent for plastics, latex carpet backing, dry film preservative, filtration papers.

Dur-O-Tron®. [Nat'l. Starch & Chem.] Polymeric pigment binders for roller and rotary screen printing.

Duroxon. [Astor Wax] Synthetic wax.

Durpeg®. [Van Den Bergh Foods] Ethoxylated esters; food emulsifier, processing aid in animal feeds; cosmetics.

Durpro® 107. [Van Den Bergh Foods] Propylene glycol mono- and diesters of fats and fatty acids with citric acid; emulsifier for cakes and cake mixes.

Dursban®. [Dow; W.A. Cleary] Chlorpyrifos; broad spectrum insecticide for ornamentals, turf, home lawns.

Durtan®. [Van Den Bergh Foods] Sorbitan esters; used in cosmetics, household prods., water softening compds., pesticides, auto polishes; fabric antistat and lubricant; food emulsifier; rust inhibitor additive; antifog in plastics; dispersant.

Dusil. [Ducey] Silicone dielectric compds.; for elec. insulation, silicone lubricants, silicone heat sinks.

Dusoran. [Solvay Duphar BV] Lanolin alcohol derivs.; emulsifier, stabilizer, softener, emollient for cosmetics, absorption bases, cleansers.

Dust Buster. [Martin Marietta Magnesia Spec.] Dust control additives.

Dustex. [Borregaard LignoTech] Lignosulfonate-based; biodegradable environmentally safe dust binder and road stabilizer.

Dustgard. [North Am. Salt] $MgCl_2$; dust suppressant.

Dusting Sulfur. [Cuproquim] 48% sulfur.

Dust Master. [Warner-Jenkinson] Dustless colors.

Dustrol. [O'Brien Industries] Dust control prods.

Dust-Set. [Mateson] Dust encapsulant resin.

Du-Ter®. [Griffin] Triphenyltin hydroxide; flowable fungicide for pecans, potatoes, sugar beets.

Dutral. [EniChem Elastomeri Srl] Hydrocarbon rubber.

Dutral®. [Ausimont] EPM/EPDM copolymer or blends; elastomer for molded goods, extrudates, insulation and cable jacketing, belting, modifying lubricating oils and polyolefins.

Duzitall. [Uncle Sam Chem.] Phenolic; germicidal cleaner.

Duz-Kleen. [Uncle Sam Chem.] Liq. detergent.

DW-Wax. [Astor Wax] Thermal expansion materials.

DY. [Ciba-Geigy/Plastics] Alkyl glycidyl ether; diluent for epoxy.

DyaFac. [Henkel] Ethoxylated esters or glycerides; emulsifier, lubricant, solubilizer for synthetic fibers, fats and oils.

Dyapol. [Yorkshire Pat-Chem] Dispers-

ants, leveling agents, retardants for disperse dyeing.

Dyasist 486. [Sandoz] High m.w. ethoxylated compd.; leveling and retarding agent for dyeing wool and blends.

Dyasulf. [Henkel] Sulfated oils or fatty acid salts; wetting agent, penetrant, dispersant, emulsifier for textiles.

Dybalube LSG. [Surpass] Dyebath lubricant for polyester and blends.

Dybild. [Apollo] Additive for textile dyeing with reactive dyes.

Dycar MBP. [Surpass] Biphenyl blend; carrier for polyester and blends.

Dycho® Antifoam D. [Dycho] Silicone antifoams for textile processing.

Dycho® Dye Assist and Lub. [Dycho] Sulfonated vegetable oil; dyeing assistant and dye dispersant.

Dycho® Liq. Detergent. [Dycho] Detergent, wetting agent, scouring agent for textile applics.

Dycho® Liq. Scour Conc. C-2A. [Dycho] Detergent, dye and bleach assistant.

Dychopen®. [Dycho] Nonionic surfactant blend; wetting penetrant, dye and bleaching assistant.

Dycho® Scour. [Dycho] Scouring agent for textiles.

Dycho® White THD (Blue). [Dycho] Peroxide stable brightener for cellulosics.

Dye Assist. [CNC Int'l.] Textile dyeing assistants.

Dyeez. [Dyetech] Lubricants, softeners for textile processing.

Dyeset®. [Sybron] Cationic; dye fixative for direct and fiber reactive dyes.

Dyeset DF. [Manufacturers Chems.] Cationic resin; dye fixative for cotton and other cellulosic fibers dyed with direct, fiber reactive, developed, or sulfur dyes.

Dyetone®. [Olin] Sodium bromate aq. sol'n.; dye oxidant for vat and sulfur dyes.

Dyetry. [Sentry] Dyeing assistants.

Dyeweld. [Sybron] Acid dye fixative.

Dykleer OL-5. [Dyetech] Blended POE amines; antiprecipitant for acid/cationic dyebaths.

Dylark®. [Arco] Styrene/maleic anhydride copolymer, some glass reinforced; engineering resin for inj. molding applics. requiring high structural rigidity, automotive parts, appliance housings and components, housewares, pkg., electronic applics.

Dylene. [Arco] Polystyrene.

Dylev CDL. [Apollo] Direct dye leveler and stabilizer for dyeing of cellulosic and polyester/cellulosic blends.

Dylite®. [Arco] Expandable polystyrene; for dynamic cushioning applics.

Dylonel. [Dylon Industries] Graphite coatings, heat transfer cements, high temp. cements.

Dylonex. [Dylon Industries] Graphite coatings, heat transfer cements, high temp. cements.

Dylonite. [Dylon Industries] Graphite coatings, heat transfer cements, high temp. cements.

Dylox. [Miles/Ag] Selective insecticide for field, vegetable, and turf crops.

Dylube. [Dylon Industries] Die casting release lubricants.

Dymax®. [Dymax] Engineering adhesive.

Dymel®. [DuPont] Aerosol propellants.

Dymerex®. [Hercules] Resin of dimeric acids; thermoplastic resin used in mfg. of protective coatings, spirit varnishes, epoxy ester resin vehicles, adhesives, and printing inks; as a soldering flux.

Dymetrol. [DuPont] Nylon monofilaments, strapping.

Dymex. [Sandoz] Modified ketones; carrier for flame retardant aramid fibers.

Dymsol®. [Henkel/Functional Prods.; Henkel-Nopco] Sulfates; stabilizer, wetting agent, coupling agent, emulsifier for polymerization, latexes, adhesives, coatings, fatliquors for tanning leather.

Dymuv 82D. [Leatex] Organic nitrogenous derivs.; leveling and migrating agent for acid dyes.

Dynacal. [Hüls Am.] Fused calcium oxide.

Dynacerin® 660. [Hüls Am.; Hüls AG] Oleyl erucate; jojoba oil substitute;

emollient for cosmetics.

Dynacet®. [Hüls Am.; Hüls AG] Acetylated monoglyceride; food emulsifier.

Dyna-Core. [Alpha Metals] Rosin core solder.

Dynaflush. [Dynaloy] Safety flushing solvent for urethanes.

Dynakoll. [Akzo] Rosin size; water repellent for paper prods.

Dynalloy. [Filterite] Metallic filter media.

Dynamar®. [3M; 3M Deutschland GmbH] Fluorochemical; mold release agent; processing aids for polymers; vulcanizing agent and cure accelerator for polyepichlorohydrin elastomers.

Dynamullit. [Hüls Am.] Fused mullit.

Dynapol. [Hüls Am.] Polyester hot melt and coating resins.

Dyna-Purge. [Shuman Plastics] Thermoplastic processing purging compds.

Dynasan®. [Hüls Am.] Glyceryl triesters; emollient and lubricant for cosmetic sticks, cakes, creams; binder, lubricant, excipient for tablets.

Dyna-San. [Huntington Labs] Food processing cleaners.

Dynasil®. [Hüls Am.; Hüls AG] Silane derivs.; coupling agent, chemical intermediate, blocking agent, release agent, lubricant, primer, reducing agent.

Dynasolve. [Dynaloy] Solvents for general cleaning, cleaning of uncured polymers, dissolving cured urethanes, silicones, anhydride epoxies, for conformal coating removal.

Dynasperse. [LignoTech] Sodium lignosulfonate; dispersant for dyes.

Dynaspinell. [Hüls Am.] Fused spinel.

Dynasylan®. [Hüls Am.; Hüls AG] Organofunctional silane compds.; bonding agent, coupling agent, chem. intermediate, blocking agent, release agent, lubricant, primer, reducing agent.

Dynatex. [Guthrie Latex] Low ammonia natural rubber latex.

Dynatherm. [Hüls Am.] Electromagnesia.

Dynazirkon. [Hüls Am.] Zirconium oxide.

Dyne. [West Agro] Dairy acid and sanitizer.

Dynemate. [West Agro] Dairy detergents.

Dynocel. [Engelhard] Alumina-based; dynamic desiccant.

Dynofoam. [Dyetech] Defoam/antifoam for dyeing.

Dypenol®. [Dexter] Penetrants, mercerizing assistants with antifoaming and detergent properties.

Dyphene. [PMC Specialties] Phenolic resins.

Dyqex®. [Georgia-Pacific] Sodium lignosulfonate; dye dispersant extender.

Dyrene®. [Bayer; Miles/Ag] Anilazine; fungicide for foliage and turf.

Dyride. [Apollo] Carrier for atmospheric and pressure becks and beams in textile dyeing.

Dyrite. [Rite Industries] Textile dyes and pigments.

Dysosoft. [Dyetech] Softener for textiles.

Dytek® A. [DuPont] 2-Methylpentamethylene diamine; epoxy curing agent; also used in polyurethanes, wet strength resins, scale and corrosion inhibitors, gasoline additives, polyamide plastics, films, adhesives, and inks.

Dytherm. [Arco] Expandable styrene copolymers.

Dytron® XL. [Advanced Elastomer Systems] Thermoplastic elastomers; electrical grades for wire and cable and other electrical applics.

E

E-. [Exxon/Tomah] Ethoxylated amines; for acid thickening, antistats, emulsification, petroleum prod. and refining, agric. adjuvants, textile processing aids, corrosion inhibition, detergent boosters, intermediates for surfactants.

E7T1. [U.S. Polymeric] Toughened epoxy prepreg system; for aerospace advanced composites.

E-0010. [Goldsmith & Eggleton] Isoprene/PBD/carbon black masterbatch.

E-12. [Key Polymer] Epoxy adhesive.

E-130, 131, 133. [Wacker Silicones] Silicone emulsions; release agents, vinyl cleaner/dressing.

E484, 486, 487, 4905, 4920, 4930. [Cosmic Plastics] Epoxy resin, glass and/or mineral filled; molding compd.; elec. and encapsulation applics.

E-3810, 3824, 8353, 8354, 8354J, 9405. [ICI Fiberite] Epoxy resin, some glass filled; molding compd., encapsulant for electronics industry.

E-9031, 9053, 9092. [ICI Polyurethanes] Imine-polyurea system; used for RIM processing for automotive exterior applics.

EA 9289, 9346, 9394. [Hysol Aerospace Prods.] Adhesive for structural, aerospace applics., potting.

Eagle Gel. [Eagle Chem] Silica gel.

Earoeel. [Croxton & Garry Ltd.] Blowing agents.

Ease Release. [George Mann] Release agent for casting and molding systems.

Eastman®. [Eastman] Antioxidants, uv absorbers, stabilizers.

Eastman® P Series. [Eastman] Polypropylene homopolymer and copolymers.

Eastobond. [Eastman] Hot-melt adhesives.

Eastobrite® OB-1. [Eastman] 2,2´-(1,2-Ethenediyl di-4,1-phenylene) bisbenzoxazole; optical brightener, fluorescent whitening agent.

Eastoflex. [Eastman] Amorphous polyolefins; for adhesives and sealants.

Eastonair. [Mateson] Resinous odor sorbent.

Eastotac. [Eastman] Hydrocarbon tackifying resins for adhesives, caulks, and sealants.

Easy-Flo. [Handy & Harman] Silver brazing alloy.

Easypoxy®. [Conap] Epoxy adhesive kit.

EB. [OxyChem] Butoxyethanol.

EB Acetate. [OxyChem] Butoxyethanol acetate.

EB1500-1AR, 3000-2, 5000-1, 7000-1. [Cuyahoga Plastics] Thermoset polyester, some glass reinforced.

Ebal. [Mitsubishi Gas] p-Ethylbenzaldehyde; additives for resins; pharmaceutical intermediate, fragrances.

E.B. Golden Glitter, Neutral Glitter. [Eastern Color & Chem.] Pearlescent material and aq. acrylic resinous binders; for decorative textile printing.

E-BR®. [Ameripol Synpol] Emulsion polybutadiene; used for treads, extruded and molded goods, footwear, rolls.

EC-25®. [Van Den Bergh Foods] Propylene glycol mono- and diesters, mono- and diglycerides, hydrog. soybean oil, lecithin, BHA, and citric acid; food emulsifier for cakes, mixes.

ECC 440, 443, 450, 453. [GE Silicones] Silicone conformal coating; for circuit assemblies.

Ecca-Clay. [Southern Clay Prods.] Ball clay.

Ecca-Tal. [Southern Clay Prods.] Talc.

Ecca-Tex. [ECC Int'l.] Kaolin; extenders for coatings.

Ecca-Tex R. [ECC Int'l.] Kaolin; extenders for rubbers.

Ecco. [Erie Foods Int'l.] Milk chemicals and proteins.

Ecco Anti-Migrant. [Eastern Color & Chem.] Antimigrants, leveling agents for textile dyeing.

Eccoblanc. [Eastern Color & Chem.] White pigment printing compds. for textile printing and dyeing.

Eccobond. [Eastern Color & Chem.] Adhesive for textiles and paperboard.

Eccobond®. [Emerson & Cuming] Epoxy, urethane, or acrylic adhesives; high performance conductive adhesives for industrial and electronic applics.

Eccobrite. [Eastern Color & Chem.] Whitening agents, fire retardants, pigment binders, anticrocks, hand modifiers for textiles.

Eccoclean. [Eastern Color & Chem.] Sulfated alcohol; detergent and emulsifier for general scouring; foam booster/stabilizer.

Eccocoat®. [Emerson & Cuming] Epoxy, urethane, or acrylic; coatings for electronics; dip and conformal coats.

Ecco Defoamer. [Eastern Color & Chem.] Silicone blends; defoamer for waste water, dyeing, printing, general processing.

Ecco Duller. [Eastern Color & Chem.] Dulling agents for textiles.

Eccodye. [Eastern Color & Chem.] Pigments for oil-phase print systems containing alkyd resins as binding agents.

Ecco Fast Binder. [Eastern Color & Chem.] Pigment binder for machine and rotary screen printing.

Ecco Finish. [Eastern Color & Chem.] Stiffening agent for fabrics.

Eccofix. [Eastern Color & Chem.] Dye fixatives for textile processing.

Eccoflameproof. [Eastern Color & Chem.] Flameproof treatment for polyester and blends.

Eccoflo®. [Emerson & Cuming] Dielectric materials.

Eccofoam® SIL. [Emerson & Cuming] RTV silicone prod.; foam for embedment of electronics subject to shock and vibration; for aerospace applications.

Eccofoam®. [Emerson & Cuming] Epoxy or urethane foams; dielectric materials.

Eccoful. [Eastern Color & Chem.] Amide ethoxylate; detergent, scouring and fulling agent for wools and blends.

Eccogard. [Eastern Color & Chem.] Resinous condensate providing flame retardancy to nylon.

Eccogel. [Eastern Color & Chem.] Protein deriv.; stiffening agent, hand modifier for fabrics.

Eccolene. [Eastern Color & Chem.] Sulfated fatty ester; wetting agent, dispersant, emulsifier, penetrant, scouring agent for textiles.

Ecco Leveler. [Eastern Color & Chem.] Anionic dye leveling agent for wool, synthetics, and blends.

Eccolube. [Eastern Color & Chem.] Lubricants for textile printing and processing.

Ecco MP®-2004. [Eastern Color & Chem.] Methylenebischlorophenol; textile fungicide, mildewproofing agent.

Ecconol. [Essential Industries] Wetting agent, detergent, emulsifier, coupling agent, thickener for detergent systems.

Ecco Pad Auxiliary 07. [Eastern Color & Chem.] Gum-type material; restricts migration of color in pigment padding.

Eccopel. [Eastern Color & Chem.] Metal salts; semidurable textile water repellent.

Ecco Polyester® Optical 525. [Eastern Color & Chem.] Benzoxazole; whitening agent for polyester fiber.

Ecco Printlube. [Eastern Color & Chem.] Lubricant for textile printing.

Eccopuff. [Eastern Color & Chem.] Printing system for paper, textiles, nonwoven, and decorative tiles in building applics.

Ecco Resin. [Eastern Color & Chem.] Acrylics or alkyds; binders for pigment pad dyeing and printing; provides soft hand and durability to laundering and drycleaning; for fabric stiffening, finishing, water repellency, soil resistance.

Ecco Rez. [Eastern Color & Chem.] Thermoset stiffing agent for textiles.

Eccoro®. [Eastern Color & Chem.]

Imidazolines; corrosion inhibitor for metals.

Eccoscour. [Eastern Color & Chem.] Detergent, wetting agent, emulsifier, scouring agent for textile and dyeing applics.

Eccoscroop. [Eastern Color & Chem.] Scrooping agent for textiles.

Eccoseal®. [Emerson & Cuming] Epoxy impregnant and casting resins.

Ecco Selbind. [Eastern Color & Chem.] Stiffening agent for fabric edging.

Eccoshield®. [Emerson & Cuming] Conductive elastomeric materials; adhesives, coatings, caulks, sealants, tapes, and gaskets for EMI/RFI shielding applics. and static dissipation and elec. ground planning.

Eccosil®. [Emerson & Cuming; Grace NV] RTV silicone prods.; adhesive, caulk, sealant, potting and embedment of circuits and heater elements.

Eccosoft. [Eastern Color & Chem.] Substantive softeners for textiles.

Eccosol. [Eastern Color & Chem.] Self-emulsifying mineral oil lubricant for fine wire drawing.

Eccosolv. [Eastern Color & Chem.] Solvent cleaner for cleaning and degreasing of machine parts.

Eccosorb®. [Emerson & Cuming] Lightweight flexible foam sheet; broadband microwave absorber.

Ecco Spersant PDQ. [Eastern Color & Chem.] Ethoxylated phenoxy condensate; dispersant for pigments.

Eccosperse. [Eastern Color & Chem.] Conc. pigment colors.

Eccospin. [Eastern Color & Chem.] Ethoxylated/propoxylated alcohol; lubricant and antistat for polypropylene fibers; spin finish for fiber processing.

Eccostat. [Eastern Color & Chem.] Antistat for textiles.

Eccoterge. [Eastern Color & Chem.] Detergent, wetting agent, dye leveling agent, scouring agent for textiles.

Eccotex. [Eastern Color & Chem.] Alkylaryl compd.; dispersant, penetrant for textile industry; wetting agent for dyestuffs.

Eccothane®. [Emerson & Cuming] Urethane casting and encapsulating compounds.

Eccotherm®. [Emerson & Cuming] Silicone or epoxy compds.; thermally conductive greases, adhesives, coatings, encapsulants.

Ecco Thickener. [Eastern Color & Chem.] Polymer for thickening pigment print formulations.

Ecco Tipping Sol'n. [Eastern Color & Chem.] For shoelace tipping.

Eccowax. [Eastern Color & Chem.] Substituted fatty amide condensate; wax conc. for softening textiles.

Eccowet®. [Eastern Color & Chem.] Sodium alkyl sulfonates; wetting agent, dispersant, penetrant for dyeing and textile applics.

Ecco White®. [Eastern Color & Chem.] Optical brighteners for polyamide, polyester, and acetate fibers and fabrics.

Ecdel®. [Eastman; Eastman Chem. Int'l. AG] Copolyester ether elastomer; for inj. molding, extrusion, blow molding, films, bags.

Eclipse. [A.E. Staley Mfg.] Industrial corn starch.

Ecoalube. [E/M Corp.] Solid lubricants.

Ecolohide. [John L. Armitage] Water-reducible coating.

Ecolozinc. [McGean-Rohco] Zinc plating processes.

Ecomeen. [W.R. Grace/Dearborn] Corrosion inhibitor.

Econofluor-2. [DuPont/Medical Prods.] Liq. scintillation counting cocktail.

Econoflux. [Alpha Metals] Liq. inorganic acid flux.

Economag. [Martin Marietta Magnesia Spec.] Magnesium oxide; feed grade.

Economix. [Vyse Gelatin] Gelatin.

Economy Flor-Dri. [Floridin] Attapulgite clay; absorbent.

Econovat. [Southern Dye & Chem.] Textile dyes and pigments.

E-Cool. [Castrol Industrial East] Synthetic polymer metalworking fluids.

Ecoset. [Ashland/Foundry Prods.] Inorganic foundry core binders.

Ecothene. [Quantum/USI] HDPE homopolymers or copolymers; post-consumer resins for blow molding, thermoforming, inj. molding, film extrusion.

ECP- Series. [Ferro/Keil] Emulsifiable chlorinated paraffins; for metal displacement applics. which leave residual protective films.

Eddi. [West Agro] Feed additive.

Edenol. [Henkel/Emery/Cospha] Phthalates or esters; emollient, cosolvent, release agent, plasticizer for lacquers, printing inks, coatings, plastics.

Edenor. [Henkel/Emery] Esters.

Edenor GMS. [Henkel Canada] Glyceryl stearate; consistency factor for creams and liq. emulsions.

Edenor PEHO. [Pulcra SA] 2-Ethylhexanol oleate; textile and multipurpose lubricant.

Edenor ST-1. [Henkel Canada] Stearic acid, triple pressed; intermediate.

Edeta®. [BASF AG] Complexing agents for cosmetics industry.

Edgar. [Engelhard] Kaolin; paper coating, filler.

Edgegum. [Chemonic Industries] Solvent or water-based polymers; provides weight to fabric edges to prevent curling.

Edgegum. [Piedmont Chem. Industries] Noncurl stiffener for edge treatment of fabrics.

Edgegum. [Sybron] Polyvinyl acetate-based polymer; used for textile selvedge for noncurling.

Edgetec. [Miles/Polyurethane] Polyurethane elastomer.

Edistir®. [Montedipe Srl] Polystyrene; for inj. molding and extrusion.

Editer. [Montedipe Srl] Reinforced ABS copolymers.

Edlon. [Edlon Prods.] Fluoropolymer prods.

Edufoam. [Edulan A/S UK] Polyurethane systems.

Edunine. [ICI Surf. UK] Softener, lubricant, antistat for textiles.

EE. [OxyChem] Ethoxyethanol.

EE Acetate. [Eastman] Ethoxyethanol acetate.

EE Acetate. [OxyChem] Ethoxyethanol acetate.

EE Solvent. [Eastman] Ethoxy ethanol.

EFA-Liquid. [Brooks Industries] Linoleic acid, linolenic acid, tocopherol; essential fatty acids for skin care prods.

Effcol. [Matsumoto Yushi-Seiyaku] Linear hydrocarbon sulfonate; antistat for warp sizing.

Eftofine. [Astor Wax] Micronized Fischer-Tropsch wax.

Efton. [Astor Wax] Synthetic wax.

EG-44. [Thiele Kaolin] Filler clay.

Eganal®. [Hoechst Celanese/Colorants & Surf.; Hoechst AG] Leveling agent, lubricant, dispersant for dyeing.

E-Gas. [Matheson Gas Prods.] Silicone tetrafluoride/oxygen blends; used in plasma etching of semiconductor materials.

EGDS-VA. [Goldschmidt] Glycol distearate; pearling agent, opacifier.

EG-ML. [Clark] Glycol laurate; coemulsifier, dispersant, defoamer, emulsifier, lubricant.

EGMS-VA. [Goldschmidt] Glycol stearate; pearling agent, opacifier.

Ekaland. [Vigar SA] Accelerator.

Ekaline®. [Sandoz Prods. Ltd.] Detergent, wetting agent, emulsifier, leveling agent, dispersant for dyeings, scouring.

Ekavyl. [Atochem Deutschland GmbH] PVC resins.

Eki U. [Eki Eeknoff Kunstof Verwerkende Industrie] Polyurethane, polyethylene, PVC, neoprene, EPDM.

Ekonol. [Carborundum] Polyester; thermally stable polymer.

E-Kote. [Acme Div.] Conductive paints.

Ektapro® EEP Solvent. [Eastman] Ethyl 3-ethoxypropionate; retarder solvent for formulating enamels, lacquers, topcoats, primers.

Ektar®. [Eastman] Engineering thermoplastic; for recreation/utility equip., transportation applics.

Ektasolve®. [Eastman] Glycol ethers; solvents for coatings, inks; coupling agent, retarder.

EKZ Series. [Katalistiks Int'l.] Fluid cracking catalysts.

Elacid Richter. [Henkel/Cospha] Hair conditioner conc., antistat; for damaged hair care prods.

Elarco. [Reusche] Ceramic colors and golds.

Elasdo. [Paniplus] Mono- and diglycerides or ethoxylates; food emulsifier.

Elastan®. [BASF AG] Polyol-isocyanate formulations; systems for sports ground surfaces, play grounds, tennis courts, artificial lawns.

Elastin CLR. [Henkel/Cospha] Elastin hydrolysate; for applic. to aging skin.

Elastinhydrolysate. [Henkel/Emery/Cospha] Hydrolyzed elastin; film former, moisturizer, stabilizer.

Elastoblend. [Ameripol Synpol] Emulsion SBR/polybutadiene blends; masterbatch for truck retreads.

Elastocarb®. [Morton Int'l.] Magnesium carbonate; filler providing flame retardancy and smoke suppression to elastomers, plastics, thermosets; for conduit, wire and cable compds., aircraft interior parts, film and sheet.

Elastocard. [Morton Int'l.] Smoke suppressant, release aid.

Elastocell. [BASF] Microcellular polyurethane suspension components.

Elastochem. [Colorite Plastics] Rubberlike compd.; for exc. resilience and low friction finish.

Elastocoat®. [BASF AG] Polyurethane; casting and coating systems; for mold making, tire filling, furniture edges, wear-resistant floor coverings.

Elastoflex®. [BASF; BASF AG] Soft polyurethane; foam systems for mattresses, dashboards, seat supports, upholstered parts.

Elastofoam®. [BASF; BASF AG] Soft integral polyurethane; foam systems for flexible integral skin foams, arm rests, steering wheels, spoilers.

Elastolen®. [BASF AG] Aromatic and aliphatic polyurethane elastomers; for textile coating, prod. of synthetic leather.

Elastolit®. [BASF; BASF AG] Hard integral polyurethane; foam and RIM systems.

Elastollan®. [BASF; BASF AG] Thermoplastic polyurethane elastomer; for inj. and blow molding and extrusion

applics., textile coating

Elastomag®. [Morton Int'l.] Magnesium oxide; chemical thickener for polyester resins; used in adhesives, fuel oil additives, as acid acceptor for specialty plastics and rubber.

Elastopal®. [Elastogran] Polyurethane elastomers; for casting of cyclones, pipe scraper discs, sieve sets, strippers, papermaking machine felts, roller coatings.

Elastopan®. [BASF; BASF AG] Thermoset polyurethane; for shoe foam systems.

Elastopor®. [BASF; BASF AG] Rigid polyurethane; for foam systems.

Elastopreg®. [Elastogran] Thermoplastic glass fiber composite; semifinished prods. of glass mat-reinforced thermoplastics for engine compartment shieldings, seat structures, carriers.

Elastorid®. [Elastogran] Wood dust-filled polypropylene sheets; carrier for inner trim on passenger cars.

Elastosil®. [Wacker Silicones] Silicone elastomers; for aerospace, appliances, automotive, elec./electronics, food handling, baby care, health care, scuba gear, fabric coatings.

Elastosoft 89. [Reilly-Whiteman] Silicone softener for textiles.

Elastosol. [Croda Chem. Ltd.] Soluble elastin and soluble collagen; conditioner for skin care prods.

Elastothene. [RCMS Polysales] Polythene for bag prod.

Elax. [Polychem Systems Srl] Polyurethane for prod. of elastomers.

Elcacid. [Int'l. Dyestuffs] Textile dyes and pigments.

Elcactive. [Int'l. Dyestuffs] Textile dyes and pigments.

Elcema®. [Degussa] Powdered cellulose; anticaking agent, tabletting aid for pharmaceutical industry.

Elcochrome. [Int'l. Dyestuffs] Textile dyes and pigments.

Elcofast. [Int'l. Dyestuffs] Textile dyes and pigments.

Elco Gear Safety 28. [Elco] Gear lubricant.

Elcoment. [Int'l. Dyestuffs] Textile dyes

and pigments.

Elcomine. [Int'l. Dyestuffs] Textile dyes and pigments.

Elcosol. [Int'l. Dyestuffs] Textile dyes and pigments.

Elcosperse. [Int'l. Dyestuffs] Textile dyes and pigments.

Elcozine. [Int'l. Dyestuffs] Textile dyes and pigments.

Elec. [Kao Corp. SA] Surfactants, antistats for plastics.

Electrafil®. [Akzo Engineering Plastics] PPO, nylon, ABS, PBT, PET, PPE, PC, PP, PE, PS, acetal, polyurethane, PES, PEEK, polyetherimide, PPS, ETFE, or polysulfone filled with carbon, stainless steel fiber, lubricants; static dissipative and conductive thermoplastics.

Electrocarb. [Ferro] Silicone carbide.

Electrodag®. [Acheson Colloids] Graphite, copper, nickel, or silver pigments with plastic binders in water or solvents; EMC shielding coatings for plastics, composites.

Electrofine. [Atochem N. Am.] Chlorinated paraffin; flame retardant.

Electromet. [Atochem N. Am.] Cleaners.

Electro-Nox. [Urban Chem.] Solder antioxidant.

Electro-Ox. [Electro Abrasives] Aluminum oxide, alumina.

Electro Pure Reagents. [Polysciences] Ultra pure reagents for electrophoresis.

Electros Colloidal Kaolin. [Charles B. Chrystal] Purified colloidal kaolin.

Electrosil. [GE Silicones] Silicone compositons; for use on electronic/elec. components.

Electrosol. [Alframine] Antistats.

Elecut. [Takemoto Oil & Fat] Alkylaryl sulfonate salt; detergent.

Elefac I-205. [Bernel] Octyldodecyl neopentanoate; emollient, SPF booster, pigment wetter and binder.

Elektroguard. [Wacker Silicones] Silicone elastomers; for potting and encapsulating electronic components.

Elektron Alloys. [Reade Mfg.] High performance magnesium alloys.

Eleminol. [Sanyo Chem. Industries] Sulfates or sulfosuccinates; emulsifier for emulsion polymerization, textile applications.

Elephant Brand. [Cominco Am.] Fertilizers.

Eleton. [Tokai Seiyu Ind.] Complex dispersion; antistatic agent for synthetic fibers.

Elexar®. [Shell] Thermoplastic elastomer; for booster cables, insulation, jacketing, and molding applics.

Elfacos®. [Akzo; Akzo Chem. BV] Polymers; consistency regulator for w/o emulsions, cosmetics; stabilizer.

Elfanol®. [Akzo; Akzo Chem. BV] Sulfosuccinates; detergent, wetting agent for cosmetics and technical applics.

Elfan®. [Akzo; Akzo Chem. BV] Sulfates, sulfonates, esters, or betaines; detergent, wetting agent for cosmetics, textiles, industrial uses; emulsifier for emulsion polymerization, agric. formulations.

Elfapur®. [Akzo; Akzo Chem. BV] Ethoxylated ethers; detergent for textile, household, personal care applics.; base material.

Elftex®. [Cabot] Oil furnace carbon black; for inks, coatings, sealants.

Elfugin. [Sandoz; Sandoz Prods. Ltd.] Antistat for textiles.

Elimina. [Marubishi Oil Chem.] Fatty alcohol sulfates or phosphates; detergent, dispersant, antistat for textiles, latex, dyestuffs, pigments, oils, waxes.

Eliminol. [H.J. Baker & Bro.] Industrial deodorizer.

Elixir. [CC Pollen] Bee pollen extract.

ELP. [Morton Int'l.] Epoxy terminated polysulfide polymers; as concrete adhesives, for chemically resistant linings and coatings, bonding to metallic substrates, modifier for other resins.

El Pinol 85. [T&R Chem.] Pine oil; for the disinfectant industry.

Eltesol®. [Albright & Wilson Am.; Albright & Wilson UK] Sulfonic acids or salts; hydrotrope, solubilizer, coupling agent, visc. modifier for detergent formulations, metal cleaning, electroplating; catalyst for foundry resins; intermediate.

Eltex. [Solvay & Cie] HDPE.

Eltex P. [Solvay & Cie] Polypropylene homopolymers and copolymers.

Elvace®. [Reichhold] VAE copolymer emulsion; adhesive base, laminating adhesive.

Elvacite®. [DuPont] Acrylic resins; for coatings, inks, adhesives, reproduction papers and films.

Elvaloy®. [DuPont; DuPont UK] Ethylene/acrylate/carbon monoxide terpolymers; resin modifier, plasticizer, impact modifier.

Elvamide®. [DuPont; DuPont UK] Nylon multipolymer resin; bonding agent, barrier coating for fabrics, structural adhesives, binder for propellants and explosives.

Elvamido. [DuPont] Nylon multipolymer rosin.

Elvanol®. [DuPont; DuPont UK] Polyvinyl alcohol; film-forming binder for adhesives, paper, coatings, textiles, films, building prods., hoses, gaskets, emulsification in emulsions and latices, additive for concrete, cement, and food, sizings.

Elvax®. [DuPont; DuPont UK] EVA copolymer; for coating and adhesive applics., specialty inks and lacquers, for blending with waxes.

EM. [OxyChem] Methoxyethanol.

EM-. [Ferro/Keil] Esters, imidazolines, alkanolamides, or phosphate esters; lubricant, sperm oil replacement, solubilizer, detergent; for floor cleaners, paint strippers, buffing compds.

Emac. [Chevron] Ethylene methyl acrylate copolymers; resins for inj. molding, blown film, compounding, and extrusion applics.

Emal. [Kao] Sodium POE alkyl sulfates; detergent, emulsifier, foam stabilizer, shampoo base, scouring agent for synthetic fibers.

Emalex. [Nihon Emulsion] Esters, ethoxylated ethers or esters, oils, alkanolamides, or triglycerides; emulsifier, cleaner, dispersant, base for cosmetics.

Emalox. [Yoshimura Oil Chem.] Potassium salt of alkyl phoshate; antistatic agent, finishing oil for dyeing of synthetic fibers, wook, silk.

Emanon. [Kao] PEG fatty acid esters; thickener for cosmetics, pigment preps.

Emargol®. [Witco/Organics] Sodium sulfoacetate of mono- and diglycerides; food processing defoamer and emulsifier; dough modifier.

Emasol. [Kao] Sorbitan esters or ethoxylates; emulsifier, solubilizer, lubricant, stabilizer for emulsion polymerization, pharmaceuticals, and cosmetics.

Emaweld®. [Ashland] Adhesives.

Emcast. [Electronic Materials] One-part epoxy resins; uv-curable flexible resin for electronic encapsulation, dipping, and adhesion applics.

Emcocel®. [Mendell] Microcrystalline cellulose NF/BP; binders, vehicles.

Emcol®. [Witco/Organics; Witco Israel; Witco SA] Surfactants, emulsifier, dispersant, wetting agent, foaming agent, detergent, antistat, conditioner, lubricant, emollient, defoamer, corrosion inhibitor for cosmetics, toiletries, household cleaners, textiles, industrial processing.

Emcompress®. [Mendell] Dibasic calcium phosphate USP/BP; binders, vehicles.

EmCon. [Fanning] Lecithin; emollient, moisturizer, emulsifier, superfatting agent, humectant, mold release agent for hair and skin care prods.

Emcosoy®. [Mendell] Soy polysaccharides; disintegrant.

Emdex®. [Mendell] Dextrates NF; binders, vehicles.

Emdithene. [Robnorganic Systems Ltd.] Polyurethane resin system.

Emerald Ram HS. [Premier Refractories] Alumina chrome plastic refractory.

Emercide®. [Henkel/Cospha] Liq. preservative system for cosmetics.

Emeressence®. [Henkel/Emery] Aromatics for fragrance and odor masking applics., as cosmetic preservatives.

Emerest®. [Henkel/Emery/Cospha; Henkel KGaA] Esters or ethoxylates; emollient, wetting agent, dispersant, conditioner, solubilizer, pearlsecent, opacifier, base for industrial lubricants, mold release agent, defoamer, flotation agent, plasticizer.

Emerlube®. [Henkel/Emery] Ethoxylated vegetable oil; lubricant for polypropylene yarns, carpet backings.

Emerox®. [Henkel/Emery; Henkel KGaA] Azelaic acid.

Emersist®. [Henkel/Textile] Dye leveler, detergent, wetting agent, emulsifier for textile applics.

Emersoft®. [Henkel/Textile] Softener, lubricant, napping assistant for textiles.

Emersol®. [Henkel/Emery] Fatty acids; cosmetics, foods, soaps, emulsifiers, chemical specialties.

Emerstat®. [Henkel/Emery] Cationic antistat for fiber lubricants.

Emerwax®. [Henkel/Emery/Cospha] Synthetic waxes; wax, emulsifier for cosmetic and pharmaceutical formulations incl. sticks, cold creams, makeup, depilatories.

Emery®. [Henkel/Emery; Henkel/ Cospha; Henkel KGaA] Emulsifier, emollient, conditioner, lubricant, moisturizer, dispersant, stabilizer, thickener, opacifier, coupling agent, cure promoter, intermediate; for pharmaceuticals, cosmetics, industrial processing, films, textiles, engine lubricants.

Emgard®. [Henkel/Emery] Dexron II/ Mercon; automatic transmission fluid; synthetic lubricant.

Emid®. [Henkel/Emery; Henkel/Cospha] Fatty alkanolamides; foam booster/stabilizer, thickener, emulsifier for shampoos, detergents, textiles.

Emigen®. [Hoechst Celanese/Colorants & Surf.; Hoechst AG] Acrylamide deriv.; prevents frosting effect during thermosol process on polyester/cellulosic fiber blend.

Emir. [Emmerich (Berlon) Ltd.] Rigid expanded polyurethane.

EMI-X®. [LNP; LNP Nederland] Polycarbonate, PPS, nylon 6 or 6/6, or thermoplastic polyester with carbon, nickel reinforcements; highly conductive composites for electromagnetic and radio frequency interference shielding applics.; for avionics housings, business machine enclosures.

Emkabase®. [Emkay] Naphthenic amino sulfonate blend; emulsifier for mineral oil.

Emkabond UR. [Emkay] Fiber reactive laminating adhesive for bonding urethane foam to fabric.

Emka® Catalyst. [Emkay] Catalyst for thermosetting resins.

Emkacide®. [Emkay] Gas fading inhibitors, mildewcide, deodorant for textiles.

Emkaclear. [Emkay] For fixing and clearing polyester dyes.

Emka® DDBSA. [Emkay] Dodecylbenzene sulfonic acid; base for compounding detergents and emulsifiers.

Emka® Defoam. [Emkay] Nonsilicone and silicone defoamers for industrial use.

Emka® Finish. [Emkay] Modified gum arabic; textile finish.

Emkafix®. [Emkay] Quaternary ammonium deriv.; dispersant, fixative for direct dyes.

Emkafol®. [Emkay] Sulfonated synthetic esters; wetting and rewetting agent.

Emkafume®. [Emkay] Nondurable gas fading inhibitor for acetate colors.

Emkagen®. [Emkay] Amino condensate and alkylaryl sulfonate blend; detergent for print washing in textile applics.

Emka® Konetex. [Emkay] Emulsified oil for use in water as a coning oil for rayon and rayon/nylon mixtures.

Emkal. [Emkay] Sulfonic acids or salts; dispersant, detergent, dyeing assistant, scouring and wetting agent, emulsifier for textile applics.

Emkalane®. [Emkay] Leveling agents, protective colloids, scouring agent for textile dyeing and scouring.

Emkalar® Base. [Emkay] Surfactant blend; emulsifier, base for dye carriers.

Emkalite. [Emkay] Wetting agent, partial stripping agent for cationic dyes from acrylic fibers.

Emkalon®. [Emkay] Quaternary or alkylamide blends; softener, antistat, lubricant for textile applics.

Emkalube®. [Emkay] Lubricant for textile processing; needle lubricant.

Emkane. [Emkay] Sulfonic acids or salts;

detergent, dyeing assistant, scouring and wetting agent, emulsifier for textile industry.

Emkanet B. [Emkay] Stiff resin finish for veils and nets.

Emkanol®. [Emkay] Mercerizing assistants, penetrants, wetting agents.

Emkanyl®. [Emkay] Nonionic leveling agent for dyeing nylon hosiery.

Emkapel®. [Emkay] Durable and semidurable water repellents.

Emkapene®. [Emkay] Pine oil penetrants, emulsifiers, scum preventatives, leveling agent, dye assistant for vat dyeing, scouring, wetting.

Emkaperm. [Emkay] Permanent stiffening finish for textiles.

Emka Permadul. [Emkay] Dulling agents for nylon, rayon, acetate fabrics.

Emkapol®. [Emkay] Soap jellies; scouring agent for textiles.

Emkapon®. [Emkay] Amide sulfonates; detergent, dyeing assistant, scouring and wetting agent, emulsifier for textile industry.

Emkapruf. [Emkay] Flame retardant for textile finishes.

Emkaron. [Emkay] Bleach, stripping agent, dyeing assistant, dispersant, retardant, stabilizer for textile dyeing.

Emkasan. [Emkay] Quaternary ammonium compd.; algicide, germicide, disinfectant, deodorant for textile and industrial use.

Emka Scroop. [Emkay] Amino complex; scroop finish for silk and synthetic fibers.

Emkasene®. [Emkay] Chelating agents.

Emkaset. [Emkay] Modified formaldehyde resin; thermoset for enhancing stiffness, crease resistance, and shrink resistance of fabrics.

Emkasize. [Emkay] Sizing agent, binder for textiles.

Emkasol. [Emkay] Sulfonates; dispersants for dyestuffs.

Emkasorb. [Emkay] Increases absorbency in textile finishing.

Emkastat®. [Emkay] Quaternary deriv.; antistat for textiles.

Emkatan. [Emkay] Antistatic softener, lubricants for textiles.

Emkatard®. [Emkay] Dispersant for vat colors.

Emkaterge®. [Emkay] Amide sulfonates; detergents, wetting agent, base for compounding graphite remover.

Emkatex®. [Emkay] Detergent, lubricant, dyeing assistant, leveling agent, retarder for textile applics.

Emkatint. [Emkay] Optical bleach and brightener for natural and synthetic textiles.

Emkatol M. [Emkay] Graphite remover for nylon lace.

Emkawate®. [Emkay] Weighter finishes.

Emkazyme. [Emkay] Mixture of amylolytic enzymes and catalysts, solubilizing resins, starches; for textile industry.

Empee® PE. [Monmouth Plastics] Polyethylenes; used in wire and cable, inj. molded parts, extruded sheet, blow molded parts, shapes, tubing, elec. applics.

Empee® PP. [Monmouth Plastics] Polypropylenes; for inj. molded parts, extruded sheet, shapes, wire insulation, elec. applics.

Empee® PS. [Monmouth Plastics] Polystyrene; for inj. molded and extruded parts.

Emphos®. [Witco/Organics; Witco Israel; Witco SA] Phosphate esters; detergent, emulsifier, lubricant, solubilizer, wetting agent, antistat, emollient, coupling agent; for personal care prods., foods, industrial detergents, metal cleaning, cutting fluids, films, textiles, emulsion polymerization, agric.

Empicol®. [Albright & Wilson Am; Albright & Wilson UK; Albright & Wilson Australia] Sulfates, sulfosuccinates, or phosphate esters; anionic detergent, foaming agent, stabilizer, wetting agent, emulsifier, dispersant, conditioner for personal care prods., pharmaceuticals, textiles, household cleaners, rubber latexes, resins.

Empicryl®. [Albright & Wilson Am.; Albright & Wilson UK] Polymethacrylates; for mfg. of polymers; plasticizer, pigment dispersant; adhesives, coatings, agric. formulations; visc. index improver for hydraulic fluids.

Empigen®. [Albright & Wilson Am.; Albright & Wilson UK] Alkanolamides, amine oxides, quaternaries, amines, or betaines; foam booster/stabilizer, visc. modifier, antistat, conditioner, intermediate, bactericide for cosmetics, cleaners, latex foams, textiles; catalyst for PU foam, resin curing agent, corrosion inhibitor, and flotation aid.

Empilan®. [Albright & Wilson Am.; Albright & Wilson UK] Ethoxylated ethers or esters, alkanolamides, or glyceryl esters; emulsifier, lubricant, foamer, superfatting agent, wetting agent, dispersant, stabilizer for latexes, cosmetics, foods, cutting oils, textile lubricants, brake fluids, agric., plastics, rubber, paper, paints, pharmaceuticals, laundry.

Empimin®. [Albright & Wilson Am.; Albright & Wilson UK] Sulfates, sulfosuccinates, sulfosuccinamates, or blends; defoamer, detergent, foaming and wetting agent, emulsifier, air entraining agent, dispersant for industrial cleaners, concrete, emulsion polymerization, plasterboard, textiles, rubber latexes.

Empiphos. [Albright & Wilson Am.; Albright & Wilson UK] Phosphate compds.; detergent builder, dispersant, stabilizer, sequestrant, peptizer, deflocculant, clarifying agent for liq. detergents, paints, rubber, water treatment.

Empiwax. [Albright & Wilson Am.; Albright & Wilson UK] Cetearyl alcohol/sodium lauryl sulfate blend; self-emulsifying wax, emulsifier for pharmaceuticals and toiletries.

Emplex. [Am. Ingredients] Sodium stearoyl lactylate; starch and protein complexing agent for bakery goods; emulsifier, conditioner for processed foods.

Empol®. [Henkel/Emery] Dimer, trimer, or polybasic acids; used for surface coatings, synthetic lubricants, polymer modification, polyamides for inks and adhesives.

Emralon®. [Acheson Colloids] Colloidal PTFE blends; dry film lubricant.

Emrite®. [Henkel/Emery] Glyceryl or sorbitan esters or ethoxylates; emulsifier for food industry; antifog for food-pkg. films.

Emsorb®. [Henkel/Emery/Cospha] Sorbitan esters or ethoxylates; coupler, emulsifier, lubricant, softener, wetting agent, solubilizer for textiles, leather, cosmetics, pharmaceuticals, household goods, paper, emulsion polymerization, metal processing.

Emtal®. [Engelhard] Talc; filler, reinforcer for thermosets, thermoplastics, thermoplastic elastomers; in caulks and sealants, dry-wall compds., adhesives, printing inks, paints, metal primers; dusting agent for tacky rubber compds.

Emthox®. [Henkel/Emery/Cospha] Ethoxylated ethers or glycerides; emollient, humectant, emulsifier, dispersant, wetting agent, penetrant, leveling agent, solubilizer for cosmetics, pharmaceuticals, soaps.

Emulamid. [Mayco Oil & Chem.] Soluble oil conc.; emulsifier, lubricant.

Emulan. [Emulan] Mink oil; emollient for skin and hair care prods.

Emulan®. [BASF AG] Ethoxylated ethers and esters; emulsifier, stabilizer for oils, waxes, polishes, metalworking, emulsion polymerization; corrosion inhibitor.

Emulbesto. [Lucas Meyer] Lecithin; emulsifier for calf milk replacement and piglet starters.

Emulbon. [Toho Chem. Industry] Borate derivs.; lubricant, plasticizer, emulsifier, dispersant, detergent, solvent.

Emuldan. [Grindsted Prods.; Grindsted Prods. Denmark] Mono and diglycerides; emulsifier and stabilizer for foods, resins, thermoplastics.

Emulfluid. [Lucas Meyer] Lecithin; for prod. of foodstuffs.

Emulgade®. [Henkel/Cospha; Henkel Canda; Henkel KGaA; Pulcra SA] Fatty alcohol blends; emulsifier, base for creams and lotions; hair conditioner, antistat.

Emulgane. [Henkel; Henkel-Nopco] Ethoxylated ethers, esters, or oils; emulsifier, dispersant, dyeing assistant.

Emulgateur. [Witco SA] Formulated emulsifier; for mineral and vegetable oils, solvents;

Emulgator. [Goldschmidt; Goldschmidt AG] Fatty alcohol blends or ethoxylates; emulsifier, stabilizer for cosmetics, pharmaceuticals.

Emulgator. [Unger Fabrikker AS] Ethoxylated alkyl phenol; used in liq. detergents; emulsifier for industrial preparations.

Emulgeant. [Ceca SA] Ethoxylated ether; surfactant.

Emulgen. [Kao] PEG aryl ethers; dispersant for pigments and dyestuffs.

Emulgit. [Zohar Detergent Factory] Blended detergents; emulsifier for pesticide formulations.

Emulmetik. [Lucas Meyer] Modified soya phospholipids; resorbable refatting agent, solubilizer, coemulsifier for skin care prods.

Emulmin. [Sanyo Chem. Industries] Ethoxylated ethers; emulsifiers, dispersants, wetting agents, detergents, thickener for cosmetics, textile printing pastes.

Emulphogene®. [Rhone-Poulenc Surf.] Ethoxylated ethers; emulsifier, detergent, dispersant, wetting agent, foam builder/stabilizer, solubilizer for latexes, textiles, corrosion inhibitors.

Emulphopal HC. [Stepan Canada] Waterless hand cleaner base.

Emulphor®. [Rhone-Poulenc Surf.] Ethoxylated ethers or esters; emulsifier, dispersant, softener, wetting agent, lubricant, stabilizer, dyeing assistant, antistat, solubilizer for cosmetics, pharmaceuticals, leather, agric., paper, textiles, plastics, corrosion inhibitors, paints, adhesives.

Emulphor ON 870, 877. [Stepan Canada] Oleth-20; surfactant.

Emulpon. [Witco SA] Ethoxylated castor oil or fatty acids; emulsifier, textile wetting agent for dyeing.

Emulpur. [Lucas Meyer] Lecithin; emulsifier for food industry.

Emulsan. [Reilly-Whiteman] PEG esters or blends; emulsifier for textile, leather, ink removal, other industrial applics.

Emulsifiant. [Seppic] Ethoxylated, sulfated alkylphenol; emulsifier for emulsion polymerization.

Emulsifier 7X. [Atsaun] Formulated blend; emulsifier, wetting agent, antistat, lubricant.

Emulsifier 30. [Pilot] Branched alkylate sulfonates.

Emulsifier 99. [Pilot] Branched alkylate sulfonic acid.

Emulsifier 632/90%. [Ethox] Modified alkyl phenol ethoxylate; low-foam surfactant, dispersant.

Emulsifier DMR. [Hoechst AG] Fatty acid polyglycol ester; emulsifier, thickener for printing of reactive dyes.

Emulsifier Four. [Exxon/Tomah] Quaternary ammonium chloride; cationic emulsifier for waxes and oils, automotive spray waxes.

Emulsifier Q. [Consos] Solvent/emulsifiers blend; solvent scour for heavily oiled and waxed textiles.

Emulsifier WHC. [Stepan] Formulated emulsifier; for waterless hand cleaners.

Emulsilac. [Witco/Humko] Dough conditioner/strengthener.

Emulsion A Oil. [CasChem] Castor oil; component of automotive and furniture polishes and cleaners.

Emulsion C-340. [Exxon/Tomah] Carnauba wax emulsion; for protective coatings on fabric, metal, wood, leather, and painted surfaces.

Emulsi-Phos. [Monsanto] Cheese emulsifier.

Emulsit. [Dai-ichi Kogyo Seiyaku] Ethoxylated alkyl phenol ethers; emulsifiers.

Emulso. [Toho Chem. Industry] Surfactants complex; floating oil disposer for oil and gas industry.

Emulsogen. [Hoechst Celanese/Colorants & Surf.; Hoechst AG] Emulsifier, corrosion inhibitor, wetting agent for metalworking fluids and lubricants.

Emulsol. [Atsaun] Ethoxylated castor oil; leveling agent, dispersant, solubilizer.

Emulson. [Auschem SpA] Ethoxylated alkylphenol or oils, sulfonates, or blends; emulsifier, dispersant, wetting agent, solubilizer for pesticides, microemulsions.

Emulsowax. [Frank B. Ross] Wax used as extender for carnauba in self-polishing floor waxes.

Emulsynt. [Van Dyk] Esters; emulsifier, thickener, emollient for creams and lotions; solubilizer for fragrances.

Emultex. [Auschem SpA] Esters or ethers; plasticizer, defoamer, softener, lubricant, emulsifier for cosmetics, textiles, agric.

Emulthin. [Lucas Meyer] Lecithin; emulsifier, dispersant, wetting agent, release agent for food industry; flour improver.

Emulvin. [Miles] Emulsifier, stabilizer, wetting agent for latexes.

Emulvis®. [C.P. Hall] PEG-150 distearate; visc. builder in cosmetics, sol. retarder for gums and resins; antiblock agent for decals; thickener in pet shampoos; melt pt. control in suppositories.

Emulzome. [Exsymol] Hydrogenated polyisobutene, stearyl heptanoate, mineral oil blend; for cosmetics.

EMU® Powder 120 FD. [BASF AG] Styrene copolymer powder; binder for paints, paper coatings, modifying cement-containing dry mortar.

Emusolvent. [ICC Industries] Emulsified ortho dichlorobenzene.

Em-U-Taine. [Emulan] Minkamidopropyl betaine.

Emvelop®. [Mendell] Hydrogenated vegetable oil NF; binders, vehicles for pharmaceutical tabletting.

Enagicol. [Lion] Betaines; mild detergent, visc. builder for cosmetics.

Enathene. [Quantum/USI] Ethylene n-butyl acrylate copolymer; for extrusion coatings, adhesives.

Encapsulated MgO. [Olin] Magnesium oxide.

Encelac®. [BASF AG] Organic and inorganic pigment formulations in nitrocellulose and dibutyl phthalate; for pigmenting nitrocellulose lacquers and nitrocellulose blend lacquers.

Enceprint. [BASF] Aq. dispersions for inks.

Enceprint®. [BASF AG] Organic and inorganic pigment formulations in nitrocellulose with plasticizers; for pigmenting flexographic and gravure inks.

Endcor. [W.R. Grace/Dearborn] Corrosion inhibitor.

Endet. [Swastik] Heavy-duty detergent powd. containing proteolytic enzyme; household detergent.

Endimal. [Int'l. Dioxcide] Oxidizing deodorant.

Endimal-C. [Pettibone Labs] Malodor neutralizer conc.

Endomine NMF. [Serobiologiques] Hydrolyzed collagen, glycerin, sodium PCA, sodium lactate, urea blend.

Endonucleine LS 2143. [Serobiologiques] Glycerin, water, propylene glycol, hydrolyzed actin, DNA, glycogen, arginine, RNA blend.

Endura. [PPG Industries] Plastic additives.

Endurolene. [Cookson Pigments] Textile dyes and pigments.

Enduronone. [Cookson Pigments] Textile dyes and pigments.

Endurophthal. [Cookson Pigments] Textile dyes and pigments.

Enduroquin. [Cookson Pigments] Textile dyes and pigments.

Endwell Prods. [Adhesives & Chem.] Shoe adhesives.

Enerade®. [Rhone-Poulenc Surf.] Alkoxylated alkylphenol formaldehyde resins.

Energize. [Uncle Sam Chem.] Fuel oil treatment, combustion catalyst.

Energizer. [Mammoth Int'l.] Humic acid; conc. liqs. for general agric. use.

Engevita. [Gist-Brocades Food Ingreds.] Dried inactive yeast.

Enhance. [DuPont/Medical Prods.] Enhancer for autoradiography.

Enhance. [Western Nutrients] Liq. humic acids.

Enlightning. [DuPont/Medical Prods.] Enhancer for autoradiography.

Enquik. [Unocal/Nitrogen Group] Herbicide, desiccant.

ENR 25, 50. [Guthrie Latex] Natural

rubber.

ENSA-6. [Hart Chem. Ltd.] Ethoxylated naphthol sulfonic acid; brightening and dispersing agent for tin plating.

Ensidom. [Henkel-Nopco] Modified fatty acids and oils; self-emulsifying wool and worsted lubricant.

Ensital. [Henkel-Nopco] Alkoxylated ethers or esters; lubricant for wool and blends.

Ensoline. [Ceca SA] Antistatic oiling of fibers.

En-Stat®. [Miller-Stephenson] Non-CFC aerosol static eliminator.

Entensify. [DuPont/Medical Prods.] Enhancer for autoradiography.

Entrapped. [Florasynth] Flavors.

Envex®. [Rogers; Cestidur Industries] Thermoplastic polyimide resins, some lubricated; for structural or insulation applics., bearings, seals, bushings.

Envirogel. [Wyo-Ben] Wyoming bentonite.

Enviro-Guard. [Southern Coatings] Epoxy, urethane, or alkyd.

Enviromelt. [Petrokem] Ice and snow melting prods.

Enviroseal. [PCR] Masonry water repellent.

Enviro-Stab 1092. [Monson Chem.] Tribasic lead sulfate, encapsulated.

Enviro-Stab 1093. [Monson Chem.] Basic white lead carbonate, encapsulated.

Enviro-Stab 1155. [Monson Chem.] Dibasic lead phthalate, encapsulated.

Enviro-Stab 1160. [Monson Chem.] Lead sulfo-phthalate blend, encapsulated.

Envirostone Gypsum Cement. [U.S. Gypsum] Cement for low-level radioactive wastes.

Envron. [Halocarbon Prods.] Non-ozone harming fluorocarbons.

EnzaBac USDA. [Amax Industrial Prods.] Liquid enzymes.

Enzeco. [Enzyme Development] Enzymes.

Enzobake. [Enzyme Development] Enzymes for baking.

Enzopharm. [Enzyme Development] Enzymes for pharmaceuticals.

Enzyami. [Alban Muller] Phyto-enzymatic complex; anti-free radical protector, cell stimulant, solar protector for cosmetic skin prods.

EP 2220 TC. [Hardman] Two-part epoxy compd.; for elec./electronic applics.

EP320002. [Epolin] uv/eb curing liq. adhesive for screen printing, roll coating, film coating applics.

EP4802-75. [Lord] Two-component adhesive for bonding thermoplastic polyolefin elastomers to textiles, leather, cured rubber.

Epal®. [Ethyl] Linear primary alcohols or blends; detergent, emulsifier intermediate.

Epalloy. [CVC Spec. Chem.] Bisphenol F and phenol epoxy novolac resins.

Epamarine. [Arista Industries] Marine lipid concs. containing omega-3 fatty acids.

Epan. [Dai-ichi Kogyo Seiyaku] PPG PEG ether; emulsifier, dispersant, lubricant, defoamer, detergent for detergents, emulsion polymerization, latexes.

EPC. [Morton Int'l.] Epoxy powd. coatings.

Epcar EPM. [BFGoodrich/Spec. Polymers] EPM rubber; for molded and extruded prods., wire insulation and jacketing, molded o-rings, brake components, sidewalls, hose, matting, automotive weatherstripping.

Epibond®. [Ciba-Geigy Plastics UK] Paste adhesive for aircraft interior applics.

Epiclon. [DIC Trading] Epoxy resins.

Epi-Cure®. [Rhone-Poulenc/Perf. Resins & Coatings] Epoxy curing agents.

Epidermin. [Henkel/Cospha] Polyvalent tissue complexes; prods. for aging and sensitive skin.

Epikote. [Chemach NV] Basic resins for paint, glue constructions.

Epikuron. [Lucas Meyer] Deoiled lecithin; dietetic supplement, choline enrichment.

Epilink®. [Akzo; Akzo Chem. BV] Polyaminoamide prods.; epoxy curing agent.

Epiphen. [Monomer-Polymer & Dajac Labs] Epoxy resins.

Epi-Rez®. [Rhone-Poulenc/Perf. Resins & Coatings] Epoxy resins or sol'ns., some halogenated; for industrial maintenance coatings, film adhesives, fiberglass-reinforced plastics, encapsulation, castings, filament winding, potting compds., chem. resist. pipe, elec. dip coatings, varnishes, solv.-free formulations, fiber finishes, textiles.

EPIstatic®. [Eagle-Picher] Tribasic lead sulfate or basic lead silicosulfate and blends; heat stabilizer for flexible and rigid PVC compds.

Epi-Tex®. [Rhone-Poulenc/Perf. Resins & Coatings] Epoxy ester resin sol'ns.; for performance primers, force-dried or low-bake applics., industrial maintenance coatings.

EPIthal. [Eagle-Picher] Basic lead phthalate; heat stabilizer for PVC wire and cable compds.

EPK. [Feldspar] Water-washed, air-floated kaolin.

Epocap®. [Hardman] Epoxy resins; for electrical applics.

Epocast®. [Ciba-Geigy Plastics UK] Adhesive for lightweight honeycomb structure reinforcement compds.

Epocel. [Pall Corp.] Epoxy resin impregnated, cellulose paper filter medium.

Epocure. [Hardman] Epoxy curing agents.

Epodil®. [Air Prods.; Pacific Anchor] Glycidyl ethers; hydrocarbon diluent for epoxy systems fortooling, elec. applics., coatings, potting, flooring.

Epolene®. [Eastman] Polyolefin resins and oxidized polyolefins; synthetic waxes used in hot-melt adhesives, coatings, floor polishes, cosmetics, inks, as paraffin modifiers, in slush and cast molding, rubber compding., vinyl compds., fruit coatings, textile lubricant/softener.

Epoleon®. [Epoleon Corp. of Am.] Odor neutralizing agents for scrubber systems, wastewaster and sludge treatment, landfill applics.

Epolite. [Hexcel] Two-part epoxy resin; surface coats, casting compds., laminating resins, potting and encapsulation, filament winding, and adhesive applications.

Epoloid. [Rowe Prods. Distribution] Epoxy base coatings.

Epo-Lux. [Steelcote Mfg.] Polyamide epoxy coatings.

Epomins®. [Aceto] Polyethylene imines; resin raw materials, crosslinkers, adhesion promoters; for paper industry.

Epon®. [Shell] Epoxy resin; protective coatings.

Epon® Curing Agent. [Shell] Epoxy curing agents.

Eponex. [Shell] Weatherable epoxy resin.

Epon® HPT®. [Shell] Epoxy matrix resin; used for advanced composites with glass, carbon, aramid or boron fibers, high-performance structural laminates and adhesives, prepregs.

Eponol. [Shell] Synthetic resin.

Eposet. [Hardman] Epoxies.

Eposir. [Sir Industriale-Gruppo Montedison] Epoxy resins.

Eposolve®. [Hardman] Clean-up solvent for cured and uncured epoxies.

Epotal® 181 D. [BASF AG] Ethylene polymer dispersion; additive to other dispersions; improves slip, blocking resistance, and hydrophobic properties of film in paper finishing, adhesives.

Epotherm. [Transene] Thermally conductive epoxy compds.

Epotuf Hardener. [Reichhold] Polyamine, polyamide, or anhydride; epoxy hardener.

Epotuf Resin. [Reichhold; Reichhold Chemie AG] Epoxy, novolac, or brominated epoxy resins; for elec. potting, encapsulating, casting, filament winding, laminating, tooling, flooring, surfacing, coatings, adhesives.

Epoweld®. [Hardman] Epoxy adhesive; for skin industry.

Epoxical. [U.S. Gypsum] Resin compds.

Epoxi-Patch®. [Dexter/Hysol] Adhesives.

Epoxol. [Am. Chem. Services] Epoxidized oils; plasticizer, acid scavenger, stabilizer for PVC compds., food pkg. materials.

Epoxybond. [Atlas Minerals & Chem.] Epoxy adhesive.

Epoxylab. [Tra-Con] Adhesive kits.

Epoxy Modifier ML. [Pacific Anchor] Diluent, plasticizer, wetting agent for epoxy systems; used for coatings, decorative flooring, exterior patching.

Epoxyprene. [Guthrie Latex] Epoxidized natural rubber.

Epsilon. [Katalistiks Int'l.] Fluid cracking catalysts.

EPsyn®. [Copolymer Rubber] EPDM terpolymer; thermoplastic rubber for wire and cable, sponge, extrudates, molded goods, hydraulic seals, hose, rubber additive.

Equex. [Procter & Gamble] Sulfates or blends; detergent, foamer, emulsifier for cosmetics, polymerization, cleaners, rug shampoos.

Equisperidin. [Burlington Bio-Medical] Horse bleeder preventative.

Equizyme. [Amano Int'l. Enzymes] Animal feed supplement for horses.

Equol. [Mackenzie Chem Works] n-Methylaminophenol sulfate; photographic developer.

Eraclear. [Enimont UK Ltd.] LLDPE.

Eraclene. [Enimont UK Ltd.] HDPE.

Eramide®. [Syn. Prods.] Erucamide; slip agents for plastic films.

Ereak PS-909. [Yoshimura Oil Chem.] Alkylamide propyl dimethyl beta-hydroxyethyl ammonium nitrate; antistatic additive to polymers, papers.

Eribate. [PMP Fermentation Prods.] Food antioxidant.

Eriochrome. [Ciba-Geigy/Dyestuffs] Textile dyes and pigments.

Erional®. [Ciba-Geigy/Dyestuffs] Aromatic sulfonic acid condensates; resist agent for nylon; fixative for acid dyes on nylon.

Erionyl. [Ciba-Geigy/Dyestuffs] Textile dyes and pigments.

Erisys. [CVC Spec. Chem.] Epoxy resins and accelerators.

Erisys 33DDS. [CVC Spec. Chem.] 3,3′-Diaminodiphenyl sulfone.

Erisys U-405. [CVC Spec. Chem.] Phenyl dimethyl urea.

Erisys U-410. [CVC Spec. Chem.] Toluene bis (dimethyl urea).

Erkantol®. [Miles/Organic Prods.] Wetting agent for cold-pad-batch dyeing.

Ernaltex. [Alframine] Textile finishing agents.

Erta. [Erta NV] High performance materials.

Erta PC. [Erta NV] Polycarbonate; engineering plastic.

Erta PEEK. [Erta NV] Polyetheretherketone.

Erta PEI. [Erta NV] Polyetherimide.

Erta PSU. [Erta NV] Polysulfone.

Ertacetal. [Erta NV] Polyacetal.

Ertalene. [Cestidur Industries] High m.w. or ultra high m.w. polyethylene.

Ertalon. [Erta NV] Nylon 6, 6/6, 11, 4/6.

Ertalyte. [Erta NV] PET.

Ervol. [Witco/Sonneborn] White mineral oil NF; emulsifier, lubricant, emollient for cosmetics, food contact applics.

ES-. [Compounding Tech.] Polyethersulfone, glass, carbon or Teflon filled.

ES-1239. [CasChem] Dimethylaminopropyl ricinoleamide benzyl chloride; emulsifier with antistatic, softening, bactericidal, wetting, dispersing properties for cosmetic, textile, agric. industries.

Esbiol. [McLaughlin Gormley King] Synthetic pyrethroid.

Escacure®. [Sartomer] Benzyl ketal or benzoin ethers; photoinitiators for composites, coatings, inks, photopolymers, electronic photoresists.

Escalol®. [Van Dyk] Sunscreens.

Escoat P-20. [Polyad] Polyvinyl octadecyl carbamate compds.; release coats, water repellents.

Escoher. [Exxon Chem. Mediterranea SpA] Synthetic waxes.

Escomer. [Deutsche Exxon Chem. GmbH] Ethylene copolymer.

Escopol®. [Exxon] Reactive polymer.

Escorene®. [Exxon; Deutsche Exxon Chem. GmbH] Polypropylene, polyethylene, EVA, or E/MA polymers; for tire industry.

Escorez®. [Exxon; Deutsche Exxon Chem. GmbH] Hydrocarbon resins; tackifying resins for tire industry.

Escort. [DuPont/Ag] Herbicide.

ESD. [RTP] Polycarbonate, glass-reinforced; flame retardant materials for static dissipative applics.

Esi-Cryl. [Emulsion Systems] Polyethylene or acrylic emulsions; for adhesives, floor finishes, general coatings, inks, lubricants, textiles, metal coatings, rust preventatives.

Esi-Det. [Emulsion Systems] Alkanolamides or phosphate esters; detergent, visc. builder for multipurpose cleaners, strippers, degreasers, hard surface cleaners, personal care and household cleaners.

Esi-Flex. [Emulsion Systems] Chemicals used in mfg. specialty coatings.

Esi-Graph. [Emulsion Systems] Acrylic emulsions or styrenated acrylics; water-based vehicles and additives for the graphic arts industry.

Esi-Terge. [Emulsion Systems] Foam builder/stabilizer, thickener, detergent, wetting agent, solubilizer for household, cosmetic, and industrial detergents.

Esokol. [Reichhold] Linseed oil.

E-Solder. [Acme Div.] Conductive adhesives.

Esperal®. [Witco/Argus] Peroxide derivs.; polymerization and crosslinking agent.

Esperase®. [Novo Nordisk] Protease; enzyme, defoamer for built liq. detergents, laundry detergents.

Espercarb®. [Witco/Argus] Peroxide derivs.; initiator, crosslinking agent for polymers.

Esperfoam®. [Witco/Argus] Peroxide derivs.; initiator for polymerization of ethylene, styrene, acrylates, curing of unsat. polyester resins; crosslinking agent for chlorinated polyethylene.

Esperox. [Witco/Argus] Organic peroxides; catalyst.

Espesilor. [Pulcra SA] Fatty alkanolamides; thickener, superfatting agent, solubilizer, emulsifier, foam stabilizer for cosmetics, detergents, lubricating oils.

Esskol. [Reichhold] Linseed oil.

Estabex®. [Akzo; Akzo Chem. BV] Epoxidized soybean oils or monooctyl ester; plasticizer, stabilizer, fungicides for PVC, plastisol, organosol, surface coatings, inks, extruded prods., chlorinated paraffins, halogenated rubbers, ethyl cellulose; processing aid; corrosion inhibitor.

Estaflex® ATC. [Akzo] Acetyl tributyl citrate; primary plasticizer for plastics incl. PP, PS, PVAc, PVB, PVC, vinyl acetate copolymers, chlorinated rubber, ethyl cellulose, nitrocellulose; suitable for coatings in contact with foodstuffs.

Estaloc. [BFGoodrich; BFGoodrich UK] Mixture of thermoplastic polyurethane, reinforcement, polymeric alloys, additives; reinforced engineering thermoplastic.

Estane®. [BFGoodrich; BFGoodrich UK] Thermoplastic polyurethane compds.; used in binders, coated fabrics, outerwear, leather finishes, adhesive blending resins, paper and foil coatings, inj. molding, automotive parts, appliance gears, extrusion of film, sheet, tubing, and shapes.

Estaram. [Unichema] Suppository bases for pharmaceuticals.

Estasan. [Unichema] Medium chain triglycerides; for personal care prods., pharmaceuticals, food additives, flavors and fragrances.

Estasol®. [Aceto] Dibasic ester; chemical for the electronics, textile, paper, and foundry industries; solvent.

Estersulf. [Climax Performance] Sulfonamide anionic ester.

Estol. [Unichema; Unichema France SA] Fatty acid esters; emulsifier, natural oil substitutes; for use in plastics, lubricants, cosmetics.

Estradur. [Estra-Kunststoff GmbH] ABS.

Estralen. [Estra-Kunststoff GmbH] Polyethylene.

Estramid. [Estra-Kunststoff GmbH] Polyamide.

Estran. [Estra-Kunststoff GmbH] SAN.

Estranat. [Estra-Kunststoff GmbH] Polycarbonate.

Estraprop. [Estra-Kunststoff GmbH] Polypropylene.

Estrasan. [Reilly-Whiteman] Methyl esters; enhances lubricity and wettability of petroleum oils for metals and other industries.

Estratil. [Resinas y Complementos Castro] Polyester resins.

Estren. [Estra-Kunststoff GmbH] Polystyrene.

Estrobond. [Eastman] Plasticizers.

Estron. [Eastman] Acetate yarn or tow.

Etadurin. [Akzo] Wet strength resin for paper industry.

Ethacure®. [Ethyl] Curing agents for epoxy and polyurethane resins; polymer modifiers.

Ethafoam. [Dow] Plastic foams.

Etha-Keratin ISO. [Brooks Industries] AMP-isostearoyl hydrolyzed keratin; cosmetics ingredient.

Ethal. [Ethox] Ethoxylated alcohols; wetting agent, intermediate, detergent, emulsifier, coupler.

Ethalin. [Gotham Ink & Color] Flexographic ink.

Ethanox®. [Ethyl] Antioxidant and stabilizer for plastic, resin, rubber, adhesives, petroleum oil, and wax.

Etha-Soy ISO. [Brooks Industries] AMP-isostearoyl hydrolyzed soy protein; for skin and hair care cosmetics.

Ethazate. [Uniroyal] Rubber accelerators; for athletic padding and insulation, latex, foam, insulated wire.

Etherdiamine 230, 400, 2000. [BASF AG] Curing agents for epoxy resins.

Ethfac. [Ethox] Phosphate esters; surfactants, detergents, emulsifiers, antistats, wetting agents.

Ethion. [FMC/Ag] Insecticide, miticide.

Ethocel. [Dow] Ethocellulose resins.

Ethoduomeen®. [Akzo; Akzo Chem. BV] Ethoxylated aliphatic diamines; emulsifier, wetting agent, dispersant, corrosion inhibitor for bitumen, waxes, paper coatings, oil recovery.

Ethoduoquad®. [Akzo] Ethoxylated diquaternary ammonium salts; antistat, emulsifier, dyeing assistant for textiles, electroplating.

Ethofat®. [Akzo] Ethoxylated fatty acids; emulsifier, detergent, wetting agent, dispersant, suspending agent, for textile, cosmetic, agric., metal and leather treating.

Ethomeen®. [Akzo; Akzo Chem. BV] Ethoxylated aliphatic amines; emulsifier, dispersant for textile processing.

Ethomid®. [Akzo; Akzo Chem. BV] Ethoxylated fatty amides; emulsifier, dispersant, detergent, dye leveling agent, lubricant for textiles, silicone finishing agents, sizing.

Ethonic. [John Campbell] Textile dyes and pigments.

Ethoquad®. [Akzo; Akzo Chem. BV] Ethoxylated quaternary ammonium chlorides; antistat, emulsifier, dyeing assistant, leveling agent, antifoam for textiles, electroplating.

Ethosperse®. [Lonza] Ethoxylated alcohols; emulsifier, humectant, spreading agent, emollient for cosmetic, pharmaceutical, and industrial uses.

Ethotal. [Witco SA] Ethoxylated fatty acid; detergent for machine washing, wetting agent for insecticides, fungicides.

Ethox. [Ethox] Ethoxylated amines, triglycerides, esters or ethers; emulsifier, dispersant, dyeing assistant, lubricant, detergent, softener. wetting agent for textiles, industrial lubes.

Ethoxamine. [Witco SA] Ethoxylated fatty amine; corrosion inhibitor, emulsifier, dispersant, antistat.

Ethoxol. [Lanaetex Prods.] Ethoxylated oleyl alcohols.

Ethoxygel. [Henkel/Organic Prods.] Ethoxlyated lanolin.

Ethoxylan®. [Henkel/Emery/Cospha] Ethoxlyated lanolin; emollient, emulsifier, dispersant, foam stabilizer, resin plasticizer for cosmetics and pharmaceuticals.

Ethoxyol®. [Henkel/Cospha] Lanolin derivs. or blends; emollient, aux. emulsifier, solubilizer, pigment wetting agent for cosmetics, liq. soaps.

Ethrel. [Rhone-Poulenc/Ag] Plant growth regulator.

Ethsorbox. [Ethox] Ethoxylated sorbitan esters; emulsifier, lubricant, softener for textiles, industrial process fluids.

Ethychem. [Union Carbide] Nonflamm. liquefied gas.

Ethyl®. [Ethyl] Alpha olefins; intermediates for biodegradable surfactants and specialty industrial chemicals.

Ethylan. [Amerchol] Lanolin ester.

Ethylan®. [Harcros UK] Alkanolamides or ethoxylated ethers, esters, or amines; detergent, wetting agent, stabilizer, emulsifier, solubilizer, antifoam, dispersant, antistat, foam booster/stabilizer, lubricant for industrial cleaners, plastics, pesticides, perfumes, cosmetics, pharmaceuticals, textiles.

Ethyl Benzene Sulfonic Acid. [Boliden Intertrade] Alkylaryl sulfonates; organic intermediate, catalyst, hydrotrope.

Ethyl Cadmate®. [R.T. Vanderbilt] Cadmium diethyldithiocarbamate; accelerator for rubbers.

Ethyl Diglyme. [Ferro/Grant] Diethylene glycol diethyl ether; solvent for electrochemistry, polymer and boron chemistry, physical processing, fuels, lubricants, textiles, pharmaceuticals, pesticides.

Ethylene Bodecandioate. [Henkel/Emery/Cospha] Musk aroma chemical.

Ethylene Brassylate. [Henkel/Emery/Cospha] Musk aroma chemical.

Ethylene Glycol Distearate VA. [Goldschmidt] Glycol distearate; opacifier and pearling agent for hair care prods.

Ethylene Glycol Monostearate VA. [Goldschmidt] Glycol stearate; opacifier, pearling agent, stabilizer, and thickener for creams and lotions.

Ethyl Ether. [Quantum/USI] Ethyl ether.

Ethylex. [A.E. Staley Mfg.] Industrial corn starch.

Ethylflo. [Ethyl] Polyalphaolefins; synthetic lubricants for automotive crankcase oils, hydraulic fluids, gear and transmission fluids, compressor lubricants, metalworking fluids, personal care items.

Ethyl Foaming Agent. [Ethyl] Alcohol ether sulfate; foaming agent for wallboard mfg., air drilling, etc.

Ethyl Glyme. [Ferro/Grant] Ethylene glycol diethyl ether; solvent for electrochemistry, polymer and boron chemistry, physical processing, fuels, lubricants, textiles, pharmaceuticals, pesticides.

Ethyl Namate®. [R.T. Vanderbilt] Sodium diethyldithocarbamate;

Ethyl Selenac®. [R.T. Vanderbilt] Selenium diethyldithiocarbamate; ultra accelerator for rubbers; vulcanizing agent.

Ethyl Tellurac®. [R.T. Vanderbilt] Tellurium diethyldithiocarbamate; ultra accelerator for rubbers.

Ethylthiurad. [Monsanto] Vulcanization accelerator.

Ethyl Tuads®. [R.T. Vanderbilt] Tetraethylthiuram disulfide; ultra accelerator for natural and synthetic rubbers; vulcanizing agent.

Ethyl Tuex. [Uniroyal] Accelerator.

Ethylux. [Westlake Plastics] Low density polyethylene.

Ethyl Zimate®. [R.T. Vanderbilt] Zinc diethyldithiocarbamate. ultra accelerator for natural and synthetic rubbers.

Etilenox. [Pulcra SA] Fatty amine deriv.; antistat for synthetic fibers.

Etingal®. [BASF AG] Antifoam for papermaking; dye carrier for leather and fur industry.

Etinox. [Aiscondel SA] PVC emulsions.

Etocas. [Croda Chem. Ltd.] Ethoxylated castor oils; solubilizer, emulsifier, lubricant, softener, leveling agent, emollient, superfatting agent, antistat, softener, detergent, humectant; used in personal care prods., fiber processing, metalworking fluids, emulsion polymerization, insecticides.

Etophen. [Zschimmer & Schwarz] Ethoxylated nonyl phenol ethers; detergents, dispersants, emulsifiers, wetting agents for household and industrial cleaners; intermediates for textile, paper, leather, and ceramic industries.

Etoxi. [Pulcra SA] Thickener, solubilizer, emollient, softener, antistat for detergent systems.

Etrofolan®. [Bayer] 2-Isopropyl-phenyl-N-methylcarbamate; insecticide effective against leafhoppers and bugs.

Eucarigid. [Van Raaijen Kunststoffen] PVC.

Eucarigid HT. [Van Raaijen Kunststoffen] Chlorinated PVC.

Eucarol. [Auschem SpA] Citrates or tartrates; nonirritating detergent, emulsifier for cosmetics.

Eucatherm. [Van Raaijen Kunststoffen] Polypropylene.

Eukesol®. [BASF AG] Binder, thickener, plasticizer for leather and fur industry, pigment finishes.

Eukesolar®. [BASF AG] Metal complex dyes; for furs, leather, cleaners, polishes, other pigment finishes.

Eulysin®. [BASF; BASF AG] Organic acids and pH regulators for textile finishing.

Eumulgin®. [Henkel/Emery/Cospha; Henkel KGaA; Pulcra SA] Ethoxylated ethers, esters, or amides; detergent, wetting agent, dispersant, emulsifier, solubilizer for paints, dishwashing, pesticides, cosmetics, floor polishes, perfumes.

Euparen®. [Bayer] Fluanid compds.; fungicide with specific action against Botrytis.

Euperlan®. [Henkel/Emery; Henkel/Cospha; Henkel Canada; Henkel KGaA; Pulcra SA] Surfactant blends; pearlescent base for shampoos, bath prods.

Eupolen®. [BASF AG] Organic and inorganic pigments with low m.w. polyethylene; for mass dyeing of polyolefins and other polymers.

Euredur. [Resinas y Complementos Castro] Hardener.

Eureka. [Atlas Refinery] Sulfated oils; emulsifier, detergent, grinding aid, carrier, plasticizer, lubricant, softener for cosmetics, textiles, coatings, pigment dispersions, fatliquors.

Eurepox. [Schering AG] Epoxy resins.

Euretek®. [Schering Berlin Polymers] Adhesion promoter for PVC plastisol.

Eurogran. [Atochem UK] PVC compds.

European Elastin. [Maybrook] Hydrolyzed elastin; substantivity agent, protective barrier, film former for skin and hair care prods.

Europrene. [EniChem Elastomeri Srl] Hydrocarbon rubbers and latexes.

Eusapon®. [BASF AG] Wetting agents, tanning assistants for leather, washing furs.

Eutanol®. [Henkel/Emery/Cospha; Henkel KGaA] Fatty alcohols; lubricant, emollient for cosmetics and pharmaceuticals.

Euthylen®. [BASF AG] Organic and inorganic pigments in polyethylene; for coloring polyolefins.

Euthylene. [BASF] Aq. dispersions for inks.

Euvinyl®. [BASF; BASF AG] Organic and inorganic pigments in VC copolymer; for mass coloring of rigid PVC, plasticized PVC, PVC pastes, VC copolymers, inks.

Euviprint®. [BASF; BASF AG] Organic and inorganic pigments in vinyl chloride copolymer; for coloring gravure inks, surface coatings for plasticized PVC and aluminum foil, other pkg. materials.

Euxyl K-400. [Calgon] Cosmetic preservative.

Evafanol®. [Nicca USA] Nonionic ether type urethane polymer; softening/antistatic agent for nylon hosiery.

Eval®. [Eval Co. of Am.] Ethylene vinyl alcohol copolymer; thermoplastic barrier resin for pkg. materials, films, bottles, sheet, tubes, coatings.

Evanacid® 3CS. [Evans Chemetics/W.R. Grace] Carboxymethyl mercaptosuccinic acid; metal chelate; metal deactivator.

Evanacure. [Evans Chemetics/W.R. Grace] Crosslinking agent for polymers.

Evangard®. [Evans Chemetics/W.R. Grace] Propionate derivs.; antioxidants, stabilizers, uv light absorbers to prevent and retard degradation.

Evanol. [Evans Chemetics/W.R. Grace] Cosmetic cream base.

Evanstab®. [Evans Chemetics/W.R. Grace] Thiodipropionate derivs.; antioxidant, stabilizers, uv light absorbers for plastics.

Evatane®. [Atochem N. Am.; Atochem UK] EVA copolymer; engineering polymers.

Evazote. [BXL Plastics Ltd.] Foam crosslinked ethylene vinyl acetate.

Event. [Am. Cyanamid/Ag] Plant growth regulator.

Everacid. [Everlight Chem. Industrial]

Textile dyes and pigments.

Everban. [McLaughlin Gormley King] Insecticide conc.

Evercide. [McLaughlin Gormley King] Insecticide conc.

Evercion. [Everlight Chem. Industrial] Textile dyes and pigments.

Evercron. [Everlight Chem. Industrial] Textile dyes and pigments.

Everdirect. [Everlight Chem. Industrial] Textile dyes and pigments.

Everflex®. [W.R. Grace/Organics] Vinyl, vinyl acetate, or vinyl acrylic emulsions; pigment binder; used in paints, paper coatings, textile treatments, caulk, mastics, roof coatings.

Evergreen. [McLaughlin Gormley King] Insecticide conc.

Evergrind. [Stewart Hall] Synthetic lubricant, coolant for metal grinding.

Everkleen. [Stewart Hall] Alkaline cleaners.

Everkut. [Stewart Hall] Synthetic lubricant and coolant for metal working.

Everlan. [Everlight Chem. Industrial] Textile dyes and pigments.

Everlube. [E/M Corp.] Solid lubricants.

Evershield. [E/M Corp.] EMI/RFI shielding coating.

Eversoft. [Specialty Chem.] Amphoteric polyethylene copolymer; durable softener/sewing and cutting lubricant for yarns, knits and piece goods.

Eversolv. [Stewart Hall] Solvent cleaners.

Everzol. [Everlight Chem. Industrial] Textile dyes and pigments.

Evolite. [Richard Daleman Ltd.] Clear and colored acrylic.

Evoprene®. [Evode Plastics Ltd.] Thermoplastic elastomer; used for molding, polymer modification, extrusion.

E Wachs BASF. [BASF AG] Montan ester wax.

EX. [Cabot Plastics Ltd.] Modified LDPE; semicondutive compd. for insulation shielding of power cables.

Exact-S. [Gaylord Chem.] Methyl sulfide; for catalyst pretreatment.

Exalin. [Gotham Ink & Color] Flexographic ink.

Exameen. [Clough] Quaternary ammonium chlorides; germicides for hard surface disinfection and sanitization; algicide for swimming pool and industrial water treatment.

Examide®. [Soluol] Dispersant, detergent, emulsifier, wetting agent, scouring agent, leveling agent, dye assistant for textiles.

EXC-33. [Releasomers] Fluorochemical copolymer; release agent for rubber, plastic, and metal industries.

Excel. [Kao/Edible Fat & Oil] Mono and/or diglyceride blends; food emulsifier; defoamer for beverages.

Excelcide. [Huge Co.] Insecticides, rodenticides, bird control prods., herbicides.

Excelerator. [E&F King] Concrete curing agent.

Excelopax. [TAM Ceramics] Zircon opacifier.

Exceparl. [Kao/Edible Fat & Oil] Esters or glycerides; emollient for cosmetics.

Exfoam. [W.R. Grace/Dearborn] Process defoamer.

Ex-It R584. [Amax Industrial Prods.] Bromacil-based; emulsifiable conc.

Exitir. [Montedipe Srl] Expandable polystyrene.

Exkin. [Hüls Am.] Antiskinning agents.

Exolit®. [Hoechst Celanese/Spec. Chem.] Ammonium polyphosphate; flame retardant for polyurethanes, epoxy, PE, PP, EVA, adhesives, paints, elastomers.

Exolite®. [Cyro Industries] Acrylic sheet.

Exolon. [Exolon-Esk] Aluminum oxide.

Expancel®. [Expancel] Expandable thermoplastic hollow microspheres; used in printing inks, paper and board, nowoven materials, body fillers, paints, explosives

Expandex®. [Uniroyal] 5-Phenyltetrazole or salts; chemical blowing agent for foaming plastics and elastomers.

Experiment Emulsion E-1614. [Rohm & Haas] Acrylic latex; vehicle for sealer finishes for flooring materials.

Explotab®. [Mendell] Sodium starch glycolate NF/BP; disintegrant for pharmaceutical tablets.

Express. [Tetra Chem.] Anhydrous calcium chloride; for oilfield applics.

Ex Rust. [Kano Labs] Rust remover.

Exsize®. [PMP Fermentation Prods.] Alpha-amylase and protease blend; enzyme for use as desizing agent in textile industry; solubilizer for starch.

Exsyfibroblastes. [Exsymol] Aids cutaneous cell regeneration and renewal of dermic connective tissue.

Exsyproteines. [Exsymol] Hydrolyzed animal elastin; used for moisturizers, oily skin cosmetic treatments, anti-aging creams.

Extan. [Lanaetex Prods.] Glyceryl or sorbitan esters.

Extendopel. [Sybron] Fluorocarbon extender.

Extendospheres. [PQ Corp.] Hollow microspheres; extenders for resin matrices, coatings, plastic parts, concrete, spray-up applics., foundry and refractory materials, adhesives, explosives, thermosets.

Extendospheres® Metalite®. [PQ Corp.] Metal-coated hollow microspheres; fillers offering functions of metal pigments and fillers with reduction in cost and weight.

Extractase. [Finnsugar Bioprods.] Pectinase; broad activity enzyme for fruit processing.

Extrakt 52. [Zschimmer & Schwarz] Surfactant blend; bath additives, hair shampoos, liq. body cleaners.

Extra Super Terra Alba. [Charles B. Chrystal] Terra alba.

Extrazine II. [DuPont/Ag] Herbicide.

Extrin. [Crompton & Knowles] Bakery flavors.

Extrudoil. [Witco] Extrusion compds., cold forming lubricants.

Extrusil. [Degussa] Calcium silicate; extender for rubber compds.

Exxal®. [Exxon; Deutsche Exxon Chem. GmbH] Higher alcohols.

Exxate®. [Exxon] Oxo-acetates.

Exxelor. [Exxon; Deutsche Exxon Chem. GmbH] Elastomer conc.; polymer modifier used with all polyolefins for specialty films/laminates, fabric coatings, medical films, disposable gloves.

Exxon Butyl. [Exxon] Isobutylene-isoprene elastomer for flat belts, coated materials, rubber-covered rolls, hose, mats, molded/extruded prods., o-rings, shock and vibration prods., inner tubes, water barrier applics., elec. goods.

Exxpar. [Exxon] Linear paraffins.

Exxsol®. [Exxon] Dearomatized aliphatic solvents.

Exzyme®. [PMP Fermentation Prods.] Protein digester.

Eymyd® Prepreg. [Ethyl] Fluorine-containing thermoplastic polyimides prepregged onto fiber or fabric; structural parts for aircraft, engines, and rockets for high temp. use.

Eymyd® Resin. [Ethyl] Polyamid acid sol'n.; forms a thermoplastic fluorinated polyimide after solv. evaporation and curing; for erosion barrier coating, thermal protective coating, tool coating, adhesive, fiber sizes.

Eypel® A Acoustic Barrier Sheet. [Ethyl] Poly(aryloxyphosphazene) elastomer; fire-resistant acoustic mass sheeting.

Eypel® A Foam. [Ethyl] Poly(aryloxyphosphazene) foam; closed-cell thermal insulation foam.

Eypel® F. [Ethyl] Polyfluoroalkoxyphosphazene compd.; fluoroelastomer for o-ring, other sealing applics.

EZA®. [Ethyl] Zeolite A; solvent, anticaking agent for detergents and desiccants.

E-Z Break. [La-Co Industries; Markal] High temp. antiseize and lubricating compds.

E-Z Cool. [Witco] Water-soluble oils for metalworking.

E-Z Melt. [Schaefer Salt & Chem.] Ice melter.

EZ Mold Lubricant. [TSE] Glycol surfactant; mold release lubricant for natural and synthetic rubber compds.

E-Z Peel. [Amax Industrial Prods.] Strippable coating.

E-Z Spread. [Rockwell Lime] Air-entraining masonry-stucco lime.

E-Z Strip. [Steelcote Mfg.] Paint remover.

E-Z-Duz-It. [Dixo] Spot/stain remover.

F

F-1, -4. [Unimin Specialty Minerals] Feldspar.

F-20. [Feldspar] Feldspar.

F 50. [GE Silicones] Phenyl silicone; for hydraulic fluid and lubrication in mechanical/ electrical applics.

F-109. [Fenocast SA] Phenolic resins.

F113. [Atochem N. Am.] Forane solvents.

F-500®, 1000, 1500, 2000, 2100, 2200, 4400. [Reheis] Aluminum hydroxide gels and powds.

F-701. [Zymet] Two-part epoxy; heat curing epoxy adhesive for bonding fiber optic components.

F-6000M. [Reheis] High density aluminum hydroxide/magnesium hydroxide blend.

FA. [Morton Int'l.] Polysulfide gum rubber.

FA. [QO Chem.] Furfuryl alcohol.

FA-1, 8, 13, 14, 21, 48. [Bacon] Epoxy resin; adhesive.

Fabex. [Vinings Industries] Couch roll corrosion inhibitor.

Fabracet. [Fabricolor] Textile dyes and pigments.

Fabracid. [Fabricolor] Textile dyes and pigments.

Fabramine. [Fabricolor] Textile dyes and pigments.

Fabricel. [Fabricolor] Textile dyes and pigments.

Fabrichrome. [Fabricolor] Textile dyes and pigments.

Fabriderm. [Fabricolor] Textile dyes and pigments.

Fabrifast. [Fabricolor] Textile dyes and pigments.

Fabrifix. [Fabricolor] Textile dyes and pigments.

Fabri-Kast® Systems. [Atochem N. Am.] Corrosion-resistant fabricated structural shapes.

Fabrilan. [Fabricolor] Textile dyes and pigments.

Fabrinyl. [Fabricolor] Textile dyes and pigments.

Fabrisol. [Fabricolor] Textile dyes and pigments.

Fabrisperse. [Fabricolor] Textile dyes and pigments.

Fabritone. [C.H. Patrick] Softener producing soft cationic-type hand.

Facegard. [CNC Int'l.] Solubilizer for acrylic fibers; under specific conditions, acts as a swelling agent to make fiber more receptive to dyes.

Facet®. [BASF AG] Quinclorac; for post-emergence control of barnyard grass and other weeds in rice.

Factice. [Am. Cyanamid] Vulcanized vegetable oil.

Fadex PAE Liq. [Sandoz] Low foaming dyebath additive improving lightfastness of milling and premetallized dyes on nylon.

Fairzine. [Fairmount] Catalyzed hydrazine.

Faktis. [Rhein Chemie] Sulfur factice; processing aid for NBR, CR, other rubbers; for molded and extruded goods, hoses, profiles, surgical and foodstuff goods, rubberized fabrics.

Falba®. [Aceto; Pfaltz & Bauer] Cosmetic and ointment base.

Famodan. [Grindsted Prods.; Grindsted Prods. Denmark] Sorbitan esters; emulsifier for foods, emulsion explosives, printing inks.

Famous. [Karlshamns] Hydrogenated soybean oil.

Fanal. [BASF] Organic pigments for textile applics.

Fanal®. [BASF AG] Color lakes of basic dyes with inorganic complex acids; pigments for inks, typewriter ribbon inks, colored pencils.

Fanchon. [Miles/Organic Prods.] Textile dyes and pigments.

Fanco Finish. [Reilly-Whiteman] Finishing agents for textiles.

Fanco Rapid Dye. [Reilly-Whiteman] Dyeing assistant for nylon.

Fancodye. [Reilly-Whiteman] Dyeing assistant for piece goods.

Fancofix. [Reilly-Whiteman] Dye fixatives for improved wash and lightfastness on acid dyed nylon.

Fancol. [Fanning] Triglycerides, lanolin derivs.; emollient, thickener, emulsifier, stabilizer, moisturizer, lubricant, plasticizer, chemical intermediate, humectant, mold release agent for pharmaceuticals, cosmetics, industrial applics.

Fancol TOIN. [Fanning] Allantoin.

Fancolene. [Reilly-Whiteman] Sequestering agents and water treatments.

Fancolev. [Reilly-Whiteman] Leveling and migrating agents for dyeing of synthetics and blends.

Fancolube. [Reilly-Whiteman] Dyebath lubricant and softener for beck or jet dyeing.

Fancor. [Fanning] Lanolin derivs.; emollient, stabilizer, emulsifier, corrosion inhibitor for personal care, pharmaceutical, and industrial applics.; plasticizer, lubricant for rubber, textiles, metalworking.

Fancorsil. [Fanning] Silicone derivs.; lubricant, gloss agents.

Fancoscour. [Reilly-Whiteman] Detergents for textile scouring.

Fancosoft. [Reilly-Whiteman] Softener for textiles.

Fancosol. [Reilly-Whiteman] Dyeing, leveling, and penetrating assistant for dyeing nylon yarn, fabrics and hosiery.

Fancowet. [Reilly-Whiteman] Wetting and rewetting agents for fibers.

Fancowhite. [Reilly-Whiteman] Optical brighteners.

Fanox N. [Exxon] Metalworking lubricant.

Fanwax. [Fanning] Fatty alcohol/alkoxylated alcohol blends; self-emulsifying wax providing emulsification for fluid systems, skin and hair lotions, antiperspirants, depilatories, creme rinses, hair dyes, bleaches.

Farez. [QO Chem.] Polyol for urethanes.

Farmacy. [Central Soya] Animal health prods.

Farmin. [Kao Corp. SA] Fatty amines; emulsifier, anticaking agent, textile additive, corrosion inhibitor, flotation reagent, softener, antistat, intermediate.

Fascat®. [Atochem N. Am.] Catalysts.

Fas/Kut. [Metal Lubricants] Cutting oils.

Fastbond®. [3M/Industrial Spec.] Contact and foam adhesives.

Fastcast. [Arnco] Two-component polyurethane elastomer; for industrial parts, rollers, skate wheels.

Fastgene. [Tokai Seiyu Ind.] Polyamine condensate; for fixing of direct dyes.

Fastogen. [DIC Trading] Phthalocyanine, quinacridone, dioxazine isoindolinone; high perfommace pigments and pigment crudes.

Fastusol®. [BASF AG] Substantive dyes; for coloring paper, textiles.

FB®. [Interox Am.] Sodium percarbonate; for textile bleaching, dye leveling, detergent enhancement.

FB-48. [U.S. Borax & Chem.] Borate; fertilizer.

FC-24, 520. [3M] Trifluoromethane sulfonic acid or salts; catalyst, reactant for polymerization of epoxies, styrenes, alkylation and acylation reactions; coatings, pharmaceuticals, explosives, dyes, and intermediates

Feedate. [Southeastern Minerals] Potassium/magnesium sulfate; feed grade.

Feedcarb. [Southeastern Minerals] 88% Sodium bicarbonate; feed grade.

Feed-K. [Southeastern Minerals] Potassium chloride; feed grade.

Feedox. [Southeastern Minerals] Magnesium oxide; feed grade.

Feedphos-24. [Southeastern Minerals] Monoammonium phosphate, nitrogen blend; feed grade.

Fekta®. [Goldschmidt] Algicide for swimming pools.

Felcobond. [Fel-Pro] Elastomer resin adhesive.

Felex 100. [Feldspar] Feldspar, 325 mesh; filler extender.

Feliderm. [Hoechst AG] Dyeing auxiliary; hydrotrope; pickling acid for leather finishing.

Felor. [DuPont] Nylon filaments.

Fel-Poxy. [Fel-Pro] Epoxy adhesives.

Fel-Protector. [Fel-Pro] Corrosion preventive coatings.

Felton. [Toho Chem. Industry] Nonionic complex; detergent for felt of paper mfg. machines.

Felzodox. [Pigment & Chem.] Zinc oxide; pigment.

Fenalan. [BASF] Aq. pigment prep.

Fenopon®. [Rhone-Poulenc Surf.] Ethoxy sulfates, cocoyl isethionates, or fatty acid taurates; surfactants.

Fermalpha. [Finnsugar Bioprods.] Alpha-amylase; enzyme for food processing.

Fermcolase®. [Genencor Int'l.] Catalase; enzyme for food processing, peroxide decomposition.

Fermcozyme®. [Finnsugar Bioprods.] Glucose oxidase; enzyme for food processing; antioxidant.

Ferment-Arom. [Hercules/FFIG] Natural flavoring ingredient.

Fermenzyme®. [Solvay Enzymes] Glucoamylase; enzyme.

Fermichamp. [Gist-Brocades Food Ingreds.] Instant active dry yeast; for vinification.

Fermipan. [Gist-Brocades Food Ingreds.] Instant active dry yeast; for baking.

Fermivin. [Gist-Brocades Food Ingreds.] Instant active dry yeast; for vinification.

Fermlipase. [Finnsugar Bioprods.] Lipase; enzyme for food processing.

Fermvertase. [Finnsugar Bioprods.] Invertase; enzyme for food processing.

Ferobestos. [Tenmat Ltd.] Fiber-reinforced phenolic.

Ferox. [Parish] Additives for gasoline and diesel fuel.

Ferralium® 255. [Haynes Int'l.] Super stainless steel.

Ferrex. [Ferro] Polypropylene, mineral-filled; for high gloss applics.

Ferric Blue. [Presperse] Ferric ammonium ferrocyanide compds.;

Ferriclear. [Eaglebrook] Acid-free hydroxylated ferric sulfate.

Ferrifloc. [Faesy & Besthoff] Antipollutant.

Ferri-Floc. [Boliden Intertrade] Coagulant for water treatment.

Ferro 1288. [Ferro] Barium/cadmium/zinc stabilizer; heat stabilizer for flexible vinyls.

Ferroblack. [Ferro-Plast Srl] Black masterbatches.

Ferrocolor. [Plastics Europe] Coloring materials.

Ferrocon. [Ferro] Conductive thermoplastic.

Ferroflex. [Ferro] TPO elastomer.

Ferroflo. [Ferro] Lubricated thermoplastic.

Ferro-Gard. [Ronco Labs] Corrosion inhibited oil base.

Ferronyl. [ISP] Dietary iron supplement.

Ferropak. [Ferro] Filled pkg. thermoplastic.

Ferro Permyl. [Ferro] Proprietary; uv absorber for polyesters, coatings.

Ferrophos. [OxyChem] Enhancer for zinc-rich primers; for weldable, conductive, or corrosion-resistant coatings.

Ferroquest. [W.R. Grace/Dearborn] Iron deposit remover.

Ferro Sil. [Kaopolite] Ferroaluminum silicate; filler for abrasion-resistant plastics; extender pigment for primers and other coatings.

Ferrosol. [W.R. Grace/Dearborn] Iron deposit remover.

Ferrostat. [Plastics Europe] Thermoplastic materials.

Ferro UV-Chek. [Ferro] Uv absorber for plastics.

Fers. [Fers SA] Phenolic resins.

Fertlstone. [Pfizer] Limestone; for agric. applics.

Fetrilon®. [BASF AG] Foliar fertilizer, micronutrient.

FF4-72. [Spartan Flame Retardants] Flame retardant.

FFA-5, -9. [Bacon] Epoxy resin; adhesive.

Fiba-Bond. [CNC Int'l.] Urea-formaldehyde resins; wet strength resins for paper industry.

Fibasol. [Finke-Farbstoffe] Liquid colors.

Fibbie. [Piemme Srl] Nylon.

Fiberfrax®. [Carborundum] Kaolin ceramic fiber; insulation.

Fibergard. [Dooley] Polymers imparting stain resist. to nylon and protein fibers.

Fiberglas®. [Owens/Corning Fiberglas] Chopped glass strand; reinforcements for plastics.

Fiberite®. [ICI Fiberite] Phenolic, epoxy, polyester, or polyimide resins; for structural aircraft exteriors, aircraft interiors, sandwich panels, aerospace structures, sporting goods.

Fiber Kal. [Kaopolite] Fibrous calcined kaolin; reinforcing fiber for plastics, sealants; filtration aid.

Fiberlay. [Chem. Processing] Textile sizing compd.

Fiberloc®. [BFGoodrich; BFGoodrich UK] Glass-reinforced vinyl composite; for appliances, business equip., construction, elec. equip., heating, ventilating and air conditioning, marine prods., plumbing and water treatment, windows.

Fibermate® AN-521. [Sybron] Textile antistat.

Fiberod®. [Polymer Composites] Thermoplastic rods, ribbons, or tapes; for pultrusion, compr. molding, braiding, laminating, filament winding, medical, recreation, aerospace, transportation, complex shapes.

Fibersil. [Goldschmidt] Self-crosslinking silicone finish for synthetic fill fibers.

Fibersize. [Nat'l. Starch & Chem.] Textile finishing agent.

Fibersoft. [Cooper] Blends of fatty amides and esters; yarn and fabric softeners.

Fibertech. [Hall Chem.] Chemicals for fiber mfg.

Fibertint. [Cooper] High m.w. polymeric water-soluble tints; fugitive liq. tints and identification crayons used to identify synthetic carpet yarns during processing.

Fibra-Cel®. [Celite] Cellulose and cotton fibers; filter aid, reinforcing fibers, functional fillers, thickener, processing aid, conditioner for thermoset and thermoplastic resins, builllding prods., grouts, pet food, asphalt mixtures, rubber goods, gaskets, sealants, braking linings, paints.

Fibral/Unix UD. [SP Systems] Reinforcements.

Fibramol®. [Hoechst Celanese/Colorants & Surf.] Softeners for textiles.

Fibran. [Nat'l. Starch & Chem.] Synthetic paper size.

Fibresinol. [Raschig AG] Phenolic resin; molding material.

Fibril. [Fibro Chem] Aliphatic hydrocarbons, spreading agents, nonionic emulsifiers blend; defoamer for jet and atmospheric dyeing equip.

Fibrite. [Shyamac Int'l.] Textile dyes and pigments.

Fibro. [Fibro Chem] Defoamers, lubricants, sequestrants, softeners, wetting agents, penetrants, resins for textile applics.

Fibrofix. [Fibro Chem] Fixative for textile dyes.

Fibrogen. [Fibro Chem] Organic peroxide stabilizers.

Fibroscour. [Fibro Chem] Scouring agents, removing soil and sizing from cotton and blends.

Fibro-Silk. [Brooks Industries] Silk; protein.

Fibrosist. [Fibro Chem] Leveling agent, dyeing assitant for textiles.

Fibrostat. [Fibro Chem] Antistat for textile use.

Fibrowet. [Fibro Chem] Wetting agent, dye assistants for textiles.

Fibrox. [Industrial Fibers] Mineral fiber filler/reinforcement; for high heat applics; visc. modifier.

Ficel®. [Schering Berlin Polymers] Azodicarbonamide; chemical blowing agent for sponge rubber, cushion vinyl flooring and fabrics, profiles/pipe, sealants, structural foam molding, wire and cable insulation, films.

Ficel® AZDN. [Schering Berlin Polymers] 2,2-Azodiisobutyronitrile; polymerization initiator.

Filamid. [Ciba-Geigy/Pigments] Dyes for polyamide fibers.

Fi-Lana. [Monsanto] Acrylic fibers.

Filester. [Ciba-Geigy/Pigments] Dyes for polyester fibers.

Fillite. [Hastings Plastics] Filler plastic.

Filmcol. [Shell] Solvents.

Filmex®. [Quantum/USI] Alcohol blends; industrial solvent, thinner for inks, chemical specialties, chemical and pharmaceutical processing, textiles.

Filmon. [Sniatechnopolimeri] Polyamide.

Filofin. [Ciba-Geigy Pigments UK] Predispersed organic and inorganic pigments for PP fibers.

Filter-Cel. [Celite] Filter aid.

Filterene. [Bond] Process, waste, potable water coagulant.

Filterfold. [Barnebey & Sutcliffe] Activated carbon; air purification adsorbers.

Filter Sil. [Unimin Specialty Minerals] Filtration media.

Filtrasol A. [Quest Int'l.] Sunscreen.

Filtrasorb®. [Calgon Carbon] Granular activated carbon; for water treatment applics.

Filtrez/Setalin. [Akzo Coatings] Rosin-based resins.

Filtros. [Akzo Coatings] Gum rosin, gum turpentine.

Fina®. [Fina Chem.] Polystyrene resins; for inj. molding, coextrusion, blow molding, extruded foam, thermoforming, lamination, film, consumer prods., medical and laboratory parts, tumblers.

Finaclear®. [Fina Chem.] Clear thermoplastic elastomers; for plastic modification.

Finagreen®. [Fina Chem.] Biodegradable oil for offshore drilling muds.

Final. [Bell Labs] Pelleted bait rodenticide.

Finamine CO. [Finetex] Cocamidopropylamine oxide.

Finapal. [Finetex] Surfactant blend; wetting and leveling agent for polyester dyeing.

Finaprene®. [Fina Chem.] Polybutadiene, styrene-butadiene, styrene isoprene, or carboxylated styrene butadiene elastomers; for footwear, tech. goods, tires, adhesives, bitumen modification, plastic modification.

Finapro®. [Fina Chem.] Polypropylene homopolymers and block copolymers; for extrusion-thermoforming, inj. molding, film extrusion, tapes, fibers, sheet, pipes, appliances, automotive, building and construction, household, and pkg. applics.

Finathene®. [Fina Chem.] High-density polyethylene homopolymers and copolymers; for blow molding, film blowing, pipe extrusion, cable jacketing.

Finazoline. [Finetex] Imidazolines; emulsifiers, wetting agents, softeners, intermediates, corrosion inhibitors for metalworking and fuel oil additives, ore flotation, cosmetics, textiles.

Findet. [Finetex] Phosphate esters; emulsifiers, wetting agents, detergents, dispersants, corrosion inhibitors for metalworking and cleaning, lubricant additives; coupling agent.

Finecat MOGL. [Finetex] Methyl hydroxycetyl gluacaminium lactate.

Fine Line. [PCR] Hexamethyldisilazane.

Finelube 4G. [Finetex] Lubricant for dyeing operations.

Finemix. [Vyse Gelatin] Gelatin.

Finesse. [DuPont/Ag] Herbicide.

Finester EH-25. [Finetex] C12-15 alkyl octanoate; emollient, solubilizer.

Finex. [Presperse] Zinc oxide or blends; powders providing feel, covering power, and skin adherance to formulations; uv absorbents, SPF boosters.

Fine Zink. [Am. MicroTrace] $ZnSO_4$ + 35.5% Zn; water soluble animal feed additive.

Finitron. [Griffin] Ethyl perfluorooctanesulfonamide.

Finizym. [Novo Nordisk] Beta-glucanase; enzyme for beer filtration.

Finnacryl. [Norplast Ltd.] Acrylic.

Finquat CT. [Finetex] Quaternium-75; hair and skin conditioner used in permanent wave sol'ns., depilatories.

Finsist. [Finetex] Quatenary ammonium compd.; retarder, dye leveling agent, compatibilizer for textiles.

Finsoft. [Finetex] Quaternium blends; hair conditioner; fabric and household softener; fluorochemical extender.

Finsolv®. [Finetex; Witco SA] Benzoate esters; nongreasy emollient for cosmetics, sunscreen, antiperspirants, deodorants; lubricant; plasticizer for hair resins, in anhyd. systems; perfume solubilizer.

Firebrake®. [U.S. Borax & Chem.] Zinc borate; flame retardant, smoke suppressant, synergist used in chlorinated polyesters, PVC plastisols, and epoxy systems.

Fire Flake. [U.S. Aluminum] Aluminum flakes.

FireGuard. [Teknor Apex] Plenum compd.; low smoke and flame compds. for cable jacketing.

Firemaster®. [Great Lakes] Halogenated phosphate esters; fire retardants for polyester thermosol applics.

Firemaster 680. [Velsicol] Bis (tribromophenoxy) ethane; flame retardant for ABS.

Firemate. [W.R. Grace/Dearborn] Fuel oil additive.

Fire PRF2. [Indspec] Two-component phenol resorcinol-formaldehyde resin; flame retardant thermoset for fabricating fume and smoke exhaust ducts, fire-retardant components.

FireShield®. [Laurel Industries] Antimony trioxide; flame retardant.

Firestone Nylon. [Firestone Textiles] Nylon 6, some glass reinforced; thermoplastic for molding, extrusion, wire jacketing.

Firing Squad. [Huntington Labs] Insecticides.

Firm-Dee. [Vitamins, Inc.] Feed supplements.

Fixanol. [ICI Am.] Resinous condensate; fixative for direct and reactive dyes.

Fixapret®. [BASF AG] Textile finishing prods. for anticrease, antishrink and easy care finishing.

Fixing Agent FP 45. [BASF AG] Organic amides/formaldehyde condensate; fixative for anionic dyes, resin size and assistants, polymer dispersions and fillers in papermaking.

Fixing Agent JP-4. [Dooley] Fixing agent for direct and reactive dyes.

Fixogene. [ICI Surf. UK] Dicyandiamide formaldehyde resinous condensate; fixing agent for dyes on cellulosic fibers.

Fixon. [Chemonic Industries] Resin complex; aftertreatments for direct, reactive, sulfur and acid dyes.

Fizul. [Finetex] Sulfosuccinates or sulfosuccinamates; emulsifier, wetting agent, dispersant, emulsifier, solubilizer, stabilizer for emulsion polymerization, foaming wallboards, floor polishes, shampoos.

FK 140, 160, 300DS, 310, 320, 320DS, 500LS. [Degussa] Silica or precipitated silica; for silicone rubber, paper and film industries; antiblocking agent for polyolefins.

FK-1517. [Chem-Tex Labs] Nonionic foam controllant for foam dyeing of carpet.

Flakeglas. [Nyco Minerals] Chemically surface-modified glass flake.

Flakeprime. [Master Builders] Epoxy primer.

Flaketar. [Master Builders] Coal tar epoxy.

Flamegard®. [Sybron] Flame retardant for textiles.

Flamenco. [Mearl] White and colored pearl pigments for cosmetics.

Flame Out. [Emco Services] Flame retardant for textiles.

Flameproof. [Apex] Flame retardants.

Flametamer-10. [OxyChem] Flame retardant for cellulosics.

Flamtard. [Alcan] Zinc compds.; flame retardant for PVC, polychloroprene, chlorosulfonated polyethylene, polypropylene, other halopolymers.

Flat-Ayd. [Daniel Prods.] Flatting agents for paints and inks.

Flatting Agent. [Degussa] Silica or surface-treated precipitated silica; for paper coating applics.

Flavorburst. [Fries & Fries] Encapsulated flavors.

Flavor-Master. [V&E Kohnstamm] Flavors, bases.

FLC-2. [Rhone-Poulenc/Perf. Resins & Coatings] Oil-sol., aliphatic hydrocarbon resin; fluid loss control additive.

Flectol®. [Monsanto] Antioxidant, preservative for CR, metal deactivator.

Fleetcol. [Rhone-Poulenc] Defoamers.

Fleetquest. [Rhone-Poulenc] Dispersant.

Flex. [CNC Int'l.] Self-crosslinking polymers; imparts hand and bulk to cellulose or synthetic fibers.

Flexamine®. [Uniroyal] Antioxidant for tires, latex, foam, molded and mechanical goods.

Flexan® 130. [Nat'l. Starch & Chem.] Sodium polystyrene sulfonate; hair fixative for setting lotions, conditioners.

Flexbind. [Hunt] Binders for textile applics.

Flexbond®. [Air Prods.] Polyvinyl acetate, vinyl acrylic, or acrylic emulsions; for adhesives, paints, caulks, ceiling tile, insulation binder, textiles, flock binder.

Flexchlor. [Witco/Argus] Chlorinated paraffin; plasticizer for use in paint, coatings, caulks, sealants, rubber, and adhesives.

Flexclad. [Goodyear] Copolyester resins; for sol'n. and hot-melt adhesives and coatings.

Flexcote. [Hunt] For sizing warps.

Flexcryl®. [Air Prods.] Acrylic emulsion; for adhesives.

Flexel 1000. [BFGoodrich/Spec. Polymers] Vinyl alloy-based thermoplastic elastomer; for insulation, cable jackets.

Flexhard Epoxy Enamel. [Southern Coatings] Polyamide epoxy.

Flexible Fyrex®. [Akzo] Phosphate blends; inorganic flame retardant for treated fabrics.

Flexipol®. [Flexible Prods.] Rigid or flexible polyurethane systems; for decorative molding, flotation, furniture, walk-in refrigeration, water heater, acoustic, pkg., toys, seats and cushioning, integral skin applics.

Flexlube. [Hunt] Lubricant for textile applics.

Flexobond. [Bacon] Two-part urethane systems; used as a fairing compd., adhesive, casting compd., or encapsulant

Flexol® Plasticizer. [Union Carbide] Plasticizers for rubber, epoxy.

Flexomer. [Union Carbide] Ethylene copolymer; for blow molding, inj. blow molding, profile and sheet extrusion, thermoforming, small parts, hose and

tubing, squeeze bottle, squeeze tube, film, laminates; blending resin.

Flexon. [Exxon] Rubber extender, process oil.

Flexoplus. [Miles/Organic Prods.] Textile dyes and pigments.

Flexopunch. [Air Prods.] Transfer additive.

Flexovoss. [Vosschemie GmbH] Polyurethane.

Flexprene. [Teknor Apex] Thermoplastic styrenics.

Flexricin®. [CasChem] Ricinoleic acid derivs.; lubricant, corrosion inhibitor, intermediate for textiles, metalworking compds.; wetting agent, wax plasticizer, and mold release agent for rubber polymers, antifoam agent, household and cosmetic applics.

Flexthane. [Air Prods.] Urethane hybrid polymer; for graphic arts, textile and vinyl coatings, leather coatings, plastic coatings, industrial coatings, film lamination.

Flexwax. [Hunt] Blended mixture of 100% saponofiables; easily desized during normal finishing operations.

Flexweld. [Imperial Adhesives] Synthetic resin adhesive.

Flexzone®. [Uniroyal] Antiozonant for latex, foam, tires, insulation jacket.

Flo-Aid. [Crowley Tar Prods.] Anticaking agent for fertilizers.

Flocculant T-9. [Toho Chem. Industry] Cationic complex; cohesion and sedimentation agent for titanium dioxide mfg.

Flo Chem. [Emco Services] Fluorochemical oil and water repellent for home furnishing industry.

Flock-Lok®. [Lord] Flexible polyurethane flock adhesive.

Floculant. [3-V] For water treatment.

Flo-Ever. [Cargill/Salt] 99.95% Sodium chloride with anticaking additive.

Floform. [R.T. Vanderbilt] Pelleted carbon black.

Flo-Fre. [Oil Dri Corp. of Am.] Anticaking agents.

Flo-Gard. [PPG Industries] Amorphous silica; absorbent, carrier, anticaking agent, flow control agent, anticaking agent for chemical formulations and processes, agric. formulations; reinforcing filler for rubber.

Flojel®. [Nat'l. Starch & Chem.] Corn starch; warp sizing.

Flomag. [Martin Marietta Magnesia Spec.] Water treatment chemicals.

Flo Mo®. [Witco/Organics] Ethoxylated alcohols, sulfonates, or formulated prods.; detergent, wetting agent, emulsifier for cleaning food handling equip., agric. formulations.

Flonac. [Kemira] Pearlescent pigments.

Flor-Add. [Kromachem Ltd.] Additives for coatings.

Floramat. [Henkel/Emery/Cospha] Ethyl-2-t-butylcyclohexylcarbonate; fragrance raw material.

Floranid®. [BASF AG] Fertilizers for horticultural crops, ornamentals, lawn, field and greenhouse crops, turf.

Florco®. [Floridin] Attapulgite clay; floor absorbent for oil, grease, water; antislip agent on floors; used by automotive, steel, transportation, commercial, pet, food and beverage, institutional, and amusements industries.

Flor-Cryl. [Kromachem Ltd.] Acrylic polymers.

Flor-Dri. [Floridin] Economical absorbent to keep floors, decks, and other surfaces cleaner, dry, and safe.

Florex®. [Floridin] Attapulgite clay; absorbent and adsorbent; for jet fuel treating, catalyst stripping, transformer oil purification, oil reclamation, min. and veg. oil purification; absorbent carrier for pesticides.

Florigel®. [Floridin] Attapulgite clay; absorbent and adsorbent.

Flor-Kleen. [Floridin] Attapulgite clay; industrial absorbent.

Flor-Plast. [Kromachem Ltd.] Saturated polyesters.

Flor-Stab. [Kromachem Ltd.] uv stabilizers.

Flotanol. [Hoechst AG] Polyglycols; flotation frothers.

Flotigam. [Hoechst AG] Fatty amine derivs.; flotation collectors, antibaking agents.

Flotigol. [Hoechst AG] Cresylic acid; flotation frother.

Flotinor. [Hoechst AG] Fatty acids; flotation collector.

Flotol. [Hoechst AG] Pine oils; flotation frothers.

Flow Agent WR-100. [Henkel] Acrylic copolymer resin; anticaking agent, flow control agent for coatings, electrodeposition.

Flowtone. [Southern Clay Prods.] Gelling and suspending agents.

Flow Wiz. [Axel Plastics Research Labs] Flow control agents.

Floxan SC-5211. [Henkel/Emery] Flocculant.

Flozene. [Witco/Sonneborn] White mineral oil.

Fluf®. [W.A. Cleary] Liq. fertilizers.

Fluftone APS. [Apollo] Softener and napping agent for natural and synthetic fibers.

Fluid Flex. [Van Den Bergh Foods] Soybean oil, fatty acid glyceryl lactates, mono and diglycerides, TBHQ; fluid shortening for cakes.

Fluidiram. [Ceca SA] Fatty amines and mixtures; anticaking agent for fertilizers.

Fluids Conc. [Mateson] Rubber and plastics deodorant.

Fluilan. [Croda Inc.; Croda Chem. Ltd.] Lanolin oil and derivs.; emulsifier, conditioner, emollient, penetrant, superfatting agent, dispersant, solubilizer, moisturizer for cosmetics, cleansers, dishwashing liqs.; plasticizer and film modifier for hair spray resins.

Fluiplast. [Scor SAS] PVC powder for coating.

Fluiscor. [Scor SAS] PVC powder for coating.

Fluo-Flux. [Alcoa] Aluminum fluoride.

Fluo HT. [Micro Powders] PTFE.

Fluolite. [ICI Am.] Fluorescent whitening agents.

Fluon®. [ICI Fluoropolymers; ICI PLC] PTFE or FEP resins.

Fluorad. [3M] Fluorinated alkyl mixtures; surfactant, wetting, spreading and leveling agent for use in electrolyte systems, coatings, alkaline cleaners, floor polishes, etchants, plating baths; emulsifier for fluoropolymer emulsions; well stimulation additive.

Fluorazine. [Surpass] Optical brightener.

Fluorel®. [3M] Fluoroelastomers; for extrusion, inj., compression, and transfer moldings, fuel line hose, seals, o-rings, calendered sheet, rolls.

Fluoresbrite Carboxylate Microspheres. [Polysciences] Fluorescent monodisperse beads with pendant carboxyl groups.

Fluorocomp®. [ICI Fluoropolymers] Filled fluoropolymers.

Fluorodyne Filter. [Pall Corp.] Hydrophilic polyvinylidene fluoride; membrane filter.

FluoroEtch. [Acton Tech.] Etchant for making fluorocarbon polymers bondable.

Fluorogold. [Furon] Teflon-based prod.

Fluoroguard. [Mfg's Chem. & Supply] Fluorochemical for oil and water resistance.

Fluorol®. [BASF AG] Polycyclic dyes with fluorescence; for coloring carburetor fuels, oils and lubricating greases.

Fluoroloc HL. [Furon] Teflon TFE resin formulation.

Fluoroloy. [Furon] Teflon TFE resin formulation.

Fluorolube®. [OxyChem] Polychlorotrifluoroethylene; lubricating grease.

Fluoromelt®. [LNP] Melt processable fluoropolymer composites.

Fluorotex®. [CNC Int'l.] Fluorochemical; oil and water repellent for outerwear.

Fluorothane. [Furon] Polyurethane resin formulation.

Fluowet. [Hoechst Celanese/Colorants & Surf.; Hoechst AG] Fluoroaliphatic ethoxylate; surfactant, intermediate.

Flushcake. [Roma Color] Pigment.

FM®. [Am. Cyanamid] Epoxy, nitrile, phenolic, or polyimide resins; foam and adhesive applics.

FM. [ICI Fiberite] Phenolic resins with fabric, glass, or cellulose reinforcements; molding compds.

F-MA 11®. [Reheis] Aluminum hydroxide/magnesium carbonate blend.

FMB. [Huntington Labs] Quaternary ammonium chlorides, fatty alkanolamides, or betaines; for formulation of disinfectants, sanitizers, fungicides, water treatment microbiocides, swimming pool algicides; preservative for pharmaceuticals.

FMB A0-8, 73-3A0. [Huntington Labs] Amine oxide.

Foamaster®. [Henkel/Emery; Henkel-Nopco] Silicone and nonsilicone types; defoamer for textiles, paints, adhesives, rubber, latexes, coatings, inks, paper/paperboard, cement, pesticides, food prods.

Foamban. [Georgia-Pacific] Defoamer for pulp and paper machine processing.

Foam Ban. [Ultra Additives] Defoamers.

Foam Block®. [Boehme Filatex] Reacted silicone dispersion; defoamer and antifoam for dyeing operations.

Foambrak. [Baker Perf. Chem.] Defoamers.

Foambreaker. [Guardian Labs] Antifoaming agent.

Foambust. [Mfg's Chem. & Supply] Defoamer and antifoam for textile use.

Foamchek. [West Agro] Nonfoaming acid for dairy use.

Foam-Coll. [Brooks Industries] Collagen derivs.; foaming proteins for skin and hair care cosmetics.

Foam-Det. [Oakite Prods.] Detergents.

Foamer. [Harcros] Anionic ether sulfate; foamer for air mist drilling, general detergents, gypsum board prod.

Foamer. [Hart Chem. Ltd.] Fatty alkanolamide sulfosuccinate; foam stabilizer for liq. detergent systems.

Foamex®. [Rhone-Poulenc/Textile & Rubber] Silicone defoamers.

Foam Fighter. [Miller Chem. & Fertilizer] Defoamer.

Foamflush. [ISP] N-Methyl-2-pyrrolidone, butyrolactone, other ingreds.; urethane remover.

Foamgard. [Rhone-Poulenc/Textile & Rubber] Nonsilicone; defoamer for carpet dyeing, dye baths, paper and pulp processing, waste treatment, print pastes.

Foam Grab. [Dacar] Adhesive.

Foamid 117. [Alzo] Cocamidopropyl dimethylamine propionate; softener, emollient for cosmetic creams and lotions, skin care prods., fingernail polish removers.

Foamink. [Foamink Tech.] Inks for foamink process.

Foamkill. [Crucible] Silicone compds.; defoamer for food, cosmetics, pharmaceutical applics., paper coatings, paints, inks, adhesives, casein, neoprene, natural latexes, water treatment.

Foam-Kon. [LNP; ICI Advanced Materials] Thermoplastic structural foam conc.; for high pressure foam systems, inj. molding compds.

Foamlite®. [General Latex & Chem.] Latex foam backing compds.

Foamole. [Van Dyk] Fatty alkanolamides or dimethylamines; emulsifier, emollient, superfatting agent, conditioner, visc. booster, foam stabilizer for hair care prods.

Foamquat IAES. [Alzo] Isostearamidopropyl ethyldimonium ethosulfate.

Foamquest. [Chemloid] Nonsilicone and silicone defoamers for textile preparation, printing, dyeing, coating, and finishing operations.

Foam-Soy C. [Brooks Industries] Sodium cocoyl hydrolyzed soy protein; for skin and hair care cosmetics.

Foam Stabilizer 5200. [Hüls AG] Sulfonate-based; foaming aid for prod. of mechanically aerated plastisols.

Foamtrol. [Ultra Additives] Antifoam/defoamer.

Foam-Trol. [Betz Industrial] Foam control agent for waste and process streams.

Foam-Wheat C. [Brooks Industries] Sodium cocoyl hydrolyzed wheat protein; mild foaming protein for shampoos, facial makeup cleansers.

Focus®. [BASF AG] Cycloxydim; post-emergence graminicide effective against annual and perennial grasses.

Fogspray. [Imperial Adhesives] Aq. spray adhesive.

Foil Flake. [U.S. Aluminum] Aluminum flakes.

Folex. [Rhone-Poulenc/Ag] Cotton defoliant.

Foliar Nitrophoska®. [BASF AG] Liq. compound fertilizer for foliar applic.

Folidol®. [Bayer] Parathion and blends; insecticide, acaricide with contact, stomach, and breathing poison action.

Folimat®. [Bayer] Omethoate; systemic acaricide and insecticide.

Foliplas. [Semiplastic Ltd.] Extruded HDPE/LDPE.

Folithion®. [Bayer] Fenitrothion; broad spectrum insecticide for controlling biting and sucking insect pests.

Fomrez®. [Witco/Organics] Foam crosslinkers, precursors for reactive diluents, chain extenders in urethane synthesis, coupling agent, catalyst, flexible cellular PU foam for textile and industrial applics., dispersing agents, coatings, adhesives, elastomers.

Fonoline®. [Witco/Sonneborn] Petrolatum USP; emollient, protective coating, binder, carrier, lubricant, moisture barrier, plasticizer, protective agent, softener for consumer use as petroleum jelly, ointments, industrial prods.

Forafac. [Atochem N. Am.] Fluorinated surfactants.

Foraflon®. [Atochem N. Am.] PVDF homopolymers; thermoplastic for extrusion and inj. molding applics. in chemical engineering, pkg., and elec. wiring fields, monofilaments, adhesives.

Foraflow Kynar. [Atochem UK] PVDF.

Foral®. [Hercules] Rosin derivs.; thermoplastic resin, tackifier for adhesives, protective and barrier coatings.

Foramousse. [Seppic] Surfactant blends; drilling assistant.

Forane. [Atochem N. Am.] Refrigerants.

Foraperle®. [Atochem N. Am.] Acrylic fluorinated resin blends; used for finishing and protection of leathers, in shoe polishes.

Forbest. [Lucas Meyer] Defoamer, leveling agent, visc. control agent, slip additive, antifloating agent, antisilking agent, gellant, corrosion inhibitors for paints, coatings, finishes, inks.

Fore®. [Rohm & Haas] Maneb blends; fungicide for turf, ornamentals.

Forestall. [ICI Am.] Soyaethyl morpholinium ethosulfate.

Forex. [Atochem N. Am.] High performance multipurpose liquefied gas; for fire extinguishing applics.

Forlan. [RITA] Lanolin, petrolatum, or cholesterol blends; emulsion stabilizer, emollient, moisturizer, solubilizer, dispersant, plasticizer for personal care prods.; solid absorption base.

Forlanit®. [Henkel/Emery/Cospha; Henkel KGaA] Phosphates; emulsifier for cosmetics and pharmaceuticals; wetting agent, flotation auxiliary for cyanide hot baths.

Forlanon. [Henkel KGaA] Sodium laureth phosphate; wetting agent, flotation auxiliary for cyanide hot baths.

Formacel-S. [DuPont] Blowing agent.

Form-A-Coat 30-200. [TACC Int'l.] One-component urethane; conforming coating resin and varnish.

Formadine. [John Campbell] Textile dyes and pigments.

Formasil®. [Union Carbide] Silicone emulsions; release agent in mfg. of medical devices, appliances, food pkg. materials.

Formcel. [Hoechst Celanese] Formaldehyde sol'ns. in alcohol.

Formetal Superfluid. [Guardian Labs] Synthetic, water-soluble metal cutting fluids.

Formion. [A. Schulman] Modified olefins and/or alloys.

Formkote. [E/M Corp.] Lubricant, mold release.

Formol 55. [BASF AG] Urea-formaldehyde precondensate; intermediate for mfg. adhesives and foundry binders.

Formrez®. [Witco] Adipate polyesters or polyether triols; used for coatings, chain extenders, cast polyurethane elastomers, microcellular urethanes, potting compds., encapsulating agents, printing rolls, coatings, adhesives.

Formula 3. [Ivar Labs] Adhesion promoter.

Formula #733. [Polymer Research Corp. of Am.] Oxazole derivs.; fluorescent whitener for textile processing.

Formvar. [Monsanto; Cairn Chem. Ltd.] Polyvinyl formal.

Foron. [Sandoz] Textile dyes and pigments.

Forprene. [Evode Plastics Ltd.] TPV's.

Fortenaxplast. [Chiorino SpA] HDPE.

Fortex®. [Petrolite/Polymers] Microcrystalline wax; used in hot-melt coatings and adhesives, paper coatings, printing inks, plastic modification (as lubricant, processing aid), as binder in ceramics, for potting, filling, and impregnating elec./electronic components, in investment casting, rubbers.

Fortifier 79. [Stewart Hall] Boiler compds. to remove oxygen.

Fortiflex®. [Solvay Polymers] HDPE; resins for inj. and blow molding of automotive, cosmetic, household and industrial goods, sheet extrusion and thermoforming, pipes, films.

Fortilene®. [Solvay Polymers] Polypropylene homopolymers and copolymers; resins for extrusion, sheet, strapping, profiles, inj. molding, caps, closures, thin-wall containers, disposable tableware, films, fibers, filaments.

Fortron®. [Hoechst Celanese/Engineering Plastics; Hoechst UK] PPS resins, some glass and/or mineral reinforced; high temp. polymer for connectors, switches, coil bobbons, caps, automotive parts, gears, industrial parts.

Fosfamide. [Henkel/Emery] Fatty amido phosphate complex; detergent, foaming and wetting agent for textiles, all-purpose, institutional and metal cleaners.

Fos-Flo. [Handy & Harman] Copper-phosphorous brazing alloy.

Fossilite. [L.A. Salomon] Diatomite.

Foster. [H.B. Fuller] Thermal insulation mastics, coatings, adhesives, and sealants.

Fosterge. [Henkel/Emery/Cospha; Henkel Canada] Phosphate derivs.; detergent intermediate; emulsifier for mineral oils and pesticides; corrosion inhibitor, dispersant, penetrant.

Fostex. [Henkel/Emery] Scale and corrosion inhibitor, dispersant, and sequestrant for water treatment.

Foulagan. [Ceca SA] Wool fulling

agents.

Foul-Up. [Mateson] Wet waste treatment.

Foundrox. [DCS Color & Supply] Red iron oxide; foundry grade.

Four Star. [McGean-Rohco] Quenching chemicals.

FPC. [OxyChem] Vinyl resins; molding, extrusion, calendering, and sol'n. resins for footwear, flooring, rigid sheet, coatings, sealants, organosols.

FR-. [Dead Sea Bromine] Bromine derivs. or nonhalogenic prods.; flame retardant for use in laminates, unsaturated polyesters, synthetic fibers, flexible polyurethane foams, plastics, rubber, textiles, adhesives, and coatings.

FR-28. [U.S. Borax & Chem.] Fire retardant for paint.

FR-D. [FMC] Phosphorus-based diol; reactive flame retardant for rigid polyurethane foam.

Fractocrete 3400. [Premier Refractories] High strength alumina castable.

Fragran. [Engelhard] Animal litter, deodorant.

Fragrance-Cap. [Ronald T. Dodge] Microencapsulated fragrance oil.

Fragrascent. [Crompton & Knowles] Fragrance bases.

Franklin. [Franklin Industrial Minerals] Calcium limestone; filler for plastics and rubber industries.

Franklin Fiber Filler. [U.S. Gypsum] Calcium sulfate whisker; fiber filler.

FRE. [Solem Industries] Alumina trihydrate; filler, flame and smoke suppressant for fiberglass-reinforced polyesters.

Free-Flo. [Cook Composites & Polymers] Polyurethane foam.

Free-Style. [Harcros] Swim pool chemicals.

Freezene. [Witco/Sonneborn] White mineral oil.

Freezgard. [North Am. Salt] $MgCl_2$; deicer.

Freezist. [A.E. Staley Mfg.] Modified tapioca starch.

Frekote. [Dexter/Frekote] Release agents for urethane elastomers and foams, rubber, plastics.

Freon. [DuPont] Fluorocarbons; refrigerants, solvents, blowing agents, dielectric gases, plastics, intermediates, aerosol propellants.

Fresh N. [Enzyme Development] Enzyme for baking.

Fresh N Fruity. [Petrokem] Deodorants.

Fric® Film. [Exxon] LDPE film; shrink bundling film for food pkg.

Frigate. [ISK Biotech] Agric. adjuvant for use with glyphosate herbicide.

Frigen. [Hoechst AG] Fluorinated hydrocarbon; solvent for removing organic and inorganic pollution.

Frigesa®. [Grünau] Hydrocolloids; thickener and stabilizer for desserts, ice cream, dressings.

Frigid-Go®. [Henkel/Emery] Moly low-temp. multipurpose synthetic greases and lubricants.

Frigidol. [G. Whitfield Richards] Extreme pressure lubricant.

Frimulsion. [Hercules/FFIG] Locust bean gum, starch, guar gum and/or gelatin blends; stabilizer and gelling agent for food systems.

Froil. [Octagon Process] Rust preventives.

Frost-Off. [Merix] Deicing fluid.

Froth-Pak Kit. [Insta-Foam Prods.] Two-component polyurethane foam kit.

Fructodan. [Grindsted Prods. Denmark] Stabilizer blends; for use in fruit ice.

Fruitfil. [A.E. Staley Mfg.] Modified tapioca starch.

Frutalone. [Hercules/FFIG] Aromatic chemical.

FSC-5. [Premier Refractories] Fused silica; castable with exc. thermal shock resistance.

FT-Wax. [Astor Wax] Synthetic wax.

Fuelsaver®. [Angus] Morpholine derivs.; preservative for diesel and other hydrocarbon fuels.

Fuel-Solv. [Betz Industrial] Fuel oil conditioner.

Fueltone. [Ampion] Fuel oil treatments.

Fulacryl®. [H.B. Fuller] Siliconized acrylic latex; caulk.

Fulton 404®. [LNP] Acetal with PTFE lubrication; thermoplastic for moving parts requiring frictional properties, bearing applics.

Fultone. [Nutex] Low foaming nonionic softener.

Fuman. [Pulcra SA] Alkylbenzene sulfonates; scouring agents for textiles.

Fumetrol. [Atochem N. Am.] Fume suppressant.

Fungamyl®. [Novo Nordisk] Fungal amylase; enzyme for baking, in starch industry, brewing.

Fungitex R. [Ciba-Geigy/Dyestuffs] Domiphen bromide; antimicrobial, fungicide for use in mouthwashes, antiseptics, cold sterilization.

Fungitrol. [Servo] Organic compd.; fungicide, bactericide for paint industry.

Funguran. [Giulini Adolfomer SA] Copper oxychloride formulations.

Furadan. [FMC/Ag] Insecticide, nematocide.

Furalac®. [Atochem N. Am.] Grouts, mortars for specialty applics.

Fura-Tone®. [Cardolite] Furan resins; impregnating resin, rubber antiozonant..

Furcarb. [QO Chem.] Furfural/furfuryl alcohol phenolic modified carbon binders.

Furnex. [Columbian Chem.] Carbon black.

Fusabond®. [DuPont] Anhydride-modified polyolefins; compatibilizers for blends and alloys; polymeric coupling agents for reinforced or recycled PE or PP; for adhesives and sealants.

Fusite. [H.B. Fuller] Two-part crosslinking urethane; adhesive.

Fusite. [Steetley Quarry Prods.] Roasted dolomite.

Fusor®. [Lord] Two-part epoxy adhesives, adhesion promoters.

Futurethane. [H.B. Fuller] Acrylic or urethane adhesives.

FW. [Degussa] Channel carbon black.

Fyarestor. [Witco/Argus] Bromochlorinated paraffin; flame retardant for textile treatment.

Fybrene. [Witco/Sonneborn] Petrolatum.

Fyran®. [Yorkshire Pat-Chem] Flame retardants for cellulosic fabrics.

Fyrebloc. [McGean-Rohco] Fire retardant plastics additive.

Fyresafe. [Nalco] Hydraulic fluids.

Fyrex®. [Akzo] Crystalline diammonium phosphate/monoammonium phosphate blend; flame retardant.

Fyrlube. [Akzo] Fire-resistant hydraulic fluids.

Fyrol®. [Akzo; Akzo Chem. BV] Phosphate derivs.; organic flame retardant for urethane foams, thermosetting and thermoplastic resins, coatings.

Fyrquel®. [Akzo] Fire-resistant hydraulic fluids.

G

G-. [ICI Spec. Chem.] Ethoxylated esters or sorbitan esters, or fatty amines; emulsifier, antistat, lubricant, softener, dispersant, wetting agent for textiles.

G-4. [Givaudan] Dichlorophene; antimicrobial, fungicide, bactericide for textiles, industrial, and veterinary applics.

G-40, -200. [Feldspar] High potash feldspar.

G-431. [Solem Industries] Alumina trihydrate; filler, flame retardant; resin extender in plastics, latex, neoprene foam systems, wire and cable insulation, vinyl wall and floor coverings.

Gafac®. [Rhone-Poulenc Surf.] Phosphate esters; emulsifier for pesticides; industrial and household detergent; emulsifier and stabilizer for emulsion polymerization; solubilizer, hydrotrope; textile antistat, lubricant, softener, dispersant, wetting agent; cosmetics ingredient.

Gafen®. [Rhone-Poulenc Surf.] Phosphate esters; surfactants.

Gaffix®. [Rhone-Poulenc Surf.] Vinylcaprolactam/PVP/dimethylaminoethyl methacrylate copolymer in ethanol; film-forming fixative resin for use in mousses, gels, glazes, hairsprays.

Gaflon. [Axial SA] PTFE.

Gafquat®. [ISP] Polyquaterniums; cationic film-forming substantive copolymers, conditioner for hair and skin care prods., antiperspirants.

Galactasol® Guar Derivs. [Aqualon] Hydroxypropyl guar; gums used as visc. increasers, flocculants, stabilizer, suspending aids.

Galaxy. [Aqualon] Guar gum and derivs.

Galaxy®. [BASF AG] Bentazon, acifluorfen; for postemergence control of annual broadleaf weeds in sobyeans and peanuts.

Galenol®. [Condea Chemie GmbH] Self-emulsifying fatty alcohols; intermediates for toiletries, cosmetics, detergents, leather and textile auxiliaries, lube oil, plastics additives, defoamers.

Galex. [Natrochem] Stabilized rosins.

Galirene. [IPE Overseas BV] Polystyrene.

Galoryl. [Tropag GmbH] Wetting agent.

Galvoline. [Dow] Magnesium anodes.

Galvomag. [Dow] Magnesium anodes.

Galvorod. [Dow] Magnesium anodes.

Gamaco®. [Georgia Marble] Calcium carbonate; extender fillers used in polyester systems, BMC/SMC, food contact applics.

Gama-Fil. [Georgia Marble] Calcium carbonate; filler for use in plastics, BMC/SMC, paint, caulks, sealants, adhesives, paper, foam urethane, modified acrylics, filled thermosets/thermoplastics, and rubber.

Gamanase. [Novo Nordisk] Galactomannase; enzyme for visc. reduction in coffee extracts.

Gama-Plas®. [Georgia Marble] Calcium carbonate; filler for use in glass fiber-reinforced polyester compds. (BMC, SMC, TMC, and wet mat molding).

Gama-Sperse®. [Georgia Marble] Calcium carbonate; filler for paint, plastics, caulks, sealants, adhesives, foam urethane, filled thermosets/thermoplastics, BMC, and rubber applics.

Gamma Series. [Katalistiks Int'l.] Fluid cracking catalysts.

Ganex®. [ISP] PVP derivs. and copolymers; moisture barrier, protective colloid, suspending aid, dyeing assistant, antiredeposition agent, dispersant for cosmetics, toiletries, coatings, polymerization, drycleaning, agric.; solubilizer for dyes.

Gantrez®. [ISP] Vinyl ether polymers and copolymers; dispersant, coupling, stabilizing agent, thickener, emulsifier, solubilizer, corrosion inhibitor, film former, antistat, tackifier, binder, used in agric., paper and textile industries, chemical processing, industrial products, detergents, cosmetics.

Garalease. [Lilly] Cellulosic lacquer; film-forming release agent.

Garbeflex. [Atochem Deutschland GmbH] Plasticizer.

Garb-O-Flakes. [Surco Prods.] Odor control granules.

Gardilene. [Albright & Wilson Am.; Albright & Wilson Australia] Sulfonates; emulsifier, dispersant, wetting and foaming agent; for polymerization, drycleaning.

Gardinol. [Albright & Wilson Am.; Albright & Wilson Australia] Sulfates; detergent, wetting agent, dispersant, emulsifier, dyeing assistant for textiles, hand cleaners, laundry prods., cosmetics.

Gardiquat. [Albright & Wilson Am.; Albright & Wilson UK] Benzalkonium chloride; bactericide, germicide for disinfectants and antiseptics for hospitals, food processing, oil drilling muds.

Gardisperse. [Albright & Wilson Am.; Albright & Wilson Australia] Sodium alkylphenoxy PEG sulfate; emulsifier, dispersant, wetting agent.

Gardol. [Apollo] Dye fixatives, fiber protectants for textiles.

Garipan®. [Boehme Filatex] Fatty acid amide deriv.; cationic softening agent for textile fibers.

Gary. [Evode Plastics Ltd.] PVC compds.

Gascodefoamer. [Gaston] Antifoam/defoamers for textile applics.

Gascofix. [Gaston] Fixative for cellulosics and blends.

Gascolev. [Gaston] Leveling agent for cellulosics and cellulosic/polyester blends.

Gascolube. [Gaston] Lubricant for textile processing.

Gascon. [Lyondell Petrochemical] General-purpose and diesel engine lubricants.

Gascopene. [Gaston] Penetrant for atmospheric systems.

Gascoreserve Salt 96. [Gaston] Mild oxidizing agent for fiber reactive dyes.

Gascoscour. [Gaston] Nonionic alkoxylate blend; scouring and soap-off agent for all fiber types.

Gascosoft. [Gaston] Softener for textiles.

Gascostend OS. [Gaston] Peroxide stabilizer for bleach baths.

Gascowet. [Gaston] Anionic alkoxylate; wetting agent for bleaching applics.

Gascowhitener. [Gaston] Stilbene sulfonates; cotton and polyester optical whiteners.

Gasket Replacer. [Hernon Mfg.] Flange gasketing and sealing.

Gatarol M30M. [Chem-Y GmbH] Myreth-3 myristate; cosmetic ingredient.

Gatodan. [Grindsted Prods.; Grindsted Prods. Denmark] Propylene glycol stearate and monoglycerides; food emulsifier.

GB TR109. [Quantum/USI] LLDPE/LDPE blend; post consumer recycled resin/virgin blend for film extrusion.

GC-44-14. [Stepan/PVO] Dipentaerythritol ester; lubricant, functional fluids for aviation, automotive, and military formulated lubricants and gas turbine engine oils.

GCA Phast. [GCA Chem.] Textile dyes and pigments.

GCP. [Betz Industrial] Biocides, additives for paint spray booths.

G-Cryl®. [Henkel] Acrylic emulsion polymers or solid resins; for inks, coatings, varnishes; as grinding vehicle.

G-Cure®. [Henkel] Acrylic resin; for paints, coatings, topcoats.

GearKote. [Witco] High visc., adherent, water-resistant gear lubricants.

Gearope. [Kano Labs] Gear and rope lubricant.

Gecet. [GE Plastics; Huntsman] Expandable engineering resin; foam bead material for use in sporting goods, automotive, furniture, building and construction, medical, shipping, materials handling applics.

Gedex. [Atochem Deutschland GmbH] Polystyrene.

Geitol. [Tokai Seiyu Ind.] Nonionic

**agents and builders; detergent, reduction cleaning agent for polyester fibers.

Gelcarin LA. [FMC/Marine Colloids] Carrageenan.

Gelcharg HP4. [Hercules] Hydroxypropyl guar; thickener, protective colloid, stabilizer.

Geleol. [Gattefosse; Gattefosse SA] Glyceryl stearate; emulsifier, stabilizer for ointments, creams, lotions.

Gelex 31. [Am. Maize Prods.] Modified amioca starch.

Gelloid. [FMC/Marine Colloids] Carrageenan; gelling agent.

Gelot 64®. [Gattefosse; Gattefosse SA] Glyceryl stearate and PEG-75 stearate; self-emulsifying base for cosmetics and pharmaceuticals.

Geloy®. [GE Plastics] Acrylic-styrene-acrylonitrile terpolymers; outdoor weatherable polymer used in the building and construction market for window and door profiles, gutters and downspouts, mobile home skirting, and capstock for siding.

Geltech Silica Powd. [Geltech] Silica; for dental composites, biomedical applics., light diffusers, polymer composites for electronic applics.

Gelucire. [Gattefosse SA] Saturated polyglycolized glycerides or hemi-synthetic glycerides; excipients for hard gelatin capsules.

Gelva. [Monsanto] Polyvinyl acetate.

Gelva Emulsion Resin TS-100. [Monsanto] Acrylates/VA copolymer.

Gelwhite GP, H. [Southern Clay Prods.] White montmorillonite.

Gelwhite MAS. [Southern Clay Prods.] Magnesium aluminum silicate.

Gemerald. [Atomergic Chemetals] Synthetic emeralds.

Gemini. [DuPont/Ag] Herbicide.

Gemtex®. [Finetex] Sulfosuccinates; surfactant, wetting agent, dye leveler for textiles, emulsion polymerization, toiletries.

Gemtone. [Mearl] Cosmetic pearl powds.

Gen III. [Lubrizol] Diesel fuel additives.

Genagen. [Hoechst Celanese/Colorants & Surf.; Hoechst AG] Ethoxylated ethers or esters; detergent, lubricant.

Genamid®. [Henkel] Amidoamine resin; epoxy curing agent; used for coatings, castings, potting, laminating, and adhesives.

Genamin. [Hoechst Celanese/Colorants & Surf.; Hoechst AG] Amines, ethoxylated amines, or quaternaries; base material, emulsifier, conditioner for hair care and industrial prods.

Genamine. [Hoechst Celanese] Ethoxylated amines; raw material for cosmetics, agric. formulations, adhesives.

Genaminox. [Hoechst Celanese/Colorants & Surf.; Hoechst AG] Alkyl dimethyl amine oxides; foaming agent, stabilizer, thickener for personal care prods.

Genapol®. [Hoechst Celanese/Colorants & Surf.; Hoechst AG] Detergent, foaming agent, lubricant, dispersant, pearlescent for cosmetics, textiles, leather, paper auxiliaries, cleaners, metalworking fluids.

Gen Carrageenan. [Hercules] Refined carrageenan; emulsifier, stabilizer, thickener, gellant for foods, pharmaceuticals, cosmetics.

Genclear®. [General Chem.] Reducing agent.

Genclor. [ICI Am.] Chlorinated PVC.

Gendriv. [Aqualon] Guar gum and derivs.; formation aid, fiber recovery and retention aid for paper industry.

Genencor® Cellulase. [Genencor Int'l.] Cellulase; enzyme for hydrolysis of cellulose.

Generol®. [Henkel/Emery/Cospha; Henkel KGaA] Soya sterol or ethoxylates; emollient, emulsifier, stabilizer, visc. modifier, solubilizer, pigment dispersant, wetting agent for cosmetics.

Genesolv®. [Allied-Signal] Fluorocarbon solvents; precision solvents for cleaning, degreasing, flux removal.

Genetron®. [Allied-Signal] Chlorine/fluorine compds.; refrigerant gas, blowing agents.

Gen Flo. [GenCorp Polymer Prods.] S/B latex; for blending.

Gen+Floc. [General Chem.] Flocculating agents.

Gen+Ion. [General Chem.] Retention/

drainage aids.

Genitron. [Schering Industrial Chem.] Chemical blowing agents, azo initiators.

Gen-Lite. [Gencorp Polymer Prods. UK] Vinyls for lightweight luggage.

Genopur. [Hoechst AG] Alkenyl dicarboxylic acid anhydride; solubilizer for heavy-duty liq. detergents.

Genotherm. [VT Plastics Ltd.] Antistatic PVC.

Gen Pac. [General Chem.] Polyaluminum hydroxychloride.

Genseke. [McIntyre] Rubber lubricants and processing aids.

Gen-Tac. [GenCorp Polymer Prods.] Styrene butadiene/vinyl pyridine latex; used for structural adhesives.

Gental®. [General Plastics] Nylon resin sol'n.; used for thread bonding, as fabric antifray for seams and edges, as stiffening agent, abrasion resistance agent.

Genton®. [General Plastics] Nylon resin dispersion; used for thread bonding, as fabric antifray for seams and edges, as stiffening agent, abrasion resistance agent.

Genu. [Hercules] Pectin; gellant for confectionary industry.

Genugel. [Hercules] Carrageenan; gellant, thickener, stabilizer, suspender for foods, pharmaceuticals, and cosmetics; water binder.

Genulacta. [Hercules] Carrageenan; gellant, thickener, stabilizer, suspender for foods, pharmaceuticals, and cosmetics; water binder.

Genuvisco. [Hercules] Carrageenan; gellant, thickener, stabilizer, suspender for foods, pharmaceuticals, and cosmetics; water binder.

Genuzan. [Hercules] Guar and xanthan gums; thickener, stabilizer used in animal feed.

Geocell. [Mearl] Low-density cellular concrete.

GeoLast®. [Advanced Elastomer Systems; Monsanto] Thermoplastic rubber; high oil resistance rubber.

Geolite. [Union Carbide] Softeners, stabilizers for urethane foam.

Geon®. [BFGoodrich; BFGoodrich UK] Vinyl resins, compds., and latexes.

Georgia Gulf. [Georgia Gulf] Vinyl compds.; for profile extrusion, molding, engineering applics.

Gepel. [GE Silicones] Silicone protective penetrant; for concrete, masonry, and dimensional stone.

Geracryl. [Cornelius Chem. Group Ltd.] Acrylic.

Geranonitril. [Henkel/Emery/Cospha] Lemon aroma chemical.

Gerbavert. [Henkel/Emery/Cospha] Fruity aroma chemical.

Germaben®. [Sutton Labs] Diazolidinyl urea blends; antimicrobial preservative for cosmetic prods.

Germall®. [Sutton Labs] Diazolidinyl urea; antimicrobial preservative for cosmetic prods.

Germul. [Ceca SA] Acrylic, vinylacrylic, or vinyl acetate resins; hand modifiers for textile finishing.

Geronol. [Rhone-Poulenc Surf.; Rhone-Poulenc Geronazzo SpA] Ethoxylate blends; emulsifier for emulsion polymerization, pesticides, cutting oils, detergent base, foamer, foam stabilizer, dispersant.

Geropon®. [Rhone-Poulenc Surf.; Rhone-Poulenc Geronazzo SpA] Sulfosuccinates, sulfosuccinamates, taurates, isethionates, or sulfonated polyester; emulsifier, wetting agent, suspending agent, dispersant, stabilizer for emulsion polymerization, pesticides, inks, textiles, paints.

Gesexel. [Atochem UK] Expandable polystyrene.

Gesil. [GE Silicones] Silicone sealant.

Getren®. [Goldschmidt AG] Silicone-free release agent for foundry and steel industry, plastics; lubricant for tire prod.

Getter. [United Desiccants-Gates] Activated carbon; for adsorption of odors.

GF. [Wilford Plastics Ltd.] PP, PVC, ABS.

G-Fill. [Feldspar] Feldspathic sand; filler extender.

GFS®. [Kelco] Xanthan gum/galactomannan blend; gum for use in food preparations; stabilizer, suspending agent, thickener.

Gilcote. [Man-Gill Chem.] Metal drawing compds.

Gilotherm. [Monsanto] Heat transfer fluid.

Gilumag. [Giulini Corp.] Magnesium hydroxide suspensions and powds.

Gistex. [Gist-Brocades Food Ingreds.] Yeast extract.

Giv-Gard DXN®. [Givaudan-Roure] 6-Acetoxy-2,4-dimethyl-m-dioxane; antimicrobial, bactericide, fungicide for industrial emulsions, fiber lubricants, spin finishes, sizes, softeners, wax and silicone emulsions, thickener sol'ns., latex binders, inks, dyestuffs.

Giv-Scents & Fresh Linen Notes. [Givaudan-Roure] Scents for bathroom, bedroom, kitchen, and dining linens and for apparel; also for masking offensive odors.

Givsorb®. [Givaudan-Roure] Industrial uv absorber, antioxidant for plastics, coatings, polishes, dyestuffs, carpet treatments.

GL 1-1010. [General Latex & Chem.] Seam and scrim adhesives; yarn splicing.

GL 301. [Central Soya] Functional soy conc.

Glacier. [Cyprus Industrial Min.] Coarse ground talc; for fertilizers and insecticides.

Glanair. [Bruce Chem.] Liq. replacement for sodium tripolyphosphate in scouring and bleaching of cotton and polyester/cotton fabrics.

Glasan. [Plastiques Obra SA] Styrene acrylonitrile.

Glascol. [Allied Colloids] Binders for inks, resins.

Glasgrip. [Ashland] Glass sealant.

Glaskyd®. [Am. Cyanamid] Glass-reinforced alkyd molding compds.

Glassclad®. [Hüls Am.] Silane surface treatments; hydrophobic treatment and lubricant for glass and ceramics.

Glass H. [FMC] Sodium hexametaphosphate.

Glas-Shot. [Cataphote] Glass bead abrasive with soft touch.

Glazamine M. [ICI Surf. UK] Stabilized hexamethylol melamine resin; thermoset for durable press finishes on cellulosic fibers , for fixation, stiffening effects on synthetic fabric.

Glazd Penta. [Vulcan Materials] Pentachlorophenol.

Glean. [DuPont/Ag] Herbicide.

Glidmint. [SCM Glidco Organics] Essential and synthetic oils.

Glidox. [SCM Glidco Organics] Pine oils, dipentene catalysts.

Glidsafe. [SCM Glidco Organics] Solvents.

Glidsol. [SCM Glidco Organics] Solvents.

Glissofluid®. [BASF AG] Aliphatic dicarboxylic acid ester; component for synthetic lubricants.

Glissolube® Grades. [BASF AG] Polyalkylene oxide derivs.; synthetic components for high-performance lubricants.

Glissopal®. [BASF AG] Intermediates and components for prod. of lubricating oil additives and lubricating oils.

Glissosafe®. [BASF AG] Hydraulic fluids for mining industry.

Glissoviscal®. [BASF AG] Polyisobutylene or diene-styrene copolymer; thickener, visc. index improver for lubricating oils.

Glitter Adhesive #5048. [Polymer Research Corp. of Am.] Aq. nonionic emulsion; thickened emulsion used for adhering glitter to fabrics.

GLOB®. [TSE] Urethane polymer with sika powds.

Globe. [Corn Prods.] Corn syrups.

Glo-Break. [Global United Industries] Glycol esters, phenolic resin oxyalkylates, or glycol epoxides; demulsifiers for crude oil and industrial emulsions.

Glo-Cor. [Global United Industries] Imidazolines or rosin oxyalkylates; corrosion inhibitor intermediate, antistat.

Glo Cryl. [Glo-Tex] Acrylic copolymers; pigment binders, hand builder.

Glo Fix. [Glo-Tex] Dye fixatives.

Glo Guard. [Glo-Tex] Fluorocarbon used as durable oil and water repellent.

Glokill. [Rhone-Poulenc Ltd.] Formulated fungicides, antimicrobials, preservatives; for water-based lubricants, toiletries and cosmetics.

Glo Lite. [Glo-Tex] Thiourea dioxide; hydro replacement for reducing vats, clearing, stripping, and cleaning equip.

Glo Lube. [Glo-Tex] Lubricant for dyes and fiber blends.

Glomax. [Dry Branch Kaolin] Calcined aluminum silicate.

Glo-Mold®. [Glo-Mold] In-place mold cleaner for preventive maintenance.

Glo-Mul. [Global United Industries] Sulfonates, sulfosuccinates, polyamides, alkanolamides, esters, or phosphate esters; emulsifier, wetting agent, detergent for oil field, agric., and industrial applics.

Glo Ox. [Glo-Tex] Oxidizing agents for bleaching, dyeing.

Glo Pel. [Glo-Tex] Water repellent, fluorochemical extender.

Glopol 461. [Rhone-Poulenc Ltd.] Acrylic quaternary ammonium polymer in IPA; antistat for mfg. of elec. conductive papers and for electrostatic printing applics.

Glo Quest. [Glo-Tex] EDTA salts; chelating agent for removal of iron from textile processing.

Glo Rez. [Glo-Tex] Polyvinyl acetate emulsion or melamine formaldehyde resin; hand modifier.

Gloria. [Witco/Sonneborn] White mineral oil USP; emollient, lubricant for food, drug, and cosmetic industries.

Glo Sil. [Glo-Tex] Silicone based softeners, water softeners, and hand modifiers.

Glo-Sperse. [Global United Industries] Quaternary amine oxyalkylate; dispersant, wetting agent, antistat, emulsifier, corrosion inhibitor.

Glo Tard. [Glo-Tex] Flame retardant for textiles.

Glo Terg. [Glo-Tex] Wetting, rewetting agent, dispersant, scouring agent, detergent for textiles.

Glo Tex. [Glo-Tex] Detergent, scouring agent for textile dyeing.

Glover. [Asarco] Lead.

Glo-Verse. [Global United Industries]

Modified polymeric amines; o/w demulsifier for crude oils and industrial emulsions.

Glo Wet Aid SR. [Glo-Tex] Sodium dioctylsulfosuccinate; wetting and rewetting agent.

GLS. [Croda Food Prods. Ltd.] Glyceryl lactostearate; used in food industry.

Gluadin®. [Henkel/Emery/Cospha; Henkel KGaA] Protein hydrolysates; additive for surfactant and skin care prods.

Glucam®. [Amerchol; Amerchol Europe] Ethoxylated or propoxylated methyl glucose ether or its esters; humectant, freezing pt. depressant, emollient, moisturizer, foam modifier, solvent, solubilizer, conditioner, lubricant, binder, plasticizer for cosmetics, pharmaceuticals;

Glucamate®. [Amerchol; Amerchol Europe] Ethoxylated methyl glucose esters; emulsifier, thickener, solubilizer for shampoos

Glucanex®. [Solvay Enzymes] Betaglucanase; enzyme.

Glucate®. [Amerchol; Amerchol Europe] Methyl glucose esters; emulsifier, conditioner, emollient, lubricant, plasticizer, pigment dispersant.

Gluconal®. [Akzo BV] Gluconate salts; bactericide, fungicide; chelating agent for alkaline media; mineral source for health foods, pharmaceuticals.

Glucopon. [Henkel/Emery] Alkyl polyglycoside; surfactant, detergent, wetting agent, surface/interfacial tension reducer, dispersant for laundry detergents, liq. cleaners, hard surface cleaners, institutional and industrial cleaners.

Glucquat 100. [Amerchol] Lauryl methyl gluceth-10 hydroxypropy dimonium chloride.

Glufil. [Agrashell] Walnut shell flour; extender/filler.

Glutarom®. [Grünau] Amino acid salt mixture on carrier; natural flavor enhancer for meat and sausage mfg.

Glutrin. [LignoTech] Calcium lignosulfonate; foundry, ceramic, and refractory binder.

Glycacil. [Lonza] Iodopropynyl butylcarbamate.

Glycerox. [Croda Chem. Ltd.] PEG glyceryl esters; solubiizer, emulsifier for cosmetics.

Glychlor®. [Lonza] 1,3-Dichloro-5,5-dimethyl hydantoin; intermediate for custom chemical synthesis, laundry bleach formulations, automatic dishwashing compds.

Glycidol Surfactant 10G. [Olin] p-Nonylphenoxy polyglycidol; surfactant for agric. emulsions, leather processing, paint, emulsion polymerization, waste paper deinking, alkaline cleaners, photographic film emulsion and coating formulations.

Glycoderm. [Henkel/Cospha] Sphingolipid liposomes with glycosaminoglycans; for dry and cracked skin care prods.; restores lipid barrier of the stratum corneum.

Glycolube®. [Lonza] Polyol ester; lubricant, antifog for plastics, PVC film.

Glycomul®. [Lonza] Sorbitan esters; emulsifier for edible, cosmetic, industrial, and pharmaceutical uses.

Glycon®. [Lonza] Fatty acids or glycerin; lubricants, defoamers, components for mfg. of food additives.

Glyconol®. [Lonza] Amide wax; synthetic wax for textiles and other applics.

Glycoserve® LAD. [Lonza] MDM hydantoin and DMDM hydantoin; household prods. preservative.

Glycosperse®. [Lonza] Ethoxylated sorbitan esters; emulsifier for food, cosmetic, pharmaceutical uses.

Glycowax®. [Lonza] Synthetic waxes; for cosmetics and other applics.

Glycox®. [Lonza] Ester wax; plastics lubricant; textile spin finish lubricant, emulsifier, antistat.

Glydant®. [Lonza] DMDM hydantoin prods.; preservative, broad spectrum antimicrobial for cosmetics and toiletries.

Glydexx. [Exxon] Glycidyl decanoate.

Glydwell. [Lunds of Bingley] Polyurethane.

Glyezin®. [BASF AG] Dye solvents and fixing assistants for textile finishing.

Glykosafe®. [BASF AG] Damping flu-

ids for engine mounts.

Glymin®. [BASF AG] Leak detector fluid for double-walled storage tanks.

Glyoxal. [Hoechst Celanese/Colorants & Surf.] Aliphatic dialdehyde; hydrogen sulfide scavengers in natural gas and sour crude prod., textiles, paper, adhesives, coatings, disinfectants, deodorants.

Glyprosol. [Brooks Industries] Yeast glycoproteins; for skin and hair care cosmetics.

Glysantin®. [BASF AG] Antifreeze for combustion engines.

Glysolube. [Morton Int'l.] External release coating for rubber.

Glytex®. [Lonza] Esters; lubricant, emulsifier, antistat for textile fiber processing.

Glythermin®. [BASF AG] Glycol-based; antifreeze thermal liqs.

GMA. [Estron] Monomer.

GMM-33. [Hefti Ltd.] Glyceryl myristate; emulsifier for pharmaceutical and cosmetic creams and bases.

GMO. [Croda Food Prods. Ltd.] Glyceryl monooleate; for use in food prods.

GMR-33. [Hefti Ltd.] Edible fatty acid glycerides; emulsifier, antifoamer for food industry.

GMS. [Croda Food Prods. Ltd.] Glyceryl monostearate; for food industry.

GMS-. [Hefti Ltd.] Polyglyceryl stearate or stearate/palmitate; emulsifier, lubricant for cosmetics, pharmaceuticals, foods.

Goal. [Rohm & Haas] Herbicide.

Gohsenal. [British Traders & Shippers] Modified polyvinyl alcohol resins.

Gohseran. [British Traders & Shippers] Modified polyvinyl alcohol resins.

Goldbond. [Lawrence Industries PLC] Silicas.

Golden Bear. [Witco] Naphthenic base oils.

Golden Dawn. [Westbrook Lanolin] Lanolin; emollient, emulsifier, ointment base, hair conditioner, lipstick binder.

Golden Fleece. [Westbrook Lanolin] Lanolin; emollient, emulsifier, ointment base, hair conditioner, lipstick binder.

Golden-Pea-Pro EN-15. [Brooks Industries] Hydrolyzed golden-pea protein; for skin and hair care cosmetics.

Golden Roast. [H.B. Taylor] Sesame seed.

Golhsefimer. [British Traders & Shippers] Modified polyvinyl alcohol resins.

Golpanol® MBS. [BASF; BASF AG] Oxidant and demetallizer for industrial cleaners.

Goltix®. [Bayer] Metamitron; selective herbicide for use in fodder and sugar beet.

Gommex. [Synthron] Textile hand modifiers, sizing agents.

Good-rite® Antioxidants. [BFGoodrich/ Spec. Polymers] Antioxidant, light stabilizer for plastics.

Good-rite® K. [BFGoodrich/Spec. Polymers] Polyacrylic acids or salts; cobuilder for laundry, dish, and other cleaners; dispersant, sequestrant.

Good-rite® Latexes. [BFGoodrich/Spec. Polymers] Vinyl pyridine or S/B latex; used for adhesives tire cord, industrial goods dips, coatings, paper saturation.

Goodtouch. [BFGoodrich/Spec. Polymers] Vinyl dispersion resin; specialty polymer for high-performance vinyl or latex disposable gloves.

Gopha-Rid. [Bell Labs] Pelleted acute rodenticide for gopher and mole control.

Gothalin. [Gotham Ink & Color] Flexographic ink.

Goulac. [Borregaard LignoTech] Calcium lignosulfonate; foundry, ceramic, and refractory binder.

GP-0098. [Georgia-Pacific] Alcohol/ fatty acid; release agent for plywood industry.

GP-090, -094. [Georgia-Pacific] Paraffin/microcrystalline wax aq. emulsion; sizing agents for composition board, gypsum board prods.

GPNS. [H. Muehlstein] Natural rubber crumb with silica partitioning agent; used for adhesives, spreading and cement applics., soft compds., sponges.

GPS. [H. Muehlstein] Natural rubber crumb with silica partitioning agent; used for adhesives, spreading and cement applics., soft compds., sponges.

Graden Butyl Oleate. [Graden] Butyl oleate; low temp. plasticizer for natural and synthetic elastomers.

Gradol. [Graden] Carboxylated polyelectrolyte salts; dispersant for paints, ceramics.

Gradonic. [Graden] Alkylolamides, ethoxylated esters or ethers; emulsifier, detergent, wetting agent, dispersant, plasticizer, stabilizer.

Grafoil. [UCAR Carbon] Flexible graphite prods.

Grainfat. [Reilly-Whiteman] Sulfated fatliquors for leather industry.

Gramet. [Ferro/Grant] Calcium stearate.

Graminon® Plus. [BASF AG] Bentazon, isoproturon, dichlorprop; for postemergence control of grasses and broadleaf weeds in winter wheat and winter barley.

Grancar. [Grant Industries] Leveling agent, dye carrier.

Grancat. [Grant Industries] Catalyst for use with silicone water repellents.

Grancoat®. [Grant Industries] Coating for textile substrates.

Grancol. [Grant Industries] Soluble animal collagen.

Granextend. [Grant Industries] Fluorochemical extender to increase oil and water repellency.

Granfresh. [Grant Industries] Odor control agents for use in dyeing, finishing.

Granguard TG. [Grant Industries] Water, oil and stain repellent for use on synthetic and cellulosic fabrics.

Granlev. [Grant Industries] Retarding and leveling agent.

Granlistin. [Grant Industries] Hydrolyzed animal elastin.

Granlou. [Grant Industries] Scouring agent for synthetics and cellulosics.

Granlube. [Grant Industries] Lubricant, dispersant, penetrant, and leveling agent for fabrics.

Granol. [ICI Am.] Seed dressings.

Granox. [ICI Am.] Seed dressings.

Granpro. [Grant Industries] Hydrolyzed collagen derivs.;

Granquat-S. [Grant Industries] Steartrimonium hydrolyzed collagen.

Granrepell. [Grant Industries] Water repellents for textiles.

Granres. [Grant Industries] Hand builder, finishing agents.

Granseq. [Grant Industries] Sequesterant for fabric processing.

Gransil®. [Grant Industries] Softener, lubricant, water repellent for textiles.

Gransoft. [Grant Industries] Softener, lubricant for textiles.

Granterge. [Grant Industries] Detergent, scouring agent, wetting agent, emulsifier for textile scouring.

Granuform. [Degussa] 91% paraformaldehyde.

Gran UP. [Sanyo Chem. Industries] Anionic/nonionic surfactants; scouring agent, soaping agent for fabrics.

Gran-U-Pel. [Martin Marietta Magnesia Spec.] Fertilizer granulation binder.

Granurea. [Terra Int'l.] Gran fertilizer urea.

Granusil. [Unimin Specialty Minerals] Silica fillers and aggregates.

Granusol. [Am. Minerals] Dispersing granular micronutrients for fertilizer blending.

Granwet. [Grant Industries] Wetting agent, dispersant, emulsifier for textiles.

Graph-I-Tite. [Carborundum] High strength, high density carbon.

GrapHsize®. [Akzo] Polyurethane emulsion; surface size and treatment for paper; water repellent.

Graphtol®. [Sandoz] Organic pigments for coloring bar soaps, gels, powders, and solids.

Grass Greenzit. [W.A. Cleary] Grass colorant.

Gravex. [Graver] Ion exchange resins.

Great Lakes BA-. [Great Lakes] Tetrabromobisphenol A compds.; flame retardants for thermoplastic and thermoset polymers.

Great Lakes BC-. [Great Lakes] Brominated organic; flame retardants for plastics.

Great Lakes BE-51. [Great Lakes] Bis (allyl ether) of tetrabromobisphenol A;

flame retardant for expandable and foamed polystyrene.

Great Lakes CD-75P®. [Great Lakes] Hexabromocyclododecane; flame retardant for thermoplastic and thermoset polymers, textile treatments, latex binders, adhesives, unsat. polyesters, and coatings.

Great Lakes DBS. [Great Lakes] Dibromostyrene; fire retardant.

Great Lakes DE-. [Great Lakes] Bromodiphenyl oxide compds.; flame retardant for thermosetting, thermoplastic, and elastomeric systems.

Great Lakes DP-45®. [Great Lakes] Tetrabromophthalate ester; fire retardant plasticizer for PVC coatings.

Great Lakes FB-72. [Great Lakes] Proprietary brominated flame retardant.

Great Lakes FF 680. [Great Lakes] Bis (tribromophenoxy) ethane; flame retardant for thermoplastic and thermoset systems.

Great Lakes FR-756. [Great Lakes] Disodium salt of tetrabromophthalate anhydride; fire retardant for wool.

Great Lakes NH-1511. [Great Lakes] Intumescent fire retardant.

Great Lakes PDBS. [Great Lakes] Poly(dibromostyrene); fire retardant.

Great Lakes PE-68. [Great Lakes] Bis (2,3-dibromopropyl ether) of tetrabromobisphenol A; flame retardant used in plastics.

Great Lakes PH-73. [Great Lakes] 2,4,6-Tribromophenol; reactive intermediate for phenol-based reactions; flame retardant, antifungal agent, or chemical intermediate.

Great Lakes PHE-65. [Great Lakes] Tribromophenol allyl ether; fire retardant.

Great Lakes PHT4. [Great Lakes] Tetrabromophthalic annhydride; flame retardant.

Great Lakes PHT4-Diol. [Great Lakes] Tetrabromophthalatediol; reactive intermediate.

Great Lakes PO-64P. [Great Lakes] Poly-dibromophenylene oxide; flame retardant.

Great Lakes SP75. [Great Lakes] Stabilized hexabromocyclododecane; fire retardant.

Great Stuff Sealant. [Insta-Foam Prods.] One-component polyurethane; sealant in aerosol can.

Gredag. [Acheson Colloids] Graphited greases.

Green Bond®. [Ashland/Foundry Prods.] Bentonite; foundry binders.

Grescofix. [Gresco Mfg.] Fixing agent for dyes.

Grescolev. [Gresco Mfg.] Leveling and compatibilizing agent for dispersed and acid dyeings on blended fibers.

Grescoscour. [Gresco Mfg.] Scouring agents for wool, cellulosics, synthetic fibers, and their blends.

Grescosoft. [Gresco Mfg.] Napping agent and softener for textiles.

Grescosperse. [Gresco Mfg.] Leveling agent, dispersant for acid and pre-metallized dyes.

Grescoterge. [Gresco Mfg.] Wetting and scouring agent for cellulosics, synthetics and their blends.

Grilamid®. [EMS-Am. Grilon; EMS-Grilon UK] Thermoplastic elastomers or nylon 12 resins; for inj. molding and extrusion applics.

Grilbond. [EMS-Grilon UK] Bonding agents.

Grilesta®. [EMS-Am. Grilon; EMS-Grilon UK] Polyester resins.

Grillocin. [RITA] Zinc ricinoleate blends; absorbs malodors from sol'ns. and surfactants.

Grilloderm L 60. [RITA] Zinc pentadecene tricarboxylate.

Grillosan DS 7911. [RITA] Disodium dihydroxyethyl sulfosuccinyl undecylenate.

Grillosol. [RITA] Ricinosuccinates.

Grilloten. [RITA] Sucrose esters; emulsifier, solubilizer.

Grilon®. [EMS-Am. Grilon; EMS-Grilon UK] Nylon resins or elastomers; resins for inj. molding, extrusion, fuel filters, medical supplies, electrical/electronic components, automotive connectors, switches, housings, monofilaments, bristles, cast blown film applics.

Grilonit. [EMS-Grilon UK] Epoxy res-

ins/hardneners.

Grilont. [EMS France SA] Polyamide 6/6.

Grilpet®. [EMS-Am. Grilon; EMS-Grilon UK] PET or PBT polyester resins; inj. molding and extrusion resins.

Griltex. [EMS-Grilon UK] Hot-melt adhesives.

Grimesolv. [Stewart Hall] Water-sol. solvent degreaser.

Grindamyl. [Grindsted Prods.] Fungal alpha-amylolytic enzyme complex; dough improvers for food industry.

Grindsted Dairy Flavors. [Grindsted Prods.] Imitation dairy flavors.

Grindsted P-Pro 2000. [Grindsted Prods.] Pea protein.

Grindsted P-Star 33. [Grindsted Prods.] Pea starch.

Grindtek. [Grindsted Prods.] Glyceryl or lactylic acid derivs. or esters; emulsifier; component in creams; lubricant, plasticizer, antistat, antifogging agent for plastics, coatings.

Gripfiller. [J & W Wegman BV] Cationic talc.

Grit-O'Cobs®. [Andersons] Corn cob meal; inert plastic extender and filler; industrial absorbent, agric. chemical carriers, livestock feed roughage.

Griton. [Toho Chem. Industry] Mineral oil/nonionic surfactant blend; water-soluble cutting oil.

Grivory. [EMS-Am. Grilon] Amorphous nylon; used for extrusion, blow molding, inj. molding, blending with other resins.

Gro-Safe. [ICI Am.] Activated carbon; for agric. applics.

Groutcide®. [Henkel] Chlorothalonil; fungicide for Portland cement grout.

Grow More. [Nat'l. Research & Chem.] Soluble fertilizer.

GRZ. [DuPont] Nylon 6/6 or 6/12, glass-reinforced; used for radiator, industrial, automotive, elec., and appliance parts.

GSD 550. [Lonza] Bromochloro-5,5-dimethyl hydantoin; intermediate for chemical synthesis, laundry bleach formulations, automatic dishwashing compds.

Guardian. [Am. Cyanamid/Ag] Insecticide cattle ear tags.

Guar Gum. [Hercules] Guar gum; thickener and stabilizer for food applics.

Guartec. [Henkel] Guar-based prods.; flotation reagent, depressant, flocculant; used in mineral processing.

Guartec. [Hercules] Strength enhancement for paper industry.

Gudy HA 10. [Filmolux] Expanded polyurethane.

Guerbitol. [Henkel KGaA] Alcohols; lubricity agent, solubilizer for metalworking, textile auxiliaries.

Gulftene. [Chevron] Alpha olefins; intermediates for biodegradable surfactants and specialty industrial chemicals.

Gull. [Petrokem] Swimming pool chemicals.

Gun-Saver. [Chem-Pak] Lubricant, rust preventative for guns.

Gusathion®. [Bayer] Azinphos compds.; insecticide and acaricide.

Guthion. [Miles/Ag] Insecticide for most major fruit pests, boll weevil in cotton and ornamental insects.

G-White. [J.M. Huber] Calcium carbonate; functional filler/extender for coatings, plastics, rubber, building prods., ceramics flux, paper fillers, adhesives, cleaning compds., and polishing agents.

H

H-25. [Pea Ridge Iron Ore] Iron oxide.

H-260. [Housmex] SBR masterbatch with high styrene resin; for blending with SBR and other polymers.

H-36. [Solem Industries] Alumina trihydrate; filler used to suppress flame and smoke in latex, vinyl, and urethane carpet-backing compds.

H7250, 7300, 7310. [Hüls Am.] Silane derivs.; blocking agent, silylating reagent for pharmaceuticals.

H7301 [Hüls Am.] Hexamethyldisilazane; adhesion promoter for photoresists.

H7310. [Hüls Am.] Hexamethyldisiloxane; low m.w. siloxanes; for cleaning, polishing, and damping media.

Hagevap. [Calgon] Copper corrosion inhibitor.

Hair Complex 20/70n. [Henkel/Cospha] Placenta extract, B vitamins, and sulfur-containing amino acids in water-alcohol medium; aq.-alcoholic lotions for regenerative hair care, oily scalps, dandruff.

Hair Complex Aquosum. [Henkel/ Cospha] Herbs and B vitamins in water-alcohol medium; for treatment of scalps with dandruff or greasiness; general hair protection prods.

Hairspray Additive S. [BASF AG] Rosin acrylate; additive for hairspray formulations.

Halar®. [Ausimont] Ethylene chlorotrifluoroethylene copolymer; melt processable fluoropolymer for extrusion, inj. molding, blow molding, rotomolding, fluidized bed or electrostatic coating processes.

Halbase. [Halstab] Lead sulfates; PVC stabilizer.

Halby. [Witco/Argus] Thiochemicals.

Halcarb. [Halstab] Lead carbonates; PVC stabilizer.

Hallco®. [C.P. Hall] Esters; plasticizers, specialties.

Hallcomid®. [C.P. Hall] Dimethyl fatty acid amides; solubilizer, solvent, dispersant, wetting agent for cosmetics, cleaners, corrosion inhibitors.

Hallcote®. [C.P. Hall] Calcium or zinc salts of carbonates or stearates; slab dip, rubber release agent.

Hal-Lub. [Halstab] Lead stearates; PVC stabilizer/lubricant.

Halobrom. [Dead Sea Bromine] 1-Bromc-3-chloro-5,5-dimethylhydantoin; broad spectrum biocide for control of algae, bacterial and fungal slimes in swimming pools and industrial water systems.

Haloflex. [ICI Resins] Vinylidene chloride/vinyl chloride/acrylic emulsions; for formulating waterborne adhesives.

Halofree. [Solem Industries] Halogen-free flame retardant for wire and cable, inj.-molded parts, extrusions, coatings, and adhesives.

Halon. [Great Lakes] Fire extinguishing agents.

Halopont. [Miles/Organic Prods.] Textile dyes and pigments.

Halox. [Hammond Lead Prods.] White non-lead inhibitive pigments.

Halphos. [Halstab] Lead phosphite, dibasic; PVC heat and light stabilizer.

Halso®. [OxyChem] Monochlorotoluene; extender, diluent, or substitute for other organic solvents; for dye carrier, fuel oil additive, sludge solvent, in paint thinners and strippers, metal parts cleaners, adhesives, agric. applics.

Halstab. [Halstab] Lead stabilizers.

Haltex. [Hitox] Alumina trihydrate; smoke suppressant and flame retardant filler for plastics, carpet backings, tub and shower stalls, wire and cable insulation, electrical uses, vinyl coated fabrics, rubber prods..

Halthal. [Halstab] Lead phthalates; PVC heat stabilizer.

Halts-Rust. [Steelcote Mfg.] Rust preventative.

Hampamide. [Evans Chemetics/W.R. Grace] Nitrilotriacetamide aq. sol'ns.

Hamp-Ene®. [W.R. Grace/Organics] EDTA or its salts; chelating agents used in photographic developers, oxidation-reduction systems, and emulsion polymerization.

Hamp-Ex®. [W.R. Grace/Organics] Pentetic acid or its salts; chelating agent; stabilizer for hydrogen peroxide in kier bleaching of textiles.

Hampfoam. [W.R. Grace/Organics] Carboxylated fatty amide sodium salt; surfactant for alkaline industrial formulations.

Hamp-Foam® 35. [W.R. Grace/Organics] Nitrogen modified anionic surfactant; high foaming agent for textile foam finishing.

Hampirex. [W.R. Grace/Organics] Chelated micronutrients.

Hamp-Iron. [W.R. Grace/Organics] Iron chelates for agric. industry.

Hamplex. [W.R. Grace/Organics] DPS; chelating agents for industrial use.

Hamp-Ol®. [W.R. Grace/Organics] HEDTA or its salts; chelating agent for control of iron, Ca, Mg.

Hamposyl®. [W.R. Grace/Organics] Sarcosine derivs.; detergent, wetting and foaming agent for hair and rug shampoos and cosmetics.

Hampshire®. [W.R. Grace/Organics] Nitrilotriacetic acid or salts or glycine derivs.; chelating agent, detergent builder for laundry detergents, specialty cleaners, water treatment, textiles, metal finishing.

Hamp-Tex. [W.R. Grace/Organics] Chelating agents for textile industry.

HAN® 857. [Exxon] Aromatic solvent.

Hanno-El. [Hanno-Werk Werner Kemper] Electrically conductive foams.

Hanno Purschaum Rapid und Super 5. [Hanno-Werk Werner Kemper] Polyurethane foam.

Hannowik. [Hanno-Werk Werner Kem-

per] Mixed fleece acrylic fibers.

Hapex. [Hastings Plastics] Epoxy components and kits.

Hapol. [Hastings Plastics] Polyester resin systems.

Har 160 Mica. [KMG Minerals] High aspect ratio wet ground mica.

Harco. [Manufacturers Chems.] Silicone defoamer.

Hardset. [Hardman] Rubber compd.

Harmony. [DuPont/Ag] Herbicide.

Haro® Chem. [Harcros] Metal soaps; stabilizer for cosmetics, oils, fats, lubricants, pharmaceuticals, paints, printing inks, plastics, rubbers, fertilizers.

Haro® Gel. [Harcros] Metal soaps.

Haroil SCO. [Graden] Sulfated castor oil; superfatting agent, emulsifier for cosmetics; plasticizer for adhesives; antisagging and antisettling agent for paints and stains.

Harol. [Graden] Polymerized naphthalene sulfonate; stabilizer, visc. depressant; dispersant for carbon black, clays; paper coating compds.

Haro® Mix. [Harcros] One-pack heat and light stabilizers and lubricants for PVC applics.

Harowan. [Harcros UK] Lubricants.

Haro® Wax. [Harcros] Lubricant blend for PVC applics.

Harshaw Antimony Oxide KR. [Atochem N. Am.] Antimony oxide; flame retardant for plastics.

Hartaine. [Hart Prods. Corp.] Betaine; detergent, emulsifier.

Hartamide. [Hart Chem. Ltd.] Fatty acid alkanolamides; coupling agent, detergent, foam and emulsion stabilizer, visc. regulator, lubricant, antistat for detergents and cosmetics.

Hartamine. [Hart Chem. Ltd.] Fatty acid imidazolines; corrosion inhibitors for oil and gas industry.

Hartasist. [Hart Chem. Ltd.] Defoamer for cementing, drilling muds; dedusting agent for coal; paraffin dispersants for oilfield applics.; cold flow improver for oil deposits; heavy-duty cleaner.

Hartasperse. [Hart Chem. Ltd.] De-inking agents.

Hartbreak. [Hart Chem. Ltd.] Phenolic

resins, polyol esters, or alkylaryl sulfonates; demulsifiers for water/oil emulsions in oil and gas industry.

Hartenol. [Hart Prods. Corp.] Sulfates; detergent, wetting agent, emulsifier, foaming agent for household detergents.

Hartex. [Hart Prods. Corp.] Surfactants and emulsifiers.

Hartex®. [Firestone Syn. Rubber] Low ammonia natural rubber latex; used for adhesives, dipped goods, foam rubber formulations.

Hartofix. [Hart Chem. Ltd.] Cationic resin; fixative for direct and reactive dyes.

Hartofol. [Hart Prods. Corp.] Dodecylbenzene sulfonates; anionic surfactants.

Hartolan. [Croda Inc.; Croda Chem. Ltd.] Lanolin alcohols; spreading agent, dispersant, stabilizer, plasticizer, o/w emulsifier and emollient for cosmetic and pharmaceutical systems.

Hartolite. [Croda Chem. Ltd.] Lanolin alcohols fraction; emulsifier, emollient, skin conditioner, moisturizing agent.

Hartolon. [Hart Chem. Ltd.] Ester blends; softener, lubricant, antistat used in textiles.

Hartomer. [Hart Chem. Ltd.] Surfactants for emulsion polymerization; antiredeposition agents, dispersants, processing aids.

Hartomul. [Hart Chem. Ltd.] Polyethylene and polyether deriv. blends; softener for use with resin finishes on lightweight fabrics; release agent for braking linings and urethane foam.

Hartonyl. [Hart Chem. Ltd.] Ethoxylated fatty amine; leveling aid, dispersant, dyeing assistant, antiprecipitant for dyestuffs, textile applics.

Hartopol. [Hart Chem. Ltd.] Polyoxyalkylene glycol; dispersant, demulsifier, emulsifier, rinse aid, defoamer for windshield washer fluids, emulsion polymerization, industrial formulations, household prods.

Hartosoft. [Hart Chem. Ltd.] Blended prod.; softener for textiles.

Hartosolve. [Hart Prods. Corp.] Emulsified solvents; solvent scour and cleaner.

Hartotrope. [Hart Chem. Ltd.] Sulfonates; hydrotrope, detergent, solubilizer, cloud pt. depressant for lt. duty and built liq. detergent systems.

Hartowet. [Hart Chem. Ltd.] Wetting agent for finishing formulations in textiles, pulp and paper processing.

Hartox. [Hart Prods. Corp.] Amine oxide; wetting agent, foaming agent, foam stabilizer.

Harvade. [Uniroyal] Plant growth regulator.

HASA. [Hernon Mfg.] Hernon anaerobic structural adhesives; for demanding applics. like audio speaker magnets, auto rearview mirrors, electric motor assembly.

Hastelloy®. [Haynes Int'l.] Corrosion-resistant and heat-resistant alloys.

Hatcol. [Hatco] Plasticizers and synthetic lubricants.

Hathane. [Hastings Plastics] Polyurethane foam.

Haynes® 242 Alloy. [Haynes Int'l.] Ni-Mo-Cr alloy; age-hardenable alloy with high temp. strength, low thermal expansion, resistance to high-temp. fluorine and fluoride environments.

Haynes Ti-3Al-2.5V. [Haynes Int'l.] Titanium alloy.

Hazy. [GE Plastics] Polypropylene film; as dielectric material for elec. insulation applics.

HB-40. [Monsanto] Partially hydrogenated terphenyl; plasticizer extender, polymer modifier, resin solvator for vinyl sheeting, films, fabric or paper coatings, vinyl protective coatings, adhesives.

HB500/1S Film. [Hercules] Saran-coated (one side) polypropylene film; for gas barrier applics., meat, cheese, and coffee pkg., lidding stocks, and pouch packages for food items, and condiments.

HBF. [Croda Food Prods. Ltd.] Edible fatty acids triglycerides; for powdered fat coatings, starch complexing agent in bread.

H.B. Fuller®. [H.B. Fuller] Adhesives.

H Chrome Green 105. [Presperse] Chromium hydroxide, bismuth oxychloride.

HCFC 123. [Allied-Signal] Dichloro-trifluoroethane.

HCFC 141b. [Allied-Signal] Dichloro-fluoroethane.

HCP. [Chemtech Industries] Copper pyrophosphate systems.

HC SynCore®. [Hysol Aerospace Prods.] Drapable low density syntactic film; for composite assemblies, closed cell core.

HC SynSkin. [Hysol Aerospace Prods.] Composite surfacing films.

HD-Eutanol®. [Henkel/Emery/Cospha] Oleyl alcohol; solubilizer for dyes and waxes, emollient.

HDO. [BASF] 1,6-Hexanediol.

HD-Ocenol®. [Henkel/Emery/Cospha; Henkel KGaA] Linear primary alcohols; intermediates for surfactant mfg., emollient, superfatting agent, carrier for cosmetics, emulsions.

HDPE. [Dow Plastics] HDPE; for inj. molding, extrusion, blow molding and sheet vacuum forming applics. incl. gas tanks, drums, large industrial parts; corrugated tubing, housewares, blown film.

Heatherlube RG. [George A. Goulston] Antistatic oil for wool and blends.

Heca III Adsorber. [Barnebey & Sutcliffe] Activated carbon; high efficiency vapor adsorber.

Hectabrite. [Am. Colloid] Sodium hectorite.

Heef 25. [Atochem N. Am.] Chromium plating.

Heico. [Heico] Chemicals.

Hei-Score. [Heico] High purity inorganic chemicals.

Hekto. [BASF] Dye.

Helastic. [KMZ Chem. Ltd.] Polyurethane dispersion.

Heliogen®. [BASF; BASF AG] Blue and green phthalocyanine pigments; for inks, paints, lacquers, plastics, artists' paints, crayons, wax chalks, textiles.

Helizarin®. [BASF AG] Pigment formulations and binders for textile dyeing and printing; crosslinking agent improving wash fastness or pigment prints.

Helmerco. [BASF] Textile dyes and pigments.

Heloxy®. [Rhone-Poulenc/Perf. Resins & Coatings] Glycidyl ethers; reactive diluent for epoxy resins; used in casting, laminating, tooling, potting, elec., adhesive, flooring, and grouting applications.

Hemicellulase Novo. [Novo Nordisk] Endopolysaccharidase.

Hemitone. [Nutex] Liq. nonionic softener and lubricant esp. for foam finishing.

Hepteen Base®. [Uniroyal] Accelerator for latex, foam, sundries; intermediate.

Herbatox®. [BASF AG] Bentazon, isoproturon, dichlorprop; for post-emergence control of grasses and broadleaf weeds in winter cereals and spring wheat.

Herbavert. [Henkel] 3,3,5-Trimethyl-cyclohexylethyl ether; fragrance raw material for shampoos, foam baths.

Herbi Spray. [Western Nutrients] Spray adjuvant for herbicides.

Hercat 627. [Hercules] Cationic dispersed rosin size; for paper industry.

Hercobond® 339. [Hercules] Water-soluble polymer for strength enhancement for paper industry.

Hercoflat® Texturing Pigments and Flatting Agent. [Aqualon] Polypropylene; used in coatings for textured, nonglare finishes.

Hercoflex® Plasticizer. [Aqualon] Pentaerythritol ester; plasticizer for PVC; used for wire insulation, government specification cable construction and high-quality plastisol formulations.

Hercolube® Synthetic Ester. [Aqualon] Polyol esters; ready-to-use compounded lubricating oils; gear oils for industrial and automotive uses.

Hercolyn®. [Hercules] Rosin derivs.; plasticizer/tackifier for lacquers, inks, adhesives, floor tiles, vinyl plastisols, artificial leather, and antifouling paints; fixative and carrier in perfumes and cosmetic preps.

Hercon®. [Hercules] Reactive sizes; for gypsum paper.

Herco® Pine Oil. [Hercules] Pine oil; for household and industrial cleaners; disinfectant, antifoam agent, textile specialties.

Hercosett®. [Hercules] Reactive polyamide-epichlorohydrin resins; wool shrinkproofing, antistat finishing; printing pretreatment.

Hercotac®. [Hercules] Modified aromatic hydrocarbon polymer; used in pressure-sensitive and hot-melt adhesives and coatings.

Hercules® 37 Series. [Hercules] Formaldehyde sol'n.; mfg. of synthetic resins used for molded articles, elec. insulation, binders for grinding wheels, plywood adhesives, varnishes, wet-strength resins for paper; chemical intermediate; hardener for proteins.

Hercules® 248, 752, 1331, X Dry Size. [Hercules] Rosin derivs.; dry size for paperboard and building prods.

Hercules® 630 Hydrocarbon Emulsion. [Hercules] Release agent for isocyanate-treated composite board.

Hercules® AS, RS, SS. [Hercules] Nitrocellulose; film-formers, bases for air-dry finishes, primers, surfacers, topcoats for transportation industry, lacquers, printing inks, adhesives, cements.

Hercules® Carbon Fabric. [Hercules] Fabric composite for structural wet layup applics., in prepregging applics.

Hercules® Carbon Fiber. [Hercules] PAN-based carbon fibers; used for weaving, prepregging, filament winding, pultrusion, and molding compds.

Hercules® Carbon Prepreg Tape. [Hercules] Prepreg tape.

Hercules® Defoamer. [Hercules] Defoamer for paper and food pkg. applics., pulp washing, waste treatment systems, coatings.

Hercules® Ester Gum. [Hercules] Glyceryl rosinates; thermoplastic resin used as softener, plasticizer in chewing gums, as clouding agent in beverages, in adhesives, inks, protective coatings; resin modifier for film-formers.

Hercules® Ethylcellulose Series. [Hercules] Ethylcellulose; film-former in coatings, hot-melts, plastic coatings, inks.

Hercules® IAD Dry-Strength Additive. [Hercules] Cellulosic polymer; increases efficiency of Kymene wet-strength resins for paper.

Hercules® Ink-Pak. [Hercules] Dye pak and acid pak for the Hercules Sizing Tester.

Hercules® PE. [Hercules] Pentaerythritols; used in prod. of alkyd resins, rosin esters, oil-modified urethane resins, drying oils, synthetic lubricants, plasticizers, intumescent paints, plastics, stabilizers for plastics, explosives.

Hercules® Release Agents. [Hercules] Release agents and creping aids for tissue and toweling machines, composite board.

Hercules® Res A. [Hercules] Resin dispersion; tackifier for acrylic-latex polymers used in label, tape, and construction adhesives.

Hercules® Resin 917. [Hercules] Cationic urea-formaldehyde resin aq. sol'n.; thermoset for modifying water-sol. polymers, imparting water resistance to films, coatings.

Hercules® Resin Emulsion 1098. [Hercules] Air entrainment agent for flotation save-alls.

Hercules® Retention Management Systems. [Hercules] Coagulant/flocculant blends for alkaline fine paper.

Hercules® Size. [Hercules] Sizing agents for paper/paperboard industry.

Hercules® Surfactant. [Hercules] PEG esters; detergent, emulsifier for industrial applics.

Hercuprene. [J-Von] Thermoplastic elastomer (PP/EPDM blends); flexible material for appliances (tubes, vibration dampers, gaskets, seals, diaphragms), automotive (bushings, tubing, hoses), building and construction, hose and tubing, connectors, sporting goods, elec. insulation, cable jacketing.

Heresite. [Heresite] Phenolic resins; thermoset for mfg. of coatings, adhesives, cements.

Herox. [DuPont] Nylon level filament; brush filaments.

Herrifin. [Paradigm Labs] Softener, lubricant, water repellent for natural and synthetic fibers.

Herriscour. [Paradigm Labs] Detergent for jet and pressure systems.

Herrisoft. [Paradigm Labs] Cationic softener for cottons and blends.

Hest MS. [Heterene] Myristyl stearate; bodying agent for creams and lotions; pearling agent, emollient; spermaceti replacement.

Het® Acid. [OxyChem] Chlorendic acid; used in unsaturated polyesters.

Hetamide. [Heterene] Fatty acid alkanolamides; foam stabilizer, emulsifier, visc. builder, detergent for car shampoos, cleaning, cosmetics, dispersants, lubricants.

Hetamine. [Heterene] Fatty amines; antistat, conditioner for hair.

Hetan. [Heterene] Sorbitan esters; emulsifier.

Hetester. [Bernel; Heterene] Esters or propoxylated esters. emollient, bodying agent, film former, pigment wetter, gloss agent, stick component for cosmetics.

Hetlan. [Heterene] Acetylated lanolin alcohols; emollient for creams and lotions.

Hetox. [Heterene] Cationics.

Hetoxalan. [Heterene] Ethoxylated lanolin.

Hetoxamate. [Heterene] Ethoxylated esters; detergent, emulsifier, lubricant, softener for cosmetics, textiles, leather, metal cleaning.

Hetoxamide. [Heterene] Ethoxylated fatty amides.

Hetoxamine. [Heterene] Ethoxylated fatty amines; emulsifier, softener, antistat, water repellent, desizing agent in agriculture, waxes, oils, textile/leather, metal cleaning.

Hetoxide. [Heterene] Ethoxylates; emulsifier, emollient, visc. control agent, lubricant, dispersant, perfume solubilizer for cosmetics, household, textile industry, metal treating and plating; intermediate.

Hetoxol. [Heterene] Alkoxylated ethers; detergent, emulsifier, leveling agent, intermediate; for personal care prods., wax, oil, textiles, scouring agents, dyes, household formulations, silicone emulsification, surfactants.

Hetphos. [Heterene] Phosphate derivs.

Hetquat S-20. [Heterene] Stearalkonium chloride.

Hetrazeen. [Heterochem] Menadione dimethylpyrimidinol bisulfite.

Hetrogen K. [Heterochem] Menadione sodium bisulfite complex.

Hetron® 92 Series. [Ashland/Composite Polymers] Fire-retardant polyester resins; for corrosion-resistant and fire retardant fiberglass equipment.

Hetron® 197. [Ashland] Halogenated thermoset polyester resins; for corrosion resistant fiberglass-reinforced plastic equipment.

Hetron® 800. [Ashland/Composite Polymers] Thermosetting furan resin; for corrosion-resistant fiberglass-reinforced plastic laminates.

Hetron® 900 Series. [Ashland/Composite Polymers] Vinyl ester resins; for corrosion-resistant reinforced thermosetting plastic equipment.

Hetsorb. [Heterene] Ethoxylated sorbitan esters; detergent, emulsifier, lubricant for cosmetics, foods.

Hetsulf. [Heterene] Sulfonates; wetting agent, emulsifier, dispersant, intermediate, detergent.

Heveagrip. [Heveatex] Resorcinol-formaldehyde latex; saturant to improve adhesion to nylon, rayon, or polyester substrates.

Heveanol. [Heveatex] Natural rubber latexes and compds.; used for adhesive dipping compds., foam compds., coatings for textiles, paper, etc.

Heveaplus. [Akrochem] Rubber latex.

Heveasyn. [Heveatex] Nitrile, PVC, polychloroprene, SBR latexes; used for dipped goods, adhesives, coatings.

Hexaphos. [FMC] Sodium hexametaphosphate.

Hexaplant Richter. [Dr. Kurt Richter; Henkel/Cospha] Water, alcohol, fennel extract, hops extract, balm mint extract, mistletoe extract, matricaria extract, yarrow extract; herbal extract as emollient for cosmetics.

Hexaryl D 60 L. [Witco SA] TEA-dodecylbenzene sulfonate; multipurpose detergent.

Hexcel. [Hercules] Thermoset phenolic, polyester, bismaleimide, or polyimide systems.

Hexcel. [Hexcel] Thermoset polyurethane compds. encapsulants; for cable plugging.

Hexcel FO 425A. [Hexcel] Petroleum solvent; general purpose cleaning and degreasing solvent for cleaning telecommunications cables and tools.

Hexcelcure. [Zeeland] Amines or diamines; epoxy curatives.

Hex-Cem. [Mooney Chems] Octoate driers; for paints, inks.

Hexetidine. [Angus] Hexetidine; antimicrobial, antifungal agent for oral hygiene prods., pharmaceuticals.

Hexol G. [Specialty Chem.] Scouring assistant, water softener.

Hexoloy. [Carborundum] Silicon carbide.

Hexomax. [UOP] Immobilized invertase enzyme.

Hexuronyl Hexosaminoglycan Sulfate. [Hepar Industries] Heparinoid.

Hexyl Jasmat®. [BASF AG] Acetyl ethyl octanoate; floral, herbel, green, jasmine-like fragrance.

HF-19. [Chem-Trend] Water-glycol type; fire-resistant hydraulic fluid providing rust and corrosion protection.

HFC 134a. [Allied-Signal] Tetrafluoroethane.

H-F Microsoy. [Howard Hall Int'l.] Powdered cellulose.

HGC 5000. [Siltech] Hair gloss prod.

Hi-Care®. [Rhone-Poulenc Surf.] Quaternary ammonium chlorides; skin conditioner for cosmetic creams and lotions.

Hi Chem. [Hill Bros. Chem] Waste water treatment chemicals.

Hicond-2000. [United Composites] Electrically conductive olefin.

HiD. [Chevron] HDPE resin; resin for pressure pipe, sheets used for toys, housewares, thermoformed parts, films for shipping bags, multiwall liners.

Hidacid. [Hilton Davis] Acid dyes for textiles.

Hidaco. [Hilton Davis] Dyes.

Hi-Ex Foam. [Henkel/Emery] Fire-fighting foam conc.

HiFax. [Himont] Polypropylene resin; resins for inj. molding, extrusion, blow molding, acoustical barrier applics.

Hi Fibe. [Hill Bros. Chem] Asbestos replacement prod.

Hi-Floc. [Gulbrandsen] Polyaluminum chloride.

Hi-Fluid. [Takemoto Oil & Fat] Alkylaryl sulfonate/formaldehyde condensates; water reducing agent for superplasticized concrete.

Hi-Foam. [Climax Performance] Patented high-sudsing detergents for shampoos and dishwashing bases.

High Purity MTBE. [Arco] Methyl t-butyl ether; extraction solvent, reaction medium in pharmaceuticals, for polymerizations; octane enhancer.

High Temperature Deodorant #4896, OS. [Andrea Aromatics] Mixture of fragrance materials; deodorant for PVC processing and finished prods.

Highlink 80®. [Hoechst Celanese/Colorants & Surf.] Aliphatic dialdehyde; hydrogen sulfide scavengers in natural gas and sour crude prod.

Highsorb 40®. [Hoechst Celanese/Colorants & Surf.] Aliphatic dialdehyde; hydrogen sulfide scavengers in natural gas and sour crude prod.

HiGlass. [Himont] Polypropylene resins, glass-reinforced; used for under-hood automotive applics., fan blades, appliances, automotive parts, elec./electronic connectors, pump housings/components, plumbing parts, athletic equip.

Hi-Grade. [Cargill] Sodium chloride.

Hilex. [Royalite Plastics Ltd.] HDPE.

Hiload. [Terra Int'l.] Liq. clay; suspension agents.

Hi Loft. [Nat'l. Starch & Chem.] Vinyl copolymer latex.

Hilon. [Himac] Nylon or polyester films; for pkg. applics.

Hiltasperse. [Hilton Davis] Color and pigment dispersions.

Hiltone. [Hilton Davis] Pigments.

Hilube. [Ruscon Plastics] Engineering plastic.

Hi-Melt. [Am. Oil & Supply] Bearing

lubricant.

Hinosan®. [Bayer] Edifenphos; non-mercurial fungicide for rice.

Hipec®. [Dow Corning] Semiconductor protective materials.

Hi-Pflex®. [Pfizer] Calcium carbonate; filler and reinforcing agent for plastics.

Hi-pHase®. [Hercules] Rosin emulsion; sizing agent.

Hi-pHorm 67. [Hercules] Rosin-based; surface treatment agent.

Hiplen. [HIP Petrohemija DP] Polyethylene blends.

Hiplex. [HIP Petrohemija DP] HDPE.

Hipnil. [HIP Petrohemija DP] PVC.

Hipochem. [High Point] Detergent, wetting agent, emulsifier, foam builder, dyeing assistant, leveling agent for textiles.

Hipocoat. [High Point] Coatings for drapery, ticking, upholstery, automotive, blinds, and nonwovens.

Hipofire. [High Point] Fire retardant treatments for coatings.

Hipofix. [High Point] Methyl methylol resin; fixing agent for direct dyes on cellulosics.

Hipofoam. [High Point] Acrylic coatings for drapery, ticking, upholstery, automotive, blinds, and nonwovens.

Hi-Point®. [Witco/Argus] MEK peroxide blends; catalyst/initiator for R.T. cures of polyesters.

Hipolon New. [High Point] Modified cationic surfactant; detergent, dye retarder, leveler.

Hipoquest. [High Point] chelating agents.

Hiposcour®. [High Point] Ethoxylates; detergent, scouring agent for textiles.

Hiposoft. [High Point] Softeners for textiles.

Hipowet. [High Point] Wetting agent, foam suppressant, deaerator for textiles.

Hipowhite. [High Point] Opticals for cotton in fabric and garment processing.

Hiprotal. [Mutchler] Soluble whey protein.

Hipten. [HIP Petrohemija DP] LDPE.

Hi-Quat. [Rhone-Poulenc/Perf. Resins & Coatings] Quaternary ammonium compds.

Hiramic. [Hitemco] Abrasion-resistant release coatings.

Hi Resin. [Toho Chem. Industry] Petroleum resins; base material for paper sizing, paints, adhesives; tackifier and softener for synthetic rubbers and adhesives; ink vehicle; pigment dispersant; water repellent.

HIRI. [PPG Industries] High index thermoset resin.

Hi-Sil®. [PPG Industries] Hydrated silica; white pigment, reinforcing filler for rubber; diluent, grinding aid, stabilizer, thixotrope, thickener, carrier; for agric., adhesives, coatings, laminating, epoxy resins, caulks, putties, sealants, pharmaceuticals, cosmetics.

Hi-Sol®. [Ashland] Naphtha; solvent.

Hisolve. [Toho Chem. Industry] Glycol ether solvents; solvents for paints, inks, adhesives, plastics, plasticizers, cleaners, paint removers, lacquer thinners, dyes, brake fluids, metal surface finishing agents; reaction solvent; separation and extraction solvent.

HiStyle 133. [Rhone-Poulenc Surf.] Polymethacrylamidopropyltrimonium chloride.

HiTec®. [Ethyl] Performance additives for formulating fuels and lubricants for automotive and industrial use.

Hi-Tex. [Cargill] 99.9% NaCl.

Hi-Tor Plus. [Huntington Labs] Germicidal detergent.

Hitox®. [Hitox] Titanium dioxide; buff colored pigment used in alkyds, acrylic urethanes, high solids systems, water reducibles, water bases, powd. coatings, inks, and adhesives.

Hi-Tri. [Dow] Solvent.

Hi-Vap. [Atochem N. Am.] Metalworking lubricants.

Hi White Clay. [Evans Clay] Filler.

HM-. [H.B. Fuller] Thermoplastic polyamide and hot-melt adhesives.

HMDS Plus. [Schumacher] Hexamethyldisilazane.

HMS. [Aceto] Homosalate.

Hobolene. [Riddle & Hobb] HDPE.

Hodag. [Hodag] Surfactants.

Hodag Antifoam F-1. [Hodag] Simethicone; antifoaming agent.

Hoe S. [Hoechst Celanese/Colorants & Surf.] Quaternaries, alcohol ethoxylates/propoxylates, EO/PO block copolymers, other surfactants; for clear conditioners and conditioning shampoos, cosmetics, plant protection.

Hoechst Wax. [Hoechst Celanese] Montan or polyolefin waxes and derivs.; water repellent for textiles.

Holcoplast. [Holland Colours UK] Plasticizer/pigment pastes.

Holcopol. [Holland Colours Apeldoorn BV] Polyol pastes.

H₂old EP-1. [ISP] Vinylpyrrolidone terpolymer.

Hollofil. [DuPont] Polyester.

Hombitan®. [Sachtleben Chemie GmbH] Titanium dioxide.

Homodan. [Grindsted Prods.; Grindsted Prods. Denmark] Emulsifiers and blends; plasticizer for margarine.

Homogenol. [Kao] Dispersant for food applics.

Homosalate®. [Aceto] Homomenthyl salicylate; cosmetics ingredient.

Homo Size 7A. [Toho Chem. Industry] Cationic complex; neutral sizing agent for paper.

Homotex. [Kao/Edible Fat & Oil] Propylene glycol or glycerine esters of fatty acids; food detergent; bacteria control.

Honol. [Takemoto Oil & Fat] Finishing oil for raylon filament/staple.

Honoralin. [Takemoto Oil & Fat] Sulfated oil; wetting and penetrating agent; dyeing auxiliary for cotton and wool.

Honostab. [Egil Sterud Agentur] Stabilizers.

Hope Line. [Hope Chem.] Chemicals.

Hopp II. [Pfizer] Modified hop extract.

Hordaflam. [Hoechst Celanese] Chlorinated paraffin; flame retardant for PS, elastomers, LDPE, unsat. polyester.

Horna. [Ciba-Geigy Pigments UK] Lead chromate yellows and molybdate oranges.

Hostacerin. [Hoechst Celanese/Colorants & Surf.; Hoechst AG] Emulsifying base, thickener, superfatting agent for cosmetics and pharmaceuticals.

Hostacor. [Hoechst Celanese/Colorants & Surf.; Hoechst AG] Corrosion inhibitor, emulsifier, and lubricant for aq. systems, drilling aids, cooling systems.

Hostadrill. [Hoechst AG] Vinylamide/vinylsulfonic acid copolymers; filtration agents.

Hostaflex. [Hoechst Celanese] PVC copolymers; for industrial coatings.

Hostaflon®. [Hoechst Celanese; Hoechst UK] PTFE, some glass, carbon, graphite or bronze filled; for extrusion, compr. molding, coatings.

Hostaflot L. [Hoechst AG] Aliphatic dithiophosphates; flotation collectors for sulfide minerals.

Hostaflot X. [Hoechst AG] Thionocarbamate; flotation collectors.

Hostaform. [Hoechst Celanese; Hoechst UK] Acetal copolymer resins, some glass or mineral filled; for inj. molding, low-friction applics.

Hostagel. [Hoechst AG] Carboxylic acid polycarboxylic acid derivs.; thickeners for oily and aq. lubricants.

Hostagliss. [Hoechst AG] Fatty acid ester; lubricating component for coolants, lubricants.

Hostalan. [Hoechst Celanese] Textile dyes and pigments.

Hostalen®. [Hoechst Celanese; Hoechst UK] Polyethylene or polypropylene resins and blends; for inj. molding, extrusion, pipe, blow molding, film.

Hostalit. [Hoechst UK] PVC.

Hostalub®. [Hoechst Celanese/Spec. Chem.] Fluoroelastomer-based processing aid for polyolefins.

Hostalux®. [Hoechst Celanese/Colorants & Surf.] Benzoxazole deriv.; optical brightener, whitening agent for polymer processing.

Hostamer. [Hoechst AG] Vinylamide/vinylsulfonic acid copolymers; for enhanced oil recovery.

Hostanox®. [Hoechst Celanese/Spec. Chem.] Antioxidant for plastics.

Hostapal. [Hoechst Celanese/Colorants & Surf.; Hoechst AG] Wetting agent, detergent, dispersant, emulsifier for textiles, emulsion polymerization.

Hostaphan. [Hoechst UK] Polyester film.

Hostaphat. [Hoechst Celanese/Colorants

& Surf.; Hoechst AG] Phosphate esters; corrosion inhibitor, lubricant, antiwear additive; emulsifier for cosmetics; antistat for textiles; hydrotrope, solubilizer.

Hostapon. [Hoechst Celanese/Colorants & Surf.; Hoechst AG] Detergent base for cosmetics, household prods., textiles, leather.

Hostapren. [Hoechst UK] Chlorinated polyethylene.

Hostaprime®. [Hoechst Celanese/Spec. Chem.] Furan dione propene polymer or maleic anhydride grafted polypropylene; coupling agent for PP, PA.

Hostapur®. [Hoechst Celanese/Colorants & Surf.; Hoechst AG] Wetting agent, detergent for textiles, cleaning agents, shampoos.

Hostarex. [Hoechst AG] Secondary and tertiary amines; anion/cation exchangers.

Hostastat®. [Hoechst Celanese/Spec. Chem.] Antistat for thermoplastics.

Hostatec. [Hoechst UK] PEEK.

Hostatex®. [Hoechst Celanese/Colorants & Surf.] Leveling carrier for dyeing.

Hostatron®. [Hoechst Celanese/Spec. Chem.] Blowing agent for PS, ABS, PE, PP.

Hostavat. [Hoechst Celanese] Textile dyes and pigments.

Hostavin®. [Hoechst Celanese/Spec. Chem.] UV absorber, light stabilizer, antioxidant for plastics.

Hostawet TDC. [Hoechst Celanese/ Colorants & Surf.] Ether carboxylate; emulsifier and wetting agent for metal cleaning formulations.

Hotbac. [DuPont] Synthetic resin.

Hotemp® Plus. [Kluber Lubrication N. Am.] High temp. chain oil for tentering and/or heat setting operations.

Houghto-Safe. [E.F. Houghton] Fire resistant hydraulic fluids.

HPC. [Aquatec Chem. Int'l.] Cooling water treatment.

HPCH Liq. [U.S. Cosmetics] Hydroxypropyl chitosan; film former for hair care prods.

HPL. [Dow] Latex.

HPP. [U.S. Polymers] High performance polymers; for coatings.

HPS. [Hernon Mfg.] Hernon porosity sealants; for sealing microporosity, vacuum impregnation.

HPX. [RITA] Lyophilized human placenta protein; contains essential amino acids; hair conditioners, creams, and lotions.

HSA. [CasChem] Hydroxystearic acid.

HSB 1900, 1903, 1904. [Mach-1 Compounding] High styrene resin/SBR fluxed blend; used in shoe soles and household goods.

HSZ. [Tosoh] Crystalline aluminosilicates; catalysts and adsorbents for refinery and petrochemical industries.

HT®. [Am. Cyanamid] Phenolic epoxy; adhesive system.

H.T. [Monsanto] Monocalcium phosphate.

HT70. [Lunds of Bingley] Elasticated nylon.

HTH. [Olin] Calcium hypochlorite; dry chlorinating agent.

HT Lakes. [Colorcon] Fine grind FD&C aluminum lakes.

HTP X-5322. [HiTech Polymers] Super tough polyamide; for tool housings, appliances, elec. connectors, sporting goods, under-the-hood automotive parts, pkg. and structural applics.

HTR-4275. [DuPont] Thermoplastic polyester elastomer; blow molding resin.

HTSA. [Hexcel] Oleyl palmitamide or stearyl erucamide; release agent providing slip, antiblocking to thermoplastics such as PP film, nylon.

HTX. [BFGoodrich/Geon Vinyl] Vinylbased alloys; engineering thermoplastics for elevated temp. performance.

Huber. [J.M. Huber/Clay Div.] Kaolin clay, carbon black, or mica; fillers for rubber and plastics industries.

Huberbrite. [J.M. Huber] Barium sulfate; pigments.

Hubercarb®. [J.M. Huber] Calcium carbonate; fillers and extenders for plastics, caulks, sealants, rubber, adhesives, paints, coatings, ceramics, paper, nonabrasive cleaners.

Huberfil®. [J.M. Huber] Silicate; paper

filler.

Hubersil®. [J.M. Huber] Hydrated silica; reinforcing filler, carrier for rubber parts incl. hose, belting, molded and extruded parts, tires, footwear, wire and cable, thermoplastic elastomers, caulks and sealants.

Hubersorb®. [J.M. Huber] Silicate; conditioning agent, carrier.

Hubron. [Hubron Ltd.] Carbon black in SAN, LDPE, PP, or PS carrier; black masterbatch for compounding, molding, film, sheet extrusion.

Humectant SD-35. [Presperse] Panthenyl ethyl ether; humectant.

Humectol. [Hoechst AG] Sulfonated fatty acid amides; wetting and leveling agent, dispersant for dyeing and printing.

Humes. [Hules Mexicanos] Nitrile elastomer or NBR/PVC blends; used for molded parts, mech. goods subjected to oils and fuels, wire and cable jackets, soles, hoses.

Humidex Tablets. [Stewart Hall] Scale and slime preventives, humidifiers.

Humifen®. [Rhone-Poulenc Surf.] Sulfosuccinates or sulfonates; surfactants.

Humiplex. [Western Nutrients] Neutralized complexed foliar fertilizers containing humic acids.

Humi Plus. [Western Nutrients] Complexed micronutrients humic base.

Humi-Sorb. [Am. Colloid] Clay desiccant.

Humus. [Humus Prods. of Am.] Conc. liq. humic acid.

Huntsman. [Huntsman] Polystyrene, expandable polystyrene, or polypropylene resins.

Hustat. [Hubron Ltd.] Carbon black in PE, PP, or PS base polymer; electrically conductive compds. for ESD control in electronic pkg. applics.

Huvasol. [Hukill] Solvents, solvent blends.

HVA-2. [DuPont] Rubber chemical.

HVP. [Hercules] Hydrolyzed vegetable protein blends; flavor enhancer.

H-White. [J.M. Huber] Calcium carbonate; functional filler extender; used in coatings, plastic and rubber fillers, building prods., ceramics flux, paper fillers, adhesives, cleaning compds., polishes.

HX- Series. [DuPont] Amorphous liq. crystal polymer, some glass or mineral-reinforced; used for chip carriers, printed circuit boards, connectors, films.

Hy-Ad HF. [Sandoz] Stabilizers and extenders for fluorochemicals; durable oil and water resistance.

Hyalite. [Richard Daleman Ltd.] Clear polystyrene.

Hyalux. [Richard Daleman Ltd.] Acrylic and polystyrene.

Hyamine®. [Lonza] Quaternary ammonium compds.; germicide, disinfectant, sanitizer, preservative for restaurant and pharmaceutical uses, textiles, starch, glue, casein.

Hybase. [Witco/Sonneborn] Sodium, barium, calcium, and magnesium petroleum sulfonates; detergent, rust inhibitor, lubricant for marine or stationary diesels.

Hybon® 2011 Rovings. [PPG Industries/ Fiber Glass Prods.] Glass roving; reinforcement for polyester and vinyl ester resin systems for pultrusion applics.

Hybond. [Pierce & Stevens] Adhesive.

Hybri-Chem. [Hybri-Chem] Polyester/ polyurethane hybrid system; for open mold casting, lamination, spray applic., resin transfer molding, tooling.

Hybrid®. [OxyChem] Vinyl prods.

Hy-Brite. [FMC] Copper pickling sol'n.

Hy-Brite Hydrogen Peroxide. [Atochem N. Am.] Oxidative pickling additive.

Hycal. [J.M. Huber] Calcined paper clay.

Hycar®. [BFGoodrich; BFGoodrich UK] Nitrile, ABS, acrylic latexes; used for adhesives, carpet backcoatings, paper coatings and saturation, pigment binding, textile coatings, leather finishes, paints, resin modification.

Hycel. [Toho Chem. Industry] Polyurethane resins; for rigid foam for thermal insulation, structural material, packing materials, bath tub, industrial refrigeration; for semirigid foam for cushioning for furniture, toys, chairs; water cutting agent, stabilizing agent; for paints, roofing, elastomers.

Hycryl. [HTS High Tech Spec.] Thickeners, additives.

Hydagen®. [Henkel/Emery/Cospha; Henkel KGaA] Ingredients for deodorants, moisturizers, other personal care prods.; softener, conditioner for hair care preps.

Hy Dense. [Mallinckrodt] Calcium or zinc stearates; release agent for PVC processing.

Hydex®. [Lonza] Sorbitol; humectants, bodying agents, moisture control agents.

Hydracol. [Gattefosse] Soluble collagen.

Hydracote. [J.M. Huber] Paper coating clay.

Hydradan. [Grindsted Prods.] Hydrated monoglycerides; dough improvers for food industry.

Hydrafil. [J.M. Huber] Paper filler clay.

Hydrafine. [J.M. Huber] Paper coating clay.

Hydragloss. [J.M. Huber] Paper coating clay.

Hydraid. [Calgon] Paper drainage and retention aid.

Hydral. [Alcoa] Hydrated alumina pigment.

Hydramatte. [J.M. Huber] Paper coating clay.

Hydra Mul. [Castrol Industrial East] Fire resistant hydraulic fluids.

Hydraphtal. [DuPont] Textile processing agent.

Hydraprint. [J.M. Huber] Paper coating clay.

Hydra Safe. [Castrol Industrial East] Fire resistant hydraulic fluids.

Hydrasperse. [J.M. Huber] Paper coating clay.

Hydraulan®. [BASF AG] Brake fluids.

Hydraway. [Monsanto] Soil and highway drainage mat.

Hydrel. [DuPont/Polymers] Polyester elastomers.

Hydrenol® D. [Henkel/Cospha; Henkel KGaA] Tallow alcohol; emollient, consistency factor for skin creams and lotions, intermediate for surfactant mfg.

Hydrepoxy. [Acme Div.] Epoxy coatings.

Hydresol®. [BASF AG] Naphthalene sulfonic or phenol-sulfonic acids/formaldehyde condensates; plasticizers for mineral binders, for improving workability of mineral-bonded mortar and concrete.

Hydrex®. [J.M. Huber] Silicates; filler for the rubber industry.

Hydrholac. [Rohm & Haas] Nitrocellulose lacquer emulsion; for leather finishing.

Hydrin®. [Zeon] Epichlorohydrin/ethylene oxide copolymers or terpolymers; elastomer for fuel pump diaphragms, pipe gaskets, fuel hose, motor mounts, rolls, adhesives, sponge goods, boots, seals, o-rings, bladders.

Hydrine. [Gattefosse; Gattefosse SA] PEG-2 stearate; stabilizer for ointments, cream lotions; opacifier for shampoos, liq. soaps.

Hydrisel. [Hydrolabs] Detergents for removal of dirt, oil and grease during peroxide bleaching, prescouring, and dyeing.

Hydrite. [ECC Int'l.] Kaolin; extender pigment for paints, plastics, rubber, adhesives, inks.

Hydritex-D. [Southern Dye & Chem.] Vat dye reducing assist.

Hydroace. [Toho Chem. Industry] Anionic/nonionic blend; emulsifier for emulsion-type fuel oils.

Hydroba-70. [Jojoba Growers & Processors] Hydrogenated jojoba wax.

Hydrocal. [U.S. Gypsum] Calcium sulfate hemihydrate.

Hydrocarb. [Omya GmbH] Calcium carbonate.

Hydrocarbon Propellant A-17. [Phillips] Butane.

Hydrocarbon Propellant A-31. [Phillips] Isobutane.

Hydrocarbon Propellant A-108. [Phillips] Propane.

Hydrocell. [Brooks Industries] Yeast protein derivs.; for skin and hair care cosmetics.

Hydro-Cem. [Mooney Chems] Drier catalysts for water systems.

Hydrocerol®. [Henley; Boehringer Ingelheim] Chemical blowing and nucleating agents for foamed plastics.

Hydroclean. [Hydrolabs] Cleaners for textile process equip.

Hydroclear. [Mammoth Int'l.] Antifog, anticondensation prod. for greenhouse use.

Hydrocol. [W. Hawley & Son Ltd.] Aq. dispersions of pigments for cement.

Hydrocoll. [Brooks Industries] Gelatin or collagen prods.; for hair and skin care cosmetics.

Hydrocolloid. [Brooks Industries] Hydrolyzed collagen.

Hydro-Cure. [Mooney Chems] Catalyst for water systems.

Hydrodarco. [Am. Norit] Activated carbon.

Hydrofix. [Hydrolabs] Fixing agent for textiles.

Hydrofol. [Procter & Gamble] Fatty acids or glycerides; used in cosmetic emulsions.

Hydrogel. [Wyo-Ben] Wyoming bentonite.

Hydrogix. [Croxton & Garry Ltd.] Impact modifiers.

Hydrokeratin. [Brooks Industries] Hydrolyzed keratin; for cosmetics.

Hydrokote®. [Karlshamns] Hydrogenated vegetable oil; specialty base used as replacement for cocoa butter in cosmetic and pharmaceutical applics.;

Hydrol. [Van Den Bergh Foods] Partially hydrogenated vegetable oils; specialty oil for vegetable dairy systems, biscuits, frying, whipped toppings, cream centers, dressings.

Hydrolac. [Rhone-Poulenc] Water-soluble lacquer.

Hydrolactin 2500. [Croda Inc.; Croda Chem. Ltd.] Milk protein; skin and hair care conditioner.

Hydrolactol 70. [Gattefosse] Glyceryl stearate, propylene glycol stearate, glyceryl isostearate, propylene glycol isostearate, oleth-25, ceteth-25; self-emulsifying base for prep. of fluid, semifluid lotions and o/w creams, formulation of mineral pigments.

Hydrolan. [Lanaetex Prods.] Hydrolyzed collagen.

Hydrolast. [Hydrolabs] Silicone elastomeric prods. imparting crease resistance to natural and synthetic fibers.

Hydrolene. [Reilly-Whiteman] Sulfated vegetable oils; for metalworking, textiles, leather, printing inks, paints, detergents.

Hydrolene H, S. [Lenox] Nonionic organic ester; hygroscopic agent and lubricant in yarn processing.

Hydrolev. [Hydrolabs] Leveling agents for cationic dyes.

Hydrolin. [Olin] Sodium hydrosulfite sol'n.; for bleaching of clay, groundwood, and thermomechanical pulp.

Hydrolith. [Magie Bros. Oil] Lithium-based greases.

Hydrolube. [AJ & JO Pilar] Fiber lubricating oils and emulsions.

Hydrolube. [Hydrolabs] Dyebath lubricant.

Hydrolyzed Mucopolysaccharides SD. [Brooks Industries] Hydrolyzed glycosaminoglycans; for skin and hair care cosmetics.

Hydro Magma. [Marine Magnesium] Magnesium hydroxide USP paste.

Hydromilk. [Brooks Industries] Hydrolyzed casein or milk protein blends; cosmetics ingredients.

Hydromix. [Omya GmbH] Masterbatches and compds.; PP/PE modifier.

Hydromond. [Croda Chem. Ltd.] Hydrolyzed almond protein; conditioner for skin and hair care prods.

Hydro Myristenol. [Dragoco] Ceteareth-6 and isopropyl myristate.

Hydron. [Hydrolabs] Dye carrier.

Hydronap. [Hydrolabs] Napping softener for textiles.

Hydropalat. [Henkel; Henkel KGaA] Polyacrylates; suspending agent for water-reducible paints.

Hydrophilol Iso. [Gattefosse SA] Propylene glycol isostearate.

Hydropower. [Hydrolabs] Scour/dye system for polyester/cotton blends.

Hydropur. [Hoechst AG] Acrylamide/acrylate polymers; visc. control agents for water-based drilling muds.

Hydroquest. [Hydrolabs] Sequestrants.

Hydrosan. [Goldschmidt] Flocculating aid for water treatment.

Hydroscour. [Hydrolabs] Solvent scour system for fabrics.

Hydrosil®. [Hüls Am.] Water-borne amino, epoxy, mercapto, methacrylate, vinyl, and vinyl-amino silanes.

Hydrosist. [Hydrolabs] Leveling agent, dispersant, dye assistant, softener, napping assistant for textiles.

Hydrosolv. [Alpha Metals] Liq. organic water-soluble flux.

Hydrosoy 2000. [Croda Inc.; Croda Chem. Ltd.] Hydrolyzed vegetable protein; conditioner for cosmetics.

Hydrosperse. [Hydrolabs] Dispersant, leveling agent for textiles.

Hydrosulfite AWC. [Henkel/Emery] Sodium hydrosulfite.

Hydrosulphit®. [BASF AG] Sodium dithionite-based; reducing bleaches for wood pulps and wood-based old paper.

Hydrothol. [Atochem N. Am.] Aquatic algicide and herbicide.

Hydrotriticum. [Croda Inc.; Croda Chem. Ltd.] Wheat protein derivs.; skin and hair conditioning agent.

Hydrovegetol C. [Gattefosse] Glyceryl stearate, propylene glycol stearate, glyceryl isostearate, propylene glycol isostearate.

Hydrox. [Cuproquim] Copper hydroxide; fungicide, bactericide, dry flowables.

Hydroxylan. [Fanning] Hydroxylated lanolin; emulsifier, conditioner.

Hydroxyprolisilane C. [Exsymol] Methylsilanol aspartate hydroxyprolinate; restructurating action on collagen, restoring action on skin elasticity; for anti-aging, stretch mark prevention, acne prevention formulations, eye creams.

Hydrumines. [Exsymol] Total plant extracts incl. both liposoluble and hydrosoluble active principles; for cosmetic and health prods.

Hyfil. [Solem Industries] Alumina trihydrate and fillers; resin extender for fiberglass-reinforced polyesters.

Hyflex. [Solem Industries] Silane surface treated alumina trihydrate.

Hyflo NS, S. [H.A. Astlett] Natural rubber with inert silica; used for adhesives, pharmaceuticals.

Hyflo Super-Cel. [Manville] Diatomaceous earth; filter aid.

Hyglyol S-26. [Nikko Chem. Co. Ltd.] Glycerin/oxybutylene copolymer stearyl ether.

Hygroplex HHG. [Henkel/Cospha] Natural moisturizing factors in hydropholic medium; for emulsified aq. and hydroalcoholic skin moisturizing cosmetics.

Hylene. [DuPont] Organic isocyanate.

Hylite LF. [Nat'l. Starch & Chem.] Glyoxal reactant; imparts permanent press in precure and postcure formulations.

Hymol. [Toho Chem. Industry] Glycol ethers; solvents for brake fluid, wet process duplicator.

Hymolon. [Hart Chem. Ltd.] Cocamide DEA; detergent, emulsifier for household prods., scouring and fulling agent for textiles.

Hymono. [Quest Int'l.] Distilled monoglycerides; for food industry.

Hyonic®. [Henkel/Emery; Henkel Canada] Alkyl phenol ethoxylates; detergent, wetting agent, emulsifier, penetrant for household and industrial cleaners, agric. chemicals, corrosion inhibitors, textiles, food applics.; intermediate.

Hypale. [Arakawa] Hydrogenated rosin and ester.

Hypalon®. [DuPont; DuPont UK] Chlorosulfonated polyethylene; elastomer used in jacketing and insulation for wire/cable, soles and heels, automotive components, coated fabrics, wh. tire sidewalls, sheet roofing, coatings, hose, linings.

Hypan. [Kingston Tech.] Acrylonitrogens copolymers; hydrogels used as thickeners, gellants, substantive agents, emulsifiers for cosmetics.

Hypan®. [Lipo] Polymers; gel-forming polymers, emulsifiers, emollients.

Hyper+Ion®. [General Chem.] Enhanced coagulants.

Hyper+Lyte. [General Chem.] Flocculating agents.

Hypermer. [ICI Am.; ICI Surf. Belgium]

Polymeric dispersants and emulsifiers.

Hyperox. [Carus] Oxidants.

Hyper+Set. [General Chem.] Neutral size bonding agent.

Hypersorb. [General Chem.] Coagulant, waste water/effluent treatment.

Hypol®. [W.R. Grace/Organics; Grace Rexolin Chem. AB] Polyurethane prepolymers; foamable hydrophilic prepolymers.

Hypowr. [Vyse Gelatin] 300 bloom gelatin.

Hypox. [Smits Neuchatel] Epoxy resin-based materials.

HyPure. [Olin] Chloric and hypochlorous acid, lithium, potassium, or sodium hypochlorites; for bleaching, organic synthesis, water treatment, hard surface cleaners.

Hy Skor. [H.B. Taylor] Artificial butter flavor, liq. and spray dried.

Hysoft®. [Rhone-Poulenc/Textile & Rubber] Fabric softeners.

Hysol®. [Dexter/Hysol] Epoxy or urethane systems; high performance structural paste and film adhesives, repair adhesives, core splice materials, and primers

Hysorb. [Celite] Adsorbent filter aids.

Hysorb. [FMC] Sodium tripolyphosphate.

Hyspin. [Castrol Industrial East] Lubricants.

Hystar®. [Lonza] Hydrogenated starch hydrolysate; lubricant, humectant for toiletries, dentifrices.

Hystrene®. [Witco/Humko] Dimer acids or fatty acids; chemical intermediate, emulsifier, corrosion inhibitor; for personal care prods., waxes, textile auxiliaries, pharmaceuticals, petroleum processing.

HyStretch. [BFGoodrich/Spec. Polymers] Acrylic elastomers; used for textile, nonwoven, adhesive bindings, coatings, and finishes, urethane blending;

Hysweet. [Celite] Solvent sweetener.

HyTemp. [Zeon] Acrylic elastomers. for hose, tubing, cable jacket, belting, and mech. goods.

HyTemp NPC-50. [Zeon] Quaternary ammonium compd.; nonpost-cure agent for use with HyTemp 4050 series elastomers.

HyTemp SR-50. [Zeon] N,N'-Diphenylurea; nonpost-cure retarder for use with HyTemp 4050 series elastomers.

Hyten®. [DuPont] Nylon 6/6 monofilament.

Hytrel®. [DuPont; DuPont UK] Thermoplastic polyester elastomer; polymers used in industrial applics. (hose, belting, flexible couplings, gears, seals, and elec. parts); sporting goods (ski equipment, ball bladders, and shoe soles); wire covering applics., automotive parts.

Hyvar. [DuPont/Ag] Herbicide.

Hyvex. [Ferro] PPS or polysulfone, some glass or carbon reinforced.

Hy-Vin. [Norsk Hydro AS] PVC compds.

Hyzeen. [Betz Industrial] Boiler feedwater conditioner.

Hyzod®. [Sheffield Plastics] Polycarbonate sheet; for aircraft seat parts, tray tables, lavatories, cargo liners, window reveals, partitions, and moldings.

I

I7860. [Hüls Am.] Isopropyldimethylchlorosilane; sterically hindered blocking agent; used in prostaglandin synthesis.

Ibbal. [Mitsubishi Gas] p-Isobutylbenzaldehyde; intermediate for pharmaceuticals, fragrances.

IBDU®. [Mitsubishi Kasei] Nitrogen fertilizer; slow release fertilizer.

Ibercarrier. [A. Harrison] Carriers for polyesters or acrylics.

Iberfix. [A. Harrison] Aliphatic aldehyde condensate; fixing agent.

Iberlene. [A. Harrison] Polyethylene emulsion; finishing agent and assistant softener.

Iberpal. [A. Harrison] Ethoxylated alkyl phenol; detergent, scouring agent.

Iberpenetrant-114. [A. Harrison] Alkylphenol polyethylene sulfonate; textile wetting agent, dispersant, foaming agent.

Iberpol Gel. [A. Harrison] Sodium aminoethyl sulfonate; detergent, textile scouring aid.

Iberpon W. [A. Harrison] Alkylaryl sodium sulfonate; detergent, textile scouring aid.

Iberquestrene. [A. Harrison] Tetrasodium EDTA; chelating agent, dyeing and finishing assistant.

Iberscour. [A. Harrison] Wetting agent, detergent, emulsifier for textile scouring.

Ibersoft. [A. Harrison] Cationic softener for most fibers.

Iberstat. [A. Harrison] Antistat for acrylics, polyester, nylon and blends.

Iberterge. [A. Harrison] Detergent, emulsifier, textile scouring agent.

Ibertex. [A. Harrison] Polyvinyl acetate emulsion; textile auxiliary.

Iberwet. [A. Harrison] Detergent, wetting and rewetting agent, sanforizing assistant for textile applics.

Ice. [Van Den Bergh Foods] Polysorbate blends; emulsifier, stabilizer for food industry; lubricant for textiles; fabric softener.

Ice Berg. [Burgess Pigment] Anhydrous aluminum silicate.

Icecap K. [Burgess Pigment] Anhydrous aluminum silicate.

ICI. [Barkley Plastics Ltd.] Polypropylene, nylon 6/6, PMMA.

Icinol. [ICI Australia] Polyoxyalkylene glycol ether; lubricant, defoamer.

ICI PEG. [ICI Am.] Polyethylene glycols.

Icomeen. [BASF] Ethoxylated fatty acid amines; wetting agent, penetrant, emulsifier, stabilizer, dispersant, antistat, lubricant, solubilizer.

Iconol. [BASF] Ethoxylated ethers or fatty acid alkanolamides; detergent, emulsifier, foam stabilizer, lubricant, penetrant, dispersant, finishing agent.

Ida-Mid. [Ideas Inc.] Alkanolamides.

Ida-Soil. [Ideas Inc.] Oil and solvent soluble corrosion inhibitor.

Ida-Sol. [Ideas Inc.] Water-soluble corrosion inhibitors.

Ida-Sperse. [Ideas Inc.] Dispersant for pigments.

Idet. [Swastik] Alkylaryl sulfonates; wetting agent, detergent, dispersant, penetrant for textiles, detergent formulation, paints.

Idox. [Ideas Inc.] Oxidized waxes.

IE-. [Innovative Engineering of Mich.] Polyurethane engineering elastomer; for rollers, pallets, prototypes, molds, fixtures, wheels, bumpers, metal forming pads.

IG-2. [Multitherm] High efficiency high temp. heat transfer fluid.

Igepal®. [Rhone-Poulenc Surf.; Rhone-Poulenc France] Ethoxylated alkyl phenyl ethers; detergent, wetting agent, emulsifier for household and industrial detergents, textiles, paper, agric., polishes.

Igepon®. [Rhone-Poulenc Surf.] Isethionates or taurates; detergent, wetting agent, dispersant for textiles, industrial detergents, paper, petroleum industry, cosmetics.

Igopas. [Igoplast-Faigle AG] Monomer casted polyamide.

II-320. [Crain] Highly conc. surfactant detergent.

Ilocut. [Castrol Industrial East] Cutting oils.

Imacol®. [Sandoz; Sandoz Prods. Ltd.] Esters or ethers; low foaming lubricant for preventing creases and abrasions in wet finishing treatments of synthetics, cellulosics, blended fabrics.

Image. [Am. Cyanamid/Ag] Herbicide.

Imbirol. [Auschem SpA] Sodium 2-ethylhexyl sulfosuccinate; detergent, wetting agent.

Imerol. [Sandoz Prods. Ltd.] Detergent, emulsifier, wetting agent, dispersant, softener, antistat for textile applics.

Imicure®. [Air Prods.; Pacific Anchor] Imidazole-based; epoxy curing agent for adhesives, composites, filament winding, inks, potting compds., powd. coatings; accelerator for anhydride curing agents.

IML-2. [DuPont] Rubber chemical.

Immergan® A. [BASF AG] Aliphatic sulfochloride; lightfast oil tanning agent for chamois; auxiliary tanning agent for soft leather.

Immunol. [Harry Miller] Meal cleaner, rust inhibitor.

Immunopure. [Pierce Chem.] Immunology/antibody prods.

Impaq. [PQ Corp.] Chromatographic adsorbent.

Impco. [Impco] Polyester resin; thermoset for impregnation of electronic components, pumps, gas and steam fittings, plastic molds, hydraulic systems, gear housing, compressors, engines, aircraft parts, builders' hardware.

Imperon. [Hoechst Celanese] Textile dyes and pigments.

Imperon Binder. [Hoechst AG] Self-crosslinking acrylate-based copolymer dispersion; for pigment dyeing.

Imperon Emulsifier 774. [Hoechst AG] Fatty acid condensate; visc. moderator for o/w emulsions in printing inks.

Imperon Softener D. [Hoechst AG] Dioctyl phthalate; softener for pigment printing.

Imperon Thickener A. [Hoechst AG] Acrylate polymer; thickener for pigment printing.

Impet®. [Hoechst Celanese/Engineering Plastics] PET polyester, some glass reinforced; for inj. molding, automotive parts, consumer electronics/appliances (motor housings, coffee makers, hair dryers), elec./electronics (connectors, terminal blocks, bobbins), furniture.

Implenal®. [BASF AG] Mixture of salts of dicarboxylic acids; finishing agent for chrome tanning, furs.

Implex. [Rohm & Haas] High impact acrylic resin molding powder.

Impregnant. [Monomer-Polymer & Dajac Labs] Acrylic polymer; electrical use.

Impriment. [Catawba-Charlab] Pigment dyestuffs.

Imprimus®. [BASF AG] Conc. pigment pastes; for sheet offset printing inks.

Imsil®. [Unimin Specialty Minerals] Microcrystalline silica; filler/extender, thickener, carrier, free flow agent, flatting agent for paints, powd. coatings, agric., cosmetics, pharmaceuticals, plastics, polishes, elec. resistors, refractory prods., insulation.

Imwitor®. [Hüls Am.; Hüls AG] Glyceryl derivs., propylene glycol, tartaric, citric, or lactic acid esters; emulsifier, dispersant, suspending agent, stabilizer, thickener for food, personal care prods., pharmaceuticals, lubricants.

Inco. [Inco Alloys Int'l.] Specialty nickel alloys and maraging steel.

Incoblend. [Americhem; Zipperling Kessler] Flexible vinyl compd. based on intrinsically conductive polymer; for applics. requiring high elec. conductivity such as EMI shielding.

Incoloy. [Inco Alloys Int'l.] Nickel-iron-chromium alloys and welding prods.

Inconel. [Inco Alloys Int'l.] Nickel-chromium alloys and welding prods.

Incrocas. [Croda Inc.] Ethoxylated castor oil; emulsifier, solubilizer, emollient, superfatting agent, lubricant for personal care prods., detergents, metalworking fluids, insecticides, herbicides, household prods.; lubricant, antistat, softener, dye leveling agent for textiles;

Incrodet TD7-C. [Croda Inc.] Trideceth-7 carboxylic acid; surfactant, foamer, neutralizer for shampoos, bubble baths, liq. soaps, skin cleansers.

Incromate. [Croda Inc.; Croda Surf. Ltd.] Dimethylamine lactates or propionates; surfactant, conditioner, foamer for personal care prods., conditioners, fabric softeners; base for cationic emulsions.

Incromectant. [Croda Inc.; Croda Surf. Ltd.] Alkanolamides or quaternary ammonium chlorides; humectant, solvent, antistat, plasticizer, moisturizer, detangling agent for shampoos, conditioners, creams and lotions.

Incromide. [Croda Inc.] Fatty acid alkanolamides; conditioner, visc. builder, foam stabilizer, emulsifier, pearling agent, opacifier for household, cosmetic, and industrial formulations.

Incromine. [Croda Inc.; Croda Surf. Ltd.] Dimethylamines; emollient, conditioner, lubricant, moisturizer, foamer, surfactant for personal care prods.

Incromine Oxide. [Croda Inc.; Croda Surf. Ltd.] Amine oxides; visc. builder, conditioner, softener, emulsifier, lubricant, wetting agent, foam stabilizer for cosmetic, household, and janitorial prods.

Incronam. [Croda Inc.; Croda Surf. Ltd.] Betaines; surfactant, emulsifier, coupling agent, visc. builder, foam booster/stabilizer, conditioner, lubricant for personal care prods., chemical specialties.

Incropol. [Croda Inc.] Fatty alcohol ethers; surfactant, emulsifier, lubricant, detergent for industrial and household prods.; coupling agent, antistat, fiber lubricant and solubilizer for personal care prods.

Incroquat. [Croda Inc.; Croda Surf. Ltd.; Croda Universal Ltd.] Quaternary ammonium compds. or blends; conditioner, foamer, antistat, emollient, dispersant for personal and hair care prods., textiles, paper, dyestuffs.

Incrosoft. [Croda Inc.] Quaternaries; fabric softener, lubricant, antistat, rewetting agent for textiles, laundry applics.

Incrosul. [Croda Inc.] Sulfosuccinates; conditioning surfactant, foaming agent for personal care prods., carpet shampoos.

Indalca. [Hercules] Gum derivs.; thickener for carpet and fabric printing dyes.

Indalloy. [Indium Corp. of Am.] Alloys.

Indanthren®. [BASF AG] Vat dyes for dyeing and printing cellulose fibers.

Indar. [Rohm & Haas] Fungicide.

Indofast. [Miles/Organic Prods.] Textile dyes and pigments.

Indopol. [Amoco] Viscous polybutenes.

Indosol®. [Sandoz] Textile dyes and pigments.

Indramic. [Barco Chem. Prods.; Industrial Raw Materials] Microcrystalline waxes.

Indrapet. [Barco Chem. Prods.] Petrolatum.

Indrawax. [Barco Chem. Prods.] Paraffin waxes.

Indrox. [Industrial Raw Materials] Oxidized microcrystalline waxes.

Induclor. [PPG Industries] Calcium hypochlorite.

Inductol®. [Ashland/Foundry Prods.] Foundry core oils.

Indulin®. [Westvaco] Kraft lignin or amine derivs.; stabilizer, dispersant, adsorbent, emulsifier, retarding agent, antistripping additive for asphalt emulsions, cements.

Indumal. [Indulor-Chemie GmbH] Additives for plastics.

Indusoil. [Westvaco] Tall oil fatty acids.

IndustraSolv. [Amax Industrial Prods.] Water-soluble degreaser.

Industrene®. [Witco/Humko] Industrial grade fatty acids; chemical intermediate, lubricant, release agent, binder; used in alkyd resins, rubber compding., water repellents, polishes, soaps, abrasives, cutting oils, candles, crayons, personal care prods., waxes, textile auxiliaries, pharmaceuticals, foods.

Industrial 24K Gold. [Technic] Potassium gold cyanide.

Industrial Masking Agents. [Givaudan-Roure] Mixture of essential oils and aromatic chemicals; masking agents for reduction of most offensive odors.

Industrial Wavingas. [Wavin Industrial Prods. Ltd.] Polyethylene.

Industrol®. [BASF] Alkoxylated esters or ethers; nonionic surfactants, emulsifier, detergent, dispersant, wetting agent.

Infrasan. [Reilly-Whiteman] Coconut alkanolamide; detergent, wetting agent.

Ingra-Tex. [Scholler] Formulated textile finishes.

Inhabitant. [Toho Chem. Industry] Special amine base; bactericides for crude oil producing and refining systems.

Inhibitant. [Toho Chem. Industry] Surfactants complex; corrosion and scale inhibitors for oil and gas field; biodice for disposal and injection water.

Inhibiteur. [Henkel-Nopco] Polyoxyalkylene condensate of a heterocyclic amine; corrosion inhibitor for drilling muds.

Inhibitor. [Zschimmer & Schwarz] Fatty acid alkanolamide; anticorrosion additive for metal cleaning and metal working; lubricant in drilling and cutting oils.

Inhibitor 60Q, 60S. [Exxon/Tomah] Corrosion inhibitor for acid cleaning applics.

Inhibitor No. 3. [Oxid Inc.] Metalworking coolant additive, vapor phase corrosion inhibitor.

Inipol. [Ceca SA] Diamines; rust inhibitor, paint and lubricant additive.

Injecta Color. [Morton Int'l. UK/Automotive] Liq. colorants for plastics.

Inklurit®. [BASF AG] Urea-formaldehyde precondensate; for mfg. petroleum prods., e.g., for firefighting.

Inkrustin®. [BASF AG] For wet and dry galvanizing, tinplating, leadcoating; as soldering machine flux in mfg. of tin cans.

Innovex. [BP Chem. Ltd.] LLDPE.

Inochrome. [ICI Am.] Textile dyes and pigments.

Inoderme. [ICI Am.] Textile dyes and pigments.

Inoset. [Ashland/Foundry Prods.] Inorganic foundry core binders.

Inplast-SPM. [Atochem UK] PVC alloys.

Inprima. [European Vinyls Corp. UK] PVC inj. molding compds.

Inscut. [Toho Chem. Industry] Paraffin-type mineral oil; water-insoluble cutting oil.

Inset. [Hart Prods. Corp.] Resin condensate; fixative for direct dyes.

Insl-Guard Epoxy Pool Coating. [Insl-X Prods.] Two-component epoxy; pool paint.

Insl-Thane Polyurethane. [Insl-X Prods.] One-component heavy-duty polyurethane.

Insl-Tile Epoxy Coating. [Insl-X Prods.] Two-part activated polyamide epoxy.

Insl-Tron Polyurethane. [Insl-X Prods.] Two-component acrylic aliphatic polyurethane.

Insol-U-25. [Georgia-Pacific] Urea formaldehyde conc.

Insta-Draw®. [Ashland/Foundry Prods.] Foundry core binders.

Instak. [Polymer Industries] Vinyl-acrylic emulsion; aq. pressure-sensitive adhesive.

Instan-set. [Development Assoc.] Polyurethane elastomer; protective coating over fiberglass-reinforced plastics, metal, PS foam prods., other substrates.

Instantbond. [Hernon Mfg.] Cyanoacrylates, ethyl and methyl types; instant bonding adhesives.

Instant Dri. [Meridian Petroleum] Absorbents.

Instant Lok. [Nat'l. Starch & Chem.] Hot-melt adhesive.

Instapak. [Plastec Iberica SA] Low density polyurethane foam.

Instapearl. [Lonza] Glycol stearate.

Insta-Seal Kit. [Insta-Foam Prods.] One-component polyurethane foam kit.

Instex. [Teknor Color] Micropulverized color conc.

Insulfas. [H.B. Fuller] Fire-retardant insulation coating for metal surfaces.

Intact Skin Care System. [Huntington Labs] Skin care prods.

Intene. [EniChem Elastomeri Srl] Hydrocarbon rubber.

Intercide®. [Akzo] Oxybisphenoxyarsine or tributyltin compds.; microbicide for PVC and other polymers, coated fabrics.

Interface. [Mason Chem. Co. Ltd.] Mold release.

Interfibe. [Interfibe] Cellulose fiber.

Interlite. [Akzo] Visc. control agent.

Interlox. [McGean-Rohco] Phosphatizing chemicals.

Interlube. [Anchor UK] Zinc fatty acids/lubricants blend; lubricant for rubber compds.

Intermediate. [Witco/Humko] Fatty acid alkanolamides; thickener, foam booster, superfatting agent.

Interox®. [Interox Am.] Textile bleaches, scouring agents.

Interplus. [Porter Int'l.] Surface-tolerant high-build polyamide-cured epoxy.

Interpol®. [Freeman] Unsaturated polyester/urethane hybrid system; for liq. inj. molding or open casting applics.

Interseal. [Porter Int'l.] High build hot-sprayed high-solids epoxy.

Intersept®. [Interface Research] Substituted ammonium salts of alkylated phosphoric acids admixed with a free alkylated phosphoric acid; broad-spectrum biostat.

Intershield. [Porter Int'l.] High build, self-priming, glass-reinforced epoxy.

Intersperse. [Akzo] Dispersants.

Intersperse. [Crompton & Knowles] Textile dyes and pigments.

Interstab®. [Akzo; Akzo Chem. BV] Stabilizer for plastisols, flexible compds., esp. hose and profile, filled systems, film and sheet; antioxidant, antistat, lubricant, release agent, chelating agent for plastics, rubber, coatings, cosmetic, metallurgy; vinyl catalysts.

Interwax. [Akzo] Lubricant for PVC.

Interwet®. [Akzo] Wetting agents, emulsifier, penetrant for coatings, adhesives, insecticides, wax dispersions, bitumen emulsions.

Intex. [EniChem Elastomeri Srl] Synthetic latexes.

Intex Scour. [Intex] Alkanolamide or phosphate; detergent, emulsifier, scouring agent, foam additive for shampoo, textile dyeing and processing.

Intol. [EniChem Elastomeri Srl] Hydrocarbon rubber.

Intrabond. [Crompton & Knowles] Textile dyes and pigments.

Intracarrier® ATM. [Crompton & Knowles] Ester/hydrocarbon deriv. blend; dyeing accelerant, carrier for polyester and triacetate.

Intrachrome. [Crompton & Knowles] Textile dyes and pigments.

Intracid. [Crompton & Knowles] Textile dyes and pigments.

Intracron®. [Crompton & Knowles] Fiber reactive dyes.

Intrafomil®. [Crompton & Knowles] Nonsilicone defoamer; for textile operations.

Intrajet. [Crompton & Knowles] Textile dyes and pigments.

Intralan®. [Crompton & Knowles] Dispersant, leveling agent, dye assistant, penetrant for textile dyeing.

Intralite. [Crompton & Knowles] Textile dyes and pigments.

Intralube®. [Crompton & Knowles] Dye assistant, oil emulsifier, lubricant, soil release agent for synthetics, natural fibers, and blends.

Intramet. [Crompton & Knowles] Textile dyes and pigments.

Intrapel. [Crompton & Knowles] Textile dyes and pigments.

Intraphasol. [Crompton & Knowles] Emulsifier, wetting agent, dye assistant, dispersant for dyeing and printing of fabrics and carpets.

Intraphor. [Henkel-Nopco] Sodium polynaphthalene sulfonate; dispersant for dyestuffs, pigments, pesticides.

Intraplast. [Crompton & Knowles] Textile dyes and pigments.

Intrapol. [Henkel; Henkel-Nopco] Solvent and alkyl polyglycol ether; foaming detergent and dispersant for textiles.

Intraquest® TA. [Crompton & Knowles] Tetrasodium EDTA; sequestering agent for textile scouring, bleaching, dyeing, wet finishing.

Intrasil. [Crompton & Knowles] Textile dyes and pigments.

Intrasoft®. [Crompton & Knowles] Fatty acid condensate; cationic softening agent for textiles.

Intrasperse. [Crompton & Knowles] Textile dyes and pigments.

Intrassist®. [Crompton & Knowles] Quaternary and ampholytic compds.; leveling agent for basic dyes on acrylics and polyester.

Intratex®. [Crompton & Knowles] Leveling agent, detergent, penetrant for textile dyeing and printing.

Intratherm. [Crompton & Knowles] Textile dyes and pigments.

Intravat. [Crompton & Knowles] Textile dyes and pigments.

Intravon®. [Crompton & Knowles] Ethylene oxide condensate; scouring and emulsifying agent, detergent, dispersant, wetting agent, penetrant for textile, household, and cosmetic applics.

Intrawet®. [Crompton & Knowles] Wetting agent, leveling agent for textile processing.

Intrawite®. [Crompton & Knowles] Fluorescent/optical brighteners for acrylic, nylon, cellulosics, wool, blends.

Intrazone. [Crompton & Knowles] Textile dyes and pigments.

Intrex. [Rhone-Poulenc Ltd.] Organic complex of phosphonic acid; hydrotrope for stabilizing nonionic surfactants in alkaline-based formulations.

Inversol. [Ferro/Keil] Complex polyglycol ester; lubricity agent for metalworking liqs.

Invertose. [Corn Prods.] High fructose corn syrups.

Iodobio 45. [Alban Muller] TEA hydroiodide; slenderizing prods.

Iodox. [Ajay] Potassium iodate.

Ionac. [Sybron] Ion exchange resins.

Ionet. [Sanyo Chem. Industries] Sorbitan or ethoxylated esters; emulsifier, scouring agent, leveling agent, penetrant, dispersant for emulsion polymerization, metal processing, personal care goods, textiles, agric.

Ionol. [Shell] Antioxidants and stabilizers for rubber, paraffin, plastics, petroleum prods.

Ionpure. [U.S. Cosmetics] Amorphous, water-soluble inorganic preservative; for cosmetics and plastic pkg.

Ionquest® 801. [Albright & Wilson Am.] Extractants.

Iosan. [West Agro] Dairy udder wash, sanitizer.

Iosol. [Crompton & Knowles] Textile dyes and pigments.

Iota. [Unimin Specialty Minerals] High purity quartz.

Iotek. [Exxon] Ionomer resins; thermoplastics for molding and extrusion.

IP-61. [Monsanto] Sodium trimetaphosphate; for detergents.

Ipelene. [IPEL Ltd.] Polypropylene biaxially oriented film.

IPP. [PPG Industries] Diisopropyl peroxydicarbonate; initiator for polymerization of unsaturated monomers.

Ircogel®. [Lubrizol] Metal complexes; thixotropic, antisag, and flow control agent for coating plastisols and organosols, fabric coatings, caulks, sealants.

Irgacolor. [Ciba-Geigy/Pigments] Mixed metal oxide inorganic pigments.

Irgacor. [Ciba-Geigy/Additives] Organic corrosion inhibitor.

Irgacure. [Ciba-Geigy/Additives] Photoinitiators.

Irgaform. [Ciba-Geigy/Additives] Photographic processing chemicals.

Irgafos®. [Ciba-Geigy/Additives] Phosphite; heat stabilizer, antioxidant for polyolefins.

Irgalan. [Ciba-Geigy/Dyestuffs] Textile dyes and pigments.

Irgalev®. [Ciba-Geigy/Dyestuffs] Diphenyl deriv.; dye assistant for nylon.

Irgalite. [Ciba-Geigy/Pigments] Organic pigment.

Irgalube®. [Ciba-Geigy/Additives] POE alkyl ester and fatty ester; extreme pressure/antiwear agents; dyeing assistant for fabrics; dye bath lubricant.

Irganox®. [Ciba-Geigy/Additives] Antioxidant, stabilizer for polymers, elastomers, coatings, food pkg. and adhesive applics., petrol. prods.

Irgapadol®. [Ciba-Geigy/Dyestuffs] Dyeing assistant.

Irgaperm. [Ciba-Geigy/Additives] Light stabilizer.

Irgarol. [Ciba-Geigy/Additives] Algicide.

Irgasan DP300. [Ciba-Geigy] Triclosan; broad-spectrum bacteriostat for deodorant prods.

Irgasol®. [Ciba-Geigy/Dyestuffs] Dye dispersant, leveling agent.

Irgasperse. [Ciba-Geigy/Pigments] Solvent and water soluble dyes.

Irgastab®. [Ciba-Geigy/Additives] Heat and light stabilizers, antioxidants for plastics.

Irgatos. [Ciba-Geigy/Additives] Process stabilizer.

Irgazin. [Ciba-Geigy/Pigments] Organic pigment.

Iridite. [Witco] Corrosion-resistant finish for metals.

Irilac. [Witco] Organic protective coatings for metals.

Iriodin. [EM Industries; E. Merck] Lead carbonate; pearl luster pigments.

Irogran. [M.T. Zouros] Thermoplastic polyurethane.

Irotyl. [Henkel/Emery/Cospha] Pungent aroma chemical.

Irox. [DCS Color & Supply] Iron oxide compd.; used with phenolic urethane foundrys and binders to eliminate subsurface porosity and expansion defects.

Irpol. [Irpen SA] Unsaturated polyester sheet and rod.

Isafil®. [Boehme Filatex] Coning oil for filament processing.

Isaflex. [ISA Industria Solventi E Adesivi Srl] Polyurethane, neoprene.

Isalatex. [ISA Industria Solventi E Adesivi Srl] Synthetic latex, natural rubber latex.

Isanol®. [BASF AG] Iso- and n-butanol; solvent for prod. of adhesives and surface coating resins.

Isaspray. [ISA Industria Solventi E Adesivi Srl] Synthetic rubber-based adhesive.

Isatin. [BASF AG] Intermediate for prod. of pharmaceuticals and dyes; stabilizer for plastic industry.

Iscolan. [Croda Chem. Ltd.] Lanolin ester; emollient, wetting and spreading agent for skin, makeup and hair care cosmetics.

Iso Beeswax. [Strahl & Pitsch] Beeswax substitute.

Isobind. [Dow] Isocyanate binders.

Isobrite. [Witco] Brighteners for plating sol'ns.

Isobutyl Niclate®. [R.T. Vanderbilt] Nickel diisobutyldithiocarbamate; antioxidant/antiozonant for protection of epichlorohydrin.

Isocal. [Vevy] Ethoxylated ether blends.

Isocreme. [Croda Chem. Ltd.] Lanolin-derived base; emollient, moisturizer, emulsifier for cosmetics and pharmaceuticals.

Isocryl. [Estron] Acrylic resin.

Isocure®. [Ashland/Foundry Prods.] Catalyzed foundry core binders.

Isodamp. [E-A-R] PVC energy absorbing material.

Isodirect. [Isochem] Textile dyes and pigments.

Isofol®. [Condea Chemie GmbH] Branched alcohols; intermediates for toiletries, cosmetics, detergents, leather and textile auxiliaries, lube oil, plastics additives, defoamers.

Iso Isostearyle WL 3196. [Gattefosse SA] Isostearyl isostearate.

Isolan. [Miles/Organic Prods.] Textile dyes and pigments.

Isolanoate. [Lanaetex Prods.] Octyl isononanoate or blends.

Isolene®. [Hardman] cis-1,4-Polyisoprene; isomerized rubber; used in adhesives, sealants, caulks, and lubricants.

Isolex. [Morton Int'l.] RFI/EMI shielding coating.

Isolite. [Ditta Salt SpA] Phenolic powder.

Isolite®. [Schenectady] Unsaturated polyester insulating varnish.

Isoloss®. [E-A-R] Polyurethane elastomers; energy absorbing material.

Isomelt®. [Schenectady] Urethane wire enamel.

Isomid®. [Schenectady] Polyester-imides; wire enamels.

Isonamid. [Dow] Thermoplastic engineering resins.

Isonate®. [Dow; Dow Ahlen] MDI; processing aid, intermediate for prod. of cast, RIM, and thermoplastic polyurethane elastomers, adhesives, binders, coatings, and sealants.

Isonel®. [Schenectady] Solvent-borne polyester insulating varnish.

Isonol. [Dow] Polyether polyol; functional polyol for polyurethane industry for industrial/consumer RIM and structural polymers, dynamic elastomers, adhesives, binders, coatings, and sealants.

Isonox®. [Schenectady] Phenols; antioxidant, thermal stabilizer for polymers and rubber systems.

Isopar®. [Exxon; Deutsche Exxon Chem. GmbH] Isoparaffins; solvents.

Isopaste®. [Ashland/Foundry Prods.] Foundry core adhesive system.

Isoplas P. [Micropol Ltd.] Crosslinkable polythene.

Isoplast. [Dow] Thermoplastic polyurethane, some glass-reinforced; amorphous engineering resin with crystalline properties; for extrusion and inj. molding applics.

Isopoxy®. [Schenectady] Water- and solvent-borne epoxy insulating varnish.

Isoprep. [Witco] Metal cleaners.

Isopropylan. [Amerchol] Isopropyl palmitate, lanolin oil blends; binder in talc and pearl powd. systems; plasticizer, emollient, moisturizer.

Isoreactive. [Isochem] Textile dyes and pigments.

Isorez. [Degen] Isophthalic alkyd.

Isoset®. [Ashland/Foundry Prods.] Emulsion polymer/isocyanate adhesive resins; foundry core binders.

Isosperse. [Isochem] Textile dyes and pigments.

Isosweet. [A.E. Staley Mfg.] High fructose corn syrup.

Isotex®. [Schenectady] Monomer for engineering plastics.

Isotron. [Atochem N. Am.] Refrigerants, aerosol propellants, blowing agents.

Isovoss. [Vosschemie GmbH] Polyurethane foams.

Isoweld®. [Schenectady] Solderable polyesterimides; for wire enamels.

Isozol. [Isochem] Textile dyes and pigments.

Ispaglas PC. [Stokvis Plastics BV] Polycarbonate.

Ispaleen H. [Stokvis Plastics BV] Polyethylene.

Ispaleen PP. [Stokvis Plastics BV] Polypropylene.

I.T. [R.T. Vanderbilt] Hydrous magnesium calcium silicate; industrial talc used as filler/extender in NR and synthetic rubbers, paints, sealants, mastics, latex compds.

Itex. [Intertex] Natural rubber.

Itz-Tuff. [BASF] Printing ink vehicle.

Iupital®. [Mitsubishi Gas] Acetal copolymer; for applics. where resistance to dimensional change, fatigue, wear, and corrosive environments is important.

Ivarbase. [Brooks Industries] Surfactant blends; cosmetics ingredients.

Ivarlan. [Brooks Industries] Lanolin or derivs.; cosmetics ingredients.

Ivex®. [CasChem] zinc undecylenate/undecylenic acid blend; antifungal powd. for foot care and personal hygiene prods.

Ixan. [Solvay & Cie] PVC.

Ixef®. [Solvay Polymers; Solvay & Cie] Polyarylamide; for shock- and vibration-absorbing parts.

Ixol. [Vevy] Polysorbates;

Ixolene. [Vevy] Sorbitan esters.

Izocyn T 80. [Organika Zachem Chem. Works] TDI isocyanate.

J

Jafaester. [Jahres Fabrikker A/S] Fatty acid esters; plasticizers, mold release agents.

JagDril. [Rhone-Poulenc/Perf. Resins & Coatings] Polymeric viscosifier for drilling operations.

Jaguar®. [Rhone-Poulenc/Perf. Resins & Coatings] Guar gums or derivs.

JAQ Powdered Quat. [Huntington Labs] Myristalkonium chloride; for formulating disinfectants, sanitizers, swimming pool algicides.

Jar-Cal. [Jarchem Industries] Food grade anhydrous calcium chloride.

Jasmacyclat. [Henkel/Emery/Cospha] Methylcyclooctylcarbonate; herbal fragrance raw material.

Jasmonan. [Henkel/Emery/Cospha] Jasmine aroma chemical.

Jasmorange®. [BASF AG] 2-Methyl-3(4-methylphenyl) propanol; fragrance material.

Jayflex®. [Exxon] Proprietary, phthalate, adipate, or trimellitate plasticizers.

Jayfloc. [Exxon] Flocculating agents.

J-Color. [Dow Corning STI] Color silicone masterbatches.

Jeffamine®. [Texaco; Texaco GmbH] Polyoxy propylene amines; epoxy modifier and curing agent.

Jeffcool. [Texaco] Industrial coolants.

Jeffersol. [Texaco] Glycol ethers; jet fuel anti-icing additive.

Jel-O-Mer. [Reichhold] Flow control agent.

Jer-Dri. [Sybron] Durable water repellents for synthetic and natural fibers.

Jerotex P. [Sybron] Methylated urea-formaldehyde resin; thermoset stiffening agent for synthetic textiles.

Jersey WT. [Sybron] Starch derivs. and crosslinkers; weighter for textile fibers.

Jet. [McGean-Rohco] Spray cleaning chemicals.

Jet Aero-Fresh. [Century Chem.] Deodorant.

Jet Amine. [Jetco] Fatty amines or diamines; emulsifier, corrosion inhibitor, ore flotation agent, dispersant, mold release, additive for lube and fuel oils; intermediate for quats, surfactants, agric. and detergent formulations.

Jet-Chlor. [Jet Inc.] Tablets for water/wastewater chlorination.

Jetco. [Jetco] Amine-based or oil-based; asphalt emulsifier, antistripping agent, coal tar modifier.

Jet Fil®. [Luzenac; R.T. Vanderbilt] Talc; filler for PP and other resin systems.

Jet Jel®. [Rhone-Poulenc/Perf. Resins & Coatings] Flocculant for settling and clarifying reserve mud pits.

Jet Quat. [Jetco] Quaternary ammonium chlorides; bactericide, textile softener, asphalt emulsifier, petroleum processing.

Jet-X. [Rockwood Systems] Fire fighting foam, high expansion foam.

JL 43155AS, 43176. [Acheson Colloids] Silver pigment/epoxy binder blends; EMC shielding coating for metals.

Jojobeads. [Jojoba Growers & Processors] Hydrogenated jojoba wax.

Jonachlor A. [Jonas Chem.] Chlorendic anhydride, high purity.

Joncryl. [S.C. Johnson Wax] Acrylic polymers; for coatings, printing inks.

Jonwax. [S.C. Johnson Wax] Wax emulsions; for coatings, printing inks.

Jonylon. [BIP Chem. Ltd.] Nylon 6, 6/6 compds.

Jordapon®. [PPG/Specialty Chem.] Cocoyl isethionates; foaming surfactant, detergent for personal care prods.; lime soap dispersant.

Jordaquat®. [PPG/Specialty Chem.] Benzalkonium chloride; bacteriostat, germicide, algicide, antistat.

Joyco. [Joy Chem.] Textile processing materials and auxiliaries.

JR-228, -228-1. [Bacon] Semiflexible adhesive systems.

J Slip. [Sybron] Modified silica; antislip agent for fibers.

J-Soft. [Sybron] Textile softener.

Juniorlan 1664. [Henkel/Organic Prods.] Lanolin wax/triolein blend; emulsifier.

Junipal. [Henkel/Emery/Cospha] Juniper-like aroma chemical.

J Wet. [Sybron] Wetting agent for flame retardant textile finishes.

JZF. [Uniroyal] Antioxidant.

K

K. [Thermal Ceramics] Insulating firebrick.

K 129. [Kaopolite] Bentonite; binder, plasticizer, flocculant, thickener, suspending agent for ceramics, liq. cleaning compds., water treatments.

KA. [Kenrich Petrochemicals] Aceto acetyl aluminates; coupling agents.

Kadel®. [Amoco] Polyketone; high performance thermoplastic for high temp. applics. in chemical processing, aviation/aerospace composites.

Kadif. [Witco UK] Sodium dodecylbenzene sulfonate; wetting and penetrating agent, detergent for industrial, institutional, and household uses.

Kagetex®. [Ore & Chem. Corp.] Natural rubber latex; for dipped goods, adhesives, carpet backing, textile coating, binding agents

Kalar®. [Hardman] Butyl rubber; used in sealants, coatings, caulks, as green enhancer for uncured butyl rubber compositions.

Kalcohl. [Kao] Fatty alcohols; raw material for plasticizers, antioxidants, ethoxylates, etc.

Kalene®. [Hardman] Butyl rubber; used in sealants, coatings, elec. encapsulating compds.

Kalex. [Hart Chem. Ltd.] EDTA and salts; chelating agents for pulp and paper, other industries.

Kalex®. [Hardman] Thermoset polyurethane compds. and adhesives; potting and encapsulating compds., adhesives.

Kalidone®. [UCIB] Potassium PCA; moisturizer for dermatological soaps, shampoos, after-sun lotions, shower gels, moisturizing creams and lotions, hair comb-out balm.

Kalixide. [Vevy] Talc and pigment blends.

Kalkup. [Malvern Minerals] Aluminum silicate, surface-modified.

Kalrez®. [DuPont] Perfluoroelastomer; finished rubber prods.

Kalsitex. [Franklin Mineral Prods.] Wet ground muscovite mica, surface-treated.

Kalthane. [Kalker] Polyurethane.

Kamar. [Finetex] Surfactant blends; emulsifier for solvents, textile scours, industrial cleaners.

Kamax. [Rohm & Haas] Acrylic-imide copolymer; amorphous thermoplastic for inj. molding, outdoor, lighting, optical applics.

Kan-Kil. [Colgate-Palmolive] Insecticides.

Kanoloid. [Kano Labs] Clear coating.

Kano Rustproof. [Kano Labs] Rust preventive.

Kantmelt. [Specialty Prods.] High-temp. grease.

Kantstik. [Specialty Prods.] Waxes; internal and external lubricants, release agents for inj. molding, extrusion, calendering, blow molding for thermoplastics and thermosets.

Kaobrite. [Thiele Kaolin] Coating clay.

Kaocal. [Thiele Kaolin] Calcined clay.

Kaocrete. [Thermal Ceramics] Kaolin castable.

Kaofill. [Thiele Kaolin] Filler clay.

Kaofine. [Thiele Kaolin] Coating clay.

Kaogloss. [Thiele Kaolin] Coating clay.

Kaokote. [Van Den Bergh Foods] Hydrog. veg. oils; coating fat and centers for food industry.

Kaola. [Van Den Bergh Foods] Hydrog. veg. oils; for food industry.

Kaolex D-6 Clay. [Evans Clay] Filler.

Kaolin. [Sachtleben Chemie GmbH] White filler with fine particles; semi-reinforcing agent for elastomers.

Kaolite. [Thermal Ceramics] Insulating castable.

Kaolloid. [Evans Clay] Hydrated aluminum silicate; reinforcing filler in NR, SR, latexes, resins, plastics; color and processing aid; used in coated materials, footwear, flooring, V-belts, rolls, belts, matting, molded and extruded goods, foam goods, o-rings, seals, sundries.

Kaolloid Clay. [Evans Clay] Filler.

Kaomax. [Van Den Bergh Foods] Hydrog. soybean oil; coating fat for confectionery.

Kaomel. [Van Den Bergh Foods] Hydrog. veg. oil; coating fat, center fat for confectionery, cosmetics, pharmaceuticals.

Kaopaque. [Dry Branch Kaolin] Aluminum silicate.

Kaopaque. [ECC Int'l.] Kaolin.

Kao-Phos. [Thermal Ceramics] Dense high alumina abrasion-resistant castable.

Kaopolite®. [Kaopolite] Ultrafine abrasives, polishing powders.

Kaoprem-E. [Van Den Bergh Foods] Partially hydrogenated soybean oil, sorbitan tristearate; coating fat, center fat for confectionery.

Kaorich. [Van Den Bergh Foods] Hydrog. veg. oil; emulsifier for food processing.

Kao-Tab. [Thermal Ceramics] Dense high alumina abrasion-resistant castable.

Kaowhite. [Thiele Kaolin] Coating clay.

Kaowool. [Thermal Ceramics] Ceramic fiber insulation.

Kapton®. [DuPont] Polyimide film, unreinforced or Teflon, carbon-filled; heat seal applics.; low dielec. film for military applics.

Kara Dye®. [Rhone-Poulenc/Textile & Rubber] Dye carrier for polyester and blends.

Karafac. [Clark] Detergents, wetting/rewetting agents for scouring, enzyme desizing, bleaching, and other textile processes.

Karalube. [Clark] Lubricants for beaming in textile industry.

Kara Lube®. [Rhone-Poulenc/Textile & Rubber] Fatty condensate; lubricant, softener, dispersant for dyestuff.

Karamide. [Clark] Fatty acid alkanolamides; detergent, emulsifier, thickener, base for floor cleaners, all-purpose cleaners.

Karapeg. [Clark] Ethoxylated esters; emulsifier, thickener, opacifier, pearlescent, dispersant, lubricant, coupling agent, solubilizer for textile processing, cosmetics, detergent systems, degreasers, paints.

Karasil NYM. [Clark] Reactive silicone softener for a soft, durable hand.

Karasist®. [Rhone-Poulenc/Textile & Rubber] Desizing, scouring, and bleaching auxiliary.

Karasoft. [Clark] Softener for textiles.

Karathane®. [Rohm & Haas] Dinitrooctylphenyl crotonate agric. fungicide/miticide for fruit, ornamentals.

Kara Wet®.. [Rhone-Poulenc/Textile & Rubber] Organic phosphate esters, sulfosuccinates; emulsifier, detergent, wetting agent for textile wet processing.

Karboresin. [Hoechst Celanese] Hydrocarbon resin; for printing inks.

Karlex. [Ferro/Engineering Thermoplastics] PC resin, unreinforced and glass or carbon-filled; PC/PET alloys; flame retardant thermoplastics.

Karmex. [DuPont/Ag] Herbicide.

Karox RTM. [Clark] Peroxide stabilizer for bleaching systems.

Kasil®. [PQ Corp.] Potassium silicate; adhesives, binders, coagulant, flocculant, dispersant, emulsifier, soil antiredeposition agent.

Kaska. [Crompton & Knowles] Conc. citrus oils.

Kasolv®. [PQ Corp.] Hydrous potassium silicate powders; adhesives, binders, surfactants, coagulants, flocculants, emulsifiers, dispersants, soil antiredeposition agents, corrosion inhibitors.

Kastone. [DuPont] Peroxygen compds.

Katapol®. [Rhone-Poulenc Surf.] Ethoxylated fatty amines; emulsifier, textile dyeing assistant, antiprecipitant, leveling agent.

Katemul. [Scher] Glycolates and gluconates; conditioner and softener for hair care prods.

Kathon®. [Rohm & Haas; Rohm & Haas France] Isothiazolinones; antimicrobials, mildewcides for textiles, metalworking fluids, industrial uses, cosmetics.

Katioran®. [BASF AG] Fatty acid and alcohol ethoxylates; emulsifier, thickener for cosmetics.

Kativo. [H.B. Fuller] Powder coatings.

Kauramin®. [BASF AG] Melamineformaldehyde based; glues for prod. of weather-resistant chipboard and plywood; impregnating resins for decorative and overlay papers.

Kauranat®. [BASF AG] Isocyanatebased; for prod. of weather-resistant chipboard.

Kauresin®. [BASF AG] Phenol-formaldehyde based; glues for prod. of weather-resistant plywood and chipboard.

Kaurit®. [BASF AG] Urea-formaldehyde based; glues for prod. of veneered board, furniture, paper impregnation; crosslinking agents for textile finishing.

Kauropal®. [BASF AG] Assistants for glue and impregnating resins.

Kayacelon. [Rite Industries] Textile dyes and pigments.

Kaycel. [Am. Fillers & Abrasives] Cellulose and cellulose compds.

Kaydol. [Witco/Sonneborn] Mineral oil USP; emollient and lubricant for cosmetics, pharmaceuticals.

Kayphobe-ABO. [Kaopolite] Kaolin clay (96%) with methyl hydrogen polysiloxane and water; for antiblocking agents, industrial coatings, inks, plastic compds.

K-Cop. [Griffin] Copper-ammonium complex; agric. fungicide.

K-Cryl. [King Industries] High solids acrylic oligomers.

KCS. [ECC Int'l.] Clay-kaolin; paper coating.

K-Cure. [King Industries] Catalysts for polymeric coatings.

KE-311. [Arakawa] Rosin ester.

Keical-Ace. [Mitsubishi Kasei] Ultra lightweight calcium silicate insulation; industrial insulator for reactor, heat exchanger, vessel, tank, pipe.

Kelacid®. [Kelco] Alginic acid; gelling agent, emulsifier, stabilizer for food, pharmaceuticals, industrial applics., paper industry.

Kelate. [Hickson Danchem] Chelating agents.

Kelate. [Tri-K Industries] EDTA and salts; chelating agent.

Kelco®. [Kelco] Sodium alginates; gelling agent, emulsifier, stabilizer for foods, pharmaceuticals, industrial use.

Kelco. [Witco] Carburizing fluids.

Kelco-Gel®. [Kelco] Gellan gum; gelling agent for use in foods, personal care prods., industrial applics.

Kelcoloid®. [Kelco] Propylene glycol alginates; gelling agent, emulsifier, stabilizer for foods, pharmaceuticals, industrial use.

Kelcosol®. [Kelco] Algin; gelling agent, emulsifier, stabilizer for foods, pharmaceuticals, industrial use.

Kelcote. [Witco] Corrosion inhibitors.

Kelcotene. [Witco] Machining oils.

Kelcut. [E.H. Kellogg] Cutting oil.

Keldax®. [DuPont] Ethylene interpolymer composition with high loadings of inorganic fillers; thermoplastic sound barrier resins for automotive, business machines, industrial equip.

Kelecin. [Reichhold] Soybean lecithins.

Kelene. [Lowenstein Dyes & Cosmetics] Sodium dihydroxyethylglycinate, trisodium HEDTA; chelating agent.

Kel-F. [3M] Fluoropolymers and elastomers; used in o-rings, seals, gaskets, paints, binders, propellants, etc.

Kelfo®. [Kelco] Xanthan gum/limestone blend; gum for animal feed; thickener, stabilizer.

Kelgin®. [Kelco] Algin and alginates; gelling agent, emulsifier, stabilizer for foods, pharmaceuticals, industrial use.

Kelgraf. [Witco] Synthetic high-temp. lubricants.

Kelig. [Borregaard LignoTech] Sodium lignosulfonate; sequestrant, dispersant, metal cleaning component, corrosion inhibitor, carrier for micronutrient metals in liq. formulations; oil well cement retarder.

Kelkut. [Witco] Cutting fluids, grinding coolants, grinding oils.

Kellin. [Reichhold] Linseed oil.

Kellogg SS2. [E.H. Kellogg] Synthetic cutting oils.

Kellox. [Reichhold] Oxidized marine oils.

Kellube. [E.H. Kellogg] Synthetic compressor oils.

Kelmar®. [Kelco] Potassium alginate; gellant, emulsifier, stabilizer in foods, industrial applics., bodying agent for creams and lotions, dental impressions.

Kelmer. [Reichhold] Marine oil.

Kelon. [Lati SpA] Polyamide 6, 6/6, or copolymers.

Kelon A Series. [Lati SpA] Nylon 66.

Kelon B Series. [Lati SpA] Nylon 6.

Kelon C Series. [Lati SpA] Nylon copolymer.

Kelpol. [Reichhold] Alkyd resin.

Kelpoxy. [Reichhold] Epoxy concs.

Kelset®. [Kelco] Sodium alginate; gelling agent, emulsifier, stabilizer in foods, pharmaceuticals, industrial use.

Kelsol. [Reichhold] Oil copolymers.

Keltan. [Copolymer Rubber] EPDM; for cable insulation, sponge articles, high hardness articles, solid profiles, hoses, sealing rings, etc.

Keltex®. [Kelco] Sodium alginate.

Kelthane. [Rohm & Haas] Agric. miticide.

Keltose®. [Kelco] Calcium alginate, ammonium alginate; gellant, emulsifier, binder, stabilizer for food and industrial applics.

Keltrol. [Reichhold] Alkyd resins.

Keltrol®. [Kelco] Xanthan gum; stabilizer for foods; thickener and emulsion stabilizer for creams and lotions; binder in toothpastes; suspending agent.

Kelvar. [Reichhold] High solids alkyds.

Kelvis®. [Kelco] Sodium alginate; gellant, emulsifier, stabilizer for food and industrial applics.

Kelzan®. [Kelco/Oil Field Prods.] Xanthan gum; flocculant, suspending agent, gellant, foam stabilizer, rheology modifier, lubricant, tackifier for in-

dustrial applics.

Kemamide®. [Witco/Humko] Fatty acid primary and secondary amides; lubricant, slip, antiblock, mold release agents for plastics, crayons, petrol. prods., asphalts, inks, metals, textiles; corrosion inhibitor; pigment grinding aid; dyestuff dispersant for paints; intermediate.

Kemamide® W Series. [Witco/Humko] Ethylene distearamide and other amides; lubricant, slip, antiblock, mold release agents for plastics, crayons, petrol. prods., asphalts, inks, metals, textiles; corrosion inhibitor; pigment grinding aid; dyestuff dispersant for paints; intermediate.

Kemamine® AS Series. [Witco/Humko] Nitrogen derivs.; antistat for polyolefins, styrenics, other plastics; lubricity aid, mold release aid, pigment dispersant.

Kemamine® BQ, Q Series. [Witco/Humko] Quaternary ammonium chlorides; antistat, textile softener, germicide, santiizer, dyeing aid, corrosion inhibitor, emulsifier, etc.

Kemamine® D Series. [Witco/Humko] Fatty acid propylene amines; gasoline detergent, bactericide, corrosion inhibitor in petrol. prod.; epoxy hardener.

Kemamine® DD, DP. [Witco/Humko] Dimer diamine; chemical intermediate, extender, crosslinking agent in polymeric systems; corrosion inhibitor.

Kemamine® P Series. [Witco/Humko] Fatty acid primary amines; emulsifier, flotation agent, dispersant, intermediate, lubricant; used in metalworking oils, as fuel oil additive; mold release for plastics and rubber

Kemamine® S Series. [Witco/Humko] Fatty acid sec. amines; grease, corrosion inhibitor; industrial use.

Kemamine® T Series. [Witco/Humko] Fatty acid tert. amines; chemical intermediate for quat. ammonium derivs., acid scavenger in petrol. prods.; epoxy hardener; catalyst.

Kematal. [Hoechst UK] Acetal copolymers.

Kemester®. [Witco/Humko] Fatty acid

esters; emollient, emulsifier for cosmetics, textiles; lubricant for leather, metalworking; carrier for agric. spray prods.; plasticizer for rubber; wetting agent; intermediate.

Kem Fluid. [Union Derivan SA] PVC lubricant.

Kemid®. [Norton Pampus GmbH] Fluoropolymer or polyetherimide film.

Kemilime. [Ash Grove Cement] Hydrated lime.

Kemilube. [Union Derivan SA] Metallic stearates.

Kemistab. [Union Derivan SA] PVC stabilizers.

Kemlex. [Ferro] Acetal, thermoplastic polyurethane resins, glass-reinforced.

Kempore®. [Uniroyal] Azodicarbonamide-based; blowing agent for foam processing.

Kemstrene®. [Witco/Humko] Refined glycerin; humectant, solvent for cosmetics, liq. soaps, inks, lubricants; intermediate.

Kemtec. [PMC Specialties] Blowing agent for plastics.

Ke-Mul®. [Georgia-Pacific] Lignosulfonate; emulsifier and stabilizer for asphalt, emulsions.

Kencolor. [Kenrich Petrochemicals] Pigments.

Kencure. [Kenrich Petrochemicals] Curing agents.

Kenflex®. [Kenrich Petrochemicals] Oligomers; plasticizer for Hypalon and neoprene.

Kenite. [Witco] Diatomite; filter aid, anticaking agents, absorbent, bulking agent, conditioner, extender pigment, filler, opacifier.

Ken Kem®. [Kenrich Petrochemicals] Cumyl phenol derivs.; modifier for epoxy, furan, and phenolic resins; epoxy cure accelerator; chemical intermediate.

Kenlastic K Series. [Kenrich Petrochemicals] Elastomeric dispersions used as accelerators, antioxidants, activators, blowing agents, vulcanizers.

Kenmag. [Kenrich Petrochemicals] Magnesium oxide dispersions.

Kenmix. [Kenrich Petrochemicals] Pastes used as accelerators, activators, blowing agents, dessicants, flame retarders, vulcanizers.

Kenplast® Series. [Kenrich Petrochemicals] Plasticizers.

Ken-React® (KR) Series. [Kenrich Petrochemicals] Titanate, zirconate, and aluminate coupling agents; also act as adhesion promoters, antioxidants, antistats, antifoamers, accelerators, blowing agents, grinding aids, hardeners, impact modifiers, internal lubricants, metal primers, process aids.

Ken-Stat. [Kenrich Petrochemicals] Antistatic agents.

Ken-Zinc. [Kenrich Petrochemicals] Zinc oxide dispersion.

KEP. [Goldsmith & Eggleton] EPM and EPDM rubbers; for wire and cable insulation, industrial parts, brake parts, etc.

Kera. [Champlain Industries] Protein hydrolysates.

Keramid. [Rhone-Poulenc UK] Polyimide resin.

Keramino. [Brooks Industries] Keratin amino acids or blends; for cosmetics.

Kerapol. [Henkel-Nopco] Coconut alkylolamide; detergent, foam booster.

Kera-Quat WKP. [Maybrook] Cocodimonium hydroxypropyl hydrolyzed keratin; substantivity agent, moisturizer, protective film-former for hair and skin care prods.

Kerasol. [Croda Inc.; Croda Chem. Ltd.] Soluble keratin; proteinic conditioner for hair and nail care prods.

Kerasol. [Henkel-Nopco] Solvent blends; detergent, kier boiling assistant; for removal of grease, tar and other impurities in textile bleaching and dyeing; cleaning agent for flame suppressors of diesel locomotives in coal and other mines.

Kera-Tein. [Maybrook] Hydrolyzed keratin or esters; protective colloid, moisturizer, substantivity agent for hair treatments for permanent waved and bleached damaged hair, skin care prods.

Kera-Tein AA. [Maybrook] Keratin amino acids; moisturizer, substantivity agent for hair and skin care prods.

Kerb. [Rohm & Haas] Herbicide.

Kerensim. [Henkel-Nopco] Alkyl polyethoxy esters and ethers; antistat, humectant, fiber treatment.

Kerimid. [Rhone-Poulenc Plastiques Tech.] Polyimide resins.

Kerinol. [Henkel-Nopco] Fatty acid alkanolamide; detergent, foam stabilizer/ thickener.

Kerobit®. [BASF AG] Phenylene diamines; antioxidant and metal deactivator for crude oil distillates.

Kerofluid®. [BASF AG] Additives to prevent ice formation in carburetors and jet engines.

Keroflux®. [BASF AG] Low m.w. wax; cold flow improvers and filtration aids for middle distillates in the mineral oil industry.

Kerokorr®. [BASF AG] Corrosion inhibitors for fuels.

Keromet®. [BASF AG] Disalicylidene-1,2-propane diamine; chelating agent and metal deactivator for fuels and lubricating oils.

Keropon®. [BASF AG] Antifouling agents for crude oil processing plants; stabilizers for storage of petroleum distillates.

Keropur®. [BASF AG] Multipurpose additives for cleaning and maintaining the inlet systems of internal combustion engines.

Kerostat®. [BASF AG] Amino carboxylates; antistatic additive for fuels.

Kessco®. [Stepan] Esters; thickener, emollient; cosmetics and pharmaceuticals; lubricant for textile and metalworking compds.

Kessco® PEG Series. [Stepan] Ethoxylated esters; nonionic surfactants for cosmetics, pharmaceuticals, food, agric., plastics; as thickeners, solubilizers, emollients, opacifiers, spreading agents, wetting agents, dispersants.

Ketac. [Am. Cyanamid] Acetone-formaldehyde; for waterproofing starch adhesives.

Ketalin. [Chattem] Chelated aluminum alcoholate.

Ketjenblack® EC, ED. [Akzo] Carbon black; electroconductive carbon black for rubber compding.

Ketjencat. [Akzo] Fluid cracking catalysts.

Ketjenfine. [Akzo] Hydrotreating catalysts.

Ketjenflex®. [Akzo] Toluene sulfonamides; plasticizer for resins, coatings, electroplating sol'ns., thermoplastics, and thermosets, nitrocellulose lacquers.

Ketjenflex® MH, MS. [Akzo] Toluene-sulfonamide/formaldehyde resin.

Ketjenlube®. [Akzo] Butanol ester of alpha olefin-dicarboxylic acid copolymers; lubricant.

Ketjensept. [R.W. Greeff] Chloramine T.

Ketjensil®. [Akzo] Sodium-aluminum silicate; reinforcing filler for silicone rubbers, disp. paints.

Ketomax®. [UOP] Immobilized glucose isomerase; enzyme converting glucose to fructose.

Kevlar®. [DuPont] Aramid; high str., lightweight, flexible material for aircraft/aerospace, boat hulls, prosthetics, footwear, sporting goods, as asbestos replacement, as reinforcement for mech. rubber goods.

Keyacid. [Keystone Aniline] Textile dyes and pigments.

Keycide®. [Witco] Tributyltin oxide; mildewcide, antimicrobial for PVAc latex paints.

Key Cool 106. [Atochem N. Am.] Cutting lubricant.

Key Epoxy. [Key Polymer] Two-component epoxy adhesives for structural applics.

Keyfast. [Keystone Aniline] Textile dyes and pigments.

Keyflon. [WJP Engineering Plastics Ltd.] Filled PTFE.

Key Gear. [Atochem N. Am.] Enclosed space reducers.

Keyphos. [Bruna Coating Ltd.] Resin phosphate.

Keyplast. [Keystone Aniline] Textile dyes and pigments.

Keysperse. [Keystone Aniline] Textile dyes and pigments.

Keystone. [Keystone Aniline] Textile dyes and pigments.

Keywax. [Key Polymer] Butyl rubber concs.

K-Flex. [King Industries] Polyester or polyurethane polyols.

KF Polymer®. [Kureha Chem. Industry] PVDF; engineering thermoplastic for corrosion resistance applics., sheet lining, powder coating, pipe, joints, pumps, valves, plates, wire insulation, piezo-electric film, fishing line, industrial yarn.

KF Series. [Shin Etsu Silicones] Silicone polymers and ethers.

Kidney Rosin. [Sovereign] Hydrocarbon rosin oil; plasticizer.

Kieralon®. [BASF; BASF AG] Surfactant blends; surfactant, detergent, wetting agent, dispersant for textile processing.

Kilfoam. [Arol Chem. Prods.] Silicone emulsion; defoamer for textile and industrial use; emulsifier for paint.

Kinel. [Rhone-Poulenc Chem. Ltd.] Polyimide thermoplastics.

King Chlor. [E&F King] Sodium hypochlorite.

King Etch. [E&F King] Aluminum etching sol'n.

King Kleen. [E&F King] Beverage and dairy cleaners.

Kingoil S-10. [Sanyo Chem. Industries] Highly sulfated fatty acid; wetting agent, penetrant, dispersant, leveling agent for bleaching, scouring and dyeing applics.

Kings Delite. [Spice King] Sweetener.

Kirnol®. [Grünau] Mono and diglycerides of fatty acids (< 60% mono content); general purpose emulsifiers for food industry.

Kisc-Direct. [Kyung-In Synthetic] Textile dyes and pigments.

Kisclon. [Kyung-In Synthetic] Textile dyes and pigments.

Kisco Leather. [Kyung-In Synthetic] Textile dyes and pigments.

Kiscocion. [Kyung-In Synthetic] Textile dyes and pigments.

Kiscozol. [Kyung-In Synthetic] Textile dyes and pigments.

Kisc-White. [Kyung-In Synthetic] Textile dyes and pigments.

Kisuma. [Kyowa Chem. Industry] Magnesium hydroxide compd.; fire retardant for thermoplastics.

Kito 40. [Van Waters & Rogers] Tributyltin complex; fungicide for leathers.

Kito 703. [Van Waters & Rogers] Ethoxylated alcohol; wetting agent, emulsifier, detergent, defoamer.

Klakup. [Malvern Minerals] Surface-modified aluminum silicate.

Klar-Aid. [W.R. Grace/Dearborn] Flocculant.

Klearall. [Octagon Process] Degreasing solvents.

Klearfac®. [BASF] Phosphate ester; surfactants; solubilzier for nonionics; emulsifier; antistat; textile processing.

Klearol. [Witco/Sonneborn] Mineral oil NF; emollient in cosmetics.

Klear-O-Line. [Stewart Hall] High pressure solvent for blocked fuel lines.

Kleen All. [Chemurgy Prods.] Liq. surfactant blends; cleaners for dyeing equip.

Kleenflo. [L.A. Salomon] Activated clay.

Kleen-Paste. [Witco Israel] Sodium alkylaryl sulfonate; raw material for detergent industry.

Kleer-Aid. [McGean-Rohco] Waste treatment chemicals.

Kleerox. [Rhone-Poulenc/Textile & Rubber] Organic stabilizers for hydrogen peroxide bleaching.

Klensorb. [Calgon Carbon] Oil and grease absorbent.

Klenzol. [Hydrolabs] Low foaming detergents for use in high turbulence equip.

Klerzyme®. [Int'l. Bio-Synthetics] Pectinase; enzyme for depectinizing fruit juices.

Kling. [ScanRoad] Asphalt antistripping additive.

Klingerflon. [Klinger] PTFE; thermoplastic for use in bearings, rings, pump caps, etc.

Kloro. [Ferro/Keil] Chlorinated paraffin; additive for soluble oils, synthetics, and semisynthetics.

Klucel®. [Aqualon] Hydroxypropylcellulose; film-former, thickener, stabilizer, suspending agent, film barrier, protective colloid for food, cosmetics, pharmaceuticals, coatings, adhesives, extrusions and moldings, paper, paint removers, encapsulation, inks.

K.L.X. [Van Den Bergh Foods] Hydrog. veg. oil. icing stabilizer.

KM. [Kerr-McGee] Chemicals.

KMD-50, -80, -100. [Ferro/Keil] N,N´-Disalicylidene-1,2-propanediamine sol'ns.; metal deactivators for gasoline, distillate fuels, and other petroleum prods.

KNA-Cumene Sulfonate 40. [Hüls AG] Potassium sodium cumene sulfonate; hydrotrope, solubilizer, coupling agent for liq. detergents.

Knit Finish WK. [Yorkshire Pat-Chem] Nonionic sewing and napping lubricant.

Knitsoft®. [Yorkshire Pat-Chem] Softeners and processing aids for knitgoods.

Knockdown. [Air Prods.] Defoamer.

Knockout. [Amax Industrial Prods.] Liq. herbicide.

Knox Out 2FM. [Atochem N. Am.] Insecticides.

K Nylon KDTs. [KD Thermoplastics Ltd.] Industrial nylon 6 and 6/6.

Kobate. [Rhone-Poulenc Geronazzo SpA] Triazine deriv.; bactericide.

Koblend. [Montedipe Srl] Styrene copolymers, modified antishock.

Kocide®. [Griffin] Copper hydroxide and mixtures; agric. fungicide wettable powds.

Kocide® Copper Sulfate. [Griffin] Copper sulfate; industrial copper, algicide, sewer treatment, wood treatment, control of tadpole shrimp in rice fields.

Kodacel. [Eastman] Cellulosic film and sheet.

Kodaflex®. [Eastman] Phthalates, adipates, trimellitates, etc.; plasticizers for coatings, vinyls, rubber.

Kodaflex® Triacetin. [Eastman] Glyceryl triacetate; plasticizer for vinyl compds., adhesives, coatings, paper.

Kodapak®. [Eastman; Eastman UK] PET polyester; for beverage bottles.

Kodar®. [Eastman; Eastman UK] Copolyester; for optical grade film and sheet.

Kodel. [Eastman] Polyester yarn and fiber.

Kodosoff. [Eastman] Polyester fiberfill.

Kofilm®. [Nat'l. Starch & Chem.] Acetylated corn starch deriv.; warp sizing, finishing.

Kohacool. [Toho Chem. Industry] Alkylether sulfosuccinate; wetting and foaming agent for nonirritating hair shampoos, bubble baths, hair conditioners, lotions.

Kohinor. [Pantasote Polymers] Vinyl compds.; for inj. molding, automotive applics., etc.

Kol Guard. [A.E. Staley Mfg.] Modified corn starch.

Kollerdur®. [Scott Bader] Polyurethane resin sol'ns.; coatings for floors, furniture, industrial finishes.

Kollidon®. [BASF AG] PVP; solubilizer, crystallization retarder, stabilizer for antibiotic suspensions; binders for tablets, dispersants.

Kollidon® CL. [BASF AG] Crospovidone; tablet-disintegrating agent, suspension stabilizer.

Kollotex. [Avebe Am.] Modified potato starches; for textile applics.

Komadur. [Norplast Ltd.] PVC.

Komeen®. [Griffin] Copper-ethylenediamine complex; aquatic herbicide for golf courses, fish ponds, potable water reservoirs.

Kompli. [Ubbink Nederland BV] Polyurethane/polyethylene.

Kongeel. [HVC] Coagulant for rubber latexes.

Konker®. [BASF AG] Vinclozolin, carbendazim; fungicide.

Konut. [Van Den Bergh Foods] Coconut oil, kosher grade; for food prods.

Kool Kut. [Castrol Industrial East] Cutting oils.

Kopanex. [Dexter] Surfactants.

Korad. [Polymer Extruded Prods.] Acrylic; durable weatherable film.

Koraid. [Kaopolite] Silica; suspension aid for pigments and abrasive particles in water-based systems.

Korantin®. [BASF AG] Corrosion inhibitors, lubricants.

Korantin® BH. [BASF AG] Butynediol;

corrosion inhibitor in acid pickles and cleaners.

Koreforte®. [BASF AG] Reinforcing resins for natural and synthetic rubbers.

Koresin®. [BASF; BASF AG; Struktol] Condensation prod. of butyl, phenyl, and acetylene; tackifier for unvulcanized rubber mixtures; adhesive additive.

Korestab®. [BASF AG] Antioxidant for natural and synthetic rubbers.

Koretack®. [Struktol; BASF AG] Octyl phenyl novolac; tackifier for natural and synthetic rubbers.

Korever®. [BASF AG] Vulcanization resins.

Korez. [Atlas Minerals & Chem.] Phenolic cement.

Korlite. [Kaopolite] Natural zeolite silicate mineral; absorbent for organic compds., sulfur dioxide gases, odors, ammonia; builder in laundry detergents; decolorizer for organic liqs., specialty filler and water treatments.

Kor-Lok. [Nat'l. Starch & Chem.] Adhesives.

Korode Kure. [Acton Tech.] Battery anti-corrosion compd.

Kortacid. [Akzo] Telomer acids.

Korthix. [Kaopolite] Bentonite; thickener, thixotropic agent for aq. paints, inks, polishes, adhesives, household prods.

Kosmos®. [Goldschmidt AG] Organotin and other compds.; catalyst for mfg. of polyurethane foam.

Kostat. [Witco] Antistats.

Kostil. [Montedipe Srl] Acrylonitrile/styrene copolymer.

Kotall. [California Prods.] Alkyd interior paints.

Kotamite®. [ECC Int'l.] Calcium carbonate; extender.

KO-Zinc. [Cuproquim] 17% Copper hydroxide.

KP-2, -4. [Ferro/Keil] Lubricity agents for plasticizing dry rust inhibitor films.

KP-140®. [FMC] Tributoxyethyl phosphate; leveling agent in floor polishes; flame retardant for plastics and rubbers.

K-Pool. [Griffin] Copper-TEA complex;

algicide for swimming pools.

KR-610. [Arakawa] Water-white rosin.

Kraton®. [Shell] Thermoplastic elastomers requiring no vulcanization; for formulating adhesives, inj. molded and extrusion prods., etc.

Kremol. [Witco/Sonneborn] White mineral oil.

K-Resin®. [Phillips; Phillips Petrol. Chem. SA/NV] Styrene-butadeine copolymers; for medical devices, housings, molded boxes, toys, displays, overcaps, novelties, housewares, pkg., thermoformed blister packs, disposable containers, extruded tubes, bottles, shrink wrap, skin pkg.

Kricinol. [Climax Performance] Potassium ricinoleate; mold release.

Krim 400. [Fabriquimica] Cetearyl alcohol, cetrimonium chloride.

Krim CH 25. [Fabriquimica] Glycol stearate, cocamide DEA, sodium laureth sulfate.

Kristalex®. [Hercules] Hydrocarbon resin derived from α-methylstyrene; thermoplastic for sealants, caulking, textile sizes, adhesives, etc.

Kristel. [Van Den Bergh Foods] Hydrog. veg. oils; processing aid for foods.

Kroil. [Kano Labs] Penetrating oil.

Krolor. [Cookson Pigments] Textile dyes and pigments.

Krolor. [DuPont] Pigments.

Kromaplast. [Ampacet] Blended dry pigments.

Kromeko. [John Campbell] Textile dyes and pigments.

Krom-Trol. [Betz Industrial] Cooling corrosion inhibitor.

Krona-Plate. [E/M Corp.] Lubricant.

Kronagold. [E/M Corp.] Lubricant.

Kronitex®. [C.P. Hall] Phosphate esters; flame retardants, plasticizers for PVC.

Kronos®. [Kronos] Titanium dioxide; pigment, opacifier; delusterant for synthetic fibers.

Krovar. [DuPont/Ag] Herbicide.

Krumbhaar. [Lawter Int'l.] Phenolic resins.

Krynac. [Polysar] NBR; used for belt covers, o-rings, seals, gaskets, hose, inj. molded and extruded goods, etc.

Kryocide. [Atochem N. Am.] Insecticides.

Krystal Klear Oils. [United Catalysts] Castor oils; dye solvent, gloss agent, emollient for lipsticks, cosmetics.

Krystaltite Film. [Allied-Signal] PVC-based formulations; thermoplastic film for shrink film applics.

Krystar. [A.E. Staley Mfg.] Crystalline fructose.

Krytox®. [DuPont] Fluorinated oils and greases, lubricants.

K Series. [Bacon] Perfluoropropylene/vinylidene fluoride copolymer; used for rubber seals, diaphragms, bellows in aircraft, missile, automotive, chemical industries.

K Series. [Cosmic Plastics] Diallyl meta phthalate resins, glass or min. reinforced; high heat resist. resins for military applics.

K-Sperse. [King Industries] Dispersants.

K-Tac. [IGI Baychem] Refined atactic polypropylene and compded. atactic polypropylene prods.

K-Tea. [Griffin] Copper-TEA complex; algicide for use on golf course, ornamental and fish ponds, potable water reservoirs.

Kucide DF. [Griffin] Copper fungicide, dry flowable.

Kumulan®. [BASF AG] Nitrothal-isopropyl, sulfur; for control of powdery mildew in apples.

Kumulus®. [BASF AG] Sulfur; for control of diseases and spider mites in fruit, vines, vegetebles, ornamentals, agric. crops.

Kunstolen. [Kunstoplast-Chemie GmbH] Polypropylene.

Kunstolen LD IHD. [Kunstoplast-Chemie GmbH] LDPE, HDPE.

Kunstonyl. [Kunstoplast-Chemie GmbH] PVC.

Kunstyrol. [Kunstoplast-Chemie GmbH] PS.

Kureton. [Kao/Edible Fat & Oil] Monoglycerides; defoaming agent for soybean curd.

Kuro Bishi®. [Mitsubishi Kasei] General use coke; for carbide, ferro alloys, copper and nickel refining.

Kurofan®. [BASF AG] PVDC copolymer; for prod. of chemically resistant, heat sealable pkg. material; binders for bonding fiber webs, textile coatings, vapor barrier coatings for construction.

Kutwell. [Exxon] Metalworking lubricant.

K-Van. [Kerr-McGee] Potassium metavanadate.

Kwik Dri. [Ashland] Aliphatic naphtha; used as solvent for rubber and latex.

Kymene®. [Hercules] Wet-strength resins for paper industry.

Kynar®. [Atochem N. Am.] PVDF, unreinforced and graphite filled; engineering thermoplastic; used for coatings.

Kynar®-Flex. [Atochem N. Am.] PVDF.

Kynodur. [Kyros] Water sol'ns. of ammonia and other dissolved gases; evaporation retardants.

Kyrovap. [Kyros] Evaporation retardants for org. liqs., monomers, and solvents.

Kytamer PC. [Amerchol Europe] Chitosan PCA; substantive humectant and conditioner for hair and skin prods.

K-Zinc. [Griffin] Zinc oxide formulation; flowable nutrient for rice seed dressing.

KZ Series. [Kenrich Petrochemicals] Zirconates; coupling agents.

L

L-7. [Am. Grease Stick] Rust preventive.

L-55R®. [Reheis] Aluminum-magnesium hydroxy carbonate; acid neutralizing agent used in prod. of polyolefin resins.

Lab-Lube. [Alvin Prods.] Penetrating oil and rust preventive.

Labrafac®. [Gattefosse; Gattefosse SA] Triglycerides; solvent for pharmaceutical creams, lotions, and gels.

Labrafil®. [Gattefosse; Gattefosse SA] Hydrophilic oils, excipients for cosmetics, pharmaceuticals.

Labrasol. [Gattefosse; Gattefosse SA] PEG-8 caprylic/capric glycerides; hydrophilic oil, excipient, solvent for pharmaceutical and cosmetic formulations.

LABS-100, 100/H.V., 100 SP. [Zohar Detergent Factory] Alkylbenzene sulfonic acid or blend; detergent intermediate.

Lacer Max. [Astor Wax] Pentaerythrityl tetrastearate, calcium stearate blend; wax used in sealers/polishes, as mold release agent.

Lacol. [Lanaetex Prods.] Isobutyl tallowate, trilinolein, laureth-2 acetate blend.

Lacolene. [Ashland] Aliphatic hydrocarbons; solvents.

Lacovyl. [Atochem UK] PVC resins.

Lacqrene. [Atochem UK] Polystyrene.

Lacqtene. [Atochem UK] LDPE, HDPE, LLDPE.

Lacqvyl. [Atochem Deutschland GmbH] PVC resins.

Lacrylic. [California Prods.] Masonry conditioner.

Lactacet. [Vevy] Cetyl lactate.

Lactarin. [FMC/Marine Colloids] Carrageenan; cold milk thickener.

Lactil®. [Goldschmidt; Goldschmidt AG] Sodium lactate, sodium PCA, hydrolyzed animal protein, fructose, urea, niacinamide, inositol, sodium benzoate, lactic acid; humectant, moisturizing agent for creams and lotions.

Lactimon. [BYK-Chemie USA] Wetting agent, dispersant.

Lactodan. [Grindsted Prods.; Grindsted Prods. Denmark] Glyceryl lactates; emulsifier for foods, cosmetics/toiletries.

Lactofil Debacterise. [Gattefosse] Lactose, casein.

Lactolan. [Serobiologiques] Hydrolyzed milk protein.

Lactol Spirits. [Unocal] Solvent.

Lactomul. [Henkel/Functional Prods.] Emulsifier for calf-milk replacer.

Lactozym. [Novo Nordisk] Lactase; enzyme for hydrolysis of lactose in dairy industry.

Laddok®. [BASF AG] Bentazon, atrazine; herbicide for selective post-emergence control of broadleaf weeds.

Ladene. [Saudi Basic Industries] HDPE, LLDPE, PVC, PS, or melamine resins.

Lafil WL 3254. [Gattefosse SA] Polyglyceryl isostearate; superfatting agent for cosmetics and pharmaceuticals.

Lagz. [WCC Industries] Adhesive.

Lakes Superior. [Warner-Jenkenson] Pigments.

Lamacit®. [Henkel Canada; Henkel KGaA] Ethoxylated glyceryl esters; emulsifier, solubilizer for cosmetics, paper applics.

Lamecerin. [Grünau] Ethoxylated lanolin deriv.; refatting agent for personal care prods.

Lamecreme®. [Henkel/Cospha; Henkel KGaA; Grünau] Surfactant blends; emulsifier for cosmetics, base for emulsions.

Lamefix. [Grünau] Accelerator for HT-fixation of polyester and triacetate.

Lameform®. [Henkel/Cospha; Henkel KGaA] Polyglyceryl ester blends; emulsifier for cosmetics.

Lamefrost®. [Grünau] Emulsifier/stabilizer blends; emulsifier and stabilizer for ice cream.

Lamegin®. [Henkel/Functional Prods.; Grünau] Mono- and diglycerides esters or tartaric acid esters; emulsifier, plasticizer, coating agent for cosmetics, foods, and drugs.

Lamemul®. [Grünau] Mono- and diglycerides; emulsifiers for baking additives; antistaling effect.

Lamephos®. [Grünau] Spray-dried emulsifier compd. based on mono- and diglycerides esterified with citric acid and phosphates on carrier; improves fat and jelly separation in sausage and meat mfg.

Lamepon®. [Henkel/Cospha; Henkel Canada; Henkel KGaA] Hydrolyzed collagen derivs.; mild surfactant, dispersant, wetting agent, stabilizer, protective colloid, detergent for personal care prods., dyestuffs.

Lamequat® L. [Henkel/Emery/Cospha; Henkel KGaA] Lauryldimonium hydroxypropyl hydrolyzed collagen; conditioner for hair and body preps.

Lamequick. [Henkel/Functional Prods.; Grünau] Emulsifier blend; food emulsifier, whipping base for toppings and aerated foods.

Lamesoft® LMG. [Henkel/Cospha; Henkel KGaA] Glyceryl laurate and TEA-coco hydrolyzed animal protein; refatting agent, thickener for foam baths, shampoos.

Lamesorb®. [Grünau] Sorbitan esters of fatty acids or polysorbates; nonionic emulsifiers for prod. of o/w and w/o emulsions.

Lamigen. [Dai-ichi Kogyo Seiyaku] Ethoxylated lanolin ethers or esters; softener, antistat for textiles, leather; emulsifier for emulsion polymerization.

Laminating Adhesive #5110. [Polymer Research Corp. of Am.] Copolymer; adhesive for textile backing, pigment and fiber binders, paper saturation.

Laminex. [Genencor Int'l.] Brewing enzymes.

Lamitex. [Protan] Thickener for printing reactive dyes on cellulose fibers.

Lampronol. [ICI Am.] Textile dyes and pigments.

LAN-401. [Nikko Chem. Co. Ltd.] Mineral oil and surfactant; coning oil for synthetic fibers.

Lanacet® 1705. [Henkel/Emery] Acetylated lanolin; emollient, superfatting agent for personal care prods.

Lanacron. [Ciba-Geigy/Dyestuffs] Textile dyes and pigments.

Lanaetex. [Lanaetex Prods.] Blends.

Lanaetex-75. [Lanaetex Prods.] Acetylated lanolin alcohol.

Lanafix. [Lenmar] Fixative for direct, fiber reactive, or acid dyes.

Lanagen. [Lanaetex Prods.] Hydrolyzed collagen.

Lanagen. [Lenmar] Antiprecipitants, compatibilizers for dyes; leveling and retarding agents.

Lanalan. [Lanaetex Prods.] Lanolin/lanolin alcohol blend.

Lanalene. [Maybrook] Lanolin blends.

Lanalox. [Maybrook] Ethoxylated lanolin compds.

Lanamine®. [Amerchol] Mixed isopropanolamines myristate; surfactant.

Lanamol. [Lenmar] Dyebath lubricant and softener.

Lanamol. [Maybrook] Mineral oil, PEG-30 lanolin, cetyl alcohol blend.

Lanapene. [Lanaetex Prods.] Isopropyl lanolate, lecithin blend.

Lanapex. [ICI Surf. UK] Wetting agent, detergent for textile scouring.

Lanaplex. [Lenmar] Liq. alkali sol'ns. for bleaching, scouring, and dyeing.

Lanapol. [Lenmar] Thickener for dyes; antimigrant, softener.

Lanapol CT. [Seppic] Isopropanolamine lanolate; cosmetic assistant.

Lan-Aqua-Sol. [Fanning] PEG-75 lanolin; emulsifier, emollient, superfatting agent, conditioner for cosmetics, pharmaceuticals, household detergents; solubilizer, wetting agent, dispersing aid.

Lanasan CL. [Sandoz] Hydrolyzed collagen.

Lanaset. [Ciba-Geigy/Dyestuffs] Textile dyes and pigments.

Lanasof. [Lenmar] Softener for cotton, synthetics, and blends.

Lanasol. [Ciba-Geigy/Dyestuffs] Textile dyes and pigments.

Lanasperse. [Lenmar] Dispersant for disperse dyes.

Lanasyn. [Sandoz] Textile dyes and pigments.

Lanatein-25. [Lanaetex Prods.] PEG-75 lanolin, hydrolyzed collagen blend.

Lanawet. [Lenmar] Wetting agent for textile use.

Lanbritol Wax N21. [Ronsheim & Moore] Cetearyl alcohol, ceteth-12, oleth-12 blend; self-emulsifying wax for pharmaceuticals, cosmetics, hair care preps.

Lancara. [Lenmar] Dye carriers.

Lancare. [Henkel-Nopco] Surfactant for shampoos, bubble baths, mild abrasive cleaners.

Lancol. [Lanaetex Prods.] Oleyl alcohol.

Lancowax. [Cray Valley Prods.] Coatings additives.

Landemul. [Harcros UK] EO/PO copolymers; demulsifier.

Lanesta. [Lanaetex Prods.] Esters.

Lanesta. [Westbrook Lanolin] Isopropyl or glyceryl lanolate; emollient, lubricant for skin care prods.

Lanethyl. [Croda Chem. Ltd.] Lanolin alcohol extract; plasticizer, film modifier for hair setting resins; emulsifier, conditioner.

Laneto. [RITA] Alkoxylated lanolin; emollient, lubricant, resin modifier, moisturizer, solubilizer, emulsifier, plasticizer for cosmetics, pharmaceuticals, detergent systems.

Lanette®. [Henkel/Emery/Cospha; Henkel Canada; Henkel KGaA] Fatty alcohols, sulfates, or blends; emollient, consistency agent, emulsifier, wetting agent for cosmetics, pharmaceuticals.

Lanette Wax. [Ronsheim & Moore] Self-emulsifying wax; for hair conditioners, creams, lotions.

Lanex. [Croda Chem. Ltd.] Lanolin alcohol extract; emollient, plasticizer, conditioner, superfatting agent, film modifier for cosmetics, hair sprays.

Lanexol. [Croda Inc.; Croda Chem. Ltd.] Alkoxylated lanolin; emollient, conditioner, superfatting agent, foam stabilizer, lubricant, emulsifier, plasticizer for cosmetics, hair spray resins.

Lanfrax®. [Henkel/Emery/Cospha] Lanolin wax derivs.; emulsifier, emollient, waxing agent for floor finishes; emollient for cosmetics; stabilizer, thickener.

Langdocyal. [Transmare BV] Phthalocyanine pigments.

Langford® Clay. [R.T. Vanderbilt] Soft kaolin; inert filler for elastomers, nonblack stocks.

Lanidox-5. [Lanaetex Prods.] PEG-6 lauramide.

Lanion. [Lanaetex Prods.] Stearalkonium chloride/lanolin derivs. blend.

Lanisolate. [Lanaetex Prods.] Isopropyl lanolate.

Lankrocell®. [Harcros] Stabilizer and mechanical foam promoter for foamed PVC applics.

Lankroflex®. [Harcros; Harcros UK] Epoxidized oils or esters; plasticizer/stabilizer for epoxy and thermoplastics.

Lankrolan SHR-3. [Harcros UK] Organic thiosulfate; shrink resist agent for wool.

Lankromark®. [Harcros; Harcros UK] uv absorbers, heat and light stabilizers, activators, antioxidants for PVC and other polymers.

Lankro Mud-Aids. [Harcros UK] General aids for drilling muds incl. mud surfactants and defoamers.

Lankro Mud Detergents. [Harcros UK] Anionic or nonionic detergents for well drilling mud applics.

Lankro Mud-Emuls. [Harcros UK] W/o and o/w emulsifiers for water-based and invert muds.

Lankromul. [Harcros UK] Emulsifier and dispersant for oil spills.

Lankropearl. [Harcros UK] Surfactant, pearlizing agent for toiletries.

Lankroplast®. [Harcros; Harcros UK] Lubricant, tackifier, visc. and rheology modifier for PVC and other formulations.

Lankropol®. [Harcros; Harcros UK] Sulfates, sulfosuccinates, or sulfosuccinamates; emulsifier, stabilizer, foaming agent, wetting agent, solubilizer for emulsion polymerization, textile, leather, metal, personal care prods., paints.

Lankrosol. [Harcros; Harcros UK] Hydrotrope, solubilizer, and visc. modifier for detergents.

Lankrostat®. [Harcros; Harcros UK] Antistats for polymers.

Lannate. [DuPont/Ag] Insecticides.

Lanobase SE. [Lanaetex Prods.] PEG-6, PEG-32, PEG-75 lanolin blend.

Lanocerin®. [Amerchol] Lanolin wax; w/o emulsifier, emollient, conditioner used in cosmetics.

Lanofoam. [Lenmar] Antifoams.

Lanogel®. [Amerchol] Ethoxylated lanolin; emollient, emulsifier, dispersant, wetting agent, solubilizer, foam stabilizer for cosmetics, personal care prods., pharmaceuticals, and facial tissues.

Lanogen 1500. [Hoechst AG] PEG mixture; ointment base, thickener.

Lanogene®. [Amerchol] Lanolin oil; emollient, moisturizer, and emulsifier.

Lanoil. [Lanaetex Prods.] Lanolin oil.

Lanoil AWS. [Lanaetex Prods.] PPG-12-PEG-50 lanolin.

Lanol. [Seppic] Alcohols or esters; emollient, ingredients for cosmetics.

Lanolene. [Maybrook] Lanolin derivs. or blends. cosmetics ingredients.

Lanolex L-40. [Nihon Emulsion] PEG-50 lanolin; reforming agent, emollient for shampoos and hair conditioners.

Lanolic Acid. [Croda Chem. Ltd.] Lanolin acid; emollient, lubricant for cosmetics.

Lanolide. [Vevy] PEG-5 pentaerythritol ether, PPG-5 pentaerythritol ether, soy sterol.

Lanolin A.C. [Lanaetex Prods.] Acetylated lanolin.

Lanolin Acetate. [Maybrook] Acetylated lanolin.

Lanolin Alcohols LO. [Solvay Duphar BV] Lanolin alcohols; emulsifier, emollient, stabilizer for tech. applics.

Lanolin Anhydrous USP. [Lanaetex Prods.] Cosmetic ingredient.

Lanolin Fatty Acids. [Solvay Duphar BV] Lanolin fatty acids; emulsifier for lubricating greases, polishes, anticorrosion compds.

Lanolol. [Lanaetex Prods.] Petrolatum, lanolin, lanolin alcohol blend.

Lanolol Distilled, Standard. [Maybrook] Lanolin alcohol.

Lanoquat®. [Henkel/Emery/Cospha; Henkel KGaA] Quaternium-33/ethyl hexanediol blends; conditioner, emulsifier, emollient for hair and skin care prods.

Lanosil. [Lanaetex Prods.] Lanolin, isopropyl lanolate.

Lanostat. [Lenmar] Antistats for textile use.

Lanosterol. [Solvay Duphar BV] Lanolin deriv.; gelling agent for cosmetics.

Lanotein AWS 30. [Fanning] Propylene glycol, hydrolyzed animal protein, PPG-12-PEG-65 lanolin oil; conditioner, film former, lubricant, humectant, and emollient for hair care prods.

Lanotex. [Auschem SpA] Ethoxylated lanolin; skin emollient and conditioner for shampoos, bubble baths, cosmetics.

Lanowax. [Lanaetex Prods.] Lanolin wax.

Lanoxal 75. [Seppic] PEG-75 lanolin; cosmetic assistant.

Lanoxide. [Lanaetex Prods.] Ethoxylated stearate.

Lanoxyl. [Witco SA] Ethoxylated lanolin; emulsifier.

Lanpol. [Croda Chem. Ltd.] Ethoxylated lanolin acids; o/w emulsifier, solvent, solubilizer, wetting agent, dispersant for cosmetics.

Lanpro. [Lanaetex Prods.] Hydrolyzed collagen blends; cosmetic ingredients.

Lanquell. [Harcros UK] Polyglycol-based; antifoam for mfg. of paper for food pkg.

Lantox. [Lanaetex Prods.] Ethoxylated lanolin. cosmetic ingredient.

Lantrol®. [Henkel/Emery/Cospha; Henkel KGaA; Pulcra SA] Lanolin oil or derivs.; emollient, emulsifier, moisturizer, dispersant, plasticizer, conditioner, vehicle for makeup, hair care prods., medicinals.

Lanycol. [Lanaetex Prods.] Ethoxylated alcohols; surfactants.

Laponite®. [Laporte/Southern Clay] Sodium magnesium silicates; carrier for active materials in horticulture and agriculture; suspending agent, antisettling aid; thickener for clays, cosmetics; antisoil for carpet and laundry prods.

Laril. [Thertec SA] Polyphenylene oxide.

Larodur®. [BASF AG] Self-crosslinking, heat-curable polyacrylate resins; for baking finishes on domestic appliances.

Laroflex®. [BASF AG] Vinyl chloride copolymers; for finishes on metal, concrete; binders for gravure inks.

Laromer®. [BASF AG] Unsaturated acrylates; reactive diluent or thinner for radiation-curable systems, finishes.

Laromid®. [BASF AG] Polyamines; hardeners for epoxy resins.

Laromin®. [BASF AG] Polyamines; hardeners for epoxy resins.

Laropal®. [BASF AG] Aldehyde and ketone resins; lightfast resins for surface coatings, pigment pastes, flexographic and gravure inks.

Larosol®. [PPG/Specialty Chem.] Leveling agents, lubricants, scouring agent, antiprecipitant, emulsifier, retarder for textiles.

Larostat®. [PPG/Specialty Chem.] Ethosulfates; antistat for synthetic fibers, fiberglass, plastics; noncorrosive mold release agent.

Larton. [Lati SpA] Polypropylene.

Larvin. [Rhone-Poulenc/Ag] Insecticide, nematicide.

L.A.S. [Gattefosse SA] PEG-8 caprylic/capric glycerides; excipient for creams, lotions; surfactant for microemulsions.

Laser®. [BASF AG] Cycloxydim; post-emergence graminicide against annual and perennial grasses.

Lasilium C. [Exsymol] Lactoyl methylsilanol elastinate; hydrating, regenerative agent for skin care and toothpaste formulations.

Laskabond. [Simlak Ltd.] Polyvinyl chloride adhesives.

Laskaflex. [Simlak Ltd.] Plasticizers.

Laskol. [Simlak Ltd.] Saturated polyesters.

Lasso Bleach. [Labbco] Sodium hypochlorite.

Lastane. [Lati SpA] Polyurethane.

Lastiflex. [Lati SpA] PVC.

Lastil. [Lati SpA] SAN.

Lastilac. [Lati SpA] ABS.

Lastirol. [Lati SpA] PS.

Lasulf. [Lati SpA] Polysulfone.

Latamid. [Lati SpA] Polyamide 6 and 6/6.

Latan. [Lati SpA] Polyacetals.

Latekoll®. [BASF AG] Polyacrylic derivs.; thickeners for polymer dispersions and latexes; for paints and textured finishes, adhesives, sealants, prod. of fiber webs.

Latene. [Lati SpA] PP homopolymers and copolymers.

Later. [Lati SpA] PBT.

Lathanol®. [Stepan; Stepan Europe] Alkyl sulfoacetates; emulsifier, wetting agent, detergent, foaming agent used in cosmetics.

Lather Lube. [Alex C. Fergusson] Special foam detergent lubricants for conveyors.

Latilon. [Lati SpA] PC.

Latimid. [Lati SpA] Polyamides.

Latol. [Climax Performance] Blown tall oil; primary emulsifier in oil muds.

Latron. [Rohm & Haas] Agric. spray adjuvant.

Laur. [Laur Silicone Rubber Compding.] Silicone elastomers; used for o-rings, roll covering, hot stamping; dispersions used as adhesives, size coatings, splicing agent.

Laural. [Ceca SA] Sulfates; foaming shampoo, emulsifier, degreaser, detergent for household prods., wool and synthetic fibers, emulsion polymerization.

Lauramide. [Zohar Detergent Factory] Fatty acid DEA; foam booster, thickener, superfatting agent.

Lauramine 20A. [Reilly-Whiteman] Anionic softener for textiles.

Lauravel SC. [Reilly-Whiteman] Nonionic softener for textiles.

Laurel. [Reilly-Whiteman] Sulfates, ethoxylated esters, or fatty acid alkanolamides; emulsifier, detergent, lubricant, softener; used in cleaners, lubricants, metalworking fluids, paint, textiles, hair preps., skin cleansers.

Laurelev. [Reilly-Whiteman] Dyeing assistant, leveling agents for acrylics, nylon.

Laurelox 12. [Reilly-Whiteman] Lauryl dimethylamine oxide; foam booster/stabilizer for light duty liqs.; nonwetting agent for textile applics.

Laurelphos. [Reilly-Whiteman] Phosphate esters; detergent, hydrotrope, deduster, scouring agent, emulsifier, rust inhibitor for cleaning compds., textiles, metalworking formulations, lubricants.

Laurelterge. [Reilly-Whiteman] Detergents for textile scouring.

Laureltex. [Reilly-Whiteman] Detergents for textile scouring; dye assistants.

Laureltex MS-1. [Reilly-Whiteman] Buffering agents for textiles.

Laureltex TS. [Reilly-Whiteman] Peroxide stabilizers for batch and continuous bleaching of textiles.

Laurelube. [Reilly-Whiteman] Napping assistants, dyebath lubricants for textile industry.

Laurene. [Vevy] Laurtrimonium chloride.

Laurex®. [Albright & Wilson Am.; Albright & Wilson UK] Primary fatty alcohols; raw material for mfg. of surfactants; lubricant; emulsion polymerization stabilizer.

Laurex®. [Uniroyal] Activator.

Lauricidin. [Lauricidin] Glyceryl laurate or blends.

Lauridit®. [Akzo BV] Fatty acid alkanolamides; detergent, emulsifier for cosmetic, household, textile applics.; corrosion inhibitor for cutting oils.

Lauroglycol. [Gattefosse] Propylene glycol laurate.

Lauropal. [Witco SA] Ethoxylated fatty alcohol; wetting agent, emulsifier, detergent for cosmetics, textiles, metal cleaning, industrial use.

Laurox®. [Akzo] Peroxides; initiator for prod. of PVC resins, acrylates, high temp. polyester cures.

Lauroxal. [Witco SA] Ethoxylated fatty alcohol; wetting agent, detergent for industrial use, textiles, metal cleaning, leather.

Laurydone®. [UCIB] Lauryl PCA; emulsifier, moisturizer for skin care prods., hair prods., toiletries, make-up.

Lavaquick®. [Boehme Filatex] Solvent/surfactant mixture; High foaming interior and exterior dye machine cleaning agent.

Lavatex®. [Boehme Filatex] Organic polyacid salt mixture; hydrogen peroxide stabilizer for silicate-free batch and continuous bleaching of cellulosics and blends.

Laventin®. [BASF AG] Ethoxylates; nonionic wetting agents, detergents for textile industry.

Lavoral. [Miles/Organic Prods.] Nonionic low temp. scouring agent.

Laxan. [Lanaetex Prods.] Polysorbates; surfactants.

LBA-133 Polyethylene. [Mitsui Petrochemicals] LDPE; clarity blown film resin.

LCA. [Bacon] Epoxy resin adhesive.

LCP-20CF/000. [Compounding Tech.] Liq. crystal polymer, carbon-filled.

LCP Pac. [LeChem] Polyanionic cellulose; for oil well drilling applics.

L.C.R.E. [Kolmar Labs] Erucyl oleate, squalane, wheat germ oil, avocado oil, C10-30 cholesterol/lanosterol esters, tocopheryl acetate, retinyl palmitate.

LDPE. [Dow] LDPE; for blown film extrusion, coatings, inj. molding.

LE. [Lubrication Engineers] Lubricants.

Lea Antistat. [Lea Mfg.] Static charge neutralizer.

Leadacid. [Leadertech Colors] Textile dyes and pigments.

Leadacryl. [Leadertech Colors] Textile dyes and pigments.

Leadirect. [Leadertech Colors] Textile dyes and pigments.

Leadisperse. [Leadertech Colors] Textile dyes and pigments.

Leadvat. [Leadertech Colors] Textile dyes and pigments.

Lea Liquabrade. [Lea Mfg.] Buffing abrasive.

Lea Liqualube. [Lea Mfg.] Lubricant.

Lea Plastishine. [Lea Mfg.] Plastics cleaner.

Leasil. [Leatex] Durable silicone elastomer for cellulosics and blends.

Leatardent. [Lea Mfg.] Tarnish preventive.

Leatex. [Leatex] Scouring agent, deaerator for dyeing, bleaching, scouring operations; textile finish.

Leatherlike. [Nat'l. Chem. & Plastics] PVC films finish.

Leather Scents. [Andrea Aromatics] Fragrances for natural and artificial leathers.

Leaxol. [Leatex] Compatibilizer and antiprecipitant.

Lebaycid®. [Bayer] Fenthion; insecticide for controlling sucking and biting pests.

Lebon. [Sanyo Chem. Industries] Betaines, glycines, imidazolines, or quaternaries; germicide, shampoo base.

Lecidan. [Grindsted Prods.; Grindsted Prods. Denmark] Monoglyceride/lecithin blends; food emulsifier.

Lecimulthin. [Lucas Meyer] Powdered lecithins; separating and emulsifying agent for wafers and dry bakery prods.

Lecitase. [Novo Nordisk] Phospholipase A-2; enzyme for food applics.

Lecithin L-Range. [Grünau] Lecithin compds.; emulsifier and coemulsifier for baking additives.

Lecithin Water Dispersible CLR. [Henkel/Cospha] Hydrolyzed soya lecithin; emollient for skin and hair care preps.

Lecithin W.D. [Troy] Lecithin prod.; wetting agent, dispersant for paint industry.

Ledate®. [R.T. Vanderbilt] Lead dithiocarbamates; accelerator for natural and polyisoprene rubbers.

Leecure B. [Leepoxy Plastics] Boron trifluoride-based; epoxy curing agents/hardeners.

Leegen. [R.T. Vanderbilt] Sulfonated petroleum prod. and mineral oil on inert carrier; rubber plasticizer.

Legend. [Rohm & Haas] Biocide.

Legoprint. [Sigma Prodotti Chimici] Binders for pigment printing.

Lektrostat®. [Dexter] Finishing agents, antistats.

Lemonal. [Huntington Labs] Water-sol. deodorant.

Lencoll. [Lensfield Prods. Ltd.] Collagen.

Lenetol. [ICI Surf. UK] Ethoxylated fatty alcohols; nonionic detergent, wetting agent for textile scouring and bleaching.

Lenolube. [Lenox] Lubricant, antistat for textiles.

Lenospin. [Lenox] Wool oil additive for processing fiber blends for improved cohesion, static control, moisture retention.

Lenostat. [Lenox] Antistat for use with wool, synthetics for all processes.

Lensol. [Lenox] Antistatic lubricant for wool, synthetics.

Lensol. [Lensfield Prods. Ltd.] Hydrolyzed collagen.

Lentex. [Lenox] Nonsoiling carpet lubricant and antistat.

Lenzing P84. [Lenzing AG] Aromatic polyimide; high performance thermoplastic used for prod. of needle felts for high temp. filtration, as reinforcement for PTFE, as compding. ingred. for inj. moldable polymers.

Lenzing PTFE. [Lenzing AG] PTFE; for multifilament yarns in sealing industry, staple fibers, weaving yarn, and sewing threads in filter industry, films in the cable industry.

Lenzing Viscose Graphite-Containing. [Lenzing AG] Viscose with 40% incorporated graphite; fibers with improved gliding behavior for braided packings, hybrid yarns.

Leocon. [Lion] Polyoxyalkylene glycol; defoamer, lubricant for industrial processing.

Leoguard. [Lion] Cellulosic resin; conditioner for hair care prods.

Leomin®. [Hoechst Celanese/Colorants & Surf.; Hoechst AG] Antistat, softener, dispersant, finishing agent for textile and plastic finishing and processing; cleaning agent.

Leonil. [Hoechst Celanese/Colorants & Surf.; Hoechst AG] Sulfonates or sulfosuccinates; wetting agent and dyeing auxiliaries for textiles.

Leophen®. [BASF AG] Wetting agent, detergent, antifoam, dispersant for textiles and fibers.

Lepton®. [BASF AG] Polymer dispersions; binders for pigments, finishes, leather dressings; pigment preparations, fillers for leather and fur industry.

Lethane. [Rohm & Haas] Synthetic organic insecticide concs.

Letocil. [Chem-Y GmbH] Fatty acid polyglycol ester; emulsifier for cosmetics.

Lets-Go. [Stewart Hall] Lubricant, penetrant.

Leucophor. [Sandoz] Fluorescent whiteners for cellulosics for laundry and cleaning prods.

Leucopure. [Sandoz] Fluorescent whiteners for synthetics and wool for laundry and cleaning prods.

Leukanol. [Rohm & Haas] Synthetic tanning agent.

Leukotan. [Rohm & Haas] Acrylic synthetic tanning agent; retanning agents for leather.

Leukotrop® W. [BASF AG] Assistant for producing white discharge prints on cellulose fibers.

Levacell. [Miles/Organic Prods.] Textile dyes and pigments.

Levaderm. [Miles/Organic Prods.] Textile dyes and pigments.

Levafix. [Miles/Organic Prods.] Textile dyes and pigments.

Levaform. [Bayer] Fatty acid deriv.; mold lubricant.

Levalin®. [Miles/Organic Prods.] Auxiliary, antimigrant, lubricant for continuous dyeing.

Levalube. [Piedmont Chem. Industries] Leveler, retarder, detergent, napping lubricant and softener.

Levapon®. [Miles/Organic Prods.] Scouring and wetting agent for textiles.

Levapren. [Bayer; Miles] EVA or VAE copolymers; synthetic rubber for tech.

moldings and extrudates, cable sheathings, insulation, cellular rubber goods, footwear soles, waterproof sheeting, adhesives, fabric proofings; impact modifier for PVC.

Levasint. [Bayer] Saponified ethylene vinyl alcohol copolymer; used as corrosion protection for metal parts used in construction and chemical industry, traffic, transport, offshore, coastal and drinking water supply installations, sports facilities, consumer and industrial goods.

Levegal®. [Miles/Organic Prods.] Leveling agent, retarder, carrier, dye transfer agent for textile applics.

Levelan. [Harcros UK] Nonionics; leveling agent, emulsifier, stabilizer, wetting agent for dyestuffs, textiles, emulsion polymers, latexes.

Levelan. [Marubishi Oil Chem.] Leveling agent for textiles, dyeing.

Levelene. [Ciba-Geigy/Dyestuffs] POE alkyl ether; rewetting and leveling agent, penetrant, stripper for textile applics.

Levelene. [Leatex] Retarding and migrating agent for dyes.

Levelol 60. [Manufacturers Chems.] Ethoxylated amine; leveling agent for acid dyeable nylons; antiprecipitant with acid and basic dyes; retarding agent.

Levelon. [Surpass] Textile dye leveling agents.

Levelox. [Yoshimura Oil Chem.] Anionic/nonionic surfactant blend; leveling agent.

Levenol. [Kao] Retarding and stripping agent, leveling agent, dispersant for dyeing, textiles.

Levilite. [Rhone-Poulenc/Fine Chem.] Amorphous silica; nonabrasive additive for dentifrice; oil absorbent for talcs and sachets; thickener for lotions.

Levn-Lite. [Monsanto] Sodium aluminum phosphate.

Levogen®. [Miles/Organic Prods.] Cationic fixative for direct and fiber reactive dyes.

Levonic. [Chemonic Industries] Blend of sulfated esters; leveling agent for use with all types of dyes and equip.

Lewisol®. [Hercules] Glyceryl rosinate; hard thermoplastic resin used in lacquers, sealers, inks, grease-resistant coatings, specialty enamels.

Lexaine. [Inolex] Betaines, sultaines; amphoteric surfactants, visc. builders, foam boosters, thickeners for detergent systems, cosmetics.

Lexamine. [Inolex] Alkyl amidopropyl dimethylamines; cationic emulsifiers, intermediates; for hair care prods.

Lexan®. [GE Plastics; GE Plastics Ltd.] Polycarbonate resin; thermoplastic resin with high impact strength, close molding tolerances, low mold shrinkage, dimensional stability, weatherability and corrosion resistance, uv stability for inj. and blow molding, extrusion, film, and foam applics.

Lexate®. [Inolex] Concentrates and blends used as conditioner, emulsifier, emollient, shampoo base, soaps, cosmetics.

Lexein®. [Inolex] Collagen derivs.; film-forming protein, hair fixative, conditioner, hair spray resin plasticizer for hair preps., makeup, skin care prods.

Lexemul®. [Inolex] Glyceryl or glycol esters or ethers; emulsifier, stabilizer, thickener, opacifier, emollient used in cosmetics and topical pharmaceuticals.

Lexgard®. [Inolex] Parabens; preservatives for cosmetics.

Lexin K. [Am. Lecithin] Lecithin; emulsifier.

Lexol®. [Inolex] Esters or blends; emollient, carrier for bath oils, topical pharmaceuticals, personal care prods.

Lexolube®. [Inolex] Esters; lubricants for textile, metalworking, coatings, inks, plastics, other industrial applics.; softeners for rubbers.

Lexomul. [Inolex] Polyethylene glycol ester.

Lexone. [DuPont/Ag] Herbicide.

Lexorez. [Inolex] Polyadipates or polyphthalates; polyol for polyurethane formulation, adhesives, flexible coatings, castable elastomers.

Lexquat®. [Inolex] Quaternary ammonium chlorides; film former, hair fixative, conditioner, emulsifier, antistat,

emollient, humectant for hair and skin prods.

LGB. [J.M. Huber] Filler and reinforcing clay.

LHS. [Emulsion Systems] Coconut oil and caustic potash.

Librel. [Allied Colloids] Chelated micronutrients.

LICA®. [Kenrich Petrochemicals] Titanates; coupling agents; adhesion promoters, antioxidants, antistats, accelerators, activators, catalysts, curatives, corrosion inhibitors, dispersants, emulsifiers, flame retardants, foamers, impact modifiers, release agents, retarders, stabilizers, etc.

Licowet. [Hoechst AG] Fluoro surfactants; wetting and flow agents for lustering emulsions.

Licristal. [EM Industries] Liq. crystals.

Lidok. [Exxon] Grease.

Light Duller NF. [Sybron] Titanium dioxide dispersion; delusterant for textile fibers.

Lightning. [Chem-Pak] Penetrant, lubricant.

Light Roast. [H.B. Taylor] Sesame seed.

Light Water. [3M] Fluorochemical; fire-fighting agent.

Light-Weld®. [Dymax] Structural adhesive.

Lignorit. [Borregaard LignoTech] Calcium lignosulfonate; extender for urea-formaldehyde resin in particle board mfg.

Lignosite®. [Georgia-Pacific] Lignosulfonates; extender, emulsifier, dispersant for adhesives, pigments, binder systems, insecticides, industrial cleaners.

Lignosol. [Borregaard LignoTech] Kraft lignin; primary dye dispersant.

Lilamac. [Berol Nobel] Amine acetate; flotation agent.

Lilamin. [Berol Nobel] Polyamine blend; bitumen adhesion agent.

Lilaminox. [Berol Nobel] Oxides; thickener, foamer for household bleaches, shampoos, hard surface cleaners.

Lilamuls. [Berol Nobel] Amines; emulsifier, adhesion agent for bitumen.

Lilion. [Sniatechnopolimeri] Polyamide monofilament.

LIM6045. [GE Silicones] Liq. two-component silicone rubber.

Limao. [British Traders & Shippers] Shellac substitute resin.

Lime-Cote. [Rockwell Lime] Hydrated lime.

Lime Resin. [Arakawa] Rosin resinate.

Lime Sulfur. [Cuproquim] 29% calcium polysulfide.

Lin-All. [Mooney Chems] Tallates; driers for paints, inks.

Linaqua. [Reichhold] Water-reducible coating resin.

Linco 5-X. [Lindley Labs] One-piece scour and leveler for sheer and support nylon pantyhose.

Lincobrite. [Lindley Labs] Optical brighteners and whiteners for synthetic and cellulosic textile fibers.

Lincol®. [Condea Chemie GmbH] Linear plasticizer alcohols; intermediates for toiletries, cosmetics, detergents, leather and textile auxiliaries, lube oil, plastics additives, defoamers.

Lincolube. [Lindley Labs] Softener and lubricant for nylon, polyester and cotton.

Lincoscour. [Lindley Labs] Scours and detergents for textile use.

Lincosoft. [Lindley Labs] Softener and lubricant for textiles.

Lincoterge. [Lindley Labs] All-purpose self-emulsifying detergent scour.

Lindax. [Lindau Chems] Epoxy cure accelerator.

Linde. [Union Carbide] Industrial gases, gas handling equip., medical gases.

Lindol. [Akzo] Flame retardant plasticizer.

Lindox. [Union Carbide] Oxygen generators.

Lindride. [Chemach NV] Accelerators.

Lindride. [Lindau Chems] Anhydride curing agent for epoxy resins.

Lindron. [Lindau Chems] Protective coating resin.

Linear. [H.B. Fuller] Reinforcing tape.

Linex®. [Griffin] Linuron; flowable herbicide for control of grasses and broadleaf weeds.

Linlev 75. [Lindley Labs] Biodegradable noncarrier system to eliminate crack marks on polyester fabrics.

Lino-Cure®. [Ashland/Foundry Prods.] Oil-based foundry core binders.

Linoil®. [Ashland/Foundry Prods.] Foundry core binders.

Linolac. [Degen] Linseed-based; polymer film former.

Linplast®. [Condea Chemie GmbH] PVC plasticizers.

Lintex. [Lindley Labs] Peroxide bleach stabilizer, carrier, cleaner, defoamer, fluorochemical extender, reducing agent, water repellent, wetting agent, or softener for textiles.

Lintexfix. [Lindley Labs] Dye fixative for direct, sulfur, or fiber reactive dyes on cotton.

Lintexlene. [Lindley Labs] Sequestrants for textile use.

Lintexlube. [Lindley Labs] Lubricant for textile use.

Lintexscour. [Lindley Labs] Low foaming scour for synthetics, cellulosics and blends.

Lintexsoft. [Lindley Labs] Napping softener and lubricant for textiles.

Lint-Gard. [CNC Int'l.] Crosslinking acrylic polymer; textile printing auxiliary.

Linwet. [Lindley Labs] Scouring agent.

Lipacide. [Rhone-Poulenc Surf.] Cosmetic ingredients; for skin preps., first aid creams, sunburn lotion, antidandruff and conditioning shampoos.

Lipal. [Aquatec Quimica SA] Esters; emollient, emulsifier, pearlescent for cosmetics.

Lipamide. [Lipo] Fatty acid alkanolamide; emulsifier, opacifier, thickener, emulsion and foam stabilizer, lubricant, humectant, antistat for personal care prods.

Lipamin®. [BASF AG] Fatliquoring agents, stabilizers for leather processing.

Lipamine SPA. [Lipo] Stearamidopropyl dimethylamine; cosmetics ingredient.

Lipcare Wax 7782. [Kahl] Microcrystalline wax, polyethylene.

Lipex. [Karlshamns] Exotic oils.

Lipitein P. [Hormel] Animal skin lipids.

Lipo. [Lipo] Glycery, ethylene glycol, or

propylene glycol esters or natural abrasives; cosmetics ingredients, spreading agents, emulsifiers, dispersants, lubricants, opacifiers, emulsion stabilizers, emollients, visc. builders.

Lipobee 102. [Lipo] Synthetic beeswax; cosmetics ingredient.

Lipo Butter. [H.B. Taylor] Natural butter flavor.

Lipocerina. [Esperis] Acetylated hydrogenated lanolin.

Lipocerite. [Vevy] Hydrogenated C12-18 triglycerides.

Lipocerite Standard. [Vevy] Cetyl palmitate, isostearyl palmitate, cetyl stearate, stearyl stearate.

Lipocire. [Gattefosse] Hydrogenated palm glycerides, hydrogenated palm kernel glycerides.

Lipocol. [Croda Chem. Ltd.] Lanolin deriv.; emollient, moisturizer, emulsifier for cosmetics, pharmaceuticals.

Lipocol. [Lipo] Fatty alcohols, ethers, ester blends or ethoxylates; emollient, spreading agent, emulsifier, dispersant, opacifier, lubricant, defoamer, solubilizer, conditioner for cosmetics.

Lipocutin®. [Henkel/Emery/Cospha; Henkel KGaA] Water, lecithin, cholesterol, dicetyl phosphate; carrier for gel and skin care preps.

Lipodan. [Grindsted Prods.; Grindsted Prods. Denmark] Distilled monoglycerides; emulsifier for foods, resins, thermoplastics.

Lipoderm®. [BASF AG] Fatliquoring agents for leather.

Lipolan. [Lion] alpha-Olefin sulfonate; emulsifier for cosmetics, emulsion polymerization; detergent base.

Lipolan. [Lipo] Lanolin derivs.; emulsifiers, solubilizers, emollients, conditioners for cosmetics, toiletries, pharmaceuticals.

Lipolase. [Novo Nordisk] Fungal lipase; for detergents, laundry powds., prespotters.

Lipomin. [Lion] Antistat for textiles, plastics; scouring agent for wool.

Lipomix. [Lion] Alpha-olefin sulfonate; deinking agent for waste paper.

Lipomulse. [Lipo] Glyceryl stearate or blends; emulsifier, emollient, opacifier, visc. builder for creams and lotions.

Lipon. [Lion] Alkylbenzene sulfonate; emulsifier.

Liponate. [Lipo] Fatty esters; emollient, thickener for personal care prods.

Liponic. [Lipo] Sorbitol or glycerin or sorbitol ethers; humectant, plasticizer, softener, lubricant for pharmaceuticals, cosmetics, adhesives, leather, and paper coatings.

Liponox. [Lion] POE alkylphenol ether; wetting agent, dispersant, penetrant, emulsifier for emulsion polymerization, textiles; dye leveling agent.

Lipopeg. [Lipo] Ethoxylated esters; spreading agent, emulsifier, dispersant, lubricant for personal care prods.

Lipo-Peptide AME 30. [Maybrook] Acetamide MEA, lauroyl hydrolyzed collagen, glycerin; moisturizer, humectant, emollient, softener, emulsifier, conditioner, bodying agent, antistat for skin and hair care prods.; foam booster for shampoos, mousses, liq. hand soaps.

Lipophos. [Lipo] Phosphate esters; coupling agent, emulsifier, wetting agent for textile scouring, emulsion polymerization.

Lipophos. [Vevy] Soybean oil, lecithin.

Lipo Polyol NC. [Lipo] Hydrogenated starch hydrolysate.

Lipoproteol. [Rhone-Poulenc Surf.] Salt of lipoaminoacid; mild additives for shampoos, facial cleansers.

Lipoquat. [Lipo] Fatty acid amide ethosulfates; antistat, conditioner, emollient, glosser, softener, latex stabilizer for personal care prods., etc.

Liporamnosan. [Vevy] Hydroxyethylcellulose.

Liporez NEP-Special. [Lipo] Trimethylpentanediol/isophthalic acid/trimellitic anhydride copolymer.

Liposiliol C. [Exsymol] Dioleyl tocopheryl methylsilanol; tissue regeneration aid for anti-aging formulations, oily cosmetics.

Liposorb. [Lipo] Sorbitol, sorbitan esters, or ethoxylates; emulsifier, lubricant, antistat, thickener.

Liposurf. [Lipo] Anionic surfactants;

cosmetics ingredients.

Lipotac. [Lion] Alkyl methyl tauride; scouring agent for cotton, wool; dispersant for pigments.

Lipotin. [Lucas Meyer] Highly polar soy lecithins; emulsifier, wetting agent, dispersant for paints, leather finishes, textiles.

Lipotol®. [Nicca USA] Soaping agent for printing.

Lipovol. [Lipo] Natural vegetable oils or specialty esters; conditioner, glosser, emollient, lubricant for cosmetics, toiletries, and pharmaceuticals.

Lipowax. [Lipo] Synthetic waxes; lubricant, emulsifier, defoamer, antitacking agent, antistat, mold release for plastics, paper, textiles, rubber, paints.

Lipoxol®. [Hüls Am.] Polyethylene glycols; surfactants for cosmetic sticks, lipsticks, deodorant sticks, shaving sticks, soap bars, powder bases, creams, pastes.

Lipozyme. [Novo Nordisk] Immobilized lipase; for ester synthesis and fat modification.

Liprot. [Fabriquimica] Hydrolyzed collagen derivs.

Liptol. [Lion] Alkyl ethoxylate; deinking agent for flotation processes.

LiquaPar® Oil. [Sutton Labs] Isopropylparaben, isobutylparaben, butylparaben; preservative for cosmetics and topical pharmaceuticals.

Liquasoap. [Lindley Labs] Scouring agent.

Liqua-Tox. [Bell Labs] Liq. diphacinone conc.; for rat and mouse control.

Liquax 488. [Astor Wax] Microcrystalline wax/zinc stearate blend; self dispersing mold release compds. for polyurethane and other plastics, liq. solv. polishes, corrosion-resistant compds.

Liquazinc AQ-90. [Witco/Organics] Zinc stearate aq. dispersion; antitack agent, lubricant for rubber; release aid and lubricant for abrasive papers.

Liquester. [Robeco] Ester C30-46 piscine oil; emollient, lubricant, moisturizer for skin prods.

Liqui-Cal. [Schaefer Salt & Chem.] Liq. calcium chloride.

Liquid Absorption Base. [Croda Inc.] Mineral oil, lanolin alcohol; emollient for liq. makeup, emulsions.

Liquid Base. [Croda Inc.] Mineral oil, lanolin alcohol; surfactant, emollient, penetrant, moisturizer, softener, emulsifier, stabilizer for personal care prods.

Liquid Code XLR. [Rhone-Poulenc/Perf. Resins & Coatings] Potassium pyroantimonate sol'n.; liq. crosslinker for guar and derivatized guar at pH 3-5.

Liquid Crystal. [Presperse] Cholesteric esters or blends; carriers for nutrients, decorative, functional, and aesthetic effects for skin care prods.

Liqui-Det. [Oakite Prods.] Detergent.

Liquid Latex. [Rhone-Poulenc/Perf. Resins & Coatings] Polyvinyl acetate; used in oil well cementing applics. for fluid loss control, rheology modification, improved bonding.

Liquidow. [Dow] Liq. calcium chloride.

Liquidox. [Liquid Carbonic] Oxygen.

Liquiflow. [Liquid Carbonic] Carbon dioxide.

Liquifluor. [DuPont/Medical Prods.] PPO-POPOP toluene conc.

Liquigel®. [Reheis] Aluminum hydroxide or blends.

Liquigel®-AM. [Reheis] Aluminum/magnesium hydroxide fluid gel.

Liquigel®-HO. [Reheis] High oxide aluminum hydroxide fluid gel.

Liquimeth. [Degussa] Liq. DL-methionine.

Liqui-Nox. [Urban Chem.] Solder antioxidant, oil.

Liqui-Nox®. [Alconox] Liq. phosphate-free detergent.

Liquipanol. [Enzyme Development] Liq. enzyme, papain prep.

Liqui Paste. [Ashland] Core paste.

Liquivac. [Mateson] Liq. sorbent.

Liquiwax. [Brooks Industries] Liquid wax emollients for skin care cosmetics.

Lissamine. [ICI Am.] Textile dyes and pigments.

Listab. [Chemson Ltd.] PVC stabilizers and additives.

Lite. [Rhone-Poulenc Basic] Sodium carbonate anhydrous.

Lite-R-Cobs®. [Andersons] Corncob

meal; inert plastic extender, absorbent, and filler for resin molding, glue, asphalt, caulking compds., rubber, industrial abrasives, agric. formulations, livestock feed roughage.

Lite Salt Mixture. [Morton Salt] 50% NaCl, 50% KCl.

Litesate. [Champlain Industries] Protein hydrolysates.

Litharge 28, 33. [Eagle-Picher] Lead oxide; acid acceptor, activator, vulcanizing agent for rubber compounding; in mfg. of dry colors, high pressure lubricants, brake linings, ceramics, glass, piezoelectric devices, chemical processes.

Lithchips. [FMC/Lithium] Lithium carbonate granules; for aluminum cell additions.

Lithene. [Revertex Ltd.] Butadiene liq. telomer; contains cyclic polymer segments which impart toughness and thermal stability to cured polymers.

Lithol®. [BASF] Organic pigments; for inks, paints, plastics, artists' colors and crayons.

Lithol® Fast. [BASF; BASF AG] Organic pigments; for inks, paints, plastics, artists' colors and crayons.

Lithopone. [Sachtleben Chemie GmbH] Zinc sulfide/barium sulfate; white pigment for thermosets, glass fiber-reinforced thermosets and thermoplastics, paper applics.

Lithospar. [FMC/Lithium] Feldspathic sand.

Lithsils. [FMC/Lithium] Soluble lithium silicates.

Litol®. [Air Prods.] Catalyst.

LK®. [Air Prods.] Nonsilicone surfactants.

LNA. [Monomer-Polymer & Dajac Labs] Leucine aminopeptidase substrate.

Lobra. [Karlshamns] Hydrogenated canola oil.

Lock-Ease. [Am. Grease Stick] Liq. graphite.

Locron. [Hoechst Celanese] Aluminum chlorohydrate.

Lo-Foam. [West Agro] Iodophor conc.

Logostat. [Morton Int'l.] Antistatic agents.

Lok-Size. [A.E. Staley Mfg.] Industrial corn starch.

LoLoss®. [Rhone-Poulenc/Perf. Resins & Coatings] Polymeric viscosifier for fluid loss control with drilled solids.

Lomar®. [Henkel/Emery; Henkel-Nopco] Naphthalene sulfonate condensates; emulsifier, leveling agent, dispersant, suspending agent, stabilizer for emulsion polymerization, dyestuff mfg., ceramics, paint, paper, rubber, agric. formulations.

Lomod®. [GE Plastics] Thermoplastic elastomer; engineering elastomer for transportation, elec./electronic, sporting goods, hose and tubing industries.

Loncizer. [DIC Trading] Polymeric plasticizer.

Loncoterge. [London Chem.] Cleaning solvent.

Londex. [DuPont/Ag] Herbicide.

London Oil. [Natrochem] Rosin oil.

Lonzaine®. [Lonza] Betaines or sultaines; amphoteric surfactant, conditioner, foamer, wetting agent for personal care prods., industrial applics.

Lonza Insta-Pearl®. [Lonza] Proprietary blend; pearlescent for hair and skin care prods.

Lonzest®. [Lonza] Esters or ethoxylated esters; emollient, penetrant, spreading agent, detergent, lubricant, dispersant, anticorrosive, emulsifier, defoamer for personal care prods., foods, textiles, petrol., insecticides, paper, agric. industries.

Loobwax. [Astor Wax] Polyethylene wax; lubricant for PVC.

Loralan-CH. [Lanaetex Prods.] Cholesterol.

Lorol®. [Henkel/Emery; Henkel KGaA] Linear primary alcohols; intermediates for surfactant mfg., components in lubricants.

Loropan. [Triantaphyllou] Fatty acid alkanolamide; foam stabilizer, emulsifier, superfatting agent, thickener for shampoos, detergents.

Lorox. [DuPont/Ag] Herbicide.

LoSilPhos. [Monsanto] Ferrophosphorus briquettes.

Lostat. [Ciba-Geigy AG] Lignosulfonate; dispersant for dyestuffs.

Lotader®. [Atochem N. Am.; Atochem UK] Ethylene-acrylic ester-glycidyl methacrylate terpolymer; for aluminum coating, plastics coating, coextrusion, pipe coating, polymer modification, hot-melt adhesives, compounds, additives.

Lo-Temp. [A.E. Staley Mfg.] Modified corn starch.

Lotrene. [Enimont UK Ltd.] VLDPE.

Lotryl®. [Atochem N. Am.] Ethylene/ acrylic ester copolymers; for films, coextrusion, gloves, sheets, coating, hot melts, compounding.

Louryl. [Triantaphyllou] TEA-alkylbenzene sulfonate; detergent for shampoos, bubble baths.

Lo-Vel®. [PPG Industries] Precipitated silica; thickener, flatting agent for coatings and lacquers.

Loving. [Kao] Sucrose ester and food additives; cleaning and peeling agents for food industry.

Low Crock. [Eastern Color & Chem.] Anticrocking agents and binders for pigment printing systems.

Low Crock. [Polymer Research Corp. of Am.] Self-crosslinking reactive prod.; used as binder for roller printing, textile applics.

Low Crock FB-1. [Yorkshire Pat-Chem] Finishing auxiliary for reduced bleeding/crocking of dyed and printed fabrics.

Lowenol. [Lowenstein Dyes & Cosmetics] Cosmetics ingredients.

Lowe's Safety. [Meridian Petroleum] Absorbents.

Lowicryl. [Lowi GmbH] Embedding media for electron microscopy.

Lowilite®. [Lowi] uv absorbers/stabilizers for polymers.

Lowinox®. [Lowi; Chemische Werke Lowi GmbH] Antioxidant for oils, fuels, foodstuffs.

Loxamid. [Henkel KGaA] Slip agents.

Loxanol. [Henkel KGaA] 12-Hydroxystearyl alcohol; for mfg. of binders for coatings industry; additive for latex decorative masonry finishes.

Loxiol. [Henkel/Functional Prods.; Henkel KGaA] Esters, acids or paraffin; lubricant, costabilizer for PVC; mold release agents, dispersants.

Loxtrex. [Enimont UK Ltd.] LLDPE.

LP®. [Morton Int'l.] Polysulfide polymers; base for sealants in construction, transportation fields and for casting compds., elec. potting compds., and fuel tank coatings.

LP-100, 200, 300, 400. [Eagle-Picher] Lead dioxide; catalyst, curing agent for polysulfide, butyl and polyisoprene rubber; oxidizer for mfg. of dyes.

LPF. [Goodyear] Butadiene/styrene latex; used in tire cord dips, spread foam, rug backing, adhesives, asphalt modification.

LPR. [Goodyear] Styrene/butadiene latex; used in carpetbacking compds. and adhesives.

LSC. [Avatar] Lecithin soya conc.

LSD. [Oakite Prods.] Detergent.

LSF-54. [Witco UK] Anionic detergent, wetting agent with high flame resistance.

LSFR. [Laurel Industries] Low smoke flame retardant.

LSP 33. [Berol Nobel] N,N-Bis (3-aminopropyl) tallow amine; intermediate, wetting agent, pigment grinding aid, dispersant, flushing agent, corrosion inhibitor for polymer industry.

Luaktin®. [BASF AG] Antiskinning agents for paints, baking finishes; improves gloss and leveling.

Lub-A-Lite. [Panef Mfg.] Dry white powdered lubricant.

Lub-A-Lock. [Panef Mfg.] Lock lubricant.

Lubar. [Lea Mfg.] Lubricant.

Lubasin®. [BASF AG] Adhesives for textile printing, furs.

Lub-A-Spray. [Panef Mfg.] Dry powd. graphite.

Lub-A-Stick. [Panef Mfg.] Silicone grease stick.

Lube-A-Tube. [G. Whitfield Richards] Drawing lubricant.

Lube Booster®. [Ferro/Keil] Lubricity additive for aluminum and ferrous metals.

Lube-Lok. [E/M Corp.] Solid lubricants.

Lubestat. [Milliken] Antistatic lubricant overspray for synthetic fibers.

Lubewax. [Chem. Processing] Textile sizing wax.

Lubewax. [Lea Mfg.] Lubricant.

Lube-Well. [G. Whitfield Richards] Cutting and drawing lubricant.

Lubit®. [Sybron] Lubricant for textile fibers.

Lubracal®. [Witco] Calcium stearate deriv.; lubricant and plasticizer for paper coatings.

Lubrajel. [Guardian Labs] Nondrying, water-sol. lubricating jelly.

Lubrajel®. [Presperse] Polyglycerylmethacrylate or blends; autoclavable nondrying water-sol. lubricant for medical and surgical use.

Lubral. [Alcoa] Surface-treated ATH.

Lubran. [Toho Chem. Industry] Blends; pour pt. depressant, dewaxing aid, visc. index improver for lubricating oils, transportation of crude oil.

Lubrazinc®. [Witco/Organics] Zinc stearate; lubricant for fiber-reinforced plastics, powd. metallurgy.

Lubrex®. [Harwick] Mold release agents, green tire lubricants, bladder treatment compds.

Lubrhophos. [Rhone-Poulenc Surf.; Rhone-Poulenc France] Phosphate esters; lubricant, rust inhibitor, emulsifier, pressure additive for lubricating and rolling oils, hydraulic, cutting and grinding fluids.

Lubri-Bond. [E/M Corp.] Solid lubricants.

Lubricant EHS. [Hoechst Celanese/ Colorants & Surf.] Fatty acid ester; heat-stable boundary lubricant for microemulsions for heavy-duty machining of aluminum.

Lubricin. [CasChem] Modified castor oil; lubricant specialties.

Lubricomp®. [LNP] PTFE-lubricated resins; internally lubricated composites for gear, cam, sliding, and bearing applics. requiring low frictional properties.

Lubricone. [Kano Labs] Silicone mold release, lubricant.

Lubri-Joint. [Markal] Gasket lubricant.

Lubrimet®. [BASF AG] PPG; lubricant, solubilizer for dyestuffs and surfactants.

Lubrinol SC. [Waco Am.] Anionic lubricant for textiles and dyebaths.

Lubrisize®. [Nat'l. Starch & Chem.] Modified starch; warp sizing.

Lubrisol. [Scher] Natural oil; textile lubricant.

Lubritab®. [Mendell] Hydrogenated vegetable oil NF; lubricant for pharmaceutical tablets.

Lubritan. [Rohm & Haas] Synthetic tanning agent.

Lubriwax. [Reilly-Whiteman] Formulated paraffin for dry lubrication of yarn and thread.

Lubrizol. [Lubrizol] Emulsifier, corrosion inhibitor, heat stabilizers for PVC coatings, films, bottles, extruded pipe; dispersant for pigments; visc. stabilizer for plastisols.

Lubro. [G. Whitfield Richards] Drawing lubricant.

Lubrol. [CNC Int'l.] Release agent for pulp and paper industry.

Lubrol. [ICI Surf. UK] Alkyl polyglycol ether; emulsifier for removing oil and grease stains caused by loom or knitting machine oil.

Lubron. [Chemonic Industries] Textile auxiliaries giving dyebath lubricity, soft, silky hand.

Lubron. [Sybron] Lubricant for textile fibers.

Lucalen®. [BASF AG] Ethylene copolymers with polar groups; adhesive agent for composites, laminating, profiles, seals, hot-setting adhesives.

Lucalor® CPVC. [Atochem N. Am.; Atochem UK] Chlorinated PVC.

Lucantin®. [BASF AG] Citranaxanthine or canthaxanthine; for pigmentation of egg yolks, skin and legs of broilers.

Lucaphos®. [BASF AG] Dicalcium phosphate; for the feed industry.

Lucarotin® 10%. [BASF AG] Betacarotene; for the feed industry.

Lucel. [MBS Plastics] POM.

Luchem. [Atochem N. Am.] Methyl methacrylate/allyl methacrylate copolymer; antishrink additive.

Lucidene. [Morton Int'l.] Flexo/gravure ink vehicles.

Lucidol. [Atochem N. Am.] Benzoyl peroxide; initiator for polymerization, curing polyester resins, elastomers, acrylics.

Lucirin® BDK. [BASF AG] uv initiator for radiation-cured finishes and putties.

Lucite®. [DuPont] Acrylic resin or sheet; offers outdoor durability, high optical properties.

Luciwax. [Morton Int'l.] Wax dispersions.

Lucky. [MBS Plastics] ABS, ASA, SAN, PS, PMMA.

Lucky. [Standard Polymers] ABS resins; used for radio housings, tape recorder housings, TV cabinets, office equip., thermos bottles, musical instruments, etc.

Lucobit®. [BASF AG] Ethylene copolymer/bitumen; for prod. of damp-proof courses in construction of buildings, as protection against corrosion for steel tanks, for improving quality of road bitumen, for prod. of inj. molded parts.

Lucolene. [Atochem Deutschland GmbH] Soft PVC compd.

Luconyl®. [BASF; BASF AG] Organic and inorganic pigment dispersions; for paints, finishes, lime washes, wallpaper inks, wood glazes.

Lucorex. [Atochem Deutschland GmbH] Hard PVC compd.

Lucovyl. [Atochem Deutschland GmbH] PVC resins.

Lucryl®. [BASF AG] Polymethylmethacrylate; thermoplastic molding and extrusion compds. for construction, motorcar, lighting engineering, household appliances, advertising displays.

Ludigol®. [BASF AG] Mild oxidizing agent for textile finishing.

Ludipress®. [BASF AG] Lactose/Kollidon based; direct tabletting auxiliary.

Ludopal®. [BASF AG] Unsaturated polyester resins; for finishes, filling compds., elastifying other grades.

Ludox®. [DuPont] Aq. colloidal silica dispersion; binder for ceramic casting.

Lufibrol®. [BASF; BASF AG] Extraction assistants for pretreatment of textile goods.

Lufilen®. [BASF AG] Organic and inorganic pigments in polyethylene; for polyolefin spin dyeing.

Lugalvan®. [BASF] For electroplating and electronics industry.

Luganil®. [BASF AG] Anionic dyes; for leather and fur dyeing, coloring cleaners.

Luhydran®. [BASF AG] Water-dilutable binders for electrocoating finishes, baking finishes, wood and plastic coatings, anticorrosion paints, printing inks, overprint varnishes.

Lumacel. [Lucky Ltd.] Textile dyes and pigments.

Lumacron. [Lucky Ltd.] Textile dyes and pigments.

Lumattin®. [BASF AG] Wax-modified silica; matting agents for paints and varnishes.

Lumicrease. [Sandoz] Textile dyes and pigments.

Lumiflon. [ICI Resins] Solvent and water-borne fluoropolymers.

Luminex. [Ikeda] Quaternium-45.

Lumiten®. [BASF AG] Complexing agents, foam suppressants, wetting agents for cements, adhesives, paints.

Lumitol®. [BASF AG] Hydroxylic polyacrylate resins; used in combination with polyisocyanates to produce polyurethane finishes with chemical and solvent resistance.

Lumo. [Zschimmer & Schwarz] Sodium alkylbenzene sulfonate; detergents.

Lumogen®. [BASF] Fluorescent yellow pigment for prod. of wax chalks, crayons, crack detectors.

Lumorol. [Zschimmer & Schwarz] Disodium laureth sulfosuccinate blends; cleansing agents.

Lumo Stabil. [Zschimmer & Schwarz] Sodium dodecylbenzene sulfonate; washing and cleansing agents, industrial cleaners.

Luna. [BASF] Organic pigments for textiles.

Luperco. [Atochem N. Am.] Organic peroxide compds.; crosslinking agent, initiator for elastomers, thermoplastic resins, for curing polyesters and acrylics.

Luperfoam. [Atochem N. Am.] t-Butylhydrazinium chloride blends; chemical blowing agent for unsaturated polyester resins.

Luperox. [Atochem N. Am.] Organic peroxide compds.; initiator for polymerization, polymer modification, thermoplastic crosslinking, curing elastomers, high-temp. cure of polyester resins.

Lupersol. [Atochem N. Am.] Organic peroxide compds.; initiator for vinyl polymerizations, curing polyester and acrylic resins; crosslinking agent.

Luphen®. [BASF AG] Polyurethane or polyesterol derivs.; for prod. of prepolymers, adhesives, sealants, coating compds.

Luphite. [Kano Labs] Graphite lubricating oil.

Lupolen®. [BASF AG] HDPE, LDPE, LLDPE, LDPE/EVA copolymer; various grades for inj. molding, extrusion, blow molding, powd. grades, compr. molding, semifinished prods.

Lupon. [MBS Plastics] Polyamide.

Lupos. [MBS Plastics] ABS GF.

Lupox. [MBS Plastics] PET.

Lupox TE. [MBS Plastics] PC/PET.

Lupoy. [MBS Plastics] PC/ABS.

Lupranat®. [BASF AG] MDI/TDI; for prod. of polyurethane flexible foam, semirigid and rigid foams, thermoplastic elastomers for furniture, automotive, construction, pkg., elec., and shoe industry.

Lupranol®. [BASF AG] Polyether polyols; for the prod. of polyurethane flexible foam, semirigid and rigid foam for the furniture, automobile, construction, pkg., elec., and shoe industry.

Lupraphen. [BASF AG] Polyester polyols; for the prod. of polyurethane, thermoplastic granules, textile coating systems, finished parts made of polyurethane and casting elastomers; for the automotive, machine and equip. construction industries.

Luprenal®. [BASF AG] Polyacrylate resins; heat-curable resins used in combination with melamine or urea resins to give hard, tough, weather-resist. and chem. resist. coatings.

Luprimol®. [BASF AG] Improves hand of pigment prints on textiles.

Luprintan®. [BASF; BASF AG] Emulsifiers and assistants for textile printing; fixing assistants for printing on synthetic fibers.

Luprintol®. [BASF] Emulsifier for textile printing.

Luprofil®. [BASF; BASF AG] Masterbatches of organic and inorganic pigments in polypropylene; for polypropylene spin dyeing.

Lupromag®. [BASF AG] Magnesium propionate; for animal feedstuffs.

Luprosil®. [BASF AG] Propionic acid or salts; preservative for feed industry.

Lurafix®. [BASF; BASF AG] Dyes for transfer printing inks.

Luramid®. [BASF AG] Metal complex dyes and acid dyes for dyeing polyamide chips in aq. liquors.

Luran®. [BASF/Engineering Plastics; BASF AG] SAN copolymer, some glass-reinforced; for inj. molding and extrusion applics.

Luran® S. [BASF AG] Acrylonitrile-styrene-acrylic ester copolymers; for inj. molding, extrusion, structural parts for outdoor use, hot water drainage pipes, patio furniture, toys.

Lurantin®. [BASF AG] Lightfast direct dyes for dyeing and printing cellulose fibers, nylon fibers, and wool.

Luranyl®. [BASF AG] Polyphenylene ether/HIPS blend; std. grades for high-precision components; glass fiber-reinforced grades for pumps and fittings; impact-resist. grades for car interior parts, instrument panels, halogen-free fire-retardant grades for elec. insulation bodies, office equip. housings.

Lurapret®. [BASF AG] Antislip agents for textile finishing.

Lurazol®. [BASF AG] Anionic dyes for leather and fur dyeing, coloring cleaners.

Luredox®. [BASF AG] Sodium dithionite derivs.; for mfg. of cleaners.

Luredur®. [BASF AG] Acrylate copolymers or modified polyacrylamides; dry strength agents for paper/paperboard industry.

Lureen. [George A. Goulston] Lubricant for dyed carpet yarn.

Luresin®. [BASF AG] Polyamido-amine-epichlorohydrin resin aq. sol'ns.; for increasing wet and dry strength of papers.

Lurol. [George A. Goulston] Antistatic coning oils, spin/process finishes for textiles.

Luron®. [BASF AG] Binders, assistants for printing inks, plating, glazing, leather finishes, fur processing.

Lurotex®. [BASF; BASF AG] Hydrophilic agent for synthetic fibers; crease inhibitor.

Lusantan®. [BASF AG] Ethoxylation prods. of p-aminobenzoate; uv absorber for sunscreen prods.

Lusep. [MBS Plastics] PPS.

Lusol. [Witco] Water-soluble metalworking fluids.

Lusolvan® FBH. [BASF AG] Diisobutyl ester of a dicarboxylic acid mixture; coalescent for aq. polymer dispersions; for paints and textured finishes.

Lusterlac. [Finetex] Delustrants for rayon, acetate, nylon, and polyester.

Lustrabrite. [Telechemische] Toluene-sulfonamide/epoxy resin or blends.

Lustralite. [Reichhold] Gloss enhancer.

Lustran® ABS. [Monsanto; Monsanto Europe] ABS resin; for inj. molding, extrusion, and thermoforming applics.

Lustran® SAN. [Monsanto; Monsanto Europe] SAN resin; for cosmetic pkg., fan blades, toys and games, business machines, interior refrigerator parts, medical parts, beverage tumblers, food containers, tableware, dinnerware.

Lustra-Pearl®. [Van Dyk] Mica, titanium dioxide; pearlescent pigments for cosmetics, makeup.

Lustrasol. [Reichhold] Acrylic/styrene resin sol'ns.

Lustre-Phos. [Monsanto] Dentifrice abrasive.

Lustrone. [Rohm & Haas] Lacquer for leather.

Lusynton®. [BASF AG] Corrosion in-

hibitor, chlorite stabilizer for textile industry.

Lutan®. [BASF AG] Basic aluminum salts; for tanning leather and fur.

Lutanol. [BASF] Polyvinyl ether.

Lutavit®. [BASF AG] Vitamin and vitamin mixtures; for feed industry, food and pharmaceutical applics.

Lutensit®. [BASF AG] Wetting agent, dispersant, solubilizer, detergent, biocide for cosmetic, industrial, and household cleaning.

Lutensol®. [BASF AG] Detergent, wetting agent, dispersant, emulsifier for household and industrial detergents, chemical processing, leather, fur, paper, paint and dye industries.

Lutetia. [ICI Am.] Textile dyes and pigments.

Lutexal®. [BASF; BASF AG] Thickeners for textile printing, fur dyeing.

Lutexan. [Burlington Bio-Medical] Natural yellow color.

Lutofan®. [BASF AG] Vinyl chloride dispersion; for prod. of laminating and heat-sealable adhesives and pkg. materials, binders for textile coating.

Lutonal®. [BASF AG] Polyvinyl ethers; soft resin for paint, ink, adhesive industries; tackifying resin.

Lutopon. [Henkel-Nopco] Linear alkylaryl sulfonate and solvent; detergent for textile processing.

Lutostat. [Henkel-Nopco] Antistats for textiles,j plastics.

Lutrizol®. [BASF AG] Dimetridazole, dimetridazole HCl; active ingredient for control of black head disease in turkeys.

Lutrol®. [BASF AG] PEG or PPG compds.; binder, solubilizer, emollient, lubricant, solvent for cosmetics, pharmaceuticals.

Lutron®. [BASF] For electroplating and electronics industry.

Luviflex®. [BASF AG] PVP copolymers; film-former for hair fixatives.

Luviform®. [BASF AG] PVM/MA copolymers or their esters; film-forming binders for hair sprays and setting lotions.

Luviquat®. [BASF AG] Polyquaterniums; substantive cationic polymer used as conditioner, film former for hair and skin care prods.

Luviset®. [BASF; BASF AG] Crotonic acid copolymers; hair fixative, film-forming agent for hair care prods.

Luviskol®. [BASF AG] PVP or PVP/VA copolymers; film-forming agent, hair fixative, thickener, protective colloid, suspending agent, dispersant for cosmetics and technical applics.

Luvitol®. [BASF; BASF AG] Synthetic oil; emollient oil component for cosmetics and pharmaceuticals.

Luwax. [BASF] Polyethylene waxes or copolymer waxes; for printing inks, paints, wax emulsions, polishes, color concs., lubricants, rubber, adhesives, cosmetics.

Luwipal®. [BASF AG] Melamine formaldehyde resins; combined with alkyd, polyacrylate, and epoxy resins to give lt.-resist. and weather-resist. baking finishes.

Luxate. [Olin] Aliphatic isocyanates or diisocyanates; urethane intermediates; for coatings, elastomers, adhesives, sealants, thermoplastics.

Luxelen. [Presperse] Titanium dioxide compds.; additive to make-up formulations providing transparency, gloss, coverage, uv absorption.

Luxematic. [JacksonLea] Liq. greaseless abrasive composition.

Luxor. [Hercules] Hydrolyzed vegetable protein blends; flavor and flavor enhancers.

Luxor. [Montedipe Srl] Polycarbonate.

Luzenac. [Luzenac; R.T. Vanderbilt] Talc; for plastics, elastomers.

LV-. [Solem Industries] Alumina trihydrate; flame-retarding filler, resin extender.

LWC. [TTC Mouldings BV] Polypropylene.

L-X. [Lubrication Engineers] Chemical supplement for diesel and gasoline fuels.

LX®- Series. [Neville] Petroleum hydrocarbon resins; used in adhesives, coatings, inks, rubber, concrete-curing compds., and caulking compds.

Lycal. [Steetley Magnesia Prods. Ltd.] Magnesium oxide or hydroxide.

Lycra. [DuPont] Spandex fiber.

Lykopon. [Rohm & Haas] Sodium hdyrosulfite; chemical reducing agent for textile vat dyeing, iron removal agent on ion exchange resins.

Lynx. [Catalyst Resources] Polypropylene catalyst.

Lyocol. [Sandoz] Dispersant for dyes and pigments.

Lyogen®. [Sandoz] Detergent, dyeing assistant, wetting agent, emulsifier, dispersant, antiprecipitant for textile fibers.

Lysmeral®. [BASF AG] 2-Methyl-3-(4-t-butylphenyl) propanol; fragrance.

Lytor. [Georgia-Pacific] Tall oil rosin.

Lytron. [Morton Int'l.] Acrylate copolymers or blends or polystyrene latexes; opacifier for cosmetics, shampoos, household and industrial detergents.

LZ Series. [Lubrizol] Antioxidant, extreme pressure additive, lubricant, tackiness agent, suspending agent for industrial applics.

M

M-25. [Pea Ridge Iron Ore] Iron oxide.

M-100. [JacksonLea; Lea Mfg.] Black phosphate coatings for steel and zinc.

M-2104, -2037. [ICI Fiberite] Melamine molding compds. with cotton or glass filler; impact compds.

M-2205. [H&S Chem.] 20% iodine conc.

M8848, 9030. [Hüls Am.] Silahydrocarbons; lubricant, wear agent.

MA. [PMC Specialties] Methyl anthranilate; industrial deodorant, aromatic.

Macaloid®. [Rheox] Magnesium aluminum silicate; suspending agent, thickener, emulsion stabilizer, rheological additive for creams and lotions.

Macco. [Witco] Phosphate-free metal cleaning compds., cleaners for machines, floor, factories; rust preventatives; alkaline cleaners; electrolytic cleaners; shearing lubricants; soluble oils for cutting, grinding.

Mackadet. [McIntyre] Surfactant blends; formulated concs. for shampoos, bubble bath, conditioners, rug shampoos, waterless hand cleaner, skin cleansers.

Mackalene. [McIntyre] Lactates or propionates; cationic conditioner, softener for hair care prods.

Mackam. [McIntyre] Amphoteric surfactants; foaming agent, conditioner, visc. builder for shampoos, cleansers, industrial and dishwash formulations.

Mackamate. [McIntyre] Sulfosuccinate; emollient surfactant.

Mackamide. [McIntyre] Fatty acid alkanolamides; foam stabilizer, thickener, conditioner, softener, emulsifier, humectant, degreaser, rust inhibitor.

Mackamine. [McIntyre] Amine oxides; conditioner, foam booster/stabilizer, visc. builder.

Mackanate. [McIntyre] Sulfosuccinates; surfactant bases for rug cleaners, shampoos, bubble baths; conditioner, wetting agent, dispersant, penetrant.

Mackanol. [McIntyre] Alcohol sulfate surfactants.

Mackazoline. [McIntyre] Imidazolines; emulsifier, corrosion inhibitor.

Mackernium. [McIntyre] Quaternary ammonium compds.; surfactants, cream rinse conditioners.

Mackester. [McIntyre] Organic esters; plasticizers, lubricants, emollients, pearlescents, emulsifiers for metalworking, textile lubricants, plastics, paper, cosmetics.

Mackinate NLP. [McIntyre] Oleamidopropyl dimethylamine propionate, palmitamidopropyl dimethylamine propionate, palmitoleamidopropyl dimethylamine propionate.

Mackine. [McIntyre] Dimethylamines or morpholines; softener; intermediate for hair conditioners.

Mackol. [McIntyre] Fatty alcohols.

Mackpearl. [McIntyre] Pearl agent for cold blends.

Mackpro. [McIntyre] Quaternized proteins; conditioner for hair and skin care prods.

Mackstat®. [McIntyre] DMDM hydantoin; broad spectrum cosmetic preservative for shampoos, skin cleansers, bath prods., creams and lotions.

Maco. [M. Argueso & Co.] Vegetable fibers.

Macol®. [PPG/Specialty Chem.] Surfactants; detergent, wetting agent, emulsifier, lubricant, defoamer, dispersant, dye leveler, gellant, antistat for cosmetics, metalworking, textile lubes, rubber, chemical intermediates, pulp/paper.

Macrobase. [Sartomer] Polystyrene/acrylate blends; dispersant for pigments; used in uv/eb curable adhesives, inks, coatings.

Macrolex. [Miles/Organic Prods.] Textile dyes and pigments.

Macromelt. [Henkel] Polydilinoleoylethylenediamide; hot-melt adhesives.

Macromer®. [Sartomer] Methacrylate-terminated polystyrene; pour pt. depressant; used in inks, coatings, sealants, glass reinforcements, adhesives, graft copolymers.

Macrospherical®. [Reheis] Aluminum chlorohydrate; antiperspirant.

Macrylic. [Ivax Industries] Self-crosslinking acrylate emulsion; shrinkage and abrasion improver.

Macrynal. [Hoechst Celanese/Fine Chem.] Acrylic resins; for paints and coatings.

Macsil. [Polymed Ltd.] Release agents.

Macwax. [Polymed Ltd.] Release agents.

Madeol. [Auschem SpA] Wetting and dispersing agent for pesticides.

Madol® 767. [Boehme Filatex] Coning oil for nylon hosiery filament.

Madurit. [Hoechst Celanese/Fine Chem.] Melamine-formaldehyde resins; for laminates.

MAE®. [General Chem.] Nitric/hydrofluoric acetic/acid blends; mixed acid etchant.

Mafco Magnasweet. [MacAndrews & Forbes] Flavor enhancer.

Mafloc®. [PPG/Specialty Chem.] Polyacrylate polymer; flocculants.

Mafo®. [PPG/Specialty Chem.] Amphoteric surfactants; detergent, dispersant, surfactant, conditioner, corrosion inhibitor, foam and visc. stabilizer, chelating agent, wetting agent, solubilizer, lubricant, emulsifier; used in personal care, dishwashing, rug and carpet cleaning applics.

Maftec®. [Mitsubishi Kasei] Alumina fiber; reinforcing agent for high temp. usage and composite materials; insulator for furnace lining.

Mafu®. [Bayer] Dichlorvos; contact, stomach, and breathing poison for control of mosquitoes, flies, cockroaches, mites, etc.

Magala®. [Akzo] Magnesium alkyls; for prod. of catalysts for polymerization of olefins or diens; alkylating agent.

Magathin. [Lucas Meyer] Lecithin deriv.; natural fat emulsifier for toffees.

Magcarb. [Marine Magnesium] Magnesium carbonate tech. powders.

MagChem®. [Martin Marietta Magnesia Spec.] Magnesium oxide; for mfg. of magnesium chemicals, construction prods., lubrication oil additives, fuel additives, oil drilling chemicals, neoprene and chlorinated polymers, sugar refining, uranium ore processing, acid neutralization, leather tanning, water treatment.

Magiesols. [Magie Bros. Oil] Petroleum solvents.

Maglite. [Marine Magnesium] Magnesium oxide tech. powders.

Magna. [Castrol Industrial East] Lubricants.

Magna. [Van Den Bergh Foods] Partially hydrogenated vegetable oils; coating fat for no-tempering coatings, centers, bakery items.

Magnabrite®. [Am. Colloid] Magnesium aluminum silicate; stabilizer, suspending agent for cosmetics, pharmaceuticals, household and industrial specialties, e.g., paints and cleaning compds.

Magnacat. [Ultra Additives] Ink, paint, and varnish drier.

Magnacide. [Baker Perf. Chem.] Biocide.

Magnacide H. [Baker Perf. Chem.] Herbicide.

Magna Concentrols. [Crompton & Knowles] Blend of natural essential oils and oleoresins in liq. form; spice flavor systems.

Magnacryl. [Beacon] Acrylic (polyrad) adhesives and uv-curing adhesives coatings.

Magnaflo. [Sybron] Replaces magnesium sulfate; used in stainblocking treatment of carpets with Stain-Free.

Magnafloat. [Prince Mfg.] Black iron oxide.

Magnamite®. [Hercules] Graphite prepreg tape for structural applics.; reinforcinga gent for weaving, prepregging, filament winding, pultrusion, molding compds.

Magnaplas. [Bay Resins] Nylon 6, barium ferrite-filled; permanent magnetic material.

Magnasoft®. [Union Carbide] Silicone softeners for textiles.

Magnesium Oxides for Neoprene. [Whittaker, Clark & Daniels] Plastics.

Magnifloc. [Am. Cyanamid] Flocculants.

Magni-Form. [Betz Industrial] Chemical oxygen scavenger for boiler condensate and feedwater.

Magnum. [Dow Plastics] ABS resin.

Magnum®. [BASF AG] Chlordazon, ethofumesate; for pre- and post-emergence control of weeds in sugar beet and fodder beet.

Magnum-White. [RMc Minerals] Magnesium hydroxide/calcium carbonate blend; fire retardant, smoke suppressant filler for PVC compds., SBR-latex formulations.

Magotex. [Kaopolite] Fused magnesium oxide; for plastics, molding compds., brake lining compds., elec. potting compds.

Magox®. [Premier Services] Magnesium oxide; adsorption, absorption agent and scavenger for plastics, rubber industries, chemical neutralization.

Mag-Plus. [Nat'l. Magnesia Chem.] High purity (97%) MgO; for processing/mfg. paper, pulp, sugar, motor/fuel oil additives, steel, aluminum, concrete, paint, leather, cellulosics, fibers, rubber, ceramics, plastics.

Maincote. [Rohm & Haas] Waterborne polymer.

Maisine. [Gattefosse; Gattefosse SA] Corn glycerides; food emulsifier.

Maki. [LiphaTech] Rodenticide.

Makon®. [Stepan; Stepan Europe] Ethoxylated ethers; detergent, emulsifier used in chemical specialties, cosmetic, agric., industrial and metal cleaners, textile, paper and petrol. industries.

Makroblend. [Bayer; Miles] PC/PET blend, some glass-reinforced; resin for inj. molding.

Makrofol®. [Bayer] Polycarbonate film; technical film.

Makrolon®. [Bayer; Miles] Polycarbon-

ate; resin for inj. molding and extrusion applics., electronics, elec. engineering, transportation, building/construction, business machine, lighting, photographic, optical equip., precision engineering, office supplies, domestic wares.

Makrolon® SF. [Bayer; Miles] Polycarbonate structural foam.

Malathion ULV. [Am. Cyanamid/Ag] Conc. insecticide.

Malco Bio. [Malco Prods.] Liq. detergent.

Malkyd. [Arakawa] Maleic modified rosin ester.

Maltrin®. [Grain Processing] Maltodextrin or corn syrup solids; used as carrier, bulking agent, for wet binding, anticaking, sweetener in foods, pharmaceuticals.

Manalube. [Manufacturers Chems.] Lubricant, antistat for textiles, paper.

Manasperse. [Manufacturers Chems.] Wetting agent, leveler, foaming agent for dyed nylon.

Manawet. [Manufacturers Chems.] Nonfoaming leveling agent for beck dyeing of polyester and nylon carpet; soaping agent and wetter for stone washing of denim.

Manex. [Griffin] Maneb/zinc coordinate; fungicide for many crops.

Manganol. [Dexter] Synthetic organic stabilizer for alkaline peroxide bleaching.

Man-Gro. [Am. MicroTrace] Mn sulfate monohydrate, 28.5% Mn; water sol. nutrient for fast plant availability in dry fertilizers.

Man-Gro AS. [Am. MicroTrace] Mn sulfate monohydrate, 28.5% Mn; highly sol. source of Mn for plant fertilizer or animal feed.

Manoxol. [Manchem] Sulfosuccinates; wetting agent for agric., pharmaceutical, cleaning, and industrial applics.

Manro. [Manro Prods. Ltd.] Surfactants; foam booster, wetting agent, thickener used in cosmetics, emulsion polymerization, cleaning prods., fire fighting foams, industrial prods.

Manromate. [Manro Prods. Ltd.] Sulfosuccinate half ester; surfactant.

Manromid. [Manro Prods. Ltd.] Fatty acid alkanolamides; foam booster/stabilizer, thickener, emulsifier, corrosion inhibitor for cosmetics, detergent systems.

Manromine. [Manro Prods. Ltd.] Modified alkanolamide; for waterless hand cleaners.

Manrosol. [Manro Prods. Ltd.] Sulfonates.

Manroteric. [Manro Prods. Ltd.] Betaines, glycinates, or propionates; amphoteric surfactants.

Manrowet. [Manro Prods. Ltd.] Dioctyl sulfosuccinate; surfactant.

Mansul. [Cuproquim] 40% Mancozeb/ 40% sulfate.

Mantrilon®. [BASF AG] Liq. manganese fertilizer; foliar fertilizer for agric. crops, vines, fruit.

Manzate. [DuPont/Ag] Fungicides.

Mapeg®. [PPG/Specialty Chem.] Esters or ethoxylated esters; emulsifier, dispersant, solubilizer, emollient, coupling agent, thickener used in cosmetics, pharmaceuticals, metalworking and fiber lubricants, emulsion polymerization, industrial applics.

Maphos®. [PPG/Specialty Chem.] Phosphate esters; lubricant with anticorrosive/antifrictional properties; textile wetting agent; hard surface detergent; dispersant, hydrotrope, solubilizer, emulsifier.

Mapico. [Columbian Chem.] Iron oxides.

Mapo®. [Aceto; Arsynco] Tris [2-(2-methyl-aziridinyl) phosphone oxide]; additive for textiles, resin mfg.

Maprenal. [Hoechst Celanese/Fine Chem.] Melamine resins; for paints, varnishes, lacquers.

Mapron®. [Mitsubishi Kasei] Soybean milk; food additive.

Maprosyl® 30. [Stepan] Sodium lauroyl sarcosinate; detergent, wetting and foaming agent for personal care and household detergent prods.

Maquat. [Mason] Quaternary ammonium chlorides; antimicrobial, germicide, disinfectant, algicide, sanitizer, deodorant.

Maquest. [Ivax Industries] Chelating agents.

Marabond 21. [Borregaard LignoTech] Lignosulfonate; low temp. oil well cement retarder.

Maracarb. [Borregaard LignoTech] Lignosulfonates; dispersant, modifier, humectant, chelating agent, stabilizer, slime control agent for dyestuffs, industrial cleaners, agric. formulations, paper mfg.

Maracell. [Borregaard LignoTech] Sodium lignosulfonate or oxylignins; dispersant, chelating agent, sludge conditioner for boiler water treatment, industrial cleaners; stabilizer; expander for lead acid batteries.

Maracon. [LignoTech] Water-reducing admixture for concrete.

Maraglas. [Acme Div.] Clear epoxy casting resins.

Maramul. [LignoTech] Lignosulfonate; emulsifier for asphalt emulsions.

Maranil. [Henkel Canada; Henkel KGaA; Pulcra SA] Dodecylbenzene sulfonates; detergent base for household and industrial cleaners.

Maranyl®. [ICI Advanced Materials] Nylon 6/6, some glass reinforced; engineering thermoplastics.

Maraset. [Acme Div.] Epoxy resins.

Marasperse. [Borregaard LignoTech] Lignosulfonates or oxylignins; dispersant, emulsion stabilizer for ceramic slurries, stucco, agric. formulations, dyestuffs, gypsum board, industrial cleaners, cement slurries, paper mfg., in mfg. of brick, tile, refractories.

Maratan. [LignoTech] Lignin; tanning agent.

Marble Dust. [ECC Int'l.] Calcium carbonate; for putties, glazes, mild abrasive compds.

Marblemite. [ECC Int'l.] Calcium carbonate; broad particle size distribution for pigment packing, cultured marble.

Mar'blend. [Georgia Marble] Calcium carbonate; single-filler system for cultured marble.

Marblewhite. [Pfizer] Ground limestone.

Marbo. [Marbo France] Mold release

agents for polyurethane.

Marbocote. [Marbo France] Mold release agents for polyester.

Marbocote TRE. [Marbo France] Mold release agents for rubber.

Marc-A-Leak. [Climax Performance] High foaming leak detector for gas lines.

Marchon®. [Albright & Wilson Am.] Catalyst in hydrogenation reactions; corrosion inhibitor; for petroleum industry.

Marc PC. [MRC Polymers] Reprocessed polycarbonate resin; for molding and extrusion applics.

Maretard. [Lenmar] Cationic retarders for basic dyes on acrylic, nylon, or polyester fibers.

Marfoam. [Lenmar] Nonsilicone defoamer for dyeing systems.

Margard® sheet. [GE Plastics] Extruded polycarbonate glazing grade sheet, uv stabilized, coated on both faces with mar-resistant silicone coating.

Marinco. [Calgon] Magnesium compds.

Marinco. [Marine Magnesium] Magnesium hydroxide tech. and USP.

Marine-Dew. [Ajinomoto] Partially deacetylated chitin; moisturizing cationic polymer for skin and hair prods.; humectant.

Marine Plasma Extract. [Brooks Industries] Complex sea plasma; moisturizer for hair and skin care cosmetics.

Mark. [Witco/Argus] Stabilizers, antioxidants, uv absorbers, chelating agents for vinyl and other plastics.

Markarry. [Ivax Industries] Carriers for dyeing polyester and blends.

Marklube. [Ivax Industries] Dyebath lubricants.

Markpel. [Ivax Industries] Water repellent for textiles.

Marksoft. [Ivax Industries] Softener for knitted, woven, pigment printed fabrics; sewing lubricant.

Marksperse. [Ivax Industries] Dispersing and sequestering agent.

Markstat®. [Witco/Argus] Quaternary ammonium chloride derivs.; antistat for plastics.

Markwet. [Ivax Industries] Wetting agent, foamer, antimigrant, peroxide stabilizer, penetrant, detergent for textiles.

Marlamid®. [Hüls Am.; Hüls AG] Fatty acid alkanolamides; foam stabilizer, thickener, superfatting agent, pearlescent for household, personal, industrial detergents.

Marlanol. [Lenmar] Detergent, wetting agent, emulsifier for dyeing and printing.

Marlasol. [Lenmar] Detergent, scouring agent, emulsifier for textiles.

Marlate®. [Kincaid Enterprises] Methoxychlor; insecticide.

Marlazin®. [Hüls Am.; Hüls AG] Fatty amine polyglycol ethers; surfactant, detergent, dyeing assistant, thickener, wetting agent for industrial cleaners, textile auxiliaries.

Marlex®. [Phillips; Phillips Petrol. Chem. SA/NV] Polyethylene or polypropylene resins; for inj., blow, or rotational molding or thermoforming of appliance parts, chemical equipment, industrial parts, containers, toys, housewares, chemical tanks, monofilament, film applics.

Marlican®. [Hüls Am.; Hüls AG] Dodecylbenzene; detergent intermediate, solubilizer.

Marlinat®. [Hüls Am.; Hüls AG] Fatty alcohol ether carboxylic acid or sulfates; surfactants, wetting agents for textile, paint, paper, cleaners, cosmetic industries.

Marlipal®. [Hüls Am.; Hüls AG] Ethoxylated ethers or esters; surfactants, dispersants, wetting agents, emulsifiers, thickeners, superfatting agents for household and industrial detergents, textiles, cosmetics.

Marlon. [Lenmar] Antifloating agent for carpet dyeing; retarder, leveling agent.

Marlon®. [Hüls Am.; Hüls AG] Alkylaryl sulfonates; detergent raw materials.

Marlophen®. [Hüls Am.; Hüls AG] Ethoxylated ethers; raw material for textile and paper auxiliaries; detergent, dispersant, wetting agent for binders.

Marlophor®. [Hüls Am.; Hüls AG]

Phosphate esters; surfactant used as base material for cleaners; detergent, wetting agent, antistat.

Marlopon®. [Hüls Am.; Hüls AG] Dodecylbenzene sulfonates; surfactant, raw material for detergents, dishwashes, personal care prods.

Marlosoft®. [Hüls Am.] Imidazolinium methosulfate; base for fabric softeners.

Marlosol®. [Hüls Am.; Hüls AG] Ethoxylated esters; raw material for finishing agents in synthetic fiber industry.

Marlowet®. [Hüls Am.; Hüls AG] Emulsifier for solvents, mineral oils, textile finishes, waxes, polishes, pesticides, leather care, metal cleaning, metalworking, mold releases, wood, paper, cosmetics.

Marlox®. [Hüls Am.; Hüls AG] PPG ethoxylated ethers; surfactants, detergents, antistats, foam controllers, wetting agents for bottle cleaning, textile auxiliaries, industrial cleaners.

Marlube. [Lenmar] Fiber spinning or winding lubricants.

Marplex. [Lenmar] Chelating agents.

Marpol. [Matsumoto Yushi-Seiyaku] Sulfates or phosphates; spinning/finishing oils for fibers.

Marquart Pigments. [Degussa] Cadmium, cobalt, or titanium pigments.

Marsorb 24. [Aceto] Benzophenone-3.

Marukarez. [Arakawa] Hydrocarbon resin.

Marvaloy. [Marval Industries] Acrylic-modified styrene; for inj. molding and extrusion of specialty medical, pharmaceutical, food, and cosmetic pkg., advertising displays, high-strength toys.

Marvanbrite. [Marlowe-Van Loan] Stilbene, oxazole, benzimidazole, or coumarin derivs.; optical brighteners for textile processing.

Marvanfix®. [Marlowe-Van Loan] Fixatives for textile processing.

Marvangel. [Marlowe-Van Loan] Sulfonated tauride; wool scour and dyeing assistant.

Marvanlev. [Marlowe-Van Loan] Leveling and migrating agent for acid dyes on nylon.

Marvanlube®. [Marlowe-Van Loan] Lubricants for textile processing.

Marvanol®. [Marlowe-Van Loan] Detergent, wetting agent, leveling agent, emulsifier, lubricant, scouring agent, dye assistant for textiles, dyeing.

Marvanquest. [Marlowe-Van Loan] Organic blend; hydrogen peroxide and bleach stabilizer; prevents calcium deposits.

Marvanscour®. [Marlowe-Van Loan] Solvents and detergents; for prescour and afterscour use.

Marvansoft. [Marlowe-Van Loan] Softeners, lubricants for textile finishing, printing.

Marvantex. [Marlowe-Van Loan] Alkaline scours; detergent; for dye stripping.

Marvelin. [Matsumoto Yushi-Seiyaku] Alkylene oxide addition prods.; dye leveling agent for wool.

Marvylan. [Limburgse Vinyl Maatschappij LVM] Suspension PVC resins.

Marvylex. [Limburgse Vinyl Maatschappij LVM] PVC-based thermoplastic elastomers.

Marvyloy. [Limburgse Vinyl Maatschappij LVM] PVC-based alloys for inj. molding.

Masil®. [PPG/Specialty Chem.] Silicone fluids and emulsions; lubricant, antistat, antifog, wetting agent, release agent, foam control agent for personal care prods., plastics, rubber, textiles, metal processing, lubricants.

Masilwax. [PPG/Specialty Chem.] Stearoxymethicone/dimethicone copolymer or blends.

Maskarome. [Synfleur] Masking agent.

Maslip®. [PPG/Specialty Chem.] Proprietary formula; lubricant base for metalworking fluids with high extreme pressure properties.

Masquodors®. [Synthron] Odor disguising agents.

Masquol®. [Synthron] Sequestering, dispersing, and antiredepositing agents.

Massa Estarinum® AM. [Hüls Am.] Trilaurin; suppository bases, carriers for pharmaceuticals.

Massa Estarinum® CM. [Hüls Am.] Hydrogenated palm glycerides, hydrogenated palm kernel glycerides.

Massive Manganese. [Chemetals] Manganese metal.

Master Bond. [Master Bond] Epoxy compds.; casting and encapsulating compd. for elec./electronic components; adhesives.

Master Color. [Ampacet] Color concs.

Master Draw. [Etna Prods.] Lubricants.

Masterflam. [VAMP Srl] Flame retardant masterbatches for thermoplastics.

Master-Foam. [Masterplast Ltd.] Blowing agent masterbatch.

Mastermix. [Harwick] Chemical dispersions.

Master-Stab. [Masterplast Ltd.] uv stabilizer masterbatch.

Master-Stat. [Masterplast Ltd.] Antistatic agents masterbatch.

Matar. [Huntington Labs] Germicidal detergent.

Mater-Bi®. [Novamont N. Am.] Thermoplastic from natural, renewable raw materials; recyclable, combustible, and biodegradable plastic for pkg. films, bags, 6-pack yokes, disposable diapers, hospital and sanitary prods.

Matexil. [ICI Surf. UK] Antifoam, neutralizer, carrier, leveling agent, dispersant, migration inhibitor, catalyst, fixing agent, crosslinking agent, retarder, stripping agent, emulsifier, softener, thickener, wetting agent for bleaching, dyeing, pigment printing, textiles.

Matrimid®. [Ciba-Geigy/Plastics] Thermoplastic polyimides; for structural composites and adhesives.

Mattina. [Mearl] Cosmetic pearl colors.

Mavan. [Dr. Madis Labs] Propenyl guaethol.

Maxacal®. [Int'l. Bio-Synthetics] Protease; enzyme which breaks down protein into water-sol. prods.; used in detergent formulations.

Maxahibit. [Climax Performance] Sodium tolyltriazole; corrosion inhibitor for nonferrous metals.

Maxaliq. [Int'l. Bio-Synthetics] Bacterial alpha-amylase; starch hydrolyzing enzyme for ethanol industry.

Maxamyl®. [Int'l. Bio-Synthetics] Alpha-amylase; enzyme for laundry detergents, starch processing.

Maxarome. [Int'l. Bio-Synthetics] Yeast extract.

Maxatase®. [Int'l. Bio-Synthetics] Protease or blends; enzyme for breakdown of proteins and starch into water-sol. prods.; used in detergent formulations.

Maxazyme®. [Int'l. Bio-Synthetics] Glucose isomerase; enzyme which catalyzes the isomeration of glucose to fructose.

Max Bond. [H.B. Fuller] Construction adhesive.

Maxcote. [Chemax] Asphalt emulsifiers.

Maxetch. [Old Bridge Chem.] Ammonium hydroxide.

Maxigard. [Drew Ind. Div.] Corrosion inhibitor.

Maxi-Gel. [A.E. Staley Mfg.] Modified corn starch.

Maxilact®. [Int'l. Bio-Synthetics] Yeast lactase; enzyme which hydrolyzes lactose to simple sugars.

Maxilene. [Dexter] Leveling agents for acid colors on nylon.

Maxilon. [Ciba-Geigy/Dyestuffs] Textile dyes and pigments.

Maximaize. [A.E. Staley Mfg.] Modified corn starch.

Maxinvert®. [Int'l. Bio-Synthetics] Invertase; enzyme used in confections and the prod. of invert syrups.

Maxithen. [Gabriel-Chemie UK Ltd.] Polymer-specific masterbatch.

Maxitol®. [Dexter] Wool lubricants, wetting agent, protective agent.

Maychlor. [Mayco Oil & Chem.] Petroleum oil additive.

Mayco Base. [Mayco Oil & Chem.] Fatty ester; lubricant and petroleum additive.

Mayfree. [Mayco Oil & Chem.] Chlorine-free EP additives.

Mayodan. [Grindsted Prods. Denmark] Stabilizer blends; food emulsifier for mayonnaise and salad dressing.

Mayoquest. [Mayo] Chelating agents.

Maypeg. [Mayco Oil & Chem.] PEG esters.

Mayphos. [Mayco Oil & Chem.] Phosphate esters; lubricant additive, surfactant.

Maypon. [Inolex] Hydrolyzed collagen derivs.; detergent for personal care prods., general purpose cleaners.

Maysol. [Mayco Oil & Chem.] Soluble oil conc.

Maysperm. [Mayco Oil & Chem.] Petroleum oil additive.

Maysyn. [Mayco Oil & Chem.] Synthetic metalworking lubricants.

May-Tein. [Maybrook] Hydrolyzed collagen derivs.; detergent, substantiviity agent, foamer, softener, moisturizer, conditioner for skin and hair care prods.; anti-irritant; emulsifier.

Mazamide®. [PPG/Specialty Chem.] Fatty acid alkanolamides; emulsifier, detergent, solubilizer, thickener, lubricant, foam builder/stabilizer used in hard surface cleaners, dishwashes, metalworking fluids, fiber and hair conditioners, dry cleaning, agric. sprays, leathers, fur, polishes; corrosion inhibitor.

Mazawax® 163R. [PPG/Specialty Chem.] Cetearyl alcohol and polysorbate 60; emulsifier for pharmaceutical and cosmetic applics.; base, emollient, thickener.

Mazawet®. [PPG/Specialty Chem.] Anionic and nonionic surfactants; wetting agent, surfactant, antifog for textile processing, emulsion polymerization, detergents, metalworking fluids, paints, inks, polishes, floor waxes, hard surface cleaning.

Mazclean. [PPG/Specialty Chem.] Terpene-based hard surface cleaner; as replacement for chlorinated solvents in some applics.

Mazeen®. [PPG/Specialty Chem.] Amines and ethoxylates; emulsifier, antistat, rewetting agent, lubricant for insecticides, textiles, lubricants, inks, and cosmetics.

Mazenate. [PPG/Specialty Chem.] Specialty fatty tertiary amine salts.

Mazide. [PPG/Specialty Chem.] Bactericides, slimicides, fungicides.

Mazol®. [PPG/Specialty Chem.] Glyceryl esters or ethoxylates; emulsifier, carrier, emollient, dispersant, plasticizer for cosmetics, pharmaceuticals, lubricants, mold releases, foods.

Mazoline. [PPG/Specialty Chem.] Fatty imidazolines.

Mazon®. [PPG/Specialty Chem.] Emulsifier, corrosion inhibitor for agric., metalworking, lubricant, emulsion polymerization, industrial cleaners, cosmetics, textiles.

Mazox®. [PPG/Specialty Chem.] Amine oxides; corrosion inhibitor, hair conditioner.

Maztreat. [PPG/Specialty Chem.] Oil and water treating compds.

Mazu®. [PPG/Specialty Chem.] Silicone prods.; defoamer, emulsifier for foods, pharmaceuticals, industrial use (inks, paints, insecticides, polymerization, adhesives, textile, paper).

Mazvap. [PPG/Specialty Chem.] Descalants.

MB 450. [Werner G. Smith] Monoethanolamine borate; corrosion inhibitor.

MBEO 1.8 Adduct. [Air Prods.] Ethoxylated acetylenic alcohol; corrosion inhibitor, leveler, brightener for electroplating.

MBS. [Transene] Moisture-resistant silicone.

MBT. [Akrochem] 2-Mercaptobenzothiazole; accelerator.

MBT®. [Uniroyal] 2-Mercaptobenzothiazole; intermediate.

MBTS. [Akrochem] Benzothiazyl disulfide; accelerator.

MC. [George Mann] Mold cleaner prods.

MC580. [GE Silicones] Silicone thermosetting molding resin.

McBloc. [McGean-Rohco] Plastic additive.

MCC. [Mississippi Chem.] Urea, ammonia, or potash.

MCI. [Melamine Chems] Melamine crystal.

M-Clene D. [OxyChem] Vapor degreasing agent.

McLube. [McLube] TFE or molybdenum disulfide lubricants and release coatings.

McNamee® Clay. [R.T. Vanderbilt] Kaolin; extender and reinforcing filler for resins.

M-C-Thin. [Lucas Meyer] Lecithin; release aid, emulsifier, wetting agent, stabilizer for food and technical applics.

MCU 2100. [Steelcote Mfg.] Moisture-cured urethanes.

MCX-A. [Mitsui Petrochemicals] Polyamide, glass-reinforced; engineering resin for automotive parts, elec./electronic components, general industrial parts.

MDA-1. [Ciba-Geigy/Additives] Oxalyl bis (benzylidenehydrazide); antioxidant.

MDD-132. [Mitsubishi Kasei] Ozone decomposition catalyst; for treatment of offensive odors from sewage disposal, night soil treatment.

MDH Series. [Mitsubishi Kasei] Hydrogenation catalyst; for mfg. of oxoalcohol.

MDK-113. [Mitsubishi Kasei] Deoxygenation catalyst; for purification of gases for electronic industry.

Mearlcrete. [Mearl] Foamed concrete.

Mearlin®. [Mearl] Titanium dioxide/mica; for pearlescent, metallic, antique, and iridescent effects in powd. coating systems.

Mearlite®. [Mearl] Bismuth oxychloride compds.; synthetic pearl pigments.

Mearlmaid. [Mearl] Guanine blends; natural pearl essence.

Mearlmica. [Mearl] Mica or blends; extender in loose powders; binder and reinforcement in lipsticks.

MEC. [Norac] Polyester resins curing agent.

Mecon MB Masterbatch. [Sydplast AB] Colors and additives for plastics.

Medialan Brands. [Hoechst Celanese/Colorants & Surf.; Hoechst AG] Fatty acid sarcosides; detergent, basic material for cosmetics, textile detergents.

Medical Fluid 360. [Dow Corning] Dimethicone.

Megapoly. [H.A. Astlett] Natural rubber latex/methyl methacrylate graft polymer; high performance industrial polymer, adhesive and bonding agent, pressure-sensitive tapes, surgical tapes, footwear, flooring, composite hoses, carpeting.

Megarad. [Dow Plastics] Polycarbonate resin.

Megasulfur. [Southern Dye & Chem.] Textile dyes and pigments.

Megatint. [Chemurgy Prods.] Polymeric fugitive tints for synthetic fibers.

Megum. [Ore & Chem. Corp.] Rubber-to-metal adhesives.

Mekon® White. [Petrolite/Polymers] Microcrystalline wax; release agent, lubricant, processing aid, binder, plasticizer, antiozonant for coatings, adhesives, inks, plastic modification, lacquers, paints, varnishes, ceramics, for potting/filling in elec./electronic components, rubber, wax size.

Mekor®. [Drew Ind. Div.] Volatile oxygen scavenger, metal passivator, corrosion inhibitor for boiler water treatment.

Mel. [CNC Int'l.] Melamine-formaldehyde resins; for textile processing.

Melaicar. [Costa Floros O.E.] Melamine-formaldehyde molding compds.

Melamag. [Dow] Metallurgical granules.

Melanol. [Seppic] Lauryl sulfate salts; base for shampoos.

Melatex AS-80. [Tokai Seiyu Ind.] Polyacrylic resin; hand modifying agent for synthetic knits.

Melax. [Axel Plastics Research Labs] Mold conditioners, sealers.

Melblend. [Metal Lubricants] Lubricating oils.

Melco. [Metal Lubricants] Lubricating oils.

Melcolene. [Metal Lubricants] Lubricating oils.

Melcolube. [Metal Lubricants] Gear oils.

Meldin®. [Furon] Proprietary polyimide resin; for bearings, thrust washers, piston rings, elec./electronic parts, gears, bearing retainers.

Melhi Resin. [Hercules] Acidic resin derived from rosin; thermoplastic used in inks, adhesives.

Melinar®. [ICI Films; ICI PLC] PET polyester; for blow molded rigid containers.

Melinex®. [ICI Films] Polyester film; for graphics applics.

Meliorian. [Ceca SA] Sodium octyl/stearyl sulfate; degreaser, ore flotation for barite and fluorspar.

Melioran 118. [Ceca SA] Sodium cetyl-stearylic alcohol sulfate; surfactant.

Melkool. [Metal Lubricants] Synthetic and semisynthetic coolants.

Melkut. [Metal Lubricants] Cutting oils.

Melmac. [Am. Cyanamid] Coating resins.

Melmex. [BIP Chem. Ltd.] Melamine-formaldehyde molding powder.

Melogel. [Nat'l. Starch & Chem.] Corn starch.

Melopas®. [Ciba-Geigy GmbH] Melamine-formaldehyde or melamine-phenolic resins; molding compds.

Melpar. [Metal Lubricants] Hydraulic oils.

Melquench. [Metal Lubricants] Quenching oils.

Melsol. [Metal Lubricants] Soluble oils.

Melsyn. [Metal Lubricants] Water-sol. hydraulic oils.

Meltac. [Metal Lubricants] Lubricants.

Meltatox®. [BASF AG] Dodemorph-acetate; for control of powdery mildew in roses and other ornamentals.

Meltran. [Metal Lubricants] Hydraulic oils.

Melurac. [Am. Cyanamid] Melamine urea-formaldehyde; wood adhesive.

Melusat. [Henkel/Emery/Cospha] Fruity aroma chemical.

Melustral. [Hoechst AG] Pigment preparations; for finishing.

Melvis. [Metal Lubricants] Lubricants.

Melvo. [Van Den Bergh Foods] Partially hydrogenated vegetable oil blend; all-purpose frying shortening.

Merbron. [Sybron] For aftercleaning of dyed polyester, cleaning fibers and equip.

Merce Assist ADB. [Sybron] Mercerization penetrant.

Mercerol®. [Sandoz; Sandoz Prods. Ltd.] Wetting agent, mercerizing penetrant for causticizing baths.

Mercury Oil. [Toho Chem. Industry] Fatty acid esters or blends; for rolling oils, drawing oils for metal working.

Merecols. [Meer] Stabilizers.

Merezan. [Meer] Xanthan gum.

Merigraph. [Hercules] Photopolymer printing plate system.

Merillite. [Pigment & Chem.] Zinc dust.

Merinol. [Takemoto Oil & Fat] Non-ionic/anionic surfactant blends; for worsted processing.

Meristami. [Alban Muller] Oak meristem extract; soothing prod.

Merix AST-1001. [Merix] uv absorber.

Merix Wash Conc. [Merix] Cleaner.

Merkyl PM-TL. [Vikon] Phenylmercury triethanol ammonium lactate; durable antibacterial agent and mildewcide for outdoor cotton textiles.

Merpol®. [DuPont] EO condensate; wetting agent, detergent, penetrant, emulsifier, leveling agent, dye assistant, antistat, emulsifier, stabilizer for textile, paper, paints, inks, medicinal ointments, antiperspirants, cosmetics.

Merquat®. [Calgon] Polyquaterniums or blends; cosmetic polymer; conditioner, antistat for hair and skin care prods.

Merse®. [Sybron] Catalyst for polyester weight reduction work.

Mersitol. [Seppic] Alkylsulfate; wetting agent for mercerizing.

Mesamoll. [Bayer] Alkyl sulfonic ester of phenol; plasticizer for PVC, rubber, and other polymers.

Mesitol®. [Miles/Organic Prods.] After-treating agent for acid dyes on nylon.

Metablen®. [Atochem N. Am.] Impact modifiers for PVC blow molding compds.

Metac. [Catalytic Prods. Int'l.] Precious metal catalysts.

Metacast. [Mereco Prods.] Casting and encapsulating resins.

Metachloron®. [Crompton & Knowles] Organo metallic compds.; color bleed assistant for direct dyes on cotton and rayon goods.

Metacure®. [Air Prods.] Organotin or ester compds.; catalyst for prod. of polyurethane coatings, adhesives, and sealants; epoxy curing agent.

Metaduct. [Mereco Prods.] Electrically conductive resins.

Metafil. [Mereco Prods.] Impregnants.

Metafloc. [Morton Int'l.] Flocculating agents.

Metafos. [Peridot Chem.] Sodium hexametaphosphate.

Metagel. [Mereco Prods.] Thixotropic dip coatings.

Metaglo. [Eastern Color & Chem.] Textile metallic print binder.

Metal Deactivator S. [Hart Chem. Ltd.] N,N´-Disalicylidene-1,2-propane diamine; copper chelating agent for refinery industry.

Metalin. [M. Dohmen USA] Textile dyes and pigments.

Metalite. [Metal Lubricants] Cleaners.

Metallic 9500. [D.J. Enterprises] Conductive additive for coatings and composites.

Metallyte. [Mitsui Petrochemicals] One side coextruded oriented polypropylene film, one side metal, heat sealable; high opacity film for laminations.

Metalon. [Chamberlain Plastics Div. Ltd.] Polyester PVC.

Metalplex. [Sullivan Chem. Coatings] Vacuum metallizing coating.

Metalplex 143. [Morton Int'l.] Metal precipitant.

Metalub. [Metal Lubricants] Drawing compds.

Metalyn 582. [Climax Performance] Pentaerythritol ester of tall oil fatty acid and polyhydric alcohols.

Metam-Fluid BASF. [BASF AG] Metam-sodium; soil fumigant for control of nematodes, soil-borne fungi, soil insects, germinating weeds.

Metanil. [Rite Industries] Textile dyes and pigments.

Metapart. [Mereco Prods.] Mold releases.

Metarin. [Lucas Meyer] Lecithin compds.; instantizer for foods.

Metasap®. [Syn. Prods.] Aluminum stearate; process aids, lubricants for polymers, specialty greases, waterproofing agents, thixotropes for paints and inks, flatting agents for lacquers, flow aids for cements, slab dips and dusting compds. for rubber.

Metaseal. [Mereco Prods.] Sealing compd.

Metasol. [Calgon] Bactericides, fungicides.

Metasol HP-500. [Marubishi Oil Chem.] Compounded prod.; heavy duty and low foaming detergent for metal cleaning.

Metastrip. [Mereco Prods.] Resin solvents and strippers.

Metasystox-R. [Miles/Ag] Insecticide for shrubs, trees, peaches, cherries, vegetable crops.

Metaterge. [Mereco Prods.] Detergents and cleaners.

Metatin. [KMZ Chem. Ltd.] Fungicides.

Met-Braze. [Degussa Metal Group] Brazing alloy containing precious metals.

Meteor. [Chem. & Pigment] Nutritional sprays.

Metglas®. [Allied-Signal] Amorphous metal alloys.

Methacrol. [DuPont] Textile lubricant.

Methar. [W.A. Cleary] Crabgrass killer.

Methasan®. [Harwick] Zinc dimethyldithiocarbamate; secondary accelerator for thiazoles and sulfenamides; fast curing accelerator for rubbers and latexes.

Methasol. [ICI Am.] Textile dyes and pigments.

Methazate®. [Uniroyal] Dithiocarbamate; accelerator for air-cured footwear, insulated wire.

Methic. [ICI Am.] Textile dyes and pigments.

Methocel. [Dow] Cellulose ethers; binders, thickeners, film formers, water retention aids, suspending agents, surfactants, lubricants, emulsifier, protective colloid for foods, cosmetics, pharmaceuticals, latex paints, construction prods., ceramics, other applics.

Meth-O-Gas. [Great Lakes] Space, soil fumigant.

Methomid 60. [Scher] PEG-50 tallow amide.

Methyl Carbitol®. [Union Carbide] Methoxydiglycol; solvent.

Methyl Cellosolve®. [Union Carbide] Methoxyethanol; solvent.

Methyl Cumate®. [R.T. Vanderbilt] Copper dimethyldithiocarbamate; accelerator for rubbers.

Methyl Dipropasol. [Union Carbide] PPG-2 methyl ether.

Methyl Ester B. [Mayco Oil & Chem.]

Fatty esters; intermediate-lubricant, wetting agent.

Methyl Ester L. [Climax Performance] Lard methyl ester.

Methyl Ester S. [Climax Performance] Soya methyl ester.

Methyl-Ethyl Tuads®. [R.T. Vanderbilt] Rubber accelerator.

Methyl Ledate. [R.T. Vanderbilt] Lead dimethyldithiocarbamate; accelerator for rubber.

Methyl Linoleate. [Hart Prods. Corp.] Fatty acid esters; base for super amides.

Methyl Namate®. [R.T. Vanderbilt] Sodium dimethyldithiocarbamate; water treatment chemical; clarification agent for wastewater from plating, photo finishing, ore beneficiation processes.

Methyl Niclate®. [R.T. Vanderbilt] Nickel dimethyldithiocarbamate; antioxidant for epichlorohydrin and peroxide vulcanized elastomers.

Methylon®. [OxyChem] Phenolic resins.

Methyl Propasol®. [Union Carbide] Methoxypropanol.

Methyl Propasol® Acetate. [Union Carbide] Propylene glycol methyl ether acetate.

Methyl Selenac®. [R.T. Vanderbilt] Selenium dimethyldithiocarbamate; vulcanizing agent, accelerator for rubber.

Methyl Tuads®. [R.T. Vanderbilt] Tetramethylthiuram disulfide; accelerator for NR and synthetic rubbers; vulcanizing agent; cure modifier for Neoprene.

Methyl Zimate®. [R.T. Vanderbilt] Zinc dimethyldithiocarbamate; accelerator for NR and synthetic rubbers.

Metlbond®. [BASF AG] Structural film adhesives for aerospace, aircraft, and industrial applics.

Met-Loy. [Degussa Metal Group] Metallurgical alloy or composite metal.

Metox. [Degussa Metal Group] Silver and anoxide of a base metal.

Metre-Grip. [Mereco Prods.] Adhesives and bonding agents.

Metron. [Reichhold] Resin impregnated overlay.

Metronic. [Atochem N. Am.] Cleaners and acid salts.

Metropad. [Yorkshire Pat-Chem] Non-resinated system for pad dyeing applics.

Metrosol. [Reilly-Whiteman] Detergents for textile scouring.

Metrotex. [Yorkshire Pat-Chem] Nonionic aq. pigment dispersions for printing.

Metso®. [PQ Corp.] Sodium metasilicate; corrosion inhibitor used in detergents to protect metal, china, glass, or ceramic prods.

Metsoak. [Atochem N. Am.] Cleaners.

Metspec. [Metalspecialties] Low-melting metals.

Meturon®. [Griffin] Fluometuron; herbicide controlling annual grasses and broadleaf weeds in cotton and sugarcane.

Metzeler. [Lermont Plastics SA] Reflective polystyrene.

Mewlon. [British Traders & Shippers] PVA fiber and filament.

Mexol. [Surpass] Powdered nonchlorine bleaching agents.

Mexpectin. [Grindsted Prods.] Pectins; suspending agents, visc. builders for beverages, cosmetics, pharmaceuticals, wound dressings.

Meyprofix 509. [Rhone-Poulenc Surf.] Vinyl acetate, isobutyl maleate, vinylneodecanoate copolymer; hair styling fixative resin.

MFF 159-B. [Siltech] Gloss additive, mild conditioner for personal care prods.

Mg-0601 T 1/8″. [M&T Harshaw] Magnesia; catalyst support.

MG-58. [Martin Marietta Magnesia Spec.] Magnesium oxide; for fertilizers.

MGH-93. [RMc Minerals] Magnesium hydroxide; dilutes or eliminates smoke produced by decomposing polymers, allows high processing temps. for olefins, nylons, etc.

MGK 264. [McLaughlin Gormley King] Insecticide synergist.

MGME-. [CVC Spec. Chem.] Magnesium methylate.

MHT. [Dow] Magnesium hydroxide.

Mibiron. [Rona] Bismuth oxychloride coated mica; pearlescent for lipsticks, pressed powders, eye makeup.

Mica. [Presperse] Mica.

Micacoat. [Nyco Minerals] Chemically surface-modified mica.

Micafil. [Filter-Media] Cryogenic vermiculite insulation.

Micatherm. [Morgan Matroc Ltd.] Inorganic thermoplastic.

Miccron. [Michigan Chrome & Chem.] Powdered resins.

Miccrosol. [Michigan Chrome & Chem.] Vinyl plastisols.

Michaelok. [David Michael] Spray-dried flavors.

Michel. [M. Michel] Alcohols or amines.

Michelene. [M. Michel] Diamine fatty acid condensate; softeners, textile finishing.

Michemcoat. [Michelman] Coatings.

Michemlube. [Michelman] Wax additives.

Michemlube. [Norwegian Talc UK Ltd.] Additive for plastic coatings.

Michemprime. [Michelman] Adhesion promoters.

Michlin. [Michlin Diazo Prods.] Insecticides.

Michtex. [David Michael] Stabilizers, emulsifiers.

Micobond. [Dexter Spec. Coatings] Primers, adhesives.

Micoflex. [Dexter Spec. Coatings] Coatings.

Micolon IV. [Dexter Spec. Coatings] Nonstick finishes.

Micothane. [Dexter Spec. Coatings] Resistant urethane finishes.

Micral®. [Solem Industries] Alumina trihydrate; smoke suppressor/flame retardant for wire and cable insulation, inj. molded polyolefins, coatings, adhesives, rubber goods, paper filler and coating, PVC, EPDM, EPR, ABS, XLPE, and compr.-molded thermosets.

Micreva. [S P & C Ltd.] Microcellular polymers EVA/PE and specialized masterbatches.

Micro-Ace. [Presperse] Talc or blends; cosmetic ingredient contributing softness, smooth feel, skin adhesion and spreadability.

Microban. [Microban Germicide] Disin-

fectant, deodorant.

Microbeads. [Cataphote] Spherical glass fillers.

Microbiotone. [Am. Labs] Hydrolyzed enzyme.

Microbloc®. [Pfizer] Surface-treated talc with proprietary coating; for film applics.

Microcat. [Bioscience] Blends of powd. activated carbon and specialized microbes; for waste treatment.

Microcel. [Sovitec SA] Hollow glass beads.

Micro-Cel®. [Celite] Calcium silicate; functional filler used as carriers, grinding aids, anticaking agent, opacifier, extender pigment, conditioner in agric. chemicals, catalyst supports, paints.

Micro-Chek 11. [Ferro] 2-n-Octyl-4-isothiazolin-3-one in plasticizers; mildewcide, antimicrobial, fungicide, preservative for PVC, polyurethane, automotive trim, awnings, pond liners, marine upholstery, outdoor furniture, interior applics.

Microcide. [Kromachem Ltd.] Biocide preservatives.

Microdip. [Atochem N. Am.] Nickel plating corrosion protection.

Microdol. [Norwegian Talc UK Ltd.] Fillers for plastics.

Micro-Dry®. [Reheis] Aluminum chlorohydrate; antiperspirant.

Microduct®. [CasChem] Maltodextrin; carrier for fragrances; emollient oils, bath prods., food additive.

Micro Emulsion 2671. [Marlowe-Van Loan] Micro-silicone emulsion; softener for textile processing; improves tear strength and elasticity.

Microfast. [Kromachem Ltd.] Pigments.

Microfine. [Astor Wax] Micronized waxes.

Microfine Bentonite. [Am. Colloid] Air-floated bentonite.

Micro-Flo. [Chem. Prods.] Barium carbonate.

MicroForm. [Hercules] Bentonite clay or acrylamide-based copolymer; coagulant for MicroForm Retention Management System.

Microken. [Witco/Inorganic Spec.] Diatomite filler.

Microlin. [Micropol Ltd.] Linear polythene powders.

Microlink. [Micropol Ltd.] Crosslinkable polyethene powder for rotational molding.

Microlite. [W.R. Grace/Construction Prods.] Vermiculite dispersions.

Microlith. [Ciba-Geigy Pigments UK] Organic pigments predispersed in vinyl or acrylic.

Microlube. [Hercules] Paraffin wax emulsion; surface treatment, lubricant for paper; improves water resistance.

Micromesh. [Mearl] Water-ground muscovite mica; filler.

Micromica. [Norwegian Talc UK Ltd.] Fillers for plastics.

Micromid. [Union Camp] Waterborne polyamide dispersions; for laminating adhesives, heat seal adhesives for pkg., release coatings.

Micromulse WIO. [Norman, Fox] Ingredient for forming stable w/o emulsions and transparent microemulsions.

Micronal. [Ferro/Transelco] Polishing agent for hard resin lenses.

Micronal®. [BASF AG] Aq. microcapsuel dispersions; for prod. of the donor face of noncarbon copying papers.

Micro-P. [D.J. Enterprises] Amorphous mineral silicate; lightweight, stable, inert filler, resin extender for aircraft, military, appliances, business machines, construction, consumer goods, elec./electronic, land transport, and marine applics.

Microperl. [Sovitec SA] Glass beads.

Microperq. [Sovitec SA] Glass-filled microspheres.

MicroPflex. [Pfizer] Surface-modified microtalc; filler.

Microplus. [Western Nutrients] Liq. plant food containing primary and secondary plant nutrients chelated with EDTA.

Micropol. [Micropol Ltd.] Polythene powders.

Micropoly. [Presperse] Polyethylene; powders improving texture and body, luster, coverage, and opacity of product formulations.

Micropor Buds. [A.E. Staley Mfg.] Low bulk density maltodextrins.

Micropro. [Micro Powders] Polypropylene wax.

Micropure® Ultra. [ISP] N-Methyl-2-pyrrolidone.

Micro Release. [Particle Dynamics] Sustained release prods.

Microseal. [E/M Corp.] Solid lubricants.

Microseal. [Himac] Laminate of paper and metallized (aluminum) heat-sealable polyester film; pastry browning wrap for microwave use.

Microsil RW. [Reilly-Whiteman] Silicone softener for textiles.

Microsized Salt. [Akzo Salt] Pulverized sodium chloride.

Microsol. [Ciba-Geigy Pigments UK] Aq. organic pigments, dispersions.

Microsphere. [Toyomenka Am.] Methacrylate polymers.

Microspin. [Hilton Davis] Pigment dispersion plastics.

Microsponge. [Advanced Polymer Systems] Styrene/DVB copolymer.

Microstat. [Micropol Ltd.] Semiconductive polyethylene powders.

Micro-Step. [Stepan; Stepan Europe] Surfactant blends; microemulsifier for pesticides.

Microsweet. [Celite] Specialty filter aid.

Microtalc. [Norwegian Talc UK Ltd.] Fillers for plastics.

Microtalc. [Pfizer] Low micron talc.

Microtex. [Guthrie Latex] Natural rubber latex.

Microthene®. [Quantum/USI] Polyolefin resins; for powder coating, rotational molding.

Microthiol Special. [Atochem N. Am.] Agrichemical.

Microtuff. [Pfizer] Surface-treated talc; filler for polyolefins.

Micro-White®. [ECC Int'l.] Calcium carbonate; general purpose fillers for paints and polymers.

Microzol. [Lubrizol] High water content fluids.

Micryl. [Michelman] Wax additives.

Midas Gold. [RBH Dispersions] Pigment dispersions.

Migassist®. [Sybron] Acid dye leveling

agent.

Mi-Gee. [Geoliquids] Methylene iodide.

Mighty Mulse. [Octagon Process] Emulsion solvents.

Miglyol®. [Hüls Am.; Hüls AG] Glycerin and/or fatty acid compds.; emollient, dispersant, lubricant, suspending agent, solubilizer, carrier/vehicle, stabilizer, and solvent for cosmetics, pharmaceuticals.

Migracrest. [Reilly-Whiteman] Fixatives for cellulosics.

Migralube. [ICI Advanced Materials] Internally lubricated molding compds.

Migralube. [Kawasaki Steel] Polycarbonate or nylon 6/10; internally lubricated thermoplastic resins for severe frictional applics.

Migranil®. [Sybron] Nonionic metallic salt; antimigrant for pigment dyeing.

Migrassist®. [Sybron] Textile leveling agents, compatibilizers.

Migratone. [Nutex] Softener for textiles.

Migregal. [Nippon Senka] Alkyl sulfate; leveling agent for fabrics.

Mikacion Dye®. [Mitsubishi Kasei] Reactive dyes.

Mikawhite®. [Mitsubishi Kasei] Fluorescent whitening agents for acetate, polyester, and polyacrylic fiber.

Miki. [U.S. Cosmetics] Jet-milled talc; for cosmetics.

Mikrokill. [Brooks Industries] Polyaminopropyl biguanide or blends.

Mil-Clean. [Baker Perf. Chem.] Liq. detergent.

Milco-150. [Milport] Sodium glucoheptanoate; chelating agent for bottle washing, metal cleaning.

Milease. [ICI Am.] Soil release agents.

Mil-Emulsifier. [Baker Perf. Chem.] Degreaser.

Milester. [Polyurethane Corp. of Am.] Polyester polyols.

Milkamino 20. [Brooks Industries] Milk amino acids; cosmetics ingredients.

Milkpro. [Ikeda] Hydrolyzed casein.

Millad®. [Milliken] Polyolefin clarifying agents.

Millamine®. [Milliken] Cycloaliphatic diamine; epoxy curing agent.

Millathane®. [TSE] Polyester or polyether urethane rubber.

Milldride®. [Milliken] Anhydrides; epoxy crosslinker; polyimide intermediate.

Millester. [Polyurethane Spec.] Hydroxyl terminated polyesters.

Millical Plus. [Pfizer] Precipitated calcium carbonate.

Millifoam. [Stepan] Ammonium ether sulfate; foamer for gypsum board.

Millothix®. [Milliken] Dibenzylidene sorbitol; thixotrope and gellant for organic systems.

Milloxane. [Polyurethane Spec.] Water-based, solvent-based and 100% solid polyurethanes.

Millsperse®. [Drew Ind. Div.] Mill water treatment.

Miloxane. [Polyurethane Corp. of Am.] Polyurethane coatings.

Milros. [Alpha Metals] Mildly activated rosin flux.

Milsize. [ICI Am.] Textile sizing agent.

Milsoft. [ICI Am.] Textile softeners.

Milstat. [ICI Am.] Textile antistats.

Milube. [George A. Goulston] Lubricants, emulsifiers, antistatic blends; antidusting agent, textile auxiliary.

Milvex. [Henkel] Nylon.

Mindel®. [Amoco; Amoco Europe] Polysulfone resins.

Mindust. [Hart Chem. Ltd.] Blend; dedusting and wetting agent for coal prep.

Minemal. [Nippon Nyukazai] Emulsifier for oils; degreaser for leather.

Mineral-Cap. [Ronald T. Dodge] Microencapsulated mineral oil.

Mineral Colloid. [Southern Clay Prods.] Purified montmorillonite.

Mineral Jelly. [Penreco] Petrolatum; preblended base for cosmetic mfg.

Mineral Seal Oil HDF-300. [Total Petroleum] Paraffin base oil.

Mineral Spray. [Western Nutrients] Micro and macronutrient blend; liq. spray and buffering agent.

Minex. [Omya GmbH] Transparent filler.

Minex. [Unimin Specialty Minerals] Sodium potassium aluminosilicate; fine ground mineral extender.

Minfil. [Steetley Minerals Ltd.] Calcium

carbonate and dolomite fillers.

Minflo. [Hercules] Water-soluble polymers; mining flotation reagents.

Minkamid 150. [Mona Industries] Minkamide DEA, mink oil.

Mink Oil W. [Cosmetochem] PEG-13 mink glycerides.

Minlon®. [DuPont; DuPont UK] Nylon resin, mineral or mineral/glass-reinforced. low cost, engineering thermoplastic.

Min-U-Gel®. [Floridin] Attapulgite clay; absorbent, adsorbent, gellant, suspending agent, stabilizer, thickener for agric. formulations, animal feeds, asphalt cutabacks, coal water slurries, paints.

Min-U-Sil. [U.S. Silica] Micron-sized silica.

Miox. [Landers-Segal Color] Micageous iron oxide.

Mipcin®. [Mitsubishi Kasei] 2-Isopropylphenyl methylcarbamate; insecticide for leafhoppers and plant hoppers on paddy rice.

Mira-Cap. [A.E. Staley Mfg.] Modified corn starch.

Miracare®. [Rhone-Poulenc Surf.] Surfactant blends; concentrates for baby shampoo, bubble bath, hand soaps, cream rinse conditioners.

Mira-Cleer. [A.E. Staley Mfg.] Modified corn starch.

Miracle Formula. [Armite Labs] Corrosion inhibitor.

Mira-Gel. [A.E. Staley Mfg.] Unmodified corn starch.

Miramid. [Industrial Polymer Services Ltd.] Polyamide 6.

Miramine®. [Rhone-Poulenc Surf.; Rhone-Poulenc France] Imidazolines; corrosion inhibitor, emulsifier, lubricant, detergent, thickener for cutting oils, acid shampoos.

Miranate®. [Rhone-Poulenc Surf.; Rhone-Poulenc France] Surfactants; detergent for shampoo systems; lime soap dispersant.

Miranol®. [Rhone-Poulenc Surf.; Rhone-Poulenc France] Amphoteric surfactant blends; detergent, wetting agent, emulsifier, solubilizer, stabilizer for shampoos, pharmaceuticals, household

and industrial uses.

Mirapol®. [Rhone-Poulenc Surf.; Rhone-Poulenc France] Polyquaterniums; polymer for hair and skin care prods.; conditioner, antistat, emollient, lubricant.

Mirapon®. [Rhone-Poulenc Surf.] Hydroxy sultaine; surfactants.

Mirasheen 202. [Rhone-Poulenc Surf.] Glycol stearate, lauramide DEA, cocamidopropyl betaine, glycerin; pearl conc. for cold blend cosmetic formulations.

Miraslip. [M-R-S Chem.] Mold release.

Mirasol. [C.J. Osborn] Alkyd, epoxy resin esters, polyesters, urethanes.

Mirasperse. [M-R-S Chem.] Pigment dispersions.

Mira-Sperse. [A.E. Staley Mfg.] Modified corn starch.

Mirastab. [M-R-S Chem.] Stabilizers.

Mirataine®. [Rhone-Poulenc Surf.; Rhone-Poulenc France] Betaines, sultaines, or propionates; detergent, visc. builder, foam booster/stabilizer, solubilizer, conditioner for shampoos, dishwashing liqs., textile, leather, industrial prods.

Miratex. [Central Soya] Textured soy flour.

Mirathen. [Anilac NV/SA] HDPE.

Mira-Thik. [A.E. Staley Mfg.] Modified corn starch.

Miravithen. [Anilac NV/SA] Ethylene vinyl acetate copolymer.

Miravon. [Rhone-Poulenc Ltd.] EO/PO adducts; rinse aids and defoamers for dishwashing and other formulations.

Mirawet®. [Rhone-Poulenc Surf.] Wetting agents for heavy-duty cleaners.

Miristamina. [Vevy] Myristamide DEA.

Miristocor. [Roche] Myristamidopropyldimethylamine phosphate.

Mirpyl. [Gattefosse] Propylene glycol myristate.

Mistabond. [MarChem] Adhesives.

Mistacoat. [MarChem] Specialty coatings.

Mistaflex. [MarChem] Plastisols and urethane elastomers.

Mistafoam. [MarChem] Urethane foams.

Mistapox. [MarChem] Epoxy compds.

Mistique Frosted Packaging Compounds. [BFGoodrich/Spec. Polymers] Vinyl molding compds.; frosted pkg. for cosmetics, toiletries, and foods.

Mistron. [Cyprus Industrial Min.] Talc; impact modifier, color enhancer, flatting agent, rheology control agent, nucleating agent, mold release agent, dusting agent for paint, plastic, rubber, pulp and paper applics.

Mitec®. [Mitsubishi Kasei] Isocyanate adducts for polyurethane finishes for wood, plastics, automotive.

Mitin®. [Ciba-Geigy/Dyestuffs] Phenylsulfonic acid deriv.; mothproofing agent.

Miton. [Astor Wax] Micronized slip agent.

Mitsubishi Carbon Black®. [Mitsubishi Kasei] Carbon black; pigments for inks, plastics, paints; conductive grades avail.

Mitsubishi Kasei GF-PET. [Mitsubishi Kasei] PET, glass filled.

Mitsubishi Kasei PBT. [Mitsubishi Kasei] PBT; for inj. molding and specialty applics. (warp resistant, heat resistant, rigid, flame retardant).

Mitsubishi Kasei PPS. [Mitsubishi Kasei] PPS; for general and electric use.

Mitsubishi Polyethylene. [Mitsubishi Kasei] HDPE, LDPE, LLDPE resins; for inj. and blow molding, extrusion, monofilament, stretched tape, pipe, film and oriented film, lamination.

Mitsubishi Polypropylene. [Mitsubishi Kasei] Polypropylene resins; for inj. molding, extrusion, inflation film.

Mitsubishi Reagents. [Mitsubishi Kasei] Reagents for moisture metering.

Mixevan. [David Michael] Vanilla sugars.

Mixxim. [Fairmount] Hindered amine light stabilizer; for polyolefins, PS, ABS, PVC, polyurethane, engineering plastics, and elastomers.

ML-. [Hefti Ltd.] Sorbitan esters or ethoxylates; emulsifier, lubricant, solubilizer, stabilizer for cosmetics, pharmaceuticals, latex stabilization, polishes, hydraulic liqs., fiber preparations, color

pastes.

MO-. [Hefti Ltd.] Sorbitan esters or ethoxylates; emulsifier for foods, cosmetics, biocides, paints, anticorrosion oils, polishes, cutting oils.

Mo-1907 T 3/16″. [M&T Harshaw] Iron molybdena; catalyst for prod. of formaldehyde.

Mobil. [Mobil/Polystyrene] Crystal or impact polystyrene; for inj. molding and extrusion.

Mobil T-125, T-400. [Mitsui Petrochemicals] C9 aromatic solvent.

Mobil T-500-100. [Mitsui Petrochemicals] Xylene range aromatic solvent.

Mobil Wax. [Mitsui Petrochemicals] Ceresin.

Moby Dick. [Werner G. Smith] Sperm oil substitutes.

MOC-. [Allied-Signal] Oxime chelants.

Mocap. [Rhone-Poulenc/Ag] Nematicide, insecticide.

M.O.D. [Gattefosse; Gattefosse SA] Octyldodecyl myristate; superfatting agent, emollient for cosmetics and pharmaceuticals.

Modacure. [Monsanto] Curing agent.

Modaflow®. [Monsanto] Resin modifier for nonaq. coatings.

Modar. [ICI Acrylics] Modified acrylic resins; for resin transfer molding, filament winding, hand lay-up, spray-up, pultrusion.

Modarez. [Protex] Acrylic copolymer; leveling agent for coatings; PVC processing aid; stabilizer for emulsion polymerization.

Modarez®. [Synthron] Thermocoagulator for acrylic emulsions; dye antimigrant; antistatic lubricant finish for fibers.

Modic. [Mitsubishi] EVA copolymer, polyethylene, or polypropylene adhesive resins.

Modicol®. [Henkel/Functional Prods.] Ethoxylated ester; coagulant, wetting agent, dispersant, stabilizer, thickener, protective colloid for latex and resin emulsions.

Modulan®. [Amerchol; Amerchol Europe] Acetylated lanolin; conditioner, emollient, softener, lubricant for cosmetics and pharmaceuticals.

Mogul®. [Cabot] Fluffy carbon black; for inks, plastics, carbon paper, ribbons, dry toners.

Moldag. [Acheson Colloids] Colloidal molybdenum disulfide.

Moldbrite. [Disco] Extender, modifier for silicone emulsions used as mold release for rubber compds.

Mold Ease. [Chem-Pak] Mold release agents.

Mold-Ease PCR. [Merix] Nonsilicone; mold lubricant for plastics, ceramic, and rubber articles.

Moldkote. [Hastings Plastics] Mold release agents.

MoldPro. [Witco/Humko] Oleo fatty deriv.; internally functional processing agents to improve mold flow, mold release, and extrusion of elastomeric compds.

Mold Wiz. [Axel Plastics Research Labs] Mold release agents.

Mold Wiz. [Centrachem AG] Mold release agents.

Mollescal®. [BASF AG] Bactericide, leather assistant, soaking and liming agent; for leather and fur industries.

Molo-Jel. [Witco/Sonneborn] Petrolatum, mineral oil.

Molol. [Witco/Sonneborn] Tech. petrolatum.

Mololin. [Clark] Detergent, wetting/rewetting agent for scouring, enzyme desizing, soaping off processes.

Molora. [BASF] Organic red pigment.

Mol Siv. [UOP] Molecular sieve adsorbent.

Molub-Alloy. [Tribol] High performance lubricants.

Molydag®. [Acheson Colloids] Molybdenum disulfide in carriers; lubricant additives.

Molydee. [Castrol Industrial East] Tapping fluid.

Molyfilm. [Kano Labs] Dry lubricant.

Molyhibit. [Climax Performance] Molybdates; corrosion inhibitor.

Molykote®. [Dow Corning; Dow Corning France SA] Silicone, molybdenum disulfide lubricant base; environmental lubricants.

Molynamel. [Lockrey] Lubricant coating.

Molysulfide. [Climax Molybdenum] Molybdenum disulfide; lubricant for automobile, mining, aerospace, railroad, metalworking, and petroleum industries.

Molyvan. [R.T. Vanderbilt] Molybdenum compds.; antiwear and EP agent, friction reducer, antioxidant for greases, engine oils, metalworking compositions, industrial and automotive lubricating oils.

Moly-White®. [Sherwin-Williams] Molybdenum compds.; corrosion inhibiting pigment for coatings.

Mona AT-1200. [Mona Industries] Amphoteric; surfactant, thickener, corrosion inhibitor for acid bowl cleaners, metal cleaners, pickling baths, petroleum processing.

Mona NF. [Mona Industries] Low foaming anionic surfactants; for use in spray, soak tank, in-place pipeline cleaners, floor scrubbing formulations.

Monacet. [Argeville] DEA-cetyl phosphate, isopropyl myristate, cetyl alcohol, glyceryl stearate SE.

Monacor. [Mona Industries] Corrosion inhibitor for slushing oils, rolling oils, metalworking lubricants, cooling towers.

Monafax. [Mona Industries] Phosphate esters; EP lubricant, corrosion inhibitor for metalworking fluids, synthetic coolants; emulsifier, coupling agent, wetting agent, antistat, dispersant for household and industrial detergents, agric., textile, paper/pulp, emulsion polymerization.

Monagum. [Mukti-Kem] Carboxymethyl polysaccharide; anionic thickener for use in printing azoid, vat and disperse colors.

Monalube. [Mona Industries] Alkanolamide; corrosion inhibitor, lubricant for metalworking fluids, fiber lubricants, textile specialties.

Monalux CAO. [Mona Industries] Cocamidopropylamine oxide.

Monamate. [Mona Industries] Sulfosuccinates; high foaming surfactants for personal care and household prods.

Monamid®. [Mona Industries] 1:1 Super fatty acid alkanolamides; foam booster/stabilizer, visc. builder, solubilizer, coupler, wetting agent for personal care, household, and industrial cleaners.

Monamide. [Zohar Detergent Factory] Coconut MEA; foam booster, thickener, and superfatting agent.

Monamilk. [Argeville] DEA-cetyl phosphate, oleyl alcohol, isopropyl myristate, PEG-2 stearate.

Monamine. [Mona Industries] 2:1 Fatty acid alkanolamides; foam booster and stabilizer, emulsifier, detergent, wetting agent, corrosion inhibitor, visc. builder, lubricant, dispersant for cosmetics, household and industrial cleaners, textiles, agric. sprays, leather and fur preparations.

Monamulse. [Mona Industries] Imidazoline, alkanolamide or alkylaryl sulfonate; emulsifier, solubilizer, dispersant, degreaser.

Monaquat. [Mona Industries] Quaternary phospholipid compds.; thickener, antistat, lubricant, softener, conditioner, corrosion inhibitor for cosmetics, household, industrial, and textile applics.

Monarch. [Ciba-Geigy/Pigments] Blue pigment for paints and plastics.

Monarch. [H.B. Fuller] Cleaners, sanitizers, lubricants.

Monarch®. [Cabot] Fluffy carbon black; for plastics, coatings.

Monastat. [Mona Industries] Antistat/cleaner for glass and plastic.

Monastral. [Ciba-Geigy/Pigments] Quinacridone pigments.

Monaterge. [Mona Industries] Imidazoline or fatty amido complexes; detergents, wetting agents.

Monateric. [Mona Industries] Amphoteric surfactants; corrosion inhibitor, detergent, emulsifier, wetting agent for metal cleaning, cutting fluids, synthetic lubricants, cosmetics, industrial detergents, drilling, cement, gypsum.

Monatex. [Henkel Chem. Ltd.] PVA.

Monatrope. [Mona Industries] Phosphate ester; hydrotrope for detergent systems.

Monawet. [Mona Industries] Sulfosuccinates; wetting agent, dispersant, emulsifier, penetrant, solubilizer for emulsion polymerization, textile, fertilizer, cosmetic, paint, food industries.

Monazoline. [Mona Industries] Imidazolines; wetting agent, emulsifier, detergent, thickener, dispersant, corrosion inhibitor, antistat, softener, bactericide used in paint, agric., and textile industries.

Monceren®. [Bayer] Pencycuron; fungicide.

Mondur. [Miles] Aromatic polyisocyanates.

Monel. [Inco Alloys Int'l.] Nickel-copper alloys and welding prods.

Monestriol. [Union Derivan SA] Fatty esters.

Monex®. [Uniroyal] Thiuram; accelerator for tires, air-cured footwear, molded soling, closed-cell sponge, insulated wire.

Monflor. [ICI Am.] Fluorocarbon surfactants.

Monitor. [Miles/Ag] Multipurpose insecticide for vegetable and field crops.

Monitor. [Monsanto] Sorbic acid/potassium sorbate.

Monnex. [ICI Am.] Fire fighting agent.

Monobed. [Rohm & Haas] Ion exchange resins.

Monocast. [Polymer Corp.] Heat-stabilized nylon; used in paper and textile, construction, mining, metalworking, and material handling industries for bearings, valve seats, seals, gears, wheels, guides, tooling fixtures, insulators, wear parts.

Mono-Coat®. [Chem-Trend; Chem-Trend A/S] Solvent-based mold release system for use on laminates, composite molding, rubber molding, mold conditioning.

Mono-Ferro. [XymaX] Single-component MIOX loaded urethane.

Monofrax. [Carborundum] Fused cast refractories.

Mono-Gard. [XymaX] Single-component coal tar epoxy replacement.

Monogen. [Dai-ichi Kogyo Seiyaku] Sodium fatty alcohol sulfate; textile scouring and washing agent; dye dispersant and leveling agent.

Monoglyme. [Ferro/Grant] Ethylene glycol dimethyl ether; solvent used in electrochemistry, polymer and boron chemistry, physical processes such as gas absorp., extraction, stabilization, industrial prods. such as fuels, lubricants, textiles, pharmaceuticals, pesticides.

Monolan. [Henkel/Emery; Henkel Canada] EO-PO block copolymers; defoamer, deduster, wetting agent for industrial processing, industrial and household cleaners, paint strippers; dispersant for pigments and pesticides.

Monolan®. [Harcros UK] EO-PO block polymer; emulsifier, stabilizer, detergent, wetting agent, antifoam for industrial and domestic detergents, emulsion polymerization, latex, detergents, synthetic lubricants, metal cleaning, resin plasticizers, textile processing.

Monolec. [Lubrication Engineers] Engine and industrial oils.

Monolex. [Clark] Preparation and bleaching detergent and scour.

Monolex. [Lubrication Engineers] Penetrating lubricant.

Monolite. [ICI Am.] Pigments.

Monomer CE. [Scott Bader] Styrene inhibited for tropical storage.

Monomid. [Steelcote Mfg.] Epoxy coatings, water-reducible epoxies.

Monomuls®. [Henkel/Cospha; Henkel KGaA; Grünau] Glycerides; emulsifier, stabilizer, dispersant, opacifier for cosmetics, foods, and pharmaceuticals.

Monoplex®. [C.P. Hall] Adipates, sebacates, tallates; monomeric plasticizers, stabilizers, lubricants for PVC.

Monopole Oil. [Henkel/Organic Prods.] Sulfated castor oil; wetting agent, penetrant, leveling agent, softener, lubricant for dyeing applics.

Monopoxy. [Hardman] Epoxy.

Monoprime. [Synair] Urethane primer.

Mono-Prime. [XymaX] Single-component universal primer urethane.

Monorywe. [Monoplas Industries Ltd.] Polystyrene.

Monoset®. [Eastman] Distilled monoglycerides; food emulsifier.

Monosiliol C. [Exsymol] Methylsilanol tri PEG-8 glyceryl cocoate; tissue regeneration aid for anti-aging cosmetic formulations.

Monosorb. [FMC] Absorptive sodium phosphate monobasic.

Monosteol. [Gattefosse SA] Propylene glycol palmito/stearate; emulsifier.

Monosulf. [Henkel/Organic Prods.] Sulfated castor oil; penetrant, emulsifier, dyeing assistant, fatliquor, plasticizer for textiles, leathers, paper coatings, glues, latex.

Monothane. [Compounding Ingredients Ltd.] Polyurethane elastomers.

Monothane. [Synair] Castable urethane.

MonoThiurad®. [Harwick] Tetramethylthiuram disulfide; accelerator for NBR and CR.

Mono-Zinc. [XymaX] Single-component moisture urethane.

Monprop. [Monoplas Industries Ltd.] Polypropylene.

Monsteol. [Gattefosse] Propylene glycol stearate; stabilizer for ointments, cream lotions.

Montacell. [Seppic] Cationic; softener for paper.

Montaclere®. [Monsanto] Styrenated phenol; nonstaining antioxidant for NR, synthetic rubber, and latexes.

Montaline. [Seppic] Surfactant blends; wetting agent; auxiliary for prod. of phosphoric acid and phosphates.

Montane. [Seppic] Sorbitan esters; emulsifier.

Montanol. [Seppic] Ceteorylglucoside; used in cosmetic and dermo-pharmaceutical prods.

Montanox. [Seppic] Ethoxylated sorbitan esters; emulsifier.

Montapol. [Seppic] Oleyl/cetyl sulfate; detergent.

Montegal. [Seppic] Ethoxylates; leveling agent for textiles.

Monteine. [Seppic] Protein derivs.; raw material for shampoos and hair conditioners.

Montelane. [Seppic] Sulfates.

Monthybase. [Gattefosse] Glycol

monostearate SE.

Monthyle. [Gattefosse; Gattefosse SA] Glycol stearate; stabilizer for ointments, creams, and lotions.

Montopol CST. [Seppic] Oleo/cetyl sulfate; detergent.

Montosol. [Pulcra SA] Sulfates; emulsifier for emulsion polymerization; foamer, wetting agent used in personal care prods., dishwashes, fire-fighting foams; solubilizer for perfumes.

Montovol. [Pulcra SA] Sulfates; detergent, wetting agent, dispersant, emulsifier, foamer for personal care prods., rug shampoos, emulsion polymerization.

Montoxyl. [Seppic] Ethoxylated propoxylated fatty alcohol; low foaming wetting agent.

Monykup. [Malvern Minerals] Surface-modified antimony oxide.

Moplen. [Enimont Iberica SA] Polypropylene.

Morad. [Morton Int'l.] Laminating adhesives.

Mor-Bind. [Morton Int'l.] Pigment binders.

Morcid. [Morlot Color & Chem.] Textile dyes and pigments.

Morco. [Penreco] Sodium sulfonates.

Morcron. [Morlot Color & Chem.] Textile dyes and pigments.

Morcryl. [Morlot Color & Chem.] Textile dyes and pigments.

Morcurd Plus. [Pfizer/Brewery & Dairy Prods.] Microbial milk clotting enzyme.

Mordantine. [Apex] Liq. antimony lactate.

Mordirect. [Morlot Color & Chem.] Textile dyes and pigments.

Morestan®. [Bayer; Miles/Ag] Chinomethinat; organic fungicide with specific action against powdery mildews; for fruits and ornamentals.

Morfast. [Morton Int'l.] Textile dyes and pigments.

Morfax®. [R.T. Vanderbilt] 4-Morpholinyl-2-benzothiazole disulfide; accelerator for NR and synthetic rubbers.

Morflex. [Morflex] Glycolates, trimellitates; plasticizers.

Morflo. [Morton Int'l.] Latex coatings.

Morillol®. [BASF AG] 1-Octene-3-ol; mushroom-like, earthy, herbaceous fragrance and flavoring.

Morlex® DEEA. [Union Carbide] N,N-Diethyl ethanolamine; corrosion control agent in boiler water and steam systems.

Morlex® DMEA. [Union Carbide] N,N-Dimethyl ethanolamine; corrosion control agent in boiler water and steam systems.

Moropol. [Moretex Chem. Prods.] Softener, hand builder, lubricant for textiles.

Moropon. [Moretex Chem. Prods.] Detergent, dispersant, penetrant, wetting agent for soaping, scouring, bleaching operations.

Morowet. [Moretex Chem. Prods.] Sodium dioctylsulfosuccinate; wetting and rewetting agent for textiles.

Morplas. [Morton Int'l.] Dyes for plastics and textiles.

Mor-Rex®. [Grain Processing] Maltodextrin; drilling mud additive.

Morsperse. [Morlot Color & Chem.] Textile dyes and pigments.

Morstik. [Morton Int'l.] Pressure-sensitive adhesives.

Morta-Lok. [Rockwell Lime] Hydrated lime; for masonry applics.

Morthane. [Morton Int'l.] Polyurethane resins; thermoplastic for inj. molding, extrusion, coatings, adhesives applics.

Morwet®. [Witco/Organics] Naphthalene sulfonates; wetting agent, dispersant for pesticides, dyestuffs.

Morwhite. [Morlot Color & Chem.] Textile dyes and pigments.

Mos Selectipur. [EM Industries] Low particulate chemicals for semiconductor prod.

Moth Gard. [Colgate-Palmolive] Insecticides.

Moth-Gas. [Colgate-Palmolive] Insecticide.

Mothjinx. [Petrokem] Moth preventatives, insecticides.

Mots. [Am. Cyanamid] EPDM accelerator.

Mould Release Agent N 32. [Chemetall

GmbH] Solvent-free mold release agent on a plastic base for the rubber and plastics industry.

Moussex®. [Synthron] Silicone and nonsilicone defoamers for finishing, bleaching, sizing, coating, dyeing, and printing.

Mowilith. [Hoechst Celanese/Fine Chem.] PVAc polymers; for coatings, adhesives, building compds., paper applics.

Mowiol. [Hoechst Celanese/Fine Chem.] Polyvinyl alcohol resins; for paper, adhesives, coatings, textiles, cosmetics, polymerization, films.

Mowital. [Hoechst Celanese/Fine Chem.] Polyvinyl butyral resins; for coatings and inks.

Mozanon®. [Mitsubishi Kasei] Sodium alginate; fungicide for tobacco crops.

MP-. [Compounding Tech.] Polyphenylene oxide, reinforced.

MP-33-F. [Hefti Ltd.] Sorbitan palmitate; softener and lubricant for polymers, emulsions, cosmetics, ointments.

M-P-A®. [Rheox] Antisettling agents, rheological additives.

MPDiol. [Arco] Glycol; used in saturated and unsaturated polyester resin systems for molding compds. and coatings.

MPL Monomer. [Monomer-Polymer & Dajac Labs] Anaerobic monomer.

M-Pyrol®. [ISP] N-Methyl-2-pyrrolidone; solvent, stabilizer for resins, fibers; used in commercial petrochemical processing, for agric. chemicals, in removal of strippable coatings, for lube oil processing.

M-Quat®. [PPG/Specialty Chem.] Fatty quaternary ammonium surfactants; conditioner, softener, emollient, emulsifier for hair and skin care prods.; antistat for cleaners.

M.R. [Witco] Hard hydrocarbon extender and plasticizer resins for rubber compding.

MRC. [GE Silicones] Hard coating.

Mrs. Ewald's Original. [Aloe Labs] Aloe vera gel.

Mr. Zip. [Am. Grease Stick] Powdered graphite.

MS-. [Hefti Ltd.] Sorbitan stearate or ethoxylates; softener and lubricant for polymers, emulsions, cosmetics, ointments.

MS-122, -136. [Miller-Stephenson] Tetrafluorethylene telomer; release agent, lubricant for molds.

MS-170. [Miller-Stephenson] 1,1,1-Trichloroethane; solvent.

MS-230 Contact Re-Nu®. [Miller-Stephenson] Tape head and contact cleaners.

MS-266 En-Stat. [Miller-Stephenson] Static eliminator.

MS-350 En-Rust. [Miller-Stephenson] Rust inhibitor.

MS- Conformal Coatings. [Miller-Stephenson] Acrylic, silicone, or urethane; conformal coatings.

MS DS. [West Agro] Iodophor conc.

MS- Freon®. [Miller-Stephenson] Trichlorotrifluoroethane or blends; solvents.

M-Soft. [Zohar Detergent Factory] Cationics; softeners for fibers, yarns, fabrics in domestic and laundry use.

MSPS K. [Eastman] Distilled monoglycerides, lecithin, polysorbate 80; food emulsifier.

MTBE. [Arco] Methyl t-butyl ether; octane enhancer for gasoline.

Muck-Up. [Mateson] Oily waste adsorbent.

Mulcoa. [CE Minerals] Aluminum silicates.

Mulgofen®. [Rhone-Poulenc Surf.] Ethoxylated fatty alcohols or castor oil.

Mullfrax. [Carborundum] Oxide refractory.

Mulsifan. [Zschimmer & Schwarz] Ethoxylated ethers or esters; emulsifier for creams and lotions, paraffins, mineral oils, waxes; solubilizer for perfumes.

Multex. [Kaopolite] Hard aluminum silicate; large particle for heavy-duty abrasives; water filtration aid.

Multibase. [MBS Plastics] PP compds. or PP/PE/PS masterbatches.

Multi-Chem. [Colorite Plastics] Flexible alloy compds. for high elongation and low temp. properties.

Multicide. [McLaughlin Gormley King] Insecticide conc.

Multicrack. [Akzo] Fluid cracking catalyst.

Multicure®. [Dymax] Engineering adhesive.

Multifect. [Genencor Int'l.] Food enzymes.

Multifex. [Pfizer] Precipitated calcium carbonate.

Multiflow. [Monsanto] Resin modifier for nonaq. coatings.

Multikup. [Malvern Minerals] Surface-modified inorganic combinations.

Multilobe. [Rohm & Haas] Opaque polymer.

Multimet®. [Haynes Int'l.] Heat-resistant alloys.

Multimix. [BASF] Organic pigments.

Multinol. [Nippon Senka] Sulfonic acid deriv.; scouring, bleaching and dyeing agent.

Multionic. [Western Water Management] Cooling water treatment.

Multiplastic. [Multiplastic SpA] Polypropylene.

Multisperse. [Western Water Management] Boiler sludge dispersant.

Multisperse CP. [BASF] Silicate; dispersant for prevention of silicate scale buildup on continuous bleaching equipment.

Multiwax®. [Witco/Sonneborn] Microcrystalline wax; plasticizer, modifier for polymeric coatings and adhesives, laminations, cleansing creams, hair pomades, pharmaceuticals, crayons.

Multiwet. [Wetronics] Liq. multiphase wetting agent.

Multizyme II. [Enzyme Development] Enzymes for animal feed and silage.

Multon. [Ubbink Nederland BV] Polyurethane/polyethylene.

Multranol. [Miles] Polyether polyols.

Multrathane. [Miles] Prepolymers and polyesters.

Murdopol Muratherm. [Industria Engineering Prods. Ltd.] UHMW PTFE.

Mureinase. [U.S. Biochemical] Multicomponent enzyme preparation for prod. of protoplasts in high yields.

Murtfeldt S. [Industria Engineering Prods. Ltd.] UHMWPE.

Musol. [Brooks Industries] Soluble yeast mucins; for skin and hair care cosmetics.

M Violet 112. [Presperse] Manganese violet, bismuthoxy chloride.

MWB Film. [Hercules] Polypropylene film; oriented film for multiwall bags.

MXP. [Malco Prods.] Oil additive.

Myacide®. [Angus] Antimicrobial for water treatment (recirculating cooling towers, process water), oilfield waterflooding operations, paper mill process water, deodorizing applics.

Myacide SP. [Inolex] 2,4-Dichlorobenzyl alcohol; antifungal agent, preservative.

Myanit. [Norwegian Talc UK Ltd.] Fillers for plastics.

Mycalex. [Mykroy/Mycalex Ceramics] Glass-bonded mica.

Mycolase®. [Int'l. Bio-Synthetics] Alpha-amylase; enzyme which converts low DE acid or modified starch hydrolysates to high maltose syrups.

MYE. [Int'l. Bio-Synthetics] Yeast extract.

Mykon®. [Sequa] Specialty softeners and lubricants for textile finishing.

Mykroy. [Mykroy/Mycalex Ceramics] Glass-bonded mica.

Mylar®. [DuPont; DuPont UK] Polyester film; transparent, strong film for pkg. foods, adhesive and extrusion laminations, vacuum metallizing process, as core wrap and insulation for telecommunications and power cable applics.

Myritol® 318. [Henkel/Emery/Cospha; Henkel KGaA] Caprylic/capric triglyceride; emollient for pharmaceuticals, cosmetics.

Myrj®. [ICI Spec. Chem.] Ethoxylated esters; surfactants, emulsifiers for cosmetics, pharmaceuticals.

Myrlene. [Gattefosse] PEG-8 myristate.

Myrol-S. [Lanaetex Prods.] Isopropyl myristate, diisopropyl sebacate.

MY Silicone. [Yoshimura Oil Chem.] Silicone-based surfactants; softener, finishing oil for textiles.

Mystic White. [U.S. Silica] Quartz aggregates.

Mytab®. [Zeeland] Myrtrimonium bromide; surfactant, emulsifier, antistat, antimicrobial for hair rinses, cosmetics, topicals.

Myvacet®. [Eastman] Acetylated monoglycerides; emulsifier, emollient, lubricant, film-former.

Myvaplex®. [Eastman] Glyceryl stearates; emulsifier for food prods.

Myvatem®. [Eastman] Diacetyl tartaric acid ester of monoglycerides; emulsifier, dispersant for food, cosmetics, pharmaceuticals.

Myvatex®. [Eastman] Blends; food emulsifier.

Myverol®. [Eastman] Glyceryl monoesters; emulsifier for foods, cosmetics.

MZO-25. [Ikeda] Zinc oxide.

N

N-30. [R.H. Carlson/Northern Labs] Suspension of fluorocarbon telomer in a nonflamm., low toxicity halocarbon; mold release agent and lubricant for casting resins and rubbers.

N-521® Biocide. [Akzo] Tetrahydro-3,5-dimethyl-2H-1,3,5-thiadiazine-2-thione; biocide for leather, paint, glue casein, starch, paper industries.

N-948® Biocide. [Akzo] Methylene bis(thiocyanate); industrial biocide.

N-1386. [Akzo] Biocides.

Na-0101 T 1/8˝. [M&T Harshaw] Sodium bifluoride; catalyst used for removal of HF.

Nabutene. [Convert SA] ABS.

Na-Butyl Monoglycol Sulfate 50. [Hüls AG] Sodium butyl monoglycol sulfate; hydrotrope for mfg. of neutral and alkaline liq. cleaners with high electrolyte content.

Nacap®. [R.T. Vanderbilt] Sodium 2-mercaptobenzothiazole; metal deactivator, corrosion inhibitor, chemical intermediate.

Nacar. [North Am. Carbon] Activated carbon prods.

Nacco. [Buffalo Color] Indigo dye.

Nacco. [Nat'l. Research & Chem.] Photographic chemicals.

Nacconol®. [Stepan; Stepan Europe] Sodium dodecylbenzene sulfonate; foamer, dispersant, wetting agent, detergent for agric., cement, dyeing, emulsion polymerization, textile, metal cleaning, metalworking, mining, paper industries.

Nacol®. [Condea Chemie GmbH] Fatty alcohol single fractions; intermediates for toiletries, cosmetics, detergents, leather and textile auxiliaries, lube oil, plastics additives, defoamers.

Nacolox®. [Condea Chemie GmbH] Fatty alcohol ethoxylates; intermediates for toiletries, cosmetics, detergents, leather and textile auxiliaries, lube oil, plastics additives, defoamers.

Nacopa®. [PMC Specialties] Monosodium-4-chlorophthalate; modifier for phthalocyanine pigments to retard crystallization; resin modifier.

Nacorr. [King Industries] Rust inhibitors for ferrous metals.

Na Cumene Sulfonate. [Hüls AG] Sodium cumene sulfonate; hydotrope, solubilizer, coupling agent for detergents and slurries.

Nacure. [King Industries] Catalysts for polymeric coatings.

Nadex®. [Nat'l. Starch & Chem.] Dextrin; textile finishing.

Nadic. [Buffalo Color] Methyl anhydride.

Nadone®. [Allied-Signal] Cyclohexanone; solvent for coating resins, printing inks, adhesives, agric. formulations.

Naetex. [Lanaetex Prods.] Cosmetics ingredients.

Nafion. [DuPont] Persulfonic membrane.

Naflon. [DuPont] Perfluorinated membranes.

Nafol®. [Condea Chemie GmbH] Fatty alcohol blends; intermediates for toiletries, cosmetics, detergents, leather and textile auxiliaries, lube oil, plastics additives, defoamers.

Naftocit®. [Ore & Chem. Corp.] Accelerator, vulcanizing agent for rubber and latex processing.

Naftolen®. [Chemetall GmbH] Plasticizers and extender oils; for natural and synthetic rubbers.

Naftolube. [Chemson GmbH] Lubricants for PVC.

Naftomix. [Chemson GmbH] PVC stabilizers and additives.

Naftonox®. [Ore & Chem. Corp.] Antioxidants for rubbers, latexes, adhesives, plastics, fuels and oils, foodstuffs and animal feeds.

Naftopast®. [Chemetall GmbH] Processing aid for rubber compounding.

Naftovin. [Ore & Chem. Corp.; Chemson GmbH] PVC stabilizers.

Naftozin® N, Spezial. [Chemetall GmbH] Stearic acid; processing aid for rubber compounding, lubricant for PVC processing.

Nagelite. [Telechemische] Toluenesulfonamide/epoxy resin or blend.

Nailsyn. [Rona] Synthetic pearl pigments for nail polish.

Nalan®. [DuPont] Polymer blends; water repellent, extender for fabrics.

Nalcast. [Nalco] Refractory materials.

Nalclean. [Nalco] Industrial cleaners.

Nalco®. [Nalco] Absorbent for diapers, agric. applics., dehydrating agent, construction materials, fire extinguishing, waste sludge stabilization; defoamer for coatings, adhesives, food processing; dispersant.

Nalcoag. [Nalco] Colloidal silicas.

Nalcolyte. [Nalco] Coagulant.

Nalcool. [Nalco] Metalworking lubricants.

Nalcote. [Nalco] Refractory materials.

Nalcotrol. [Nalco] Drift control agents.

Nalidone®. [UCIB] Sodium PCA; moisturizer for dermatological soaps, shampoos, after-sun lotions, shower gels, moisturizing creams and lotions, hair comb-out balm.

Nalkylene®. [Vista] Alkylbenzene sulfonates; surfactant intermediates.

Nalprep. [Nalco] Corrosion inhibitors.

Nalquat. [Nalco] Polyquaternaries.

Nalram. [Nalco] Refractory ramming mix.

Naltene. [Convert SA] LDPE.

Nalzin. [Rheox] Zinc compds.; anticorrosive pigment.

Namate®. [R.T. Vanderbilt] Sodium dithiocarbamates; accelerators for latexes.

Nanofuture. [Induchem AG] Caprylic/capric triglyceride blends.

Nanospheres 100 Blends. [Exsymol] Porous polymeric particles blended with various ingredients; provides con-

trolled release of actives for cosmetic and health prods.

Nansa®. [Albright & Wilson Am.; Albright & Wilson UK] Surfactant intermediates, foamers, emulsifier, dispersant, wetting agent for domestic and industrial detergents, agric., leather, paper and pulp, textile, metal, coatings, polymerization industries.

Nap-All. [Mooney Chems] Naphthenates; driers for paints, inks.

Napclor. [Chapman] Sapstain and mold control.

Napcocrest. [Reilly-Whiteman] Napping assistants for textile industry.

Naphid. [Hewchem] Crude naphthenic acid.

Naphthanil. [Cookson Pigments] Textile dyes and pigments.

Naphthanil. [DuPont] Diazo pigments.

Naphthanilide. [Yorkshire Pat-Chem] Textile dyes and pigments.

Naphtholite. [Unocal] Solvent.

Naplube. [Dooley] Reactive silicone softeners for cotton and blends.

Napocrest. [Reilly-Whiteman] Napping assistant for synthetic fibers.

Nappar. [Deutsche Exxon Chem. GmbH] Solvent.

Napsoft. [CNC Int'l.] Textile napping softener.

Naptal. [Napex Ltd.] Industrial acetal.

Naptex. [Degen] Lithographic vehicle.

Narcarb. [North Am. Refractories] Resin bonded alumina silicone carbide plastics and ram mixes.

Nargosol. [Nutex] Synthetic weighter for knits and woven fabrics.

Narlex. [Hart Chem. Ltd.] Dispersant, superplasticizer for pigments, clays, particulates, fillers.

Narphos. [North Am. Refractories] Phosphate bonded alumina plastics and ram mixes.

NAS®. [Novacor] Methyl methacrylate styrene copolymers; for medical applics. , meter covers, brush blocks, drafting instruments, housewares, displays, outdoor, clarity, and automotive applics.

Na-Sul. [King Industries] Petroleum corrosion inhibitor prods.

Nasuna®. [Henkel Canada; Henkel KGaA] Vinyl pyrrolidone/vinyl acetate copolymer; fixing agent, basic material for hair setting lotions, sprays.

Natac®. [Whitney & Oettler] Resin acids-amine resin soaps blend; tackifier for rubber; molding aid; inhibits bloom in NR and NR/SBR blends.

National 78-1898. [Nat'l. Starch & Chem.] Distarch phosphate.

National Brand. [Nat'l. Ammonia] Anhydrous ammonia.

National Starch 28-4979. [Nat'l. Starch & Chem.] Acrylates/octylacrylamide copolymer; hair spray polymer.

Natipide. [Am. Lecithin; Nattermann Phospholipid] Lecithin blends.

Na Toluene Sulfonate. [Hüls AG] Sodium toluene sulfonate; hydotrope, solubilizer, coupling agent for detergents and slurries.

Natrex. [Henkel-Nopco] Detergent, foamer, wetting agent, emulsifier for textiles.

Natro-Cel. [Natrochem] Dispersed rubber chemicals.

Natro-Cure. [Natrochem] Poly allyl monomers.

Natro Flex. [Natrochem] Plasticizers.

Natro-Flex IOT. [Harwick] Isooctyl tallate; plasticizer.

Natro-Rez. [Natrochem] Coumarone-indene resins.

Natrosol®. [Aqualon] Hydroxyethyl cellulose or derivs.; thickener, protective colloid, suspending aid, binder for cosmetics, paints, emulsion polymerization, cements, paper.

Natro Tar. [Natrochem] Pure pine tar.

Natrovis®. [Aqualon] Hydroxypropyl hydroxyethylcellulose; water-sol. polymer, additive to hydraulic binder and latex-modified building materials.

Natsyn®. [Goodyear] Polyisoprene; elastomer for adhesives, footwear, sponge prods., tires, molded and mech. goods.

Naturalgel. [Wyo-Ben] Wyoming bentonite.

Natural Sterols. [Universal Labs] Sterols.

Naturechem®. [CasChem] Esters; emollient, emulsifier, opacifier, thickener, dispersant, stabilizer for cosmetics, household prods.

Nature-Sol. [Brulin] Environmentally safe cleaners.

Naturol. [Lanaetex Prods.] Mink oil.

Naturon. [Rona] Guanine blends; pearlescent in nail polish, lotions.

Naugalube®. [Uniroyal] Alkylated secondary aromatic amine; antioxidant for automatic transmission fluids, turbine oils, synthetic jet turbine lubricants; petroleum additives.

Naugard®. [Uniroyal; Uniroyal Chem. Ltd.] Antioxidant, metal deactivator, processing aid, polymerization inhibitor, stabilizer for adhesives, thermoplastic and thermoset polymers; styrene inhibitors.

Naugawhite. [Uniroyal] Phenolic/bisphenolic; nondiscoloring antioxidant for thermoplastics, hot-melt adhesives, molded and mechanical goods.

Naugex®. [Uniroyal] Accelerators.

Nautolex. [Gencorp Polymer Prods. UK] Marine vinyls.

Navinon. [Sunbelt] Textile dyes and pigments.

Naxchem. [Ruetgers-Nease] Surfactant blends; detergent, foam booster/stabilizer, dispersant, emulsifier for shampoos, carpet shampoos, bubble baths, dishwashing and laundry detergents, textile processing, pigment dispersions, metal processing.

Naxel. [Ruetgers-Nease] Dodecylbenzene sulfonic acid or salts; detergent intermediate, foaming agent, wetting agent, emulsifier for household and industrial detergents, textile scours, carwashes, drycleaning systems, cosmetics.

Naxide 1230. [Ruetgers-Nease] Lauryl dimethylamine oxide; detergent, foam booster, stabilizer for cosmetics, hand cleaners, light duty detergents, textiles, lubricants, paper coatings.

Naxoco. [Convert SA] Acrylic or ABS.

Naxol. [Convert SA] Resins for paints.

Naxol®. [Allied-Signal] Cyclohexanol; solvent for fats, oils, resins; coupling agent.

Naxolate. [Ruetgers-Nease] Sodium lauryl sulfates; detergent, wetting agent, foamer, emulsifier, dispersant for cosmetic and household prods., rug shampoos, toothpastes, acid soap processing, pigment dispersion, compounded detergent formulations, textile wool scouring.

Naxolene. [Convert SA] Polystyrene.

Naxonac. [Ruetgers-Nease] Phosphate esters; emulsifier, lubricant, hydrotrope, wetting agent, stabilizer for heavy duty and household detergents, emulsion polymerization, solvent degreasers, paint and wax strippers, electrolytic cleaners.

Naxonate®. [Ruetgers-Nease] Sulfonates; hydrotrope, stabilizer, solubilizer for detergents, inks, electroplating baths, dyestuffs, polymers.

Naxonic. [Ruetgers-Nease] Nonoxynol series; detergent, emulsifier, stabilizer, solubilizer, penetrant, wetting agent, dispersant for textiles, insecticides, cosmetics, industrial and household detergents, latex, polymers, waxes, polishes, bottlewashing compds.; demulsifier for crude petroleum.

Naxonol. [Ruetgers-Nease] Coconut acid diethanolamide; foaming and thickening agent, emulsifier, detergent for shampoos and detergents.

NB. [Angus] 2-Nitro-1-butanol; chemical and pharmaceutical intermediate, in tire cord adhesives, as formaldehyde release agents, deodorants, antimicrobials.

NBC. [DuPont] Rubber chemical.

NC. [Ennar Latex] NR latex; for extruded thread, carpet backing, latex foam, dipped goods, coatings, medical applications, textiles, adhesives.

NC-4. [Feldspar] 200 mesh soda feldspar.

NC-300. [Norton Chem. Process Prods.] No$_x$ removal catalyst.

NCC. [North Coast Compounders] Polyurethane alloys; for profile and sheet extrusion, inj. molding, calendering, wire/cable, etc.

ND-. [Premier Malt Prods.] Malted barley extract or blends; syrup for food applics.

NE. [Angus] Nitroethane; intermediate, stabilizer for halogenated solvs., as fuels, explosives, and solvs. for coatings or industrial processes.

Neacid. [Degussa] Pickling agent for cleaning of dental alloys.

Near-A-Lard. [G. Whitfield Richards] Cutting and drawing lubricant.

Neatsan D. [Reilly-Whiteman] Sulfated oil; emulsifiable oil for processing leather.

Nebony®. [Neville] Aromatic petroleum hydrocarbon resin; used in wire/cable coatings, rubber cements, mech. and molded goods, tires, caulking compds.

Neccobond. [Northern Prods.] Aq. adhesives.

Neccoplex. [Northern Prods.] Chromium complex.

Necomar/Necolin. [Ashland-Südchemie-Kernfest GmbH] Copolymer resins.

Necon. [Alzo] Esters; cosmetic ingredients.

Necowel. [Ashland-Südchemie-Kernfest GmbH] Water-soluble resins.

Neelacid. [Spectra Dyestuffs] Textile dyes and pigments.

Neelactive. [Spectra Dyestuffs] Textile dyes and pigments.

Neelasolve. [Spectra Dyestuffs] Textile dyes and pigments.

Neelasperse. [Spectra Dyestuffs] Textile dyes and pigments.

Neeldirect. [Spectra Dyestuffs] Textile dyes and pigments.

Nekal®. [Rhone-Poulenc Surf.] Sulfonates or sulfosuccinates; emulsifier, wetting agent, dispersant, stabilizer, foamer for textiles, paints, pesticides, dyes, latex emulsions, emulsion polymerization, industrial use.

Nekanil®. [BASF AG] Alkyl phenol ethoxylate; detergent, wetting agent, dispersant for textiles.

Nemacur®. [Bayer; Miles/Ag] Fenamiphos; systemic nematicide for cotton, soybeans, peanuts, turf.

Neobee®. [Stepan/PVO; Stepan Europe] Naturally derived oils; diluents, vehicles, cosolvents, emollients, solubilizers, spreading agents, penetrants, clouding agents for food, cosmetics, pharmaceuticals.

Neobon. [Atlas Minerals & Chem.] Neoprene coating.

Neobor. [Borax Consolidated Ltd.] Sodium borate.

Neocation. [Nikko Chem. Co. Ltd.] Quaternary ammonium salt; surfactant, leveling agent for cationic dyes.

Neo-Cem 250. [Mooney Chems] Drier for high solids resin paints.

Neochel. [Atochem N. Am.] Copper and bronze plating additive.

NeoCryl. [ICI Resins] Acrylic resins and emulsions; used in construction compounds, carpet shampoos, coatings, polishes, consumer equipment, inks, as vehicle for wood, plastics, concrete, metal, and wall board.

Neocrystal. [Nicca USA] Chelating and dispersing agent for desizing, scouring, bleaching operations.

Neodene®. [Shell] Alpha olefins; intermediate for surfactants and specialty industrial chemicals.

Neodol®. [Shell] Alcohols and ethoxylates; detergent intermediate, emulsifier, wetting agent, dispersant, solubilizer.

Neodye. [Reilly-Whiteman] One-bath scour/dye/finish chemicals for hosiery.

Neo-Fat. [Akzo] Fatty acids.

Neofix. [Nicca USA] Dye fixative.

Neoflex®. [Shell] Alcohols; for plasticizers.

Neol. [BASF] Neopentyl glycol.

Neolan. [Ciba-Geigy/Dyestuffs] Textile dyes and pigments.

Neolisal. [Seppic] Ethoxylated fatty amine; stripping agent.

Neomodelon 100. [Toho Chem. Industry] Cationic complex; neutral sizing agent for paper.

Neonite®. [Ciba-Geigy GmbH] Epoxy molding compd., long glass fiber-reinforced.

Neopac. [ICI Resins] Water-borne urethane acrylic copolymers.

Neopelex. [Kao] Alkylbenzene sulfonic acid or salts; detergent base.

Neopen®. [BASF AG] Metal complex dyes; for inks, ball pen pastes, copying toners.

Neopolen®. [BASF AG] Polyethylene or polypropylene foams; for cushioning and resilience applics. in pkg., building trade, maritime, automotive, sport, recreation, leisure industries.

Neopon. [Witco SA] Lauryl sulfate salts; detergent, base for cosmetics.

Neoprene. [DuPont; DuPont UK] Polychloroprene; versatile synthetic rubber for wire/cable; industrial hose and belting; automotive gaskets, seals, hose; tire sidewalls; molded and extruded goods; cellular prods.; adhesives, sealants, and protective coatings; for shoe, home crafts, sporting goods.

Neoprotex. [Nikko Chem. Co. Ltd.] Methylol amide surfactant; softener, water repellent for resin treatment.

Neorad. [ICI Resins] Radiation-curable resins.

Neorate. [Nikko Chem. Co. Ltd.] Alkyl phosphate; penetrant for polyester fibers.

Neorate®. [Nicca USA] Peroxide bleach stabilizer.

Neo Rez. [ICI Resins] Polyurethane resins and emulsions.

Neoscoa. [Toho Chem. Industry] Detergent, scouring agent, wetting agent, deinking agent for textiles, metal cleaning, paper processing.

Neosol. [Shell] Solvent alcohols.

Neosolve. [M.S. Paisner] Anionic and nonionic wetting agents.

Neosolvent NSG. [Toho Chem. Industry] Polypropylene polyethylene glycol monomethyl ether; solvents for brake fluid.

Neosorb. [Roquette] Sorbitol.

Neospinol. [Toho Chem. Industry] Blends; dispersant for sulfur in rayon processing.

Neosyl. [Crosfield] Silicon dioxide.

Neotac. [ICI Resins] Water-borne adhesive polymers.

Neotan. [Fabriquimica] Salicylates.

Neotex. [Columbian Chem.] Carbon black.

Neothene. [Geoliquids] Solvent for geoscience work.

Neoweld. [Imperial Adhesives] Neoprene adhesive latex.

Neozapon®. [BASF; BASF AG] Metal complex dyes; for prod. of flexographic and gravure inks.

NEPD. [Angus] 2-Nitro-2-ethyl-1,3-propanediol; chemical and pharmaceutical intermediate, in tire cord adhesives, as formaldehyde release agents, deodorants, antimicrobials.

Nepol®. [Neste Composite Materials] Polypropylene, glass-reinforced; for inj. molded parts and prods. requiring high stiffness and good impact resistance.

Nepoxide. [Atlas Minerals & Chem.] Epoxy coating.

Nepton® EXT. [Yorkshire Pat-Chem] Durable water repellent and fluorocarbon extender.

Neptun®. [BASF; BASF AG] Dye bases; for prod. of ball pen pastes, typewriter ribbon inks.

Neral. [BASF AG] Z-3,7-Dimethyl-2,6-octadiene-1-al; lemon-like, green fragrance and flavoring.

Nercolan F. [Rhone-Poulenc Ltd.] Phenolic blend of fungicides and surfactants; broad spectrum fungicide.

Nerolidol. [BASF AG] Z,E-3,7,11-Trimethyl-1,6,10-dodecatriene-3-ol; sweetly floral, green, woody fragrance and flavor.

Nervan CP. [Rhone-Poulenc Ltd.] Disodium methylene dinaphthalene disulfonate; dispersant.

Nervflex. [Tramico SA] High resilient polyurethane foam.

Nesatol. [Vevy] C10-18 triglycerides.

Neto. [A.E. Staley Mfg.] High maltose corn syrup.

Neu Mag. [Zinkan Enterprises] Magnesium hydroxide.

Neuphor®. [Hercules] Dispersed rosin size; sizing agent for paper/paperboard mfg.

Neustrene®. [Witco/Humko] Hydrogenated oils or glycerides; textile lubricant, pharmaceutical intermediate, emulsifier, mold release agent, buffing compd.

Neutrafilm. [Betz Industrial] Condensate corrosion inhibitor.

Neutraflux. [Transene] Self-neutralizing organic flux.

Neutral Degras. [Fanning] Fatty esters of wool grease; lubricant, plasticizer for wire drawing compds., slushing and cutting oils, rust preventative, adhesives; textiles; inhibits crystallization of wax components.

Neutrameen. [Betz Industrial] Neutralizing amine.

Neutrase®. [Novo Nordisk] Protease; enzyme for breakdown of proteinaceous matter into peptides and amino acids; used in brewing industry.

Neutrex. [A-Aroma Tech] Odor counteractants.

Neu-Tri. [Dow] Solvent.

Neutrigan®. [BASF AG] Alkaline salts and salt mixtures; basifying, deacidifying, and masking agents for chrome leather, furs.

Neutroda. [Stewart Hall] Fuel oil and petroleum deodorant.

Neutrol 9. [Reilly-Whiteman] Dyeing assistant for nylon.

Neutrol® TE. [BASF; BASF AG] Tetrahydroxypropyl ethylenediamine; neutralizing agent for cosmetics.

Neutronyx® 656. [Stepan] Nonoxynol-11; detergent, dispersant, wetting agent.

Nevastain®. [Neville] Phenolic compds.; nonstaining antioxidants used in mastic adhesives, rubber goods such as mech. and molded goods, caulking compds., cement, and antiskinning agents.

Nevastane. [Atochem N. Am.] Lubricating grease.

Nevchem®. [Neville] Alkylated petroleum hydrocarbon resins; used in adhesives, coatings, inks, rubber, concrete-curing compds., caulking compds.

Never-Seez. [Bostik Div./Emhart] Antiseizing and lubricating compds.

Nevex®. [Neville] Modified hydrocarbon resin; used in adhesives, coatings, rubber cements, mechanical and molded goods, tires.

Nevillac®. [Neville] Hydroxy modified resin; used in adhesives, epoxy coatings, rubber cements, antiskinning agents.

Nevoxy®. [Neville] Hydroxy modified resin; used in adhesives, epoxy coatings, rubber cements, antiskinning agents.

Nevpene®. [Neville] Modified hydrocarbon resin; tackifier; used in adhesives, coatings, rubber, and caulking compds.

Nevroz®. [Neville] Modified hydrocarbon resin; replacement resins for rosin derivatives, and in inks.

Nevtac®. [Neville] Synthetic polyterpene resins; tackifier for adhesives, coatings, rubber prods., concrete-curing compds., and caulking compds.

Newark. [Newark Chem.] Hot-melt adhesive.

Newbon®. [Nicca USA] Leveling agent for dyeings.

Newcol. [Nippon Nyukazai] Ethoxylated ethers or esters; emulsifier, stabilizer, dispersant, antistat, corrosion inhibitor, wetting agent, lubricant, detergent, penetrant for emulsion polymerization, dyes, detergents, agric. chemicals, machine oils, textiles.

Newdamp Balancing Fluid. [Bacon] Blends of halogenated alkylaryl hydrocarbons.

New Econa 200 CH. [Kao/Edible Fat & Oil] Hydrogenated tallow glycerides, hydrogenated coconut oil.

Newkalgen. [Takemoto Oil & Fat] Blends; emulsifier for agric. formulations.

Newlon. [Takemoto Oil & Fat] Polyglycol ether; raw wool scouring agent.

Newpol. [Sanyo Chem. Industries] EO/PO block copolymer; base material for household and industrial detergents; plasticizer, antistat for phenol resins; emulsifier for agric. pesticides and emulsion polymerization; pigment and pitch dispersant.

Ney 380. [Ney Prods.] Fluxless aluminum solder.

Neylo. [Ney Prods.] Low temp. white metal alloys.

NF 600. [Davidson Metals] Unsupported free film coated with a permanent rubber base adhesive and self-wound on a release liner; for lamination applics. requiring a permanent pressure-sensitive adhesive.

NFB. [Monsanto] Phosphoric acids for aluminum brightening.

NH-. [Compounding Tech.] Nylon 11, glass-reinforced.

N-Hance. [Aqualon] Guar hydroxypropyltrimonium chloride; substantive polymer for hair and skin care prods.

Ni-. [M&T Harshaw] Nickel compds.; catalysts for slurries, hydrogenation, hydrogenolysis, and amination reactions.

NI-. [Compounding Tech.] Nylon 6/10, glass-reinforced.

Niacet®. [Niacet] Calcium acetate, potassium acetate, sodium acetate, sodium diacetate, calcium propionate, or sodium propionate; chemicals for industrial, bakery, cheese, feed and grain applics.

Niaproof®. [Niacet] Sulfates; detergent, wetting agent, penetrant, emulsifier used in metal cleaning, electroplating, photo chemicals, adhesives, emulsion polymerization, adhesives, metal finishing, leather treating.

Niax® Catalyst C-224. [Union Carbide] Catalyst for molded rigid foams incorporating both fluorocarbon and water blown formulations utilizing polymeric MDI isocyanates;

Niax® PWB-1200. [Union Carbide] Polyether polyol; used to mfg. waterborne polyurethane coatings and adhesives.

Niblend. [Technopol GmbH] Technical plastic blends.

Nicca Fi-None®. [Nicca USA] Flame retardant for textiles.

Nicca Silicone®. [Nicca USA] Silicone softener for textiles.

Nicca Sunsolt. [Nicca USA] Dispersant, leveling agent for dyeing textiles.

Niccatex. [Nicca USA] Leveling agent, antimigrant, color enhancer for dyeing operations.

Nicepole®. [Nicca USA] Antistat for polyester textured fabrics.

Nickel-Glo. [JacksonLea] Barrel nickel plating brightener.

Nickel-Lume. [Atochem N. Am.] Nickel plating.

Niclate. [R.T. Vanderbilt] Rubber antioxidant.

Nicron. [Montana Talc] Platy talc; low oil absorption filler for paints, plastisols, and coatings.

Nidaba 3. [Vevy] Lauramide MIPA.

Nidaba 318. [Vevy] DEA-hydrolyzed lecithin.

Nihon Polyglyceryl-10 Distearate. [Nihon Surfactant] Polyglyceryl-10 distearate.

Nikkol. [Nikko Chem. Co. Ltd.] Surfactants for cosmetics, pharmaceuticals, industrial applics.

Nikkol Aquasome. [Nikko Chem. Co. Ltd.] Sodium lauroyl methylamino propionate; shampoo and facial cleanser base, foaming agent, detergent.

Nikkol Estepearl. [Nikko Chem. Co. Ltd.] Glycol distearates; pearlescent for shampoos, rinses, hair conditioners.

Nikkol Sarcosinate OH. [Nikko Chem. Co. Ltd.] Oleoyl sarcosine; corrosion inhibitor.

Niklad. [Witco] Industrial chemicals; for sensitizing, catalysts, accelerators, etchants, nickel compds., photoresists.

Nilac. [Technopol GmbH] ABS.

Nilfom. [Am. Emulsions] Nonsilicone defoamers for textiles.

Nilo VON. [Sandoz Prods. Ltd.] EO sulfonate; emulsifier for mineral oils.

Nimate. [Albright & Wilson Am.] Nickel sulfamate; for nickel electroforming and plating for aerospace and electronics applics.

Nimco®. [Henkel/Emery] Lanolin derivs.; emulsifier, stabilizer, emollient for o/w systems, creams, lipsticks.

Nimcolan®. [Henkel/Emery] Absorption base; emollient, emulsifier.

Nimlesterol®. [Henkel/Emery] Blends; hypoallergenic emulsifier, emollient for skin care prods.

Nimonic. [Inco Alloys Int'l.] Nickel-chromium alloys.

Ninate®. [Stepan; Stepan Europe] Alkylbenzene sulfonates; detergent, emulsifier for pesticides and other formulations.

Ninol®. [Stepan; Stepan Canada; Stepan Europe] Fatty acid alkanolamide; detergent for floor cleaners, wax strippers, alkali cleaners, sanitizers; emulsifier for industrial lubricants; thickener for personal care prods.; foam booster/stabilizer.

Ninox. [Stepan Europe] Fatty amine oxides; thickener, foam booster/stabilizer, detergency enhancer, antistat for scale-removing liqs., cleaning foams.

Nin-Sol. [Pierce Chem.] Ninhydrin sol'ns.

Niosan. [Industrias Nioco SA] Polyamide 11.

Niosi. [Industrias Nioco SA] Polyamide 6.

Niox. [Pulcra SA] Ethoxylated ethers or esters; dispersant, emulsifier, spreading agent, defoamer, lubricant, solubilizer for cosmetics, detergents, resins, textiles, leather, pesticides, agric., metals; intermediate.

Nipabenzyl. [Nipa Labs] Benzylparaben; color developing agent for heat-sensitive recording papers.

Nipabutyl. [Nipa Labs] Butylparaben or salts; preservatives for foodstuffs, pharmaceuticals, cosmetics.

Nipacide® BCP. [Nipa Labs] o-Benzyl-p-chlorophenol; germicide in disinfectant and cleaner prods. for hospitals, schools, homes, public facilities.

Nipacide® BK. [Nipa Labs] Hexahydro-1,3,5-tris (2-hydroxyethyl)-s-triazine; preservative for cutting fluids and coolants.

Nipacide® MX. [Nipa Labs] p-Chloro-m-xylenol; for disinfectant, algicide, slimicide, water treatment, and pesticide formulations.

Nipacide® OPP. [Nipa Labs] o-Phenyl phenol; preservative, disinfectant for detergents, cooling lubricants, adhesives, paper, citrus fruits, polishes, soap so'ns.; leather and textile finishing agent.

Nipacide® PTAP. [Nipa Labs] p-t-Amyl phenol; disinfectant for industrial use.

Nipacombin. [Nipa Labs] Preservatives for foodstuffs, pharmaceuticals, cosmetics.

Nipa Esters. [Nipa Labs] Esters of p-hydroxybenzoate or their sodium salts; preservatives for foods, pharmaceuticals, medicinals, cosmetics, toiletries, industrial prods.

Nipagin. [Nipa Labs] Methylparaben or salts; preservatives for foodstuffs, pharmaceuticals, cosmetics.

Nipa GMPA. [Nipa Labs] Glyceryl p-aminobenzoate.

Nipaguard DMDMH. [Nipa Labs] DMDM hydantoin; preservative, antimicrobial for cosmetics.

Nipanox®. [Nipa Labs] Liq. antioxidant mixtures.

Nipanox® BHT. [Nipa Labs] BHT.

Nipantiox 1-F. [Nipa Labs] BHA.

Nipa PABA. [Nipa Labs] p-Aminobenzoate.

NiPar. [Angus] Nitropropane; intermediate, solvent for inks and coatings; esp. for NC, chlorinated rubber, vinyl, epoxy, acrylic, PU, polyamide systems; automotive finishes.

Nipar-640®. [Angus] Nitroparaffin blend; specialty solvent additive.

Nipasept. [Nipa Labs] Preservatives for foodstuffs, pharmaceuticals, cosmetics.

Nipasol. [Nipa Labs] Propylparaben or salts; preservatives for foodstuffs, pharmaceuticals, cosmetics.

Nipastat®. [Nipa Labs] Methylparaben, butylparaben, ethylparaben, and propylparaben; preservative offering rapid-kill of common spoilage microbes.

Ni-Plex. [Atochem N. Am.] Nickel stripper.

Nipocera. [Frank B. Ross] Japan wax substitute.

Nipol. [Stepan] Block polymers; emulsifier component for agric. formulations.

Nipol®. [Zeon] Butadiene-acrylonitrile copolymer or NBR/PVC blends; used for hose, sheet packing, molded and extruded goods, oil seals, sundries, adhesives, plastic modification, o-rings, coated materials.

Nipol® AR. [Zeon] Acrylic rubber; used for seals, o-rings, gaskets, automotive parts, etc.

Nipolon. [Tosoh] Polyethylene resins; used for blow molding, blown film, extrusion pipe, extrusion coating and laminating, inj. molding, filament.

Niramid. [Technopol GmbH] Polyamide

6 and 6/6 resins.

Nirex. [Leatex] Low foam scouring agent for natural and synthetic fibers.

Nirez®. [Arizona] Terpene-phenolic resin; specialty ink solvents.

Nirion. [Technopol GmbH] Polycarbonate.

Ni-Rod. [Inco Alloys Int'l.] Welding prods. for welding cast irons.

Niroform. [Technopol GmbH] Acetal homopolymers and copolymers.

Nirox 600. [New Riverside Ochre] Micronized natural brown pigment.

Nissan Amine. [Nippon Oils & Fats] Amines; intermediate for cationic surfactants; anticorrosive agent, germicide, wetting agent, mold release agent, softener, emulsifier, dispersant, intermediate used in textiles, water treatment, concrete, asphalt, agriculture, ceramics.

Nissan Anon. [Nippon Oils & Fats] Betaines or glycines; amphoteric antistat, softener, germicide in foods, cosmetics, and cleaning industries, asphalt antistripping agent, textile treatment, water treatment; extraction aid in fermentation;

Nissan Asphasol. [Nippon Oils & Fats] Asphalt emulsifiers.

Nissan Cation. [Nippon Oils & Fats] Quatenaries or other cationics; germicide in water treatment, petrol., paper, foods and textile industries, antistat in plastics, pulp and paper industries, rinse agent, pigment coating agent, dispersant, coagulant, softener.

Nissan Chlorpearl. [Nippon Oils & Fats] Nonionic/cationic blend; drycleaning soap for 1,1,1-trichloroethane type solvent.

Nissan Diapon. [Nippon Oils & Fats] Taurates; detergent, emulsifier, dyeing auxiliary, scouring agent for textiles, hair dyes.

Nissan Disfoam. [Nippon Oils & Fats] Polyalkylene glycol derivs.; defoamers for fermentation applics., pulp, synthetics.

Nissan Dispanol. [Nippon Oils & Fats] Ethoxylated ethers; emulsifier and detergent for animal, cellulosic, and synthetic fibers.

Nissan Elegan S-100. [Nippon Oils & Fats] Antistat for plastics.

Nissan Monogly. [Nippon Oils & Fats] Glyceryl stearate; emulsifier for cosmetics and pharmaceuticals.

Nissan New Elegan. [Nippon Oils & Fats] Cationic; antistats for soft and rigid PVC.

Nissan Newrex. [Nippon Oils & Fats] Alkylbenzene sulfonate; detergent.

Nissan New Royal P. [Nippon Oils & Fats] Nonionic/cationic blend; drycleaning soap for perchlorethylene type solvent.

Nissan Nonion. [Nippon Oils & Fats] Ethoxylated ethers or esters; emulsifier, dispersant, detergent, wetting agent, thickener for cosmetic, pharmaceutical, food, textile, metalworking, agric., inks, paints, industrial applics., polymerization;

Nissan Nymeen. [Nippon Oils & Fats] Amines; wetting agent, pigment dispersant, emulsifier, acid cleaner additive for metals, textile antistatic agent and auxiliary.

Nissan Nymide. [Nippon Oils & Fats] POE alkylamide; modifier for soaps.

Nissan Ohsen. [Nippon Oils & Fats] Alkylbenzene sulfonate; general purpose detergent and emulsifier.

Nissan Panacete. [Nippon Oils & Fats] Triglyceride; diluent for perfumes; raw material for pharmaceuticals and specialty foods.

Nissan PEG. [Nippon Oils & Fats] PEGs; emulsifier, wetting agent, plasticizer for cosmetics, pharmaceuticals, other uses.

Nissan Persoft. [Nippon Oils & Fats] Detergent, emulsifier, wetting agent, dyeing assistant; base for liq. detergent; dispersant for pulp pitch; lubricant for bottle cleaning processing; textile processing.

Nissan Plonon. [Nippon Oils & Fats] POE-POP ether; emulsifier, solubilizer, dispersant, detergent, antifoaming agent, wetting agent used in metal cleaning, emulsion polymerization, fermentation, paper industries.

Nissan Polystar. [Nippon Oils & Fats]

Sodium carboxylate; dispersant for dyes, pigments, clay, agric. chemicals.

Nissan Rapisol. [Nippon Oils & Fats] Sodium dioctyl sulfosuccinate. wetting agent, polymerization agent for PVC, dyeing auxiliary.

Nissan Softer. [Nippon Oils & Fats] Amide amine; softening and finishing agent for textiles.

Nissan Soft Osen. [Nippon Oils & Fats] Linear alkylbenzene sulfonate; detergent, emulsifier.

Nissan Stafoam. [Nippon Oils & Fats] Fatty acid alkanolamide; thickener, detergent, foam stabilizer for cosmetics, textiles, detergents.

Nissan Sunalpha. [Nippon Oils & Fats] Alpha-sulfonated fatty acid ester blend, sodium salt; detergent for household and industrial cleaning.

Nissan Sunamide. [Nippon Oils & Fats] Fatty alkylolamide ether sulfate; base for shampoos and dishwashing detergents.

Nissan Sunbase. [Nippon Oils & Fats] Sodium fatty acid ester sulfonate; lime soap dispersant; nonphosphate detergent base; builder for household detergents; emulsifier, dispersant.

Nissan Sun Flora. [Nippon Oils & Fats] Nonionic/cationic blend; drycleanig soap for petroleum type solvent.

Nissan Tertiary Amine. [Nippon Oils & Fats] Tertiary amines; intermediate for various surfactants, visc. index improver for lubricating oil, curing catalyst for epoxy resin, corrosion inhibitor, germicide.

Nissan Trax. [Nippon Oils & Fats] POE alkylaryl ether sulfate; penetrant, emulsifier, detergent, wetting agent for emulsion polymerization, textiles, resins.

Nissan Unilube MB-38. [Nippon Oils & Fats] PPG-3 butyl ether.

Nissan Unisafe. [Nippon Oils & Fats] Amine oxides; foam stabilizer, detergent.

Nissan Unister. [Nippon Oils & Fats] Neopentyl polyol fatty acid ester; lubricating oil, plasticizer, cosmetics base.

Niterox. [Chilean Nitrate] Refined sodium nitrate; for industrial applics.

Nitran. [Zipp Industries] Dry fertilizer.

Nitrane. [W.R. Grace/Organics] Nitroparaffin solvents; industrial chemicals;

Nitrene. [Henkel/Emery] Cocamide DEA; thickener, foamer, detergent, wetting agent, emulsifier for industrial and household cleaners.

Nitriflex EPDM. [A. Schulman] EPDM; for wire and cable insulation, automotive parts, hoses, seals, extruded and molded goods, weatherstripping.

Nitriflex N. [A. Schulman] NBR; used for gaskets, o-rings, cements, adhesives, oil seals, hose, automotive parts, extruded prods., compr. moldings.

Nitrobac-Rx. [Polybac] Bacterial deammonifying culture.

Nitro BT. [Monomer-Polymer & Dajac Labs] Nitro blue tetrazolium.

Nitro Fast. [Sandoz] Specialty dyes for coloring automobile, shoe, and furniture polishes, oils, waxes, and solvents.

Nitrofuel. [Angus] Racing fuel.

Nitrophos®. [BASF AG] Nitrogen/phosphate complex fertilizer.

Nitrophoska®. [BASF AG] Complex nitrogen/phosphate fertilizer; for agric., horticultural crops.

Nitropore®. [Uniroyal] 4,4′-Oxybis (benzenesulfonhydrazide); nitrogen-releasing blowing agent for elastomers, thermoplastics, rubber-resin blends.

Nitrorace. [W.R. Grace/Organics] Liq. fuel for internal combustion engines; also for clothing.

Nivionplast. [Enimont UK Ltd.] Polyamide.

Nixox®. [Air Prods.] Catalyst.

Nix Stix. [Dwight Prods.] Mold releases for plastics and rubbers.

NJ. [Compounding Tech.] Nylon 12, glass-reinforced.

NK. [Medical Chem.] Heparin anticoagulant.

NK Guard®. [Nicca USA] Fluorocarbon water/oil repellent for natural and synthetic fibers and their blends.

NL-. [Compounding Tech.] Nylon 6/12, glass or carbon-reinforced.

NM. [Angus] 2-Nitro-2-methyl-1-propanol; intermediate, stabilizer for

halogenated solvs., as fuels, explosives, and solvs. for coatings or industrial processes.

NMP. [Angus] 2-Nitro-2-methyl-1-propanol; chemical intermediate, formaldehyde donor for adhesives and foundry resins, textile reactant; reduces formaldehyde on finished cloth.

NMP. [ISP] Methylpyrrolidone; solvent.

NN-. [Compounding Tech.] Nylon 6/6, glass, bronze or carbon-reinforced, some lubricated.

No. 1 White. [ECC Int'l.] Calcium carbonate; general purpose.

No. 22 Barytes. [Cyprus Industrial Min.] Ground natural unbleached barite.

Nobestos. [Lydall] Asbestos-free chloroprene, nitrile, or acrylic.

Noblecoat. [Engelhard] Insoluble anodes.

Nobs. [Am. Cyanamid] Rubber accelerators.

Nochek 4607. [C.P. Hall] Microcrystalline wax.

No-Cy. [Witco] Electroplating sol'ns. and additives.

Nodor SPR. [Ivax Industries] Odor eliminator for textile dyeing and finishing.

No Flame. [Flame Control Coatings] Fire and flame retardant paints, varnishes, chemicals, and coatings.

Nofoam JM. [Ivax Industries] Nonsilicone jet dyeing defoamer.

Nofome. [Surpass] Silicone or nonsilicone defoamers for textile wet processing, dyeing, bleaching, scouring, finishing, sizing operations.

Nofome. [Sybron] Defoamer for textiles.

Noform. [Finetex] Odor control and masking agents.

Noigen. [Dai-ichi Kogyo Seiyaku] Ethoxylated ethers; emulsifier.

Noiox. [Pulcra SA] Ethoxylated esters or ethers; lubricant, emulsifier, antistat, solubilizer, defoamer, spreading agent, wetting agent for inks, plastics, textiles, cosmetics, pharmaceuticals.

Nolibond®. [Rhone-Poulenc] Polyamide-imide resin; high-temperature adhesive.

Nolicoat. [Rhone-Poulenc] Polyamide-imide resin; as protective varnish, for coatings and paints; provides self-lubricating, high heat resistant, and decorative properties.

Nomofome CO. [Chemurgy Prods.] Silicone emulsion; defoamer for most textile applics.

Nonal. [Toho Chem. Industry] Alkyl phenol ethoxylate; detergent, penetrant, emulsifier, scouring agent, pitch dispersant.

Nonarox. [Seppic] Alkylphenyl polyethoxy ether; detergent, wetting agent.

Nonasol. [Hart Chem. Ltd.] Sulfate or sulfonates; detergent, dyeing assistant.

Nonatell. [Shell] Proprietary surfactants; deinking surfactant for use on xerographic finish, groundwood finish, in pulp deresination; digester process aids for kraft and sulfite processes.

Nonex. [Hart Chem. Ltd.] Ethoxylated esters; emulsifier, lubricant, solubilizer, softener for industrial and textile oils.

Non Flam. [Imperial Adhesives] Neoprene N/F.

Non Flame. [Flame Control Coatings] Fire and flame retardant paints, varnishes, chemicals, and coatings.

Nonicol. [Atsaun] Ethoxylated alkylphenol; emulsifier, wetting agent.

Nonionic. [Hodag] Ethoxylated alkylaryl ethers; detergent, wetting agent for insecticides, other formulations.

Nonipol. [Sanyo Chem. Industries] Ethoxylated ethers; detergent, penetrant, wetting agent, emulsifier, dispersant for agric., textile, emulsion polymerization, household cleaners.

Nonipol. [Toho Chem. Industry] Surfactant/mineral oil blends; spinning oil for wool.

Nonox. [ICI Am.] Antioxidants.

Nonox®. [Akzo] 2,2´-Methylenebis [4-methyl-6-(1-methyl-cyclohexyl) phenol]; antioxidant for plastics.

Nonseparal. [Witco] Calcium greases.

Nonstix 84. [Nutex] Lubricants to prevent pick-off in processing.

Nonychosine. [Exsymol] Acrylic polyester/methacrylic polyester/methionin blend; hardener and normalizer for nail growth.

No-Ox-Id. [Sanchem] Compounded formulas for control of corrosion by chemical and mechanical action.

Nopafoam. [Pillo Pak] Polyethylene foam.

Nopalcol. [Henkel/Emery] Ethoxylated fatty esters; emulsifier, plasticizer, lubricant, dispersant, wetting agent, binder, thickener for cosmetics, dry cleaning, leather, textile industries.

Nopco®. [Henkel/Emery; Henkel-Nopco] Sulfated surfactants; wetting agent, defoamer, emulsifier, dispersant, detergent for scouring, latexes, paints, adhesives, textiles, paper, metal treatment.

Nopcocastor. [Henkel/Functional Prods.] Sulfated castor oil; emulsifier, superfatting agent for cosmetics.

Nopcochex RA. [Henkel/Emery] Fatty amide condensate; oil-soluble corrosion inhibitor.

Nopcocide®. [Henkel; Henkel-Nopco] Mildewcide, fungicide, antimicrobial for paints and coatings.

Nopco® Colorsperse. [Henkel/Coating Chem.; Henkel-Nopco] Dispersant, wetting agent for pigments.

Nopcogen. [Henkel/Emery; Henkel-Nopco] Detergent, wetting agent, emulsifier, softener used in textile, asphalt, paper industries.

Nopcolan. [Henkel] Organic thiosulfate; shrink resistor for wool.

Nopcosant®. [Henkel/Coating Chem.; Henkel-Nopco] Dispersant for latex systems, paint, paper, textile, leather coating compds.

Nopcosperse®. [Henkel/Emery; Henkel-Nopco] Dispersant for pigments, paints, pesticides.

Nopcostat OT-85. [Henkel/Emery] Sulfated vegetable oil.

Nopcosulf. [Henkel/Emery; Henkel-Nopco] Sulfated compds.; softener, plasticizer for cotton goods, finishing starches, gums; furniture polish base.

Nopcote. [Henkel-Nopco] Calcium stearate; lubricant for paper coatings.

Nopcowax. [Henkel] Ethylene bis-stearamide; synthetic wax, binder, thickener for latex formulation, coatings, adhesives; used in powd. metallurgy as internal lubricant.

Noram. [Ceca SA] Amines; synthesis

intermediate, anticaking agent, flotation, antistripping for road making, soil stabilization; auxs. for fuel additives, rust inhibition, paint; chemical intermediate for quats., betaines, amine oxides.

Noramac. [Ceca SA] Alkyl amine acetates; flotation agent, bactericide, emulsifier, anticaking agent, soil stabilizer, flocculant, corrosion inhibitor.

Noramer. [Norsohaas SA] Acrylic/sulfonate copolymers; corrosion inhibitor booster, dispersant for water treatment, oil recovery.

Noramium. [Ceca SA] Quaternary ammonium chlorides; bactericide, fungicide, demulsifier, textile softener, hair conditioner for cosmetics, paints, antibiotic mfg.

Noramox. [Ceca SA] Ethoxylated fatty amines; emulsifier, drying assistant, rust inhibitor; antistat for ABS, PS; metal treatment.

Norane®. [Sequa] Water repellants and fluorochemical extenders.

Norasol. [Norsohaas SA] Acrylic copolymers; detergent addtives, textile auxiliaries, rinse aids, dishwashing.

Norbloc. [Noramco] uv stabilizer.

Norbond. [Norton Materiaux Avances] Structural adhesives.

Norcast. [R.H. Carlson/Northern Labs] Epoxy resins or systems; for casting, encapsulation, potting, dipping, tooling, adhesives, coatings.

Norcure. [R.H. Carlson/Northern Labs] Curing agent, epoxy hardener.

Nordcell. [Nordchem SpA] Rigid expanded PVC granules.

Nordel®. [DuPont; DuPont UK] EPDM; used in conveyor belts, automotive and industrial molded parts, rolls, industrial and steam hose, elec. insulation, cellular prods., hose, sheeting, tape, weatherstripping.

Nord-HT. [Nordchem SpA] Heat-resistant CPVC.

Nordvil. [Nordchem SpA] Rigid PVC granules.

Norfox®. [Norman, Fox] Surfactants, soaps, lubricants, specialties; base for shampoos, soaps, industrial and agric.

wetting agents and detergents; defoamer, gellant, stabilizer, lubricant for cosmetics, paper, textiles, leather, metalworking.

Norfox Hercules Conc. [Norman, Fox] Built detergent blend; for heavy duty cleaning, truck and auto washing, degreasing.

Norit. [Am. Norit; R.W. Greeff] Activated carbon.

Norkool. [Union Carbide] Industrial prods.

Norland Hi Pure Liq. Gelatin. [Norland Prods.] Fish gelatin aq. sol'n.

Norland High Tack Fish Glue. [Norland Prods.] Water-sol. fish glue.

Norland Optical Adhesive. [Norland Prods.] uv-curing adhesive for assembling glass lenses.

Norlig. [Borregaard LignoTech] Lignosulfonates; dispersant, binder, resin extender, soil and dust stabilizer, pelletizing of coal and charcoal, ceramic additive.

Nor-Mer. [R.H. Carlson/Northern Labs] Epoxy system; uv-curable adhesive.

Normount. [Norton Perf. Plastics; Norton Materiaux Avances] Sealant.

Norm-X. [Poly Research] Conc. standards.

Norox. [Norac] Organic peroxide initiators.

Norpar®. [Exxon] Normal paraffin solvent.

Norpel. [Northern Prods.] Chromium complex.

Norpol. [Jotun GmbH] Unsaturated polyester resins.

Norpolefin. [Statoil UK Ltd.] Polypropylene or polyethylene.

Norprene. [Norton Perf. Plastics] Thermoplastic elastomer.

Norseal. [Norton Pampus GmbH] PVC/PVR sealing foams.

Norsil. [R.H. Carlson/Northern Labs] Silicone rubber, compds., and greases.

Norsolene. [C.P. Hall] Hydrocarbon resins.

Norsophen®. [Norold Composites] Phenolic resins and compds.; composites used for aerospace, construction, mass transit, mining, and automotive applics.

Norsorex®. [Atochem N. Am.; Atochem Deutschland GmbH] Polynorbornene or blends; synthetic elastomer used for damping compds.

Nortuff. [Quantum/USI] Polyethylene or polypropylene resins; engineering resin for inj. molding, extrusion.

Norust. [Ceca SA] Fatty amines blend; corrosion inhibitor, acid passivator.

Norvan. [Artilabo SA] Polyethylene.

Norvic. [Marlin Chem. Ltd.] PVC resins.

Norvinyl. [Norsk Hydro AS] Suspension PVC.

Noryl®. [GE Plastics; GE Plastics Ltd.] Modified polyphenylene oxide resins; engineering resin for inj. molding, extrusion, structural foam; used for computers, business equip., automotive, elec., electronics, construction, telecommunications, appliances, and other industries.

No-Seize. [J.C. Whitlam Mfg.] Antiseize lubricant and thread sealant.

No Shock. [Monsanto] Antistatic fiber.

Nosifeed 40. [Mitsubishi Kasei] Feed additive, growth promoter for swine and poultry.

Nosiheptide. [Mitsubishi Kasei] Feed additive, growth promoter for swine and poultry.

Nosil A-57. [Vikon] Metallo organic chelant blend; organic stabilizer for peroxide bleach baths permitting omission of silicate.

No Slip. [Petrokem] Ice and snow melting prods.

Nostat SA. [Ivax Industries] Ethoxylated amine; antistat agent for synthetics.

No-Swab. [E/M Corp.] Glass mold release coating.

Nouryflex. [Akzo] Phthalate ester; plasticizer for lacquers and coatings.

Nouryset®. [Akzo] Monomer; used for optical applics.

Nova. [Rohm & Haas] Fungicides.

Nova. [Rowe Prods. Distribution] Novalac/epoxy prods.

Nova. [Ubbink Nederland BV] Polyurethane/polyethylene.

Novablend®. [Novatec Plastics & Chem.] PVC compd; for inj. molding, extruded profiles, bottles, business ma-

chines, housings, elec./telecommunications industry and fittings.

Novacarb. [Nova Polymers] Carbon black; black conc. and impact modifier for ABS and PVC.

Novacite. [Malvern Minerals] Microcrystalline silicon dioxide.

Nova Flav. [Champlain Industries] Autolyzed primary yeast extracts.

NovaFlo. [Georgia-Pacific] Liq. rosin size.

Novafloc. [Lewis Industrial Prods.] Nylon fiber.

Novagel. [Whittaker, Clark & Daniels] Magnesium hydroxide paste.

Novakup. [Malvern Minerals] Silane-treated microcrystalline silicon dioxide.

Novalar. [Nova Polymers] Impact modifier for ABS, PVC, PC, polyurethanes, epoxies, PBT, acrylics.

Novalast. [Nova Polymers] Low-cost specialty cured thermoplastic elastomers.

Novalene. [Nova Polymers] Impact modifier; upgrades properties of polystyrene, polyethylene, polypropylene.

Noval Mineral Products. [Alcan] Alumina (60%), magnesium oxide (11%), aluminum nitride (17%), aluminum (3%); used for fused grains, cement, glass, foundry and steel mixes, refractories.

Novaloy®. [Novatec Plastics & Chem.] ABS/PVC alloy; engineering alloy for inj. molding, profile and sheet extrusion; used for appliance housings, elec. boxes, communications, lawn and garden applics., recreational prods.

Novalyte. [Aldoa] Electroplating brighteners.

Novamid®. [Mitsubishi Kasei] Nylon 6, some glass or mineral reinforced; for inj. molding, extrusion, monofilament, film applics.

Novamin®. [Condea Chemie GmbH] Modifiers, crosslinking agents, chain extenders for polyurethane prod.; additives for gasoline and diesel fuel.

Novamyl. [Novo Nordisk] Carbohydrase blend; for baking applics.

Novapol®. [Novacor Ltd.] Polyethylene resins; for food, pharmaceutical, consumer and industrial pkg., woven fabrics, paperboard containers, paper coatings, profile extrusions, sheet, tape, foam, film, liners, blends, household, industrial and automotive chemical bottles.

Novaprint. [3-V; Sigma Prodotti Chimici] Polyacrylic acid polymer; thickener, rheological additive for textile printing applics.

Novarex®. [Mitsubishi Kasei] Polycarbonate, some glass reinforced; for inj. molding, blow molding, extrusion.

Nova Seal. [Huntington Labs] Water-based concrete seal.

Novaset. [Ashland/Foundry Prods.] Air setting foundry binders.

Novasil. [Engelhard] Anticaking agent for animal feed addition.

NovaSize. [Georgia-Pacific] Tall oil rosin; internal sizing agent.

NovaSperse. [Georgia-Pacific] Rosin aq. dispersion; size.

Novata®. [Henkel/Cospha; Henkel KGaA] Cocoglycerides; suppository bases.

Novate. [Novachem] Rubber accelerators.

Novatec. [Poliaden Petroquimica SA] HDPE.

Novatec®-AP. [Mitsubishi Kasei] Extrudable adhesive polyolefin resins; for coextrusion, film, bottle, tube, sheet, barrier and thermoplastic bonding.

Novathane. [Ashland/Foundry Prods.] Foundry core binders.

Novawet®. [Novachem] Wetting agents, emulsifiers for fibers.

Novax. [Novachem] Rubber accelerators.

Novazone. [Uniroyal] Antiozonants for tires, molded and mechanical goods.

No Vein Compound. [DCS Color & Supply] Iron oxide compd.; used to eliminate expansion defects in castings.

Novel®. [Vista] Ethoxylated alcohol; surfactant, wetting agent, emulsifier for detergent formulation.

Novex. [BP Chem. Ltd.] LDPE.

Noviplast. [Novkabel] PVC compd.

Novoblanc A. [Sigma Prodotti Chimici] Titanium dioxide dispersion; white pig-

ment for pigment printing.

Novocoll®. [Pierce & Stevens] Isocyanate-terminated polyester urethane adhesive; laminating adhesive for transparent pkg. films, foils, metallized surfaces.

Novocolor. [Novosystems Farben und Additive GmbH] Liq. colorants.

Novodur®. [Bayer] ABS, some glass-reinforced; for inj. molding, extrusion, electroplating.

Novofoam. [Novosystems Farben und Additive GmbH] Blowing agent dispersions.

Novogard. [Novosystems Farben und Additive GmbH] Flame retardant dispersions.

Novogel® ST. [Rhone-Poulenc] Aluminum stearate/mineral oil blend; stabilizer for emulsion systems; gelling agent, waterproofing agent, lubricant; improves pigment adhesion and slip.

Novogen®. [Condea Chemie GmbH] Phenolic and melamine resins; bonding resins for wood processing, abrasives, foundry industry.

Novol. [Croda Inc.; Croda Chem. Ltd.] Oleyl alcohol; emollient, emulsion stabilizer, superfatting agent, pigment suspending aid, used in cosmetics, personal care prods.

Novolen®. [BASF AG] Polypropylene; for inj. molding and extrusion, automotive parts, elec. engineering, pkg., pharmaceuticals, medical technology, fibers, slit film yarns, monofilaments.

Novolube. [Novosystems Farben und Additive GmbH] Lubricants, additives for plastics.

Novon. [Novon] Specialty polymers from renewable sources (corn starch, potato starch); decomposable material for food pkg., personal and health care, other consumer prods. and pkg.

Novor. [Akrochem] Urethane vulcanizing system; accelerator.

Novostab. [Novosystems Farben und Additive GmbH] Stabilizer dispersions.

Novostat. [Novosystems Farben und Additive GmbH] Antistatic materials.

Novox. [Novosystems Farben und Additive GmbH] Antioxidant dispersions.

Novozym 188. [Novo Nordisk] Liq. cellobiase; used in conjunction with cellulase when complete hydrolysis of cellulose to glucose is required.

Noxamine. [Ceca SA] Cationics; foamer, antistat, bactericide, emulsifier, corrosion inhibitor.

Noxamium. [Ceca SA] Quaternary ammonium derivs.; antistatic, bactericidal emulsifier.

NP-10, -25. [Neville] Aromatic plasticizer; used in adhesives, rubber, caulks.

NP-55. [Hefti Ltd.] Ethoxylated nonylphenyl ethers; wetting agent, detergent component, emulsifier for agric., pigments, textiles, industrial and household cleaners, paper, leather.

NPG® Glycol. [Eastman] Neopentyl glycol; resin intermediate.

NRC 76. [NRC] Tantalum mill prods.

NSA-17. [Sanyo Chem. Industries] Proprietary; additive for heavy-duty powd. detergents.

NSC Esbrid. [Thermofil] Ceramic fiber/glass fiber-reinforced nylon; for replacement for die cast metal parts used in business machines, home appliances, and automotive applics.

NSIL. [GE Silicones] Silicone sealant.

N® Sodium Silicate. [PQ Corp.] Sodium silicate; corrosion inhibitor for water systems; used for metals in industrial plants, textile mills, laundries, office bldgs., municipalities, oil refineries.

N-Sol. [Mississippi Chem.] Nitrogen sol'n.

NT-. [Compounding Tech.] Toughened nylon 6/6, glass or mineral-reinforced.

NTA. [Monsanto] Sodium NTA; sequestering agent.

Nucap. [J.M. Huber] Treated reinforcing clay.

Nuchar. [Westvaco] Active carbon.

Nu Cop. [Cuproquim] 50% copper oxychloride.

Nucrel®. [DuPont; DuPont UK] Thermoplastic ethylene methacrylic acid copolymers; for extrusion into sheet, film, and coatings, for inj. molding, blow molding, thermoforming; blendable with other materials for adhesives and sealants.

Nufilm. [Miller Chem. & Fertilizer] Spreader-sticker, residue control agent.

Nu-Flex-Glu. [Industrial Adhesives] Asbestos encapsulants.

Nu-Iron. [Boliden Intertrade] Oxalate for iron deficiencies in horticulture and agriculture.

Nukem No. 21C. [Sauereisen Cements] Furan resin mortar/grout.

Nulok. [J.M. Huber] Treated reinforcing clay.

Numel. [IGI Baychem] Structural hotmelt adhesives for bonding polyolefins to wood, glass, metal, other polyolefin and plastic substrates.

Nuocure. [Hüls Am.] Polyurethane catalyst.

Nuodex. [Servo] Organic mercury compd.; fungicide, bactericide for paints; preservative for aq. systems.

Nuoplaz®. [Hüls Am.] Proprietary; plasticizer for PVC.

Nuosept C. [Hüls Am.] Polymethoxy bicyclic oxazolidine.

Nuosperse 657. [Hüls Am.] Dispersing agents.

Nuostab®. [BASF AG] Heat stabilizers for rigid and plasticized PVC, internal and external applics., PVC foams.

Nupetal. [Nutex] Wetting agents.

Nupol. [Cook Composites & Polymers] Thermosetting acrylic resins.

Nusoft. [First Preference Prod.] Fabric softener.

Nusoft. [Nutex] Liq. quaternary softener.

Nusol JW-2. [Nutex] Nonionic ethoxylates; low foaming scour and emulsifier.

Nustar. [A.E. Staley Mfg.] Modified tapioca starch.

Nuto H. [Exxon] Hydraulic oil.

Nutrabond. [Nutex] Durable hand builders.

Nutralok. [Nutex] Antislip textile finishes.

Nutramix. [Western Nutrients] Complex rice mix with zinc lignin base.

Nutraphos. [Uniroyal] Foliar nutrient.

Nutraplex. [Western Nutrients] Complexed chelated metal salt micronutrient sol'n. with lignin base; for agricultural use.

Nutrapon. [Clough] Sulfates or blends; anionic surfactants for personal care prods. and detergents.

Nutrex PG. [Fabriquimica] Glycosaminoglycans.

Nutrex PV. [Fabriquimica] Hydrolyzed corn starch.

Nutrex RT. [Fabriquimica] Water, spleen extract.

Nutrex TM. [Fabriquimica] Water, thymus hydrolysate.

Nutricol. [FMC/Marine Colloids] Konjac flour; gelling agent, thickener.

Nutrifos. [Monsanto] Meat curing ingredients.

Nutrigent. [Lubrication Engineers] Industrial detergent.

Nutrilan®. [Henkel/Emery/Cospha; Henkel KGaA] Hydrolyzed keratin, elastin, or collagen; for surfactant and skin and hair care preps.

Nutri Leaf. [Miller Chem. & Fertilizer] Soluble fertilizer.

Nutrilife®. [Grünau] Enzyme-protein compds.; synergistic additive with emulsifiers for baking process.

Nutriloid. [TIC Gums] Nutritonal functional soluble dietary fiber.

Nutri-Mag. [FMC/Marine Colloids] High purity (59.1% Mg) MgO; nutritional source of palatable magnesium for animal feed supplements, fertilizers.

Nutrimalt®. [Grünau] Malt extract in powder form; taste and volume improver for baked goods.

Nutrio Pea Prods. [Grindsted Prods.] Fiber; dough improvers for food industry.

Nutrisoft®. [Grünau] Distilled monoglyceride; water-dispersible emulsifier for baking additives and confectionery; antistaling effect.

Nutrol. [Clough] Surfactants; foaming agent, detergent, stabilizer, wetting agent, emulsifier, dispersant for cosmetics, household and industrial cleaners, polymerization, pesticides.

Nutrox®. [Atochem N. Am./Textiles] Scouring agent for fabrics.

Nuts N' Bolts. [Hernon Mfg.] Adhesives for threaded parts.

Nuts Off. [Spray Prods.] Penetrant.

Nuva. [Hoechst AG] Fluorine derivs.; water and oil repellent finishing agents for textiles, leather.

Nuvis. [Hüls Am.] Bodying agent for paints.

NY-. [Compounding Tech.] Nylon 6, glass or carbon-reinforced.

Nyacol®. [PQ Corp.] Colloidal silicas as binder for pkg. prods., antislip agent, retention aid, and coating for paper applic.; antimony pentoxide compds. as flame retardant.

Nyad® Wollastonite. [Nyco Minerals] Wollastonite.

Nybex. [Ferro/Engineering Thermoplastics] Nylon 6, 6/6, or 6/12, some glass, carbon, or mineral-reinforced.

Nycoa Nylon. [Nylon Corp. of Am.] Nylon 6, some plasticized, nucleated, impact-modified; for extrusion, inj. molding.

Nydur. [Miles] Polyamide 6 resins; for automotive, consumer, and industrial/mechanical applics.

Nydye. [Crompton & Knowles] Textile dyes and pigments.

Nykon®. [Rheox] Tetra alkyl or trialkyl aryl ammonium smectites with corrosion inhibitor; gellant, rheological additive, corrosion inhibitor.

Ny-Kon®. [LNP] Nylon 6, 6/6, 6/10, or 6/12 with molybdenum disulfide lubricant; internally lubricated molding compds.

Nylafix. [Apex] Dye fixing agents.

Nylanthrene. [Crompton & Knowles] Textile dyes and pigments.

Nylar. [McLaughlin Gormley King] Insect growth regulator.

Nylasan. [Akzo] Acrylic emulsion; reduces and prevents scale and sludge in digesters, evaporators, heat exchanges, water, lines, and cooling towers.

Nylatron®. [DSM] Nylon 6, 6/6, or 6/12 resins, some glass or mineral-reinforced, molybdenum disulfide-lubricated; for inj. molding applics.

Nylatron Ertalon. [Plastiques Obra SA] Polyamides.

Nylfix. [Lindley Labs] Antimigratory fixing agent for nylon.

Nylofixan®. [Sandoz] Fixing agent and reserving agent.

Nyloid. [Rhone-Poulenc] Warp sizes for nylon.

Nyloil. [Granby Plastics Ltd.] Oil-filled nylon.

Nylok®. [J.M. Huber] Amino-functional calcined clay; filler, reinforcement for polyamide resin systems.

Nylomine. [ICI Am.] Dyes.

Nylomine Assistant DN. [ICI Colors] Surfactant blend; dyeing assistant, leveling agent for dyeing fibers.

Nylon N-012, SI-N. [Presperse] Nylon 6/6, methicone blends; surface-treated nylon powders for use in anhydrous systems, emulsions, powdered formulations; provides lubricious feel for cosmetics.

Nylon Resist NCO. [Eastern Color & Chem.] Dye resist for nylon.

Nylopak. [Dow] Plastic film.

Nylosan®. [Sandoz] Specialty dyes for coloring aq. media, fertilizers.

Nyloset Finish. [Scher] Amide resin; nylon builder and softener.

Nylox. [Hardman] Nylon adhesive.

Nysel. [M&T Harshaw] Nickel; catalyst for hydrogenation of oils.

Nysist. [Eastern Color & Chem.] Leveling agent, dye assistant, retarder for nylon.

NYsyn®. [Copolymer Rubber] Acrylonitrile-butadiene copolymer; for industrial and automotive hose and seals, cable jackets, footwear, wire and cable compds., belts, rolls, molded and extruded mech. goods.

NYsynblak®. [Copolymer Rubber] Acrylonitrile-butadiene copolymer black masterbatch; used for extruded and molded goods.

Nytal®. [R.T. Vanderbilt] Magnesium silicate; filler, reinforcing extender for art pottery, ceramic casting slips.

Nytek 10. [Maag Agrochem.] Solubilized copper-8-quinolate; industrial fungicide.

Nytek 10 WP. [Maag Agrochem.] Solubilized copper-8-quinolate and water repellents; industrial fungicide.

Nytek 645. [Maag Agrochem.] Industrial fungicide, penetrant sealer, and water repellent.

Nytek GD. [Maag Agrochem.] Solubilized copper-8-quinolate; sapstain control agent.

Nytek WD. [Maag Agrochem.] Solubilized copper-8-quinolate; water-dispersible industrial fungicide.

Nyzist. [Eastern Color & Chem.] Organic sulfonate; nylon dye resist.

NZ® Series. [Kenrich Petrochemicals] Zirconates; coupling agents; adhesion promoters, antioxidants, antistats, accelerators, activators, catalysts, curatives, corrosion inhibitors, dispersants, emulsifiers, flame retardants, foamers, impact modifiers, release agents, retarders, stabilizers, etc.

O

O®. [PQ Corp.] Sodium silicate; liq. alkali stabilizer for pad-batch dyeing.

O9810, 9816. [Hüls Am.] Low m.w. siloxanes; for cleaning, polishing, and damping media.

OA. [Witco/Argus] Chlorinated paraffin or esters, sulfurized additives, or diphenylamines; antioxidants, additives for indusrial oil and grease, coolant or water-based systems, pour pt. depressants, processing aids.

OAA. [Nat'l. Starch & Chem.] Tert. octyl acrylamide.

Oakite Drycid. [Oakite Prods.] Descalant.

Oases®. [Air Prods.] Waste water treatment systems.

Oasis. [U.S. Cosmetics] Polyacrylate/lecithin; moisturizing surface treatment for cosmetics.

OA Waxes BASF. [BASF AG] Oxidized polyethylene waxes; for wax emulsions.

Obanol. [Toho Chem. Industry] Polyoctyl polyamino ethyl glycine and POE alkylphenol ether; germicide, disinfectant, deodorant, fungicidal cleaning aid.

Obax. [A.E. Staley Mfg.] Industrial corn starch.

Obazoline. [Toho Chem. Industry] Imidazoline derivs.; antistat, softener, base material for hair rinses, textile fibers.

O'B Floc. [O'Brien Industries] Polymers for water clarification.

O'B Hibit. [O'Brien Industries] Acid corrosion inhibitors.

O'B Kool. [O'Brien Industries] Cooling water treatments.

O'B No Foam. [O'Brien Industries] Defoamers.

OBTS. [Akrochem] N-Oxydiethylene-2-benzothiazole sulfenamide; primary accelerator for natural, SBR, nitrile, and other general-purpose rubbers.

OC200, 1000. [O&C] Litharge.

OC300. [O&C] Battery oxide.

OC400. [O&C] Red lead.

Ocenol. [Henkel/Emery/Cospha] Ethoxylated alcohol; emulsifier, lubricant aid for metalworking fluids.

Ocetox. [Witco SA] Ethoxylated fatty alcohol; emulsifier, detergent.

OCI. [Olin] Chloro triazinetrione compds.; dry chlorinator, sanitizer for food and beverage industries, swimming pool stabilizer, in dry bleaches for laundries, dishwashing compds., scouring powds., as intermediates in industrial applics.

Ocropon. [Organic Dyestuffs] Wetting and scouring agent.

OCS-M. [Drew Ind. Div.] Corrosion inhibitor.

OctaBoost® 620. [Akzo] Alumina controlled fluid cracking catalysts.

Octafilm. [Betz Industrial] Filming amine.

Octamine®. [Uniroyal] Amine antioxidant; for tires, air-cured footwear, molded soling, closed-cell sponge, athletic padding and insulation, insulated wire, molded and mechanical goods.

Octapol. [Sanyo Chem. Industries] Ethoxylated octylphenyl ether; base material for detergents.

Octaprotein. [Vevy] Soy protein.

Octaron. [Seppic] POE nonylphenol ether sulfated, sodium salt; detergent.

Octoate® Z. [R.T. Vanderbilt] Zinc 2-ethylhexoate; activator for natural and synthetic rubbers.

Octocure. [Tiarco] Zinc dithiocarbamates or mercaptobenzothiazoles; accelerators for latex and rubber.

Octoguard. [Tiarco] Antimony trioxide or antimony trioxide/decabromodiphenyloxide blends; flame retardant dispersions for water-based polymers, latex adhesives, binders, coatings, foams.

Octoil. [Octagon Process] Rust preventives.

Octojet. [Tiarco] Carbon black dispersions.

Octolite. [Tiarco] Antioxidant emulsions and dispersions for latex compds., rubber, natural and synthetic adhesives, foams.

Octomer. [Tiarco] Maleates; plasticizers for vinyl resins, in PVC and vinyl acetate polymerization reactions, latex paints; as intermediates for surfactants.

Octopirox. [Hoechst AG] Piroctone olamine; antidandruff agent.

Octopol. [Tiarco] Liq. dithiocarbamates; polymerization shortstop for SBR, CR, and natural rubber; precipitant for heavy metals in wastewater treatment; natural rubber latex preservative.

Octosol. [Tiarco] Sulfosuccinamates, sulfosuccinates, or sulfates; emulsifier, stabilizer, frothing aid, foaming agent, dispersant for industrial applics., emulsion polymerization, mfg. of carboxylated latexes.

Octosperse. [Tiarco] Silicone emulsion defoamers.

Octotint. [Tiarco] White or color dispersions.

Octowax. [Tiarco] Wax emulsions; for use in paper and wood coating, sizing, textile lubrication, as processing aid in latex foams, as antiozonant.

Octowet. [Tiarco] Sodium dioctyl sulfosuccinate; wetting agent, emulsifier, penetrant for textile processing, agric, mining, paper, printing, other industrial applics.

Octyl Dimethyl PABA. [Nat'l. Starch & Chem.] Octyl dimethyl p-aminobenzoate.

Odomaster. [Surco Prods.] Odor control systems.

Odorgon. [Stewart Hall] Fuel oil and petroleum deodorant.

Odorsorb. [Advance Chem] Chemical deodorant liq.

Odo'Zone. [Mateson] Odor control agents.

OFA 6. [Vista] Ore flotation agent.

Ofax®. [Stepan] Special blends; high foaming surfactants for oilfield applics.

Oftanol®. [Bayer; Miles/Ag] Isofenphos; insecticide effective against soil insects and leaf-eating plants.

Ogtac. [Chem-Y GmbH] Glycidyl trimethyl ammonium chloride; intermediate for surfactants, starch modifier; used for paper, textile and cosmetic industry.

OHlan®. [Amerchol; Amerchol Europe] Hydroxylated lanolin; emulsifier, stabilizer, pigment wetting and dispersing agent, emollient and conditioner in personal care prods.

Ohmex-AG. [Transene] Thermosetting silver for ohmic bonding.

Ohso. [Steetley Quarry Prods.] Pulverized limestone.

Oilamid. [Licharz GmbH Nylontechnik] Polyamide 6 with oil.

Oildag. [Acheson Colloids] Colloidal graphite in refined naphthenic petroleum oil; lubricant additives.

Oil-Dri. [Oil Dri Corp. of Am.] Oils and greases.

Oilfos. [Monsanto] Glassy sodium phosphate; drilling mud conditioner.

Oil Gard. [Gard] Oil additive.

Oil-Perge. [W.R. Grace/Dearborn] Microbicide.

Oil-Treet. [Schaefer Tech.] Fuel oil conditioner.

Ointment Base. [Penreco] White petrolatum USP; ointment base for eye and skin medications; carriers for medical materials.

Okemcoat. [Oakite Prods.] Coatings.

Okerins. [Astor Wax] Antiozonant waxes for rubber.

Okirol. [Ina-Oki] General-purpose and impact grade polystyrenes.

Okirol E. [Ina-Oki] Expandable polystyrene.

Okisan. [Ina-Oki] ABS.

Okiten. [Ina-Oki] LDPE.

Okstan. [Bärlocher GmbH] Organotin stabilizers for plastics.

Olamin®. [Henkel Canada; Henkel KgaA] Surfactant blend; wetting agent, protectant for permanent waves.

Olapon. [Reilly-Whiteman] Detergents

for textile scouring.

Old Reliable. [Peninsula Copper Industries] Cupric oxide.

Oleine. [Ceca SA] Oleic acid; surfactant.

Oleo-Coll. [Brooks Industries] Hydrolyzed collagen blends; for skin and hair care cosmetics.

Oleo-Keratin ISO. [Brooks Industries] AMP-isostearoyl hydrolyzed keratin, isostearic acid, myristyl myristate, isopropyl palmitate; cosmetics ingredient.

Ole-Ole. [ISK Biotech] Agric. fungicide.

Oleo-Soy C. [Brooks Industries] Cocohydrolyzed soy protein; for hair and skin care cosmetics.

Oleosterines. [Exsymol] Plant extracts containing liposoluble active principles; for cosmetic and health prods.

Olepal. [Gattefosse; Gattefosse SA] Ethoxylated esters; solvent, emulsifier for cosmetics, pharmaceuticals.

Olew. [Bruce Chem.] Fiber processing finishes for overspray use on textiles.

Olicat®. [Alox] Polyisobutylene in mineral oil; tackiness agent, lubricant in lubricating oils and greases.

Olicine. [Gattefosse; Gattefosse SA] Peanut glycerides; food emulsifier.

Oligoidine. [Vevy] Metal salts of aspartic acid.

Oligomex® N. [Boehme Filatex] Quaternary ammonium compd.; dyeing machine cleaner.

Olitex 75. [Reilly-Whiteman] Sulfated olive oil; lubricant.

Olympic. [Cyprus Industrial Min.] High purity talc; for cosmetics applics.

Omacide®. [Olin] Zinc pyrithione blends; antimicrobial for PVC systems.

Omadine® MDS. [Olin] Bispyrithione, magnesium sulfate; antidandruff agent for nonalkaline hair care prods.; antimicrobial agent for Gram-negative and Gram-positive bacteria; fungicide.

Omaha. [Asarco] Lead.

Omega. [Aquatec Chem. Int'l.] Cooling tower compds.

Omega. [Katalistiks Int'l.] Fluid cracking catalysts.

Omega. [Omega] Antistats, flame retardants, penetrants, scouring agents, softeners, stabilizers for textile processing.

Omite. [Uniroyal] Miticide.

Omnifil. [J.M. Huber] Paper filler clay.

Omni-Fix. [Omni/Ajax] Hazardous waste detoxifier.

Omni-Kap. [Omni/Ajax] Hazardous waste solidifier.

Omni-Trap. [Omni/Ajax] Hazardous waste filter media.

Omo. [Lever Bros.] Soap base.

OMTS. [Akrochem] N-Oxydiethylene-2-benzothiazole-sulfenamide; delayed-action accelerator for SBR, NR, and nitrile rubbers.

Omyacarb. [Omya] Wet and dry ground natural calcium carbonate.

Omyalene G200. [Omya SA] 88% calcium carbonate masterbatch.

Omyalite. [Omya GmbH] Calcium carbonate.

Once. [Hardman] Hand cleanser for removal of resins, adhesives, epoxies, gasket cement, paint, pitch, polyester, silicones, tars, urethanes, carbon black, dirt, grease, grime, inks, putty.

Oncor. [Rheox] Basic lead silico chromate; anticorrosive pigments.

Oneida Polish. [Malco Prods.] Silver, brass, and metal polish.

One Touch. [GE Silicones] Silicone sealant.

Onguard. [Drew Ind. Div.] Water treatment controller.

Onifine. [Otsuka] Azodicarbonamide blends; blowing agent for prod. of cross-linked PVC foam of closed-cell structures; cushioning and absorbing materials; insulation, water flotation items; sealant in motor and construction industries.

Onymyrrhe. [Alban Muller] Biological nail regenerator.

Onynex. [Merck] Ascorbyl palmitate, citric acid, glyceryl stearate blends.

Onyxide® 200. [Stepan] Hexahydro-1.3.5-tris (2-hydroxethyl)-s-triazine; preservative for sol. cutting fluids and coolants.

Onyx Premier. [Alcan] Alumina trihydrate; filler for cultured onyx.

OP-100, OP-200. [RTD Chem.] Sodium stearate.

OP-2000. [BASF] PPG-26 oleate; cosmetic emollient, lubricant, defoamer, visc. control agent, dispersant, spreading agent for personal care prods.

Opacicoat. [ECC Int'l.] Calcium carbonate; for polymers.

Opacimite®. [ECC Int'l.] Calcium carbonate; for paper industry.

Opacitex. [ECC Int'l.] Calcined clay-kaolin.

Opacoat. [Colorcon] Colored film coating.

Opadry. [Colorcon] Complete film coating conc.; for pharmaceuticals.

Opalux. [Colorcon] Coating conc.; for pharmaceuticals.

Opalwax. [CasChem] Hydrogenated castor oil.

Opaspray. [Colorcon] Pigment dispersion film coating.

Opatint. [Colorcon] FD&C aluminum lake conc.; for foods.

Opazil®. [BASF AG] Activated bentonite deriv.; for absorption of interfering substances in papermaking.

Opex®. [Uniroyal] Chemical blowing agents.

OPK-1000. [RTD Chem.] Potassium stearate.

OPPalyte®. [Mitsui Petrochemicals] Oriented polypropylene film; high opacity films for use as a single web or in laminations.

Oppanol®. [BASF; BASF AG] Polyisobutylene; for adhesives and sealants, elec. insulating oils, bases for chewing gums, for prod. of damp-proof courses containing fillers in construction industry.

Oppasin®. [BASF AG] Inorganic and organic pigment concs.; for coloring rubber compds. and sol'ns.; optical brighteners for textile fibers.

Opssalit. [Opssa Resinas Sinteticas] Phenolic molding compds.

Opssalkyd. [Opssa Resinas Sinteticas] Unsaturated polyester molding compds.

Opssamin. [Opssa Resinas Sinteticas] Melamine-phenol molding compds.

Opssapol. [Opssa Resinas Sinteticas] Unsaturated molding compds.

Optal. [Henkel Chem. Ltd.] Casein.

Optec. [Atomergic Chemetals] Crystal-growing grades of chemicals.

Optene-B. [Neste Composite Materials] Barrier polymer; adhesion and barrier materials used in multilayer pkg.

Optiblanc. [3-V; Sigma Prodotti Chimici] Additives for detergents, textiles, paper, plastics; optical brightener for fibers.

Optical Whitener. [Polymer Research Corp. of Am.] Polyester, nylon acetate whitener.

Optichrome. [Lawter Int'l.] Fluorescent color bases.

Opticite. [Dow] Label films.

Opticlean®. [Solvay Enzymes] Bacterial protease.

Opti-Color. [Opticolor] Visc. regulator.

Optigard. [Dow Corning] Optical fiber coating.

Optigel. [United Catalysts] Bentonite, or smectite prods.; thixotrope for aq. cosmetic creams and lotions.

Optima. [Van Den Bergh Foods] Hydrogenated vegetable oils; center fats for confectionery, ice cream coatings.

Optimase®. [Solvay Enzymes] Protease or protease amylase; enzymes.

Optimix. [BASF] Flush colors.

Optinol. [Yorkshire Pat-Chem] Carriers for dyeing polyester.

Optiwhite. [Burgess Pigment] Anhydrous aluminum silicate.

Optiwite. [CNC Int'l.] Optical whites for textile processing.

Oracet. [Ciba-Geigy/Pigments] Solvent-soluble dyes for plastics.

Oraid® Liq. [Dycho] Chelating complex; boil-off and scouring agent for synthetic fibers; dye assistant.

Oramide. [Seppic] Alkanolamides or ethoxylated fatty amides; cosmetic additives.

Oramix. [Seppic] Glucosides or sarcosinates; cosmetic ingredients.

Orapol. [Seppic] Fatty acid alkanolamides; detergent.

Orapret WTNB 25. [Ceca SA] Alkylaryl sulfate and mineral oil; surfactant.

Orasol. [Ciba-Geigy/Pigments] Solvent-soluble dyes for paints.

Orco. [Organic Dyestuffs] Textile dyes, pigments, accelerators, antifoams, antimigrants, scours, catalysts, dye assistants and carriers, sequestrants, flame retardants, levelers, lubricants, binders, water repellents, cleaners, optical brighteners.

Orcoacid. [Organic Dyestuffs] Textile dyes and pigments.

Orcobrite WL. [Organic Dyestuffs] Organic brightener and color intensifier for wool.

Orcoceylene. [Organic Dyestuffs] Detergent, emulsifier, wetting agent, kier boiling assistant, bleaching aid, dispersant, dye assistant.

Orcochrome. [Organic Dyestuffs] Textile dyes and pigments.

Orcocil. [Organic Dyestuffs] Gas fade inhibitor; textile dyes and pigments.

Orcocilacron. [Organic Dyestuffs] Textile dyes and pigments.

Orcocryl. [Organic Dyestuffs] Acrylic copolymer emulsion; textile resin finish with firm hand.

Orcofix. [Organic Dyestuffs] Dye fixing agents.

Orcoform. [Organic Dyestuffs] Textile dyes and pigments.

Orcofresh. [Organic Dyestuffs] Odor control agent for formaldehyde resins in textile finishing plants.

Orcogal ST. [Organic Dyestuffs] Stripping assistant for vat, sulfur, and direct dyes; dispersant in soaping prints.

Orcolan. [Organic Dyestuffs] Textile dyes and pigments.

Orcolite. [Organic Dyestuffs] Thiourea deriv.; reducing agent.

Orcolitefast. [Organic Dyestuffs] Textile dyes and pigments.

Orcomine. [Organic Dyestuffs] Textile dyes and pigments.

Orcomol. [Organic Dyestuffs] Fulling assistant.

Orconol. [Organic Dyestuffs] Detergent for greasy soils; dye retardant; emulsifier, dispersant.

Orcopal. [Organic Dyestuffs] Scouring and fulling agent for wool.

Orcoset. [Organic Dyestuffs] Polyvinyl acetate emulsion; semidurable hand building finish for cellulosics and synthetics.

Orcosil. [Organic Dyestuffs] Low crock additive.

Orcosoft. [Organic Dyestuffs] Softeners for textile finishing, dyebaths.

Orcoterge. [Organic Dyestuffs] Detergent, emulsifier for fibers.

Orcotol. [Organic Dyestuffs] Kier boiling assistant for vat dyed or printed materials.

Orcowet. [Organic Dyestuffs] Wetting and dispersing agent for dyeing.

Orcozine. [Organic Dyestuffs] Textile dyes and pigments.

Orcozine Retarder. [Organic Dyestuffs] Retarders for use in dyeing of acrylic fibers with basic dyes.

Orel. [DuPont] Tapered polyester filaments.

Oremet Titanium. [Oregon Metallurgical] Titanium ingots, mill prods., castings, and sponge.

Orevac®. [Atochem N. Am.; Atochem UK] Polyethylene, propylene, or EVA terpolymers; for blown film coextrusion, cast film coextrusion, extrusion coting, skin pkg. adhesive, adhesive emulsions.

Orgamide. [Atochem UK] Nylon 6.

Organex. [Certified Processing] Caffeine anhydrous USP.

Organiclear. [Standard Tar Prod.] Wood preservative.

Organosilane. [Degussa] For penetrating sealers for concrete and wood.

Orgasol. [Atochem N. Am.; Atochem Deutschland GmbH] Nylon; engineering polymers.

Orgasol. [Lipo] Nylon 12.

Orgater. [Atochem UK] PBT.

Orgozon. [Clough] Phosphate esters; detergent, emulsifier.

Oriole. [Warner-Graham Ltd.] Shellac, lacquer solvents, water repellents, polyurethane, lacquer, alcohols.

Orion. [Ubbink Nederland BV] Polyurethane/polyester.

Orlene. [Calgon] Paper chemical.

Ormagel. [Assessa-Industria] Algae extract/sorbitol blends.

Oroglas. [Rohm & Haas Europe] Polymethylmethacrylate.

Oromid. [Alfordshire Ltd.] Nylon 6/66.

Oronal. [Seppic] Sodium sulfates or blends.

Oropon. [Rohm & Haas] Protease; enzymes for bating hides before tanning.

Orotan. [Rohm & Haas] Tanning agent.

Oroxolate Powd. [Organic Dyestuffs] Reducing agent for stripping dyes.

Or/Scrub. [Huntington Labs] Antimicrobial skin cleanser.

Ortegol®. [Goldschmidt] Crosslinking additive for prod. of polyurethane slabstock and foams.

Orthochrom. [Rohm & Haas] Nitrocellulose lacquers; for leather finishing.

Ortho-Clear. [Rohm & Haas] Lacquers for leather finishing.

Ortho Danitol®. [Chevron] Fenpropathrin; pyrethorid with insecticidal and miticidal activity for use on ornamentals, fruits, cotton, and other field and veg. crops.

Ortholite. [Rohm & Haas] Vinyl lacquers; for leather finishing.

Ortho Orthocide®. [Chevron] cis-N-[(Trichloromethyl) thio]-4-cyclohexene-1,2-dicarboximide; fungicide for control of diseases on a variety of fruit and vegetables.

Orthophen®. [Atochem N. Am.] Amylphenol; intermediate for chemical specialties; also in mfg. of photographic chemicals, oil demulsifiers, phenolic resins, agric. surfactants, antiskinning agents.

Ortho Prunit. [Chevron] (E)-1-(p-Chlorophenyl)-4,4-dimethyl-2-(1,2,4-triazol-1-yl)-1-penten-3-ol;plant growth retardant for wide variety of native and ornamental trees.

Ortho Select. [Chevron] Selective postemergence grass herbicide.

Ortho Spotless. [Chevron] Diniconazole; systemic, sterol-inhibiting fungicide for use on peanuts, apples, grapes, small grains, etc.

Ortho Sumagic. [Chevron] Uniconizole; plant growth retardant for container-grown ornamentals.

Ortolan®. [BASF AG] 1:2 metal complex dyes; for dyeing and printing of wool and nylon fibers.

Orvus. [Procter & Gamble] Ammonium laureth sulfate, cocamide MEA, SD alcohol 40-B.

Orzan®. [Rayonier] Lignosulfonates; binder, dispersant, emulsifier, suspending agent for asphalt, gypsum board, skins, leather, metallurgy, briquetting, resins, adhesives, pesticides, water treatment, ore flotation.

Orzol. [Witco/Sonneborn] White mineral oil USP; emollient, lubricant for food, drug, and cosmetic industry.

OS-2. [Werner G. Smith] Emulsifiable ester; for buffing, lapping, drawing, grinding compds.; rust preventative.

OS-59. [Monsanto] Dielectric heat transfer liq.

OS-124. [Monsanto] High temp. industrial functional fluid.

Osage. [Lyondell Petrochemical] Gas engine oils.

Osimol Grunau. [Grünau] Sulfonic acid, sulfonates, amide amine, or ethylene oxide derivs.; leveling agent, dispersant, migrating agent, protective colloid for dyeing applics.

OSO®. [Hitox] Red and yellow iron oxides; synthetic iron oxides for coatings and plastics; natural red iron oxide for highly loaded coatings, primers, industrial maintenance finishes, exterior applics., building prods., rubber, plastics.

O2 Sorb. [United Desiccants-Gates] Oxygen scavenger packets.

Ospin. [Tokai Seiyu Ind.] Detergent, leveling agent, migrator, retarder for dyeing.

Ospol. [Tokai Seiyu Ind.] Nonionic; scouring and washing agent.

OSR-7348. [Crain] Oil spill remover.

Ottalume. [Ferro] Fluorescent zinc oxide; uv stabilizer, opacifier for plastics and paints.

Ottasept®. [Ferro] Chloroxylenol; antimicrobial, preservative, disinfectant for industrial, chemical, and cosmetic uses incl. adhesives, shoe polishes, printing inks, cutting fluids, shampoos, medical powds., antiseptics, sanitizing soap, cosmetics.

Oust. [DuPont/Ag] Herbicide.

Outrite®. [Lord] Mold release agents for

rubber, plastics and metals.

Ovazyme. [Finnsugar Bioprods.] Glucose oxidase; enzyme for food processing.

Ovucire. [Gattefosse SA] Hemi-synthetic glycerides; excipients for suppositories.

OW-1®. [Air Prods.] Secondary acetylenic alcohol; corrosion inhibitor for oil well acidizing, steel pickling, and electroplating.

O-W Lubekote. [Chem. Processing] Textile overwax.

Oxaban®. [Angus] Oxazolidines; antibacterial, preservative for cosmetics.

Oxamin LO. [ICI Australia] Lauryl dimethylamine oxide; detergent, food foamer and foam stabilizer for personal care prods. and detergent formulations.

Oxetal. [Zschimmer & Schwarz] Ethoxylated fatty alcohol ethers; detergent, dispersant, emulsifier, wetting agent used in detergents and cleaners for household and industry; aux. agent for textile, paper and leather industry.

Oxi-Chek. [Ferro] Hindered phenol; antioxidants for rubber, plastics, latexes, adhesives.

Oxidan. [3-V; Sigma Prodotti Chimici] Isocyanuric acid or salts; disinfectant, sanitizer for hospitals, breweries, dairy, laundry bleaches, automatic dishwashing, swimming pool, water treatments.

Oxine. [Surpass] Peroxygen bleaching compd. for bleaching cotton and blends.

Oxitol. [Shell] Ethylene glycol monoethyl ether; solvent.

Oxitone. [Nutex] Liq. cationic softener/lubricant for sanforizing, napping, and sanding operations.

Oxone. [DuPont] Monopersulfate compd.

Ox-Out. [Chemclean] Stainless steel scale and oxide remover.

Oxsol. [OxyChem] Nonflamm. solvent.

Oxy. [OxyChem] PVC resins; for blending, adhesives, flooring, coatings, records, color concs., automotive applics., inj. molded bottles and pipe fittings, calendered and extruded films, rigid foams, pipes, profiles, wire and cable, rotational castings.

Oxyblend. [OxyChem] PVC compds.

Oxybloc. [Western Water Management] Boiler water oxygen scavenger.

Oxycat. [Met-Pro] Air pollution control systems.

Oxychlor. [Int'l. Dioxcide] Chlorine dioxide generating systems.

Oxycop 8L. [Cuproquim] 8% Copper ammonium carbonate.

Oxycop 8LS. [Cuproquim] 8% Copper ammonium carbonate, 5% sulfur.

Oxycop Dry 5. [Cuproquim] 50% Copper oxychloride sulfate.

Oxyfume. [Union Carbide] Sterilant.

Oxygo. [Genencor Int'l.] Glucose oxidase; enzymes.

Oxymag. [Premier Services] Magnesium oxide for oxychloride cements.

Oxypol-11. [Gattefosse] Octoxynol-11.

Oxypon. [Zschimmer & Schwarz] Ethoxylated esters; superfatting agent, solubilizer for cosmetics.

Oxypruf. [Olin] Alkoxylated hydrazines or dimethylpyrazoles; corrosion inhibitor in functional fluids.

Oxysiv. [Union Carbide] Oxygen generator.

Oxyvet. [Henkel/Emery/Cospha] Animal aroma chemical.

Oysterical. [J.W.S. Delavau] Directly compressible oyster shell calcium carbonate.

Ozokerite. [Astor Wax] Ozokerite.

Ozone Benign. [Foamtek] Non-CFC urethane foam system.

P

P-. [Bacon] Epoxy compds.; potting compds.

P®-10 Acid. [CasChem] Ricinoleic acid; lubricant, rust inhibitor, base for cutting oils, grease, soaps, resin plasticizers, ethoxylated derivs.

P-60-10, P-60-20 Cold Molding Compounds. [Perma-Flex Mold] Two-component polysulfide systems; molding compds.

P-80-K. [Unimin Specialty Minerals] Mica.

P0114. [Hüls Am.] Phenethyltristrimethylsiloxysilane; compatibilizer for silicones and hydrocarbons.

P 210-D. [Novacor] Acrylic copolymer; PVC processing aid.

P 506. [Novacor] Styrene methyl methacrylate copolymer; for appliances, displays, medical devices, toys, office accessories.

P-0620 T 1/8″. [M&T Harshaw] Phosphoric acid on alumina; catalyst.

P0820. [Hüls Am.] Propyltris (trimethylsiloxy)silane; low visc. fluids with unique solvent properties.

PA-. [Bay Resins] Nylon 6 or 6/6, some glass or mineral reinforced, lubricated, impact modified; molding compds.

PA-. [Exxon/Tomah] Amines.

PA-14 Acetate. [Exxon/Tomah] Cationic salt; gellant, wetting agent for clays, fillers, fibers in organic coatings and sealants.

PA-57, PA-80. [Akrochem] Lightly cross-linked natural rubber with oil extender; processing aid for extrusions, calendering, and open steam curing.

PA-59-997, PA-59-999. [Polymer Applics.] Alkylphenol-acetylene reaction prod.; used in tread splice cements, tire compds.

Pabco. [Air Prods. Nederland BV] Polyurethane additives, catalyst.

Pace. [Olin] Conc. chlorinating agent.

Pacific Sea Kelp Glycolic Extract B-1063. [Bell Flavors & Fragrances] Pacific sea kelp extract; mineral supplement, lustering agent for hair care cosmetics.

Pacifix. [Pacific Dyes & Chems.] Textile dyes and pigments.

Pacisol. [Pacific Dyes & Chems.] Textile dyes and pigments.

Pacivat. [Pacific Dyes & Chems.] Textile dyes and pigments.

Pacol. [Witco/Sonneborn] Tech. white mineral oil.

Pactor Fume Up & At'Em Insecticide. [Uncle Sam Chem.] Water-based insecticide for compactors.

Pad-1. [Vevy] Magnesium carbonate, aluminum silicate.

Padding Emulsion. [Yorkshire Pat-Chem] Pigment binders for textiles.

Padimate O. [Lipo] Octyl dimethyl PABA.

Padyecron. [Pacific Dyes & Chems.] Textile dyes and pigments.

PAE®. [General Chem.] Phosphoric/nitric/acetic/water blends; phosphoric acid etchants.

Pah-Nol. [Houghton Chem. Corp.] Antifreeze for windshield washers, gas line.

Paint Kill. [Calgon] Polymeric detackifying agent.

Pak-N-Foam. [Cook Composites & Polymers] Polyurethane foam for pkg.

Palamid®. [BASF AG] Masterbatches of organic and inorganic dyes in polyamide; for polyamide spin dyeing.

Palamoll®. [BASF AG] Plasticizers for plasticized PVC prods. resistant to oil, gasoline, and bitumen.

Palanil®. [BASF; BASF AG] Disperse dyes for polyester, acetate, triacetate, nylon, polyacrylonitrile, PVC, polyvinyl alcohol fibers.

Palanthrene. [BASF] Textile dyes and pigments.

Palapreg®. [BASF AG] Unsaturated polyester resins; SMC/BMC resins for elec., automotive, building, and sanitary industries.

Palatal®. [BASF AG] Unsaturated polyester and vinyl ester resins; for building sector, containers, pipes, silos, boat building, motor vehicles, elec., chem., mech. engineering, space and air travel, consumer goods, furniture, concrete, marble, fillers, buttons, pourable sealing compds. and laminates.

Palatase. [Novo Nordisk] Fungal esterase/lipase; enzyme for hydrolysis of fats.

Palatex®. [BASF; BASF AG] Assistants to prevent creasing in textile dyeing.

Palatin®. [BASF; BASF AG] Fast dyes for coloring cleaners, textile dyeing and printing.

Palatinol®. [BASF] Phthalates, adipates, or trimellitates; plasticizer for PVC, nitrocellulose dyes, rubber, plastisols.

Pale. [CasChem] Polymerized castor oil; plasticizer, lubricant, penetrant, wetting agent, dispersant, coupling solvent, adhesion promoter for cellulose lacquers, inks, adhesives, polish, caulks, leather dressing, hydraulic fluids, rubber compding., gasket cement.

Palegal®. [BASF; BASF AG] Leveling agent for dyeing polyester fibers with disperse dyes.

Palenine. [Catawba-Charlab] Leveling and stripping agent.

Paliofast. [BASF] Pigments.

Paliogen®. [BASF] High quality organic pigments for paints, plastics, tin plate inks, artists paints and crayons.

Paliotan®. [BASF] Co-finish pigments for industrial finishes, polymer dispersions.

Paliotol®. [BASF] Plastic metal complex pigments. for specialty surface coatings, coloring plastics, artists' paints and crayons, tin plate inks.

Pallamerse. [Technic] Palladium barrier layer.

Pallaspeed. [Technic] Palladium.

Pallcell. [Pall Corp.] Cellulose; pleated, high area disposable filters.

Palliag. [Degussa] Palladium alloy; for dental applics.

Pallinal®. [BASF AG] Metiram, nitrothal-isopropyl; for control of powdery mildew in fruit, vegetables, hops, ornamentals.

Pallitop®. [BASF AG] Nitrothal-isopropyl blends; for control of powdery mildew in apples.

Pallsorb. [Pall Corp.] Absorbent filter cartridge.

Palmabeads. [Hoffmann-La Roche] Dry vitamin A palmitate beadlets.

Palma-Sperse. [Hoffmann-La Roche] Dry vitamin A palmitate beadlets.

Palmsal. [Leatex] Cationic softener for textiles.

Palomar. [Miles/Organic Prods.] Textile dyes and pigments.

Palusol®. [BASF AG] Glass fiber/hydrated sodium silicate fireproof sheet; for construction materials.

Pamak 4. [Hercules] Tall oil acid.

Pam Masterbatches. [Pamplast ApS] Masterbatches and concs.

Pamolyn® 100. [Hercules] Oleic acid; detergent intermediate, emulsifier, fiber lubricant, textile processing aid, defoamer, emulsion breaker.

Pampus. [Rias A/S] PTFE.

Panacea. [CC Pollen] Bee propolis tablets.

Panalane®. [Amoco] Hydrogenated polyisobutene; for cosmetics applics.

Panalite. [Paniplus] Mono- and diglycerides or derivs.; food emulsifier.

Panatex. [Paniplus] Distilled monoglycerides; emulsifier system.

Pancogene® S. [Gattefosse; Gattefosse SA] Soluble collagen; humectant for cosmetics.

Pangel. [Omya GmbH] Sepiolithe.

Pan-L-Bond. [W.J. Ruscoe] Structural adhesive.

Panodan. [Grindsted Prods.; Grindsted Prods. Denmark] Monoglyceride diacetyl tartaric acid ester; food emulsifier.

Panol. [Enzyme Development] Enzyme, papain preparation.

Pan-O-Lite. [Monsanto] Leavening

agent.

Panph. [Pfaltz & Bauer] pH indicator.

Pansil. [Omya GmbH] Sepiolithe.

Pansorb. [Colgate-Palmolive] Ingredient in medicine.

Pantalast®. [Pantasote Polymers] VC-EVA copolymers; engineered materials.

Pantex. [Texo] Alkylaryl sulfonate; for institutional hand washing.

Panther Creek. [Am. Colloid] Southern bentonite.

Pantoxyl®. [Condea Chemie GmbH] Phenolic and melamine resins; bonding resins for wood processing, abrasives, foundry industry.

Panzyme. [Burlington Bio-Medical] Animal feed ingredient.

Papain P-100. [Finnsugar Bioprods.] Protease; enzyme for food processing.

Paperaid. [Polypure] Polymers for pulp and paper process.

Papi. [Dow] Polymeric MDI; for polyurethane industry for adhesives, binders, coatings, sealants, molded elastomers, structural foam molding, appliances, rigid and flexible foams, industrial/consumer RIM and structural polymers.

Pa-Quel. [Monsanto] Bicomponent acrylic fibers.

Parabar. [Exxon] Oxidation inhibitors.

Parabis. [Dow] Bisphenol A; resin intermediate used to produce PC and polysulfone engineering thermoplastic resins.

Parabolix®. [Merix] Degreaser, destaticizer for electronics, parabolic light fixtures.

Parachlor. [DuPont] Red pigment.

Paraclor. [Uniroyal] Chlorinated polyethylene elastomer; used in wire, cable, hose, mech. goods, extruded, inj. molded, and calendered goods, sheet, sponge, and for blending.

Paracol. [Nippon Nyukazai] Alkyl phosphate; emulsifier, anticorrosion agent, emulsion breaker for insecticides.

Paracol®. [Hercules] Paraffin or microcrystalline wax emulsion or wax-rosin emulsion; sizing agent for paper/paperboard mfg.

Paracril®. [Uniroyal; Uniroyal Chem. Ltd.] NBR or NBR/PVC fluxed blends; for inj. and trasnfer molding, extrusion, blending, hose, seals, o-rings, adhesives, automotive parts, shoe soles, rolls, wire and cable jacketing, hose tubes and covers, sponge, belting, gasketing, military applics.

Paradene®. [Neville] Coumarone-indene resin; used in adhesives, coatings for aluminum, citrus fruit, emulsion, epoxy, industrial, marine, paper, traffic, wire & cable, rubber (cements, mechanical and molded goods, tires), and caulking compds.

Paradet R329. [Amax Industrial Prods.] Super strength industrial detergents.

Paradize. [Petrokem] Insecticides.

Paradyne. [Exxon] Distillate-fuel improvers.

Paraffin Wax Emulsifier. [Croda Chem. Ltd.] Ethoxylated fatty alcohols; paraffin wax emulsifier.

Parafflex. [IGI Boler] Paraffin wax blends.

Paraflow. [Exxon] Additives.

Para-Flux. [C.P. Hall] Saturated polymerized petroleum hydrocarbon.

Paragon. [J.M. Huber] Filler clay.

Paragon. [McIntyre] Propylene glycol, DMDM hydantoin, paraben blends; dual action preservative for cosmetics and toiletries.

Paralan. [Croda Chem. Ltd.] Tech. lanolin; emulsifier.

Paralgol. [Ceca SA] Algicide for water treatment.

Paraloid®. [Rohm & Haas; Rohm & Haas Europe] Acrylic-based polymers; impact modifier, toughener, blending additive, gloss reducer, processing aid.

Paramel. [Am. Cyanamid] Paper resins.

Paramins. [Exxon] Fuel and oil additives.

Paramount. [Van Den Bergh Foods] Hydrogenated vegetable oils blends; center fat or coating fat for confectionery, icings, cosmetics, pharmaceuticals.

Paramul®. [Bernel] Surfactants; emulsifier, pearlizing agent, opacifier.

Paranol K. [Finetex] Surfactants for use as detergents, wetting agents, emulsifiers, and scours.

Paranox. [Exxon] Detergent inhibitors.

Parapel® HC. [Bernel] Linoleamidopropyl ethyldimonium ethosulfate, dimethyl lauramine isostearate; emulsifier, hair conditioner, fragrance solubilizer for hair care prods.

Parapel® LAM-1000. [Bernel] Lactamide MEA; hair conditioner, humectant.

Parapel® LIS. [Bernel] Dimethyl lauramine isostearate; shampoo and conditioner additive; emollient.

Paraplast. [Hexcel] Inorganic powders; used to cast mandrels or cores used in prod. of hollow plastic articles.

Paraplex®. [C.P. Hall] Sebacates, adipates, polyester, or epoxidized soybean oil; plasticizers/stabilizers for PVC, NC, CAP, chlorinated rubber.

Paraplex®. [Rohm & Haas] Sebacic acid plasticizing coating resin.

Parasepts. [Kalama] Preservatives.

Parasoft. [Paradigm Labs] Softener for textile finishes with soft, silky hand.

Parasonarl Mark II. [Nippon Chem. Co. Ltd.] Ethyl urocanate.

Paratac. [Exxon] Tackiness improvers.

Para Toluene Sulfonic Acid. [Boliden Intertrade] Catalyst, organic intermediate, hydrotrope.

Paratone. [Exxon] Viscosity index improvers.

Par® Clay. [R.T. Vanderbilt] Hard kaolin; reinforcer and inert filler for elastomers, nonblack compds.

Parco®. [Thibaut & Walker] Vinyl acrylic emulsions; used in paints.

Parcroloid. [Rowe Prods. Distribution] Chlorinated rubber base coatings.

Parcryl®. [Thibaut & Walker] Acrylic emulsions; used in paints, caulks, roof and tennis-court coatings.

Parel®. [Zeon] Propylene oxide/allyl glycidyl ether copolymer; elastomer for dynamic mounts and high flex applics.

Parez. [Am. Cyanamid] Paper resins.

Paricin®. [CasChem] Hydroxystearates or acetoxystearates; wax modifier, emollient, mold release, lubricant, slip additive.

Paris. [ICI Am.] Textile dyes, pigments.

Paroil®. [Dover] Chlorinated paraffin; flame retardants.

Parol. [Penreco] Mineral oil, tech.; used for animal feed dedusting, food pkg. materials, meat pkg., household cleaners and polishes.

Parolite®. [Henkel/Textile] Zinc formaldehyde sulfoxylate; stripping agent for nylon and wool.

Paron. [Rudd] Chlorinated rubber paint.

Parsol®. [Bernel] UV absorber, sunscreen agent.

Part-Ease. [Am. Grease Stick] Lubricant.

Parteze. [Kano Labs] Silicone mold release.

Particulo. [General Chem.] Low particle chemicals.

Particu-Lo® LTM. [General Chem.] Ultra pure low particle grade chemicals for electronic applics.

Partsprep Degreaser. [ISP] N-Methyl-2-pyrrolidone and surfactant.

Parvan. [Exxon] Paraffin.

PAS-60. [Igoplast-Faigle AG] Polyamide 6.

PAS-80 MO. [Igoplast-Faigle AG] Polyamide 6/6 and molybdenum disulfide.

PAS-80 X. [Igoplast-Faigle AG] Polyamide 6/6 and polyethylene.

Pasco. [Pigment & Chem.] Zinc oxide pigment.

Paselli. [Avebe Am.] Potato starches for food industry.

PAS-L. [Igoplast-Faigle AG] POM.

PAS-LX. [Igoplast-Faigle AG] POM and PE.

PAS-PE 10. [Igoplast-Faigle AG] UHMWPE.

Passavant. [Zimpro Passavant Environmental Systems] Clarifier, filter press.

Paste Wiz. [Axel Plastics Research Labs] Mold release agent.

Patco. [C.H. Patrick] Textile dyes and pigments.

Patco® 3. [Am. Ingredients] Sodium stearoyl lactylate/calcium stearoyl-2-lactylate blend; conditioner/softener; starch and protein complexing agent for bakery prods.

Patcofix. [C.H. Patrick] Polymeric fixing agent.

Patcolev. [C.H. Patrick] Leveling agent for textile dyeings.

Patcorez. [C.H. Patrick] Durable press resin for fabrics.

Patcosperse. [C.H. Patrick] Textile dyes and pigments.

Patcosul. [C.H. Patrick] Textile dyes and pigments.

Patcote®. [Am. Ingredients] Silicone and nonsilicone emulsions; defoamers for coatings, resins.

Patcovat. [C.H. Patrick] Textile dyes and pigments.

Patfix. [Yorkshire Pat-Chem] Fixing agent for fiber reactive and direct dyes.

Pat-Fix. [Yorkshire Pat-Chem] Fixative for fiber reactive and direct dyes, prints.

Path. [Solem Industries] Titanium dioxide; extender pigment.

Patilon®. [Atochem N. Am.] Engineering polymers.

Patinal. [EM Industries] Vacuum deposition chemicals.

Pationic. [RITA] Lactylates; surfactant, emulsifier, detergent, conditioner, visc. builder, emollient, moisturizer, protein complexer for cosmetics; perfume solubilizer.

Pationic®. [Am. Ingredients] Esters; mold release agent, antistat, process aid for inj. molding and extrusion, lubricant, melt flow enhancement, pigment wetter for color concs.; catalyst scavenger.

Patlac. [Am. Ingredients] Lactic acid or derivs.; surfactant, emollient, humectant, stabilizer, pH buffer for cosmetics.

Pat-Lube. [Yorkshire Pat-Chem] Lubricants for jet and beck dyeing.

Patogen. [Yorkshire Pat-Chem] Wetting and scouring agent, foam booster, detergent, emulsifier for textiles.

Patoran®. [BASF AG] Metobromuron; for pre-emergence weed control in potatoes, soybeans, tobacco, tomatoes.

Pat-Quest CS. [Yorkshire Pat-Chem] Chelating/sequestering agent for textile bleaching and scouring.

Patsoft. [Yorkshire Pat-Chem] Softener for natural and synthetic fabrics and blends.

Pat-Soft. [Yorkshire Pat-Chem] Softeners, lubricants, finishing auxiliaries, antiozonants for textiles.

Pat-Wet. [Yorkshire Pat-Chem] Alkyl phosphate; wetting and scouring agent for dyeing.

Pave. [Morton Int'l.] Complex amide amine; additive for asphalt, coatings; wetting and bonding aid.

Paxon®. [Paxon Polymer] HDPE; resins for blow molding, inj. molding, rotational molding, extrusion, films, structural foam.

Paxwax. [Nat'l. Wax] Microcrystalline waxes.

PBN. [Elco] Metalworking additive.

PBO-8. [Prentiss Drug & Chem.] Emulsifiable piperonyl butoxide conc.

PBT-. [Bay Resins] PBT polyester, some glass reinforced, impact modified, flame retarded.

PBZ-. [Zeon] Hydrogenated nitrile rubber/PVC blends; elastomer for fuel hose, fuel diaphragms.

PC-. [Bay Resins] Polycarbonate, carbon reinforced; molding compd. for elec./electronic, business machines applics.

PC-. [Compounding Tech.] Polycarbonate, some carbon or glass reinforced, lubricated; molding compds.

PC-35. [Paniplus] Mono- and diglyceride hydrate; food emulsifier.

PC072, 075, 085. [Hüls Am.] Platinum catalysts.

PC-1244, -1344. [Monsanto] Ethyl acrylate/2-ethylhexylacrylate copolymer; defoamer in nonaq. hydrocarbon and solv. systems, lubricating oils.

PCA. [Polyurethane Corp. of Am.] Polyurethane prepolymers or polymers.

PCL. [Dragoco] Esters and blends; cosmetic ingredients.

PCL. [Velsicol] Hexachlorocyclopentadiene.

PCMX. [Ferro] Fungicide or biocide.

PCQ. [Bell Labs] Meal bait rodenticide; anticoagulant for rats and mice.

PCR. [PCR] Fine chemicals.

Pd-0803 E 1/16 ˝. [M&T Harshaw] Palladium; catalyst for hydrogenation of alkenes in aromatic streams.

PDDS. [Molecular Rearrangement] Paradiazo diphenylamine sulfate.

PDI. [Pigment Dispersions] Plastics additives.

PDM. [ECC Int'l.] Kaolin; paper filling clay.

PDO-40. [Mantrose-Haeuser] Ink varnish/vehicle.

PDX. [LNP] Nylon 6/10 or 11, PEEK, PPS, PC, ABS, PEI, PC, or PBT, some metal filled; conductive attenuating composites effectively shielding electromagnetic and/or radio frequency interference; used in avionics housings, business machine enclosures, and other electronic devices.

PE. [Chevron] Polyethylene resins; for extrusion, film, coating, lamination

PE-25. [Washington Penn Plastics] MDPE; extrusion grade resin.

PE 100 228. [Zipperling Kessler] Halogen-free flame-retardant polyethylene compd.; for extrusion.

Peacock. [Geo. Pfau's Sons] Animal oils, lard oils, mink oils, neatsfoot oils, tallows, tallow oils, stearine, blown oils, fatty acids, oleo stearine.

Pea Pro-Tein. [Maybrook] Hydrolyzed pea protein; substantivity agent, filmformer, moisturizer for skin and hair care prods.

Pearex-L®. [Solvay Enzymes] Pectinase.

Pearl I, II, III. [Presperse] Bismuth oxychloride; pearl pigments.

Pearlex. [Clough] Proprietary blend; pearlizing/opacifying agent for shampoos, bubble baths, liq. hand soaps.

Pearl-Glo®. [Van Dyk] Bismuth oxychloride; pearlescent pigment for cosmetics.

Pearl Luster. [Hoechst Celanese] Glycol distearate.

Pearl Supreme, Super Supreme. [Presperse] Bismuth oxychloride or blends.

Pearsall. [Witco] Aluminum chloride.

Pebax®. [Atochem N. Am.; Atochem UK] Polyether block amide polymers. extrusion/molding grade polymer used in sport, automobile, elec./electronics, medical, agriculture, and mechanical handling industries, adhesives.

Pebble Lime. [Rockwell Lime] Dolomitic pebble lime.

Peceol. [Gattefosse] Glyceryl oleate.

Peceol Iso. [Gattefosse; Gattefosse SA] Glyceryl isostearate; emulsifier, superfatting agent, pigment dispersant, additive for lipsticks.

Pecogel. [Phoenix] PVP/ester blends.

Pecosil. [Phoenix] Silicones.

Pectinex. [Novo Nordisk] Pectinase; pectin-decomposing enzymes for food, wine, and citrus processing.

Pectinol®. [Genencor Int'l.] Pectinase; enzyme for juice, wine industry.

Peerless®. [R.T. Vanderbilt] Kaolin clay; filler used in adhesives, wallboard, paint, paper, fertilizer, roofing, crayons, soaps, pharmaceuticals, ceramics, sanitaryware, artware, floor tile, porcelain, refractories.

PEG Esters. [Stepan] Emulsifiers, emollients, opacifiers, spreading agents, wetting agents, dispersants.

Pegnol. [Toho Chem. Industry] Ethoxylated alkyl ethers, esters, or amines; emulsifier, dispersant, antistat, thickener, dispersant, leveling agent.

Pegol®. [Rhone-Poulenc Surf.] Alkoxylated glycol; defoamer, dispersant, wetting agent, emulsifier, demulsifier, leveling agent and detergent.

Pegosperse®. [Lonza] Polyethylene or ethylene glycol fatty acid esters; emulsifier, stabilizer, emollient, lubricant, thickener, wetting agent, plasticizer, and pigment dispersant in pharmaceuticals and cosmetics.

Peladow. [Dow; Schaefer Salt] Calcium chloride.

Pelaspan. [Dow] Expandable polystyrene.

Pelemol. [Phoenix] Esters; cosmetics ingredients.

Pelex. [Kao] Sulfonates or sulfosuccinates; wetting agent, penetrant, emulsifier, foam controller for latex, emulsion polymerization, textile processing.

Pellethane®. [Dow] Polyurethane elastomers; for automotive, health care (tubing, catheter components, transdermal patches) applics., in film or sheets as bladders, liners, or belting; in fabricated prods. for casters, athletic shoes, bushings.

Pelplus. [Martin Marietta Magnesia Spec.] Pellet binder for animal and aquaculture feed.

Pelron. [Pelron] Polyurethane prods.

Pels. [PPG Industries] Caustic soda beads.

Peltek. [Hydrolabs] Fluoropolymer emulsions; durable fabric and fiber finishes offering water and oil repellency.

Peltex. [LignoTech] Ferro-chromium lignosulfate; wetting agent, emulsifier, dispersant, oil well drilling mud thinner, conditioner.

Pel-Unite. [Oil Dri Corp. of Am.] Binders.

Pemublend. [Plastenterprise Pemu] PP.

Pemuflon. [Plastenterprise Pemu] PTFE.

Pemulen® TR-1. [BFGoodrich/Spec. Polymers] Acrylates/C10-30 alkyl acrylate crosspolymer; emulsifier for creams, lotions, waterproof sunscreens, fragrance prods.

Pemuplasten. [Plastenterprise Pemu] PP, HDPE.

Pemusil. [Plastenterprise Pemu] Silicone rubber.

Pemutherm BG. [Plastenterprise Pemu] Polyamide.

Penacolite®. [Indspec] Resorcinol-formaldehyde resin; adhesives for tire and rubber industry; for dry bonding, dipping, special compounding.

Penelox®. [Atochem N. Am./Textiles] Peroxide bleaching system activator.

Penephite. [Kano Labs] Graphite penetrating and lubricating oils.

Penestrol. [Tokai Seiyu Ind.] Secondary alcohol ethoxylate; penetrant, wetting agent, bleaching assistant.

Peneteck. [Penreco] Mineral oil, tech.; emollient.

Penetize. [Penetone] Disinfectant cleaner, deodorizer, fungicide, virucide.

Penetone DC. [Penetone] Quaternary-based; disinfectant cleaner, deodorizer, fungicide, sanitizer.

Penetone Formula 861. [Penetone] Fuel oil additive.

Penetral NA 20. [Ceca SA] Low-foaming wetting agent for mercerizing.

Pene-Tri-A-Nol. [Scholler] Wetting and penetrating agents for wet-out, scour, and dyebaths.

Penetrol 2-EHS. [Clark] Ammonium salt of 2-ethlhexylsulfate; wetting agent.

Penetron OT-30. [Hart Prods. Corp.] 2-Ethylhexyl sulfosuccinate; penetrant, wetting agent.

Penford Gums. [Penford Prods.] Ethylated corn starches.

Pengloss. [Penford Prods.] Starch.

Pennad 150. [Atochem N. Am.] Diethylaminoethanol; intermediate, emulsifier, catalyst in urethane foams, curing agent, corrosion inhibitor for boiler water.

Pennant. [Lyondell Petrochemical] Gear oils.

Penncap-M. [Atochem N. Am.] Insecticides.

Pennchem® Mortar. [Atochem N. Am.] Two-component, silica-filled chemical-resistant, vinyl ester resin-based mortar; mortar for construction in pulp/paper, chemical process, food and beverage plant industries.

Pennco. [Penn Color] Color pigment dispersions.

Penncoat® 101 Membrane. [Atochem N. Am.] Three-component chemical-resistant asphaltic barrier for concrete substrates.

Penncozeb. [Atochem N. Am.] Fungicide.

Penn-Drake. [Penreco] Oils, greases.

Pennex. [Exxon] Metalworking lubricant.

Pennfloat®. [Atochem N. Am.] Mercaptan; intermediate.

Pennguard®. [Atochem N. Am.] Inorganic foamed borosilicate glass block; protectant for metals, concrete and FRP substrates from acid corrosion; provides thermal insulation and energy conservation.

Pennlon®. [Furon] Filled PTFE; bearing material.

Penn Mar Kote. [Dryden Oil] Solid lubricants.

Pennodorant®. [Atochem N. Am.] Tetrahydrothiophene; odorant for natural gas to permit detection of leaks.

Pennstop®. [Atochem N. Am.] N,N-Diethylhydroxylamine; free radical scavenger for rubber industry as an emulsion polymerization inhibitor; vapor phase inhibitor for olefin or styrene monomer recovery systems.

Penntrowel. [Atochem N. Am.] Epoxy surfacers.

Pennwalt 4P®. [Atochem N. Am.] Tert. dodecyl mercaptan; modifier in polymerization reactions, esp. for SBR and NBR.

Pennwalt n-Dodecyl Mercaptan. [Atochem N. Am.] n-Dodecyl mercaptan; modifier in polymerization reactions, esp. for SBR; reducing initiator and chain transfer agent.

Pennzone. [Atochem N. Am.] Thioureas; rubber additives.

Penreco. [Penreco] Petrolatum, petroleum distillates, or hydrocarbon solvents; emollient, base, carrier, binder, processing aid for rubber, polishes, corrosion preventatives, lubricants, inks, solder pastes, agric. sprays, fruit and veg. processing, cleaners.

Pensil®. [GE Silicones] Silicone foam or rubber sealant; for crack and void filling, fire spread prevention.

Penstick. [Eni-Trade Grafiske AS] Static clinging vinyl.

Penta. [Chapman] Wood preservative.

Pentacat. [Akzo] Alcoholysis catalyst.

Pentacite. [Arizona] Resin ester.

Pentacite. [Reichhold] Synthetic resin.

Pentagen. [Pentapharm AG] Glycogen.

Pental. [Hercules] Glycerol ester of tall oil rosin; thermoplastic resin for furniture lacquers, sanding sealers, inks, low-bake enamels, grease-resistant coatings.

Pentalan. [Croda Chem. Ltd.] Pentaerythritol tetraester; corrosion protectives, greases.

Pentalyn®. [Hercules; Hercules Europe] Rosin derivs.; thermoplastic resin for inks, polishes, food pkg. and processing, adhesives, specialty coatings; binder, leveling agent, tackifier.

Pentaphen®. [Atochem N. Am.] Phenolic compositions; intermediate for chemical specialties; in germicidal formulations; also in mfg. of photographic chemicals, oil demulsifiers, phenolic resins, agric. surfactants, antiskinning agents.

Pentaprint. [Eni-Trade Grafiske AS] Rigid PVC.

Pentaquest. [Clough] DTPPA, pentasodium pentetate or DTPAA; chelating agents.

Pentech. [Owens/Corning Fiberglas] High efficiency polyester resin for filled systems.

Pentek-600. [Pentek] Self-stripping polymers for removing rust, oxides, surface contamination and radioisotopes.

Pentex®. [Rhone-Poulenc Surf.] Sulfonates or sulfosuccinates; wetting agent, detergent, penetrant, emulsifier, dispersant for textiles, industrial processing, leather, paper.

Pentine. [Clough] Dodecylbenzene sulfonic acid or salts; emulsifier for degreasers, drycleaning, agric. formulations.

Pentol. [Witco/Sonneborn] White mineral oil.

PentoXone. [Shell] 4-Methoxy-4-methyl-pentanone-2; high boiling solvent used in prep. of NC, acrylic, vinyl, and cellulose acetate butyrate lacquers, alkyd and epoxy enamels, and urethane and thermosetting acrylic coatings; used in cleaning compds.

Pentrone. [Rhone-Poulenc Ltd.] Disodium alkoxysulfosuccinate deriv.; used for water-based systems.

Penwet. [Manufacturers Chems.] Wetting agent, penetrant for dyeing, bleaching operations.

Peorol. [Toho Chem. Industry] Surfactants complex; for use in water flooding for enhanced oil recovery.

Pep®. [Air Prods.] Polyester promoter.

Pepol. [Toho Chem. Industry] Ethoxylated fatty alcohols; detergent, wetting agent, penetrant.

Pep Set®. [Ashland/Foundry Prods.] Catalyzed foundry core binders.

Peptein®. [Hormel] Hydrolyzed animal or wheat proteins; conditioner for cosmetics.

Peptidyl. [Serobiologiques] Glycerin,

water, serum protein, glycogen.

Peptizer. [C.P. Hall] Oil/sulfonate blends.

Pepton. [Am. Cyanamid] Plasticizers for rubber.

Peracit. [Perstorp Chemitec] Phenolic resins.

Peral. [Raschig] Additive used as bonding agent for coatings.

Peralube. [Perstorp Chemitec] Cutting fluids.

Peramin. [Perstorp Chemitec] Amino resins.

Peramit. [Pulcra SA] Ammonium lauryl sulfate; for light duty detergents, hand cleaning pastes.

Peran. [Perstorp Chemitec] Epoxy resin for flooring.

Peranat. [Henkel/Emery/Cospha] Fruity, pear-like aroma chemical.

Perapret®. [BASF; BASF AG] Fillers, stiffeners, leveling agents for textile finishing and dyeing.

Perbunan. [Bayer; Miles] NBR or NBR/PVC fluxed blends; used in tech. moldings, seals, sleeves, diaphragms, bellows, valves, vibration dampers, footwear soles; roll covers, printing blankets, belting, fabric proofings, hoses, cable jackets, brake linings, sponge rubber, adhesives, profiles.

Perc. [Ferro] Porcelain enamel powder coatings.

Perchem®. [Akzo] Montmorillonite or hectorite clays; rheological additives, gellant, antisettling agent, viscosifier.

Perclene. [OxyChem] Perchlorethylene; transformer, dry cleaning, vapor degreasing grades.

Percol. [Allied Colloids] Flocculants.

Peregal®. [ISP] PVP polymer; dispersant, stripping assistant, detergent, leveling agent, suspending agent for dyeing, paper, textiles.

Perenol. [Henkel; Henkel KGaA] Polymers; defoamer, anticaking agent for coatings and paints.

Perfecta®. [Witco/Sonneborn] Petrolatum; emollient for cosmetics, pharmaceuticals.

Perfectamyl. [Avebe Am.] Potato starches; for textile, food industries.

Perfection. [Rhone-Poulenc Basic] Sodium acid pyrophosphate; food grade leavening acid.

Perfekthion®. [BASF AG] Dimethoate; systemic insecticide for control of sucking and biting insects.

Perflex. [Van Den Bergh Foods] Partially hydrogenated vegetable oil; emulsified shortening for cakes, yeast-leavened prods.

PerforMax®. [Drew Ind. Div.] Corrosion inhibitor, cooling water treatment.

Pergan. [Newgate Simms Ltd.] Organic peroxides.

Pergascript. [Ciba-Geigy/Dyestuffs] Color former.

Pergasol. [Ciba-Geigy/Dyestuffs] Direct dye for textiles.

Periclase. [Martin Marietta Magnesia Spec.] Magnesium oxide.

Perkacit®. [Akzo] Accelerators for rubber.

Perkadox®. [Akzo; Akzo Chem. BV] Peroxide derivs.; initiators, crosslinking agents.

Perkalink®. [Akzo] Cyanurates or methacrylates; co-agent to improve efficiency of peroxide-induced cross-linking of rubber; sensitizer for radiation-cured compds.

Perkasil®. [Akzo] Precipitated silicas; reinforcing filer for rubber industry.

Perlankrol®. [Harcros UK] Surfactants; for personal care prods., textiles, household and industrial cleaners, emulsion polymerization.

Perlatum. [IGI Petroleum] Petrolatum.

Perlex. [Rhone-Poulenc] Bismuth compds.; pearlescent for makeup, emulsions, dry systems.

Perlextra. [Rhone-Poulenc] Bismuth compds.; pearlescent for makeup, emulsions.

Perlglanz-Konzentrat. [Goldschmidt AG] Cocamide DEA; pearlescent, opacifier, detergent for personal care prods.

Perlglanzmittel. [Zschimmer & Schwarz] Blends; pearlescent, opacifier for personal care prods.

Perlite King. [Filter-Media] Cryogenic perlite insulation.

Perlit® SE. [Miles/Organic Prods.] Aftertreating agent for improving crockfastness of dyeings on synthetic fibers.

Perltex. [Solico-Southwest Vermiculite] Paint filler, texturizer.

Perm-A-Clor. [Detrex] Trichloroethylene.

Permacol. [Betz Industrial] Filming amine.

Permacor. [Arnco] Two-component urethane; liq. urethane for repair of pneumatic tires.

Permacurate. [Hoffmann-La Roche] Sodium ascorbate; meat grade.

Permafin. [Nat'l. Starch & Chem.] Softener, lubricant for textile processing.

Permaflex. [Goldschmidt] Amino-modified polysiloxane; softener for fabrics.

Perma-Flex. [Perma-Flex Mold] Polysulfide or silicone systems; for making elastic molds and patterns, for precision casting, plaster casting and foundry pattern material.

Perma-Flo. [A.E. Staley Mfg.] Modified corn starch.

Permaflon. [Permali Gloucester Ltd.] PTFE.

Permafresh®. [Sequa] Glyoxal-based thermosetting resins; for durable press textile applics.

Permagrip. [Imperial Adhesives] Oil and heat resistant solvent cement.

Perm-A-Kleen. [Detrex] Perchloroethylene.

Perma Kleer®. [Rhone-Poulenc/Textile & Rubber] Chelating agents.

Permalease. [George Mann] Mold release agents.

Permalene. [Rhone-Poulenc/Textile & Rubber] Amine-coconut oil condensate; detergent, dispersant, dye leveler, lubricant.

Permalite. [Grefco] Perlite.

Permaloid. [Rhone-Poulenc] Textile sizes.

Permalose. [ICI Surf. UK] Hydrophilic copolymer aq. dispersion; soil-release/antistatic agent for polyester fabric treatment.

Permalux. [DuPont] Neoprene accelerator.

Perma-Mold. [Brulin] Release agents.

Permanax. [Akzo] Antioxidants and antiozonants for rubbers.

Permanent Encapsulant. [Hexcel] Urethane compd.; castable elastomer for encapsulating telecommunications cable.

Permapol®. [Products Research & Chem.] Polythioether polymers; adhesion promoter, processing aid, reactive plasticizer; coreactant to form polyurethanes or polyester.

Permasep. [DuPont] Permeators.

Perma-Slik. [E/M Corp.] Solid lubricants.

Permasint. [Plastic Coatings Ltd.] Polypropylene powder coatings.

Permasist® CG. [Rhone-Poulenc/Textile & Rubber] Liq. replacement for sodium hexametaphosphate to overcome dissolving and caking problems.

Perma-Tac. [ScanRoad] Asphalt anti-stripping additive.

Permatox. [Chapman] Wood-treating chemical.

Permax. [Yoshimura Oil Chem.] Antistats for synthetic fibers.

Permel®. [Am. Cyanamid] Textile resins.

Perm-Ethane DG. [Detrex] Chlorinated solvent.

Permethyl®. [Presperse] Polyisobutene, isohexadecane, isoeicosane, or isododecane; cosolubilizer for nonhydrocarbon materials, solvent, plasticizer; for makeup, sun care and skin prods., cleansers.

Permnasoft. [Manufacturers Chems.] Softener for denims.

Permyl®. [Ferro/Bedford] Proprietary; light stabilizer, uv absorber for polymers (fiberglass-polyester corrugated outdoor panels; PVC film for outdoor use; extruded garden hose; cellulosic eyeglass frames).

Peronal. [Seppic] Anionic surfactant blend; textile detergent.

Perone. [DuPont] Hydrogen peroxide.

Peropal®. [Bayer] Azocyclotin; acaricide for fruits, vegetables, cotton, other crops.

Perowhite. [Hydrolabs] Textile auxiliary.

Peroxal. [Seppic] Anionic/nonionic surfactants; textile detergent.

Peroxide Stabilizer. [Hoechst AG] Sodium polycarboxylate; stabilizer for hydrogen peroxide bleaching.

Peroxidol. [Reichhold] Epoxidized soya or tall oil; plasticizer, stabilizer.

Peroximon®. [Akrochem] Peroxides; vulcanizing and crosslinking agents for rubber industry.

Perrindo. [Miles/Organic Prods.] Textile dyes and pigments.

Per Sec. [Vulcan Materials] Perchloroethylene.

Persellig. [Ceca SA] Leveling agent for dispersed dyes on polyamide and polyester.

Persistol®. [BASF AG] Hydrophobic and oleophobic agents for textile finishing, furs.

Persoftal®. [Miles/Organic Prods.] Antistat, lubricant, softener for textiles.

Perspex. [Mooney Plastics Ltd.] Acrylic.

Perthane. [Rohm & Haas] Synthetic (polychor) organic insecticide.

Perylene. [Sun Chem. Corp.] Textile dyes and pigments.

Pestilizer. [Stepan] Phosphate esters; emulsifiers for liq. fertilizers.

Petcat. [Laurel Industries] Antimony trioxide.

Petlon. [Miles] PET polyester, glass and/or mineral-reinforced; thermoplastic for inj. molding of automotive, elec/electronic, and mechanical industrial components.

Petra®. [Allied-Signal] PET polyester, glass and/or mineral-reinforced.

Petrac®. [Syn. Prods.] Stearic acid, metallic stearates, amides, or waxes; lubricant, activator, dispersant, plasticizer, mold release agent, emulsifier, slip agent, thickener, anticaking agent for plastics, syn. lubricants, bar soaps, cosmetics, rubbers, polishes, paints, inks, waterproofing.

Petrex. [Hercules] Thermal-curing alkyd resin; for inks.

Petro®. [Witco/Organics] Alkyl naphthalene sulfonates; hydrotrope, surfactant, germicide, detergent, dispersant, wetting agent for agric., industrial and household cleaners, plating baths, dyestuffs.

Petrobac. [Polybac] Adapted bacterial hydrocarbon degrader.

Petrocote. [Petrolite/Polymers] Anticorrosion coating.

Petrodarco. [Am. Norit] Activated carbon.

Petroflux. [Witco] Process oils, extender oils.

Petrolatum RPB. [Witco] Petrolatum, tech.

Petrolig. [LignoTech] Drilling mud dispersant.

Petrolig ERA. [LignoTech] Enhanced oil recovery agents.

Petrolite®. [Petrolite/Polymers] Synthetic waxes; used in coatings, adhesives, paper, printing inks, plastic modification (as lubricant and processing aid), lacquers, paints, varnishes, as binder in ceramics, for potting elec./electronic components, in investment castings, textiles, cosmetics.

Petromiser. [Stewart Hall] Fuel oil additives for soot removal.

Petromix. [Witco/Sonneborn] Petroleum sulfonate; emulsifier, base for cutting oils, solvent degreasers.

Petronate. [Witco/Sonneborn] Petroleum sulfonate salts; detergent, rust inhibitor, wetting agent, emulsifier, dispersant, rust preventative for lube and industrial oils, fuels, greases, drycleaning soaps, textile processing.

Petronauba®. [Petrolite/Polymers] Oxidized microcrystalline wax; for polishes, emulsions; carnauba substitute.

Petro-Rez. [Akrochem] Aromatic hydrocarbon resin; plasticizer, tackifier, processing aid, reinforcing extender for rubber compounding, resin modification, adhesives, inks.

Petro-Rez. [Lawter Int'l.] Hydrocarbon resin.

Petrosan. [Reilly-Whiteman] Methyl esters; metal lubricants.

Petroscale. [Witco/Organics] Organic polyphosphonate; chelating agent for industrial water treatment.

Petrosolve®. [Witco/Organics] Aromatic solvent; additive for carburetory and soak cleaners.

Petrostep. [Stepan] Petroleum sulfonates; for enhanced oil recovery operations.

Petrosul®. [Penreco] Sodium petroleum sulfonate; surfactant, corrosion inhibitor; as motor and fuel oil additives, rustproofing formulations, drycleaning solvents, leather processing, printing inks, oil well drilling fluids, ore flotation, metalworking fluids.

Petrothene®. [Quantum/USI] Polyethylene or polypropylene resins; for inj., rotational, and blow molding, sheet and profile extrusion, vacuumforming, thermoforming, film, wire and cable, industrial containers, closures, toys, foamed parts, pipe, food pkg., housewares, adhesives, coatings.

Petrowet®. [DuPont] Sodium alkyl sulfonate; wetting agent, dispersant, detergent, penetrant for petrol., metalworking, textile, and paper industries; industrial formulations and cleaning.

Pevikon. [Norsk Hydro Plast Oy; Norsk Hydro AS] PVC resin emulsion.

Pexite® Wood Rosin. [Hercules] Refined natural resins; thermoplastic acidic resin for paper sizes, soaps, emulsifiers, and greases, and for mfg. of syn. resins for adhesives, paints, varnishes, food-pkg. and -processing.

Pexol®. [Hercules] Rosin sizes for paper industry.

PF-. [Compounding Tech.] Polysulfone, some carbon or glass reinforced.

P-Fibre. [Grindsted Prods.] Fibers; for food industry.

Pfinyl. [Pfizer] Surface-treated ground limestone.

P-Flakes. [Karlshamns] Hydrogenated palm oil.

P&G Amide No. 27. [Procter & Gamble] Cocamide MEA; foam stabilizer, thickener; foam booster and visc. builder for detergents and shampoos.

PGE-. [Hefti Ltd.] PEG esters; emulsifier, plasticizer, solubilizer, antistat for cosmetics, pharmaceuticals, paper, deinking, plastics, textiles, agric.

PGM-Mica. [Presperse] Mica; cosmetics ingredients offering softly lusterous to brilliantly reflective properties.

PGMS 70. [Croda Food Prods. Ltd.] Propylene glycol stearate; for whipped desserts, whipped toppings, cake mixes and shortenings.

PG No. 4. [Hart Prods. Corp.] PEG 400 stearate; emulsifier, base for textile softeners.

Phaase 25. [Ivar Labs] Ice nucleation catalyst.

Phantolid. [Hercules/FFIG] Patented aromatic.

Pharmaceutical Lanolin Anhydrous USP. [Croda Inc.] Lanolin.

Pharmasorb. [Engelhard] Attapulgite clay; adsorbent for pharmaceuticals.

Pharmatinic. [Pharmachem Labs] Iron dextran.

Pharmolin. [Engelhard] Kaolin; for pharmaceutical applics.

Phase Alpha®. [Ashland] Resins for automotive and truck body composite fabrication.

Phase II®. [Ashland] SMC resins for automotive and truck body composite fabrication.

Phenobac. [Polybac] Adapted bacterial hydrocarbon degrader. for pharmaceutical applics.

Phenodur. [Hoechst Celanese] Phenolic resins for coatings, laminates and molding.

Phenofoam. [Lewcott] Phenol formaldehyde insulating foam.

Phenoline. [Carboline] Phenolic.

Phenonip. [Nipa Labs] Phenoxyethanol; preservative system.

Phenothiazine LVT. [ICI Polymer Additives] Phenothiazine aq. dispersion; stabilizer and chain stopping agent for acrylic polymers.

Phenoweld. [Hardman] Phenolic adhesives.

Phenoxetol. [Nipa Labs] Phenoxyethanol; antimicrobial preservative for cosmetics and pharmaceuticals.

Phenrez®. [Hercules] Petroleum hydrocarbon thermoplastic resin; used in adhesives, coatings, inks, molding compositions; binders for foundry core sand and fibrous prods.

Phenyl Sulphonate. [Hoechst Celanese;

Hoechst AG] Calcium alkylaryl sulfonate; base material for mfg. of biocide emulsifiers.

Pheonate 3 DSA. [Phoenix] PEG-3 distearate.

PHG. [Cargill] 99.95% NaCl.

Philacid. [United Coconut Chem.] Fatty acids; intermediate for mfg. of detergents, toilet soaps, cosmetics, amines, esters for use as lubricants, emollients, flavors, and perfumes.

Philcohol. [United Coconut Chem.] Fatty alcohols; intermediate for mfg. of detergents; consistency agent, emollient.

Philsavon. [United Coconut Chem.] Coconut fatty acids.

Phloroglucinol. [Schering Berlin Polymers] 1,3,5-Trihydroxybenzene; intermediate for pharmaceuticals; coupler in dyeline printing; photographic chemicals.

Phobotex®. [Ciba-Geigy/Dyestuffs] Water repellent for textiles; extender for fluorochemical finishes.

Phoenamid. [Phoenix] Fatty acid alkanolamides.

Phoenix PS-2766. [Phoenix] Behenoxy dimethicone.

Phoenoxol. [Phoenix] Cetearyl alcohol/ethoxylated cetearyl ether blends.

Phorwite. [Miles/Organic Prods.] Textile dyes and pigments.

Phos-Ad 100. [Chemron] Tributyl phosphate; defoamer for oilfield applics.

Phosal. [Nattermann Phospholipid] Lecithin blends; prod. of liposomes for dermatology and cosmetics; for creams, shampoos, bath oils, soap additive, hair rinses.

Phosaver. [Grain Processing] Phosphate mining chemical.

Phosbrite® 172. [Albright & Wilson Am.] Proprietary metal finishing additive; used to chemically polish aluminum.

Phos-Chek. [Monsanto; Monsanto Europe] Flame retardants.

Phosfac. [Rhone-Poulenc/Textile & Rubber] Phosphate ester; emulsifier, penetrant, antistat, wetting agent, detergent, lubricant, corrosion inhibitor and dispersant used in liq. detergent formulations, textiles, metalworking fluids, pigments.

Phosfetal. [Zschimmer & Schwarz] Phosphoric acid ester; cleansing agents.

Phosflex. [Akzo] Phosphates; flame retardant plasticizers for plastics and rubber.

Phosokresol. [Hoechst AG] Aromatic dithiophosphates; flotation collectors for sulfide minerals.

Phosphac. [Synthron] Wetting agents.

Phosphanol. [Toho Chem. Industry] Phosphate esters; emulsifiers, antistats, solubilizers for spinning oils, textiles, cosmetics, polymerization, lubricants.

Phosphate Ester 123. [Hoechst Celanese] Phosphoric acid ester; corrosion inhibitor, lubricant.

Phosphite-1. [Ciba-Geigy/Additives] Di-t-butylphenylphosphonite condensation prod. with biphenyl; antioxidant.

PhosPho. [Fanning] Phospholipids or hydroxylated or hydrogenated phospholipids; surfactant, suspending agent, emulsifier, solubilizer, superfatting agent for nail polishes, hair and skin conditioning prods., face creams

Phosphogel. [Pierce Chem.] Affinity chromatography gel for phosphorylated proteins, peptides, and amino acids.

Phospholan® [Harcros UK] Phosphate esters; surfactant, hydrotrope, wetting agent, detergent, dispersant, coupling agent, emulsifier for textiles, drycleaning, industrial and domestic cleaners, agric., metal cleaning, emulsion polymerization.

Phospholan KPE4. [Henkel/Emery] Potassium arylethoxy phosphate.

Phospholipid. [Mona Industries] Alkyl amidopropyl PG-dimonium chloride phosphates; synthetic phospholipids for use in cosmetic preparations.

Phospholipon. [Am. Lecithin; Nattermann Phospholipid] Lecithin prods.; emulsifier, producer of liposomes for dermatology and cosmetics.

Phosphoteric® T-C6. [Mona Industries] Substituted carboxylated cocoimidazoline organophosphate; surfactant, hydrotrope for cleaning formulations.

Photine. [Ronsheim & Moore] Dimorpholine type brightener; fluorescent brightener for washing powds.

Photochromic. [Molecular Rearrangement] Dyes.

Photocure 51. [Aceto] Benzil dimethyl ketal; uv photoinitiator.

Photoglaze. [Lord] Uv-cure coatings.

Photomer®. [Harcros UK] Acrylic derivs.; radiation-curing resins for inks, varnishes, glass lamination, oriented polypropylene film lamination, coatings for plastics, blister pack adhesives.

Photomet. [Atochem N. Am.] PCB coating.

Phreeguard. [Calgon] Cooling water treatment.

Phthalavin. [Colorcon] Enteric coating polymer.

Phthalopal®. [BASF AG] Oil-free phthalate resins; for spirit-based and nitrocellulose finishes; in hydrolyzed form as binders for aq. flexographic inks and ballpen inks.

Phylderm® Filatov. [Gattefosse] Human placental protein; biological additive.

Phytaxyl. [Serobiologiques] SD alcohol 39-C, barley extract, propylene glycol, polysorbate 20.

Phyt'iod. [Alban Muller] Ethiodized oil; slenderizing prods.

Phytoresinoids. [Bio-Botanica] Highly conc. botanical extracts.

PI. [Compounding Tech.] Polyetherimide, some glass or carbon reinforced.

Pibiter. [Enimont UK Ltd.; Montedipe Srl] PBT.

Picaltal®. [BASF AG] Mixture of aromatic sulfonic acids; complex-forming acid in chrome tanning.

Piccantase. [Int'l. Bio-Synthetics] Lipase; enzyme.

Picco®. [Hercules] Aromatic hydrocarbon resins; thermoplastic tackifier in elastomer-based sealants, adhesives, inks; as extender, processing and reinforcing agent in rubber compd. and extruding applics.

Piccodiene®. [Hercules] Aliphatic hydrocarbon resin; thermoplastic with balanced tack, adhesive, and cohesive props. when blended with elastomers.

Piccofyn®. [Hercules] Terpene hydrocarbon resin; used as tackifiers in adhesives and rubber compding.; in laminating agents, plastics modification, paints, varnishes, and printing inks.

Piccolastic®. [Hercules] Hydrocarbon resin derived from styrene monomer; thermpolastic resin, softener, plasticizer in hot-melt compositions, adhesives, coatings, and rubber compd.,food-pkg. and processing operations.

Piccolyte®. [Hercules] Terpene hydrocarbon resin; thermoplastic resin, tackifier, modifier in adhesives, coatings, laminations, rubber, wax, sealants, inks, medical goods; waterproofing resin for textile sizing.

Piccomer®. [Hercules] Aromatic hydrocarbon resins; thermoplastic resin, vehicle, tackifier, softener, plasticizer, reinforcer for paints, varnishes, coatings, inks, adhesives; modifier for rubber and wax.

Picconol®. [Hercules] Aliphatic hydrocarbon resin emulsions; used in combination with other aq. thermoplastic and/or elastomeric systems to produce coatings, paints, and adhesives; tackifiers for natural and syn. rubber systems; waterproof finishes for paper, textiles, and textile backings.

Piccopale®. [Hercules] Aliphatic hydrocarbon resin emulsions; tackifier, binder for rubber, adhesives, construction materials; saturant/waterproofing agent for paper, textiles; replacements for gloss oils in paints, varnishes, and coatings.

Piccotac®. [Hercules] Aliphatic hydrocarbon resin emulsions; tackifier, binder for rubber, adhesives; saturant/waterproofing agent for paper, textiles; modifier for waxes.

Piccotex®. [Hercules] Vinyltoluene/alpha-methylstyrene copolymer resins; thermoplastic used in hot-melts, adhesive systems, rubber compding., specialty paint and lacquer formulations.

Piccovar®. [Hercules] Hydrocarbon resins; plasticizer, softener and tackifier in rubber compd., pressure sensitive applic., coatings, and adhesive systems;

food processing operations.

Pickelene. [Atochem N. Am.] Acid salts.

Pidolidone®. [UCIB] PCA; cellular penetration vector for amino acids or mineral salts; for skin and hair care formulations.

Pierce. [La-Co Industries] Rust inhibitor.

Pierce. [Markal] Aerosol penetrant.

Piersolve. [Pierce Chem.] Highly purified 2-methoxyethanol; solvent.

Pigmosol. [BASF] Dry pigment dispersions.

Pillmaster. [Sybron] Acrylic-based; durable anti-pilling agent for textiles.

Pilot® SXS. [Pilot] Sodium xylene sulfonate; coupling agent, hydrotrope, solubiizer, dipersant, solvent, stabilizer for liq. cleaners, organic polymers and dyestuffs, petrol. industry, pulping, animal glues.

Pil-Trol. [Monsanto] Low pill acrylic fibers.

Pinamine. [Consos] Pine oil/surfactant blends; scour and detergent for textiles.

Pineotrene. [Reilly-Whiteman] Detergents for textile scouring.

Pinnacle. [DuPont/Ag] Herbicide.

Pioloform. [Wacker Chem.; Wacker Chemie GmbH] Polyvinyl butyral.

Pioneer. [Dry Branch Kaolin] Kaolin.

Pioneer. [Theodor Leonhard Wax] Pure refined yellow beeswax.

Pionier PLW. [Hansen & Rosenthal] Mineral oil, polyethylene.

Piosolv. [Akzo] Solvent blends for cleaning and deinking.

Piotron. [Akzo] Nonionic surfactants; for rewetting, deinking, and dispersing.

Pioze 40. [Nippon Oils & Fats] PEG-75.

Pipetite. [La-Co Industries] Pipe thread compd.

Pirimor. [ICI Am.] Insecticide.

Piror. [Union Carbide] Slimicides for paper industry.

Pittabs. [PPG Industries] Calcium hypochlorite.

Pittclor. [PPG Industries] Calcium hypochlorite; algicide, antimicrobial, fungicide used to chlorinate swimming pools and as a disinfectant for cleaning pool parts, locker room floors; used for municipal water treatment, sewage

treatment, in textile and paper mills, tanneries.

Pix®. [BASF AG] Mepiquat chloride; bioregulator for reduction of undesired vegetative growth of cotton, better boil retention, earlier maturity; improves yield and market quality of garlic and onions.

PK. [Compounding Tech.] PEEK, some glass or carbon reinforced.

PKP. [Transene] Purified kodak photoresists.

PL-. [Chem-Trend] Lubricating oil blends; plunger lubricants.

PL-71. [Crain] Corrosion inhibitor.

Placentaliquid. [Henkel/Cospha] Extract of unborn bovine placentas; for skin toning, anti-aging skin prods., hair loss protection prods.

Placonite. [Peterson] Industrial maintenance coatings.

Planell Oil. [Brooks Industries] Squalene, squalane, glycolipids, phytosterol, and tocopherol; natural plant emollients for skin care cosmetics.

Planifoline. [Florasynth] Imitation vanilla.

Plantaren. [Henkel/Cospha; Henkel KGaA] Fatty alcohol glycoside; solubilizer, wetting agent for dishwashing agents, laundry detergents, cleaners.

Plant Exsyliposomes. [Exsymol] Plant unsaponifiables/glycerophospholipids blend; for cosmetics.

Plantvax. [Uniroyal] Fungicide.

Plasblak®. [Cabot Plastics Ltd.] Carbon or furnace black masterbatches with polyethylene, EVA, polypropylene, polystyrene, or SAN carriers; black masterbatch for molding and extrusion of articles for pkg., electronic equip. cases, radios, TVS, domestic appliances, pipe and sheet, film, cable sheathing.

Plas-Chek. [Ferro/Bedford] Epoxidized soybean or linseed oils; stabilizer, plasticizer for PVC compds.

Plascize. [Goo Chem.] Acrylates/ diacetoneacrylamide copolymers or derivs.

Plasdeg. [Cabot Plastics Ltd.] LDPE/ starch-based masterbatch; degradation promoter imparting controlled degradability to polyolefin films and moldings.

Plasdone®. [ISP] PVP; excipient, tablet binder, coating agent, solubilizer, stabilizer, protective colloid, vehicle for pharmaceuticals.

Plasfalt. [Kao Corp. SA] Amine-based; for slurry seal additives.

Plasgrey®. [Cabot Plastics Ltd.] LDPE/titanium dioxide/carbon black masterbatch; light gray masterbatch used for coloring LDPE, HDPE, and PP; applications incl. films (garbage bags, heavy-duty sacks), inj. molded parts and pipes.

Plaskon. [Plaskon Electronic Materials] Epoxy molding compds.; for the encapsulation of semiconductor devices.

Plaslube®. [Akzo Engineering Plastics] Acetal, nylon 6, 6/6, 6/12, PC, or PPS resins lubricated with PTFE, silicone, molybdenum disulfide, or carbon.; internally lubricated engineering thermoplastics.

Plaspylene. [Plastribution Ltd.] Polypropylene.

Plastaplex. [Sullivan Chem. Coatings] Polystyrene coating.

Plastazote. [BXL Plastics Ltd.] Foamed crosslinked polyethylene.

Plasteccol. [Eclipse Colours Ltd.] DOP dispersions.

Plastech. [Cabot Plastics Ltd.] Polypropylene, polyethylene, or polystyrene resins, some glass or mineral filled; for inj. molded parts for automotive industry, e.g., dashboards, door moldings, lamp housings, under-the-hood applics. such as fans, fan housings, tech. parts, garden furniture.

Plasthall®. [C.P. Hall] Tallates, phthalates, glutarates, adipates, azelates, sebacates, trimellitates, other esters; plasticizers for rubber, adhesives; lubricant, softener, penetrant for textiles, caulk, printing ink, surface coatings.

Plasthene. [Plastribution Ltd.] Polyethylene.

Plasticizer 9. [BASF AG] Glycerol ether alcohol; plasticizer for naphtha-resistant and benzene-resistant nitro-lacquers, sheet gelatin, aq. shellac sol'ns.

Plasticizer REO. [Akrochem] Paraffinic/naphthenic process oils and petroleum sulfonate mixture; plasticizer, process aid for synthetic and natural rubber compds.

Plastiflex. [Plastics & Chem.] Plasticizers and halogenated hydrocarbons.

Plastifoam. [Bubble & Foam Industries NV] Expanded polyethylene.

Plastigel®. [Plasticolors] Metal oxides or hydroxides in vehicles; thickener in the prod. of thickenable molding compds. (polyester SMC, BMC, TMC).

Plastigen® G. [BASF AG] Carbamide resin; lightfast, nonhydrolyzable resin for nitrocellulose, chlorinated rubber finishes.

Plasti-Grit Clear-Cut. [Composition Materials] Thermoset plastic for dry stripping composites and thin aluminum media.

Plasti-Grit Flour. [Composition Materials] Crosslinked thermoset powder; reinforcing agent, filler, extender for plastics.

Plastilease. [Lilly] Silicone, PVAL, or polyolefin; release agents.

Plastilit® 3060. [BASF AG] Polypropylene glycol alkylphenyl ether; plasticizer for polymer dispersions in paint, construction, chemical, adhesive, and sealant industries.

Plastilube. [Plastics & Chem.] Specialty surfactants and lubricants.

Plastisan. [3-V] Plastic additives.

Plastislip. [Dow Corning France SA] Special lubricants for plastics.

Plastisorb. [Plasticolors] Uv and weather-resistant colorants, additives.

Plastisperse®. [Plasticolors] Granular, dustless colorants for polymers, thermoplastic elastomers, bulk molding compds.

Plasto. [Crompton & Knowles] Textile dyes and pigments.

Plastoflex®. [Akzo] Epoxidized soybean oils; plasticizer/stabilizer for PVC.

Plastogen. [King Industries] Rubber plasticizers.

Plastogen. [R.T. Vanderbilt] Plasticizer, softener for elastomers, latex.

Plastol®. [BASF AG] Low visc. polymer plasticizers for PVC.

Plastolein®. [Henkel/Emery] Plasticizers for PVC, elastomers, food pkg. films, plastisols, calendered fabrics, appliance cords, gasketing, wall coverings, automotive constructions, elec. insulation and tape, adhesives.

Plastomag®. [Morton Int'l.] Oil-dispersed magnesium oxide; chemical thickener for polyester resins; anticaking agent; used in syn. rubber compounding, adhesives, fuel oil additives, and as acid acceptor for specialty plastics.

Plastomoll®. [BASF AG] Adipic acid esters; plasticizers for plasticized PVC, nitro lacquers.

Plastone®. [Harwick] Plasticizer, peptizer, processing aid.

Plastopal®. [BASF AG] Urea-formaldehyde resins, butanol or methanol etherified; binders for baking finishes, acid-curable coatings, film coatings, nitrocellulose finishes, epoxy finishes.

Plastriox. [Mines De La Lucette] Flame retardants.

Plastyrene. [Plastribution Ltd.] Polystyrene.

Plasvita®. [Hüls AG] Methylene casein; tablet disintegration agent for pharmaceuticals.

Plaswite®. [Cabot Plastics Ltd.] Titanium dioxide masterbatch in polyethylene, polypropylene, or polystyrene carriers; some with lithopone or calcium carbonate as extenders; white masterbatch for blown and cast film, inj. molding, blow molding of drug, cosmetic, and food containers.

Platabond. [Atochem Deutschland GmbH] EVA hot melt.

Platamid®. [Atochem N. Am.; Atochem UK] Hot melt adhesive produced from a copolyamide base; for extrusion of fusible film, monofilament, netting, web, and multifilament; for thin film fusible coatings; for fusion bonding of textiles, leather, wood, glass, and metals; automotive interiors, belts, color concs., rainwear, shoes, wire coatings.

Plathen. [Atochem Deutschland GmbH] PE hot melt.

Platherm®. [Atochem N. Am.; Atochem UK] Copolyester hot melt; engineering polymers.

Plathuran. [Atochem UK] Polyurethane powders.

Platone. [Am. Labs] Hydrolyzed enzyme.

Plenco. [Plenco Plastics Eng.] Melamine, phenolic, or alkyd compds., some mineral, cellulose filled; molding compds.

Plenex. [Vista] Thermoelastic polymer.

Plex. [Vikon] Chelating agents for textile scouring, peroxide bleaching.

Plexar®. [Quantum/USI] EVA or polyethylene-based resin adhesives; tie-layer resins for coextruded pkg.

Plexene®. [Sybron] Chelating agents for dyeing, peroxide bleaching; sequestering agent for Ca and Mg under hard water conditions.

Plexiglas®. [Rohm & Haas; Rohm GmbH] Acrylic resins; inj. molding and extrusion materials, plastic sheet.

Pliabrac. [Merrand] Phosphates; plasticizers for rubbers and plastics, some flame retardant.

Pliabrac®. [Albright & Wilson Am.] Antiwear and extreme pressure additives for lubricants; plasticizers.

Pliobond®. [Ashland] Adhesives.

Pliocord®. [Goodyear] Vinyl pyridine-styrene-butadiene latex; used for cord adhesion in tires, conveyor belts, hose.

Plioflex®. [Goodyear] SBR; used in general purpose masterbatches, shoe soles, household goods, carpet underlay, sponge prods.

Pliogrip®. [Ashland; Ashland-Südchemie-Kernfest GmbH] Polyurethane-based structural adhesives.

Pliolite®. [Goodyear; Goodyear Europe] Styrene/acrylate, styrene/butadiene, vinyl-toluene/acrylate, or vinyl-toluene/butadiene emulsions; latexes used in spread foam, molded foam, scatter rug, backsizing, and adhesives.

Plion. [Vikon] Lubricant and softener for cottons, synthetics, blends.

Plio-Nail®. [Ashland] Adhesives.

Plio-Seam®. [Ashland] Adhesives.

Pliovic®. [Goodyear] Vinyl resins; for

plastisols, organosols, blending applications, dipped goods, coatings for awnings, rainwear, wall coverings, foams for garments, upholstery, and carpet backings, weatherstripping, toys, novelties.

Plioway®. [Goodyear] Acrylic resin.

Pluracol®. [BASF; BASF AG] Polyoxyalkylene polyols, PEGs, PPGs polyols for foams, blown elastomers, molding, crosslinking agent; PEGs and PPGs as intermediate, coupling agent, thickener, lubricant, mold release, defoamer, conditioner, antistat, sizing agent, dispersant.

Pluradot. [BASF] Polyoxyalkylene glycols; emulsifier, solubilizer, deduster, wetting agent, demulsifier, detergent, dispersant; surfactant for hard surface cleaning, machine dishwashing.

Pluradyne. [BASF] Oilfield chemicals.

Plurafac®. [BASF; BASF AG] Oxyalkylated alcohols; detergent, dispersant, wetting agent, emulsifier, defoamer, deduster used in domestic and commercial cleaning formulations, agric. adjuvants.

Pluraflo®. [BASF] Nonionic surfactant blends; dispersant, wetting agents for pesticides.

Plurasafe. [BASF] Fire resistant hydraulic fluid.

Pluriol®. [BASF AG] PEGs, PPGs, or PO/EO block polymers; emulsifier, solubilizer, humectant, binder, dispersant, plasticizer, softener, mold release, lubricant, defoamer in cosmetics, dyestuffs, inks, textiles, coatings, paper, adhesives, metalworking, rubber and latex, agric., emulsion polymerization.

Plurol. [Gattefosse; Gattefosse SA] Polyglyceryl esters; emulsifiers.

Pluronic®. [BASF; BASF AG] PO/EO block copolymers; emulsifier, wetting agent, binder, stabilizer, plasticizer, lubricant, solubilizer, dispersant, visc. control agent, defoamer, intermediate for detergents, textiles, cosmetics, pharmaceuticals, pulp, paper, petroleum, agric., water treatment.

Plyamul®. [Reichhold] Vinyl acetate emulsions; adhesives.

Plyocite. [Reichhold] Resin-impregnated paper.

Plyophen®. [OxyChem] Phenolic resins.

Ply-Pro 25. [Aqualon] Tissue-converting adhesive for towel stock.

Plysolene PIB. [Plysolene Ltd.] Polyisobutylene.

Plystran® C. [Seydel-Woolley] Carboxymethylcellulose-based; dry blend warp size for blends and cotton.

Plysurf. [Dai-ichi Kogyo Seiyaku] Phosphate esters; antistats, emulsifiers, detergents, dispersants for agric., metals, emulsion polymerization, pigments, detergent systems.

PMAS. [W.A. Cleary] Phenyl mercuric acetate; fungicide.

PME. [Vevy] PPG-5 pentaerythritol ether.

PMF® Fiber. [Pall Corp.] Calcium-alumino silicate; filler-reinforcement.

PMM. [Pall Corp.] Pleated meta membrane filters.

PMP. [PMP Fermentation Prods.] Gluconic acid or derivs.; dequestrant.

PMS-33. [Hefti Ltd.] Propylene glycol stearate; emulsifier, solubilizer for food, cosmetics.

Poast®. [BASF AG] Sethoxydim; postemergence graminicide against annual and perennial grasses.

Pocan®. [Bayer] PBT or PET polyester, some glass or mineral reinforced; engineering thermoplastic for automotive, consumer, industrial/mech., elec./electronic markets.

Poem-LS-90. [Riken Vitamin Oil] Sodium cocomonoglyceride sulfate.

Poem-S-105. [Riken Vitamin Oil] PEG-5 glyceryl stearate.

Pogol. [Hart Chem. Ltd.] PEGs; lubricant, solubilizer, antistat, softener, dispersant, emulsifier, humectant, mold release, lubricant, plasticizer, tablet binder for rubber, agric., paper coatings, textiles, metals, pharmaceuticals, cosmetics.

Poise. [Engelhard] Mineral pet litter.

Poison-Pruff. [Atomergic Chemetals] Bitter aversive agent.

Poiz. [Kao] Polymeric carboxylate; dispersant for pigments and fillers.

Pokonobe. [Pokonobe Industries] Vegetable oils.

Polacure®. [Air Prods.] Diamine curative.

Polaktyny. [Organika Zachem Chem. Works] Reactive dyes.

Polamine®. [Air Prods.] Oligomeric diamines; curing agents for elastomer, coatings, and adhesives applics.

Polane Systems. [FSW Coatings Ltd.] Two-part polyurethane.

Polar. [Ciba-Geigy/Dyestuffs] Textile dyes and pigments.

Polargel®. [Am. Colloid] Bentonite; thickener, gellant, suspending agent for cosmetics, pharmaceuticals, household and industrial specialties, e.g., paints and cleaning compds.

Polarin®. [Grünau] Glycerol and glycerol derivs.; solvents, humectants for food industry.

Polaris Gums. [Penford Prods.] Ethylated potato starches.

Polarite. [Am. Colloid] Sodium bentonite; suspending agent, gellant for household and industrial specialties.

Polarite. [ECC Int'l.] Surface-treated minerals; low-profile additives for low and zero-shrink DMC and BMC moldings, high gloss prods.

Polarlink R. [ECC Int'l.] Surface-modified kaolins; for rubber industry.

Polarose. [Polarome Mfg.] Rose type synthetic.

Polarsan. [Polarome Mfg.] Sandalwood substitute.

Polathane®. [Air Prods.] Polyurethane prepolymers.

Polawax®. [Croda Inc.; Croda Surf. Ltd.] Emulsifying wax NF; emulsifier, thickener, opacifier, suspending agent, stabilizer.

Polcarb. [ECC Int'l.] Calcium carbonates.

Polectron® 430. [ISP] Styrene/PVP copolymer; binder and adhesive for wood, cotton, paper, glass, flour, concrete; stabilizer and opacifier; laundry processing; stabilizer for detergents; textile and paper coatings, latex rug backings, floor wax emulsions, and cosmetics.

Polefine. [Takemoto Oil & Fat] Alkyl naphthalene sulfonate/formaldehyde condensates; reducing agent for concrete.

Polestar 400. [ECC Int'l.] Kaolin; for coatings.

Polexan. [Domingo Pascual Carbo SA] Foam polyethylene.

Pol-E-Z. [Calgon] Emulsion polymers.

Polidan. [Padanaplast SpA] Shrink polyethylene.

Polidene®. [Scott Bader] Vinylidene chloride copolymers; binder, base for paper coatings, fire-retardant coatings, adhesives, textile fibers, pigment printing, paints, caulks.

Polidux ABS. [Aiscondel SA] ABS.

Polidux EPS. [Aiscondel SA] Expandable polystyrene.

Polidux PS. [Aiscondel SA] Crystal or high impact polystyrene.

Polidux SAN. [Aiscondel SA] SAN copolymer.

Poligen®. [BASF AG] Polyethylene emulsions; for dry-bright floor polishes, release coats.

Poliglicoleum. [Vevy] Ricinoleth-40.

Polirol. [Auschem SpA] Ethoxylates or sulfate derivs.; emulsifier for emulsion polymerization.

Politol®. [Westvaco] Sodium lignate; protein coagulants in purification of fats and oils.

Polity PS-1900. [Lion] Polystyrene sulfonic acid, sodium salt; dispersant for pigments, dyes, ceramics, clay; emulsion stabilizer.

Pollopas. [Hüls Am.] Urea-formaldehyde molding compds.

Pol-Nu. [Chapman] Wood preservative.

Polurene. [Sapici SpA] Polyurethane resin for varnishes.

Polwinity. [Organika Zachem Chem. Works] PVC plasticized compds.

Polyaldo®. [Lonza] Polyglyceryl fatty acid esters; emulsifier, emollient, lubricant for cosmetics, toiletries, pharmaceuticals, and household speciality prods.

Polyaminon. [Nippon Chem. Co. Ltd.] Ethyl or diethyl glutamate and aspartate blends.

Polyanthrene®. [Crompton & Knowles]

Cationic condensate; fixing agent for direct dyes on cellulosic fibers.

Polybatch. [A. Schulman] Additive concentrates.

Poly bd®. [Atochem N. Am.; Atochem UK] Butadiene polymer; functional liq. polymers for elec. applics., polyurethane and polyester polymerization, ester or ether derivatives.

Polybead Microparticles. [Polysciences] Monodisperse latex beads; used in medical research.

Polybinder. [Morton Int'l.] Binder for pigment printing.

Polyblak. [A. Schulman] Black concentrates.

Polybond®. [BP Performance Polymers] Polypropylene or polyethylene compounds; thermoplastic for use as chemical coupling agent, compatibilizing agent, metal adhesive and nucleating agent; impact modifier; additive in polymer alloys, for adhesion to metals and polar polymers.

Polybond® 90. [Morton Int'l.] Carbohydrate ether; finishing agent for textiles.

Polybor. [U.S. Borax & Chem.; Borax Consolidated Ltd.] Sodium octaborate; for fire retardant treatment of lumber or other cellulosic materials.

Polybut. [British Traders & Shippers] Polybutene.

Poly C4M. [Crowley Tar Prods.] Olefinic polymer oils, grease, caulk, seal compd.

Polycarb. [W.R. Grace/Dearborn] Boiler compd.

Polycare® 133. [Rhone-Poulenc Surf.] Polymethacrylamidopropyltrimonium chloride; cationic conditioner for hair care systems; provides soft hold.

Polycat®. [Air Prods.; Air Prods. Nederland BV] Tert. amine or its acid salts; catalyst for polyurethane coatings.

Poly-Chill. [Olin] Propylene glycol.

Polychol. [Croda Inc.; Croda Chem. Ltd.] Ethoxylated lanolin alcohols; emulsifier, dispersant, solubilizer, emollient and gelling agent for cosmetics and pharmaceuticals.

Poly-Cide. [O'Brien Industries] Liq. industrial odor control.

Polycin. [CasChem] Urethane polyol.

Polycizer®. [Harwick] Rubber and plastics plasticizers.

Polyclad. [Carboline] Vinyl protective coating.

Polyclar®. [ISP] Polyvinylpolypyrrolidones; stabilizer for beverage clarification; adsorbent in thin-layer and column chromatography.

Poly Clay. [Burgess Pigment] Hydrous aluminum silicate.

Polyclear. [Hoechst UK] PET.

Polyclear®. [Henkel/Textile] Clearing agent for dyed textile goods.

Poly-Clear. [Transene] Casting polymer for metallurgical specimens.

Polyco. [Rohm & Haas] Vinyl acetate or vinyl acetate acrylic polymers; emulsion for paints, paper coatings; pigment binder for paper and paperboard coating; fiberglass sizing.

Polycol. [Atochem UK] PVC compds.

Polycolor. [A. Schulman] Pigment concentrates.

Polycomp®. [ICI Fluoropolymers] Polymer-filled PTFE composites; for high wear applics.

Polycomplex A-11. [Guardian Labs] Oil microdispersant, solubiizer.

Polycon. [Witco/Humko] Sorbitan esters; food emulsifiers.

Poly-Cone. [Olin] Silicone emulsion; mold release agent for rubber.

Polycotton. [Crompton & Knowles] Textile dyes and pigments.

Polycoupler. [CNC Int'l.] Copolymer; textile resin fixative.

Poly-Coupler. [Mateson] Resin copolymer.

Polycron. [Keystone Aniline] Textile dyes and pigments.

Polycryl®. [Morton Int'l.] Textile finishes, binders.

Poly-Cure. [Hernon Mfg.] Uv-cured conformal coating systems for printed circuit board and electronic assembly.

Poly-Cure. [Mooney Chems] Polyester catalyst.

Poly-DAC 40. [Rhone-Poulenc] Polyquaternium-6.

Polydex. [Dexter Spec. Coatings] Fabric

coating.

Polydis®. [Struktol] Blends; binder, coupling agent between polymer systems and fillers; processing aid, blending and dispersing agent, wetting agent; lubricant for plastics.

Poly DNB®. [Lord] Poly-p-dinitrosobenzene wax dispersion; conditioner for butyl rubber.

Polydon. [Industrial Polymers UK] Flexible PVC.

Polydyol. [Eastern Color & Chem.] Dye carrier for polyester dyeings.

Polydyolev [Eastern Color & Chem.] Organic sulfonate; nylon dye resist.

Polyelph. [W.R. Grace/Dearborn] Cooling water treatment.

Poly-Emulsion. [Chem. Corp. of Am.] Polyethylene emulsions.

Polyene. [Meridionale des Plastiques SA] HDPE.

Polyester 1606, N-95. [Chemron] Emulsion breaker for oilfield applics.

Polyfil. [Polykemi AB] Filled PP.

Polyfil®. [J.M. Huber] White, anhydrous organofunctional pigment; filler and reinforcing clays for EPR, crosslinked polyethylene.

Polyfilm®. [Morton Int'l.] Warp size for polyester or nylon filaments.

Polyfin. [Polysciences] Embedding wax.

Polyfine®. [Advanced Web Prods.] Thermoplastic resins; economical substitute for polyacetal resin in applics. where friction and wear props. are required; for VTR tape reel, noiseless gear, bearing surface.

Polyflam. [A. Schulman] Thermoplastic flame retardant compds.

Polyflo. [Indol Chem.] Polyethylene dispersions.

Polyflo. [Magruder Color] Polyethylene color concs.

Polyflo. [UOP] Fuel oil additive, antifoulant.

Poly-Floc. [Betz Industrial] Coagulant aid.

Polyflow. [Morton Int'l.] Flow additive for pigment printing.

Polyfluo. [Micro Powders] PTFE/polyethylene blends.

Polyfol. [Polyfol Klepsch & Co. GmbH]

Oriented polystyrene film.

Polyfon®. [Westvaco] Sodium lignosulfonate; dispersant for industrial applications, agric. chemicals, dyestuffs, ceramics.

Polyfort. [A. Schulman] Polypropylenes, polyethylenes.

Poly-G®. [Olin] PEGs, polyether polyols, or polyalkylene glycol derivs.; PEGs as intermediate, carrier in cosmetics, pharmaceuticals, textiles, rubber mold releases, printing inks, dyes, metalworking fluids, etc.; polyols for adhesives, caulks, sealants, coatings, castable elastomers, tire fill, potting compds.

Polygard®. [Uniroyal] Stabilizer, antioxidant for air-cured footwear.

Poly-Gee. [Geoliquids] Sodium polytungstate.

Polygel. [3-V; Sigma Prodotti Chimici] Carboxyvinyl polymers; stabilizers for plastic additives, detergents, car waxes and polishes, scouring creams.

Polygel Hydraulic Elastomers. [Wacker Silicones] Silicone rubber compds.

Polygloss. [J.M. Huber] Paint grade clay.

Polyglucadyne. [Brooks Industries] Polyglucan; enhances natural defense mechanisms of skin against aging, uv effects; for skin care prods.

Polyglycol. [Dow Europe] Polypropylene glycol; antifoaming agent, coupler for pottery, ceramics, latex, emulsion paints, agric. formulations, lubricant base.

Polyglycol. [Hoechst Celanese/Colorants & Surf.] Random EO/PO copolymers; textile and fiber chemicals.

Polygon. [Goldschmidt] Stripping agent for permanent silicone textile treatments.

Polygrade. [Ampacet] Degradable masterbatches.

Polyhall®. [Rhone-Poulenc/Perf. Resins & Coatings] Polyacrylamide.

Polylac. [Polykemi AB] ABS.

Polylan®. [Amerchol; Amerchol Europe] Oleyl linoleate, lanolin linoleate; emollient, dispersant, conditioner, penetrant, lubricant, cosolv., dye solubilizer used in cosmetics and pharmaceuticals.

Polylev #745. [Polymer Research Corp. of Am.] Hydrozyethylated compds.; dispersant, penetrant, and leveling agent for dyestuffs.

Polylite. [Uniroyal] Antioxidant for molded soling, closed-cell sponge, latex, foam, molded and mechanical goods.

Polylite®. [Reichhold; Reichhold Chemie AG] Thermoset polyester resins; casting resin producing a Corian look-alike prod.

Polylube. [Hart Chem. Ltd.] Polyoxyalkylene derivs.; lubricant, antistat for fibers.

Poly-Lube. [Hillcrest Labs] Wire cable pulling lubricant.

Polylube #745. [Polymer Research Corp. of Am.] Hydroxyethylated compds.; dispersant, penetrant, and leveling agent for dyestuffs.

Polylube-B. [Polymer Research Corp. of Am.] Pigment softener.

Polyman. [A. Schulman] ABS/PVC alloys; for high temp. applics.

Polymaron. [Arakawa] Coating agent for paper.

Polymate. [W.R. Grace/Dearborn] Boiler/cooling water compds.

Polymatte. [J.M. Huber] Structured clay; for paint mfg.

Polymax. [Chemax] Metal finishing additives.

Polymeg. [QO Chem.] Polytetramethylene ether glycols.

Polymekon®. [Petrolite/Polymers] Microcrystalline wax; antislip and antimar agent for paint and ink industries.

Polymel #7. [Frank B. Ross] Modified polyethylene wax; low m.w. wax incorporated into rubber batches for mold release properties.

Polymer Additive GH. [BASF] Polyoxyisobutylene/methylene urea copolymer.

Polymer C. [Crowley Tar Prods.] Microcrystalline wax replacement.

Polymer SBOCP. [Brooks Industries] Cocamidopropyl dimethylammonium C8-16 isoalkylsuccinyl lactoglobulin sulfonate.

Polymeric Acid. [Henkel] Crude polymeric fatty acid; neutralizer for oil-sol. corrosion inhibitor formulations.

Polymex. [PSG Group Ltd.] Color-coded polyester.

Polymica. [Franklin Industrial Minerals] Wet processed muscovite mica; for coatings and polymer applics.

Polymid. [Lawter Int'l.] Polyamides.

Polymin®. [BASF AG] Based on polyacrylamide, modified polyethylene imines; retention, dewatering, and flocculating agents.

Polymoist® Mask. [Henkel/Emery/Cospha; Henkel KGaA] Collagen fiber material; moisturizing effect for skin; humectant.

Polymon. [ICI Am.] Plastic pigments.

Polyotic. [Am. Cyanamid/Ag] Tetracycline.

Polyox®. [Union Carbide] PEGs; thickener.

Poly-Pale®. [Hercules] Rosin derivs.; pale thermoplastic resin for lacquers, varnishes, adhesives, driers, synthetic resins, ink vehicles, floor tile, rubber compds., solder fluxes, and wax modification; tackifier.

Polypeg. [Olin] PEG fatty ester; visc. control agent, dispersant, surfactant.

Polypenco Acetal. [Polymer Corp.] Acetal copolymer; used in bearings, gears, antifriction parts, elec. components, etc.

Polypenco Cast Acrylic. [Polymer Corp.] Cast acrylic rod; optical clarity resin for displays, signs, furniture components, lenses, elec./electronic parts.

Polypenco Nylon 101. [Polymer Corp.] Nylon 6/6; for food processing, machinery, electronics, military and other industries for bearings, bushings, valve seats, seals, rollers, gears, insulators, fasteners, liners, tooling fixtures, forming dies, etc.

Polypenco Polycarbonate. [Polymer Corp.] Polycarbonate; thermoplastic for stand-off insulators, coil forms, optical and transparent structural components.

Polypenco Q200.5. [Polymer Corp.] Crosslinked polystyrene; rigid insulating material for UHF, VHF, and microwave insulators, communication and electronic equipment.

Polypeptide. [Inolex] Hydrolyzed collagen; surfactant for personal care prods.

Polypeptide. [Maybrook] Hydrolyzed collagen; surfactant for personal care prods.

Polypet. [Crowley Tar Prods.] Rubber plasticizer.

Polyphase. [Troy] Industrial fungicide.

Polyphor T. [Monomer-Polymer & Dajac Labs] Scintillation plastics.

Polyphos®. [Olin] Sodium hexametaphosphate; for municipal and industrial process and potable water treatment.

Polyplasdone®. [ISP] Polyvinylpolypyrrolidone or crospovidone; dry binder, disintegrant, excipient for pharmaceutical tablets; adsorbent in thinlayer chromatography; suspension aid, complexing agent.

Polyplate. [J.M. Huber/Clay Div.] Kaolin clay; for paints.

Polyplex. [Sullivan Chem. Coatings] Acrylic coating.

Polypol 19. [Crowley Tar Prods.] Low visc. amorphous polypropylene.

Poly-Prep. [O'Brien Industries] Dry internal boiler water treatment.

Polypro 5000. [Hormel] Hydrolyzed collagen.

Polyproducts. [Polyproducts] Epoxy and polyester resin systems.

Poly Pross. [Disco] Formulated process aid for natural and synthetic rubber processing.

Polypur. [A. Schulman] Polyurethane compds.

Polyquart®. [Henkel/Emery/Cospha; Henkel KGaA] Ethoxylated polyamines; antistat, softener, conditioner, hair fixative for personal care prods., hair care prods.

Polyquest. [CNC Int'l.] Sequesterants for textile processing.

Polyquest. [W.R. Grace/Dearborn] Boiler compds.

Polyrad. [Polymer Industries] Uv curable reactive coatings; lacquer coating for graphic arts applics.

Polyrad®. [Hercules] Ethoxylated amine; corrosion inhibitor, detergent,

wetting agent, emulsifier for petrol. processing equipment; inhibits HCl in industrial and household cleaners.

Polyram. [Ceca SA] Polyamines; emulsifier for bitumen; antistripping agent for road making.

Polyram®. [BASF AG] Metiram; for control of fungus diseases in fruits, hops, etc.

Polyrene. [J. Gaillon SA] Polystyrene.

Polyrez®. [OxyChem] Phenolic resins.

Polyron®. [Hoechst Celanese/Colorants & Surf.] Activator, stabilizer for bleaching, kier boiling assistant.

Polysalt. [BASF AG] Based on salts of polycarboxylic acids; dispersants for extenders, papercoating pigments; for stabilizing coating mixtures and slurries; grinding assistants for chalk.

Polysan. [Reilly-Whiteman] Nonionic oils.

Polysar. [Polysar] Isobutylene/isoprene copolymers, EPDM, or EPM rubbers; rubber for use in tire inner liners, pharmaceutical closures, mechanical goods, gaskets, seals, hose, profiles, insulation, chewing gum base, tank linings, sponge, adhesives, sealants.

Polysar®. [Novacor] Crystal or impact polystyrenes; used in inj. foam, inj. blow molding, extrusion, dinnerware, biaxially oriented sheet, profiles, housewares, cosmetic and medical molding, consumer electronics, molded pkg., toys, and as blending resin.

Polysar EPDM. [Miles/Polysar Rubber] EPDM polymers; for inj. molding, hose and profile extrusion, sponge and cast curing, roofing membranes, black sidewalls.

Polysat. [Polysat] Styrene-acrylate copolymers or hydrocarbon resins.

Polyscent. [Rotuba Extruders] Fragranced thermoplastic.

Polyscents. [Andrea Aromatics] Deodorants and fragrances for plastics.

Polyscour. [Ivax Industries] Surfactant, wetting agent for scouring operations.

Polyseed. [Polybac] Seed inoculum.

Polysene®. [Nat'l. Starch & Chem.] Chelating agents.

Polyset. [CNC Int'l.] Acrylic triazine

resin; stiffening agent for polyester fabrics.

Polyset. [Morton Int'l.] Epoxy molding compds.

Polysilicate. [DuPont] Polysilicate; binder for refractory, ceramic, metal, inorg. fiber, catalyst support, inorg. paint systems, zinc coatings for marine and industrial applic.; glass surface modifier.

Polysilk. [Micro Powders] Polyethylene wax.

Polysize® 524. [Morton Int'l.] Acrylic resin; warp sizing additive for spun yarns.

Polysoft. [CNC Int'l.] Antistatic agent, lubricant, and softener for the textile industry.

Polysoft B. [Sybron] Polyethylene; textile softener.

Polysoftener®. [Boehme Filatex] Softener/yarn lubricant and oligomer binder.

Polysol. [ICI Surf. UK] Polyvinyl alcohol sol'n.; finishing agent for cellulose fabrics.

Poly-Solv®. [Olin] Ethylene and propylene glycol ether solvents; for brake fluids, hard-surf. cleaners, leather dyeing, paints, coatings, printing inks, textile vat dyeing and printing, adhesives, antifreeze, floor waxes/polishes, insect repellents; solubilizer for dyes; plasticizer.

Polyspend. [U.S. Cosmetics] Polyethylene; surface treatment for cosmetics.

Polysphere 3000 SP. [Presperse] Polystyrene, squalane; binder for pressed powder formulations; lubricious, lusterous, high-grade filler.

Polyspin. [Hart Chem. Ltd.] Polyoxyalkylene derivs. or blends; lubricant for filament polypropylene.

Polystab. [Harcros UK] Alkoxylates; solvent, hydrotrope, solubilizer for polyurethane foam mfg.

Polystat. [A. Schulman] Black polymer with electrical conductivity.

Polystat Agent #5033. [Polymer Research Corp. of Am.] Phosphate ester; antistat for plastics.

Polystate C. [Gattefosse; Gattefosse SA] PEG-6 stearate; base for cosmetic lotions.

Polystay. [Goodyear] Anilino-phenyl methacrylamide; antioxidant for emulsion polymers, rubbers.

Polystep®. [Stepan; Stepan Europe] Sulfonates, sulfates, or nonoxynols; emulsifier, surfactant, dispersant, stabilizer for emulsion polymerization, floor polishes, coatings.

Polystron. [Arakawa] Additive for paper industry.

Polystyrene. [Novacor Ltd.] Crystal or impact polystyrene; for medical molding, housewares, sheet glazing, coextrusion, oriented sheet, inj. blow molding, inj. foam, thermoformed drinkware, rigid pkg., audio cassettes, cosmetic molding, electronic pkg., labware, toys, closures; blending resin.

Polystyrol. [BASF PLC] General purpose and high impact polystyrene.

Polysweet. [Guardian Labs] Low calorie sweeteners.

Polysynlane. [Nippon Oils & Fats] Hydrogenated polyisobutene.

Polysystems. [Olin] Urethane foam chemical systems; for insulation, flotation, molding.

Polytac. [Crowley Chem.] Sealant, adhesive.

Polytal. [Whittaker, Clark & Daniels] Filler for polyolefins.

PolyTalc. [Pfizer] Surface-modified platy talc; for polymer applics.

Polytard®. [Westvaco] Lignin sulfonic acid derivs.; additive for masonry cement.

Poly TDP 2000. [Eastman] Thiodipropionate polyester; antioxidant for polyolefins.

Polyter. [Meridionale des Plastiques SA] LDPE.

Polyterge. [CNC Int'l.] Surfactant; textile printing auxiliary which removes and suspends residual disperse dyes on polyester.

Poly-Tergent®. [Olin] Alkoxylated linear alcohol or EO/PO block polymers; surfactant, wetting agent, defoamer, emulsifier, dedusting dispersant for household and industrial detergents, textiles, paper industries.

Poly-Tex. [Rhone-Poulenc/Perf. Resins & Coatings] Acrylic resin; for use in thermosetting industrial finishes, automotive topcoats, coil coating, appliance finishes, metal decorating,, maintenance coatings.

Polytex Resin. [Estron] Toluenesulfonamide/epoxy resin.

Polythane. [Hexagon Enterprises/Chem. Components] Urethane foam systems and coatings.

Polytrap®. [Dow Corning] Acrylates copolymer or blends; adsorptive powder for control of fluid delivery; for makeup, sun care prods., skin care prods., antiperspirants, perfumes, pressed powds., facial cleansers.

Polytreat. [Aquatec Chem. Int'l.] Boiler compds.

Polytriox. [Mines De La Lucette] Masterbatches.

Polytrol. [W.R. Grace/Dearborn] Boiler compd.

Polytron. [BFGoodrich/Spec. Polymers] Static dissipative alloys.

Polytrope. [A. Schulman] Thermoplastic elastomer compds.

Polytrope. [Rheox] Modified montmorillonite clay; rheological additive, thickener for unsaturated polyester laminating resins.

Polytung. [Degen] Thermolized tung oil.

Polyval. [Ceca SA] Vat dye stripper, direct dye clearing auxiliary.

Polyvin. [A. Schulman] Flexible and semirigid PVC compds.

Polyvinate. [Rowe Prods. Distribution] Polyvinyl acetate coating.

Polyviol. [Wacker Chemie GmbH] Polyvinyl alcohol.

Polywax®. [Petrolite/Polymers] Polyethylenes; release agent; modifier for paraffin waxes, plastics; component in hot-melt coatings, adhesives, chewing gum base; lubricant in plastics and rubber processing, elec. insulation, powd. coatings, printing inks, textiles; antiblocking agent.

Polywet®. [Uniroyal] Polyfunctional oligomer salts; emulsifier for emulsion polymerization; dispersant for miner-

als, pigments, fillers in paints, latexes, coatings, adhesives, paper and paperboard; boiler water treatment.

Polywhite. [J.M. Huber] Titanium dioxide; extender for paints.

Poly-Zole® AZDN. [Olin] 2,2′-Azobisisobutyronitrile initiator for vinyl, acrylic, or other polymerization.

POM. [Climax Performance] Molybdenum trioxide; corrosion inhibitor for cooling water systems.

Pomoco. [Piedmont Chem. Industries] Penetrant, detergent, stabilizer, leveler, antimgrant, dye assistant.

Pomofix. [Piedmont Chem. Industries] Fixing agent for dyes.

Pomoflex. [Piedmont Chem. Industries] Polymeric textile auxiliaries.

Pomoguard. [Piedmont Chem. Industries] Fluorochemicals for oil, water and stain repellency on textiles.

Pomoleine. [Piedmont Chem. Industries] Emulsifier, sequestrant, suspending agent for textile processing.

Pomolev. [Piedmont Chem. Industries] Wetting agent, leveler for dyes.

Pomolube. [Piedmont Chem. Industries] Lubricant, softener for natural and synthetic fibers and fabrics.

Pomoscour. [Piedmont Chem. Industries] Detergent, scouring agent.

Pomosoft. [Piedmont Chem. Industries] Softener for textiles.

Pomosol. [Piedmont Chem. Industries] Penetrant, detergent for textile use.

Pomosolv. [Piedmont Chem. Industries] Solvent scour for synthetic fabrics.

Pomosperse. [Piedmont Chem. Industries] Sequestrant and suspending agent.

Pomovol. [Piedmont Chem. Industries] Dye carriers.

Ponolith. [Miles/Organic Prods.] Textile dyes and pigments.

Pontamine. [Miles/Organic Prods.] Textile dyes and pigments.

Pop-All. [Van Den Bergh Foods] Partially hydrogenated vegetable oil; for popping and seasoning corn, snack foods.

Porocel. [Engelhard] Activated bauxite; adsorbents.

Porofor. [Bayer; Miles] Azodicarbonamide, hydrazides, or other chemical blowing agents; for prod. of plastic foams, rubber goods.

Poron®. [Rogers] Cellular polyurethane; used for gaskets, seals, vibration mounts, motor mounts, RF shielding, PCB cushions, spacers, foam-backed tapes, athletic padding in automotive, elec./electronic industries.

Porosponge. [Advanced Polymer Systems] Acrylates copolymer.

Porox. [Ferro/Refractories] Alumina ceramic adsorbents.

Portadamp. [E-A-R] Portable extensional damping material.

Posidyne. [Pall Corp.] Nylon 6/6 positive zeta potential high area filter elements; for microbial and particulate removal.

Post-4. [Rheox] Castor oil complex deriv.; thixotrope, gellant, post additive for trade sales and industrial finishes.

Potato-Pro EN-15. [Brooks Industries] Hydrolyzed potato protein; for skin and hair care cosmetics.

Poten. [Bullen Chem.] Cleaner-degreaser.

Pot-Sil. [Rhone-Poulenc Basic; Crosfield] Potassium silicate.

Pounce. [FMC/Ag] Insecticide.

Pourgel. [DynaGel] Liq. gelatin.

Powderlink. [Am. Cyanamid] Powder coating resin.

Powdex. [Graver] Powdered ion exchange resins.

Powdurablue. [PMC Specialties] Toner for printing inks.

Powerclear. [Hydrolabs] Reduction clearing agent for afterscouring of dyed polyester.

Power-Det. [Oakite Prods.] Cleaner.

Powergizer. [Mammoth Int'l.] Liq. mixed fertilizers containing humic acids.

Powershield. [Lubrizol] Fuel additive for protection of engines using leaded gasoline.

Powertrace Micronutrients. [Mammoth Int'l.] Micronutrients for agric. formulations.

PP-. [Compounding Tech.] Polypropyl-ene, some glass or calcium carbonate-reinforced.

PP-. [Washington Penn Plastics] Polypropylene homopolymers or copolymers, some mineral or glass-reinforced; for automotive parts, furniture, toys, appliances, lawn equip., pump housings, instrument panels, housewares, closures and containers for foods/drugs.

PP 100 205/03 FH-VP. [Zipperling Kessler] Halogen-free flame-retardant compd.; for film extrusion.

PPG Perchlor. [PPG Industries] Perchlorethylene; solvent for vapor degreasing, cold cleaning, or drycleaning applics.

PPG Trichlor. [PPG Industries] Trichlorethylene; solvent for vapor degreasing, cold cleaning, cleaning electronic components, synthesizing chemicals.

P-Pro. [Grindsted Prods.] Proteins; for food industry.

PPX-. [Compounding Tech.] Polypropylene, highly chemically coupled, glass reinforced.

Praepagen. [Hoechst AG] Dialkyl-dimethyl ammonium chlorides; raw material for formulation of fabric softeners, carwash drain aids.

Praestol. [Stockhausen] Water treatment polymers.

Pran. [Fabriquimica] Hydrolyzed protein derivs.

Preact. [Mineral Research & Development] Galvanizing flux.

Preadd. [Premix OY] Additive masterbatches.

Preamasoft C. [Specialty Chem.] Amphoteric or polyamido quaternary; softener for half-hose and dyed goods.

Preblack. [Premix OY] Black masterbatches.

Preblend. [Premix OY] Polymer blends.

Prebond. [Premix OY] Coupling agents.

Prechem. [Ivax Industries] Wetting agent, dyeing assistant, kier boiling assistant, scouring agent, detergent, peroxide bleach stabilizer for textiles.

Precifac ATO. [Gattefosse SA] Cetyl palmitate; tabletting agents.

Precirol. [Gattefosse; Gattefosse SA]

Glyceryl esters; additive for tablets, binder, lubricant

Precol. [Premix OY] Color masterbatches.

Pree®. [BASF AG] Metazachlor; for control of grasses and broadleaf weeds in maize.

Pre-Elec. [Premix OY] Conductive compd.

Prefera®. [Henkel/Emery/Cospha; Grünau] Sodium or calcium stearyl-2-lacrylates; emulsifiers for improvement of fermentation tolerance, volume and texture of yeast-raised baked goods; antistaling effect; dough conditioner.

Prefill. [Premix OY] Filled compd.

Prejel. [Avebe Am.] Potato starches; for food industry.

Prelete. [Dow] Defluxer solvent.

Premax Modules. [Premier Refractories] Ceramic fiber modules.

Premier. [ECC Int'l.] Kaolin; paper coating clay.

Premier Flake Salt. [Cargill] 99.9% NaCl.

Premier Maltzymes. [Premier Malt Prods.] Amylase; enzyme which converts gelatinized starch to dextrins and small amts. of low m.w. carbohydrates; useful for prod. of alcohol from ground corn.

Premise. [West Agro] Disinfectant.

Premolac. [Crompton & Knowles] Cultured butter flavor.

Premose Syrup. [Premier Malt Prods.] 80% corn, 20% malted barley; syrup with sweet but mild malt flavor; used in ice cream prods., bakery goods, nutrient for distilled vinegars, etc.

Premovan. [Crompton & Knowles] Cultured butter-vanilla flavor.

Prenol. [BASF AG] 3-Methyl-2-butene-1-ol; fresh, herbal, green, fruity fragrance and flavoring.

Prentox. [Prentiss Drug & Chem.] Pesticides.

Prep. [Accurate Chem. & Scientific] Cesium chloride; for laboratory use in ultra-centrifugation.

Prep. [Rhone-Poulenc/Ag] Plant growth regulator.

Prepetal. [Zschimmer & Schwarz] Fatty alcohol polyalkylene; emulsifier, detergent, wetting agent, antifoaming additive.

Prepfos. [Olin] Sodium tripolyphosphate; for food applics.

PrepRite Coating Remover. [ISP] N-Methyl-2-pyrrolidone, butyrolactone, and other ingreds.

Prep-Sol. [DuPont] Solvent.

Preserval. [Laserson & Sabetay] Parabens; preservatives.

Preserve-X. [Poly Research] Water preservatives.

Preserv-O-Sote. [Crowley Chem.] Wood preservatives.

Prespersion. [Syn. Prods.] Nonpolymeric dispersions of rubber chemicals.

Press-Aid. [Presperse] Synthetic wax or blends with corn gluten protein; binder.

Prestogen®. [BASF AG] Stabilizers for peroxide bleaching in textile industry.

Pre-Tectg. [Calgon] Boiler water treatment.

Pretreat SS-10. [Yorkshire Pat-Chem] Pretreatment for pigment garment dyeing.

Prevail. [Dow Plastics] Thermoplastic polyurethane/ABS blend; engineering thermoplastic for bumpers and fairings on commercial trucks, on snowmobiles, ATVs and campers, and in the automotive market.

Preventol. [Akzo] Flocculation control agent.

Preventol. [Polymer Research Corp. of Am.] Chlorophene, benzylhemiformal, or dichlorophene.

Prevex®. [GE Plastics] Polyphenylene ether resins; for inj. molding, extrusion, structural foam molding for pumps, housings, small appliances, lawn care tools, power tools, industrial devices, sheet profile, business machine parts, elec. enclosures and connectors.

Preview. [DuPont/Ag] Herbicide.

Prevox. [Kano Labs] Rust preventative.

Prewhite. [Premix OY] White masterbatches.

Priacetin. [Unichema] Acetate esters; used in cigarette filter tips, concrete additives, foundry auxiliaries.

Priadit. [Unichema] Polymer modifiers, antistats.

Priamid. [Unichema] Alkanolamides.

Pri Bond. [Sullivan Chem. Coatings] Metal prep. coating.

Pricat. [Hart Chem. Ltd.] Hydrogenation catalysts.

Pricat. [Unichema] Nickel catalyst; for hydrogenation of edible oils and fats, industrial fats and fatty acids.

Pricerine. [Unichema] Glycerin; for pharmaceuticals, surface coating resins, nitration prods., tobacco, emulsifiers, cosmetics, esters, food additives.

Prifac. [Unichema] Tallow acids; chemical intermediate for surfactants, stabilizers, detergents, fabric softeners, soaps, buffing formulations.

Prifat. [Unichema] Industrial triglycerides; for catalyst mfg., leather, waxes, textile auxiliaries, cosmetics, fatty acid derivs., and plastics.

Prifrac. [Unichema; Unichema France SA] Fractionated fatty acids; intermediate for esters, surfactant prods., synthetic lubes, substituted glycerides for cosmetics, leather and rubber treatment, paint drier, textile auxiliaries.

Primabond 2. [Nutex] Durable hand builder.

Primacor. [Dow] Adhesive copolymers; for blown film, extrusion coatings, adhesive extrusion laminates; for flexible pkg. structures, coatings, inks, adhesives.

Primafast®. [Gresco Mfg.] Alpha amylase; enzyme.

Primal. [Rohm & Haas] Resin emulsions and pigment dispersions; for leather finishing.

Primallor. [Degussa] Gold alloy for dental applic.

Primapel®. [Rohm & Haas] Acrylic carboxylic copolymer; carpet soil retardant.

Primarol. [Henkel Canada] Alcohols; extender and solvent for dyes and fragrance oils; intermediate for prod. of comps. for applics. in lubricants, emulsifiers, metal processing, and textiles.

Primasol®. [BASF; BASF AG] Wetting agents, detergent, stabilizer, and textile auxiliary.

Primatone. [Hilton Davis] Pigment dispersions plastics.

Primax. [Air Prods.] Modified UHMW-PE; used in cast polyurethane formulations to impart increased abrasion resist., reduced coeff. of friction, improved tear str., reduced part dens.

Primazin®. [BASF; BASF AG] Reactive dyes or fixing agent for printing and dyeing cellulose fibers.

Prime. [Fries & Fries] Meat and savory flavors.

Primef. [Solvay & Cie] Polyphenylene sulfide.

Primene. [Rohm & Haas] Tert-alkyl primary amine.

Primex. [Lawter Int'l.] Ink varnish quick set vehicle.

Primid. [Rohm & Haas] Aq. pigment dispersions; for leather coloring.

Priminox. [Rohm & Haas] Ethoxylated tertiary alkylamine.

Primipel. [Rohm & Haas] Synthetic tanning agent.

Primojel. [Avebe Am.] Potato starch; tablet disintegrant.

Primojel. [Generichem] Sodium starch glycolate.

Primorol 1511. [Henkel/Emery] C24-26-28 branched chain alcohol.

Prinlin. [Pierce & Stevens] Kraton dispersions.

Printac. [Toho Chem. Industry] Anionic complex; leveling and penetrating agent for acrylic fabric printing.

Printex. [Degussa] Carbon or furnace blacks; for electrostatic powder toners, uv inks, black coating systems.

Printlok. [Catawba-Charlab] Vinyl-acrylic or acrylic pigment printing binder.

Printogen® HD. [Sandoz] Weakly cationic glycol ethers; carrier for superheated steaming of polyester and Qiana printed with disperse dyes.

Printol. [3-V] For textiles.

Printrite. [BFGoodrich/Spec. Polymers] Binders, defoamers, printing additives.

Printsolve. [ISP] N-Methyl-2-pyrrolidone, dipropylene glycol methyl ether; ink remover.

Prinza®. [Grünau] Guar gum; thickener and stabilizer for food industry.

Priol. [Unichema] Polyester polyols.

Priolene. [Unichema] Oleic acids; intermediates for ethoxylates, esters, nitrogen derivs., surfactants; used in lubricants, metalworking fluids, personal care prods.

Priolube. [Unichema] Esters; synthetic lubricant bases for engine, industrial, and metalworking lubricants, greases, additives.

Priosorine. [Unichema] Isostearic acid or isostearyl alcohol; emulsifier for esters/soaps; textile softener, antistat; anticorrosion additive; for skin care emulsions, sunscreen prods.

Priowax. [Auschem SpA] PEGs; lubricant, carrier, plasticizer, solubilizer for industrial applics.

Priplast. [Unichema; Unichema France SA] Azelates or epoxizied oleate; plasticizers for PVC, PVDC, PVAc, cellulosics, specialty rubbers, nitrocellulose, polyvinylbutyral films; polyester polyols for polyurethane foam, coatings, and thermoplastic elastomers.

Priplus. [Unichema] Additives for animal feed supplements.

Pripol. [Unichema; Unichema France SA] Dimer or trimer acids; modifier for nylon, polyester fibers; used in polyamide for hot melt adhesives, thermographic inks, urethane elastomers, industrial lubricants, fuel additives, surface coating resins, spin finishes.

Pripure. [Unichema] High purity dimer acids or dimer derivs.; for personal care prods.

Prisavon. [Unichema] Soap bases; for preparation of specialty soaps, toilet soap, household soap bars, soap powders, industrial and liq. soaps.

Prism. [Miles] Polyurethane RIM systems.

Prisorine. [Unichema] Isostearic acids; used in lubricant and fuel additives and cosmetics.

Prist. [PPG Industries] Aviation fuel additive.

Pristene. [UOP] Mixed tocopherols or blends or rosemary extract; natural food grade antioxidants.

Pristerene. [Unichema; Unichema France SA] Stearic or palmitic acids; intermediate for ethoxylates, esters, nitrogen derivs., personal prod. formulations, soap, detergents, stabilizers, candles, paper chemicals, rubber, inorganic coatings, textile auxiliaries.

Pro 18. [Pea Ridge Iron Ore] Iron oxide pellets.

Proaid. [Akrochem] Processing and dispersing aid, homogenizing agent, softener for rubbers, unsaturated polymers; peptizing agent for natural and polyisoprene rubber.

Proban®. [Albright & Wilson Am.] Flame retardants for textiles.

Procas. [Croda Chem. Ltd.] Propoxylated hydrogenated castor oil; emollient for lipsticks.

Procetyl. [Croda Inc.; Croda Chem. Ltd.] PPG cetyl ether compds.; emollient, coupler, cosolvent, plasticizer, superfatting, wetting and spreading agent, penetrant, lubricant in cosmetics and personal care prods.

Prochem. [Prochem] Fluorocarbon; fluorochemical finishes imparting oil and water resistance to fibers.

Prochem. [Protameen] Hydrolyzed collagen or lecithin blends.

Procion. [ICI Am.] Dyes for textiles.

Proclean. [Prochem] Solvent cleaner, degreaser, machine cleaner.

Proco R394. [Amax Industrial Prods.] Wax-based rust preventative.

Procoal. [Tokai Seiyu Ind.] Nonionic surfactant; deairing, wetting and penetrating agent.

Procol. [Protameen] Ethoxylated fatty alcohol ethers.

Procom. [ICI PLC] Polypropylene compds.

Procon. [Central Soya] Soy protein conc.

Procond. [United Composites] Electrically conductive polypropylene copolymer; for extrusion.

Procostat. [Process Control Co.] Antistatic cleaning fluid.

Pro-Cote. [Protein Tech. Int'l.] Paper coating pigment binder.

Prodag. [Acheson Colloids] Graphite in

water dispersion; lubricant additive.

Prodew. [Ajinomoto] Sodium lactate, sodium PCA, sorbitol, hydrolyzed animal protein, proline; formulated moisturizer for cosmetics, soaps, hair care prods.; humectant.

Prodotto T 8455. [Auschem SpA] Nonionic blend; thickener for hydrochloride acid sol'ns.

Prodox. [PMC Specialties] Phenolic antioxidants.

Product 21, 21LF. [Reilly-Whiteman] Detergents for textile scouring.

Product 98. [Heterene] Polysorbate 80, cetyl acetate, and acetylated lanolin alcohol.

Product BCO, DDN. [DuPont] Betaines; wetting agent, detergent, emulsifier, dispersant, surfactant; dyeing applics.; softener for textiles; leveling and rewetting agent in the paper industry; dyeing assistant and degreaser in the leather industry; antistat on plastic films.

Product PW. [CNC Int'l.] Print wash auxiliary for use on nylon printed with acid dyes.

Product VN-11. [Henkel/Emery/Cospha] Ethoxylated oleyl alcohol; emulsifier for metalworking fluids.

Produkt. [Croda Chem. Ltd.] Succinic acid derivs.; corrosion inhibitor.

Produkt. [Zschimmer & Schwarz] Anticorrosion agent, spreading auxiliary, solubilizer, emollient, detergent, opacifier for cosmetics, cleaners.

Produkt S-17-20. [Chem-Y GmbH] PEG-25 glyceryl stearate.

Produkt W 37194. [Stockhausen] Acrylamidopropyltrimonium chloride/acrylates copolymer.

Profan. [Sanyo Chem. Industries] Fatty acid alkanolamide; foam stabilizer and thickener for shampoo.

Pro-fax®. [Himont] Polypropylene homopolymers or copolymers; for inj. molding, compr. molded sheet, ram extrusion, thermoforming, film, fiber, automotive, hospital and institutional ware, housewares, closures, toys, containers, appliances, chemical process equip., slit tape, twine, totes, trays, and furniture.

Profloc. [Vinings Industries] Polyacrylamide.

Progacyl®. [Rhone-Poulenc/Textile & Rubber] Guar gums or derivs.; thickener, suspending agent for personal care prods.

Progalan. [Rhone-Poulenc/Textile & Rubber] Ammonium alkylaryl ether sulfate; foaming and scouring agent for textiles, detergent formulations.

Progallin®. [Nipa Labs] Esters of gallic acid; antioxidants for cosmetics.

Progasol®. [Rhone-Poulenc/Textile & Rubber] Fatty alkanolamide or sulfonate; wetting, leveling, and scouring agent, emulsifier for textiles, detergents.

Pro Gen. [Fleming Labs] Arsanilic acid.

Pro Grip. [Fel-Pro] Cyanoacrylate adhesive.

Project® 70 Stainless Type 316. [Carpenter Tech.] Molybdenum bearing austenitic steel with increased percentages of nickel; alloy for paper pulp handling equip., process equip. for producing photographic chemicals, inks, rayon, rubber, textile bleaches and dyestuffs, high temp. equip.

Proklene. [Stoner] Cleaner and degreaser.

Prolagen C. [Proalan SA] Hydrolyzed collagen.

Prolagen I. [Hormel] Hydrolyzed collagen or derivs.

Pro-Lak. [H.J. Baker & Bro.] Protein supplement.

Pro-Lan. [Lanaetex Prods.] Sorbitol, TEA-coco-hydrolyzed animal protein.

Prolase®. [Int'l. Bio-Synthetics] Protase; protein digestive enzyme for pharmaceuticals.

Prolastine. [Proalan SA] Hydrolyzed animal elastin.

Prolube. [Prochem] Lubricant for textiles, dyebath.

Promax. [Central Soya] Functional soy protein conc.

Promaxon. [Nyco Minerals] Synthetic hydrated calcium silicate.

Prometol. [Viobin] Conc. wheat germ oil.

Promine. [Central Soya] Functional soy conc.

Promocaf. [Central Soya] Soy conc.

Promodan. [Grindsted Prods.] Propylene glycol esters; emulsifier for foods, cosmetics.

Promosoy. [Central Soya] Soy protein.

Promotor 301. [Akzo] Metal compd., hydrocarbon solvent; accelerator.

Promount. [Stoner] Conc. liq. tire mounting lubricant.

Promulgen®. [Amerchol] Fatty alcohol condensates; gelling agent, emulsifier, emollient, and stabilizer for cosmetics and pharmaceuticals.

Promyr. [Amerchol] Isopropyl myristate; emollient and solvent for cosmetics, toiletries, makeups.

Promyristyl PM-3. [Croda Inc.; Croda Chem. Ltd.] PPG-3 myristyl ether; emollient for clear analgesic, deodorant, and fragrance sticks.

Pronal. [Toho Chem. Industry] Nonionic complex; defoamer for paper, latex, paint, food, drilling mud, natural gas, fertilizer, petrochemical, and fermentation industries.

Pronova. [Protan] Sodium hyaluronate; skin moisturizer for personal care; hemostatic agent.

Propacyl. [Rhone-Poulenc/Textile & Rubber] Guar deriv.; thickener, suspending agent.

Propadex. [Dependable Extrusions Ltd.] Polypropylene.

Propafilm®. [ICI Films] Oriented polypropylene films.

Propagen. [Hoechst AG] Quaternary ammonium salts; textile softener, antistat.

Propal. [Amerchol] Isopropyl palmitate; emollient and solvent for cosmetics, toiletries, makeups.

Propanil 60 DF. [Terra Int'l.] Dry flowable herbicide.

Propaply. [ICI Films] Oriented polypropylene films.

Propasol. [Union Carbide] Butoxypropanol; industrial solvent.

Propathene. [ICI PLC] Polypropylene.

Propetal. [Zschimmer & Schwarz] Fatty alcohol polyalkylene glycol ether; detergents, wetting agents, emulsifiers, antifoams, intermediates for prep. of detergent systems.

Propiofan®. [BASF] Polyvinyl propionate; binder for paints, textured finishes, composite thermal insulation, concrete coatings, textile coatings; modifier for cement mortar; raw material for building, laminate, and pkg. adhesives.

Proplast. [Aquatec Quimica SA] Esters; lubricant, antistat for plastics.

Propomeen®. [Akzo] Propoxylated amines.

Proponite. [Borden] Oriented PP film.

Propoquad®. [Akzo] Propoxylated quaternary ammonium compds.

Propoxyol® 1695. [Henkel/Emery/Cospha; Henkel KGaA] PPG-5 lanolin wax glyceride; emollient, stabilizer, and pigment dispersant for anhydrous makeups; cosmetic additive.

Propsolv. [Quantum/USI] Solvent blends; Solvent for shellac, chemical specialties, latex coagulants.

Propylan. [Lankro Polychem AB] Polyether polyols.

Propylene Phenoxetol. [Nipa Labs] Phenoxy isopropanol.

Propylex. [Royalite Plastics Ltd.] Polypropylene.

Propyltex. [Micro Powders] Polypropylene texturizing agent.

Propylux. [Westlake Plastics] Polypropylene.

Propyl Zithate®. [R.T. Vanderbilt] Zinc isopropyl xanthate; accelerator.

Prosil®. [PCR] Organosilane derivs.; coupling agents.

Prosoft. [Prochem] Softener for textiles.

Prosol. [Hart Chem. Ltd.] Mineral oil blends; lubricant for wool processing.

Pro-Solv. [Anderson Labs] Specially denatured alcohol.

Prospin. [Prochem] Water-white polypropylene spin finish.

Prostat. [Prochem] Antistat for textiles.

Prostearyl 15. [Croda Inc.; Croda Chem. Ltd.] PPG-15 stearyl ether; emollient, lubricant for cosmetics; coupler for fragrances.

Protachem. [Protameen] Surfactants;

cosmetics ingredients.

Protalan. [Protameen] Lanolin derivs.; cosmetics ingredients.

Protamate. [Protameen] Ethoxylated esters; cosmetics ingredients.

Protamide. [Protameen] Fatty acid alkanolamides; cosmetics ingredients.

Protamine. [Nat'l. Starch & Chem.] Amide-amine fatty deriv.; emulsifier, softener, lubricant base.

Protaphos. [Protameen] Phosphates; cosmetics ingredients.

Protapon. [Protameen] Sodium methyl alkyl taurate.

Protaquat 2HT-75. [Protameen] Distearyldimonium chloride.

Protasorb. [Protameen] Polysorbates.

Protastat P-211. [Protameen] Methylparaben, propylparaben, potassium sorbate.

Protavic. [Protex] Epoxy molding compd. for electronic component encapsulation.

Protect. [Pro Chem Chemicals] Flame retardant chemicals.

Protectein. [Hormel] Propyltrimonium hydrolyzed animal protein; substantive quaternary reducing irritation potential of surfactants; for skin and hair prods.

Protecto. [Meridian Petroleum] Absorbents.

Protectol®. [BASF AG] Biocidal surfactants, preservatives, disinfectants for cleaning, water treatment, other industrial applics.

Protector. [Vyse Gelatin] 300 bloom gelatin.

Protectorite. [Octagon Process] Phosphating compd.

Prote-Fix®. [Synthron] Fixing agent for dyeing.

Prote-Gal®. [Synthron] Oxidation inhibitor, leveling agent, antimigrant additive for dyeing and printing.

Protegin®. [Goldschmidt; Goldschmidt AG] Petrolatum/ozokerite blends; emulsifier, emollient, absorption base for cosmetics and pharmaceuticals.

Pro-Tein. [Maybrook] Hydrolyzed collagen derivs.; cationic substantivity agent, anti-irritant for hair prods., facial toners, antiperspirants, after shaves.

Prote-Nyl. [Synthron] Fire retardants.

Proteodermin. [Henkel/Cospha] Soluble proteoglycans; for aging or sun-damaged skin care preps.

Proteolene. [Vevy] Hydrolyzed animal protein.

Proteosilane C. [Exsymol] Methylsilanol elastinate; tissue regeneration aid for skin care creams, anti-aging creams, stretch mark prevention formulations.

Prote-Pon. [Protex] Alkyl ether phosphoric acids or potassium phosphates; wetting agent, detergent, hydrotrope, rust inhibitor, EP agent for detergent formulations, metal cleaners, hard surface cleaners, textile scours, metal lubricants, dry cleaning soap, textile lubricants, emulsion polymerization, pesticides.;

Prote-Pon®. [Synthron] Detergent, wetting agent.

Proteric. [Protameen] Sodium or disodium cocoamphoacetates or blends.

Protesine. [Synthron] Paper chemicals.

Prote-Sol. [Synthron] Wetting agent, dispersant, stabilizer for dyeing, pigments, bleaching.

Prote-sorb. [Protex] Sorbitan esters; surfactants, emulsifiers for petroleum oils and solvents, vegetable oils, waxes, silicones; for agric., cosmetic, leather, metalworking, and textile industries.

Pro-Tex. [Griffin] Maneb (32.63%) and triphenyltin hydroxide (4.72%) sol'n.; flowable fungicide for potatoes and sugar beets.

Prothane. [Europa Plastics BV] Polyurethane raw materials and prods.

Protoferm. [Finnsugar Bioprods.] Protease; enzyme for food processing; meat tenderizer.

Protol. [Witco/Sonneborn] White mineral oil USP; emollient, lubricant for cosmetics.

Proto-Lan. [Maybrook] Surfactant blends; emollient, moisturizer, emulsifier, conditioner, antistat for hair care prods., creams, lotions, hand and face soaps.

Protolube. [Nat'l. Starch & Chem.] Lubricant, softener, scrooping agent for textiles.

Proton. [Spice King] Protein.

Protopet®. [Witco/Sonneborn] Petrolatum USP; carrier, lubricant, emollient, moisture barrier, protective agent, softener for cosmetics, pharmaceutical ointment, industrial applics.

Protorez. [Nat'l. Starch & Chem.] Glyoxal reactant for textile operations.

Protosil. [Nat'l. Starch & Chem.] Nonyellowing silicone for white knit and woven goods.

Protostat. [Nat'l. Starch & Chem.] Antistat, softener, lubricant for textiles.

Protowet. [Nat'l. Starch & Chem.] Surfactants; wetting agent, scouring agent, detergent, emulsifier, penetrant, dyeing assistant, stabilizer, leveling agent for textiles.

Protox. [Protameen] Ethoxylated alkyl amines; cosmetics ingredients.

Protulines. [Exsymol] Hydrolyzed vegetable proteins; moisturizers for skin care prods., oily cosmetics, regeneration and skin treatments, anti-aging and anti-wrinkle creams.

Prove. [Fabriquimica] Hydrolyzed vegetable proteins.

Proventin®. [Henkel/Textile] Antioxidant for nylon fabrics.

Provol. [Croda Inc.] Propoxylated oleyl alcohols; emollient, superfatting agent, lubricant, pigment dispersant, coupler for personal care prods.

Prowl®. [Am. Cyanamid/Ag] Pendimethalin; emulsifiable conc. herbicide for control of annual grasses and broadleaf weeds in field crops.

Prox. [Kano Labs] Rust preventative.

Prox®. [Synthron] Textile resins.

Prox E. [Protex] Epoxidized surfactants for water-based epoxy systems.

Prox-E. [Synthron] Epoxidized emulsifier.

Prox-Amine. [Synthron] Softener, lubricant for textiles.

Proxel CRL. [ICI Biocides] 1,2-Benzisothiazolin-3-one sol'n.; microbiostat preservative.

Proxmelt. [Pierce & Stevens] Hot melt adhesive.

Prox-onic. [Protex] Ethoxylated ethers, esters, or amines; surfactants for industrial and cosmetic applics.

Proxycarb. [Riverside Prods.] Sodium percarbonate.

Prozate. [Champlain Industries] Protein hydrolsates.

Prozine. [Am. Cyanamid/Ag] Herbicide.

Prozone. [Halocarbon Prods.] Non-ozone harming fluorocarbons.

Pruv. [Mendell] Sodium stearyl fumarate NF; lubricant for pharmaceutical tablets.

Pryfon. [Miles/Ag] Insecticide for treatment of soil for control of subterranean termites.

Prym®. [Sequa] Specialty soil release agents for textile finishing.

PS. [Hüls Am.] Silicone fluids, emulsions, or fluorosilicones.

PS-. [Compounding Tech.] PBT polyester, some glass reinforced.

PSA. [Hart Chem. Ltd.] Phenol sulfonic acid; electrolyte for tin plating; catalyst for resins.

PSD. [Hüls Am.] Siloxanes; diffusion pump fluids.

PSE-. [GE Silicones] Silicone rubber; for insulation, jacketing.

Pseudocollagen. [Brooks Industries] Plant pseudocollagen; forms moisture retentive films on skin; for cosmetic creams and lotions.

P-Star. [Grindsted Prods.] Starch; for food industry.

PT-0602. [Astor Wax] Synthetic wax; filler/binder in cosmetic sticks.

Ptal. [Mitsubishi Gas] p-Tolualdehyde; additive for resins; intermediate for pharmaceuticals, fragrances.

PT Color. [Mitsubishi Kasei] Disperse dyes for transfer printing method.

PTFE-. [Presperse] PTFE or blends; binder.

PTN®. [Novo Nordisk] Pancreatic trypsins; enzyme for bating in leather industry.

PTSA 70. [Hart Chem. Ltd.] p-Toluene sulfonic acid; catalyst for resins.

PTZ®. [ICI Polymer Additives] Phenothiazine; antioxidant, monomer stabilizer.

PU-100. [European Master Batch] Pigment disp. for polyurethane coating.

Pueblo. [Asarco] Litharge.

Pulmix. [Chemithon] Detergent agglomerator.

Pulpex® Thermal Bonding Pulps. [Hercules] Polyolefin pulp; absorbent pad binder used in feminine hygiene prods., disposable diapers, dressings.

Pulsar. [Olin] Calcium hypochlorite; dry chlorinating agent for industrial sanitation treatment, food and beverage plants, sewage disposal, agric. applics.

Pulse. [Dow] Engineering thermoplastic.

Pulvi-Lan. [Lanaetex Prods.] Lanolin oil, calcium silicate.

Punctilious®. [Quantum/USI] Specially denatured ethyl alcohol; solvent, thinner, extraction media for cosmetics, pharmaceuticals, detergents.

Purac Powder H. [Purac Am.] Lactic acid powder.

Pural®. [Condea Chemie GmbH] Alumina; for prod. of catalysts for petroleum refining, vehicle pollution control, chemical processes.

Puralox®. [Condea Chemie GmbH] Alumina; for prod. of catalysts for petroleum refining, vehicle pollution control, chemical processes.

Purasolv. [Purac Am.] Biodegradable and toxicologically safe solvents.

PUR-BD, -HD, -HE, -HR, -HTT. [Allrim SA] Continuous high temp. polyurethane.

Purdenz. [Accurate Chem. & Scientific] Iodine composition; for laboratory use as a centrifugation medium.

Purecat. [Hall Chem.] Chemicals for catalyst mfg.

P.U.R.E.-CMC. [Perma-Flex Mold] Polyurethane elastomers; cold molding compds.

Pureco®. [Karlshamns] Coconut oils or blends.

Pure-Dent®. [Grain Processing] Corn starch; binder, diluent, absorbent, disintegrant for pharmaceuticals, tablets.

Pure-Flo. [Oil Dri Corp. of Am.] Bleaching clay.

Purelast. [Polymer Systems Corp.] Urethane elastomer systems.

Puremist. [Chemol] Crop oil conc.

Pure Sil. [Unimin Specialty Minerals] Low iron sand.

Puress 997. [United Coconut Chem.] Glycerin 99.5% USP; solubilizer, humectant for toothpaste, creams and lotions, tobacco.

Puretan. [E&F King] Chrome tanning sol'ns.

Puretronic. [Hall Chem.] Chemicals for electronics prod.

Purette. [Vikon] Durable textile bacteriostats, fungistats.

Purex. [Morton Salt] Vacuum pan salt.

Pur-Fect Tool®. [Ciba-Geigy] Polyurethane casting system; used for assembly jigs, backfilling, core boxes, foundry gates and risers, laminating molds, mold cores, pattern plates, prototypes.

Purflo. [Purflo DTL SA] Polyester and polyethylene.

Purgitol. [Harry Miller] Metal working cleaner.

Purifloc. [Dow] Flocculant.

Purilan. [Ubbink Nederland BV] Polyurethane/polyester.

Purity® 21. [Nat'l. Starch & Chem.] Corn starch; binder, filler, and disintegrant for cosmetic and pharmaceutical formulations.

Pur-Oba®. [Goldschmidt AG] Natural wax ester; for cosmetics.

Purrcil. [Bruce Chem.] Replacement for sodium silicate and sodium hydroxide in peroxide bleaching.

Pursuit. [Am. Cyanamid/Ag] Herbicide.

Purtalc. [Charles B. Chrystal] Talc.

Purton. [Zschimmer & Schwarz] Fatty acid alkanolamide; foam stabilizer, thickener, superfatting agent for cosmetics, cleaners.

Purzaust®. [Allied-Signal] Catalysts for air purification.

Pusher. [Dow] Oil field prods.

Puxol. [Pulcra SA] Dodecylbenzene sulfonic acid or salts; detergent intermediate; scouring and wetting agent, detergent for formulation of liq. detergents, personal care prods., car washing shampoo, textiles; emulsifier for emulsion polymerization reactions; pigment dispersant

PVC 1195. [Air Prods.] PVC homopolymer resin; for rigid inj. molding, calendering, sheet extrusion, and blow molding; used for pipe fittings, film and sheet, bottles, plasticized applics.

PVC Deodorant. [Andrea Aromatics] Mixture of fragrance materials; deodorant for PVC processing and finished prods.

PVO 44-0. [Pacific Anchor] Water-based emulsion of conjugated oil.

PVP. [H&S Chem.] Iodine prods.

PVP-Iodine. [BASF] PVP-iodine.

PVP K-. [ISP] PVP; binder, stabilizer, protective colloid, carrier, film-former for cosmetics, adhesives, detergents, coatings, paper, ink, textiles, printing, antifreeze, agric. formulations.

PVP/VA. [ISP] PVP/VA copolymer; film-former for cosmetics, protective masks and bandages, antitarnish coating for metals, shoe polishes, inks, plastics.

PX-. [Aristech] Phthalates, adipates, maleates, or trimellitates; reagent grade plasticizers.

PX-. [Compounding Tech.] Polyphenylene ether, some carbon or glass filled.

Pycal. [ICI Am.] POE aryl ether; plasticizers.

Pycoacid. [Pylam Prods.] Textile dyes and pigments.

Pydrin. [DuPont/Ag] Insecticide.

Pyracur® FL. [BASF AG] Chloridazon, metalochlor; pre-emergence herbicide for control of grasses.

Pyradex®. [BASF AG] Chloridazon, triallate; pre-plant incorporated herbicide for control of broadleaf weeds and grasses in sugar and fodder beet.

Pyradur®. [BASF AG] Chloridazon, metalochlor; pre-emergence herbicide for control of grasses and broadleaf weeds in sugar beet and fodder beet.

Pyramin®. [BASF AG] Chloridazon; for pre- and post-emergence control of weeds in sugar beet, fodder beet, Swiss chard, some ornamentals.

Py-Ran. [Monsanto] Anhydrous monocalcium phosphate; for baking applics.

Pyratex. [Bayer] Butadiene-styrene-2-vinyl-pyridine latex; for rubber-fabric bonding (tire cord, V-belt, conveyor belting, etc.).

Pyrax®. [R.T. Vanderbilt] Pyrophyllite; inert filler, extender, diluent, carrier, anticaking agent for rubbers, agric. toxicants, paints, plastics, cosmetics, pharmaceuticals, animal feeds, ceramics, refractories.

Pyrazol. [Sandoz] Textile dyes and pigments.

Pyridine 1°. [Nepera] Pyridine.

Pyroban®. [CNC Int'l.] Flame retardants for paper and textiles.

Pyrobloc. [McGean-Rohco] Fire retardant plastic additive.

Pyrocat. [Mackenzie Chem Works] Combustion catalyst.

Pyro-Chek®. [Ferro/Keil] Brominated polystyrene; flame retardant for plastics.

Pyrocide. [McLaughlin Gormley King] Insecticide conc.

Pyrocrete. [Carboline] Paint-on fireproofing.

Pyrofine. [Atochem N. Am.] Ceramics.

Pyrogallol. [Schering Berlin Polymers] 1,2,3-Trihydroxybenzene; chemical intermediate for electronics; photographic chemicals.

Pyro-Gen. [GenCorp Polymer Prods.] Vinylidene chloride latexes.

Pyrolube. [Kano Labs] High temp. lubricant.

Pyrolux. [Auralux] Flame retardants for textiles.

Pyro-Mag. [Nat'l. Magnesia Chem.] Magnesium oxide; for refractories and ceramics.

Pyromet®. [Carpenter Tech.] Nickel-base alloy; for heat shields, furnace hardware, gas turbine engine ducting, combustion liners, chemical plant hardware, seawater applics.

Pyron. [Chemonic Industries] Brominated phosphonated amines and nitrogen phosphorus salts; flame retardants for textiles.

Pyron. [Pyron Corp.] Hydrogen-reduced iron powder.

Pyronate. [Witco/Sonneborn] Sodium petroleum sulfonates; wetting agent, dispersant; oil and froth flotation.

Pyronil. [Atochem N. Am.] Flame retardants.

Pyronyl. [Prentiss Drug & Chem.] Synergized pyrethrum conc.

Pyrophobe. [Surpass] Durable flame retardant.

Pyrosan. [Reilly-Whiteman] Flame retardant for polyester, wool, blends.

Pyroscat. [Premier Refractories] Fire retardant cement.

Pyroset®. [Am. Cyanamid; Cyanamid BV] Flame retardants for textiles.

Pyroshield. [Lubrication Engineers] Open gear lubricant.

Pyroter. [Ajinomoto; Nihon Emulsion] PCA derivs.; emulsifier, solubilizer, moisturizer, dispersant, and thickener used in personal care prods.

Pyrotol®. [Air Prods.] Catalyst.

Pyrovatex®. [Ciba-Geigy/Dyestuffs] Flame retardant cotton.

Q

Q-. [Exxon/Tomah] Quaternaries.

Q-1300, -1301. [Wako Chem. USA] N-Nitrosophenylhydroxylamine salts; analytical reagents, radical polymerization inhibitor, chelating agent, antioxidant, agric. chemicals, germicides, fungicides.

QA-555. [Floridin] Attapulgite clay; absorbent, adsorbent.

Q-Broxin. [Georgia-Pacific] Oil drilling mud additive.

QC-8800. [Quantum Composites] Vinyl ester-based sheet molding compd.; for compression molding of components requiring high structural strength.

Q-Cel®. [PQ Corp.; Omya GmbH] Inorganic silicate microspheres; extender/filler for plastics, fiberglass-reinforced plastics, cultured marble, cast polyester furniture and decorative parts, bowling ball cores, cast urethane and epoxy systems, autobody repair fillers, marine putties, PVC plastisol compds.

Q-Cide. [Huntington Labs] Germicidal detergent.

QDO®. [Lord] p-Quinone dioxime; vulcanizing agent for synthetic elastomers.

QO-33-F. [Hefti Ltd.] Sorbitan sesquioleate; emulsifier for cosmetic emulsions.

QO® Furan. [QO Chem.] Furan; chemical intermediate in mfg. of herbicides, pharmaceuticals, plastics, and fine chemicals.

QO® Furcarb. [QO Chem.] Modified furan-phenolic resins; for molding or extrusion; used in bonding carbon, graphite, basic refractory grains, silicon carbide, sand, and other aggregates to form shaped articles.

QO® Furfural. [QO Chem.] 2-Furaldehyde; chemical intermediate; solvent used in petrol. lubricating oil, gas oil, and diesel fuel; extractive distillation of C4 and C5 hydrocarbons for the mfg. of syn. rubber.

QO® Furfuryl Alcohol (FA®). [QO Chem.] Furfuryl alcohol; used in the prod. of foundry sand binders and corrosion-resistant resins; intermediate for esterification and etherification; impregnating sol'n. and carbon binder.

QO® Polymeg®. [QO Chem.] Polytetramethylene ether glycol polyol; used in urethane elastomers, fibers, coatings, and adhesives, in the prod. of high-performance thermoset and thermoplastic elastomers, elastomeric polyesters, as polyester modifiers.

QO® Quacorr® Resin/Catalyst Systems. [QO Chem.] Furfuryl alcohol-based resins and liq. catalyst; for laminates, fiberglass-reinforced plastic equip. with outstanding corrosion resistance, low flame spread, low smoke emission.

QO® Tetrahydrofuran. [QO Chem.] Tetrahydrofuran; industrial solvent.

QO® Tetrahydrofurfuryl Alcohol (THFA®). [QO Chem.] Tetrahydrofurfuryl alcohol; solvent and carrier for pesticides, paper processing; chemical intermediate.

Quab. [Degussa] Cationizing reagent.

Quabond®. [Rhone-Poulenc/Textile & Rubber] Polyvinyl acetate emulsion; textile size and finish; paper coating.

Quacorr. [QO Chem.] Furfuryl alcohol; laminating resin.

Quadefome. [Rhone-Poulenc/Textile & Rubber] Modified silicone; defoamers for fet and beck dyeing.

Quadrafos. [Peridot Chem.] Sodium tetraphosphate.

Quadrafos®. [Marlowe-Van Loan] Polyphosphate; water softener.

Quadralube. [Manufacturers Chems.] Fiber processing aid and lubricant for synthetics.

Quadrastat. [Manufacturers Chems.] Fiber processing aid, antistat for synthetics.

Quadrilan®. [Harcros UK] Antistat, germicide, surfactant, emulsifier.

Quadrol®. [BASF; BASF AG] Tetrahydroxypropyl ethylenediamine; polyol; chelating agent; intermediate used in resins, emulsifiers, surfactants, pharmaceuticals, herbicides, fungicides, insecticides, adhesives, and plasticizers; neutralizing agent for detergents.

Quaefoam 2462. [Rhone-Poulenc/Textile & Rubber] Nonsilicone effluent defoamer for textile mill waste water.

Qualiflon. [Wills Engineered Polymers Ltd.] Fluorocarbon materials.

Quamectant AM-50. [Brooks Industries] 6-(N-Acetylamino)-4-oxahexyltrimonium chloride; emollient, humectant, moisturizer for hair and skin care cosmetics.

Quanto. [Huntington Labs] Germicidal detergent.

Quantum. [Hernon Mfg.] Cyanoacrylates; performance instant bonding adhesives.

Quantum Performance Films. [Quantum/USI] Polypropylene film; biaxially oriented and metallized films.

Quar-A-Poxy. [H.B. Fuller] Epoxy adhesive.

Quartamin. [Kao Corp. SA] Quaternary ammonium chlorides; emulsifier, softener, corrosion inhibitor, sanitizing agent, antistat for personal care prods., textiles.

Quartermate. [West Agro] Dairy teat dip.

Quartzil. [Unimin Specialty Minerals] High purity quartz sand.

Quassa II. [Penco of Lyndhurst] Bittering agent.

Quat-Coll. [Brooks Industries] Quaternary collagen or gelatin derivs.; for hair and skin care cosmetics.

Quat DS. [Arsynco] Quaternary ammonium compd.

Quaternary O. [Ciba-Geigy] Quaternary oleyl imidazoline; detergent, wetting agent, penetrant, antistat used in

acids, solvs., germicides, fungistats, polishes, acid cleaners and corrosion inhibitor formulations, ore flotation, asphalt wetting.

Quatex S. [Lanaetex Prods.] Steartrimonium hydrolyzed animal protein.

Quat Keratin. [Brooks Industries] Quaternary keratin derivs.

Quat-Keratin WKP. [Brooks Industries] Cocodimonium hydroxypropyl hydrolyzed keratin; cosmetics ingredient.

Quat-Pro. [Maybrook] Hydrolyzed protein derivs.; cationic substantivity agent, moisturizer, film-former for hair and skin care prods.

Quatracid. [McGean-Rohco] Pickling salts.

Quat-Silk QTM-10. [Brooks Industries] Hydroxypropyltrimonium hydrolyzed silk.

Quat-Soy. [Brooks Industries] Hydrolyzed soy protein prods.; for hair and skin care cosmetics.

Quatrene. [Henkel/Emery] Amidoamine; wetting agent, demulsifier, and corrosion inhibitor for petrol. prod.

Quatrex. [Chemron] Quaternary ammonium chlorides or blends; surfactant, hair conditioner and softener for cosmetics; corrosion inhibitor intermediate.

Quatrex. [Dow] Epoxy resin; electronic grades.

Quatrisoft Polymer LM-200. [Amerchol; Amerchol Europe] Polyquaternium-24; stabilizer, thickener, conditioner for hair and skin prods.

Quat-Veg Q-30. [Brooks Industries] Hydroxypropyltrimonium hydrolyzed vegetable protein; for skin and hair care cosmetics.

Quat-Wheat. [Brooks Industries] Hydrolyzed wheat protein prods.; substantivity agent for hair care prods.; shampoo ingredient.

Quebracho. [Climax Performance] Wood-derived tannin; for leather tanning and water treatment applics.

Quell-Oil. [Harcros UK] Oil spill dispersant.

Quenzine. [Witco] Degreasers, heat treating salts, tempering oils, blackening oils and salts, quenching oils.

Querton. [Berol Nobel] Quaternary ammonium chlorides; surfactants.

Quesfloc. [Ques Industries] Waste water treatment flocculants.

Questal. [Clough] Sodium EDTA or HEDTA; chelating agents.

Quester. [Surpass] Sequestering agent for bleach and scouring baths.

Questex. [Rhone-Poulenc Basic] EDTA or tetrasodium salts; chelating agents.

Questric Acid 5286. [Clough] EDTA; chelating agent.

Quickcure. [Cray Valley Prods.] Polyamide resin.

Quickset. [CNC Int'l.] Accelerator/catalyst for fixing thermoplastic resins, soil-release textile finishes.

Quickset®. [Witco/Argus] MEK peroxide; initiator for curing polyester resin.

Quicksperse®. [BASF AG] Conc. pigment pastes in special varnishes for web offset inks.

Quik-Freeze®. [Miller-Stephenson] Freezing and fault isolation prods.

Quikote. [Morton Int'l.] External release coating for rubber.

Quikwet. [Hydrolabs] Wetting and scouring agents for continuous dyeing and printing operations.

Quilon. [DuPont] Chrome complex; water repellent, release agent.

Quimipol. [Quimigal-Quimica] Ethoxylated ethers; emulsifier, foam booster/stabilizer, thickener, wetting agent, detergent, solubilizer, dispersant, antistat, defoamer, for cosmetics, household detergents, textiles, plastics, metal cleaning, agric.

Quincat. [Enterprise Chem. Corp. Ltd.] Catalyst formulations.

Quindex. [Hüls Am.] Fungicide.

Quindo. [Miles/Organic Prods.] Textile dyes and pigments.

Quinplex. [Lubrication Engineers] Food grade lubricants.

Quinta-Pro Conc. [Maybrook] Hydrolyzed collagen, triethonium hydrolyzed collagen ethosulfate, cationic collagen polypeptides, hydrolyzed keratin, collagen amino acids; protein blend for hair and skin care prods.

Quintox. [Bell Labs] Cholecalciferol; pelleted bait rodenticide.

Quso®. [Degussa] Precipitated silica; thickener, suspending agent, defoamer for cosmetics.

Qwiksalt. [North Am. Salt] Sodium chloride deicer.

R

R-60 Z-5. [Werner G. Smith] Fish oil fractions; for marine lubricants, metalworking oils, rust preventatives, scavenger.

R-100. [Akzo Salt] Sodium chloride; for use in chemical industries.

R-1000. [Reheis] Aluminum hydroxide.

R3124 Ester. [Reilly-Whiteman] PEG 600 ditallate; multi-purpose emulsifier.

RA-061. [Himont] Ethylene-propylene copolymer; EPM rubber used as impact modifier and raw material

RAC-100. [Tokai Seiyu Ind.] Polyamine deriv.; lubricating and softening agent.

Racoflame. [Paradigm Labs] Flame retardant for cottons, rayon, and blends.

Racon. [Atochem N. Am.] Refrigerants.

Racumin®. [Bayer] Coumatetralyl; for control of rats and mice.

Rad-Cure. [Rad-Cure] Uv-curable, heat-sealable adhesive coatings.

Radel®. [Amoco; Amoco Europe] Poly-arylsulfone resins; for medical devices, chem. processing, elec./electronic, food pkg., aviation applics.

Radia®. [Fina Chem.] Hydrogenated triglycerides or fatty esters; lubricant, chemical intermediate, plasticizer, emollient, solvent for plastics, cosmetics, pharmaceuticals, textiles, leather, paper, metalworking, coatings.

Radiacid®. [Fina Chem.] Fatty acids; lubricants for PVC.

Radiaflot®. [Fina Chem.] Deinking chemicals for waste paper recycling.

Radialube®. [Fina Chem.] Biodegradable base oils for lubricants.

Radiamac®. [Fina Chem.] Alkyl amine acetate; flotation reagent, anticaking aid, corrosion inhibitor, emulsifier; for fertilizers.

Radiamine®. [Fina Chem.] Fatty amines or diamines; corrosion inhibitor, pigment dispersant, emulsifier, lubricant, mold release for cosmetics, mineral flotation, rubber, textiles, chemical synthesis, fertilizers, road construction; antistat and antifog additive for plastic foils.

Radiamuls®. [Fina Chem.] Sorbitan or glyceryl esters or ethoxylates or monoglycerides; food emulsifier, dispersant, antistaling agent, spray drying aid, solubilizer, dryness improver, whipping aid; used for food and feed industries.

Radianol. [Fina Chem.] Fatty alcohols; lubricant, plasticizer, solubilizer, food additive.

Radiaquat®. [Fina Chem.] Quaternary ammonium chlorides; softener, bactericide for detergents, textiles.

Radiastar®. [Fina Chem.] Calcium, magnesium or aluminum stearates; anticaking agent for powdered food prods.

Radiasurf®. [Fina Chem.] Surface-active fatty esters; chemical intermediate, emulsifier, detergent, lubricant, wetting agent, corrosion inhibitor for chemical synthesis, cosmetics, pharmaceuticals, cleaning prods., textiles, metalworking, paints, inks, plastics.

Radiex JK. [Leatex] Wetting and rewetting agent.

Radiflam. [Radici Novacips SpA] Nylon; molding and extrusion compds.

Radilon. [Radici Novacips SpA] Polyamide 6 and 6/6

Radox. [Mateson] Latex moisture barrier.

Rainbow. [IMC Fertilizer] Premium granulated fertilizers.

Rak®. [BASF AG] Pheromone; pheromones of moth varieties used in mating disruption.

Raku-Pox. [Rampf Giessharzsysteme GmbH] Epoxy resins.

Raku-Pur. [Rampf Giessharzsysteme GmbH] Polyurethane resins.

Rakusol®. [BASF AG] Organic and inorganic pigments in paraffin oil and glycerol ester; for coloring plastics.

Rally. [Rohm & Haas] Fungicides.

Ralox. [Raschig] Antioxidants for the plastics and rubber industries.

Ralufon. [Raschig] Surfactants.

Ralulac. [Raschig] Additive for lacquers.

Ralupol. [Raschig AG] Polyester resin; molding material.

Raluquin. [Raschig] Antioxidant for animal feed, plastics.

Ramak. [North Am. Refractories] Pitch bonded alumina silicon carbide plastics and ram mixes.

Ramapo. [DuPont] Lightfast pigments.

Ramasit®. [BASF; BASF AG] Hydrophobic agents for textile finishing.

Ramol. [Witco/Sonneborn] White mineral oil.

Ram Polymer 110. [Atramax] Ammonium acrylate; crosslinkable dispersing agent for pigments used in textiles; vehicle for flexo inks.

Ramtite 60. [Premier Refractories] High alumina heat setting plastic.

Ranbow. [Taiwan Surf.] 80% copper triethanolamine compd.; dual action algicide for swimming pools.

Raneoff®. [Eastern Color & Chem.] Silicone; durable water repellent for textiles.

Raney. [W.R. Grace/Davison] Nickel and nickel compound catalysts.

Ranthane. [Randolph Prods.] Polyurethane.

Rapi-Cure. [ISP] Vinyl ethers; reactive diluent.

Rapidase®. [Int'l. Bio-Synthetics] Amylase; enzyme for textile starch desizing.

Rapidase® C80. [Int'l. Bio-Synthetics] Pectinase; enzyme.

Rapidblend. [Anchor UK] Zinc oxide, magnesium oxide, and liq. alkylated diphenylamine antioxidant; vulcanizing/antidegradant system for polychloroprene.

Rapid Clean. [Aquatec Chem. Int'l.] Scale removing compds.

Rapid Scour®. [Int'l. Bio-Synthetics] Scouring agent for cotton, blends, and synthetics.

Rapid Set. [Mereco Prods.] Cyanacrylate adhesive.

Rapid-Solv®. [Int'l. Bio-Synthetics] Solvent scour.

Rapid-Wash®. [Int'l. Bio-Synthetics] Nonionic detergent for textile applics.

Rapidyne. [West Agro] Dairy germicide.

Raplan. [API Applicazion Plastiche Industriali] Thermoplastic rubber.

Rapyel Plus. [Rayonier] Sodium lignosulfonate powder.

Rareox. [W.R. Grace/Davison] Polishing prod.

Ratstaurant. [LiphaTech] Rodenticide.

Raven. [Columbian Chem.] Carbon black.

Ravinil. [European Vinyls Corp. GmbH] PVC.

Rawhide. [Rohm & Haas] Herbicides.

Rax Powder. [Prentiss Drug & Chem.] Warfarin concs.

Ray. [Asarco] Copper.

Raybinder. [Rayonier] 50% Sodium lignosulfonate sol'n.

Raycafix. [ICI Am.] Dye fixing agents.

Raycalev. [ICI Am.] Dye leveling agents, retarding agents.

Raycalube. [ICI Am.] Lubricant, antistat, and oil scavenger.

Raycapene. [ICI Am.] Penetrant, wetting and prescouring agents.

Raycapol. [ICI Am.] Detergent, wetting agent, emulsifier, dispersant for textile applics.

Raycasalt. [ICI Am.] Dyebath additive and stabilizer.

Raycaset. [ICI Am.] Acrylic binding emulsions; enhance crockfastness on pigment, vat and sulfur dyes.

Raycasoft. [ICI Am.] Softener for textile fibers.

Raycasperse. [ICI Am.] Dispersing and leveling agent for dyeing of synthetic fibers.

Raycassist. [ICI Am.] Defoamer for dye operations; antimigrant, thickener for printing operations.

Raycastrip. [ICI Am.] Stripping agent for textiles.

Raycatex. [ICI Am.] Carriers for synthetic fibers.

Raychelate. [ICI Am.] Sequestrant.

Ray Krome. [Rayonier] Chrome lignosulfonates; dispersant for drilling mud.

Raylasses. [Rayonier] 50% Sodium lignosulfonate sol'n.

Raylig. [Rayonier] Sodium lignosulfate; dispersant, suspending agent, deflocculation and visc. control in water dispersions of solids.

Raymix. [Rayonier] Sodium lignosulfate; dispersant for concrete admixtures.

Rayolan®. [Boehme Filatex] Finishing agent for lubrication of sewing threads.

RC 7. [Releasomers] Fluorocarbon; mold release agent and lubricant for silicone rubber.

RC-32. [Thiele Kaolin] Rubber clay.

RD-. [Ciba-Geigy/Plastics] Glycidyl ethers; diluent for epoxy.

RD-5078. [Akzo] Antistat for detergent/antistat/softener formulations.

RD Heparin. [Hepar Industries] Low m.w. heparin.

RDP. [Chem-Trend] Multipurpose lubricant and penetrant.

RDX. [Rhone-Poulenc] Epoxy resin.

REA-I-1. [Firestone Syn. Rubber] Vinyl compd.; semirigid extrusion compd.

Reach®. [Reheis] Aluminum chlorohydrates or aluminum-zirconium complexes; antiperspirant.

Reach-All. [Armite Labs] Silicone oiler.

ReAct. [Hernon Mfg.] Two-component acrylic adhesives; structural adhesives.

Reactant M-22, MRF. [Sybron] Reactants for textile processing.

Reactex. [Ivax Industries] Reactant, stabilizer for textile finishes.

Reacticryl. [Glo-Tex] Acrylic copolymers; pigment binders, hand modifiers, low crocks.

Reactifix. [Glo-Tex] Nonformaldehyde cationic dye fixatives.

Reacti-Gel. [Pierce Chem.] N,N´-Carbonyldiimidazole activated supports.

Reactint®. [Milliken] Reactive polymeric colorants for urethanes.

Reactisol. [Glo-Tex] Durable press reactant.

Reactmel. [Ivax Industries] Melamine thermosetting resin for finishing cellulosic, nylon, or polyester fabric.

Reactodye. [Ivax Industries] Mild oxidant to maximize brilliancy of reactive color.

Reactofix. [Ivax Industries] Dye fixatives.

Reactomer. [Morton Int'l.] Acrylic monomers.

Reactosil. [Ivax Industries] Silicones for textile finishing, softening, antislip, lubrication.

Reactosoft. [Ivax Industries] Softener, lubricant for textiles.

Reade. [Reade Advanced Materials] Metal, ceramic, and intermetallic materials.

Reademm. [Reade Advanced Materials] Metal, ceramic, and intermetallic materials.

Reakt. [BASF; BASF AG] Colorless dye intermediates for noncarbon copying paper.

Realox. [Alcoa] Low soda, calcined aluminas.

Reax®. [Westvaco] Lignosulfonates; dispersant, wetting agent, air entrainment aid, plasticizer, grinding aid, suspending agent, chelating agent for micronutrient formulations, pesticides, fertilizers, cement, dyestuffs.

Reclamite. [Witco/Golden Bear] Asphalt rejuvenator.

Recodan. [Grindsted Prods.; Grindsted Prods. Denmark] Emulsifier, heat stabilizer for milk prods.

Rec-Oil. [Betz Industrial] Prods. for separating soluble oils from wastewater.

Recov. [Covan Ltd.] Specially formulated sodium borohydride used in metal finishing and recovery processes.

Red #10. [Laur Silicone Rubber Compounding] 66.67% iron oxide pigment in silicone rubber base.

Red 139. [Presperse] Iron oxides, bismuth oxychloride.

Redball. [Int'l. Sulphur] Sulfur.

Redcote. [Synair] Moldmakers compd.

Red Diamond. [Liquid Carbonic] Industrial gases.

Red Hot Pellets. [Schaefer Salt & Chem.] Pelleted calcium chloride.

Redicote. [Akzo; Akzo Chem. BV] Cationic asphalt emulsifiers.

Redihop. [Pfizer] Modified hop extract.

Redimag. [Nat'l. Refractories & Minerals] Agric. liming material with high acid neutralizing effect.

Redimix. [Harwick] Chemical dispersions.

Rediset. [Akzo] Highway chemicals.

Redisol. [A.E. Staley Mfg.] Modified tapioca starch.

Red PP. [Cabot Plastics Ltd.] Polypropylene copolymer; permanently dissipative compd. for inj. molding.

Reduce®-150. [Am. Ingredients] Blend of sodium stearoyl lactylate, calcium sulfate, and sodium sulfite; conditioner for foods.

Reduce-IMM. [Pierce Chem.] Immobilized reductant for proteins.

Reducing Scour OEM. [Reilly-Whiteman] Clearing agents for dyestuff clearing on textiles.

Reductone®. [Olin] Sodium hydrosulfite; for continuous vat dyeing and afterscouring operations.

Redux®. [Ciba-Geigy Plastics UK] Epoxy adhesive.

ReedLite. [Reed Plastics] Heavy metal-free color concs.; for automotive, electronic, mechanical industries.

Reed Stock Concs. [Reed Plastics] Color and additive concs. for plastics.

Reel. [Abso-Clean Industries] Conc. detergents.

Reenterable Encapsulant. [CasChem] Polyurethane system.

Refinex®. [Floridin] RVM attapulgite; contact adsorbent for reclaiming motor oils.

Refractory Sheet. [Zircar Prods.] Ceramic fiber reinforced alumina prods., high temp. thermal insulation.

Refrax. [Carborundum] Refractory.

Refresh Tablets. [Stewart Hall] Slime preventives, humidifers.

Regal. [Chlorinators] Gas chlorinator and gas sulfonator for water and wastewater treatment.

Regal. [Electro Science Labs] Reinforced glass/alumina multilayer dielectric materials.

Regal®. [Cabot] Oil furnace carbon black; for inks, coatings, plastics, paper.

Regane. [Ivax Industries] Weighting agent for textile processing.

Regent 12XX. [Rhone-Poulenc Basic] Monocalcium phosphate monohydrate.

Reginol. [Hart Prods. Corp.] Surfactant blend; scour in dyeing.

Regisil. [Regis] Silylating reagent.

Regitant. [Tokai Seiyu Ind.] Phenol deriv.; resisting agent for textiles.

Regulus. [Advanced Web Prods.] Polyimide thermoplastic film; heat-resistant film for wire and cable, thermoplastic composites, pressure-sensitive adhesive tapes, primary insulation.

Rehatuin. [Intergen] Fetal bovine serum.

Reheptar. [Tech. Chem.] Food supplement for animals and humans.

Rehydragel®. [Reheis] Aluminum hydroxide; protein adsorbent.

Rehydrol® II. [Reheis] Aluminum chlorohydrate/propylene glycol complex; antiperspirant.

Reillydipipamine. [Reilly Industries] Polymer intermediate.

Reillypdhp. [Reilly Industries] Polymer intermediate.

Relaxer Conc. [Brooks Industries] Cetearyl alcohol/lanolin deriv. blends; hair relaxer formulation.

Release 161, 166DF. [CNC Int'l.] Release agents for pulp and paper industry.

Release Agent. [Akzo] Wax mixtures; release agent for molded prods.

Release Agent No. 2. [Scott Bader] Polyvinyl alcohol sol'n.; release.

Release and Paint. [Stoner] Paintable release agent.

Releasil. [Dow Corning France SA] Release agent.

Releez. [Alzo] Methyl oleate, methyl stearate, methyl palmitate, methyl laurate, methyl myristate; asphalt release agent.

Reli-O-Bond. [W.J. Ruscoe] Rubber cement.

Relpel. [Reliance Chem. Prods.] Durable silicone water repellent for cotton, wool, synthetic fabrics.

Relugan®. [BASF AG] Resin or aldehyde tanning agents; for leather and fur industries.

Remanol. [Kempen] Collagen protein hydrolysate; component for shampoos.

Rematard. [Hoechst AG] Quaternary ammonium compds.; retarders for dyeing of acrylic fibers with carbonic dyes.

Remazoln. [Hoechst Celanese] Textile dyes and pigments.

Remazol Salt FD. [Hoechst Celanese; Hoechst AG] Sodium chlorinated carboxylate; printing auxiliary for dyes.

Remcopal. [Ceca SA] Ethoxylated ethers or esters; emulsifier, degreaser, antistat, detergent, intermediate, stabilizer, solubilizer, retarder, wetting agent, dispersant for emulsion polymerization, detergents.

Remol®. [Hoechst Celanese/Colorants & Surf.; Hoechst AG] Aromatic hydrocarbon compds.; carriers for dyeing of polyester fibers.

Remolgan. [Hoechst AG] Oxethylate prods.; degreasing agents for leather.

Renacit. [Miles/Polysar Rubber] Pentachlorothiophenol prods.; peptizing agent for rubber industry.

Ren:C:O-Thane. [Ciba-Geigy] Polyurethane elastomer systems.

Renex®. [ICI Am.; ICI Surf. Belgium] Ethoxylated ethers or esters; detergent, wetting agent, emulsifier, solvent, dispersant for domestic and industrial cleaners, metal cleaning, textile scouring, paints.

Renite. [Renite] Special lubricants, release agents, swabbing compds., coatings.

Rennilase®. [Novo Nordisk] Milk-cotting enzymes for cheese prod.

Renosol. [Renosol] Plastisol formulations and polyurethane systems.

Reocor. [Ciba-Geigy/Additives] Carboxylic acid deriv.; corrosion inhibitor for lubricants.

Reodorants. [Givaudan-Roure] Mixture of essential oils and aromatic chemicals; masking agents for formaldehyde, latex coatings, finishes, polyurethane latex foam backing, solvent-based print pastes.

Reogen. [King Industries] Rubber plasticizer.

Reomet®. [Ciba-Geigy/Additives] Substituted benzotriazole deriv.; copper corrosion inhibitor for metalworking fluids; optical brightener.

Repak. [PPG Industries] Calcium hypochlorite.

Repel-O-Tex. [Rhone-Poulenc/Textile & Rubber] Water repellent for fabrics, fluorochemical extender.

Re-Pneu. [Arnco] Two-component urethane; used to permanently flatproof moderate to high pressure pneumatic tires.

Reposa. [Commercial Quimica Insular SA] Polyester resin.

Reproxal®. [Condea Chemie GmbH] PVC plasticizers.

Repsol. [Commercial Quimica Insular SA] HDPE, LDPE, polypropylene.

Rescue. [Uniroyal] Herbicide.

Rescue. [UOP] Multipurpose gasoline additive.

Reserve® Salt Flake. [Ciba-Geigy/Dyestuffs] Sodium m-nitrobenzene sulfonate; stabilizer for dyeing of fibers; assistant in discharge printing.

Reservol® F, P. [Yorkshire Pat-Chem] Sodium m-nitrobenzene sulfonate; organic oxidizing agents for prevention of streaks on discharged printed fabrics.

Resi-Bond. [Georgia-Pacific] Wood products resins.

Resicure. [Atochem N. Am.] Epoxy resins, curing agents, accelerators.

Residol Plus. [West Chem. Prods.] Cockroach insecticide.

Resi-Flake. [Georgia-Pacific] Phenolic resins.

Resiflow. [Estron] Paint additive.

Resiglas. [Fibreglass Evercoat] Polyester laminating resin.

Resi-Grow. [Georgia-Pacific] Liq. resins for fertilizer.

Resi-Lam. [Georgia-Pacific] Saturating resins for paper laminating.

Resi-Mat. [Georgia-Pacific] Resins for roofing mat.

Resimelt. [H.B. Fuller] Polyamide; hot-melt adhesive.

Resimene®. [Monsanto] Hexamethoxymethylmelamine; condensation agent for resorcinol-type bonding systems.

Resi-Mix. [Georgia-Pacific] Ready-to-use adhesives for plywood.

Resin 164. [Hexcel] Polyether polyurethane compd.; for moisture blocks and pressure dams in paper, pulp, and plastic insulated telecommunications cable.

Resin 731. [Hercules] Rosin derivs.; thermoplastic for hot-melt adhesives and coatings for paper and paperboard substrates, as tackifier and processing aid for rubber-based adhesives and molding compds.

Resin 15940. [Akzo] Allyl styrene/benzoate copolymer.

Resin QR. [Rohm & Haas] Methyl methacrylate copolymer or polyvinyl imidazolinium acetate; thermoplastic for barrier coatings, as modifier.

Resina. [Interensco NV] Polyols.

Resinex®. [Harwick] Dark thermoplastic hydrocarbon resin.

Resinoid. [Resinoid Engineering] Phenolic resin, some glass or fabric reinforced; thermoset for inj., transfer, and compression molding.

Resinol. [Raschig AG] Phenolic resin molding material.

Resinol. [Resina Chemie BV] Rigid foam polyol compd.

Resinoplas. [Atochem UK] PVC compds.

Resin Rinse. [Nalco] Ion exchange maintenance program.

Resinset Insolubilizers. [Harborchem] Paper coating insolubilizer; used in polyvinyl alcohol formulations as crosslinking agents.

Resi-Patch. [Georgia-Pacific] Plywood patching compd.

Resi-Seal. [Georgia-Pacific] Edge sealer.

Resi-Set. [Georgia-Pacific] Industrial resins.

Resi-Shell. [Georgia-Pacific] Liquid foundry resins.

Resist. [Obron Atlantic] Bronze powder.

Resist-Aid. [Transene] Promoter for adhesion of photoresists.

Resistat. [Composition Materials] Grits impregnated with antistats.

Resistone WR. [Rhone-Poulenc Ltd.] Quaternary ammonium compd.; hydrophobe in wax and wash car treatments; antistat and dispersant for paint and pigmented prods.

Resistox. [SCM Metal Prods.] Copper powder.

Resitex. [Estron] Sulfonamid resin.

Resi-Vat. [Georgia-Pacific] Log vat treatment.

Resiweld®. [H.B. Fuller] Epoxy; for adhesive, sealant, concrete patching applics.

Resoflex. [Cambridge Industries Co. of Am.] Plasticizers.

Resogen®. [Crompton & Knowles] Aminoplast precondensate; fixative for dyes on cellulosics.

Resoltex®. [Condea Chemie GmbH] Phenolic and melamine resins; bonding resins for wood processing, abrasives, foundry industry.

Resolvyl. [ICI Surf. UK] Vinyl resin aq. dispersion; finishing agent.

Resomer. [Boehringer Ingelheim KG] Biodegradable polymers.

Resopol TLS 40. [Rewo GmbH] TEA-lauryl sulfate.

Res-O-Sperse. [Dover] Water-dispersed chlorinated paraffins.

Response. [Central Soya] Textured soy protein conc.

Respumit. [Miles/Organic Prods.] Non-silicone defoamer for textiles.

RestEasy. [BASF] Polyurethane slab foam.

Resydrol. [Hoechst Celanese] Water-reducible polymers for coatings and inks.

Resyn®. [Nat'l. Starch & Chem.] Vinyl acetate/crotonic acid polymers; hair fixative; uv absorber for hair and skin care prods.

Retain. [Dow Plastics] Postconsumer recycle content plastics.

Retarder. [Eastern Color & Chem.] Dyeing assistants, retarders for textiles.

Retarder A. [Crompton & Knowles] Quaternary ammonium deriv.; retarder for level dyeing on acrylics.

Retarder A, AK, BA, BAX, PX, SAFE, SAX. [Akrochem] Retarding agents for rubbers.

Retarder ESEN®. [Uniroyal] Additive for air-cured footwear, molded soling, closed-cell sponge.

Retarder N. [Hart Prods. Corp.] Dimethyl lauryl benzyl ammonium chloride; retarders in dyeing, antistatic agents.

Retarder V-48. [Sybron] Retarder, leveling agent for cationic dyes.

Retardine®. [Henkel/Textile] Retarder/ leveling agent for direct, sulfur, and vat dyes on cellulosics.

Retardit A. [Eastern Color & Chem.] Cationic organic condensate; retarder and leveling agent for textiles.

Retardol®. [Albright & Wilson Am.] Flame retardants for pulp and paper industry.

Retardsol. [Unocal] Solvent.

Retelan. [Reter Srl] ABS resin.

Reten®. [Hercules] Acrylamide-based polymers; flocculant, retention aid, thickener, suspending agent, leveling agent, film-former, antistat, crosslinking agent for pulp and paper, hair treatment, adhesives.

Retentol RM. [Zschimmer & Schwarz] Quaternary ammonium compd.; disinfectant.

Reticusol. [Croda Inc.; Croda Chem. Ltd.] Hydrolyzed reticulin; moisturizer, conditioner for skin care prods.

Retilox®. [Akrochem] Peroxides; curing and crosslinking agents for rubber industry.

Retrocure® G. [Akrochem] Tolyltriazole; retarder for rubber industry.

Revacryl. [Harlow Chem. Co. Ltd.] Acrylic, styrene/acrylic and ester copolymer dispersions.

Revatol® S/SP. [Sandoz] Sodium m-nitrobenzene sulfonate; kier boiling assistant for vat dyed materials.

Reveal. [UVP] Leak detection system.

Revertex. [Diversified Compounders] High solids natural rubber latex; used for carpet backing, adhesives, cement and asphalt additives.

Revultex. [Diversified Compounders] Precured natural latex; used for dipped goods, balloons, gloves, catheters, molding compds.

Reward. [Lever Bros.] Soap base.

Rewo-Amid L203. [Rewo GmbH] Lauramide MEA.

Rewocid®. [Rewo GmbH] Fatty acid alkanalamides; lubricant, detergent, emulsifier, fungicide, bactericide, foam builder/stabilizer for toiletries.

Rewocor. [Rewo GmbH] Corrosion inhibitors for metalworking fluids.

Rewocoros. [Rewo GmbH] Corrosion inhibitors for metalworking fluids.

Rewoderm®. [Rewo GmbH] Ethoxylated glyceryl esters or sulfosuccinates; emulsifier, mild surfactant, thickener, superfatting agent, solubilizer for cosmetics.

Rewolan®. [Rewo GmbH] Lanolin derivs.; superfatting agent, moisturizer, emollient for personal care prods.

Rewolub. [Rewo GmbH] Lubricant for metalworking fluids, synthetic cooling oils, textile auxiliaries.

Rewomat. [Rewo GmbH] Sulfosuccinamides; detergent, foaming agent/ stabilizer, emulsifier, solubilizer, dispersant for cosmetics, emulsion polymerization, latexes.

Rewomid®. [Rewo GmbH] Fatty acid alkanolamide; detergent, foam booster/ stabilizer, emulsifier, thickener, lubricant, superfatting agent, corrosion inhibitor for cosmetics, metalworking, detergent systems.

Rewomine. [Rewo GmbH] Imidazolines; corrosion inhibitor, emulsifier, penetrant, wetting agent; used in leather and metalworking industry, paint and dyes, for carbonization baths.

Rewominox. [Rewo GmbH] Alkyl amine oxide; foam booster, antistat for personal care prods.

Rewominoxid. [Rewo GmbH] Alkyl amine oxide; foam booster, emulsifier, softener, antistat, conditioner for personal care prods.

Rewomul. [Rewo GmbH] Esters or ethoxylated ethers; emulsifier for cosmetics.

Rewopal®. [Rewo GmbH] Ethoxylates; wetting agent, dispersant, emulsifier, solubilizer, detergent, coupler, solvent, thickener, pearlescent for cosmetics, pharmaceuticals, electroplating baths, emulsion polymerization, insecticides, metalworking fluids, textiles, paper.

Rewophat. [Rewo GmbH] Ethoxylated ether phosphate; corrosion inhibitor, emulsifier, dispersant, wetting agent, antistat, hydrotrope, solubilizer for industrial cleaners.

Rewopol®. [Rewo GmbH] Anionic surfactants; foaming agent, emulsifier, raw material for personal care prods., detergents, emulsion polymerization, paper, textile, paint, and dye industries.

Rewopon®. [Rewo GmbH] Betaines or imidazolines; detergent, foam booster/stabilizer, wetting agent, corrosion inhibitor, emulsifier for personal care prods., cleaning agents, coatings.

Rewoquat. [Rewo GmbH] Quaternary compds.; disinfectant, antistat, fabric and hair softener, bacteriostat, fungicide.

Reworyl®. [Rewo GmbH] Sulfonic acids or salts; hydrotrope, catalyst, raw material; for detergent systems, polymerization, textiles, leather tanning.

Rewotein. [Rewo GmbH] Coco hydrolyzed animal protein salts.

Rewoteric. [Rewo GmbH] Amphoteric surfactants; for personal care prods., industrial cleaners, pickling baths

Rewowax. [Rewo GmbH] Emulsifier for cosmetics; spermaceti wax replacement; gloss emulsion for floors, varnishes, plastics.

Rexan. [Dexter] Fatty esters and sulfates; low foam leveling agent for disperse dyes.

Rexax. [Dexter] Auxiliary for dyeign with reactive dyes.

Rexene®. [Rexene Prods.] Polypropylene or polyethylene polymers; for inj. or blow molding, extrusion coating, laminating, profiles, sheet, wire and cable, automotive parts, construction prods., medical equip. parts, rigid containers, food pkg., laboratory ware, medicine vials, hypodermic syringes, kitchen ware.

Rexfoam. [Graden] Defoamers for paints, inks, chemical processing, pulp and paper mfg., adhesive formulations, water treatment.

Rexobase®. [Emkay] Emulsifier, deter-

gent, scouring agent.

Rexobond®. [Emkay] Resin finish.

Rexoclean®. [Emkay] Detergent, scouring agent, dye carrier.

Rexodull®. [Emkay] Delustrant for textile applics.

Rexofos. [Emkay] Water conditioner and softener.

Rexogel. [Emkay] Gelatin base; stiff finish for textiles.

Rexogum®. [Emkay] Gum resin mixture; weighter and body builder for textile finishing.

Rexoil. [Viobin] Wheat germ oil vitamin fortified.

Rexol. [Hart Chem. Ltd.] Ethoxylated ethers; emulsifier, dyeing assistant, antistat, defoamer, wetting agent, detergent, dispersant, solubilizer, stabilizer for textiles, detergent systems, agric., paper, paints, coatings, leather, polymerizations.

Rexole®. [Emkay] Oils; plasticizer, softener, lubricant, scroop finish, sizing agent for textile processing.

Rexolene. [Emkay] Sulfonate; dyeing assistant, dispersant for acetate dyes.

Rexoloid®. [Emkay] Resin finish.

Rexolube®. [Emkay] Diazotizer, lubricant, chafe eliminator for textiles.

Rexonic. [Hart Chem. Ltd.] Alcohol ethoxylates; detergent, wetting agent, emulsifier, intermediate, dispersant for cosmetics, industrial detergents, textiles, pulp and paper, paint, pesticides, drycleaning, metal cleaning, leather.

Rexonit D. [Emkay] Protein resin mixture; heavy finish for rayon and acetate fabrics.

Rexopal. [Hart Chem. Ltd.] Alkoxylated alcohols; low foaming emulsifier, wetting agent, scouring agent for textiles.

Rexopene®. [Emkay] Sodium alkylaryl sulfonate; wetting agent, leveling agent, penetrant, scouring assistant.

Rexophos. [Hart Chem. Ltd.] Phosphate esters; detergent, emulsifier.

Rexopon®. [Emkay] Detergent, scouring agents.

Rexoscour®. [Emkay] Blend of soaps, solvents, fatty acid. detergent, kier boil assistant, fulling agent, desizing agent in continuous machines.

Rexoslip®. [Emkay] Nonslip and weighter for textiles.

Rexosolve®. [Emkay] Sulfonated oil blends; detergent, scouring assistant, oil and grease remover, petroleum solvents.

Rexowax®. [Emkay] Wax blends; for textile finishes, paper mfg.; binders and lubricants for starch and gelatin formulations.

Rexowet®. [Emkay] Sulfated surfactants; wetting agent, penetrant, dyeing assistant, leveling agent for textiles.

Rextac. [Rexene Prods.] Amorphous poly alpha olefins.

Rez®. [CNC Int'l.] Urea-formaldehyde resins; textile auxiliary for crease resistance, shrinkage control, hand appeal, textile finishes.

Rezal®. [Reheis] Aluminum zirconium pentachlorohydrate or tetrachlorohydrex-glycine.

Rezax. [Chem. Processing] Textile sizing compd.

Rezcat. [CNC Int'l.] Polyfunctional aziridine prods.; crosslinkers for polymeric systems used for textile finishing.

Rezcote. [Dacar] Coatings.

Rezista. [A.E. Staley Mfg.] Modified corn starch.

Rez-N-Bond. [Schwartz] Plastic bonding solvents.

Rez-N-Dye. [Schwartz] Colors for plastics.

Rez-N-Glue. [Schwartz] Adhesives.

Rez-N-Lac. [Schwartz] Lacquers.

Rezolin. [Hexcel] Polyurethane compds.

Rez-O-Sperse®. [Dover] Chlorinated paraffin; flame retardants.

Rez-Set CI. [CNC Int'l.] Modified urea-formaldehyde resin.

Rezthane®. [CNC Int'l] Polyurethane emulsion coatings.

RG Lecithin. [Central Soya] Lecithin.

Rhenalkote®. [Condea Chemie GmbH] Alkyd resins, polyesters, or acrylates; coating resins for paints, plasters, varnishes, adhesives, and sealants.

Rhenalyd®. [Condea Chemie GmbH] Alkyd resins, polyesters, or acrylates; coating resins for paints, plasters, varnishes, adhesives, and sealants.

Rhenital®. [Condea Chemie GmbH] Phenolic and melamine resins; bonding resins for wood processing, abrasives, foundry industry.

Rhenoblend. [Rhein Chemie] Chloroprene, fluoro rubber, or NBR/PVC blends; for tech. molded goods.

Rhenocure. [Rhein Chemie] Accelerator, crosslinking agent, curing agent for rubber.

Rhenodiv. [Rhein Chemie] Release agents for rubber compounding and extrusion.

Rhenofit. [Rhein Chemie] Blow promoter for cellular rubber; filler/activator, crosslining agent for rubber.

Rhenomag. [Rhein Chemie] Magnesium oxide; acid acceptor and vulcanization activator for rubber goods, tech. molded and extruded articles, adhesives based on CR, erasers, chlorinated paraffins.

Rhenopor. [Rhein Chemie] Sodium hydrogen carbonate with dispersants; blowing agent for cellular and micro-cellular rubber articles.

Rhenosin. [Rhein Chemie] Thermoplastic resin; softener, homogenizer for rubber goods.

Rhenosorb. [Rhein Chemie] Calcium oxide; desiccant for seals, molded and extruded rubber goods, conveyor belts.

Rhenovin. [Rhein Chemie] Antioxidant, plasticizer, vulcanizing agent, accelerator, activator.

Rheocin. [United Catalysts] Trihydroxystearin; thixotropes, thickening agents for paints and coatings.

Rheodol. [Kao] Sorbitan esters or ethoxylates; emulsifier, dispersant, stabilizer for cosmetics, pharmaceuticals, polymerization, inks, paints.

Rheolate. [Rheox] Acrylates copolymer; thickener, gellant, rheological additive for coatings.

Rheopearl KL. [Chiba] Dextrin palmitate.

Rheothik Polymer. [Henkel/Emery] Polysulfonic acid; thickener, suspending agent, slip agent for lubricants, acid and alkaline systems.

Rheotix. [United Catalysts] Antisettling and antisag agent for paints.

Rheotol. [R.T. Vanderbilt] Polymerized alkyl phosphate; dispersant, wetting agent, leveling agent for pigmented coatings.

Rhodacal. [Rhone-Poulenc Surf.; Rhone-Poulenc France] Sulfonates; emulsifier, detergent, foaming agent, wetting agent, penetrant, leveling agent, solubilizer for cosmetics, agric., polymerization, leather finishing, textiles, dyestuffs, paper.

Rhodafac®. [Rhone-Poulenc Surf.; Rhone-Poulenc France] Phosphate esters; detergent, emulsifier, visc. builder, stabilizer, antistat, lubricant, softener, coupling agent, corrosion inhibitor for creams and lotions, industrial cleaners, emulsion polymerization, textiles, drycleaning, agric.

Rhodameen®. [Rhone-Poulenc Surf.; Rhone-Poulenc France] Ethoxylated fatty amines; wetting agent, penetrant, emulsifier, stabilizer, dispersant, antistat, lubricant, textile dyeing assistant, antiprecipitant, stripping agent, leveling agent, corrosion inhibitor, scouring agent.

Rhodamox. [Rhone-Poulenc Surf.] Dimethylamine oxides; foaming agent, foam stabilizer, thickener, emollient for cosmetics, toiletries, fine fabric detergents.

Rhodapex. [Rhone-Poulenc Surf.; Rhone-Poulenc France] Sulfates; high foaming surfactant, air entraining agent, frothing agent for cosmetics, concrete, gypsum wallboard, built detergents; lime soap dispersant; emulsifier for emulsion polymerization; antistat.

Rhodapon. [Rhone-Poulenc Surf.; Rhone-Poulenc France] Sulfates; wetting agent, emulsifier, detergent for cosmetics, emulsion polymerization, rug shampoo, latex stabilization.

Rhodaquat. [Rhone-Poulenc France] Alkyl dimethyl benzyl ammonium chlorides; cationic emulsifier, dispersant, bactericide for institutional disinfectant cleaners, swimming pool algicides, toiletries, medicated soaps; dye leveling and retardant in textile processing.

Rhodasurf®. [Rhone-Poulenc Surf.; Rhone-Poulenc France] Ethoxylated alcohols; detergent, foamer, solubilizer, stabilizer, emollient, wetting agent, penetrant for detergent formulations, cosmetics, textiles, metal cleaning, emulsion polymerization.

Rhodazol. [Hoechst Celanese] Textile dyes and pigments.

Rho-D-Ban. [Florasynth] Industrial deodorant.

Rhodeftal. [Rhone-Poulenc Plastiques Tech.] Polyimides-amides.

Rhodialux. [Rhone-Poulenc] Benzophenones.

Rhodiasurf. [Rhone-Poulenc] Ethoxylates.

Rhodigel®. [Rhone-Poulenc Surf.] Xanthan gum; emulsion stabilizer, suspending agent, thickener for cosmetics, pharmaceuticals.

Rhodium Sulphate TP. [Technic] Rhodium.

Rhodocap. [Rhone-Poulenc] Cyclodextrin.

Rhodopol®. [Rhone-Poulenc Surf.] Xanthan gum; for agric. suspension concs.

Rhodorsil® Antifoam, Oils. [Rhone-Poulenc Surf.] Silicone derivs.

Rhodorsil® RS. [Rhone-Poulenc] Silicone rubber.

Rhonite. [Rohm & Haas] Resins for textile finishes.

Rho-Perc. [Rho-Chem] Chlorinated vapor degreasing solvent.

Rhoplex®. [Rohm & Haas] Acrylic emulsions; for coatings, adhesives.

Rhoplex Multilobe 200. [Rohm & Haas] 100% Acrylic binder; for exterior flat paints, sheen paints, opaque stains.

Rhotex. [Rohm & Haas] Synthetic resin finishes, textile sizing, thickening agent.

Rho-Thane. [Rho-Chem] Chlorinated vapor degreasing solvent.

Rho-Tri. [Rho-Chem] Chlorinated vapor degreasing solvent.

Rho-Tron. [Rho-Chem] Fluorinated solvents.

Rhozyme®. [Genencor Int'l.] Amylase, pentosanase-hexonase, or protease; enzymes for starch liquefaction, food processing, desizing textiles.

Riacryl. [Rias A/S] PMMA.

RIA CS. [Olin] Modified urea; cure accelerator and activator.

Riag-PC. [Industriplas Ltd.] Polycarbonate.

Riblene. [Enimont UK Ltd.] LDPE.

Ribotide. [Takeda USA] Flavor enhancer.

Ricaccel. [Ricon Resins] Cure rate accelerator for chloroprene elastomers.

Rice-Pro EN-20. [Brooks Industries] Hydrolyzed rice protein; for skin and hair care cosmetics.

Rice Pro-Tein BK. [Maybrook] Hydrolyzed rice protein; film-former, substantivity agent, moisturizer, anti-irritant for skin and hair care prods.; ingredient in oil absorbent formulations.

Richaid. [Witco] Rubber processing aids.

Rich Cure. [Witco] Rubber accelerators.

Ricinion. [Gattefosse; Gattefosse SA] PEG-33 castor oil; solvent, emulsifier for cosmetics, pharmaceuticals.

Ricino Viscoil. [Vevy] PEG-25 castor oil.

Ricketson Mineral Colors. [DCS Color & Supply] Iron oxide pigments; for mortar and cements.

Rico. [Rite Industries] Textile dyes and pigments.

Ricoamide. [Rite Industries] Textile dyes and pigments.

Ricobond. [Ricon Resins] Reactive adhesive promoter for compounding with elastomers.

Ricoderm. [Rite Industries] Textile dyes and pigments.

Ricofast. [Rite Industries] Textile dyes and pigments.

Ricolan. [Rite Industries] Textile dyes and pigments.

Ricon. [Ricon Resins] Polybutadiene polymers; thermoset for potting, molding compds. and castings, rubber modifiers, wire coating, laminates, coatings.

Ricoroof. [Ricon Resins] Two-part R.T. curing elastomer; for outdoor applics. such as roofing, flashing repair, parking garage concrete sealant.

Ricoseal. [Ricon Resins] Two-part elastomer; sealant, potting compd.

Ricosolve. [Rite Industries] Textile dyes and pigments.

Ricotuff. [Ricon Resins] Anhydride/epoxy system; structural adhesive for polyolefins, elec. applics., epoxy toughener.

Ricsan. [Ricsan] Metalworking oils.

Ric-Syn Wax. [United Catalysts] Hydrogenated castor oil.

Ridacto®. [Kenrich Petrochemicals] Amine; activator.

Ridafoam. [PPG/Specialty Chem.] Surfactant blend; textile dyeing and finishing.

Ridall-Zinc. [LiphaTech] Rodenticide.

Rid-O-Germ. [Uncle Sam Chem.] Pine disinfectant.

Ridstone. [West Agro] Dairy low foam acid.

Riescour D. [Riechem/Mt. Vernon Mills] Surfactant/wetting agent blend for desizing, alkaline scour and peroxide bleaching.

Riesoft S. [Riechem/Mt. Vernon Mills] Softener for cottons.

Rigidex. [BP Chem. Ltd.] HDPE.

Rigidite®. [BASF AG] Advanced composites based on reactive resins and glass, carbon, or aramid fibers.

Rigipore. [BP Chem. Ltd.] Expandable polystyrene.

Rilanit. [Henkel/Functional Prods.; Henkel KGaA] Esters; lubricant for metalworking oils, drawing oils, greases, motor oils, extrusion of ceramics, fatting agents in textile and leather auxiliaries.

Rilsan®. [Atochem N. Am.; Atochem UK] Nylon 11; for extrusion, inj. molding, powder coatings.

Rilsoft III. [Rhone-Poulenc/Textile & Rubber] Softener, antistat for pad or exhaust applic. to cotton or polyester/cotton knits.

Rimflex®. [Syn. Rubber Tech.] Thermoplastic elastomer; for extrusion and inj. molding.

Rimlease. [Dexter/Frekote] Release coating.

RIMline®. [ICI Polyurethanes] Polyurethane or polyurea systems; for RIM processing.

Rimthane. [Dow] Urethane elastomer; for RIM processing.

Ringdex. [Sanraku] Cyclodextrin.

Rins. [Merix] Antistat.

Rinsite. [Oakite Prods.] Rinsing aid.

Rioklen. [Auschem SpA] Ethoxylated nonylphenol ethers; emulsifiers, detergents, wetting agents.

Ritacetyl®. [RITA] Acetylated lanolin; superfatting agent for soaps, shampoos; film-former.

Ritachol®. [RITA] Surfactant blends; emulsifier for cosmetics, pharmaceuticals; liq. absorption base.

Ritacholesterol. [RITA] Cholesterol.

Ritacyl. [RITA] Cetyl acetate.

Ritaderm®. [RITA] Petrolatum, lanolin, sodium PCA, polysorbate 85; emollient, moisturizer, lubricant, base for cosmetics.

Ritahydrox. [RITA] Hydroxylated lanolin; emulsifier, hypoallergenic emollient.

Ritalafa®. [RITA] Lanolin acid or ethoxylates; film-former, emollient; rewetting of makeup preparations.

Ritalan®. [RITA] Lanolin oil derivs.; emulsifier, moisturizer, emollient, blending agent, penetrant.

Ritaloe. [RITA] Aloe vera gel.

Ritamide. [Rit-Chem] Polyamide resins.

Ritapeg. [RITA] Ethoxylated fatty esters; thickener, emulsifier.

Ritapro. [RITA] Fatty alcohol/ethoxylated ether blends; emulsifier for personal care prods.

Ritaromines. [Rit-Chem] Aromatic amine epoxy curing agents.

Ritasol Base. [RITA] Isopropyl lanolate or blends; emollient, spreading agent, film former for lip prods.

Ritasynt IP. [RITA] Glycol stearate; foam booster, thickener, opacifier.

Ritawax. [RITA] Lanolin alcohol or derivs.; emollient, lubricant, moisturizer, penetrant, solubilizer, dispersant, plasticizer for personal care prods.

Rit-Cizer 8. [Rit-Chem] N-Ethyl o,p toluenesulfonamide.

Riteflex®. [Hoechst Celanese/Engineering Plastics; Hoechst UK] Thermoplastic polyester elastomer; for blow molding, inj. molding, or extrusion, polymer modification, hydraulic tubing, wire coatings, bellows, seals, dust covers.

Riteflex® BP. [Hoechst Celanese] Thermoplastic polyester elastomer alloy; for demanding applics., e.g., automobile fascia, subject to abuse in service.

Rite Reactive. [Rite Industries] Textile dyes and pigments.

Ritoform. [Rias A/S] ABS.

Ritoleth. [RITA] Ethoxylated oleyl ethers; emulsifier, solubilizer.

Ritolite. [Rit-Chem] Plasticizer resin.

RIX 80482. [Rhone-Poulenc] Resole phenolic aq. dispersion; used for adhesives, fiber finishes, and binders for filter media and abrasives.

RIX 90149, 90911. [Rhone-Poulenc/Perf. Resins & Coatings] Modified cycloaliphatic amine adduct; curing agent for epoxy resins; for industrial floor toppings, high build glaze, sealer or gel coatings, general purpose castings and encapsulations.

RJ-100. [Monsanto] Styrene allyl alcohol.

R J Resins. [Cairn Chem. Ltd.] Styrene allyl alcohol.

R-MA®. [Reheis] Aluminum hydroxide, magnesium carbonate blends.

RO-8, 9, 20. [Rogers Anti-Static] External coatings or internal antistats for plastics.

RO-40. [Georgia Marble] Ground calcium carbonate; filler for asphalt, putty, ceramic material, foamed compds.

RO A1. [Rogers Anti-Static] Antistatic coatings for plastics.

Roach Defense. [In-Cide Tech.] Insecticide.

Roach Stoppers. [Burlington Bio-Medical] Blatticide.

Robane®. [Robeco] Squalane NF; moisturizer, emollient, lubricant, humectant for skin cosmetics and pharmaceuticals.

Robeco-DNA. [Robeco] DNA or its salts.

Robecote. [Robeco] Shark liver oil; emollient for cosmetics, dermatologicals; skin protectant.

Robeyl. [Robeco] Squalene, hydrogenated shark liver oil; emollient.

Robond. [Rohm & Haas] Waterborne adhesive emulsions; for construction, pkg., pressure-sensitive applics.

Robuoy. [Robeco] Pristane; lubricant, buoyant; adjunct to oceanographic research and arctic lubrication.

Rock Dust. [Georgia Marble] Ground calcium carbonate; filler for coal mine dusting.

Rocket Release. [Stoner] Food grade paintable release agent.

Rocket® Ultra. [BASF AG] Tridemorph, fenpropiomorph; systemic fungicide for control of cereal diseases.

Rockwood. [Rockwood Systems] Fire fighting foams and wetting liquids.

Rocoat. [Hoffmann-La Roche] Coated B-complex vitamins.

Rocrolite. [Rowe Prods. Distribution] Alkyd coating.

Rocryl. [Rohm & Haas] Specialty monomers.

Rocsol. [Croda Chem. Ltd.] Oxidized hydrocarbon wax; for household and industrial polishes, as PVC processing aid.

Rodent Cake. [Bell Labs] All-weather rodenticide (anticoagulant for rats and mice).

Rodform. [R.T. Vanderbilt] Rubber accelerators.

Rodine. [Parker & Amchem] Corrosion inhibitors.

Rodip. [McGean-Rohco] Chromate conversion chemicals.

Rodo. [R.T. Vanderbilt] Essential oil blends; aromatic odors.

Rodol. [Lowenstein Dyes & Cosmetics]

Rohagit. [Rohm GmbH] Acrylates copolymer.

Rokleen. [McGean-Rohco] Soak/electrocleaning chemicals.

RokLok®. [Flexible Prods.] Polyurethane binders; for control of water and broken strata in mines, tunnels, shafts, and stream sealing applics.

Rokon. [R.T. Vanderbilt] 2-Mercaptobenzothiazole; metal deactivator, copper corrosion inhibitor for fuels, indus-

trial lubricants, automotive chemicals, and industrial cleaners; rubber accelerator.

Rol. [Fabriquimica] Ethoxylated esters or glycol, glyceryl, or propylene glycol esters.

Rolamet. [Auschem SpA] Cocamine polyglycol ether; emulsifier, codetergent for industrial use.

Rolamid. [Auschem SpA] Cocamide DEA; thickener, foam booster, superfatting agent for toiletries.

Rolfat. [Auschem SpA] Ethoxylated esters; emulsifier, codetergent, thickener for industrial use.

Rolfor. [Auschem SpA] Ethoxylates; emulsifier, detergent, solubilizer for industrial use.

Rolox. [Hardman] Epoxies.

Rolpon. [Auschem SpA] Sulfates, sulfosuccinates, or carboxylates; raw material for detergent toiletry preparations.

Roma. [Roma Color] Pigments.

Romicron. [Rohm & Haas] Ultra filtration system chemical.

Romie. [Tokai Seiyu Ind.] Polyquaternary ammonium salt; softener for acrylic and other synthetic fibers.

Romilat. [Henkel/Emery/Cospha] Herbal aroma chemical.

Rondis. [Arakawa] Disproportionated rosin and soap.

Ronex MP. [Exxon] Multipurpose grease.

Ronfalin. [DSM UK Ltd.] ABS resins; for inj. molding of housings and components of photographic equip., office machines, elec. household appliances, home computers, video game computers.

Rongal®. [BASF AG] Reducing agents for dyeing cellulose fibers with vat dyes in textile finishing.

Rongalit® C. [BASF AG] Sodium hydroxymethane sulfonate; reducing and discharge agent for textile printing.

Ronilan®. [BASF AG] Vinclozolin; contact fungicide for use in vines, fruit, etc.

Ropaque. [Rohm & Haas] Opaque polymer.

Ro-pel. [Burlington Bio-Medical] Rodent, pest and vermin repellent.

Ropet. [Rohm & Haas] Molding resin.

Roprepp. [McGean-Rohco] Alkaline derusting chemicals.

Rosbif. [Champlain Industries] Protein hydrolysates.

Rose Ether Phenoxyethanol. [Henkel/ Emery/Cospha] Rose aroma chemical.

Rose Mitcham. [A.M. Todd] Oil of peppermint.

Roseville. [Roseville Charcoal & Mfg.] Granular charcoal.

Rosheene. [McGean-Rohco] Bright dipping chemicals.

Rosinal. [Crowley Chem.] Softener, tackifier for rubber.

Rosintene. [Crowley Chem.] Terpene-rosin polymer.

Rosite®. [Rostone] Thermoset polyester, glass reinforced; bulk and sheet molding compds.

Ross. [Frank B. Ross] Synthetic waxes.

Ross Chem. [Ross Chem.] Surfactants, emulsifiers, dispersants.

Ross Emulsifier. [Ross Chem.] Ethoxylates; low foaming surfactant, emulsifier.

Ross Thix. [Frank B. Ross] Proprietary wax; thickener for aq. systems.

Ross Wax. [Frank B. Ross] Synthetic waxes.

Rostrip. [McGean-Rohco] Metal stripping chemicals.

Rotax®. [R.T. Vanderbilt] 2-Mercaptobenzothiazole; accelerator for rubber; corrosion inhibitor.

Rotoflex. [Bowdene-Redcar Ltd.] Polyethylene powder.

Rotolan®. [RITA] Lanolin deriv.; pigment dispersant for inks.

Rotopol. [Bowdene-Redcar Ltd.] Polyethylene powder.

Rotuba. [Rotuba Plastics] Cellulose acetate; for molding and extruding applics.

Rouilux. [Irpen SA] Glass fiber-reinforced polystyrene.

Rovace. [Rohm & Haas] Vinyl acetate or vinyl acrylic emulsions; for joint cement, carpet back size, acoustical coatings, textile sizing, corrugated adhesives.

Rovel. [Dow] Weatherable polymer.

Rovimat. [Chomarat et Cie, Les Fils D'Auguste] Woven roving/glass mat combination.

Rovinap. [Chomarat et Cie, Les Fils D'Auguste] Roving/mat combination.

Rovral. [Rhone-Poulenc/Ag] Fungicide.

Rowalid. [Rowa GmbH] Color masterbatches.

Rowalid PP. [Rowa GmbH] Pigment preps.

Rowaset. [Rowa GmbH] Paint additive.

Roxanthin. [Hoffmann-La Roche] Canthaxanthin beadlets.

Royal. [Corn Prods.] Glucose liq.

Royal. [Uniroyal Adhesives & Sealants] Roofing adhesives and sealants.

Royal MH 30, Slo-Gro. [Uniroyal] Plant growth regulator.

Royal-T. [Corn Prods.] Nonagglomerated dextrose/maltodextrin mixture.

Royalac®. [Uniroyal] Rubber accelerator for insulated wire, molded and mechanical goods.

Royalcast®. [Uniroyal] Low pressure castable polymers.

Royalene®. [Uniroyal; Uniroyal Chem. Ltd.] EPDM terpolymers; used for wire and cable, sponge, automotive weatherstrip, mechanical goods, coated fabrics, cellular goods, tire sidewalls, blending.

Royalite. [Royalite Plastics Ltd.] ABS, PS, PVC, ABS/PVC, PP, or thermoplastic rubber.

Royaltherm®. [Uniroyal] Silicone-modified EPDM elastomers; for automotive parts, wire and cable, architectural gaskets, tubing, food containers, appliance wire, molded elec. goods, laminating, coil coating, sheet sponge, pipe insulation.

Royaltuf. [Uniroyal; Uniroyal Chem. Ltd.] Modified EPDM polymer blends; impact modifier for thermoplastics.

Roycevat. [Passaic Color & Chem.] Textile dyes and pigments.

Rozol. [LiphaTech] Rodenticide.

RP2. [RapidPurge] Nonabrasive purging compd. for thermoplastic processing equipment.

RPP. [Ferro] Polypropylene, chemically coupled, glass reinforced.

RR 5. [Releasomers] Semipermanent

mold release agent for thermoset rubber and plastics.

RR Zinc Oxide. [Akrochem] Zinc oxide; activator for rubber.

RS-55-40. [Hefti Ltd.] PEG-40 stearate; emulsifier for cosmetics, pharmaceuticals, silicone oil, cooling emulsions; glass surface finishing agent.

RS Nitrocellulose. [Hercules] Nitrocellulose; film-former for lacquers, inks, adhesives, nail polishes, and protective and decorative coating applics.

RT-3. [Disco] Process aid, tackifier for elastomers.

RT/Duroid® M. [Rogers] PTFE, glass-reinforced; wear-resistant material for seats, seals, bearings for chemical and food processing equip., backup rings in hydraulic service, rub strips in B-1B bomber, microwave dielectric, thermal insulation, clutch face gaskets.

RTF. [GE Silicones] Silicone foam or rubber.

RTM. [Transene] Room temp. metallizing for alumina.

RTP. [RTP] Polypropylene, PBT, PET, PPS, PES, polyurethane, nylon 6/6, 6/10, 6/12, 11, PC, polyphthalamide, PS, SAN, ABS, polyethylene, acetal, or liq. crystal polymers, some glass, carbon, or mineral reinforced; thermoplastics.

RTV. [GE Silicones] RTV silicone; for adhesives, encapsulating applics.

RTX 366. [Rhone-Poulenc] Supercooled liq. cyanate ester monomer; for large aircraft composite structures, microwave antennas, radomes, spaceware, and structural or electronic grade adhesives.

Rubber Calk. [Products Research & Chem.] Construction sealant.

Rubberknit. [Richmond Oil Soap & Chem.] Synthetic lubricant.

Rubbermakers Sulfur. [Akrochem] Sulfur; for rubber industry applics.

Rubiflex Polyurethane Systems. [ICI Polyurethanes] Formulated systems for molded flexible foam.

Rubilene. [Lyondell Petrochemical] General purpose lubricants.

Rubinate®. [ICI Polyurethanes] Isocyanates; used in the mfg. of high performance elastomeric materials incl. thermoplastics, cast elastomers, sealants, coatings, adhesives, and encapsulants.

Ruco. [Occidental Chem. Europe] Saturated polyesters.

Rucoflex®. [Ruco Polymer] Polyester polyols; for polyurethanes, sol'n. laminating adhesives, sol'n. coatings, prepolymers, thermoplastic elastomers, and one-shot castables.

Rucoplex. [Ruco Polymer] Low profile additives for SMC.

Rucote®. [Ruco Polymer] Polyester resins for powder coatings.

Rucothane. [Ruco Polymer] Polyurethane latex; used as frothable interlayer/adhesive, as saturant and binder for nonwovens, and as blending resin for other latexes.

Rudol®. [Witco/Sonneborn] White mineral oil NF; binder, carrier, conditioner, defoamer, dispersant, extender, heat transfer agent, lubricant, moisture barrier, plasticizer, protective agent, and/or softener in adhesives, agric., chemicals, cleaning, cosmetics, food, pkg., plastics, and textiles.

Ruducite. [HVC] Reducing agent for chromium reduction in plating baths, waste and water treatments, dechlorination, pulp/paper deinking.

Rueterg. [Finetex] Alkylaryl sulfonates; detergent base for household and industrial applics.

Rulan. [DuPont] Flame-retardant plastic.

Rulon®. [Furon] PTFE compds.; bearing materials.

Runox. [Toho Chem. Industry] Condensed naphthalene sulfonate; dispersant for dyestuffs and pigments; detergents, rust preventives for metal surfaces; rust and paint remover.

Rust-Ban. [Exxon] Rust preventives, protective coatings.

Rust Buster. [Armite Labs] Penetrant.

RustChek. [E.H. Kellogg] Rust preventive.

Rustclean. [Octagon Process] Rust removers.

Rust-Foe. [Witco] Corrosion-resistant coating for metals.

Rustilo. [Castrol Industrial East] Rust

preventives.

Rustripper. [Oakite Prods.] Rust and scale remover.

Rustshield. [Octagon Process] Phosphating compd.

Rust Veto. [E.F. Houghton] Rust preventives.

Rust-X. [Crain] For removal of scale and soil from metal surfaces.

Ruthenium TP. [Technic] Platinum.

RWL. [Morton Int'l.] Floor polish latexes.

RX®. [Rogers] Diallyl phthalate or epoxy, some glass or mineral reinforced; engineering thermoset molding materials.

Rychem®. [Reilly-Whiteman] Polyglycol ester; emulsifier, lubricant, softener, rewetting agents for textile, leather and other industrial uses.

Ryco. [Reilly-Whiteman] Defoamers.

Rycofax®. [Reilly-Whiteman] Emulsifiers, wetting agents, softeners, debonding aids, surfactants.

Rycolube. [Reilly-Whiteman] Release agents, dyebath lubricants.

Rycomid. [Reilly-Whiteman] Alkanolamides; detergents, surfactants, foam stabilizer.

Ryflon. [Marnic PLC] PTFE prods.

Rylex. [Ferro/Bedford] Vulcanizing and curing agents.

Rylex NBC. [DuPont] Nickel dibutyldi-

thiocarbamate; uv absorber, stabilizer for polyolefins.

Rynite®. [DuPont; DuPont UK] PET or PBT resins; for elec./electronic, automotive, appliance housings, water pump housings, structural housings, encapsulation.

Ryno-Tar. [Southern Coatings] Coal tar epoxy.

Ryno-Thane. [Southern Coatings] Urethane enamel.

Ryolex Perlite. [Silbrico] Insulation.

Ryoto Sugar Ester. [Mitsubishi Kasei] Sucrose esters; emulsifier, softener, conditioner, aerating agent for foods.

Rytol. [Witco/Sonneborn] White mineral oil.

Ryton®. [Phillips; Phillips Petrol. Chem. SA/NV] PPS resins; high performance engineering thermoplastic for connectors, encapsulation caps, automotive and electronic parts, appliance components, pump housings and impellers, valves, cams, underhood parts, transistors, resistor networks.

Ryuron. [Tosoh] Ethylene-vinyl chloride, PVC or blends, or EVA/PVC polymers; for rigid and foamed profiles, flexible tubes, film, sheet, shoes and boots, crash pads, inj. molding, wire and cable insulation, pipe, dip and spray coatings, castings.

S

S-25. [Arcadian] 25% nitrogen sol'ns. containing sulfur.

S-60 RVM. [Floridin] Attapulgite clay; absorbent, adsorbent.

S160 Beads and Powder. [Degussa] Carbon black.

S-201, 507, 701. [La Roche Chem.] Claus catalyst.

S-210. [Procter & Gamble] Soya fatty acid; intermediate for mfg. of soaps, amides, esters, surfactant and non-surfactant applics.

Sabithane. [Rohm & Haas] Insecticides, fungicides, disinfectants.

Sabre. [Dow Plastics] PC/polyester resin; engineering resin.

Sachtolen. [Sachtleben Chemie GmbH] Filler masterbatches.

Sachtolith®. [Sachtleben Chemie GmbH; Ore & Chem.] Zinc sulfide; white pigment for thermoplastics, thermosets, elastomers, textile fibers, paper, sealants, lubricants.

Saci. [Witco/Sonneborn] Rust preventive concs.

Sadolin. [Perrite Plastic Compds.] Colored masterbatches.

Saduren®. [BASF AG] Synthetic resin sol'ns., melamine-formaldehyde precondensates; binders for bonding fiber webs, esp. glass fiber webs.

Saf. [ECC Int'l.] Kaolin; airfloated filling clay.

Safacid. [Jahres Fabrikker A/S] Fatty acids.

Safanol. [Sanyo Chem. Industries] Surfactants; softener for synthetic fibers, cotton and cotton/synthetic blends.

Safegard. [Sanchem] Water-based anticorrosive emulsion to prevent rust of metal compds.

Safegard CC. [Sanchem] Aluminum conversion coating.

Safe N Dri. [Oil Dri Corp. of Am.] Oil and grease.

SafeSolv. [Amax Industrial Prods.] Nonconductive degreasing solvent.

Safester A-75. [Lipo] Ethyl linoleate.

Safetol. [Halocarbon Prods.] Inert, nonflamm. hydraulic fluids.

Safe-T-Quench. [Castrol Industrial East] Synthetic quenchants.

Safety-Cool. [Castrol Industrial East] Synthetic grinding coolants.

Safety-Draw. [Castrol Industrial East] Drawing compds.

Safety First. [Amax Industrial Prods.] Citrus-based degreaser.

Safety-Lube®. [Chem-Trend] Blends; die lubricant for aluminum castings.

Safety Solvent 378. [Amax Industrial Prods.] Dielectric degreaser.

Safezone®. [Miller-Stephenson] Cleaning solvent and flux remover.

Saflex. [Monsanto] Polyvinyl butyral film.

Safoam. [Reedy Int'l.] Chemical endothermic nucleating and blowing agent for thermoplastics.

Saf-T-Sol. [Crowley Chem.] Insecticide solvent, industrial cleaning compd.

SAG®. [Union Carbide] Dimethicone and silica; antifoams.

S.A.G. [Midwest Rubber Reclaiming] Fiber reclaimed from tires; used for reinforcement in mud flaps, truck bed liners.

Sagtex®. [Union Carbide] Silicone antifoams for textiles.

Saisir. [Bruce Chem.] Solubilizing agent, stabilizers for boil out, bleach operations.

Salabon. [Takemoto Oil & Fat] Alkyl polyamine ethyl glycine hydrochloride; germicide.

Salacos. [Nisshin Oil Mills] Esters.

Saladizer. [TIC Gums] Blended stabilizers for emulsions, salad dressings.

Salcare. [Allied Colloids] Polyquaterniums or steareth-10 allyl ether/acrylates copolymer.

Salfax. [Chem-Y GmbH] Fatty acid and cationics; hair conditioner.

Sali. [Zircar Prods.] Aluminum oxide fibrous ceramic thermal insulation.

Salioca. [Am. Maize Prods.] Waxy blend.

Salt 100, 200, 300 M. [Tokai Seiyu Ind.] High polymeric active agent/solvent blend; dyeing assistant for dyeing wool at low temps.

Saltplex. [Aqua Process] Removes or prevents depositon due to ammonia chloride salts or other inorganic salts.

Salut®. [BASF AG] Chlorpyrifos, dimethoate; insectide.

Salute. [Miles/Ag] Herbicide for soybeans.

Saluthion®. [BASF AG] Chlorpyrifos, dimethoate; insectide.

Samaron Thickener N. [Hoechst AG] Acrylic acid based high polymer; thickener for printing with nonionic dispersing dyes on polyester and triacetate fibers.

Sampaque. [J.M. Huber] Structured clay for paper applics.

Sanac. [Karlshamns] Amphoteric surfactants; detergents, wetting and foaming agents, solubilizers.

Sana Prep. [Huntington Labs] Povidone iodine; surgical prep.

Sana Scrub. [Huntington Labs] Povidone iodine; surgical scrub.

Sancure. [Sanncor Industries] Water-based urethane resins.

Sandacid®. [Sandoz] Buffer, dispersant for dyeings.

Sandene®. [Sandoz] Cationic polymer for Vagabond styling effects.

Sandet. [Sanyo Chem. Industries] Sulfates or sulfonates; detergent base, emulsifier, wetting agent, penetrant, for emulsion polymerization, shampoos, bleaching detergents.

Sandin. [Sandoz] Sulfonated hydrocarbon; antistat for PVC, talc, etc.

Sandobet SC. [Sandoz] Cocamidopropyl hydroxy sultaine; mild surfactant for cosmetics, toiletries.

Sandoclean®. [Sandoz] Cleaning agent for achieving uniform hydrophilic finish on cotton and blends.

Sandocorin. [Sandoz] Phosphate esters; film-former, corrosion inhibitor.

Sandocryl®. [Sandoz] Specialty dyes for coloring aq. media.

Sandoderm. [Sandoz] Textile dyes and pigments.

Sandofix®. [Sandoz] Dye fixatives.

Sandoflam. [Sandoz UK] Flame retardants.

Sandogen®. [Sandoz] Dye leveling agent; antiprecipitant for dyeing nylon carpet.

Sandolan®. [Sandoz] Specialty dyes for coloring aq. media, fertilizers.

Sandolite. [Sandoz] Uv absorber for exhaust dyeing.

Sandolube®. [Sandoz Prods. Ltd.] Aliphatic hydrocarbons aq. dispersion; enhances sewing properties of knitwear and textiles.

Sandopan®. [Sandoz; Sandoz Prods. Ltd.] Surfactants; detergent, emulsifier, wetting agent, scouring agent, for cosmetics, household, industrial use.

Sandoperm FE. [Sandoz] Dimethiconol and amodimethicone; finish for pad applic. on woven and knitted cellulosic and cellulosic/synthetic blends.

Sandoperme® ME. [Sandoz] Amino functional silicone softener; for continuous and exhaust applics.

Sandopur. [Sandoz Prods. Ltd.] Ethoxylated ether; washing-off agent for textiles.

Sandopure®. [Sandoz] Print washing detergent, washing off agent for dyeings.

Sandorin. [Sandoz UK] Organic pigments.

Sandospace® HPB. [Sandoz] Quaternary amine; reactive agent for increasing affinity of cellulosic fibers for cellulosic dyes.

Sandostab P-EPQ. [Sandoz] [Tetrakis (2,4-di-tert-butylphenyl) 4,4-biphenylenediphosphonite]; processing stabilizer, antioxidant for polymers.

Sandoteric. [Sandoz] Amphohydroxypropyl sulfonates; extremely mild surfactant producing synergistic visc. increase with alkyl sulfates; weak ampholyte.

Sandotex® A. [Sandoz] Stearamidoethyl ethanolamine phosphate, cetoleth-24; antistat for synthetic pile carpets.

Sandothrene. [Sandoz] Textile dyes and pigments.

Sandotor HV. [Sandoz] Permanent hydrophilic finish for nylon.

Sandoxylate®. [Sandoz] Alkoxylated alcohols, fatty acids, amines, alkyl phenols; wetting agent, dispersant, surfactant for household and industrial applics.

Sandoz. [Sandoz] Surfactants.

Sandozin®. [Sandoz; Sandoz Prods. Ltd.] Wetting agent, detergent for textiles.

Sandozol®. [Sandoz; Sandoz Prods. Ltd.] Sulfonated oil; for finishing and sanforizing; textile dyes and pigments.

Sanduvor. [Sandoz] Oxalanilide deriv.; uv absorber for films and coatings.

Sanfix. [Sanyo Chem. Industries] Cationic resins; fixing agent for direct and reactive dyes.

Sanfloc. [Sanyo Chem. Industries] High m.w. polymers; flocculants for water treatment.

Sanfodex. [Dexter] Wetting agents.

Sani Fect. [Alex C. Fergusson] Nascent chlorine producing system.

Saniflame. [Sanitized Inc.] Flame retardant.

Sanikleen. [West Chem. Prods.] Quaternary disinfectant.

Sanimal. [Nippon Nyukazai] Emulsifier for pesticides and insecticides.

Sani-Matic System. [Surco Prods.] Odor control system.

Sanisol. [Kao Corp. SA] Benzalkonium chloride; disinfectant, sanitizer for pharmaceuticals, industrial and cooling water treatment.

Sanitary 1700. [GE Silicones] Silicone sealant.

Sanitized®. [Sanitized; Sanitized AG] Bacteriostat, antimicrobial treatments for textiles.

Sanleaf. [Sanyo Chem. Industries] Nonionic/cationic surfactant; yarn finishing agent for cheese-oiling applic.

Sanmorin. [Sanyo Chem. Industries]

Penetrant, wetting agent for textile processing.

Sanogran®. [Sandoz] Pigments for coloring gels, powders, and solids.

Sanperox®. [Sandoz] Stabilizer with sequestering and dispersing properties; for peroxide bleaching of cotton and blends.

Sanres. [Sanncor Industries] Solvent-based urethane resins.

Sansil. [PPG Industries] Inorganic paper chemicals.

Sansilic 11. [Ceca SA] Antifoamer for fiber processing.

Sanstat. [Sanyo Chem. Industries] External antistat for plastics, polyester fibers.

Santac. [Santech] High m.w. polybutene tackifier in LLDPE; cling additive for stretch-cling film.

Santechem 21-21. [Santech] Azodicarbonamide conc.; blowing agent for polyolefins, extruded and inj. molded structural foam.

Santechem Grey F.R. P.E. Conc. [Santech] Halogen compd. and antimony oxide in LLDPE; flame retardant conc. for polyethylene.

Santechem Solid Color PE Concs. [Santech] Pigments dispersed in LDPE.

Santicizer. [Monsanto; Monsanto Europe] Plasticizers for resins.

Santochlor. [Monsanto] p-Dichlorobenzene.

Santocure®. [Monsanto] Benzothiazole derivs.; accelerator for sulfur-curable elastomers.

Santoflex®. [Monsanto] Phenylene diamines or blends; antiozonant, antiflex cracking agent for rubber.

Santogard® PVI. [Monsanto] N-(Cyclohexylthio) phthalimide; prevulcanization inhibitor.

Santolink. [Monsanto] Crosslinking agent.

Santone®. [Van Den Bergh Foods] Polyglyceryl esters; food emulsifiers.

Santonox. [Monsanto] Nonrubber antioxidant.

Santoprene®. [Advanced Elastomer Systems] Thermoplastic rubber.

Santoquin. [Monsanto] Ethoxyquin; antioxidant for fats and rendered prods.

Santosol. [Monsanto] Polyphenyl deriv.; hydraulic fluid.

Santotrac. [Monsanto] Synthetic hydrocarbon.

Santovac. [Monsanto] Vacuum diffusion pump fluid.

Santovar® A. [Monsanto] 2,5-Di (t-amyl) hydroquinone; staining antioxidant in noncuring applics., adhesives; NBR stabilizer.

Santowax. [Monsanto] Mixed isomeric terphenyls.

Santoweb®. [Monsanto] Treated cellulose fibers; reinforcing agent for rubbers, plastics.

Santowhite®. [Monsanto] Antioxidant for rubber, synthetics, latexes.

Sanuril 115 Tablets. [Eltech Int'l.] 70% Calcium hypochlorite tablets; used to disinfect wastewater.

Sanwax. [Sanyo Chem. Industries] Low m.w. polyethylene; lubricants and pigment dispersants for plastics.

Sanwet. [Hoechst Celanese; Hoechst AG] Sodium polyacrylate starch.

Sanwet. [Sanyo Chem. Industries] Starch grafted polyacrylate; absorbent polymers for diapers, sanitary napkins, soil conditioners.

Sanylene. [Sandoz UK] Masterbatch for PP fiber.

Sanyo Levelon. [Sanyo Chem. Industries] Naphthalene sulfonate; dispersant, water reducer for concrete molding.

Sanyo PEG Series. [Sanyo Chem. Industries] PEGs; intermediates for surfactants, plasticizers, lubricants, etc.

Sapogenat T Brands. [Hoechst AG] Ethoxylated tributyl phenol; surface active basic material.

Saporin®. [BASF AG] Feed industry additive.

Sapstain Control Chemical NP-1. [Kop-Coat] Specially formulated sapstain chemicals.

Saran. [Dow] Vinylidene chloride polymers; extrusion resins for containers, film, sheet, barrier applics.

Saranex. [Dow] Plastic films; barrier films for ostomy appliances, trans-

dermal drug delivery systems, tissue culture pouches, medical and drug pkg., instrument pkg.

Sarapoll. [Nat'l. Wax] Blended coatings.

Sarbox®. [Sartomer] Resins with acrylate, anhydride, and carboxyl functionality; for photoimaging, inks, metal coatings, plastic coatings, sealants, adhesives.

Saret®. [Sartomer] Acrylic crosslinking agents.

Sar Gel®. [Sartomer] White paste indicating presence of water bottoms in storage tanks.

Sarkosine. [Hoechst Celanese] Sarcosinates.

Sarkosyl®. [Ciba-Geigy AG] Sarcosine derivs.; detergent, corrosion inhibitor, foam booster/stabilizer, wetting agent, lubricant, emulsifier for dentifrices, personal care, and household cleaning prods., pharmaceuticals, metal processing and finishing, metalworking and cutting oils.

Sarlink. [DSM] Thermoplastic elastomers; for hose, tubing, coated fabrics, weatherstripping, diaphragms, gaskets, seals, extruded sheet, ducting, belts, trays, tank linings, sports grips, medical stoppers, o-rings, furniture parts, connectors.

Sarmaflam. [Sandoz UK; Sarma SpA] Flame retardant masterbatch.

Sarmastab. [Sarma SpA] Light stabilizer.

Sarmatron. [Sarma SpA] Expanding agent for polyolefins.

Sartomer. [Sartomer] Intermediates for urethanes, polyesters, allyls, prepolymers; plasticizer, fire retardant for films, coatings.

Satexlan 20. [Croda Inc.] PEG-20 hydrogenated lanolin; emulsifier, emollient, thickener, perfume solubilizer.

Satialgine H8. [Mendell] Alginic acid NF; disintegrant for pharmaceutical tabletting.

Satina. [Van Den Bergh Foods] Hydrogenated vegetable oils; center fats for confectionery.

Satintone®. [Engelhard; Lawrence Industries PLC] Calcined aluminum silicate; reinforcing extender for rubber and polymer systems.

Satulan. [Croda Inc.; Croda Chem. Ltd.] Hydrogenated lanolin; emulsifier, emollient for personal care prods.

Saturseal. [Crowley Tar Prods.] Joint sealant.

Savan. [Kerr-McGee] Sodium ammonium decavanadate.

Savey. [DuPont/Ag] Miticide.

Savinase®. [Novo Nordisk] Bacterial protease; enzyme, defoamer for built liq. laundry detergents.

Savinyl®. [Sandoz] Specialty dyes for coloring adhesives, oils, waxes, solvents, wood stains.

Savondol GP-9. [Nippon Oils & Fats] PPG-8 polyglyceryl-2 ether.

Savopine. [West Chem. Prods.] Disinfectant, cleaner, deodorant.

Savosellig. [Ceca SA] For soaping fiber reactive dyes on cotton.

Sav-Ox®. [Olin] Hydrazine sol'ns.; corrosion protector for industrial boilers.

Sawaclean. [Lion] Alpha-olefin sulfonate; milling and scouring agent for feathers, wool, raw wool.

Saytex®. [Ethyl] Brominated flame retardants for plastics.

SB-. [Solem Industries] Alumina trihydrate; filler with flame retarding/ smoke suppressing properties.

SB522/1S Film. [Hercules] Balanced oriented polypropylene film; for lightweight food pkg.

SC-10, 12, 17. [Bacon] Filled silicone-based resins; thermally conductive polymers.

SC-53. [ECC Int'l.] Calcium carbonate; coarse ground fillers.

Scalebreaker. [Kamco Ltd.] Descaling pumps and chemicals.

Scaleclean F. [Stewart Hall] Scale and deposit inhibitor for potable water systems.

Scale-Cleen. [W.R. Grace/Dearborn] Cleaner.

Scale-Off. [Ambersil Ltd.] Water scale remover.

Scale-Prep. [Crown Tech.] Oxide scale conditioner.

Scalite. [Jersey Ind. Chem. Prod.] Com-

plete boiler water treatment.

Scarab. [BIP Chem. Ltd.] Urea-formaldehyde molding powder.

Scavenger 1200. [Western Nutrients] Hydrogen sulfide scavenger for oil drilling muds.

Scav-Ex®. [Hercules] Synthetic pulp; fibrous processing aid for toweling, gypsum paper, boxboard, and fine paper mfg.

Scav-Ox. [Olin] Hydrazine sol'n.

SCC. [Solvit] All-purpose liq. cleaner conc.

ScentCap. [M-CAP Tech. Int'l.] Microencapsulated fragrance mineral oils.

Scentenal. [Firmenich] Fragrance ingredient.

Scentinel. [Phillips] Gas and LPG odorants.

Scepter®. [Am. Cyanamid/Ag] Ammonium salt of imazaquin; herbicide for soybean crops.

Schercamox. [Scher] Amine oxides; conditioner, detergent, wetting agent, antistat, softener, foam booster, stabilizer, emollient.

Schercemol. [Scher] Esters; emollient, emulsifier, opacifier, stabilizer, thickener for personal care prods.

Schercoat. [Scher] Substantive poly emulsion; lubricant for glass containers.

Schercodine. [Scher] Dimethylamines; emulsifier, intermediate, conditioner, visc. builder, softener for hair and skin prods.

Scherco Finish. [Scher] Resin dispersion; lubricant for textiles.

Schercomid. [Scher] Fatty acid alkanolamides; foam booster/stabilizer, thickener, emulsifier, solubilizer, humectant, suspending agent, for cosmetic and detergent applics.

Schercomul. [Scher] Emulsifier, detergent.

Schercophos. [Scher] Complex phosphate; detergent builder, water conditioner for textile industry.

Schercopol. [Scher] Sulfosuccinates; wetting agents, solubilizers, softeners, surface tension depressants for personal care products, home and industrial cleaners.

Schercopon. [Scher] Ethoxylated sulfosuccinate; drycleaning detergent, wetting agent.

Schercoquat. [Scher] Quaternary ammonium compds.; emulsifier, conditioner, bactericide for hair care prods.

Schercotaine. [Scher] Betaines; mild surfactants, detergents, wetting agents, foamers, cloud pt. depressants for cosmetics, toiletries.

Schercoterge. [Scher] Ethoxylated amide; detergent, wetting and textile scouring agent, emulsifier, dye assistant.

Schercoteric. [Scher] Imidazolinium amphoterics; surfactants for cosmetic and industrial cleaners.

Schercowet DOS-70. [Scher] Sodium dioctyl sulfosuccinate; emulsifier for emulsion polymerization; wetting agent.

Schercozoline. [Scher] Imidazolines; surfactant, antistat, dispersant, wetting agent, emulsifier, microbicide used in acid and emulsion cleaning, cleaners, polishes, surf. treatment, textile and leather processing, agriculture and cosmetic intermediate.

Scheroba Oil. [Scher] Isostearyl erucate and erucyl erucate; jojoba oil substitute; for cosmetic formulations.

Schlichte®. [BASF AG] Sizes for cellulose and synthetic fibers.

Schulamid. [A. Schulman] Nylon compds. or alloys.

Schulman. [Comco Plastics Ltd.] Masterbatch additives.

Scin Sol. [Monomer-Polymer & Dajac Labs] Scintillation solvent.

Sclair. [D M & C Watering BV] Polyethylene.

Sclairfilm®. [DuPont Canada] Polyolefin films.

SCL Conc. [Elco] Lubricant additive.

Sclomo. [Ferro/Keil] Sulfur chlorinated bases; extreme pressure agents for threading and tapping operations, cutting and grinding oils.

Scolefin. [Anilac NV/SA] HDPE.

Scorguard. [Western Water Management] Organic cooling water scale and

corrosion inhibitor.

Scotch. [3M] Adhesives.

Scotchban. [3M] Water/oil/grease treatment for paper.

Scotch-Clad. [3M] Coatings.

Scotchgard®. [3M] Water/oil repellent treatment for fabrics.

Scotchkote®. [3M] Epoxy coatings; corrosion protective resins, electrical coating.

Scotchlite. [3M] Glass bubbles; engineered fillers for industrial applics.

Scotch-Weld®. [3M] Thermosetting structural adhesives.

Scotex. [Protan] Thickener for printing reactive dyes on cellulose fibers.

Scourol. [Kao] Ethoxylated ether or blends; textile processing assistant, scouring agent, desizing agent, bleach assistant.

Scouron. [Chemonic Industries] Emulsifier for removal of lubricants and spinning oils.

Scourwhite. [Aerochem] Blended surfactants and fluorescent dye; one-bath scouring and brightening agent for nylon.

Scovinyl. [Anilac NV/SA] PVC.

Scripset. [Akzo] Surface sizing agent.

Scripset®. [Monsanto] Styrene/maleic anhydride copolymer; emulsifier, binder, sizing agent, visc. modifier, stabilizer; starch modifier; pigment dispersant, protective colloid, sizing, coating, water-paint calsomines, adhesives, printing, paints.

Scroop B. [CNC Int'l.] Ethylene oxide condensate; antistat, fiber lubricant, scroop agent for textiles.

Scrubidine. [Medical Chem.] Povidone iodine scrub.

Scrub-N-Shine. [Advance Chem] Neutral detergent.

Scrub-Stat IV. [Huntington Labs] Antimicrobial sol'n.

SCS 40. [Witco SA] Sodium cumenesulfonate.

SD-. [Reilly-Whiteman] Fatty acid alkanolamides; detergent, emulsifier, corrosion inhibitor, visc. builder for hard surface cleaners, laundry prods., textile scouring, metalworking fluids, wax stripping, shampoos.

SD-35. [Presperse] Panthenyl ethyl ether.

SD-9101, 9104. [Polysar] PC/modified acrylic alloy.

SDL-1. [Akzo] Synthetic engine lubricant.

SDM. [ICI Am.] Organo nitrates.

SE-. [GE Silicones] Silicone rubber compds.; for compr., transfer, and inj. molding, extrusion, and calendering, o-rings, cylinder liner seals, hydraulic seals, valve gaskets, tubing, belting, shock mounts, rolls, closed cell sponge, coated fabrics.

Seabond. [Hilton Davis] Pigments and dispersions.

Seaco. [Southeastern Adhesives] Urea-formaldehyde resins and emulsion adhesives.

SeaGel. [FMC/Marine Colloids] Locust bean gum; milk stabilizer, gelling agent.

SeaKem. [FMC/Marine Colloids] Carrageenan; milk thickener, stabilizer, gelling agent.

Sea-Klear. [Vanson Chem.] Chitosan sol'n.; for clarification of pool, spa, and hot tubs and to treat waste waters.

Sealmore. [Armite Labs] Sealing pipe joint.

Seal-Tyte. [J.C. Whitlam Mfg.] Boiler compd.

Seal-Unyte. [J.C. Whitlam Mfg.] Pipe joint thread and gasket compd.

SeaPlaque. [FMC/Marine Colloids] Agarose; gelling media for research or clinical diagnostic tests.

Seapol. [Sanyo Chem. Industries] Surfactants/solvent blend; low toxicity oil spill remover.

SeaSep. [FMC/Marine Colloids] Agarose.

Seasoft. [Nutex] Cationic softener for textiles.

SeaSpen. [FMC/Marine Colloids] Carrageenan; suspending agent.

Seatone. [Hilton Davis] Pigments and dispersions.

Sebamid. [Industrial Polymer Services Ltd.] Polyamide 6.

Sebase. [Westbrook Lanolin] Ethoxylated lanolin blend; emollient, lubricant,

visc. stabilizer for cosmetics.

Sebopessina. [Vevy] Polyisoprene, soy sterol.

Sebosan. [Stockhausen] Cationic softeners for textiles.

Seboside. [Vevy] Soy sterol, PPG-5 pentaerythritol ether, PEG-5 pentaerythritol ether.

Secco. [Southeastern Chem.] Antifoams, cleaners, wetting agents, fixing agents, oil and water repellents, leveling agents, chelating agents, lubricants, flame retardants, scouring agents for textiles.

Secco Clay. [Southeastern Clay] Hard rubber clay.

Secolat. [Stepan Europe] Alkyl disodium sulfosuccinamate; foamer for latex emulsions; emulsifier for emulsion polymerization.

Secomine. [Stepan Europe] Ethoxylated alkylamine; lubricant, wetting agent, antistat, emulsifier.

Secosol®. [Stepan Europe] Sulfosuccinates; mild foamer, dispersant, wetting/rewetting agent for cosmetics, cleaners, textiles.

Secoster®. [Stepan Europe] Esters; lubricant, antistat, emulsifier, dispersant, solubilizer, emollient, pearlescent for cosmetics, cutting oils, pesticides.

Secosyl. [Stepan Europe] Sodium lauroyl sarcosinate; detergent, foaming agent, base, anticorrosion additive for rug shampoos, personal care prods.

Securon® 668. [Henkel/Textile] Chelating agent.

Sedamon. [Henkel/Emery/Cospha] Jasmine aroma chemical.

Sedaplant Richter. [Dr. Kurt Richter; Henkel/Cospha] Polyvalent herbal extract blend; emollient.

Sedefos 75®. [Gattefosse; Gattefosse SA] Glycol stearate, PEG-2 stearate, and trilaureth-4 phosphate; self-emulsifying base for cosmetics and pharmaceuticals.

Sedipol®. [BASF AG] Higher aliphatic alcohols; antifoams for industrial and sewage applics.

Sedipur®. [BASF AG] High m.w. anionic, cationic, or nonionic polymers;

flocculants for industrial and communal water treatment, sewage sludge treatment, mining sol'ns.

Sedoran FF. [Sanyo Chem. Industries] Base materials for detergents.

SE/Duroid. [Lydall] Cellulose fiber/ polymer sheet material; fiber board used as elec. barrier insulation and low-cost structural material.

Seeclear. [Vyse Gelatin] 275 Bloom gelatin.

Seeley Flavors. [Champlain Industries] Butter and dairy flavors.

Seenox 412S. [Witco/Argus] Pentaerythrityl tetrakis (beta-laurylthiopropionate).

SEF. [Monsanto] Modacrylic fibers.

Selar®. [DuPont; DuPont UK] Melt-extrudable barrier polymers.

Select. [Oil Dri Corp. of Am.] Gellant.

Select-A-Sorb. [R.T. Vanderbilt] Hydrous magnesium silicate; reinforcing filler for rubber, wire and cable stocks.

Selecthane. [Xenox] Thermoplastic urethane elastomer.

Selectipur. [EM Industries] Chemical for semiconductor industry.

Select-Unyte. [J.C. Whitlam Mfg.] Pipe-joint compd.

Selene. [Polychem Systems Srl] Epoxides for elec. use.

Selexol. [Union Carbide] Gas treating solvent.

Selexsorb. [Alcoa] Selective adsorbents.

Self Locker. [Hernon Mfg.] Preapplied anaerobic thread locking adhesives.

Self Seeler. [Hernon Mfg.] Preapplied pipe thread sealants.

Selin® O. [Grünau] Trioleate; basic oil for release agents, bread cutting oil.

Sellifix. [Ceca SA] Fixative for fiber reactive dyes.

Sellig. [Ceca SA] Antifoam, bleach stabilizer, corrosion inhibitor, wetting agent, detergent, leveling agent for textile treatment.

Selligor. [Ceca SA] For washing out of acidic dyes.

Sellogen. [Henkel/Emery; Henkel-Nopco] Sulfates or sulfonates; wetting agent, detergent, dispersant, emulsifier for textiles, agric., household and in-dustrial cleaners, fire control, paint strippers, emulsion and suspension polymerization aids, latex paints.

SEM-. [Harcros] Silicone emulsion; lubricant, release agent, emulsifier for plastics, rubber, metalworking, printing, textile softeners, cosmetics.

Semacylase. [Novo Nordisk] Pen-v-acylase; enzyme for mfg. of penicillin intermediates.

Semi-Base I. [Ferro/Keil] Semisynthetic conc. base for lubricants; corrosion inhibitor.

Semirit. [Hüls Am.] Aluminum oxide semifriable partial glyceride blend of natural saturated vegetable.

Semtol®. [Witco/Sonneborn] White mineral oil, tech.; binder, carrier, conditioner, defoamer, dispersant, extender, heat transfer agent, lubricant, moisture barrier, plasticizer, protective agent, and/or softener in adhesives, agric., chemicals, cleaning, cosmetics, food, pkg., plastics, and textiles.

Sencor®. [Bayer; Miles/Ag] Metribuzin; selective herbicide for weed control in potatoes, tomatoes, soybean, turf, etc.

Senka Antifoam. [Nippon Senka] Silicone emulsion; antifoaming agent.

Senocryl. [Senova Kunststoffe GmbH] PMMA.

Senodur. [Senova Kunststoffe GmbH] PVC.

Senolen. [Senova Kunststoffe GmbH] PE, PP.

Senolux. [Senova Kunststoffe GmbH] PC.

Senosan. [Senova Kunststoffe GmbH] PS.

Sensi Tech. [A-Aroma Tech.] Odor counteractant/fragrance combinations.

Sentry. [Union Carbide] Simethicone; antifoam; antiflatulent in OTC drugs; in food-grade emulsifiers for antacid prods.

Seodol. [Nihon Emulsion] Ethoxylated octyldodecyl ethers.

Sepabase®. [BASF AG] EO/PO adduct; for dehydration of crude oil emulsions and removal of residual salts in petroleum prod.

Sepabeads®. [Mitsubishi Kasei] Highporous type hydrophilic polymers; for industrial purification of protein and enzyme by chromatography.

Sepacid®. [BASF AG] Biocides for oilfield applics., crude oil prod.

Sepaclear®. [BASF AG] Removal of residual oil from water in refineries and in crude oil prod.

Sepacorr®. [BASF AG] Nitrogen-containing condensate; corrosion inhibitor for crude oil production and refining equipment.

Sepaflood®. [BASF AG] Water-soluble polymers; for tertiary crude oil prod.

Sepaflux®. [BASF AG] Pour pt. and/or visc. depressants for crude oils and residual oils.

Sepakoll®. [BASF AG] Protective colloids for drilling fluids and borehole cementing.

Sepapar® P. [BASF AG] Paraffin inhibitor.

Separan. [Dow] Flocculants.

Separex®. [Air Prods.] Membrane systems.

Separol®. [BASF AG] EO/PO adduct; demulsifier for crude oil emulsions.

Sepascale®. [BASF AG] Scale inhibitors for petroleum prod. and processing.

Sepasolv® MPE. [BASF AG] Purification of natural gas.

Sepawet®. [BASF AG] Surfactants for petroleum prod. and for pipeline transport.

Sepicide. [Seppic] Antimicrobial preservatives for cosmetics.

Sepigel. [Seppic] Polyacrylamide, C13-14 isoparaffin, laureth-7; thickening agent.

Sepiogel. [Floridin] Sepiolite mineral; gelling and suspending agent for drilling muds.

Sepisol. [Alcan Rubber & Chem.] Textile dyes and pigments.

SEQ. [Vikon] Chelating agents for heavy metal ions.

Seqlene®. [Pfanstiehl Labs] Sodium glucoheptonate or blends; sequestrant for metal applics.; scavenger for antioxidants, bactericides; bottle washing, alkaline cleaning, textile applics.

Sequabond®. [Sequa] Acrylic and acrylic copolymer emulsions; for water-based inks and overprint varnishes.

Sequapel®. [Sequa] Water/oil repellents and water repellents.

Sequasoft®. [Sequa] Softener for textiles.

Sequenase. [U.S. Biochemical] T7 DNA polymerase; enzyme for dideoxy DNA sequencing.

Sequestrene®. [Ciba-Geigy/Dyestuffs] EDTA or salts; chelating agent for water softening, chemical cleaning, processing of textile, paper, and leather, in metal treatment, and for syn. rubber, photographic developer baths, personal care prods., animal feeds.

Sequestrox. [Transene] Etchant adjuncts for silicon devices.

Sequex. [Sequa] High performance starch crosslinkers for nonwoven bonding, roofing mat, adhesive, and agric. film applics.

Ser-Ad. [Servo] Surfactants; wetting agent, emulsifier, thickener for paint industry.

Serdas. [Servo] Surfactants; defoamer for paints, paper, sugar, textiles, leather.

Serdet. [Servo] Sulfates, sulfonic acid or derivs.; foamer, detergent base, emulsifier, wetting agent, latex stabilizer, spreading agent, dispersant for personal care, household detergents, emulsion polymerization.

Serdolamide. [Servo] Fatty acid alkanolamides; foam stabilizer, refatting agent, visc. modifier, corrosion inhibitor for cutting oils, detergents, personal care prods.

Serdox. [Servo] Ethoxylates; detergent, wetting agent, foamer, emulsifier, stabilizer, intermediate, dispersant for textiles, leather, household and industrial cleaners, metal cleaning.

Serenasol. [Nippon Senka] Dye solubilizer.

Serfene. [Morton Int'l.] Polyvinylidene chloride latex.

Sergene. [General Plastics] Nylon resin sol'n.

Serica. [Tokai Seiyu Ind.] Refined vegetable, mineral oil, and modified fatty alcohol blends; soaking agent for raw silk.

Sericite. [Presperse] Mica or blends.

Sericite WL. [Ikeda] Mica, lauroyl lysine, lecithin; softness and smooth feel, skin adhesion and spreadability for skin formulations; filler.

Sericron. [Pfizer] Extender pigment.

Series #34. [Syn. Surfaces] Patented instant grab adhesives; for factory bonding of automotive parts and panel assemblies, outdoor installation of roofing, sport surfaces, marine and construction materials under adverse conditions.

Series #98. [Syn. Surfaces] Patented latent tack adhesives; for factory bonding of automotive parts and panel assemblies, outdoor installation of roofing, sport surfaces, marine and construction materials under adverse conditions.

Series 7000. [Martin Marietta Magnesia Spec.] Fuel oil additives.

Serilene. [Yorkshire Pat-Chem] Disperse dyes for textile printing.

Serilube. [Yorkshire Pat-Chem] Lubricants for jet dyeing various blended and 100% cotton fabrics.

Seriplast. [Yorkshire Pat-Chem] Disperse dyes for textile printing.

Seripol. [Yorkshire Pat-Chem] Dispersants, leveling agents, retardants for disperse dyeing.

Seriprint. [Yorkshire Pat-Chem] Disperse dyes for textile printing with synthetic printing thickeners.

Serisol. [Yorkshire Pat-Chem] Disperse dyes for acetate.

Sermul. [Servo] Surfactants; emulsifier, stabilizer for emulsion polymerization.

Sermulen. [Servo] Oleyl polyglycol ether; coemulsifier for mineral and vegetable oils.

Serpol. [Servo] Polyacrylates; pigment dispersant for paper, paints.

Serumpro EN-10. [Brooks Industries] Serum protein; cosmetics ingredients.

Servamine. [Servo] Antistat, lubricant, germicide, disinfectant, sanitizer, emulsifier, corrosion inhibitor, intermediate.

Serviblock. [Chemloid] Printing auxiliaries improving pigment print coverage on dyed fabrics.

Serviclean. [Chemloid] Surface cleaners for removal of dyes, chemicals, greases, and soils.

Servigum. [Chemloid] Thickeners for printing vat, disperse, acid, fiber reactive, azoic and cationic dyes.

Servil®. [Grünau] Wax esters; release agents for confectionery and baked goods.

Servilev. [Chemloid] Reserving and leveling agents for direct, acid, and premetallized dyes.

Servilite. [Chemloid] Dyebath bleaching agents and stabilizers for hydrogen peroxide bleaching and desizing processes.

Servilok. [Chemloid] Pigment printing binders for metallic and pearl pigments.

Serviprint. [Chemloid] Printing assistants for pigments, vat, disperse, fiber reactive, azoic, and acid dyes.

Servirox. [Servo] Ethoxylated castor oil; used in textiles, leather, pharmaceuticals, paper, and cosmetics industries.

Servisene. [Chemloid] Stabilizers, sequesterants, assistants for hydrogen peroxide bleaching and desizing processes.

Serviset. [Chemloid] Hand builders, crosslinking agents.

Servisoft. [Chemloid] Softeners for pigment printing, napping, and finishing.

Servisperse. [Chemloid] Dispersants for dyes and pigments.

Servistat. [Chemloid] Antistats for natural, synthetic, and blended fabrics.

Servit®. [Grünau] Citric acid esters of mono- and diglycerides of fatty acids with other emulsifiers; emulsifiers for mfg. of instant dry yeast.

Servitak. [Chemloid] Adhesives for use when printing pigments and dyes on natural, synthetic, and blended fabrics.

Servitène. [Serviplast] Recycled polyethylene.

Servithix. [Chemloid] Auxiliary thickeners.

Servitol. [Chemloid] Wetting agents, dispersants, scouring compds. for desizing, bleaching, mercerizing, finishing, afterscouring dyed and printed fabrics.

Serviwhite. [Chemloid] Pigment white

dispersions for special effects printing.

Serviwick. [Chemloid] Printing auxiliaries.

Servo. [Servo] Quaternaries; softener, bactericide for shampoos, dyestuffs, oil industry.

Servoxyl. [Servo] Phosphate esters; wetting agent, detergent, emulsifier, antistat for personal care prods., drycleaning, metal cleaners, pesticides, textiles.

Serwet. [Servo] Sodium di-2-ethylhexyl sulfosuccinate; wetting agent, dispersant for pigments.

SES. [Transene] Silicone elastomer; for semiconductors.

Sesolvan® L. [BASF AG] Dye solvent for textile dyeing.

Setacin. [Zschimmer & Schwarz] Sulfosuccinates; detergents for cosmetics, cleaning agents.

Setamol®. [BASF AG] Dispersants for dye dispersions used for vat dyes for synthetic and cellulose fibers.

Setan. [Steac] Drilling mud conditioners.

Setsit®. [R.T. Vanderbilt] Activated dithiocarbamate; accelerator for rubber latexes.

Sett®. [Grünau] Hydrogenated triglycerides; visc. enhancer, hardstock for margarine, coating agent.

Sevin. [Rhone-Poulenc/Ag; W.A. Cleary] Carbaryl; insecticide for control of turf insects, fire ants, and ticks.

Sevinil. [Commercial Quimica Insular SA] PVC.

Sexadecyl Alcohol. [Lanaetex Prods.] Isostearyl alcohol, mineral oil; cosmetics ingredient.

Seyco®. [Seydel-Woolley] Sizes, binders, lubricants for textile applics.

Seycofilm®. [Seydel-Woolley] Dry polyester size for use with starch, PVA and blends to improve adhesion, elongation, and weaving efficiency.

SF-. [Compounding Tech.] PPS, some carbon or glass reinforced; used for precision electro-mech. components, corrosive chem. and fluid handling equip.

SF. [GE Silicones] Silicone derivs.; mold release agent, lubricant, flow control agent, emollient for coatings, rubber, plastic, hydraulic and lubricating fluids, dielectric coolant, textiles, skin and hair cosmetics, metals, damping, defoaming applics.

SF/Duroid Series. [Rogers] Polyester fiber/epoxy resin sheet materials; for electrical insulating.

S-Flakes. [Karlshamns] Hydrogenated soybean oil.

SFR 100. [GE Silicones] Silicone fluid; flame retardant for polyolefins.

SGP 502S Absorbent Polymer. [Henkel] Corn starch/acrylamide/sodium acrylate copolymer.

SGS. [NRC] Tantalum mill prods.

Shalex. [Chem-Lig Int'l. Industries] Potassium lignosulfonate.

Shamrock. [Erie Foods Int'l.] Milk chemicals and proteins.

Shamrock. [Witco] Water-soluble chemical emulsion for metalworking.

Shantoplast. [Shanti GmbH] PP, PE, cellulose acetate, cellulose acetate butyrate, cellulose propionate, PC, PS, polyurethane.

Sharpmold. [Sharp Chem.] Foundry sand additive.

Shawinigan Acetylene Black. [Chevron] Carbon black prod.

Shealoe. [Terry Labs] Shea butter/aloe blend.

Shearlon®. [Nicca USA] Napping agent for synthetic and cellulosic fibers and their blends.

Shear Magic. [Witco] Polyurea grease; shear-stable extreme pressure agent.

Shebu. [RITA] Shea butter.

Shell Cyclo Sol. [Shell] Aromatic hydrocarbon solvent; for dry cleaning, coatings, automotive and chemical specialties.

Shellflex. [Shell] Rubber process oils.

Shellmax. [Shell] Wax prods.

Shell Mineral Spirits. [Shell] Mineral spirits; solvent for dry cleaning, coatings, automotive and chemical specialties.

Shell Polybutylene. [Shell] Polybutylene resin; thermoplastic resin for polymer blending, film, pipe applics.

Shell Polypropylene. [Shell] Polypropylene homopolymers or random copolymers; for inj. molding, fiber spinning, stretch fiber and tape, film extrusion, blow molding, sheet and pressure forming applics.

Shell Sol. [Shell] Hydrocarbon solvents; for drycleaning, coatings, automotive, and chemical specialties.

Shellsolve. [Shell] Solvent.

Shell Toluene. [Shell] Toluene; solvent for drycleaning, coatings, automotive, and chemical specialties.

Shell Tolu-Sol. [Shell] Aliphatic naphtha solvent; for drycleaning, coatings, automotive, and chemical specialties.

Shell VM&P Naphtha. [Shell] Aliphatic naphtha solvent; for drycleaning, coatings, automotive, and chemical specialties.

Shellwax. [Shell] Paraffin and microcrystalline waxes; binders, coatings, moisture protectant, adhesives.

Shell Xylene. [Shell] Xylene; solvent for drycleaning, coatings, automotive, and chemical specialties.

Sherbrite. [PMC Specialties] Sodium saccharin; industrial grade brightener for nickel plating.

Sherex. [Sherex] Mining reagents.

Sher-Guard. [PMC Specialties] Corrosion inhibitor for cellulose insulation.

Shield Carpet Care. [Huntington Labs] Carpet cleaners.

Shieldex. [W.R. Grace/Davison] Nontoxic anticorrosion pigment.

Shinju. [Mearl] Synthetic pearl pigments.

Shinol. [Chemtech Industries] Electroplating bath agent.

Ship Shape. [ISP] N-Methyl-2-pyrrolidone, butyrolactone, deceth-6, fragrance; resin cleaner.

Shiro Bishi®. [Mitsubishi Kasei] Foundry coke; for malleable and general grade cast iron.

Shokusen SE. [Mitsubishi Kasei] Sugar ester, propylene glycol, ethanol, sodium citrate; detergent for foods.

Shoo. [Petrokem] Insecticide.

Shreeactive. [Trade Search Assoc.] Textile dyes and pigments.

Shreezol. [Trade Search Assoc.] Textile dyes and pigments.

Shur-Coal. [Sherex] Coal process chemicals.

Si 69. [Degussa] Sulfur grade organosilane; crosslinking and coupling agent for rubber.

Si 264. [Degussa AG; Struktol] 3-Thiocyanatopropyltriethoxy silane; reinforcing agent for rubbers.

Siamp-Cedap. [Polyemballages SA] PS and complexes for thermoforming.

Sica Antimony Trioxide. [Elders Exsud Ltd.] Antimony trioxide.

Siccatol®. [Akzo; Centrachem AG] Fatty acid-based; paint drier.

Sico®. [BASF; BASF AG] Predominantly azo pigments; for baking and air-drying finishes, inks, coloring plastics, textiles.

Sicocab®. [BASF AG] Organic and inorganic pigments in cellulose acetobutyrate vehicle; for metallic paint, wood stains.

Sicocer®. [BASF AG] Decorative colors for ceramics, tiles, enamel, glass, sanitaryware, porcelain, underglaze.

Sicodop®. [BASF AG] Organic and inorganic pigments in DOP; for coloring plasticized PVC.

Sicofast. [BASF] Organic pigments.

Sicoflush®. [BASF AG] Organic and inorganic pigments in carriers; for industrial finishes, artists' paints, wood stains and protectants.

Sicolen®. [BASF AG] Organic and inorganic pigments in polyethylene; for coloring polyethylene hollow articles, inj. molding and extrusion goods, cable sheathing.

Sicolub®. [BASF AG] PVC lubricants.

Sicomet. [Henkel Chem. Ltd.] Cyanoacrylate.

Sicomet®. [BASF AG] Anionic, cationic, fat-soluble dyes, organic pigments, inorganic pigments, or pigment concs.; colorants for cosmetics.

Sicomin®. [BASF] Inorganic chromate pigments; for paints, surface coatings, coloring plastics, flexographic or gravure inks, laminated paper coloring.

Sicomix®. [BASF AG] Pigment combinations for surface coatings.

Sicopal®. [BASF] Inorganic chromate pigments; for industrial finishes, coloring plastics.

Sicopharm®. [BASF AG] Soluble dyes or pigments; for pharmaceutical preparations.

Sicoplast®. [BASF AG] Predispersed pigment mixtures; for mass coloring of thermoplastics.

Sicopos®. [BASF AG] Organic and inorganic pigment concs.; for coloring PET.

Sicopurol®. [BASF AG] Organic and inorganic pigment concs. in ester polyol; for coloring polyurethane foams.

Sicopur®. [BASF AG] High purity iron oxides; pigments for paper refiners.

Sicor®. [BASF] Anticorrosion pigments.

Sicorin®. [BASF] Anticorrosion pigments.

Sicostab®. [BASF AG] Heat stabilizers for rigid and plasticized PVC, internal and external applics., PVC foams.

Sicostyren®. [BASF AG] Organic and inorganic pigment concs. in polystyrene; for coloring polystyrene and styrene copolymers.

Sicotan®. [BASF] Inorganic pigments; for surface coatings, plastics with high processing temps.

Sicotherm®. [BASF; BASF AG] Cadmium sulfide/zinc sulfide or cadmium sulfide/selenide mixed crystals; pigments for coloring plastics and paints.

Sicotrans®. [BASF; BASF AG] Synthetic iron oxide pigments; for high quality paint systems esp. metallics, for coloring plastics.

Sicoversal®. [BASF AG] Organic and inorganic pigment concs.; for coloring thermoplastics.

Sicovinyl®. [BASF AG] Organic and inorganic pigment concs. in plasticized PVC; for mass coloring of plasticized PVC, PVC cable sheathing.

Sicovit®. [BASF AG] Soluble colorants for foodstuffs.

Sicovoss. [Vosschemie GmbH] Silicone construction resins.

Sidamil. [UCB NV Filmsektor] Polyamide, PET, PVC complexes.

Sidanyl. [UCB NV Filmsektor] Polyamide films.

Sident. [Degussa] Silica; thickener for toothpastes.

Sigilgomma. [Mario Lombardini Srl] Thermoplastic rubber.

Sigilplast. [Mario Lombardini Srl] PVC.

Sigma. [Katalistiks Int'l.] Fluid cracking catalysts.

Sigma-Plus. [Katalistiks Int'l.] Fluid cracking catalysts.

Sigma-Strip. [Urban Chem.] Stripper.

Silacryl. [Rowe Prods. Distribution] Silicone-acrylic base coating.

Silacto. [Kenrich Petrochemicals] Amine accelerator; activator.

Silamide DCA-100. [Siltech] Conditioner for personal care prods.

Silamine. [Siltech] Cationic conditioners, gloss agents, softeners for personal care prods.

Silanol Exsyliposomes. [Exsymol] Silanol/glycerophospholipids blend; for cosmetics.

Silaprene®. [Uniroyal] Industrial adhesives and sealants based on chloroprene, PVC, acrylic, nitrile, natural rubber, SBR, PC, ABS, urethane, or epoxy or blends.

Silasorb. [Celite] Calcium silicate; adsorbent, specialty filter aid.

Silastic®. [Dow Corning; Dow Corning France SA] Silicone rubber compd.; for compr., transfer, or inj. molding, calendering, extrusion, and blending.

Silastomer. [Hernon Mfg.] RTV silicones; flange gasketing sealants.

Silbione Antifoam. [Rhone-Poulenc] Simethicone; antifoam.

Silbione Oils. [Rhone-Poulenc] Silicone derivs.

Silbond®. [Akzo] Silicates; intermediate for binders for inorganic zinc coatings, investment casting molds, cores, ceramic shapes, coatings.

Silcat® R. [Union Carbide] Vinylsilane; crosslinking agent for polyethylene modification.

Sil-Cell. [Norwegian Talc UK Ltd.] Low density fillers.

Sil-Cell®. [Silbrico] Glass microcellular filler (silicon dioxide, aluminum oxide, potassium oxide); resin extender.

Silcolease. [ICI Am.] Silicone; release agents, antifoams.

Sil-Co-Sil. [U.S. Silica] Ground silica.

Silcron. [SCM] Fine particle silica gel.

Silene®. [PPG Industries] Hydrated silica; reinforcing pigment for rubber.

Silenka. [Centrachem AG] Rovings.

Sil-Fin. [High Point] Silicone emulsions or elastomers; hand modifier, softener, water repellent finish for textiles.

Silflake. [Handy & Harman] Silver flake.

Silflex A22. [Rhone-Poulenc] Sodium/ TEA-lauroyl animal collagen amino acids.

Sil-Fos. [Handy & Harman] Silver brazing alloy.

Sil-Free. [Sybron] Additive to prevent silicate deposit in peroxide/silicate bleach systems.

Silgan. [Wacker Silicones] Silicone rubbers.

Silglaze®. [GE Silicones] Silicone sealant.

Sil-Glyde. [Am. Grease Stick] Lubricant.

Silgrip. [GE Silicones] Silicone adhesive.

Silhydrate C. [Exsymol] Copper PCA methylsilanol; moisturizer for skin care and anti-aging formulations.

Silica FK. [Degussa] Silica; filler for silicone rubbers.

Silicate Cluster 102. [Olin] Tris (tributoxysiloxy) methyl silane; functional fluid for use in specialty defoaming, pressure-sensitive adhesives, heat-transfer fluid, lubricants, mold release, high-performance hydraulics, dielectric coolants.

Silicoat. [Transene] Silicone coating; for PC assemblies.

Silicone AF. [Harcros] Silicone compds.; emulsifier, antifoamer for food processing, chemical processing (adhesive, ink, and soap mfg., latex and starch processing), textiles and paper, leather finishing, metal working.

Silicone Antifoam Emulsion. [Wacker Chemie GmbH; Wacker Silicones] Silicone emulsion; antifoam, processing aid for pharmaceuticals, foods, cosmetics, fermenation, mfg. of plastics.

Silicone Compounds SWS-. [Wacker Silicones] Silicone compds.; for water repellents, sealants, and coatings for elec./electronic use; lubricants for rubber and plastic; release agents for rubber and plastic molding; damping media.

Silicone Defoamer #5037. [Polymer Research Corp. of Am.] Silicone emulsion; defoamer for fermentation applications.

Silicone Emulsion. [Akrochem] Silicone emulsions; mold release agent for tire and mechanical goods, wire and cable, plastics.

Silicone Emulsions E-, SWS-. [Wacker Silicones] Silicone emulsions; for mold release, furniture polishes, textile lubricants and softeners, cosmetic formulations.

Silicone Fluid. [Akrochem] Dimethylpolysiloxane; release and slip agents for tire and mechanical goods, fan belts, o-rings, floor mats, hose, toys, shoe heels, floor tile, bath mats, wire and cable applics.; additives in rubber and plastics for water-repellent treatments.

Silicone Fluids SWS-. [Wacker Silicones] Silicone fluids; for dielectric coolants, brake fluids, lubricants, auto care prods., release agent, heat transfer, aerosols, damping media, household prods., antifoams, cosmetics, fluid power transmission, precision aircraft instruments, shock absorbers.

Silicone L-31, L-45. [Union Carbide] Methicone and dimethicone respectively.

Silicone Release Agent #5038. [Polymer Research Corp. of Am.] Dimethicone emulsion; release agent for rubber and plastics.

Silicone Systems. [R.H. Carlson/Northern Labs] Silicone systems; adhesive sealants.

Silicopearl SR. [Tokai Seiyu Ind.] Silicone, neutral oil, nonionic blend; lubricant for synthetic fibers.

Siligen®. [BASF; BASF AG] Plasticizers, glazing agents, and silicone elastomers; additives for improving ease of sewing, scuff and tearing resistance of textiles.

Silikoftal®. [Tego] Silicone-modified polyester resin; binders for high gloss coatings, one-coat systems, decorated coatings.

Silikophen®. [Tego GmbH] Phenylmethyl polysiloxane resins; binders for anticorrosion coatings, one-coat coatings.

Silikroil. [Kano Labs] Kroil plus silicone.

Silikup. [Malvern Minerals] Surface-modified silica.

Siliphos. [Giulini Corp.] Slowly soluble water treatment compd.

Siliporite. [Atochem N. Am.] Molecular sieve prods.

Silkos. [Franklin Mineral Prods.] Wet ground muscovite mica.

Silk Pro-Tein. [Maybrook] Hydrolyzed silk; moisturizer, substantivity agent, protective barrier for elegant hair and skin care prods., soaps, shave preps.

Silk Protein Complex. [Croda Chem. Ltd.] Hydrolyzed silk protein; conditioner for hair and skin care prods.

Silksoft®. [Sybron] Textile softener.

Sillikolloid. [Hoffmann Min.] Quartz-kaolinite; filler.

Sillitin. [Hoffmann Min.] Quartz-kaolinite; filler.

Sillum. [D.J. Enterprises] Alumina silicate; filler.

Silmar®. [BP Chem. Inc.] Thermoset polyester resins; for synthetic marble, laminating applics.

Silmate. [GE Silicones] Silicone rubber.

Silmax. [C.P. Hall] Mandrel release fluid.

Silmod. [Union Carbide] Silicon-modified polyether polymers; for elastomers, sealants, adhesives, potting compds.

Silock. [Calgon] Boiler water treatment.

Silogram. [A. Margolis & Sons] Lubricants, oils, greases, solvents.

Silopren. [Bayer] Silicone rubber; used in seals and other technical moldings and extrudates, roll covers, conveyor belting, cable and wire insulations, pharmaceutical goods.

Siloxane SWS-03314. [Wacker Silicones] Volatile silicone fluid; slip agent, lubricant, release agent, emollient for personal care prods.; plasticizer for hair spray resins; in aerosol-based antiperspirants.

Siloxan Tego®. [Goldschmidt AG] Organic modified polysiloxane; mold release agent for rubber molding.

Siloxide Etchant. [Transene] Selective etchant for deposited silicon dioxide.

Silo-Zyme. [Sybron/Biochemical] Ensilage improver.

Silplus. [GE Silicones] Silicone elastomer.

Silpowder. [Handy & Harman] Silver powder.

Silproof. [SCM] Fine particle silica gel; for chillproofing beer.

Silpruf®. [GE Silicones] Silicone sealant.

Silquat. [Siltech] Cationic conditioners for personal care prods.

Silsoft. [Chemonic Industries] Silicone softeners for fabric finishing.

Silsoft. [Union Carbide] Water-white silicone microemulsion conc.

Silsolv LS. [Vikon] Alkaline blend of chelating agents; for hydrogen peroxide bleach baths.

Siltek® GR, M, M Super, PL. [Petrolite/Polymers] Polyethylene.

Siltek® L Polymer. [Petrolite/Polymers] Ethylene/propylene copolymer.

Siltem®. [GE Plastics] Silicone/polyetherimide copolymer; for combustion corrosion-sensitive applics., cable insulation.

Siltex. [Kaopolite] Fused silica; for plastics, elec. molding and potting compds., epoxy compds., sealants.

Siltex. [Nutex] Silicone softeners for pigment prints and fabrics.

SILTherm. [Industrial Dielectrics] Thermoset polyester; for bulk and sheet molding compds.

Siltouch®. [Yorkshire Pat-Chem] Silicone emulsions; softeners for textiles.

Siluminite. [Tenmat Ltd.] Fiber-reinforced phenolic.

Silvatol®. [Ciba-Geigy/Dyestuffs] Scouring agent, wetting agent, detergent, stabilizer for textiles, peroxide bleaching.

Silver Bond. [Unimin Specialty Minerals] Ground silica; filler, flatting agent for paints, mastics, adhesives, buffing compds., elec. epoxy compds.

Silverline. [Montana Talc] Platy talcs;

for paints, industrial coatings, rubber applics.

Silver-Lume. [Atochem N. Am.] Silver plating.

Silvet. [Silberline Ltd.] Masterbatch.

Silwax. [Siltech] Silicone waxes; for personal care prods.

Silwet®. [Union Carbide] Dimethicone copolyol; surfactant, dispersant, emulsifier, leveling agent, antifog, lubricant, antiblock, slip additive for adhesives, agric., automotive, coatings, printing inks, textiles, household specialties, cutting fluids, petrol. extraction, paper, plastics, rubber.

Simazine 90 DF. [Terra Int'l.] Dry flowable herbicide.

Simchin. [RITA] Jojoba oil; moisturizer, emollient, conditioner for skin and hair care prods.

Simco. [Simco BV] Antistat.

Simulsol. [Seppic] Ethoxylates; emulsifier, solubilizer.

Sinbar. [DuPont/Ag] Herbicide.

Sinkral. [Enimont UK Ltd.] ABS.

Sinochem. [Sino-Japan] Esters; emulsifier, lubricant.

Sinocol. [Sino-Japan] Surfactants; emulsifier, spreading and sticking agent, penetrant for agric. formulations, phosphate compds.

Sinol. [Sino-Japan] Surfactants; degreaser, emulsifier, leveling agent for leather, paraffin.

Sinonate. [Sino-Japan] Sulfates, sulfonates, or sulfosuccinates; emulsifier, wetting agent, penetrant, dispersant, solubilizer, detergent, foaming agent, antistat, anticorrosive agent for emulsion polymerization.

Sinopol. [Sino-Japan] Ethoxylated esters, ethers, amines; emulsifier, dispersant, suspending agent, solubilizer, stabilizer, defoamer, corrosion inhibitor, extreme pressure additive, lubricant, mold release, dyeing auxiliary for emulsion polymerization, textiles, etc.

Sinoponic. [Sino-Japan] EO/PO ethers; low foaming detergent and wetting agents for phosphatizing bath; emulsifiers for acrylic polymers.

Sinotex. [Sino-Japan] Quaternary ammonium compds.; softener, conditioner, antistat for fabrics, hair prods.

Sintimid. [Sintimid Hochleistungkunststoffe GmbH] Polyimide.

Sinvet. [Enimont UK Ltd.] PC.

Siokal. [Norwegian Talc UK Ltd.] Fillers for plastics.

Sioplas. [Dow Corning] Crosslinking technology for polyolefins.

Sipernat®. [Degussa] Silica; adsorbent, anticaking and free-flow agents, carrier; antiblock for films.

Sipex®. [Rhone-Poulenc Surf.] Sulfates; surfactant, wetting agent, dispersant, emulsifier, detergent, foamer, flotation agent for personal care and industrial applics.

Sipon®. [Rhone-Poulenc Surf.] Lauryl sulfate salts; detergent, emulsifier, foaming agent, solubilizer, penetrant for cosmetics, fabrics, polymerization.

Siponate®. [Rhone-Poulenc Surf.] Sulfonates; detergent, emulsifier, solubilizer, visc. builder for emulsion polymerization, personal care prods., paints, textiles, metal cleaning, agric., household and industrial cleaning, oil recovery.

Siponic®. [Rhone-Poulenc Surf.] Ethoxylated ethers; detergent, emulsifier, wetting agent, emollient, thickener, conditioner, solubilizer for cosmetics, pharmaceuticals, household and industrial cleaners, metal cleaning, rubber, dyeing, textiles, emulsion polymerization, paper, agric., waxes, polishes.

Sipothix. [Rhone-Poulenc Surf.] Acrylate copolymers; thickening agent for liq. detergents, shampoos, and cosmetics.

Siral. [Siral-Kunststoffwerk Siebauer] Polypropylene.

Siral®. [Condea Chemie GmbH] Silica-doped alumina.

Siralchid. [Sir Industriale-Gruppo Montedison] Alkyd resins.

Sirales. [Sir Industriale-Gruppo Montedison] Saturated polyester resins.

Siralox®. [Condea Chemie GmbH] Silica-doped alumina.

Siramin. [Sir Industriale-Gruppo Montedison] Aminic resins.

Sirester. [Sir Industriale-Gruppo Monte-dison] Unsaturated polyester resins.

Sirius. [Miles/Organic Prods.] Textile dyes and pigments.

Sirrix. [Sandoz] Boiling off assistant with high sequestering ability.

Sisalite Lotonite. [Ditta Salt SpA] Phenolic powder.

Sisellig. [Ceca SA] Antislip, anti-agglomerant for fibers.

Sisthane. [Rohm & Haas] Fungicide.

Sitostene. [Vevy] Soy sterol.

Sitren®. [Goldschmidt] Modified siloxanes; water repellent for treatment of mineral fiber insulation, plasterboard.

Size 777S. [Aqualon] Size used with alum in paperboard and insulation board.

Size CB. [BASF] Water-soluble polyacrylate; sizing agent for cellulosic fibers and blends.

Sizepine. [Arakawa] Sizing agent for paper.

Sizing Wax PA, PT, SM. [Hüls AG] Polyethylene glycols; raw material for prod. of water-sol. sizing auxiliaries; antistats and dyeing auxiliaries.

SK-7101 (Aquadene), 7532, 7542. [Stiles-Kem] Sequestering agent, corrosion inhibitor for potable use.

SK-9881, 9882. [Stiles-Kem] Scale and corrosion inhibitor for process cooling water.

Skamex. [DuPont] Metallurgical additive.

Skane® M-8. [Rohm & Haas] 2-n-Octyl-4-isothiazolin-3-one; mildewcide for paints.

Skellite®. [Texaco] Solvent used as stove and lamp fuel.

SK Fert. [Kao Corp. SA] Amine-based; anticaking agent for fertilizers.

SK Flot. [Kao Corp. SA] Amine-based; mineral flotation agent.

Skid. [Stoner] Rust breaker, lubricant.

Skino. [Mooney Chems] Antiskinning agent.

Skinotan. [Zschimmer & Schwarz] Polysiloxane polyglycol ether; shampoo additive.

Skliro. [Croda Inc.; Croda Chem. Ltd.] Lanolin acid; emollient, emulsifier for water-repellent films, makeup.

Skybond. [Monsanto] Polyimide.

Skydrol. [Monsanto] Hydraulic fluid.

Skyprene®. [Tosoh] Polychloroprene; used for wire and cable jackets, automotive parts, general industrial parts, sponge and construction material, extruded and calendered prods., hose, adhesives, belts, rolls, wet suits.

Skytone. [Hilton Davis] Textile pigments.

SLA. [Acheson Colloids] Colloidal molybdenum disulfide, graphite, or PTFE; lubricant additive for gear oils, engine oils, chain lubricants, aerosols, machine oils, greases.

Slab Dip AC699. [Ayer's Cliff] Calcium carbonate pigmented powder with dispersants; rubber slab dipping system for rubber industry.

SLC-1000. [Glastic] Glass-reinforced polyester compd.; low cost BMC compds.

SLCC-D. [Drew Ind. Div.] Ammonium hydroxide sol'n.; corrosion inhibitor for steam generating systems.

Slick. [Morton Int'l./Automotive & Ind. Finishes] Paint additive.

Slick Slide. [Graphite Prods.] Friction-reducing lubricants.

Slide®. [Percy Harms] Mold releases, lubricants, cleaners, rust preventives.

Slik. [Ash Grove Cement] Hydrated lime.

Slikwik. [Andersons] Processed corncob absorbent for organic and nonorganic liquids and sludges.

Slime-Trol. [Betz Industrial] Slime control agent.

Slimex. [E.F. Houghton] Slimicides.

Slip-Ayd. [Daniel Prods.] Predispersed polyethylenes and waxes for paints and inks.

Slip-Eze®. [Syn. Prods.] Oleamide; slip agents for plastic films.

Slipicone. [Dow Corning] Silicone lubricant.

Slip Plate. [Superior Graphite] Graphite-based lubricating paint.

Slip-Quick®. [Syn. Prods.] Patented blend; slip agent for plastic films.

Slipspray. [DuPont] Dry lubricant.

Sludge Buster. [Martin Marietta Magnesia Spec.] Asphaltene control fuel additive.

Sludgtrol. [W.R. Grace/Dearborn] Boiler compd.

SM. [GE Silicones] Silicone derivs.; emulsions for textile softeners, furniture, vinyl, and auto polishes, mold release for rubber, plastics, foundry applications, particle treatment, hair mousse, printing.

SMA®. [Atochem N. Am.; Atochem UK] Styrene/maleic anhydride copolymer; soil release, dispersant for carpet shampoos and mfg., paints, inks, paper coatings, commercial laundries, in emulsion polymerization; leveling resin for floor polishes.

Smash. [Panef Mfg.] Penetrating oil, rust solvent.

S-Maz®. [PPG/Specialty Chem.] Sorbitan esters; solubilizer, emulsifier, dispersant, lubricant, antistat, softener, process defoamer, opacifier, suspending agent, coupler for lubricants, coolants, cosmetics, food formulations, industrial oils, and household prods.

Smith Lime Flour. [Smith Lime Flour] Hydrated lime.

Smithol. [Werner G. Smith] Lubricants for metalworking, leather treating.

Smokebloc. [McGean-Rohco] Fire retardant plastic additive.

Smoothar. [Yoshimura Oil Chem.] Surfactant/oil blends; coning oil, lubricant for textiles.

Smoothie. [Abso-Clean Industries] Fabric softeners, laundry soaps.

Smooth-On. [Smooth-On] Epoxy adhesives, casting compds.

SMQ. [Indium Corp. of Am.] Pastes, powders, fluxes.

SMR CV, L, WF, XL. [Akrochem] Rubbers.

SMT Adhesive 995. [Dymax] Adhesive for surface mount attachment.

SN-. [Compounding Tech.] SAN, glass reinforced.

Snac-Kote. [Van Den Bergh Foods] Partially hydrogenated vegetable oil; coating fat for bakery coatings.

Snap. [Witco] Antipit agent for electroplating sol'ns.

Snelstrip. [Pilzecker's Industrie BV] Polyurethane and PVC foam.

Sniamid. [Nylon Corp. of Am.; Sniatecnopolimeri] Nylon 6 or 6/6, some glass or mineral filled; for inj. molding, film extrusion, extrusion of sheet, rod, and tube, carburetor filters, brake fluid reservoirs, monofilament, friction and wear applics., automotive grills, hub caps, power tool housings.

Sno Boy. [Harcros] Ammonia, bleach.

Snobrite. [Evans Clay] Hydrated aluminum silicate; reinforcing filler in NR, SR, latexes, resins, plastics; color and processing aid; used in coated materials, footwear, flooring, V-belts, rolls, belts, matting, molded and extruded goods, foam goods, o-rings, seals, sundries.

Snofil. [Evans Clay] Filler.

Snomelt. [Standard Tar Prod.] De-icing chemicals.

Snowcal. [Croxton & Garry Ltd.] Chalk whiting fillers.

Snow Fine. [Allied-Signal] Sodium sesquicarbonate.

Snowflake. [Ash Grove Cement] Hydrated lime; for food applics.

Snowflake. [ECC Int'l.] Calcium carbonate; pigment for protective coatings, rubber, plastics, caulks, glazing compounds, mastics, filled systems.

Snow Flake. [Allied-Signal] Sodium sesquicarbonate.

Snow Floss. [Celite] Mineral filler.

Snowfort. [Croxton & Garry Ltd.] Coated and uncoated wollastonite.

Snow Fresh. [Monsanto] Antibrowning agent.

Snowmaster. [Schaefer Salt & Chem.] Calcium chloride.

Snowtack. [Eka Nobel] Tackifier resin dispersions; for pressure-sensitive applics.

Snow-Tex. [U.S. Silica] Calcined clay.

Snow White. [Unimin Specialty Minerals] Wet ground mica.

Snow White Filler. [U.S. Gypsum] Fine gypsum (anhydrite) filler.

Snow White Petrolatum. [Stevenson Bros.] Petrolatum.

Soal. [La Roche Chem.] Liq. sodium

aluminate.

Soarblen. [British Traders & Shippers] EVA copolymer.

Soarflex. [British Traders & Shippers] EVA copolymer.

Soarnol. [British Traders & Shippers] Vinyl alcohol/ethylene copolymers.

Sobalg. [Grindsted Prods.; Grindsted Prods. Denmark] Alginates; suspending agent, visc. builder for beverages, bandage materials, cosmetics, dental industry, paper, pharmaceuticals, textiles, water purification, welding rods, wound dressings.

Sobral. [Scott Bader] Alkyd, acrylic, acrylamide, or polyurethane blends; for industrial finishes, domestic appliances, laboratory furniture, inks, plasticizing.

Socci®. [Morton Int'l.] Antimicrobial for cordage, rope, textiles, pharmaceuticals; mold release.

Sochamine. [Witco SA] Detergent, wetting agent, foamer, softener, antistat for shampoos, textiles.

Soda Ash Briquettes. [HVC] Soda ash tablets for acid neutralization.

Sodaphos®. [FMC] Sodium tetrametaphosphate; water conditioner for dyeing, printing, scouring, kier boiling, delustering operations, prevention of lime soap deposits.

Sodium Aluminate. [U.S. Aluminate] Sodium aluminate granules.

Sodium Chlorite. [Olin] Sodium chlorite; slimicide for paper mill systems; algicide in cooling towers; used in the electronics industry; as intermediate.

Sodium Octyl Sulfate Powder. [Henkel/Emery/Cospha] Sodium octyl sulfate; wetting agent for cleaning prods., reprographic coating baths, electroplating.

Sodium Omadine. [Olin] Sodium 2-pyridienthione; industrial microbiostat, chelating agent; for aq. metal coolant and cutting fluids, latex emulsion, inks, fiber lubricants.

Sodyeco®. [Sandoz] Antimigrants, defoamers, levelers, penetrants, lubricants, softeners, chelating agents, dyeing assistants.

Sodyecron. [Sandoz] Textile dyes and pigments.

Sodyefac®. [Sandoz] Wetting and rewetting agent for textiles.

Sodyefide®. [Sandoz] Reducing agent for sulfur dyes.

Sodyefresh. [Sandoz] Odor masking agent for dyebaths.

Sodyelube®. [Sandoz] Low foaming agent for dyeing machines; prevents and reduces creases and crack marks on synthetic fabrics.

Sodyesul. [Sandoz] Textile dyes and pigments.

Sodyevat. [Sandoz] Textile dyes and pigments.

Sofbon. [Takemoto Oil & Fat] Quaternary ammonium compd.; softeners for textiles.

Sofnon. [Toho Chem. Industry] Imidazolines; softeners for textiles, glass fibers.

Sof'N-Soil. [U.S. Gypsum] Lawn, garden gypsum.

Sofprene. [Softer SpA] Thermoplastic rubber (SBS-based).

Soft. [CNC Int'l.] Synthetic softeners for textile finishes.

Softal. [Tokai Seiyu Ind.] Polyamine or amide; softener, lubricant for synthetic fibers.

Soft Detergent. [Lion] Sulfonates; detergent, emulsifier, dispersant for emulsion polymerization.

Softech. [Dyetech] Soil, oil and water repellent finishes for carpets and textile fabrics.

Softener. [Polymer Research Corp. of Am.] Softener for textiles.

Softener 77, 636D, C. [Reilly-Whiteman] Softeners for textiles.

Softener J.B. [Dycho] Nitrogen-bearing fatty base; orlon and synthetic fiber softener.

Softenol®. [Hüls Am.] Fatty acid triglyceride or glyceryl stearate; emulsifier, lubricant, release agent, emollient, antistat for food processing equip., textile sizing, cutting oils, mfg. and processing of plastics, textiles, leather, oils and polishing materials; compression aid for tablets.

Softex. [Kao] Polyamide deriv.; textile softener.

Softex. [Leatex] Softener for textiles.

Softigen®. [Hüls Am.; Hüls AG] Glyceryl derivs.; emollient, refatting agent, wetting agent, solubilizer, ointment base, moisturizer, stabilizer for personal care prods., pharmaceuticals.

Softisan®. [Hüls Am.; Hüls AG] Triglycerides of saturated fatty acids; moisturizer, cream base, emollient.

Softlon. [Yoshimura Oil Chem.] Polyamide surfactants or blends; softener, lubricant for synthetic fibers before knitting.

Softnap. [Reilly-Whiteman] Napping assistants for textile industry.

Softol. [Dexter] Softener for textiles.

Softonic. [Chemonic Industries] Wax emulsions to achieve desired hand on natural or synthetic fabrics.

Soft Touch. [Paniplus] Hydrated mono- and diglycerides, stearoyl lactylate, and polysorbate 60; emulsifier system.

Soft Touch. [Yorkshire Pat-Chem] Silicone enhanced softeners for textiles.

Softyne. [Hart Prods. Corp.] Amido-fatty quaternary; textile softening agents.

Soi Bact. [Societa Ital. Emulsionanti] Hexahydrotriazine; bactericide for pesticides.

Soil Buster. [Mammoth Int'l.] Soil conditioner, penetrant; breaks up clay soils.

Soilife. [La Roche Industries] Mixed fertilizers.

Soi Mul. [Societa Ital. Emulsionanti] Sodium petroleum sulfonate/rust inhibitor/solvent blends; emulsifier, rust inhibitor for metalworking industries.

Soitem. [Societa Ital. Emulsionanti] Emulsifier, dispersant, suspending agent for pesticide formulations.

S'OK! [Shamrock Spec.] Penetrating oil.

Sokalan®. [BASF; BASF AG] Polyacrylic acid or salts, PVP, or maleic copolymers; dispersant, antiredeposition inhibitor for detergents, water softening.

Solamidine. [John Campbell] Textile dyes and pigments.

Solamine. [Seppic] Quaternary or imidazoline; softener for fibers.

Solan. [Croda Inc.; Croda Chem. Ltd.] Ethoxylated lanolin; hydrophilic emollient, emulsifier, conditioner, thickener, superfatting agent, foam stabilizer, plasticizer, humectant for personal care prods.; fragrance solubilizer.

Soland. [Climax Performance] Oil field chemicals.

Solangel 401. [Croda Inc.] PEG-75 lanolin; emulsifier, humectant for soap.

Solanos. [Seppic] Softener.

Sol-Aqua-Fast. [Crompton & Knowles] Textile dyes and pigments.

Solar. [BASF] Textile dyes and pigments.

Solar. [DuPont] Barrier resins.

So-Lara. [Monsanto] Dyed acrylic fibers.

Solarchem® O. [CasChem] Octyl dimethyl PABA; uv absorber for suncare and cosmetic prods.

Solar Soap Powder. [Guelph Soap] Soap used for blending hand soaps or laundry prods.

Solef®. [Solvay Polymers; Solvay & Cie] PVDF homopolymer or copolymer; for equip. for chemical, petrochemical, metal, pharmaceutical, food, and nuclear industries, paper/pulp, metal surface coatings, elec./electronics, films for building and automobile industries, tech. parts.

Solegal®. [Hoechst Celanese/Colorants & Surf.; Hoechst AG] Alkyl phenyl polyglycol ester; emulsifier for textile printing.

Solem ATH. [Solem Industries] Alumina trihdyrate; filler providing translucency and flame and smoke suppression to cultured onyx.

Sole-Mulse B. [Hodag] Modified ethoxylates; emulsifier for emulsion degreasers, industrial cleaners.

Sole-Onic CDS. [Hodag] Diethylene glycol laurate; defoamer, emulsifier for emulsion paints, industrial emulsions.

Sole-Terge. [Hodag] Sulfosuccinate or sulfate; detergent, emulsifier, wetting agent, penetrant for personal care prods., industrial applics.

Solflex®. [Goodyear] Styrene/butadiene vinyl sol'n.; for tire formulations, molded goods.

Solidegal. [Hoechst AG] Amine-based;

leveling agent for vat dyes.

Solidester. [Robeco] Ester-hydrogenated piscine oil; emollient, lubricant, moisturizer for skin prods.

Solidokoll®. [Hoechst Celanese/Colorants & Surf.; Hoechst AG] Polyacrylamide; auxiliaries to inhibit migration of dyes during intermediate drying.

Solidur®. [Solidur; Solidur Kunststoffwerk] UHMWPE; for molding, extrusion, sheet, rod, tube, profiles, wear applics., conveyor equip., linings for coal chutes.

Solimide® Foam. [Ethyl] Flexible polyimide foam; for thermal and acoustical insulation applics.

Solka-Floc®. [Mendell; James River; Grefco] Cellulose; filter aids and fillers.

Sollagen®. [Hormel] Soluble animal collagen; humectant for cosemtics.

Solobond. [Soluol] Adhesives.

Solocod G. [Reilly-Whiteman] Sulfated fish oil; emulsifiable oil for processing leather.

Solon Conc. [Eastern Color & Chem.] Tetrasodium EDTA; chelating agent.

Solophenyl. [Ciba-Geigy/Dyestuffs] Textile dyes and pigments.

Solopol. [Stockhausen] Stabilzier for peroxide bleaching, retarding agent for cationic dyes.

Soloron. [Rona] Mica/titanium dioxide/tin oxide blends; pearlescents for cosmetics.

Solpon 4488. [Boehme Filatex] Low foaming wetting, scouring, and emulsifying agent for textiles.

Solprene. [Housmex] Polybutadiene or SBR; elastomers for tires, molded, extruded, and calendered goods, microcellular sponge prods., v-belts, belt covers, hose, tubes, soles, and heel; processing aid, modifier for plastics, adhesives, caulks, sealants.

Solricin®. [CasChem] Ricinoleic acid soaps; emulsifier, mild germicide, lubricant, foam stabilizer in foamed rubber, for cosmetics and household applics.

Sol-Speedi-Dri. [Engelhard] Attapulgite clays; floor absorbents used in metal-

working and automotive industries, animal barns, and butcher shops.

Soltite. [Blowmocan Ltd.] Solvent-resistant HDPE.

Solton. [Toho Chem. Industry] Surfactant/EP agent blends; water-soluble cutting oil.

Soltrol. [Phillips] Isoparaffins.

Solubilisant Gamma. [Gattefosse; Gattefosse SA] Ocotxynol-11/polysorbate 20 blends.

Solubilizer L-76. [Lipo] Trideceth-12, laureth-12.

Solubond®. [Soluol] Urethane laminating adhesive.

Solubor. [U.S. Borax & Chem.] Soluble plant food borate.

Solu-Coll. [Brooks Industries] Collagen derivs. or blends; for hair and skin care cosmetics.

Solucote®. [Soluol] Urethane coatings.

Sol-U-Gro. [Miller Chem. & Fertilizer] Soluble fertilizer.

Soluhoba. [Jojoba Growers & Processors] Alkoxylated jojoba oil.

Solukast. [Soluol] Urethane prepolymer or elastomer.

Solulac. [Grain Processing] Animal feed additive.

Solulan®. [Amerchol; Amerchol Europe] Lanolin derivs.; emulsifier, wetting agent, dispersant, lubricant, pearlescent, spreading agent, plasticizer, emollient, solvent, conditioner, foam stabilizer for cosmetics, detergent systems, pharmaceuticals, waxes, polishes, leather treatment.

Solu-Lastin. [Brooks Industries] Hydrolyzed animal elastin; for hair and skin care cosmetics.

Solu-Mar. [Brooks Industries] Hydrolyzed collagen; film former and moisturizer for skin and hair care prods.

Solumin F. [Rhone-Poulenc Ltd.] Ether sulfate; surfactant for water-based systems.

Solumine®. [Soluol] Substantive softener for textiles.

Soluol OM. [Kao Corp. SA] Quatenary ammonium salt; antistat for acrylic fibers.

Soluphor® P. [BASF AG] Pyrrolidone-

2; solvent for veterinary medicine.

Solupon®. [Soluol] Textile softeners.

Solupret. [Hoechst AG] Silicon sol'ns.; water repellent finishing.

Sol-U-Pro. [DynaGel] Protein hydrolysates.

Soluscour. [Soluol] Scouring agents.

Solusil®. [Soluol] Fabric lubricant.

Solu-Silk. [Brooks Industries] Hydrolyzed silk or silk amino acids.

Solusoft NK, WA, WL. [Hoechst AG] Silicone elastomer; finishing agent for textiles.

Solusoft®. [Soluol] Textile softener.

Solusol®. [Am. Cyanamid] Dioctyl sodium sulfosuccinate; wetting agent for cosmetics.

Solu-Soy. [Brooks Industries] Hydrolyzed soy protein; for skin and hair care cosmetics.

Sol-U-Tein. [Fanning] Albumen or hydrolyzed soy protein; binder, coagulant for pharmaceuticals and personal care prods.; dye mordant in textiles, adhesives, veneers, sizing and making papers; gilding leather; book binding; food applic.

Solutene TER. [ICI Surf. UK] Self-emulsifiable dichlorobenzene; carrier for dyeing polyester fibers and blends with disperse dyes.

Solu-Tofu. [Brooks Industries] Hydrolyzed soy protein.

Solutol® HS 15. [BASF AG] PEG-660 hydroxystearate; solvent for injection sol'ns.

Solu-Veg. [Brooks Industries] Hydrolyzed vegetable protein or blends; for hair and skin care cosmetics.

Soluvit Richter. [Dr. Kurt Richter; Henkel/Cospha] Multivitamin-herbal complex; vitamin treatment for skin and hair care prods.

Solva. [Giulini Corp.] Special additives for processed cheese.

Solvable. [DuPont/Medical Prods.] Aq. tissue and gel solubilizer.

Solvar. [Lawter Int'l.] Ink varnish visc. control.

Solvatex. [Leatex] Penetrant, stabilizer, and sequestrant for peroxide bleaching of cotton and cotton blends.

Solvawax. [Frank B. Ross] Waxes for mfg. of solvent, liq. and paste polishes, leather finishes, lubricating sticks.

Solvenol®. [Hercules Europe] Mixed terpenes; hydrocarbon solvent for industrial cleaners, deodorants.

Solvenon®. [BASF AG] Solvents, diluents for adhesives, paints, resins, dyes, surface coatings, cleaning agents.

Solvent 111. [Vulcan Materials] Trichloroethane.

Solvent GC. [BASF AG] Ethylene glycol acetate mixt.; solvent for printing inks and rubber stamp inks.

Solvent PM. [BASF; BASF AG] Methoxypropanol; solvent for cosmetic industry.

Solvent Scour. [Hart Chem. Ltd.] Alkanolamides and alkyl phenol ethoxylate; scouring agent.

Solvent-Scour. [ICI Surf. UK] Chlorinated solvent/surfactant blend; general purpose scouring agent for all fibers.

Solvesperse. [Hilton Davis] Pigment dispersions.

Solvic. [Solvay & Cie] PVC.

Solvite. [Leatex] Low foaming solvent scours.

Solvitose. [Avebe Am.] Potato starches; for textiles.

Solvocine®. [BASF] Print paste assist for discharge printing.

Solvoil. [Reilly-Whiteman] Solvent fatliquors for leather industry.

Solvolan. [Smits Neuchatel] Polyurethane resin-based material.

Solvonic. [Chemonic Industries] Solvents and detergents for grease and oil removal from fabrics.

Solwax. [Van Schuppen] Alkoxylated lanolin or cholesterol derivs.; emulsifier, foam stabilizer, softener, conditioner, emollient.

Somepon T25. [Seppic] Sodium methyl cocoyl taurate.

Sonic-Solve. [London Chem.] Defluxing solvent.

Sonite. [Smooth-On] Mold release agents.

Sonojell®. [Witco/Sonneborn] Petrolatum blends; emollient bases for cleansing creams, cosmetics.

Sonolube. [Witco/Sonneborn] Food-grade lubricant.

Sonostat®. [Henkel/Textile] Softener for natural and synthetic fibers and blends.; napping aid; in resin finishing.

Sontex. [Witco/Sonneborn] White mineral oil.

Sonwood. [Sondex AB] PVC and cellulosic fibers.

Soponol. [Surpass] Detergent, emulsifier, scouring agent for textiles.

SOP®. [Floridin] Fuller's earth.

Sopralub. [Henkel-Nopco] Polyoxyalkylene fatty glycerides; softener, plasticizer for textile finishes and coatings.

Soprofor. [Rhone-Poulenc Geronazzo SpA] Emulsifier, detergent, wetting agent, dispersant, suspending agent, anticaking agent, intermediate, antistat, lubricant, softener.

Soprol VR.50. [Rhone-Poulenc] PEG-200 glyceryl stearate.

Soprophor®. [Rhone-Poulenc Surf.; Rhone-Poulenc Geronazzo SpA] Alkoxylated alcohols, EO/PO block polymers, sulfonates, or ethoxylates; surfactant, wetting and suspending agents for degreasers, acid pickling compds., pesticides.

Sorba. [Croda Chem. Ltd.] Lanolin; emollient for cosmetics, pharmaceuticals.

Sorban. [Witco SA] Sorbitan ester; emulsifier, antifoamer, corrosion inhibitor.

Sorbanox. [Witco SA] Ethoxylated sorbitan ester; detergent, emulsifier, solubilizer.

Sorbaset. [Omni/Ajax] Hazardous waste solidifier, alkali neutralizing solidifier.

Sorbasolv. [Omni/Ajax] Oil absorbent.

Sorbaspray. [Uniroyal] Foliar nutrient.

Sorbassist. [Paradigm Labs] Wetting and rewetting agent for scouring, boiling-off, bleaching, dyeing, and finishing operations.

Sorbax. [Chemax] Sorbitan esters or ethoxylates; emulsifier, solubilizer for perfumes, flavors, cosmetics, textile and metal lubricant industries.

Sorbelite. [Mendell] Sorbitol NF; tabletting aid for pharmaceuticals.

Sorbeth. [Croda Surf. Ltd.] Ethoxylated sorbitol esters; emulsifier, dispersant for agric. chemicals and the oil industry.

Sorbilene. [Auschem SpA] Ethoxylated sorbitan esters; emulsifier, solubilizer for cosmetics, pharmaceuticals, agric. formulations.

Sorbirol. [Auschem SpA] Sorbitan esters; emulsifier for cosmetics, pharmaceuticals, agric. formulations.

Sorbit. [Ciba-Geigy AG] Alkyl naphthalene sulfonate; wetting agent, foamer, emulsifier, detergent, germicides, textiles.

Sorb-It. [United Desiccants-Gates] Silica gel-based desiccant; for adsorption of moisture.

Sorbo. [ICI Am.] Sorbitol USP.

Sorbo-Cel. [Celite] Filter aid.

Sorbon. [Toho Chem. Industry] Sorbitan esters or ethoxylates; emulsifier, dispersant, solubilizer, detergents for cosmetics.

Sorbothane®. [Sorbothane; Leyland & Birmingham Rubber] Polyurethane material; shock and vibration absorbent for sports, medical, engineering, electronics, acoustics, and computer applics.

Sorbsil. [Crosfield] Silica gel.

Sorgen. [Dai-ichi Kogyo Seiyaku] Sorbitan esters or ethoxylates; emulsifier, antifoamer for foods, cosmetics, pharmaceuticals.

Soricinol. [Climax Performance] Sodium ricinoleate; mold release.

Soromin®. [BASF AG] Sizing agents, assistants for synthetic fibers, primary spinning.

Sorpol. [Toho Chem. Industry] Surfactant blends; emulsifier for pesticides.

So/San 30M. [Stepan] Softener/sanitizer for fabrics.

Sotex. [Morton Int'l.] Fatty acid esters; nonionic surfactants; dispersant, wetting agent for paints, coatings, dyes, inks; stabilizer in vinyl plastisols.

Sovatex. [Standard Chem. UK] Sodium dodecylbenzene sulfonate and fatty alcohol ethoxylate blend; detergent, wetting agent, and dye assistant for textile processes.

Sovermol. [Henkel KGaA] Unsaponifiable hydroxyl component for high-quality polyurethane finishes.

Soyabits. [Central Soya] Soy grits.

Soyafluff. [Central Soya] Soy flour.

Soyalac. [Degen] Soya-based polymer.

Soy-Amino Quat L/O. [Brooks Industries] Lauryloleylmethylamine soy amine acids; for skin and hair care cosmetics.

Soyarich. [Central Soya] Soy flour and grits.

Soy Polymer. [Protein Tech. Int'l.] Paper coating and industrial polymer.

Soy-Quat C. [Maybrook] Cocodimonium hydroxypropyl hydrolyzed soy protein; cationic substantivity agent, conditioner, moisturizer for hair and skin care prods.

Soy Solinox. [Reichhold] Soybean oil plasticizers.

Soy-Tein NL. [Maybrook] Hydrolyzed soy protein; moisturizer, protective film-former, substantivity agent for hair and skin care prods.; improves tensile strength of hair; mitigates damage due to bleaching, perming, hot combing.

SP. [Chevron] Ethylene-methyl acrylate copolymer; for film, coating, laminating.

SP-. [Schenectady] Phenolic resins; crosslinking agent, tackifier, plasticizer for rubber compds., adhesives, and sealants.

SP-500. [Kobo] Nylon 12.

SP-731, 732, 734, 735. [Specialty Prods.] Acetates/amine blend; line flushing and clean-up solvent for adhesives, resins, urethane foam.

Spac. [Mitsubishi Kasei] Sodium propionate; mold inhibitor for animal feed.

Span®. [ICI Spec. Chem.; ICI Surf. Belgium] Sorbitan esters; emulsifier, stabilizer, thickener, lubricant, softener, antistat for foods, pharmaceuticals, cosmetics, cleaning compds., textiles.

Spanscour®. [CNC Int'l.; Reilly-Whiteman] Gas fading inhibitor, surfactant for scouring nylon/lycra, spandex.

Spark-L®. [Solvay Enzymes] Pectinase; enzyme for depectinization and pulp washing.

Sparklon. [Morton Int'l.] Gloss coatings for paper.

Spartan. [Spartan Flame Retardants] Flame retardant chemicals.

Spartan EP. [Exxon] Extreme pressure lubricant.

Spartcide®. [Mitsubishi Kasei] 2-3-Dichloro-N-4-fluorophenylmaleimide; fungicide for apple fruit spot, melanose and scab of citrus, coffee berry disease, pink disease on rubber.

Spauldite®. [Spaulding Composites] Composite laminates.

Spaulrad®. [Spaulding Composites] Woven glass fiber polyimide epoxy lamintes; for elec. insulation, magnet construction.

Specflex. [Dow; Dow Ahlen] Polyurethane molded system.

Special Bio. [Accurate Chem. & Scientific] Cesium chloride.

Special Black. [Degussa] Carbon or furnace black; for uv inks.

Special Fat. [Hüls Am.] Hydrogenated fats or triesters.

Special Oil. [Hüls Am.] Glyceryl triesters; tracer additive to other fats such as dairy butter.; release agent, lubricant in cosmetic emulsions, lipsticks, hair care prods.; pigment dispersant for sticks and liners.

Spectone. [DCS Color & Supply] Liq. iron oxide pigments; used in mortar and cement coloring.

Spectra. [Spectra Colors] Textile dyes and pigments.

Spectrablend. [Warner-Jenkinson] Dry blended pharmaceutical pigmented coatings.

Spectracid. [Spectra Colors] Textile dyes and pigments.

Spectracoat. [Warner-Jenkinson] Conc. dispersions.

Spectracote. [Flexible Prods.] Multiuse coatings and adhesives for athletic items, refrigeration prods., architectural panels, lamination, encapsulation, textiles, concrete.

Spectradiazo. [Spectra Colors] Textile dyes and pigments.

Spectradyne® G. [Lonza] Chlorhexidine gluconate; antimicrobial for pharmaceuticals, hospital disinfectants, veterinary prods., antiplaque dental prods.

Spectrafil. [Specmat Ltd.] Highly filled plastics.

Spectra Fine. [Spectra Polymer] Granular color concs.

Spectraflo. [Ferro/Plastic Colorants] Liq. colorants.

Spectrafoam. [Specmat Ltd.] Lightweight rigid foam.

Spectralite. [DSM Resins UK Ltd.] Pigment dispersions.

Spectraliz. [Spectra Colors] Textile dyes and pigments.

Spectramine. [Spectra Colors] Textile dyes and pigments.

Spectra-Pearl. [Van Dyk] Colored titanium dioxide/mica; pearlescent for cosmetics.

Spectraresorcine. [Spectra Colors] Textile dyes and pigments.

Spectrasol. [Spectra Colors] Textile dyes and pigments.

Spectrasorb. [Specmat Ltd.] Radiation absorbers.

Spectra-Sorb®. [Am. Cyanamid] Uv absorbers for cosmetics, sunscreens, pharmaceuticals.

Spectraspray. [Warner-Jenkinson] Conc. color and protective coatings for pharmaceuticals.

Spectratech® CM. [Quantum/USI] Uv absorber/inhibitor, antiblock, antistat, antioxidant, optical brighter for film, sheet, pkg., molded items.

Spectratech® Color Concs. [Quantum/USI] Color concs. in polymer carriers.

Spectratech® FM. [Quantum/USI] EVA, LDPE, or PS foam concs.; for structural foam, extrusion, inj. moldings, wire and cable, blown film, business machine housings.

Spectratech® FR. [Quantum/USI] HDPE/EVA compd. with nonhalogenated flame retardant; flame retardant compd. for extruded sheet for lamination to aluminum and plywood.

Spectratech® PM. [Quantum/USI] Processing aid for LLDPE and HDPE extrusion.

Spectrazine. [Spectra Colors] Textile dyes and pigments.

Spectrazurine. [Spectra Colors] Textile dyes and pigments.

Spectrim. [Dow; Dow Ahlen] Urethane systems; for structural RIM applics. for automotive, consumer, and industrial goods, dynamic elastomers, load-bearing applics.

Spectro Acid. [Spectro Color & Chem.] Textile dyes and pigments.

Spectro Basic. [Spectro Color & Chem.] Textile dyes and pigments.

Spectro Direct. [Spectro Color & Chem.] Textile dyes and pigments.

Spectro Disperse. [Spectro Color & Chem.] Textile dyes and pigments.

Spectro Reactive. [Spectro Color & Chem.] Textile dyes and pigments.

Spectrum. [Spectrum Chem. Mfg.] Laboratory reagents.

Speed Bonder #312. [Loctite] Structural adhesive.

Speedcure. [Aceto] Uv photoinitiators.

Speedye. [Yorkshire Pat-Chem] Pigment colors for exhaust dyeing of fabric or garments.

Spenkel. [Reichhold] Urethane polyol.

Spenlite. [Reichhold] Urethane polyol.

Spensol. [Reichhold] Urethane polyol.

Spermwax. [Robeco] Cetyl esters wax NF; synthetic spermaceti used in toiletries, cosmetics, dermatologicals as stiffening agent, slip and visc. aid.

Sperse Polymer IV. [Atramax] Ammonium polycarboxylate; dispersant for pigments used in textiles; vehicle for flexo inks.

Speswhite. [ECC Int'l.] China clay.

Spezyme. [Genencor Int'l.] Amylase; enzyme for textile desizing, food processing, starch liquefaction.

Spherica. [Ikeda] Silica or silica/hyaluronic acid blend.

Sphericel. [Potters Industries] Hollow glass spheres; filler for plastic compds., molded parts; yields weight reduction for finished parts.

Spheriglass®. [Potters Industries] Solid glass spheres; additive for thermoplastic and thermosetting resin systems; enhance processing; for automotive, chemical, electronic, industrial, engineering, and photographic industries.

Spherititan. [Ikeda] Titanium dioxide.

Spheron. [Presperse] Silica or blends; spherical microbeads for powders, anhydrous systems, emulsions; oil absorbent; carriers for sunscreens, fragrances, emollients.

Sphingoceryl. [Serobiologiques] Octyldodecanol/phospholipids/glycosphingolipids blends.

Sphingolipid. [Nikko Chem. Co. Ltd.] Sphingolipids.

Spink-Gel Bentonite. [H.C. Spinks Clay] Wyoming bentonite.

Spinks 211. [H.C. Spinks Clay] Sodium polyacrylate.

Spinks Blend Clay. [H.C. Spinks Clay] Ball clay.

Spinomar NaSS. [Tosoh] Sodium p-styrenesulfonate; dyeing assistant, fiber modifier, emulsifier for emulsion polymerization, flocculant and scale inhibitor, dispersant for cosmetics, antistat for paper, fibers, plastics; ion exchange resin; photo chemicals, pharmaceutical, artificial bio-membranes.

Spinrite. [Lenox] Cohesive agent, lubricant, antistat for textiles.

Spinzit. [Nat'l. Starch & Chem.] Lubricant, antistat for carpet processing.

Spiroflor. [Henkel] 3-Ethyl-2,4-dioxaspiro (5.5) undec-8-ene; natural sweet odor fragrance raw material.

Spongolit®. [Henkel/Emery/Cospha; Grünau] Esterified glycerides; aerating emulsifiers for Madeira and sponge cakes.

Sponto®. [Witco/Organics; Witco Israel; Witco SA] Surfactants; emulsifier for agric. formulations.

Spotleak®. [Atochem N. Am.] Butyl mercaptan blends; odorant for natural gas to permit detection of leaks.

Spotrete. [W.A. Cleary] Thiram fungicide and animal repellent.

Spra-Seal. [Rowe Prods. Distribution] Epoxy mastic coating.

SprayCore®. [Omega] Sprayable composite systems; print barriers, sprayable cores, bonding adhesives, lamination systems.

Spray Graph. [Am. Resin] Dry film graphite lubricant and release agent.

Spraymet. [Atochem N. Am.] Cleaners.

Sprayset®. [Witco/Argus] MEK peroxide; catalyst for polyester curing.

Spraythane. [Enodim SA] PU foam for caulking buildings.

Spraywax 660-A Conc. [Finetex] Proprietary imidazolinium; for industrial car wash sprays.

Spreading Agent ET0672. [Croda Chem. Ltd.] Complex nonionic alkoxylate; surface spreading agent for oils used in bath oils, for mineral oils used for mosquito control.

Spritz. [Merix] Antifog liquid.

Spud-Nic®. [Aceto] Chloropropham; herbicide.

Squadron. [Am. Cyanamid/Ag] Herbicide.

SR. [GE Silicones] Silicone conformal coating.

SR-. [Sartomer] Acrylates; comonomer to produce crosslinked polymers; crosslinking and curing agent used in emulsion polymerization, castings, ion exchange resins, rubber compd., plastisols, coatings, fibers, papers, and other fabrications.

SR-7475. [Firestone Syn. Rubber] Butadiene-styrene copolymer rubber; blendable modifier for thermoplastic resins and asphalt.

SRF. [Schenectady] Resorcinol formaldehyde preformed resin sol'n.; bonding agent used in rubber compds. to improve adhesion; resorcinol donor in resorcinol-formaldehyde latex dips.

SS. [Acheson Colloids] Nickel and/or silver pigments with polymeric binders; EMC shielding coating for composites/aluminum; protects sensitive electronic equip.

SS. [GE Silicones] Silicone blends; water repellent, release agent, emollient for rubber, plastics, cosmetics, polishes, textile, leather.

SS-07/SE-9034. [GE Silicones] Silicone rubber; strand sealant compd.

S-Safe 2010. [Nippon Oils & Fats] PPG-20-decyltetradeceth-10.

SSE. [Transene] Self-catalyzed silicone elastomer.

SSR. [Jersey Ind. Chem. Prod.] Cooling

water treatment.

S.S.T.® Sump Saver Tablets. [Angus] Tris (hydroxymethyl) nitromethane; antibacterial agent, preservative for metalworking fluids.

ST. [Morton Int'l.] Polysulfide gum rubber.

Sta-B12. [Hoffmann-La Roche] Vitamin B12. Visc. suppressant, antisettling agent, processing aid.

Stabar. [ICI Films] PEEK and polyether sulfone films.

Stabicol®. [Allied Colloids] Stabilizer, developer for peroxide bleaching.

Stabicote. [Hoffmann-La Roche] Vitamin B12.

Stabil-9. [Monsanto] Food leavening agent.

Stabileze. [ISP] Crosslinked Gantrez with 1,9-decadiene.

Stabilisal S Liq. [Hoechst AG] Polysulfide sol'n.; stabilizer for sulfur dyeing.

Stabilite 75. [Apollo] Organic peroxide stabilizer designed for continuous bleaching, cold pad batch and regular batch systems.

Stabilizer 9-A [Givaudan-Roure] Mono and diisopropylated m- and p- cresols; antioxidant for fatty acids and their derivs. used in antistats, fiber lubricants, stearic acid, oleic acid, vegetable and animal oils, paraffin.

Stabilizer 128. [Yorkshire Pat-Chem] Silicate-free organic stabilizer; for textile bleach baths.

Stabilizer 1097. [Miles/Polysar Rubber] Acid chloride in butyl acetate; stabilizer used to extend the pot life of the bonding agent system/plastisol mixt.

Stabilizer 2013-P®. [TSE] Carbodiimide; activator for vulcanizates based on millable polyurethane.

Stabilizer AWN, CS, SIFA. [Sandoz Prods. Ltd.] Stabilizing agents for hydrogen peroxide bleaching baths.

Stabilizer C, I-FF. [Bayer] Phenyl stabilizers.

Stabilizer CB. [ICI Surf. UK] Sodium polycarboxylate mixture; stabilizer for hydrogen peroxide bleaching.

Stabilizer NS. [Marlowe-Van Loan] Organic/inorganic; peroxide stabilizer.

Stabilizer T. [Grünau] Protein fatty acid condensate; stabilizer for bleaching.

Stabilo. [Vyse Gelatin] 125 Bloom gelatin.

Stabilon®. [Ciba-Geigy/Textiles] Stabilizers for bleaching operations.

Stabilor. [Degussa] Gold alloy; for dental applics.

Stabilox. [Henkel KGaA] Stabilizer/lubricant masterbatches.

Stabiol. [Henkel KGaA] Heat stabilizers for PVC processing.

Stabiram. [Ceca SA] Emulsifiers for bitumen.

Stabland®. [Karlshmans] Fractionated partially hydrogenated soybean oil; specialty oil.

Stabolec C. [Amico] Lecithin, propylene glycol, citric acid, t-butyl hydroquinone.

Staburags®. [Kluber Lubrication N. Am.] Grease for bearings.

Sta-Clad. [Reichhold] Synthetic resin.

Staclipse. [A.E. Staley Mfg.] Industrial corn starch.

Staflex. [Reichhold] Esters; plasticizers.

Stafoam. [Nippon Oils & Fats] Fatty acid alkanolamides; detergent, foam booster and stabilizer, softener, antistat, rust inhibitor, penetrant for shampoo, cosmetics, dishwashing, laundry detergent, drycleaning, textile softener aux., and metal cleaning.

Sta-Form. [Georgia-Pacific] Urea-formaldehyde resin; for industrial use.

Stag Epoxide Resins. [Stag Polymers & Sealants Ltd.] Liq. epoxides.

Stainaway. [Am. Emulsions] Stain blocker for nylon 6/6 and 6 carpet fibers.

Stain-Free. [Sybron] Stain resist auxiliary for nylon carpet.

Stainguard. [Am. Emulsions] Dye fixative for acid dyes in nylon.

Staleydex. [A.E. Staley Mfg.] Liq./granular dextrose.

Sta-Lok. [A.E. Staley Mfg.] Industrial corn starch.

Stam. [Rohm & Haas] Selective herbicide.

Stamere. [Meer] Carrageenan prods.

Stamid. [Clough] Fatty acid alkanolamide; foam stabilizer, emulsifier, thickener for household, industrial, and cosmetic formulations.

Sta-Mist. [A.E. Staley Mfg.] Modified corn starch.

Stamulan HD. [DSM Deutschland GmbH] HDPE.

Stamylan. [DSM UK Ltd.] LDPE, HDPE, PP, UHMWPE.

Stanclere®. [Akzo] Tin complexes; stabilizers for PVC, urethane catalysts.

Standamid®. [Henkel/Emery/Cospha] Fatty acid alkanolamides; foam booster and stabilizer, superfatting agent, thickener, wetting agent, detergency builder.

Standamox. [Henkel/Emery/Cospha; Pulcra SA] Amine oxides; wetting agent, foam builder/stabilizer, thickener, emollient, conditioner, softener, lubricant for cosmetic, household, and industrial prods.

Standamul®. [Henkel/Emery/Cospha] Emulsifier, solubilizer, emollient, base for cosmetics, pharmaceuticals, food prods., paper, textile, leather, household cleaning.

Standapol®. [Henkel/Emery/Cospha; Henkel KGaA; Pulcra SA] Surfactants; detergent, foamer, thickener, conditioner, antistat, emollient for personal care, household, and industrial applics.

Standapon® 4149 Conc. [Henkel/Textile] Scouring agent/detergent for continuous scouring, desizing, bleaching, and pot dyeing.

Standard Super-Cel. [Celite] Diatomaceous earth; filter aid.

Stanlev®. [Henkel/Textile] Leveling agent for acid and disperse dyed nylon.

Stan-Mag®. [Harwick] Magnesium oxide powders, rubber and plastic dispersions.

Stan-Mask. [Harwick] Reodorants.

Stannine. [Rhone-Poulenc Ltd.] Thiourea-based blends; inhibitors for use in pickling and cleaning ferrous metals.

Stannochlor. [Atochem N. Am.] Plating compds.

Stannolume. [Atochem N. Am.] Tin plating.

Stanno-Plus. [Goldschmidt] Stabilizer for metal salt sol'ns.

Stanol. [Henkel-Nopco] Organic sulfates/nonionics blend; scouring and fulling agent for wool; shampoo base; bubble baths.

Stanomerse. [Technic] Immersion tin.

Stantex®. [Henkel/Emery; Henkel Canada] Ester sulfate or detergent/solvent blends; wetting and rewetting agent for dyeing and finishing; detergents for textiles; jet car wax agent.

Stan-Tone®. [Harwick] Pigments and pigment dispersions.

Sta-Nut EE. [Van Den Bergh Foods] Partially hydrogenated cottonseed oil, fatty acid mono and diglycerides; peanut butter stabilizer.

Stanyl®. [DSM; DSM UK] Nylon 4/6, some glass or mineral reinforced; for electrical applics.

Sta-O-Paque. [A.E. Staley Mfg.] Modified corn starch.

Staph-O-Cide. [Crain] General cleaner and germicide.

Stapron. [DSM; DSM UK] Styrene/maleic anhydride copolymer, rubbermodified, some glass reinforced; amorphous engineering plastic.

Star. [Procter & Gamble] Glycerin USP; humectant.

Star®. [PQ Corp.] Sodium silicate; liq. alkaline bleaching aid and kier boiling agent.

Staramide. [Ferro Enamel Espanola SA] Polyamide composites.

Starblast. [DuPont] Abrasive sand.

Starburst Dendrimers. [Polysciences] Natural product extractions.

Star C. [Ferro Enamel Espanola SA] Carbon fiber-reinforced engineering plastics.

Starch 1500. [Colorcon] Tablet and capsule excipient.

Star-Dri. [A.E. Staley Mfg.] Maltodextrins or corn syrup solids.

Starflam. [Ferro Enamel Espanola SA] Flame retardant engineering plastics.

Starfol®. [Sherex] Difatty esters; emollient for cosmetics.

Starglas. [Ferro Enamel Espanola SA] Reinforced engineering plastics.

Star L. [Ferro Enamel Espanola SA] Lubricated engineering plastics.

Starplex®. [Am. Ingredients] Monoglyceride blends; food emulsifier, stabilizer; starch complexing agent.

Starpol. [A.E. Staley Mfg.] Polymerizable starch.

Starpylen. [Ferro Enamel Espanola SA] Reinforced PP.

Star Rov®. [Manville] Fiber glass roving; for filament winding, pultrusion, weaving.

Star Stran. [Schuller] Continuous filament glass strand; reinforcement for polypropylene, PPS, PEI.

Start Up. [Western Nutrients] Chelated metal salts of Zn, Mn, Mg, Ca, Fe for use with starter fertilizers citric base.

Start-Up. [Vanson Chem.] Sol'n. for clarifying plaster pools on initial water fill.

Starwax®. [Petrolite/Polymers] Hard microcrystalline wax; for coatings, adhesives, paper, printing inks, plastic modification (as lubricant and processing aid), lacquers, paints, varnishes, as binder in ceramics, for potting/impregnant in elec./electronic components, rubber, cosmetics.

Sta-Slim. [A.E. Staley Mfg.] Modified corn starch.

Stasoft. [Reilly-Whiteman] Sulfated neatsfoot or fish oils; fatliquor for leathers.

Sta-Soft. [Reichhold] Polyurethane resin.

Sta-Sorb. [A.E. Staley Mfg.] Absorbent starch.

Sta-Tac. [Reichhold] Synthetic resin.

Sta-Tac®. [Arizona] Mixed olefin hydrocarbon resin; for adhesives.

Statex. [Columbian Chem.] Carbon black.

Statexan. [Miles/Polysar Rubber] Sulfonated aliphatic hydrocarbon; internal antistat or external coating for PVC, PS.

Staticide®. [ACL] Water-based topical antistat coating.

Statik-Blok® FDA-3. [Amstat Industries] Industrial antistat for plastics, food contact applics.

Stat-Kon®. [LNP; LNP Nederland] ABS, PE, PC, polyetherimide, polysulfone, PES, acetal, PEEK, PP, PPS, nylon 6, 6/6, 6/10, PBT, or PBO, some with carbon or stainless steel reinforcement; statically dissipative thermoplastic composites for protection against electrostatic discharge damage; used in electronic pkg. systems and functional components.

Statoil. [Statoil Petrokemi AB; MBS Plastics] LDPE, HDPE, PP compds.

Statonic. [Chemonic Industries] Antistat for synthetic fibers.

Statran. [Interpolymer] Antistatic agent.

Stat-Rite®. [BFGoodrich; BFGoodrich UK] Static dissipative polymers.

Stature. [Dow] Static control additive.

Stauffer N-1386®. [Akzo] Bis (trichloromethyl) sulfone; industrial biocide, slimicide, perservative for paper/paperboard, adhesives, latexes, secondary oil well recovery.

Stauffer NR-1. [Rhone-Poulenc Basic] Reagent.

Staybelite®. [Hercules] Rosin derivs. or esters; thermoplastic resin for adhesives; tackifier, plasticizer, processing aid for rubbers, laminations, barrier coatings, chlorinated rubber finishes; modifier for film-formers, elastomers, waxes; softener/plasticizer for chewing gum.

Stayclean. [Merix] Destaticizer cleaner.

Stayco. [A.E. Staley Mfg.] Industrial corn starch.

Staysize. [A.E. Staley Mfg.] Industrial corn starch.

Stealim. [L.A. Salomon] Talc; anticaking agent for animal feed.

Steamaster 79. [Stewart Hall] Boiler compds. to prevent corrosion and scale.

Steamfilm FG. [Drew Ind. Div.] Octadecylamine aq. emulsion; corrosion inhibitor for boiler water treatment.

Stearal. [Amerchol] Stearyl alcohol; emollient, emulsifier, texturizer.

Stearalchol. [Lanaetex Prods.] Mineral oil/lanolin alcohol blend.

Stedbac®. [Zeeland] Stearalkonium chloride; hair conditioner, emulsifier.

Steelast. [Steelcote Mfg.] Vinyl prods.

Steelgard. [Harry Miller] Rust preventive.

Steel Grip. [Alpha Metals] Stainless steel

inorganic acid flux.

Steelhide. [Etna Prods.] Rust preventives/corrosion inhibitors.

Steelskin. [Atochem/Wire Mill Prods.] Metalworking compds.

Stellar. [Cyprus Industrial Min.] Talc; fine grind, high brightness for plastics applics., polyolefin films.

Stellite®. [Haynes Int'l.] Cobalt-based alloys; wear-resistant alloys.

Stenol®. [Henkel/Emery; Henkel KGaA] Linear primary alcohol; intermediate for surfactant mfg.

Steol®. [Stepan; Stepan Canada; Stepan Europe] Sodium or ammonium laureth sulfates; detergent, emulsifier, foamer, dispersant, wetting agent for personal care prods., cleaning prods., textile mill applics., emulsion polymerization.

Stepan. [Stepan] Methyl esters; intermediate for mfg. of detergents, emulsifiers, wetting agents, stabilizers, lubricants, plasticizers, resins, and textile specialties.

Stepanate. [Stepan] Sulfonates; hydrotrope, solubilizer, coupler for detergent systems.

Stepanflo. [Stepan] Alpha olefin sulfonates; anionic surfactants for enhanced oil recovery operations.

Stepanflote®. [Stepan] Sodium alkyl ether sulfate; flotation reagent.

Stepanfoam. [Stepan] Polyurethane resins and foam systems.

Stepanform. [Stepan] Anionic-nonionic blends; emulsifier for vegetable oils in non-edible applics.; foaming agent for cellular concrete, drilling.

Stepanhold®. [Stepan; Stepan Europe] PVP copolymers; hair fixatives.

Stepan-Mild®. [Stepan; Stepan Europe] Sulfosuccinates; for shampoos, hand soaps, bubble baths, baby prods., dishwashing detergents.

Stepanol®. [Stepan; Stepan Canada; Stepan Europe] Alkyl sulfates; detergent, foamer, wetting and suspending agent for personal care prods., household, metal, and industrial cleaners; fruit washing; insecticides; textile and leather processing; pharmaceuticals.

Stepanon CG. [Stepan] Amphoteric surfactant; brine tolerant foamer for acid or alkaline media.

Stepan Pearl Range. [Stepan Europe] Surfactant blends; satining, opacifying, or pearlescing agent.

Stepanpol. [Stepan] Urethane polyols.

Stepanquat®. [Stepan; Stepan Europe] Quaternary methoxysulfates; conditioner, antistat for personal care prods.

Stepan Tab-2. [Stepan] Di(hydrogenated) tallow phthalic acid amide; emulsifier, suspending agent for conditioning and anti-dandruff shampoos.

Stepantan®. [Stepan] Alkylaryl sulfonic acid or salts or sodium naphthalene formaldehyde sulfonate; emulsifier, dispersant, wetting agent.

Stepantex. [Stepan; Stepan Europe] Fatty quaternary methosulfate; softeners for textile and home laundry use.

Stepfac®. [Stepan; Stepan Europe] Ethoxylated nonylphenol phosphate; compatibility agent for liq. fertilizers.

Step-Flow. [Stepan] Dispersants and surfactants for aq. flowable agric. formulations.

Steposol®. [Stepan] Ammonium ether sulfate; for gypsum board, cellular concrete, air drilling, foam cleaners.

Stepsperse. [Stepan; Stepan Europe] Surfactants blend; dispersant for flowables and dry flowable agric. formulations.

Stepwet. [Stepan; Stepan Europe] Sodium dodecylbenzene sulfonate; wetting agent, dispersant for pesticides.

Steraffine. [Laserson & Sabetay] Stearyl alcohol.

Steralchol. [Lanaetex Prods.] Multisterol base.

Steramine. [Henkel-Nopco] Softener, antistat, lubricant, finishing agent for textiles.

Stereon®. [Firestone Syn. Rubber] SBR copolymer; modifier for thermoplastic resins.

Steri-Det. [Oakite Prods.] Germicidal detergent.

SteriLine. [Montana Talc] Platy talcs; high purity talcs meeting USP, CTFA, and European Pharmacopeia specs.; for cosmetics, antiperspirants, dusting powders, pharmaceutical excipients.

Sterling. [Cabot] Oil furnace carbon black.

Stero. [Climax Performance] Lanolin or wool grease.

Sterol. [Auschem SpA] Esters or glycerides; superfatting agent, emulsifier, thickener, solubilizer, pearling agent for cosmetics and pharmaceuticals.

Sterotex®. [Karlshamns] Hydrogenated vegetable oils; lubricants, binders for pharmaceutical tableting, pressed powders.

Sterox®. [Monsanto] Ethoxylated alkylphenol; detergent, penetrant, intermediate for textile, paper, metal cleaning, leather processing, household and industrial cleaners.

Sterpon. [Convert SA] Unsaturated polyester resins and gelcoats.

Stetalc Artic Mist. [Steetley Minerals Ltd.] Talc fillers.

Stevens. [Stevens Elastomerics] Polyurethane film; for headphone ear pads, wheelchair pads, IV pressure infusers, inflatable splints, artificial heart components, scuba divers buoyancy compensators, diaphragms, cable jacketing, pkg., fabric laminates, adhesive tapes, noise/vibration dampers.

Stiffener DSC. [Anchor UK] Amine salt on inert filler; thickener.

St. John's Wort Oil CLR. [Henkel/Cospha] Vitamin E carrier with natural tocopherol sin soya oil medium; for general skin care and improvement in skin tone.

Stokopol LO. [Stockhausen] Sodium C12-18 alcohols sulfate.

Stomp. [Am. Cyanamid/Ag] Herbicide.

Stonelite. [Steetley Quarry Prods.] Pulverized limestone.

Stoner. [Stoner] Rust and corrosion preventives, mold release for plastic, rubber.

Stop-It! [Omni/Ajax] Nonselective absorbent.

Stop-Pit. [W.R. Grace/Dearborn] Corrosion inhibitor.

Storm®. [BASF AG] Bentazon, acifluorfen; for postemergence control of broadleaf weeds in soybeans and peanuts.

Stralpitz. [Strahl & Pitsch] Wax blends.

Stratabed. [Rohm & Haas] Ion exchange resin; for stratified bed systems.

Stratos®. [BASF AG] Cycloxydim; post-emergence graminicide against annual and perennial grasses.

Strip-N-Stick®. [CHR Industries/Furon] Silicone tape; pressure-sensitive tape for gasketing, vibration damping, and thermal insulation use.

Striptron Stripper. [Dow] Inhibited methylene chloride with additives; solvent stripper for dry film photoresist and screen inks.

Stroblite. [Stroblite] Luminous and fluorescent prods.

Strodex®. [Dexter] Polyphosphoric ester acid anhydride or salts; detergents, emulsifier, dispersant, wetting agent, stabilizer for paints, pigment grinding.

Structovis®. [Kluber Lubrication N. Am.] Oil for tentering or heat-setting operations.

Struktol®. [Struktol] Resin blends; plasticizer, processing agent, homogenizing agent for elastomer blend compds.

Struktol® Activator 73. [Struktol] Mixture of zinc salts of aliphatic and aromatic carboxylic acids; vulcanization activator for natural rubber.

ST Wetting Agent. [Solem Industries] Visc. suppressant, antisettling agent, processing aid.

Stycast®. [Emerson & Cuming; Grace NV] Epoxy resins or urethane rubbers; casting resins for impregnation, potting.

Stygene. [Chemfax] Polynuclear aromatic polymers; resins for rubber compding., joint cements, plastic compds., protective coatings, fiber board, inks, epoxy potting compds., adhesives, insecticides, briquettes, floor tile, etc.; soft grades as rubber plasticizers; hard grades as rubber extenders.

Stymer. [Monsanto] Synthetic resin; textile sizes.

Stypol®. [Cook Composites & Polymers] Polyester resins; for synthetic marble casting applics., automotive and appliance parts, reinforced plastic prods., impregnation of elec. coils, cores, and windings.

Styra Clear. [Westlake Plastics] Polystyrene.

Styrene XL-8035. [Dow] HIPS.

Styresol. [Reichhold] Styrenated resin sol'n.

Styrid. [Specialty Prods.] Woven glass fiber polyimide epoxy lamintes; styrene emissions reducer for fiberglass molding.

Styrochrom®. [BASF AG] Polystyrene-based batches; antistatic finishing for consumer goods; surface improvement for housings and tech. parts.

Styrocolor®. [BASF AG] Colored molded shapes; for pkg., displays, reusable containers, transport pallets.

Styrodur®. [BASF AG] Extruded rigid polystyrene foam; for thermal insulation of roofs, floors, walls, as frost protection of subsoils below roads, aircraft runways.

Styrofan®. [BASF AG] Styrene-based polymer dispersions; gloss binders for paper, paper coatings, overprint varnishes; raw material for laminating adhesives; binders for bonding fiber webs.

Styrofill®. [BASF AG] Loose padding and filling material for cushioning goods during transport.

Styrofoam. [Dow] Extruded rigid polystyrene foam; for residential sheathing.

Styrolit®. [BASF AG] For prod. of formed foam plastic parts, for full mold casting, for thin-walled formed parts.

Styrolux®. [BASF AG] Styrene/butadiene block copolymer; for inj. molding, extrusion, thermoforming, blow molding, pkg. material, domestic goods, toys, tech. parts, medical applics.

Styron. [Dow Plastics] Polystyrene; general-purpose, impact, and structural foam resins.

Styronal®. [BASF AG] Butadiene/styrene polymer dispersions; binders for paper and board coating.

Styroplus®. [BASF AG] Styrene/butadiene copolymer; for sealable pkg. film, lids for packing containers.

Styropor®. [BASF; BASF AG] Expandable polystyrene; insulating material.

Styrothane. [Futura Coatings] Aromatic urethane coating; for protection of expandable polystyrene, Styrofoam, plywood, urethane and phenolic foam boardstocks.

Stysolac AW. [Kane Int'l.] Plasticizer.

Styvex. [Ferro/Engineering Thermoplastics] Polystyrene, SAN, ABS, or PPO resins, some glass or carbon reinforced; thermoplastics.

SU 3. [Releasomers] Release agent for cast or molded urethanes.

Sublaprint. [Keystone Aniline] Textile dyes and pigments.

Sublimed Blue Lead. [Eagle-Picher] Basic blue lead sulfate; lubricating/friction aid for mfg. of brake linings, clutch facings, high-pressure greases; rust inhibitive pigment for structural steel.

Suconox®. [Zeeland] Processing aids and antioxidants for plastics.

Sucro Ester. [Gattefosse; Gattefosse SA] Saccharose esters; food emulsifier.

Sudan®. [BASF; BASF AG] Oil- and fat-soluble dyes; for marking and coloring mineral oil prods., e.g., engine fuels, heating oil, lubricating greases, shoe polishes, floor polishes and waxes.

Sufatol. [Standard Chem. UK] Sulfated fatty alcohol or blends; scouring, wetting agent, softener, dye assistant, detergent for textiles.

Sugartab®. [Mendell] Sucrose; tablet base, vehicle.

Sulcoidal. [UPI] Colloidal sulfur.

Sulfads®. [R.T. Vanderbilt] Dipentamethylene thiuram tetrasulfide; accelerator, vulcanizing agent for natural and synthetic rubbers.

Sulfadye. [Clark] Reducing agent for sulfur dyes.

Sulfa-Hitech®. [Atochem N. Am.] Solvent used to dissolve sulfur in the prod. of sour gas wells, in sour-gas pipelines, and in refinery and chemical plant flowlines.

Sulfan. [PVS] Stabilized sulfur trioxide.

Sulfasan®. [Monsanto] 4,4′-Dithiomorpholine; vulcanizing agent, crosslinking agent for elastomers; sulfur donor.

Sulfetal. [Zschimmer & Schwarz] Sulfates; detergent, wetting agent, flotation agent for cosmetics, industrial and metal cleaners.

Sulfidal. [Hüls Am.] Colloidal sulfur.

Sulfidal. [Pettibone Labs] Colloidal sulfur; for acne, hair and skin treatment.

Sulf-N 45. [Allied-Signal] Ammonium sulfate fertilizer.

Sulfochem. [Chemron] Sulfates; detergent, foamer for shampoos, bubble baths, cleansers, drilling fluids.

Sulfocos 2B. [Cosmetochem] PEG-6 isolauryl thioether.

Sulfodur. [Colores Hispania SA] Pigments resistant to sulfur dioxide.

Sulfolane W. [Shell] Tetramethylene sulfone; solvent for extraction of benzene, toluene, other aromatic hydrocarbons from oil refinery streams.

Sulfole. [Phillips] Tertiary mercaptans.

Sulfolink. [Pierce Chem.] Sulfhydryl reactive gel; for affinity chromatography.

Sul-fon-ate. [Boliden Intertrade] Sulfonates; wetting agent, antifoam, corrosion inhibitor, coupler, solubilizer, emulsifier for cement, food, commercial laundry, cosmetics, fertilizers, insecticides, leather, paper, petroleum, and rubber processing, metal cleaning, electroplating, pickling.

Sulfonated Castor Oil. [Nat'l. Starch & Chem.] Fatty glyceride sulfate; emulsifier, lubricant, dyeing assistant, dye dispersant and leveler.

Sulfonated GTO. [Nat'l. Starch & Chem.] Fatty glycerine sulfate; emulsifier, dyeing assistant, dye dispersant and leveler.

Sulfonated Red Oil. [Nat'l. Starch & Chem.] Sulfated fatty prod.; dyeing assistant for cellulosics or acid colors.

Sulfonex. [Estron] Sulfonamid resin.

Sulfonic. [Boliden Intertrade] Sulfonic acids; detergent, penetrant, intermediate for textiles, industrial cleaning.

Sulfonic Acid LS. [Hart Chem. Ltd.] Linear alkylbenzene sulfonic acid; intermediate for detergent formulations.

Sulfopon®. [Henkel Canada; Henkel KGaA; Pulcra SA] Sulfates; detergent, wetting agent, emulsifier, foamer for personal care prods., emulsion polymerization, detergent systems.

Sulfosil P-491. [Witco] Sodium silicate/ sodium sulfate-based builder; detergent booster for household and industrial cleaning.

Sulfostat. [Zschimmer & Schwarz] Amido alkylamine acetate; antistat for cosmetics.

Sulfotex. [Henkel/Emery/Cospha; Henkel Canada] Sulfates or sulfosuccinates; wetting agent, detergent, emulsifier, foamer, dispersant, hydrotrope, solubilizer for personal care prods., detergent systems, carpet backing, fire fighting foams, polymerization, food processing, textiles, inks.

Sulfox. [Hickson Danchem] Oxidizing agent for sulfur and vat dyes.

Sulframin. [Witco; Witco SA] Alkylaryl sulfonic acid or salts; penetrant, lubricant, dispersant, detergent, wetting agent, antistat for household and industrial cleaning, personal care prods., textiles, petroleum industry, polymerization.

Sulframine Acid B. [Witco SA] Dodecylbenzene sulfonic acid.

Sulftech®. [General Chem.] Sodium sulfite.

Sulfur 6 Flowable. [Cuproquim] 52% Sulfur.

SulfuSorb. [Calgon Carbon] Activated carbon; for vapor phase applics.

Sullux. [Sullivan Chem. Coatings] Polyester coatings.

Sul-Perm®. [Ferro/Keil] Sulfur bases; lubricant, extreme pressure agent for gear oils, cutting oils, industrial gear lubricants, greases.

Sulphonated Lorol. [Ronsheim & Moore] Sulfates; base for hair and carpet shampoos.

Sulphonic Acid LS. [Hart Chem. Ltd.] Alkylbenzene sulfonic acid; base for detergents.

Sul-Po-Mag. [IMC Fertilizer] Potassium magnesium sulfate.

Sultafon. [Stockhausen] Wetting agents for textiles.

Sumifix. [Blackman Uhler] Textile dyes and pigments.

Sumine®. [Zeeland] Aromatic amines.

Sumiplast. [GCA Chem.] Solvent dye for textiles.

Summit Brand. [Mammoth Int'l.] Emulsified paraffin-based petroleum oils; for crop spraying with herbicides.

Sumquat®. [Zeeland] Aromatic quaternary ammonium salts.

Sunaptol. [ICI Surf. UK] Ethoxylated ethers, esters, or oils; detergent, emulsifier, dispersant, leveling agent for desizing, scouring, bleaching, dyeing.

Sunbond. [Sequa] Vinyl acetate and copolymer latexes; for paper coating applics.

Sunbrite. [Sun Chem. Enterprises] Azo red orange yellow; ink grade pigments.

Suncryl®. [Sequa] Vinyl acetate and copolymer latexes; for textiles and specialties.

Sun Espol G-318. [Taiyo Kagaku] Triisostearin.

Sunfast. [Sun Chem. Enterprises] High performance pigments.

Suniso. [Witco/Sonneborn] Refrigeration oil.

Sunkem. [Sequa] Polyethylene emulsions; lubricants for paper coatings.

Sunkote. [Sequa] Calcium stearate dispersions; lubricants for paper coatings.

Sunlife. [Nicca USA] Uv absorber, dye resist agent for textiles.

Sunmorl. [Nicca USA] Scouring and penetrating agent, leveling agent, dispersant for textiles.

Sunnol. [Lion] Sulfates; detergent, foamer, dyeing assistant for textiles, detergent and personal care prods., emulsion polymerization.

Sunny Safe. [Dai-ichi Kogyo Seiyaku] Sucrose ester; cleansing agent for foods.

Sunolite®. [Witco] Petroleum-derived wax; antisunchecking agent, antiozonant, lubricant, processing aid for rubber goods, PVC.

Sunolox®. [Atochem N. Am./Textiles] Peroxide bleaching system stabilizer.

Sunproof®. [Uniroyal] Protective wax.

Sunproofing Wax. [Frank B. Ross] Complex hydrocarbon mixture; antiozonant, anticracking, sunchecking wax for rubber.

Sunrez. [Sequa] Water-soluble starch insolubilizers for fine paper coating.

SunShade. [Santech] ZnO/synergist system; uv stabilizer for thermoplastics.

Sunsize. [Sequa] Specialty sizing agents for paper applics.

Sunsoflon. [Nikko Chem. Co. Ltd.; Nicca USA] Polyamide or imidazoline; surfactant, softener, lubricant for textiles.

Sunsoft 601. [Taiyo Kagaku] Diglyceryl stearate malate.

Sunsolt. [Nikko Chem. Co. Ltd.] Alkylaryl ester; dispersant, leveling agent for textiles.

Sunsperse. [Sun Chem. Corp.] Easily dispersible pigments.

Sunveil. [Ikeda] Titanium dioxide.

Sunyl®. [SVO Enterprises] Surfactant, emulsifier, dispersant, plasticizer, penetrant, lubricant for textile processing.

Suparamin. [Toho Chem. Industry] Cationic complex; wet strength resin for paper industry.

Supatwin. [Wavin Industrial Prods. Ltd.] PVC.

Supec®. [GE Plastics] PPS crystalline polymer; high performance resin for industrial, elec./electronic, and aircraft applics.

Super. [Mearl] Cosmetic pearl colors.

Super 45, 49. [Texasgulf] Super phosphoric acid.

Superac. [E.L. Puskas] Accelerators.

Superactiv. [E.L. Puskas] Activators.

Superadoplast. [Ceca SA] Softener for textile finishing.

Superaid. [E.L. Puskas] Dispersants, processing aids, internal lubricants.

Superaid. [Grefco] Diatomite; filter aid.

Super Alkyd®. [Thibaut & Walker] Alkyd resin; for enamels, primers, structural steel coatings, traffic paints.

Superanox. [E.L. Puskas] Antioxidants.

Superanoz. [E.L. Puskas] Antiozonants.

Super-Beckacite. [Reichhold] Substituted phenol-formaldehyde resin.

Super-Beckamine. [Reichhold; Reichhold Chemie AG] Melamine-formaldehyde resins.

Super-Beckosol. [Reichhold] Synthetic resin.

Superbond. [E.L. Puskas] Adhesives.

Superbond®. [Grain Processing] High-performance carrier for bonding high ring crush or recycled medium and liner.

Superbonder #495. [Loctite] Cyanoacrylate adhesive.

Supercadoplast. [Ceca SA] Fiber softeners.

Supercarrier. [Pierce Chem.] Immune modulator for enhanced antibody response.

Supercel. [E.L. Puskas] Blowing agents.

Superchlon. [British Traders & Shippers] Chlorinated rubbers, polyethylene, or polypropylene.

Super-Chlor. [Harcros] Bleach.

Superclear. [Vyse Gelatin] 300 Bloom gelatin.

Superclear®. [Henkel/Textile] Polysaccharide colloids; natural gum for textile dyeing and printing.

Supercoat®. [ECC Int'l.] Surface-treated, ultrafine ground calcium carbonate; for paper, plastics.

Super Cobalt. [Ultra Additives] Ink, paint and varnish driers.

Supercol®. [Aqualon] Guar gum.

Supercond. [E.L. Puskas] Conditioners.

Supercore®. [Grain Processing] Modified corn starch; for stucco slurries, wallboard.

Super Corona. [Croda Inc.] Refined anhydrous lanolin USP; superfatting emollient, emulsifier for cosmetics, pharmaceuticals.

Supercure. [E.L. Puskas] Curing and vulcanizing agents.

Superdrymix. [E.L. Puskas] Powdered form of liquids.

Superdye. [Reilly-Whiteman] Dyeing assistant for nylon.

Superedge. [Castrol Industrial East] Water-soluble cutting fluids and coolants.

Superelease. [E.L. Puskas] Release agents.

Super Ester. [Arakawa] Modified rosin ester.

Super Filmeen. [W.R. Grace/Dearborn] Corrosion inhibitor.

Super Fine. [Celite] Mineral filler.

Superfine Lanolin. [Fanning] Lanolin; superfatting emollient.

Superfine Lanolin Anhydrous USP. [Croda Inc.] Lanolin; superfatting emollient.

Super Fine Zink. [Am. MicroTrace] Zinc sulfate monohydrate + 35.5% Zinc; for foliar sprays.

Super Flake®. [General Chem.] Calcium chloride.

Superflex. [Amco] Two-component urethane polymer; flatproofing system for low-pressure tires.

Superflex. [R.T. Vanderbilt] Rubber antioxidant.

Superfloc. [Am. Cyanamid; Cyanamid BV] Polyacrylamide; flocculants.

Super Floss. [Celite] Diatomaceous earth; mineral filler.

Supergel. [E.L. Puskas] Gelled dispersions.

Supergrip. [Bostik Div./Emhart] Adhesives.

Super Hartolan. [Croda Inc.; Croda Chem. Ltd.] Lanolin alcohol; spreading agent, dispersant, stabilizer, plasticizer, emulsifier, and emollient for cosmetics and pharmaceuticals.

Super High Grade. [Eagle-Picher] Lead sulfide; friction additive in clutch facings, disc brake pads, railroad brake shoes; as nonreactive dense material for oil-drilling muds.

Super I Alumina. [Universal Scientific] More active grade alumina.

Superior. [Akzo Salt] Salt for use in chemical industries.

Superkleen C. [CNC Int'l.] Polymer dispersion; imparts hydrophilic and soil release properties to 100% polyester fabrics.

Super Kleenite. [West Agro] Dairy manual detergent.

Superla. [Vyse Gelatin] 175 Bloom gelatin.

Superla Mineral Oil. [Amoco Lubricants] Mineral oil.

Superlite. [R.T. Vanderbilt] Rubber antioxidant.

Superlitefast. [Crompton & Knowles] Textile dyes and pigments.

Superloid®. [Kelco] Refined ammonium alginate; gum used as gelling agent, thickener, emulsifier, film-forming agent, suspending agent, and stabilizer in food, pharmaceutical, and industrial applics.,paper, textiles.

Supermite®. [ECC Int'l.] Calcium carbonate; for paints, coatings.

Supermontaline SLT65. [Seppic] Dioctyl sodium sulfosuccinate; wetting agent, emulsifier.

Supernaltene. [Convert SA] HDPE.

Supernat. [Degussa] Precipitated silica.

Super Nevtac® 99. [Neville] Polyterpene resins; tackifying resin for polyolefins, rubber, solvent and hot melt-based adheseives.

Super-Nyl. [Flexello Castors (Sales) Ltd.] Polyurethane and nylon.

Supernylite. [Crompton & Knowles] Textile dyes and pigments.

Superohm. [A. Schulman] Crosslinked EPDM compds. in pellet form.

Superol. [Procter & Gamble] Glycerin USP; humectant.

Superox. [Reichhold] Peroxide derivs.; catalyst, polymerization initiator for polyester, vinyl ester resins, vinyl monomers.

Superpax. [TAM Ceramics] Zircon opacifier.

Superpep. [E.L. Puskas] Peptizers.

Super-Pflex®. [Pfizer] Surface-modified, precipitated calcium carbonate; reinforcing filler for PVC.

Superplast. [E.L. Puskas] Plasticizers.

Superpol. [E.L. Puskas] Polymers.

Superpolystate. [Gattefosse SA] Self-emulsifying polyglycol stearate; base for pharmaceutical and cosmetic lotions.

Super Prill. [Arcadian] Fertilizer grade urea; large prills.

Super Pro 5A. [Inolex] TEA coco-hydrolyzed collagen; detergent, conditioner, emulsifier, moisturizer, foamer for personal care prods.

Super Protek-Sorb. [W.R. Grace/Davison] Desiccant.

Super Rainbow. [IMC Fertilizer] Premium granulated fertilizers.

Super Ream. [H.B. Fuller] Chlorinated detergent.

Super Refined. [Croda Inc.; Croda Surf. Ltd.] Natural reinfed oils; emollients, lubricants, softeners for cosmetics.

Super-Sat. [RITA] Hydrogenated lano-

lin or ethoxylates; emollient, emulsifier, plasticizer for cosmetics, pharmaceuticals.

Super Sericite SS-88. [U.S. Cosmetics] Sericite mineral; provides creamy feel to cosmetic formulations.

Super-Set. [Acme Resin] Acid setting no-bake binders.

Super Slash. [Alex C. Fergusson] Liq. sanitizer cleaner.

Superslip. [E.L. Puskas] Slip agents.

Superslip. [Micro Powders] Slip agent.

Supersoft. [CNC Int'l.] Silicone emulsions; synthetic softeners for textiles.

Supersoft OES. [Reilly-Whiteman] Cationic softener for textiles.

Supersol ICS. [Sybron] Solvent scour for textiles.

Super Solan Flaked. [Croda Inc.] PEG-75 lanolin; emollient, conditioner, superfatting agent, solubilizer.

Supersolv. [E.L. Puskas] Solvents.

Super-Solv. [Uncle Sam Chem.] Emulsion degreaser.

Superspension. [E.L. Puskas] Water dispersions.

Superspersion. [E.L. Puskas] Elastomer dispersions.

Super Sta-Tac®. [Arizona] Specialty hydrocarbon resin; for adhesives, sealants; tackifier for polymers.

Super Sterol Ester. [Croda Inc.; Croda Chem. Ltd.] C10-30 cholestrol/lanosterol esters; emollient, lubricant, and moisturizer for dry skin, cosmetics, pharmaceuticals.

Superstif. [E.L. Puskas] Stiffening agents.

Superstik. [Evode Spec. Adhesives Ltd.] Cyanoacrylates.

Super-Strip. [Uncle Sam Chem.] Ammoniated stripper.

Super-Strip. [Urban Chem.] Insulation stripper.

Super Strip 100. [Reilly-Whiteman] Clearing agents for dyestuff clearing on textiles.

Super Sublimed White Lead 41. [Eagle-Picher] Basic sulfate white lead; white pigment, ingred. in the mfg. of bank notes and printing inks; in brake linings, clutch facings, and high-pressure lubricants.

Supersul. [Cuproquim] 80% sulfur.

Supersurf. [Am. Emulsions] Wetting agent, penetrant, leveler for textiles.

Supertac. [Crowley Tar Prods.] High visc. amorphous polypropylene.

Supertack. [E.L. Puskas] Tackifiers.

Supertak. [Bostik Div./Emhart] Aerosol adhesives.

Supertak. [Kano Labs] Gummed tape adhesifier.

Supertard. [E.L. Puskas] Retarders.

Super Tel ZN. [Am. MicroTrace] Zinc sulfate monohydrate + 35.5% zinc; for foliar sprays.

Supertex. [Vyse Gelatin] 225 Bloom gelatin.

Super-Thane. [Flexello Castors (Sales) Ltd.] Polyurethane.

Super Thane®. [Thibaut & Walker] Oil-modified urethanes; for varnish applications, gym floor finishes, industrial enamels, traffic markers.

Super Tin® 4L. [Griffin] Triphenyltin hydroxide; flowable fungicide for pecans, potatoes, sugar beets.

Supertone. [E.L. Puskas] Colors, dispersed colors.

Supervan. [David Michael] Vanillin replacements.

Super Vilex. [Atomergic Chemetals] Denatonium saccharide; bittering aversive agent.

SuperWash DZ. [DeeZee] Detergent blend; textile scouring agent and laundry detergent.

Super Westone. [West Chem. Prods.] Dust control treatment.

Superwet. [Degen] Resin modified alkyd.

Super Wet. [RITA] Ethoxylated fatty acid and ether; wetting agent and emulsifier for pesticides; soil penetrant.

Super Wet. [W.A. Cleary] Nonionic wetting agent for penetration to compacted soil and thatch, promoting soil drainage and pesticide or fertilizer efficiency.

Superwet BOE. [General Chem.] Etchants.

Superwetmix. [E.L. Puskas] Wetted powders.

Superwhip. [Vyse Gelatin] 250 Bloom gelatin.

Superwool. [Thermal Ceramics] Mineral wool insulation.

Supoweiss. [Unichema] Suppository bases for pharmaceuticals.

Suppocire. [Gattefosse] Hydrogenated palm glycerides, hydrogenated palm kernel glycerides; excipients for suppositories.

Supra EF. [Cyprus Industrial Min.] Talc; for cosmetics, creams and lotions, antiperspirants, dusting and pressed powds.

Supraene®. [Robeco] Purified squalene; natural emollient.

Suprafino. [Cyprus Industrial Min.] Talc; for cosmetics, antiperspirants, aerosols, soaps.

Supragil®. [Rhone-Poulenc Surf.; Rhone-Poulenc Geronazzo SpA] Sodium methyl naphthalene sulfonate; dispersant, suspending agent for pesticides.

Supramica. [Mykroy/Mycalex Ceramics] Glass-bonded mica.

Suprapal®. [BASF AG] Styrene copolymers; for roadmarking paints, gravure and flexographic inks, paper coatings, zinc dust primers.

Suprapur. [EM Industries] High purity chemicals.

Suprasec. [ICI Am.] Urethane curing agents.

Sup-R-Conc. [Hilton Davis] Pigments and colorants; for aq. systems, inks, coatings.

Sup-R-Cryl. [Hilton Davis] Acrylic flushed colorants; for industrial coatings.

Suprel. [Vista] SVA engineered thermoplastic.

Supreme. [Champlain Industries] Protein hydrolysates.

Supreme Green. [La Roche Industries] Mixed fertilizers.

Supreme USP. [Cyprus Industrial Min.] Talc.

Suprex. [J.M. Huber] Filler and reinforcing clay.

Suprmix DBEEA. [C.P. Hall] Dibutoxyethoxyethyl adipate.

Supronic. [Rhone-Poulenc Ltd.] Ethylene/propylene oxide adduct; foam control agent for food industry.

Supronyl. [Hoechst UK Films] Polyamide film.

Supro-Tein. [Maybrook] Hydrolyzed proteins/sorbitol blends; mild surfactant, emulsifier, solubilizer, moisturizer for cosmetics.

Suprovac. [Hoechst UK Films] Polyamide-polyethylene laminate films.

Surchlor. [Surpass] Sodium hypochlorite sol'n.; bleaching agent for full whites on cotton, rayon, and other cellulosic fibers.

Surcopur®. [Bayer] Propanil; herbicide for control of weeds in rice crops.

Surcotech. [Surco Prods.] Odor control systems.

Surebond. [Evode Spec. Adhesives Ltd.] Polyurethane.

Surecat. [Engelhard] Sulfur recovery catalysts.

Sure-Curd. [Pfizer/Brewery & Dairy Prods.] Microbial milk clotting enzyme.

Surefloc. [Schaefer Tech.] Wastewater treatment.

Surelease. [Colorcon] Ethylcellulose-based sustained release coating.

Sureseal. [Evode Spec. Adhesives Ltd.] Sealants.

Sure Sol. [Koch Chem./Muskegon] Aromatic solvents.

Suresperse®. [Drew Ind. Div.] Antifoulant, dispersant for industrial air washers; disperses oil, lint, biological matter, mud, textile fibers, dust, deposits, and debris.

Sure-Step. [Schaefer Salt & Chem.] Chip calcium chloride.

Surestik. [Evode Spec. Adhesives Ltd.] Solvent and water-based adhesives.

Suretex®. [Drew Ind. Div.] Static control agent for air washer treatments in textile mills.

Surett. [Exxon] Lubricating oil.

Surfac®. [Sherex Polymers] Corrosion inhibitor, ore flotation agent, surfactant for metal processing.

Surface Shield. [XymaX] Mold sealers and release agents.

Surfact-Amps. [Pierce Chem.] Highly purified detergents.

Surfactant AR 150. [Hercules] Ethoxylated rosin ester; emulsifier, low foaming detergent for food industry.

Surfactant WK. [DuPont] Spray additive.

Surfacto. [Vapor Blast Mfg.] Wetting agent.

Surfactol®. [CasChem] Surfactants; wetting agent, dispersant, wax plasticizer, mold release agent, antifoamer for textiles, leather, paints, household, cosmetics, dyeing, tanning, finishing, sizing, cutting and sol. oils.

Surfadone. [ISP] Pyrrolidone derivs.; nonionic surfactants.

Surfagene. [Chem-Y GmbH] Phosphates or sulfosuccinates; detergents, wetting agents, emulsifiers for cosmetics, detergent formulations, pesticides.

Surfam. [Sherex Polymers] Amines; corrosion inhibitor, emulsifier, lubricant, flotation agent, intermediate for textiles, gasoline and fuel additive, floor finishes, paints, metalworking, agric., water treatment.

Surfaron. [Synthron] Wetting agents.

Surfax. [Aquatec Quimica SA] Betaines, sulfates, sulfosuccinates, phosphate esters, sulfosuccinamates, or blends; detergent, dispersant, foamer, wetting agent for cosmetics, household and industrial cleaners, toothpastes, emulsion polymerization, rubber and plastics, agric., ore flotation.

Surfax. [E.F. Houghton] Wetting and rewetting agents.

Surfine. [Finetex] Carboxylates; detergent, wetting agent, dispersant, solubilizer, coupler, emulsifier for household and cosmetic formulations.

Surfix. [Surpass] Resin fixative.

Surflo®. [Exxon] Surfactants; dispersant, detergent, wetting agent, foamer for drilling operations.

Surfonic®. [Texaco] Ethoxylates; emulsifier, wetting agent, dry cleaning detergent, penetrant, solubilizer, lime soap dispersant, antifoamer used in agric., cosmetics, industrial cleaners, ceramics, concrete, film developing, emulsion polymerization, latexes.

Surfynol®. [Air Prods.] Diols or glycols; defoamer, dispersant, wetting agent, solubilizer for paints, inks, rinse aids, coatings, adhesives, dyestuffs, cements, metalworking fluids, latex dipping and paper coatings, agric. formulations.

Surgemaster. [Stewart Hall] Boiler water defoamer.

Surgex. [Stewart Hall] Boiler water defoamer.

Surgicide. [Medical Chem.] Povidone iodine sol'n.

Surlyn®. [DuPont; DuPont UK] Ionomer resins; extrudable resin for flexible pkg., in coextrusions, laminations, golf ball and bowling pin covers, automotive exterior body trim, footwear components, wire and cable insulation, ski boots, metal coating, glazing.

Surmax®. [Chemax] Alkaline stable surfactants; for formulating alkaline detergent concs.

Surpassol. [Climax Performance] Esters in a hydrocarbon base; paper mill defoamer; emulsifier for oil field drilling.

Surpawite. [Surpass] Peroxygen compd.; for bleaching, scouring, boil-off for textiles.

Sur-Wet®. [Pacific Anchor] Epoxy curing agents.

Suspend-Ayd. [Daniel Prods.] Suspending agent for paints and inks.

Suspendite SDT. [Shieldalloy Metallurgical] Fining abrasive additives.

Sustain. [PPG Industries] Calcium hypochlorite.

Sustamid. [Röchling Sustaplast KG] Nylon.

Sustane®. [UOP] Preservatives, antioxidants, stabilizers for food, flavors, cosmetics, vitamins, oils, waxes, essential oils, tallow, sausage, chewing gum base, shortening, lard, food pkg. materials, potatoes, and cereals.

Sustarin. [Röchling Sustaplast KG] Acetal.

Susteel. [Tosoh] PPS; engineering plastic for elec. and electronic parts (switch bases, relay components, connectors, coil bobbins), appliance components, automotive applics.

Sustilan® N. [Miles/Organic Prods.] Agent to prevent reduction of acid, direct and selected disperse dyes.

Sustonat. [Röchling Sustaplast KG] PC.

Sutro. [ICI Am.] Industrial polyols; humectants, plasticizers, sequesterants.

Suttocide® A. [Sutton Labs] Sodium hydroxymethylglycinate; antimicrobial preservative for cosmetics.

Swanic. [Swastik] Surfactants; foam booster/stabilizer, wetting agent, emulsifier, detergent for cosmetics, detergent systems, textiles, concrete.

Swanol. [Nikko Chem. Co. Ltd.] Detergent, wetting agent, dispersant.

Swascol. [Swastik] Sulfates; detergent, wetting agent, foamer, emulsifier for personal care prods., liq. detergents, emulsion polymerization.

Swastik Detergent Powder. [Swastik] Sodium alkylaryl sulfonate with alkaline builders, STPP; detergent powder.

Sway. [Swastik] Sodium alkylaryl sulfonate with STPP, alkaline builders, etc.; heavy-duty detergent for cottons.

Swedstab. [KMZ Chem. Ltd.] PVC stabilizers.

Sweetose. [A.E. Staley Mfg.] High conversion corn syrup.

Sweet-Pea. [Mateson] Acid spill treatment.

Sweetrex®. [Mendell] Dextrose, fructose, maltose, isomaltose blends; directly compressible chewable tablet base.

Sweetzyme®. [Novo Nordisk] Immobilized glucose isomerase; enzyme for conversion of glucose to fructose.

Swelltite. [Am. Colloid] Below grade waterproofing.

SWS. [Morton Int'l.] Industrial aq. reducing agent.

SWS. [Wacker Silicones] Silicone elastomers and compounds.

SX. [AEI Compds.] Polymers/catalyst masterbatches; thermosetting systems for film, inj. and blow molding, extrusion, calendering, pipe, flooring materials.

SXS 40. [Witco SA] Sodium xylenesulfonate.

Syenex. [Elkem A/S Nefelin] Nepheline syenite.

Sykanol. [Henkel KGaA] Sulfated castor oil; wetting agent for bath preps.

Sylfat. [Arizona] Tall oil fatty acids.

Sylgard®. [Dow Corning; Dow Corning France SA] Two-part silicone system; electrical/electronic insulating resin.

Sylodent. [W.R. Grace/Davison] Synthetic hydrated silica.

Syl-Off. [Dow Corning] Paper release coating.

Syloid®. [W.R. Grace/Davison] Silica gels; bonding agent for adhesives.

Sylomer. [Getzner Chemie GmbH] Elastic polyurethane material.

Sylon. [3M] Fluorosilicones.

Sylosiv. [W.R. Grace/Davison] Zeolite powders.

Sylox. [W.R. Grace/Davison] Micronsized silica.

Syltherm®. [Dow Corning] Heat transfer liquids.

Sylvadym. [Arizona] Dimer acids.

Sylvalite®. [Arizona] Rosin resins.

Sylvaros. [Arizona] Tall oil rosin.

Sylvatac®. [Arizona] Modified tall oil rosin; adhesive tackifiers.

Sylvatal. [Arizona] Distilled tall oils.

Symalit GMT. [Symalit; MBS Plastics] Glass mat reinforced thermoplastics; for automotive bumper systems, front end, under-hood, underbody panels.

Synaceti. [Werner G. Smith] Cetyl esters; synthetic spermaceti.

Synasol. [Union Carbide] Industrial solvents.

Synatone. [Nutex] Liq. cationic softener esp. for foam finishing.

Syncal. [PMC Specialties] Saccharin or salts; sweetening agents for foods, toothpastes, cosmetics, pharmaceuticals.

Syn-Chek. [Ferro/Keil] Chlorinated lubricity and extreme pressure additive; for metalworking lubricants

Syncrolube. [Croda Universal Ltd.; Croda Surf. Ltd.] Internal and external lubricants and processing aids.

Syncrowax. [Croda Inc.; Croda Surf. Ltd.] Synthetic waxes or esters; emulsifier, emollient, opacifier, lubricant, suspending agent, stabilizer, thickener.

Syndralubric. [Castrol Industrial East] Fire resistant hydraulic fluids.

Synesstic. [Exxon] Compressor lubricant/industrial oil.

Synex. [King Industries] Liquid ion exchange reagents for metal extraction.

Syn Fac®. [Milliken] Polyols; emulsifier, dispersant, intermediate, lubricant, reactive diluent for paints, coatings, textiles, insecticides

Syn Grind. [Castrol Industrial East] Grinding lubricants.

Synhappret BAP. [Miles/Organic Prods.] Polyurethane shrinkproof finish on wool.

Synkad®. [Ferro/Keil] Borate or carboxylate salts; corrosion inhibitor fur synthetic cutting, drawing, and grinding fluids.

Syn Kut. [Castrol Industrial East] Cutting oils.

Syn Lube. [Milliken] Ethoxylated glyceride ester; emulsifier, lubricant for textiles.

Syn Mould. [Castrol Industrial East] Ceramic mold release agents.

Syno. [Sidney Springer] Antimigrants, antibleeds, antislips, pigment binders, lubricants, cleaners, dye fixatives, penetrants, delustrants, softeners for textiles.

Syn-O-Ad. [Akzo] Phosphites; antioxidant and antiwear agent in gear and transmission oils; stabilizers and metal deactivators.

Synocron. [Sidney Springer] Textile dyes and pigments.

Synocure. [Cray Valley Prods.] Acrylic resins.

Synodirect. [Sidney Springer] Textile dyes and pigments.

Synolac. [Cray Valley Prods.] Alkyd resins.

Synolec. [Lubrication Engineers] Synthetic lubricants.

Synolite. [DSM Resins UK Ltd.] Polyester resin.

Synolube. [Nat'l. Starch & Chem.] Softeners.

Synomulsion. [Sidney Springer] Hand builder.

Synopel. [Sidney Springer] Water repellent.

Synopen. [Sidney Springer] Wetting agent, lubricant, dye leveler.

Synopon. [Sidney Springer] Detergents.

Synoquart. [Aquatec Quimica SA] Fatty acid alkanolamides; detergent, wetting agent, thickener, foam booster/stabilizer, superfatting agent for cosmetic and household prods.

Synoreact. [Sidney Springer] Textile dyes and pigments.

Synoreactant. [Sidney Springer] Fabric stabilizer.

Synotex. [DSM Resins UK Ltd.] Cyclized rubber and other rubber derivs.

Synotol. [Aquatec Quimica SA] Fatty acid alkanolamides or amine oxides; foam stabilizer, thickener, superfatting agent for cosmetic and household preps.

Synox 5LT. [Neville] 2,2′-Methylene bis (4-methyl-6-t-butylphenol); nonstaining antioxidant for natural, S/B, BR, CR, and polyisoprene rubbers.

Synperonic. [ICI Am.; ICI PLC] Ethoxylated ethers or EO/PO copolymers; detergent, foamer, emulsifier, wetting agent, solubilizer for personal care and household prods., textiles, agric., emulsion polymerization.

Synpol. [Ameripol Synpol] Styrene/butadiene rubber.

Synpro®. [Syn. Prods.] Esters; additives, mold release agents.

Synprol Alcohol. [ICI Am.] Tridecyl alcohol; intermediate.

Synprolam. [ICI PLC] Emulsifier, anticaking agent, sanitizer, biocide, corrosion inhibitor, antistat, conditioner, lubricant for personal care prods., cleaners, textiles, plastics, rubber.

Synpron®. [Syn. Prods.] Antimony mercaptide, dibutyltin dilaurate, or metallic stearates; PVC heat stabilizer, lubricant, process aid.

Synpro-Ware. [Syn. Prods.] Rubber chemical dispersions; processing aids.

Syntaryl. [Witco SA] Blend; multipurpose detergents, bases for liq. detergents.

Syntase®. [Rhone-Poulenc Surf.] Benzophenones; uv absorbers for plastics, cosmetics, resins, paints, varnishes, and lacquers.

Syntemp. [Lubrication Engineers] Synthetic lubricants.

Syntens. [Hefti Ltd.] Ethoxylated alcohols or EO/PO adducts; wetting agent, detergent base material, emulsifier for household and industrial cleaning, textiles, paper, leather.

Syntergent®. [Henkel/Organic Prods.; Henkel-Nopco] Fatty amido condensate; wetting agent, detergent for textiles, leather, paper, metalworking.

Syntesqual. [Vevy] Polyisoprene.

Syntex®. [Rhone-Poulenc/Perf. Resins & Coatings] Modified phthalic alkyd resin; used as a grinding liq. in the formulation of tinting colors.

Synthabond. [Piedmont Chem. Industries] Reserving and resist agents for nylon carpet.

Synthacryl. [Hoechst Celanese] Acrylic resins; for paints and coatings.

Synthalen. [3-V; Sigma Prodotti Chimici] Carbomers; thickener, suspending agent for cosmetics.

Synthalube. [Piedmont Chem. Industries] Anionic/polymer blend protecting knits and woven fabrics during preparation and dyeing.

Synthamica. [Mykroy/Mycalex Ceramics] Synthetic mica.

Synthapal®. [Boehme Filatex] High temp. dyeing assistant for polyester yarns and blends.

Syntharome. [Florasynth] Flavors.

Synthaset®. [Piedmont Chem. Industries] One-piece blends of resin, catalyst, softener, penetrant for fabric processing.

Synthasil. [Piedmont Chem. Industries] Silicone emulsions; softener for printing, hand modifier for textiles.

Synthawhite. [Piedmont Chem. Industries] Sodium percarbonate-based compd.; for bleaching of cotton and blends.

Synthe-Copal®. [Arizona] Thermoplastic hydrocarbon resin; used for web offset news inks and vehicles.

Synthemul®. [Reichhold/Emulsion Polymers] Acrylic, styrene-acrylic, or vinyl emulsion polymers; for industrial coatings.

Synthesize. [Abco Industries] Warp sizing agents for spun yarns.

Syntheso®. [Kluber Lubrication N. Am.] Calender oils for finishing and wet processing equip.

Synthionic. [Witco SA] Block polymer; wetting agents, low foaming detergents, rinse aids, defoamers.

Syntholube. [Lenox] Lubricant, antistat for polyester/rayon blends.

Synthopel. [Synthron] Water repellents.

Synthospin. [Lenox] Antistat spinning lubricant for acrylics, modacrylics, polyesters, nylon, and wool.

Synthrapol. [ICI Am.] Ethoxylates; wetting and rewetting agent, detergent, emulsifier, penetrant, dyeing assistant for textile processing.

Syntilo. [Castrol Industrial East] Synthetic metalworking fluids.

Syntofor. [Witco SA] Polyglycol fatty ester; emulsifier for textiles and cosmetics.

Synton®. [Uniroyal] Polyalphaolefins; synthetic lubricants.

Syntophos. [Witco SA] Phosphate ester; wetting agent, detergent for textiles.

Syntopon. [Witco SA] Ethoxylated alkyl phenol ethers; wetting agent, detergent, emulsifier for textiles, leather, metal cleaning, degreasing, latexes, perfume, insecticides.

Syntran. [Interpolymer] Acrylate copolymer blends.

Synwax. [Reilly-Whiteman] Synthetic wax; for leather finishing, polishes for furniture and floors, in wax molding, carbon paper coating.

Systanat. [Anilac NV/SA] MDI/TDI for polyurethanes.

System. [CasChem] Polyurethane systems.

Systhane. [Rohm & Haas] Fungicides.

Systol. [Anilac NV/SA] Polyether and polyester for polyurethanes.

T

T-9. [Harcros; Werner G. Smith] Ethoxylated tallate; low foam surfactant.

T-11, 18, 20, 22. [Procter & Gamble] Tallow or hydrogenated tallow acids; intermediates for mfg. of soaps, amides, esters, alcoholamides, surfactant and nonsurfactant applics.

T-600, 1000, 3000, 4000. [Olin] PPG triol; plasticizer, chemical intermediate for resins, brake fluids, rigid and flexible urethane foams, nonfoam urethane coatings, adhesives, elastomers, sealants and caulks;

T1000, 2000, 3000, 4000 Series. [Hüls Am.] Silane derivs.; blocking agents, silylation reagent.

T1750, 1920, 1928. [Hüls Am.] Alkoxysilanes; dielectric fluids, heat exchange applics., solar panels, airborne radar.

TA-1618. [Procter & Gamble] Cetearyl alcohol.

Tabs. [British Traders & Shippers] Biodegradable solvent.

Tactix. [Dow Plastics] Thermoset performance polymers for aerospace applics.

Tagat®. [Goldschmidt; Goldschmidt AG] Ethoxylated glyceryl esters; preparation of o/w emulsions; solubilizer for flavors, perfumes, vitamin oils; dispersant and antistat.

TAHP-80. [Witco/Argus] 80% t-Amyl hydroperoxide sol'n.

Taiacryl. [T&T Industries] Textile dyes and pigments.

Taicron. [T&T Industries] Textile dyes and pigments.

Taifix. [T&T Industries] Textile dyes and pigments.

Taipol. [Goldsmith & Eggleton] Sol'n. polybutadiene; for tires, retreads, footwear, belting, golf balls, mech. goods.

Tairus. [T&T Industries] Textile dyes and pigments.

Taitalac. [Regent Chem. Ltd.] ABS and SAN.

Tak. [Kano Labs] Gummed tape adhesifier.

Takanal. [Ikeda; Tri-K Industries] Quaternium 51.

Taka-Sweet®. [Solvay Enzymes] Glucose isomerase; enzyme for prod. of fructose syrups from glucose.

Taka-Therm®. [Solvay Enzymes] Alpha-amylase; emzyme for starch liquefaction and textile desizing.

Taktene. [Polysar] Sol'n. polybutadiene; for tires, footwear, belting, hose, floor tile, golf ball centers, molded and extruded goods, masterbatches; impact modifier.

Talc MS. [Presperse] Talc; for cosmetics.

Talcoseptic C. [Vevy] Talc, phenoxyethanol, methylparaben, ethylparaben, propylparaben, butylparaben.

Talcron. [Pfizer] Talc.

Tallates, K. [Murphy-Phoenix] Potassium tallate; detergent, emulsifier for petroleum and agric. prods.

Tallene. [Westvaco] Tall oil pitch.

Tallex. [Westvaco] Distilled abietic acid; tall oil pitch.

Tallopol. [Stockhausen] Antistat for conductive carpet backing.

Tallow Amine, Diamine, Tetramine, Triamine. [Exxon/Tomah] Tallow amine, diamine, tetramine, triamine; ore flotation agent, emulsifier, corrosion inhibitor

Tally® 100 plus. [Van Den Bergh Foods] Glyceryl stearate, PEG-20 glyceryl stearate, hydrog. soybean oil; emulsifier, dough strengthener and crumb softener for breads; textile lubricant and softener.

Talstar. [FMC/Ag] Insecticide, miticide.

T.A.M. [Exsymol] Thenoyl methionine;

organic sulfur source for scalp and hair treatments.

Tamanol. [Arakawa] Phenolic resin.

Tamanori. [Arakawa] Adhesive for paperboard and gummed tape.

Tamaron®. [Bayer] Methamidophos; insecticide and acaricide.

Tamasof. [Arakawa] Rubber softener.

Tamol®. [BASF AG] Naphthalene sulfonates or carboxylates; dispersants for pigments, dyestuffs, carbon black; grinding aid, stabilizer, solubilizer; tanning agent for leather.

Tamol®. [Rohm & Haas] Naphthalene sulfonates, carboxylates, or copolymers; dispersants for paints, dyes, pigments, cement.

Tamolan®. [BASF AG] Synthetic laking agent for alkaline flexographic inks.

Tamsil. [Unimin Specialty Minerals] Silica; filler for coatings, adhesives.

Tanabond. [Sybron] Durable, pressure-sensitive screen printing table adhesive.

Tanabron. [Sybron] Detergent, wetting agent for textile processing.

Tanaclean. [Sybron] Replacement for caustic/hydro in reduction cleaning, machine cleaning.

Tanafresh HFO. [Sybron] Odor masking agent.

Tanalev®. [Sybron] Direct dye leveler, retardant.

Tanalon®. [Sybron] Dye carrier.

Tanalube®. [Sybron] Lubricant for textile fibers.

Tanapal®. [Sybron] Phosphate ester; leveling agent for disperse dyes.

Tanapel 54. [Sybron] High m.w. thermoset; water repellents for textiles; fluorocarbon extender.

Tanapon®. [Sybron] Phosphate ester; detergent, wetting agent for alkaline cleaning compds., pigments, adhesives, textiles.

Tanapure®. [Sybron] Acid dye leveling agent.

Tanaquad. [Sybron] Quaternary ammonium compd.; wetting agent, bactericide, sanitizer, dye retardant, antistat.

Tanasoft®. [Sybron] Textile softener.

Tanassist®. [Sybron] Dye migrator for pressure equip.

Tanastat®. [Sybron] Textile antistat.

Tanaterge®. [Sybron] Phosphate deriv.; detergent for industrial cleaners; desizing agent for textiles.

Tanatex® Nostick. [Sybron] Specialty additive to prevent polymer buildup on dry cans or equip.

Tanavol®. [Sybron] Textile dye carriers.

Tanawet®. [Sybron] Wetting agent for textile finishes.

Tancobind. [Evode-Tanner Industries] Pigment binders.

Tancobond. [Evode-Tanner Industries] Blanket adhesive for textile printing.

Tancosene. [Evode-Tanner Industries] Chelating agent.

Tancosoft. [Evode-Tanner Industries] Softeners, antistats, napping assistant for textiles.

Tancotard. [Evode-Tanner Industries] Fire retardants for textiles.

Tancowet. [Evode-Tanner Industries] Wetting agent, penetrant for textiles.

Tandem. [Witco/Humko] Mono and diglyceride derivs.; food emulsifier, conditioner, softener.

Tannex®. [Sybron] Organic stabilizer/chelate for textile bleaching.

Tannochrome. [AJ & JO Pilar] Specialty chrome chemicals.

Tarene. [Harwick] Pine tar replacements.

Target. [Ashland] Electronic chemicals.

Tari. [Giulini Corp.] Special additives for processed meats.

Tarset. [Porter Int'l.] Coal tar epoxy coating.

Tauranol. [Finetex] Isethionates, taurates; detergent, foamer, dispersant, conditioner for personal care, pharmaceutical prods., textile processing.

TBAB. [Hexcel] Tetrabutyl ammonium bromide.

TBC. [Croda Surf. Ltd.] Tributyl citrate; plasticizer for vinyl and cellulose resins.

TBHP-70. [Witco/Argus] t-Butyl hydroperoxide sol'n.; initiator.

TBTO. [Atochem N. Am.] Bioactive chemical.

TC-100. [CHR Industries/Furon] Thermally conductive silicone elastomer preform.

TC-1005, 1010, 1010T. [Procter & Gamble] Tallow/coconut fatty acids; intermediates for mfg. of soaps, amides, esters, alcoholamides, surfactant and nonsurfactant applics.

T-Carb. [Harcros] Calcium carbonates.

TCC. [Monsanto] 3,4,4′-Trichlorocarbanilide; bacteriostat for bar soaps.

T-Chlor. [Thatcher] 12% Sodium hypochlorite sol'n.

T-Det®. [Harcros] Ethoxylated ethers; emulsifier, detergent, wetting agent, dispersant, solubilizer for textiles, detergent formulations, agric., household prods., leather, metal processing, paper, wax, polish, rubber, polymerization, water treatment, paint, petroleum processing.

Teban®. [Boehme Filatex] Leveling agent for acid dyes on nylon.

Tebol 99. [Arco] t-Butyl alcohol; solvent, cosolvent, compatibilizer, coupling agent, processing aid for pharmaceuticals, personal care prods., aq. coatings and adhesives, agric. formulations, polymer processing, cleaners/disinfectants.

Tebolan®. [Boehme Filatex] Anti-creasing agent and lubricant for synthetic fabric processing.

Tecbond. [Raffi & Swanson] Acrylic adhesives.

Tecfil. [Filtec Ltd.] Hollow ceramic microspheres.

Technical Hydrox. [Cuproquim] 59% Copper hydroxide.

Techni-Gold HS. [Technic] Gold.

Techni-Rhodium. [Technic] Rhodium.

Techni-Silver. [Technic] Silver.

Technistrip. [Technic] Gold strips.

Technyl. [Rhone-Poulenc Plastiques Tech.] Polyamide 6/6, 6, 6/10, copolymers for inj. molding.

Tech Pet. [Witco] Petrolatum tech.

Techster. [Rhone-Poulenc Plastiques Tech.] PBT and PET polyesters.

Tech Tin. [Technic] Immersion tin.

Techwet. [Dyetech] POE carboxylate; wetting agent, penetrant for continuous or rope bleaching of cotton.

Tecnoflon®. [Ausimont] Fluoroelastomer polymer; for inj., transfer, or compr. molding, extrusion, blending, o-rings, valve stem seals, gaskets, bellows, expansion joints, fuel hoses, custom molded goods, as processing aid.

Tecnoprene. [Montedipe Srl] Reinforced PP.

Tecnopro. [Mario Lombardini Srl] PP and EPDM.

Tecnovil. [Vulcaflex SpA] Rigid PVC for thermoforming.

Teco-Sil. [CE Minerals] Fused minerals and chemicals.

Tecperl. [Filtec Ltd.] Solid microspheres.

Tecpol. [Raffi & Swanson] Vinyl type polymeric emulsions; fabric finishes.

Tecpril. [Filtec Ltd.] Lightweight ceramic aggregate.

Tecquinol®. [Eastman] Hydroquinone; antioxidant for latexes, fats, oils, monomers, polyester resins.

Tecrothene. [Rotec Chem. Ltd.] Polymeric materials, predominantly polyethylene.

Tecsol®. [Eastman] Denatured alcohol; solvents for heat transfer printing inks.

Tectilon. [Ciba-Geigy/Dyestuffs] Textile dyes and pigments.

Teda. [Tosoh] Amine or metal-based catalysts; for polyurethanes.

Tedimon. [Montedipe Srl] Isocyanates.

Tedion V18®. [Solvay Duphar BV] Tetradifon; acaricide for control of spider mites.

Tedlar®. [DuPont] Polyvinyl fluoride film; for transportation industry (airline interiors, autobody trim), building prods. (wall coverings, insulation cover), decorative applics., release film.

Tedur®. [Bayer; Miles] PPS, glass and/ or mineral reinforced; for inj. molding, embedding, conductive, reflective, extrusion applics.

Teepol. [Shell] Sulfonates or sulfates; detergent, emulsifier, solubilizer for dishwashing, hard surface, germicidal, and general cleaners.

Teflon®. [DuPont; DuPont UK] PTFE, PFA or FEP resins; fluoropolymers for powder coatings, wire and cable insulation, general coating, impregnation, gaskets, extruded tubing, pipe, molded goods, chemical linings, unsintered tapes, labware, melt-processable, oil and water repellent applics.

Tefose®. [Gattefosse; Gattefosse SA] Ethoxylated esters or blends; self-emulsifying base for cosmetics, pharmaceuticals.

Tefsin®. [Advanced Elastomer Systems] Thermoplastic elastomers; for general purpose, food, and medical applics.

Tefzel®. [DuPont; DuPont UK] Ethylene/tetrafluoroethylene copolymer; melt processable high performance thermoplastic for extrusion, inj. molding, elec. applics., mechanical applics., wire and cable insulation, and in chemical service.

Tegacid®. [Goldschmidt] Glyceryl stearate blends; self-emulsifying base for creams and lotions.

Tegamine®. [Goldschmidt] Alkyl amidopropyl dimethylamine; conditioner for hair care and bath prods.

Tegamin® Oxide WS-35. [Goldschmidt] Cocamidopropylamine oxide; surfactant, foamer, visc. builder for hair and skin cleansing.

Tegiloxan®. [Goldschmidt; Goldschmidt AG] Methylsilicone oils; antifoams for rubber and plastics, lubricant for tire prod.; additive for polishes.

Tegin®. [Goldschmidt; Goldschmidt AG] Esters; emulsifier, emollient, plasticizer, stabilizer, antistat, lubricant, solubilizer for cosmetics, pharmaceuticals, foodstuffs, dyestuffs.

Teginacid®. [Goldschmidt; Goldschmidt AG] Glyceryl esters or blends; emulsifier for cosmetic creams and lotions, acid and salt resistant emulsions.

TegMeR®. [C.P. Hall] Ethoxylated esters; lubricant, plasticizer for aluminum can, rubber, textile industries.

Tego® Airex. [Tego] Silicone polymers; deaerator.

Tego® Amid. [Goldschmidt] Alkyl amidopropyl dimethylamine; emulsifier, conditioner for cosmetics.

Tego® Amid S 18. [Goldschmidt AG] Stearamidopropyl dimethylamine; cationic emulsifier for creams and lotions; conditioner for hair care prods.

Tegoamin®. [Goldschmidt] Amines; catalyst, activator for mfg. of polyurethane foams and elastomers.

Tego® Antiflamm® N. [Goldschmidt] Flame retardant for polyurethane foams.

Tego® Antifoam [Goldschmidt] Antifoam for waste water treatment.

Tego® Betaine. [Goldschmidt; Goldschmidt AG] Betaines; amphoteric surfactants, foam stabilizer, visc. builder for cosmetics, baby prods., dishwashes.

Tego® Care. [Goldschmidt; Goldschmidt AG] Nonionic blends; emulsifier for cosmetic creams.

Tegochrome® 22. [Goldschmidt] 2,2-Ethylene dithiodiethanol; organic intermediate for producing color developers in photography industry.

Tegocoll®. [Goldschmidt] One and two-component polyurethane adhesives.

Tegocolor®. [Goldschmidt] Pigment dispersions in a polyether polyol; for coloring polyether polyurethane foam.

Tego® Dispers. [Tego] Polycarboxylic acid derivs.; wetting agent, dispersant for pigments; prevents flooding; antisagging agent.

Tegodont®. [Goldschmidt] Chlorinating agent for water treatment.

Tego® Effect. [Tego] Polyurethane-based sol'n.; thickener for aq. polymer dispersions.

Tego® EL-HA-CE. [Goldschmidt] Chlorinated agent for water treatment.

Tego® Emulsion. [Goldschmidt AG] Silicone and nonsilicone emulsions; aq. release agents.

Tego® Flow. [Tego] Acrylic or siloxane polymers; flow and leveling agents for aq. and solvent systems.

Tego® Foamex. [Tego] Siloxane copolymers; defoamer for water-based emulsion paints, aq. systems, printing inks.

Tego® Glide. [Tego] Siloxane copolymers; mar resistant and flow additive for water and solvent-based paints, printing inks; deaerator for epoxy resins.

Tego® Hammer 300000. [Tego] Methyl silicone oil; hammer finish additive for solvent-based paints.

Tego® IMR®. [Goldschmidt] Internal mold release agent for demolding polyurethane RIM formulations.

Tegomuls®. [Goldschmidt; Goldschmidt AG] Glyceryl esters; emulsifier, defoamer for food industry.

Tego®-Pearl. [Goldschmidt; Goldschmidt AG] Surfactant blends; pearlescent and opacifiers for hair care prods.

Tego® Phobe. [Tego] Fluorinated or silicone resins; water-repellent additive for paints, printing inks, leather.

Tegopren®. [Goldschmidt] Organo modified siloxanes; surfactants used as antistats, wetting and leveling agents, emulsifiers, dispersants, and for the improvement of lubricity; additive for polishes.

Tegosil®. [Goldschmidt] Silicone-based; aerosol for release of inj. molded plastic and rubber moldings.

Tego® Silicone. [Goldschmidt; Goldschmidt AG] Silicone acrylate; for release coatings.

Tegosipon®. [Goldschmidt] Silicone; antifoam for mfg. of stomach and intestinal preps.

Tegosivin®. [Goldschmidt; Goldschmidt AG] Modified siloxanes; protectant for concrete.

Tegosoft. [Goldschmidt; Goldschmidt AG] Isononanoates; emollient, superfatting agent, solvent for cosmetics.

Tegostab®. [Goldschmidt] Silicone surfactants; stabilizer for polyurethane foams.

Tego® Stannous Oxalate. [Goldschmidt] Tin (II) oxalate; catalyst for stannous esterification.

Tegotain. [Goldschmidt AG] Amine oxides or betaines; foam booster, antistat, thickener, mild amphoteric surfactant.

Tegotens. [Goldschmidt AG] Solubilizers, emulsifers, wetting agents, antistat, surface coating for expanded polystyrene beads containing a propellant.

Tegotrenn®. [Goldschmidt] Release agents for polyurethane foams, RIM systems, etc.

Tego® Wet. [Tego] Siloxane surfactants; wetting agent for solvent and aq.

systems.

Teinowax. [Lanaetex Prods.] Cetearyl alcohol, polysorbate 60, PEG-150 stearate, steareth-20.

Tekalen. [Terbrack Kunststoff GmbH] UHMWPE, HDPE.

Tekstim 8504. [Exxon/Tomah] Corrosion inhibitor for acid cleaning applics.

Tektamer®. [Calgon] Methyldibromo glutaronitrile or blends; antimicrobial preservative for cosmetics.

Tek Tan. [Van Waters & Rogers] Sodium naphthalene sulfonate; dispersant, tanning agent, bleaching agent for leather.

Tek-Wet. [Van Waters & Rogers] Ethoxylates; wetting and emulsifying agent, dispersant for leather processing.

Telar. [DuPont/Ag] Herbicide.

Telcar. [Teknor Apex] Thermoplastic olefins.

Telcon. [Alvin Prods.] Dry lubricant, antistick agent.

Telene. [BFGoodrich/Spec. Polymers] Engineered resin systems; RIM and thermoplastics.

Telkanol. [Dexter] Leveling agents, retarders, migrating aid, solubilizer for dyes.

Telloy. [R.T. Vanderbilt] Tellurium; vulcanizing agent.

Tellurac. [R.T. Vanderbilt] Rubber accelerator.

Telon. [Miles/Organic Prods.] Textile dyes and pigments.

Telprene. [Teknor Apex] Dynamic vulcanizates.

Telura. [Exxon] Process oil.

Tel ZN. [Am. MicroTrace] Zinc sulfate monohydrate + 35.5% zinc; for prod. of quality zinc ammonium complex.

Temac. [Nichimen Italia SpA] Polyacetate resin.

Temasept. [Hexcel] Salicylanilide; antimicrobial for resins, latexes, plastics.

Tembind. [Temfibre] Ammonium lignosulfonate; asphalt emulsifier and stabilizer.

Temik. [Rhone-Poulenc/Ag] Insecticide, nematicide.

Tempo. [Miles/Ag] Insecticide for lawn

and ornamentals.

Temprite. [BFGoodrich; BFGoodrich UK] Chlorinated PVC and low combustility plastics.

Tempstran. [Celite] Glass microfiber.

Temsperse. [Temfibre] Lignosulfonates; dyestuff dispersant, water reducer, slurry thinner.

Tenac. [Pan Polymers Ltd.] Acetal.

Tenase®. [Solvay Enzymes] Alpha-amylase; enzyme for starch liquefaction.

Tenax. [Westvaco] Tall oil fatty acid, maleated; intermediate.

Ten-Cem. [Mooney Chems] Neodecanoate driers for paints, inks.

Tenderfil. [A.E. Staley Mfg.] Modified tapioca starch.

Tender-Jel. [A.E. Staley Mfg.] Modified corn starch.

Tenite®. [Eastman; Eastman Chem. Int'l. AG] Cellulosic, polyester, or polyolefin plastics.

Tenlo®. [Henkel/Coating Chem.] Grinding aid, dispersant for nonaq. coating systems.

Tenn-Cop 5E. [Boliden Intertrade] Copper salts of fatty and rosin acids; agric. fungicide.

Tenn-White. [Manufacturers Chems.] Optical brighteners for textiles.

Tenox®. [Eastman] Antioxidants, stabilizers for foods, plastics, rubber.

Tensagex DLM. [ICI Am.; ICI PLC] Sodium trideceth sulfate; for liq. detergent blends, shampoos, bubble baths.

Tensami. [Alban Muller] Xanthan gum blends; natural emulsifiers.

Tensianol. [ICI PLC] Surfactant blends; raw material for mfg. of alkali-free soap bars.

Tensiofix. [OmniChem NV] Surfactant blends; wetting agent, dispersant, emulsifier for pesticides, agrochemical oil formulations.

Tensopol. [ICI PLC] Sulfates; surfactants for detergents, toiletries, emulsion polymerization, pigment dispersion, latex foam, pharmaceuticals, toothpaste.

Tensuccin. [ICI PLC] Sulfosuccinate; mild surfactant for shampoos, bubble baths, shower gels, dermatological preparations.

Tenterlube. [Ivax Industries] Lubricant for tenter frame chains.

Tephal Grunau. [Chem-Y BV] Protein fatty acid condensate; washing, milling, and wetting agent for textiles.

Tequat. [Auschem SpA] Quaternaries; germicide, softener, antistat, disinfectant, algicide.

Teracol. [DuPont] Polyether glycol.

Teramine. [West Chem. Prods.] Disinfectant.

Teraprint. [Ciba-Geigy/Dyestuffs] Textile dyes and pigments.

Terasil. [Ciba-Geigy/Dyestuffs] Textile dyes and pigments.

Terathane®. [DuPont] Polytetramethylene ether glycol; soft segment in polyurethane resins.

Terblend®. [BASF AG] ASA/PC blend; thermoplastic polymer for inj. molding of car instrument covers, tail light assemblies, etc., housings for small appliances, transformer housings, switchgear for house wiring, meter housings.

Teresstic. [Exxon] Lubricating oil.

Terg-A-Zyme®. [Alconox] Alkylaryl sulfonate, lauryl alcohol sulfate, phosphate, carbonate, and protease enzyme; detergent, wetting agent, sequestrant for hospitals, laboratories, dairies.

Tergelan® 1790. [Henkel/Emery] Mixed isopropanolamines myristate, mixed isopropanolamines lanolate; detergent, foam builder for personal care prods.

Tergenol. [Hart Chem. Ltd.]. Sodium oleyl methyl taurates or blends; textile scouring agent, detergent, dye leveling agent, emulsifier, dispersant for laundry powders.

Tergitol®. [Union Carbide] Ethoxylated ethers; detergent, emulsifier, wetting agent, defoamer, spreading agent, dyeing aid, leveling agent for textiles, household and industrial cleaners, paper/pulp, agric. formulations.

Tergon. [C.H. Patrick] Penetrant, emulsifier, detergent for textile processing.

Tergum. [Rowa GmbH] Sticky resins.

Teric. [ICI Australia] Ethoxylated ethers, esters, amines; emulsifier, defoamer, wetting agent, dispersant, stabilizer, antistat, lubricant, detergent for household and industrial cleaners, textile, paper, leather, metal cleaning, agric., coatings, ceramic, polishes.

Terlon. [Lawter Int'l.] Ink, varnish, and pigment wetting agent.

Terluran®. [BASF AG] ABS polymers; for inj. molding, extrusion, electroplating.

Terlux®. [BASF/Engineering Plastics] Clear ABS.

Termamyl®. [Novo Nordisk] Alpha-amylase; heat-stable enzyme for starch liquefaction, for laundry and dishwash detergents, alcohol, brewing, and textile industries.

Termanto. [Polimex SpA] Expanded polymers.

Termodur. [Colores Hispania SA] Heat-stable lead chrome pigment.

Ternil B. [Montedipe Srl] Polyamide 6.

Terpal®. [BASF AG] Chlormequat chloride or mepiquat chloride blends; bioregulator.

Terpanol. [Crowley Chem.] Dipentene substitute.

Terphane®. [Rhone-Poulenc/Film Div.; Rhone-Poulenc UK] Polyester film; for food pkg., magnetic tape, elec. cable insulation, graphic arts.

Terpinoxo. [Crowley Tar Prods.] Oxygen-former.

Terra Alba. [U.S. Gypsum] Fine gypsum filler.

Terraclor. [Uniroyal] Fungicide.

Terracur® P. [Bayer] Fensulfothion; nematocide and insecticide.

Terra-Dry. [Terry Labs] Freeze-dried aloe vera gel (powder).

Terra-Green Soil Conditioner. [Oil Dri Corp. of Am.] Soil conditioner.

Terra-Green Top Dressing. [Oil Dri Corp. of Am.] Soil conditioner.

Terran. [Terra Int'l.] Ammonium nitrate prill.

Terraneb SP. [Kincaid Enterprises] Turf fungicide.

Terra Nitrogen Sol'ns. [Terra Int'l.] Nonpressure nitrogen sol'ns.

Terrazole. [Uniroyal] Fungicide.

Terrea. [Terra Int'l.] Feed urea.

Terr-O-Cide. [Great Lakes] Multipurpose soil biocide.

Terr-O-Gas. [Great Lakes] Soil fumigant.

Terr-O-Gel. [Great Lakes] Soil fumigant.

Tersan. [DuPont/Ag] Turf fungicides.

Tertac. [Rowa GmbH] Sticky resins.

Tesal. [Gattefosse; Gattefosse SA] Propylene glycol stearate SE; self-emulsifying base for cosmetic and pharmaceutical ointments, creams, and lotions.

Tescol. [Allied Colloids] Sizing agents.

Tessilite. [Ditta Salt SpA] Phenolic powder.

Tetracarrier. [Reilly-Whiteman] Carriers, swelling agents for disperse dyeing of polyester.

Tetradecene-1. [Ethyl] C14 alpha olefins; intermediate for surfactants and industrial chemicals.

Tetradur. [Tetradur Kunststoff-Produktion GmbH] Unsaturated polyester molding compds.; for BMC.

Tetraglyme. [Ferro/Grant] Tetraethylene glycol dimethyl ether; solvent for electrochemistry, polymer and boron chemistry; processes such as gas absorp., extraction, stabilization; used in industrial prods. such as fuels, lubricants, textiles, pharmaceuticals, pesticides.

Tetralene LL. [Reilly-Whiteman] Dyeing assistant for cotton.

Tetralev. [Reilly-Whiteman] Dyeing assistant for polyester.

Tetralin®. [DuPont] 1,2,3,4-Tetrahydronaphthalene; solvent for oils, resins, waxes, rubber, asphalt; kier boiling assistant.

Tetralizer CA. [Reilly-Whiteman] Dyeing assistant for acrylics.

Tetralon. [Allied Colloids] Sequestrants.

Tetraloy®. [LNP] Reinforced fluoropolymers; for chemical and wear-resistant applics., seals, rings, bearings.

Tetralube. [Reilly-Whiteman] Dyebath lubricants for textiles.

Tetranol. [Sandoz] Sodium butyl oleate; wetting and leveling agent for dyeing processes; emulsifier, detergent for scouring.

Tetranyl. [Kao Corp. SA] Alkyl dimethyl benzyl ammonium chlorides; disinfectant, sanitizer for pharmacueticals, industrial and cooling water treatments.

Tetraquest. [Reilly-Whiteman] EDTA and NTA-based; sequestrants for water conditioning where dyeing problems could occur.

Tetra San. [Alex C. Fergusson] Liq. germicide algicide.

Tetrasil MAL. [Reilly-Whiteman] Silicone softener for textiles.

Tetrasperse. [Reilly-Whiteman] Dyeing assistant for polyester.

Tetrassist. [Reilly-Whiteman] Dyeing assistant for polyester.

Tetrastat JS. [Reilly-Whiteman] Antistat for textiles.

Tetraterge. [Reilly-Whiteman] Detergents for textile scouring.

Tetrathal. [Monsanto] Tetrachlorophthalic anhydride; flame retardants.

Tetrawet DWN. [Reilly-Whiteman] Wetting agents for cellulosics and blends.

Tetrone. [DuPont] Rubber accelerator.

Tetronic®. [BASF] EO/PO block copolymers; emulsifier, thickener, wetting agent, dispersant, solubilizer, stabilizer, antistat for cosmetics, pharmaceuticals, petroleum, detergents, molding powders, metal treatment, emulsion polymerization, paints, cutting fluids, rubber vulcanization.

Tettolight. [Filmolux] Expanded polystyrene.

Tettopor. [Filmolux] Expanded PVC.

Tettoren. [Filmolux] Polystyrene.

Tewax. [Auschem SpA] Ethoxylated esters, ethers or blends; lipophilic component for cosmetics, hair dye, pharmaceutical preps.

Texacar®. [Texaco; Texaco UK] Ethylene or propylene carbonate; solvents for polymers, selective or extractive solvent applics.

Texacat®. [Texaco; Texaco GmbH] Amines; catalysts for polyurethane foam mfg.

Texadd. [Texaco GmbH] Additives for polyurethane industry.

Texadril. [Henkel/Emery/Cospha] EO/PO block copolymer; low foaming wetting agent, coemulsifier.

Texal-L. [Lanaetex Prods.] Isopropyl myristate, myristyl alcohol.

Texalon. [Texapol] Nylon 6, 6/6, 6/10, or 6/12 resins, some glass filled; for extrusion, inj. molding, automotive applics., mechanical parts and housings, fasteners, tubes, wire and cable jacketing, kitchenware.

Texamid®. [Henkel KGaA] Sodium alginate; thickener for cosmetics, pharmaceuticals (toothpastes, gels, tableting auxiliary).

Texamin. [Henkel/Emery/Cospha; Henkel KGaA] Emulsifier for metalworking fluids; corrosion inhibitor.

Texamine. [Zohar Detergent Factory] Alkanolamide and ethanolamine alkylbenzene sulfonate; raw material for detergent mfg.

Texanol® Ester Alcohol. [Eastman] 2,2,4-Trimethyl-1,3-pentanediol mono-isobutyrate; solvent, coalescing agent, defoamer in coatings, inks, latexes, drilling muds; chemical intermediate.

Texaphor. [Henkel Canada; Henkel KGaA] Suspending agent for paints.

Texapol. [Texapol] Acetal or PBT resins.

Texapon®. [Henkel/Emery/Cospha; Henkel/Functional Prods.; Henkel KGaA; Pulcra SA] Sulfates or sulfosuccinates; surfactant, wetting agent, foamer, visc. builder, solubilizer, detergent for cosmetics, pharmaceuticals, fire fighting foams, carpet shampoos, dishwashing.

Texapret®. [BASF AG] Fillers and stiffeners for textile finishing.

Texaquart. [Henkel KGaA] Additives to increase the conductivity of electrostatic or powder spray coatings.

Texatein. [Lanaetex Prods.] Coco hydrolyzed animal protein salts.

Texchem II. [Franklin Industrial Minerals] NSF approved filler for plastics.

Texflo 40. [Nat'l. Starch & Chem.] Stabilized corn starch; warp sizing.

Texicote®. [Scott Bader] PVAc, VA/acrylic, polyester, or styrene emulsions; binder, bonding agent for adhesives, paints, textile processing, wallpaper.

Texicryl®. [Scott Bader] Acrylic or styrene-acrylic emulsions; binder for surface coatings, adhesives, polishes, sealants, fabric backing, paper impregnant, paper clay coatings, grouts.

Texigel®. [Scott Bader] Aq. polyacrylate gels; thickening agents for natural and synthetic latexes, e.g., in carpet-backing.

Texin. [Henkel Canada; Henkel KGaA] Sodium sulfosuccinates; wetting agent, detergent, emulsifier for household and industrial cleaners, hand soaps, glass, metal cleaners, dust repellents, firefighting foams.

Texin. [Miles] Thermoplastic polyurethane; for inj. molding, extrusion (hose, tubing, profiles, wire and cable, film and sheet), blow molding.

Texipol. [Scott Bader] Acrylamide copolymer emulsion; thickener for adhesives, pigment printing, carpet backing compositions.

Texi TD. [CNC Int'l.] Replacement for sodium hydrosulfite for reduction clearing, stripping dyes and equip. cleaning.

Texlin®. [Texaco; Texaco GmbH] Polyethylenepolyamines; for agric., asphalt, chelating agents, ion exchange resins, paper, textiles, epoxy curing agents, petroleum prod. additives.

Tex Lube. [Leatex] Low foaming lubricant for bleaching and dyeing of cotton and blends.

Texnol. [Nippon Nyukazai] Quaternary; antistat, emulsifier, fungicide, softener.

Texo. [Texo] Sulfonates or ethoxylates; cleaner, wetting agent, mold release agent.

Tex₂O. [Texo] Water treatment prods.

Texofor. [Rhone-Poulenc Ltd.] Ethoxylated alcohols; wetting agent, emulsifier for cosmetics, toiletries.

Texogum. [Mukti-Kem] Finishing agent, adhesive for textile applics.

Texox®. [Texaco; Texaco UK] PEGs, PPGs or EO/PO derivs.; intermediates, lubricants, plasticizers, solvents, coupling agents, frothing agents, heat transfer fluids, defoaming agents; used in solder reflow applics., boiler defoaming, ore flotation, inks, dyes.

Tex-Phlo. [FMC] Liq. alkaline sol'n.

Texport®. [Nicca USA] Defoaming/

penetrating agent for dyeing and bleaching.

Tex-Sil®. [Chem. Prods.] Sodium silicate sol'ns.; bleaching assistants, silk weighing, boiling off agent.

Texsolve®. [Texaco] Solvents for paint, coatings, rubber compding.

Textamin AT 1, 2, 3. [Henkel/Emery] Amino heptyl triazole.

Textamine. [Henkel/Emery] Amines or imidazolines; corrosion inhibitor, foamer, thickener, emulsifier, dispersant, ore flotation agent, intermediate, lubricant for inks, asphalt, agric., metalworking fluids, fuel additives.

Textamine Carbon Detegent K. [Henkel/Emery] Fatty nitrogen compd.; removes carbon deposits from aircraft, diesels, metal surfaces.

Textamine Oxide. [Henkel] Amine oxides; wetting agent, foamer, conditioner, thickener for brines, household and industrial cleaners.

Textamine Polymer. [Henkel] Amines; corrosion inhibitor, wetting agent, emulsion breaker.

Tex-Tel®. [Atochem N. Am./Textiles] Fluorochemical oil and water repellent.

Texthane. [Morton Int'l.] Polyurethane sol'ns.; textile coating and adhesive prods.

Textile Resin 2309 Conc., NF-U. [BASF] Crosslinking agents for resin finishing of textiles.

Textile Spirits. [Unocal] Solvent.

Textile Wax W. [BASF] Wax-like substance for addition to sizing and finishing liquors for textile processing.

Textol. [Zohar Detergent Factory] Ethanolamine alkylbenzene sulfonate; raw material for detergent mfg.

Textone®. [Olin] Sodium chlorite or blends; source of chlorine dioxide for oxidizing applics. incl. bleaching of natural foliage, upgrading of fats and oils, stripping dyestuffs from textiles, pulp bleaching, copper etching.

Textreat. [Texaco] Gas-treating chemical.

Textura. [Vyse Gelatin] Gelatin.

Textured Procon. [Central Soya] Textured soy protein conc.

Tex-Wet. [Intex] Wetting agents, disperant, emulsifier, foam stabilizer, solubilizer for textiles, dyeing.

Texzyme®. [PMP Fermentation Prods.] Desizing agent.

Thancat®. [Condea Chemie GmbH] Amine catalysts; for polyurethane prod.

Thanecure®. [TSE] Zinc chloride/ benzothiazyl disulfide; vulcanization activator for sulfur-curable millable urethane elastomers.

Thanol®. [Arco] Polyols; for prod. of urethane foams, carpet cushioning, rigid foams for building insulation, nonfoam urethanes for sealants, adhesives, elastomers, industrial and consumer coatings.

The Enhancer. [Dow Plastics] Carpet backing.

The Natural Resource. [Warner-Jenkinson] Natural food colors.

Theophyllisilane C. [Exsymol] Methylsilanol theophyllinacetate alginate; for slimming and anti-aging formulations.

Therban. [Bayer; Miles] Saturated hydrocarbon ACN copolymer; specialty elastomer for power transmission belting, hoses, membranes, seals, gaskets, bellows, linings, extruded profiles for automotive, industrial, and oil field applics.

Thermaclean. [Cook Composites & Polymers] Cleaning conc.

Thermact®. [Olin] Amine; catalyst.

Thermacure®. [Cook Composites & Polymers] Dicumyl peroxide; vulcanizing agent or polymerizing catalyst in rubber or plastics.

Thermaflo. [Evode Plastics Ltd.] PVC compds.

Thermagel. [Polymer Research Corp. of Am.] Dye migration retardant.

Thermagloss. [Michelman] Coatings.

Thermally Conductive Adhesive 991. [Dymax] Filled thermally conductive adhesive for mounting heat-sensitive electronic components.

Thermalux®. [Westlake Plastics] Polysulfone film.

Thermasil. [Transene] Thermally conductive silicone adhesive.

Thermax®. [Cancarb; R.T. Vanderbilt] Thermal carbon black; reinforcer for rubber industry.

Therm-Chek. [Ferro/Bedford] Light and heat stabilizers for PVC, plastisols, organosols, rigid formulations, calendered, molded, and extruded plasticized compositions.

Thermica. [Mykroy/Mycalex Ceramics] Synthetic mica.

Thermid®. [Nat'l. Starch & Chem.] Polyimides; thermosetting resin for laminating, composite prepregs, molding, coatings, adhesives (structural and electronic).

Therminol®. [Monsanto] Heat transfer fluids.

Thermocal B. [ICI Australia] Corrosion inhibiting heat transfer fluid for industrial cooling and refrigeration systems and moderate high temp. heat exchange systems.

Thermocoat. [GE] Lubricants.

Thermocomp®. [LNP; LNP Nederland] ABS, SAN, PS, PC, polyetherimide, HDPE, polysulfone, nylon 6, 6/6, 6/10, 6/12, 11, 12, polyurethane, PBT, PES, acetal, PEEK, PP, PPS, or PPO resins, some glass, carbon or lubricant filled; reinforced thermoplastics.

Thermogrip. [Bostik Div./Emhart] Adhesives.

Thermoguard®. [Atochem/Plastics Additives] Antimony oxide or brominated compds.; flame retardants.

Thermolin®. [Olin] Flame retardants for thermoplastics and thermosets for construction, furniture, transportation industries.

Thermolite. [Atochem N. Am.] Organotins; PVC stabilizers.

Therm-O-Lok. [Imperial Adhesives] Hot-melt adhesive.

Thermoplast®. [BASF; BASF AG] Dyes for mass dyeing of thermoplastic and thermosetting plastics, cellulose derivs.

Thermoscreen. [Imperial Adhesives] Silk screening adhesive.

Thermo-Seal. [H.B. Fuller] Hot butyl sealant.

Thermoset. [Thermoset Plastics] Epoxy, silicone, or polyurethane compds.; for adhesive, casting, tooling, laminating, conformal coatings, potting, moldmaking, elec./electronic insulation applics.

Thermo-Toe. [Imperial Adhesives] Hot melt adhesive; for shoes.

Thermozyme 70. [Novo Nordisk] Thermostable starch desizing enzyme.

The Solution. [Streett Industries] Windshield washer solvent containing methanol.

THFA. [QO Chem.] Tetrahydrofurfuryl alcohol.

Thiate®. [R.T. Vanderbilt] Thioureas; accelerator for rubber.

Thickener #5004. [Polymer Research Corp. of Am.] Carboxyl-containing emulsion; thickener.

Thimet®. [Am. Cyanamid/Ag] Phorate; soil and systemic insecticide for veg., field, and forage crops.

Thin-N-thik. [A.E. Staley Mfg.] Modified corn starch.

Thiocaulk. [Steelcote Mfg.] Polysulfide sealing compd.

Thiodan. [FMC/Ag] Insecticide.

Thiofide®. [Monsanto] 2,2´-Dithiobis (benzothiazole); accelerator for sulfur-curable elastomers.

Thiokol®. [Morton Int'l.] Polysulfides; for caulk/sealant, building, aviation, automotive, leather goods, adhesives, potting and encapsulating, molds, gaskets, barrier coatings, insulating glass, marine industries.

Thioset®. [Evans Chemetics/W.R. Grace] Ethanolamine sulfite.

Thiostat® B. [Uniroyal] Sodium dimethyl dithiocarbamate; bactericide, fungicide for industrial use; slimicide for paper mfg.

Thiostop®. [Uniroyal] Additive for polymer mfg.

Thiotan®. [Sandoz] Selective displacement agents for acid and premetallized dyes in continuous dyeing of nylon and wool carpet.

Thiotax®. [Monsanto] 2-Mercaptobenzothiazole; accelerator for sulfur-curable elastomers.

Thiovanic®. [Evans Chemetics/W.R. Grace] Thioglycolic acid.

Thiovanol®. [Evans Chemetics/W.R. Grace] Thioglycerin; stabilizer for acrylonitrile polymers; crosslinking agent for coatings; accelerator, reducing agent; used in hair waving and straightening, hair dyes, depilatories, textiles, furs, pharmaceuticals, surfactants.

Thiovite. [Goodpasture] Ammonium thiosulfate.

Thiox. [Hydrolabs] Thiourea dioxide; reducing agent for clearing, stripping, print washing, machine cleaning.

Thiurad®. [Harwick] Tetramethylthiuram disulfide; accelerator, sulfur donor.

Thiuram C. [Miles] Tetramethyl thiuram disulfide.

Thixatrol. [Rheox] Rheological additives, thixotropes, gellants.

Thixcin® [Rheox] Stabilizer, thickener, thixotrope, suspending agent for coatings.

Thixon®. [Morton Int'l.] Vulcanizable bonding system; adhesive for bonding elastomers.

Thixseal. [Rheox] Castor oil deriv.; thixotrope, gellant for solv.-based systems; sag and slump control agent for caulks, sealants, and mastics.

Thorcat. [Thor; Polymed Ltd.] Catalysts.

Thornel® Carbon Fiber. [Amoco] Carbon fibers.

Thorowet. [Clough] Sodium dioctyl sulfosuccinate; wetting and rewetting agent.

THQ. [Eastman] Toluhydroquinone; antioxidant for unsaturated polyesters.

Thunderbolt. [Huntington Labs] Degreaser/cleaner.

T-Hydro® Sol'n. [Arco] t-Butyl hydroperoxide; polymerization initiator; coinitiator in polyethylene processes; finishing catalyst in emulsion polymerizations and unsat. polyester resin thermosets; intermediate for peroxygen derivs.

Ti-0720 T 1/8´´. [M&T Harshaw] Titania; catalyst.

Ticaloid. [TIC Gums] Blended stabilizers for suspension, emulsification, thickening and coating.

Ticalose. [TIC Gums] Cellulose gum.

Ti-Code 12. [Titanium Metals] Corrosion resistant titanium alloy.

Ticon. [TAM Ceramics] Dielectric materials.

Tideguard. [Ameron Protective Coatings] Sprayable protective marine cladding.

Tiger Aero Air-Entraining. [Rockwell Lime] Masons hydrate type S.

Tiger Jiffi-Soak. [Rockwell Lime] Finish hydrate type S.

Tiger Morta-Mate. [Rockwell Lime] Masons hydrated lime type S.

Tildenet. [Kerrypak] HDPE.

Tile-Cote. [California Prods.] Polyamide epoxy coatings.

Tim-Bor. [U.S. Borax & Chem.] Timber preservative.

Timica. [Mearl] Titanium dioxide/mica blends; for frosted/iridescent effects in lipsticks, cosmetics.

Timiron®. [Rona] Titanium dioxide-coated mica; pearlescent for lipsticks, pressed powders, makeup.

Timonox. [Cookson Minerals Ltd.] Antimony oxide.

Timpreg. [Chapman] Wood preservative.

Tinamul®. [Hüls Am.] Partial glycerides from natural fat raw materials; food emulsifier for mfg. of Tahina and Halva.

Tinegal®. [Ciba-Geigy/Dyestuffs] Leveling agent, retarding agent, dyebath stabilizer, surfactant, emulsifier for textiles.

Tinofix®. [Ciba-Geigy/Dyestuffs] Dye fixative.

Tinopal®. [Ciba-Geigy] Fluorescent whitening agent, optical brightener for laundry detergents.

Tint-Ayd. [Daniel Prods.] Tinting colorants for paints and inks.

Tinuvin®. [Ciba-Geigy/Additives] Uv absorber, light stabilizer for plastics, coatings.

Tiodize. [Tiodize] Titanium anodize.

Tiofine. [TDF Tiofine BV] Titanium dioxide.

Tioga Adhesion Promoter. [Tioga Coatings] Adhesion promoters for polypropylene, TPO, TPR.

Tiolon. [Tiodize] PTFE dry film lubricants and mold release.

Tiolube. [Tiodize] High pressure dry film lubricants.

Tiona®. [SCM] Titanium dioxide or blends; for use in coatings, plastics, paper, and rubber.

Tioxide. [Tioxide Group PLC] Titanium dioxide pigment.

Tipelin. [Tiszai Vegyi Kombinát] HDPE.

Tipolen. [Tiszai Vegyi Kombinát] LDPE.

Tipplen. [Tiszai Vegyi Kombinát] PP homopolymers and copolymers.

Ti-Pure®. [DuPont] Titanium dioxide; for adhesives.

Tisolate-BS. [Lanaetex Prods.] Isobutyl stearate, octyl isononanoate.

Ti-Sphere. [Presperse] Spherical titanium dioxide or blends; cosmetic ingredients, sunscreens.

Titadine. [Titan] Sulfonic acid derivs.; wetting agent, emulsifier for textiles, leather, paper, household detergents.

Titan Castor No. 75. [Titan] Sulfonated castor oil; emulsifier for textile processing.

Titan Decitrene. [Titan] Alkylated aromatic sulfonate; wetting agent; textile, leather, paper, household detergents.

Titanium Dioxide 110. [Presperse] Titanium dioxide, bismuthoxychloride.

Titanium Dioxide P25. [Degussa] Titanium dioxide.

Titankup. [Malvern Minerals] Surface-modified titanium dioxide.

Titanole. [Titan] Alkylated aryl sodium sulfonate; wetting agent; textile, leather, paper, household detergents.

Titanox. [Kronos] Obsolete tradename for titanium dioxide pigments; superceded by Kronos®.

Titanterge. [Titan] Detergent, wetting agent for textile processing.

Titapene Conc. [Titan] Sodium sulfosuccinate; wetting agent for textile processing.

Titazole. [Titan] Sulfonates; detergent for textile processing.

Tivar®. [Poly-Hi Menasha] UHMWPE; used in paper industry (pulping and finishing), textile industry (loom components), chemical industry (process equip.), food and beverage industry, mining and metals processing, transportation, consumer goods, industrial parts, porous molds.

Tixogel. [United Catalysts] Bentonite deriv. blends.; thixotrope, suspension aid for cosmetics, nail lacquers, antiperspirants, paints, coatings.

Tixosil. [Rhone-Poulenc] Hydrated silica; flatting agents, rheology control agents for coatings, adhesives, plastics, paper, food, cosmetics, toothpaste, and pharmaceutical applics.

Tizox. [Ferro/Transelco] Polishing compds.

TL-. [ICI Advanced Materials] Fluoropolymer-based lubricants.

TL 4190. [Mace Adhesives & Coatings] Water-based polyester film laminating adhesive.

TL Brand. [Theodor Leonhard Wax] Pure white beeswax.

T-Maz®. [PPG/Specialty Chem.] Ethoxylated sorbitan esters; emulsifier, solubilizer, wetting agent, antistat, stabilizer, dispersant, visc. modifier, suspending agent used in the food, cosmetic, drug, textile and metalworking industries.

TM/ETD. [Akrochem] 60% Tetramethyl thiuram disulfide/40% tetraethyl thiuram disulfide; rubber accelerator.

TMI (Meta). [Am. Cyanamid] Unsaturated aliphatic isocyanate.

TMPD® Glycol. [Eastman] 2,2,4-Trimethyl-1,3-pentanediol.

TMT 15, 55. [Degussa] Heavy metal precipitants.

TMTM. [Akrochem] Tetramethyl thiuram monosulfide; accelerator.

T-Mulz®. [Harcros] Phosphate ester; emulsifier, solvent degreaser, textile lubricant, agric. formulations, cleaners.

TMXDI (Meta). [Am. Cyanamid] Aliphatic isocyanate.

TNA-5/CMF. [Akzo Salt] Textile dyeing reagent salt.

TN Cleaner S. [Tokai Seiyu Ind.] Nonionic blend; scouring and washing agent for synthetic fibers and cotton.

TO-. [Hefti Ltd.] Sorbitan trioleate or ethoxylates; emulsifier, antifoam, solubilizer for cosmetics, pharmaceuticals, cattle feed, textiles, biocides, paints, plastics, leather, fur, wood preservation, furniture polishes, various tech. applics.

TOF. [Merrand] Tri 2-ethylhexyl phosphate; plasticizer for NBR, NBR/PVC, CR, CPE elastomers, PVC, PVAc, cellulosics, PS, ABS.

Toffix®. [Hüls Am.] Special hard fat; for mfg. of caramels and chewing sweets.

Tohol. [Toho Chem. Industry] Fatty acid DEA; foam stabilizer, thickener for personal care prods.

Toho Me-PEG Series. [Toho Chem. Industry] Methoxy polyethylene glycols; base material for surfactant, synthetic resin, plasticizer, lubricating industries; wetting, softening, penetrating, lubricating, and cleaning agent for textile, paper, ink, pigment and other industries.

Toho PEG Series. [Toho Chem. Industry] Polyethylene glycols; base material for surfactant, synthetic resin, plasticizer, lubricating industries; wetting, softening, penetrating, lubricating, and cleaning agent for textile, paper, ink, pigment and other industries.

Toho Quench T-3. [Toho Chem. Industry] Anionic poly soap; water-soluble quenching oil.

Toho Salt. [Toho Chem. Industry] Surfactant blends; solubilizer, dispersant, leveling agent for dyestuffs, polyester fibers.

Toi-De-Fresh. [Century Chem.] Deodorant.

Toku Bishi®. [Mitsubishi Kasei] Foundry coke; for ductile and high quality cast iron.

Tokuthion®. [Bayer] Prothiofos; insecticide.

Tolcide MBT. [Albright & Wilson Am.] Methylenebis thiocyanate; biocide for water treatment, paper, antifoulant paint, leather, timber preservation.

Tolex. [Gencorp Polymer Prods.] UK] Vinyls for graphic arts.

Tolplaz TBEP. [Albright & Wilson Am.] Tributoxyethyl phosphate; plasticizer for rubber.

Tolu-Sol. [Shell] Lacquer diluent.

Tomah. [Exxon/Tomah] Emulsifiers, corrosion inhibitor, lubricant, antistat, flotation reagent for textile processing, oilfield, plastics.

Tonalid. [Hercules/FFIG] Patented aromatic.

Tone®. [Union Carbide] Caprolactone-based polyols, diluents, and monomers; for coatings, adhesives, and urethane elastomers.

Tone® Polymer P-767. [Union Carbide] ε-Caprolactone homopolymer; as thermoplastic replacement for plaster orthopedic casts; as powd. coating or extrusion coating for shoe counters; as mold release agent for engineering polymers; pigment dispersant for pigment masterbatch prep.

Tonerclean. [Nippon Nyukazai] Nonionic blend; deinking agent for reclaimed paper prod.

Tonox®. [Uniroyal] Aromatic amine; curing agent, fortifier for epoxy resins; for molded and mechanical goods.

Tonsil. [L.A. Salomon] Activated clay; adsorbent for bleaching oils, fats.

Toolcast. [Am. Resin] Casting, bonding resin.

Topanex. [ICI Polymer Additives] Uv absorber, stabilizer, antioxidant.

Topanol®. [ICI Polymer Additives; ICI Surf. Belgium] Antioxidants, antiozonants for polymers.

Topcithin. [Lucas Meyer] Soya lecithin; flow control agent.

Top Flake. [Morton Salt] Flake salt grades.

Top-Flo Evap Salt. [Cargill] 99.85% NaCl with anticaking additive.

Topmulgat. [Lucas Meyer] Natural nonionic stabilizer blends; structure improver, consistency stabilizer for soups, sauces, dressings, mayonnaise.

Topp-Kote. [Urban Chem.] Evaporation suppressant.

Topsin M. [Atochem N. Am.] Fungicide.

Tor. [Huntington Labs] Germicidal detergent.

Torlon®. [Amoco; Amoco Europe] Poly(amide-imide) resin, graphite, glass or carbon reinforced; high-strength resins for connectors, switches, bearings, switches, relays.

Tornac. [Polysar] Hydrogenated nitrile rubber; for automotive (timing belts, fuel hose, water pump seals, engine gaskets), oil well (rotary seals, o-rings), rocket fuel bladders, industrial/business machine rolls, geothermal seals and liners, conveyor belting, medical and pharmaceutical parts.

Tosete. [Stoner] Liq. tire mounting lubricant.

Toso-CSM®. [Tosoh] Chlorosulfonated polyethylene elastomer; for elec. cables, rubber coatings, life boats, life jackets, windbreakers, escalator handrails, paints, coatings, coated fabrics, floor tiles, hoses, rolls, machine parts.

Totablan. [Ceca SA] Hydrogen peroxide bleaching stabilizer, wetting agent.

Totalide. [Hercules/FFIG] Aromatic chemical.

Tough Gel. [Dow] Ion exchange resins.

Toval. [DuPont] Polyester filaments.

Tower Brand. [Lonza] Camphor.

Tower Heavy Duty Alkaline Cleaner. [Tower Chem.] Highly alkaline cleaner for spray applics.

Tower Safety Solvent. [Tower Chem.] Nonflamm. degreasing solvent.

Tower Treat. [Schaefer Tech.] Cooling water algicide.

Toxanon. [Sanyo Chem. Industries] Surfactant; emulsifier, dispersant for insecticides, herbicides, pesticides.

Toximul. [Stepan; Stepan Europe] Sulfonate/nonionic blends; pesticide emulsifiers.

Toyocat®. [Tosoh] Amine catalyst for polyurethane foam.

TP. [Morton Int'l.] Plasticizers.

TP-2. [Laurel Industries] Dry antimony pentoxide.

TP 511, 519, 520, 710, 711. [AEI Compds.] Thermoplastic compds.; for extrusion, inj. molding, cable insulation and sheathing.

TP-L. [Laurel Industries] Colloidal antimony pentoxide.

TPR. [Monsanto] Thermoplastic rubber.

TPX-80CNI. [Compounding Tech.]

Polymethylolpentane, carbonyl iron powder.

TR-2 Powders. [Dycho] Built detergent/dispersant blend; assistant in removing transfers and/or tars and gums; processing assistant in turbid waters.

TR14 Surfacers. [Atochem N. Am.] Surfacers.

Tra-Bond. [Tra-Con] Epoxy resins; for laminating, bonding, metal repair, adhesives, electronic insulating, fiber optic applics.

Tra-Cast. [Tra-Con] Epoxy casting systems.

Tracel. [Rowa GmbH] Blowing agent.

Tra-Duct. [Tra-Con] Conductive nickel or silver epoxy adhesives.

Traffaid 30 B. [Borregaard LignoTech] Barium-calcium lignosulfonate; antiscumming agent for brick and tile mfg.

Traflam. [Rowa GmbH] Plastic additive.

Tragum. [Rowa GmbH] Sticky resins.

Tramisol. [Am. Cyanamid/Ag] Levamisole.

Transcutol. [Gattefosse SA] Ethyldiglycol; solvent for active ingredients in pharmaceutical preps.; cosurfactant for microemulsions.

Transeal. [Transene] Silicone adhesive sealant.

Transepoxy. [Transene] Self-catalyzed epoxy systems.

Transetch. [Transene] Selective etchant.

Transferin®. [Boehme Filatex] Dispersant for disperse and vat dyes.

Trans Gard. [Gard] Transmission additive.

Transjojoba. [Jojoba Growers & Processors] Transisomerized jojoba oil.

Translastic. [Transene] Silicone elastomer systems.

Translink. [Lawrence Industries PLC] Silane-treated clay.

Translink®. [Engelhard] Calcined aluminum silicate; reinforcing extender for plastics and rubber.

Trans-Oxide. [Hilton Davis] Iron oxide pigments.

Transpafill. [Degussa] Nonreactive transparent filler for printing inks.

Transpalene®. [Neste Composite Materials] Polypropylene; for high clarity and gloss pkg. applics.

Transpalite SS. [Stanley Plastics Ltd.] High grade crosslinked acrylic material.

Transport Plus. [Nalco] Boiler water treatment.

Transtext. [Coated Specialities Ltd.] Cellulose acetate.

Trapex. [Rowa GmbH] Wax.

Trap-FM. [Petrokem] Insecticides.

Tra-Prime. [Tra-Con] Primers.

Trap-Stik. [LiphaTech] Rodenticide.

Trapylen. [Rowa GmbH] Binder.

Trasar Technology. [Nalco] Cooling water treatment.

Trasiv. [Rowa GmbH] Aq. adhesive.

Trastab. [Rowa GmbH] Plastic additive.

Trastan. [Climax Performance] Tanning extracts.

Trastan LS. [Climax Performance] Low sugar calcium lignosulfonate powder.

Trastatic. [Rowa GmbH] Plastic additive.

Travel-Jon. [Century Chem.] Deodorant.

Trawax. [Rowa GmbH] Wax.

Traxol. [Ameripol Synpol] Special extender for improved traction.

Traytuf. [Goodyear; Goodyear Europe] PET polyester; for high purity high strength pkg. with optimum transparency.

T & R Brand. [T&R Chem.] Turpentine; solvents for paints.

TRC-. [Calgon] Cooling water treatment.

Trefsin®. [Advanced Elastomer Systems] Thermoplastic rubber; low gas permeability rubber for medical, industrial and consumer goods.

Trem® LF-40. [Henkel/Coating Chem.] Sodium alkylallyl sulfosuccinate; polymerizable surfactant, solubilizer.

Trenbest 500. [Lucas Meyer] Trimethyl hydroxyethyl ammonium ester of a carboxyglycerol phosphoric acid; mold release agent for plastics and caoutchouc industry.

Trenchcoat. [Dow] Polyolefin films.

Trendwatch. [Drew Ind. Div.] Water treatment controller.

Trennspray Tego®. [Goldschmidt] Silicone-free aerosol for release of inj. molded plastic and rubber moldings.

Trespaphan. [Hoechst UK Films] Biaxially oriented PP film.

Tretinoin. [Hoffmann-La Roche] Retinoic acid.

Triabon®. [BASF AG] Special complex fertilizer for substrates, greenhouse crops.

Triadine®. [Olin] Hexahydro-1,3,5-tris (2-hydroxyethyl)-5-triazine or blends; microbiostat for metalworking fluids.

Triagran®. [BASF AG] Bentazon, dichlorprop, MCPA; post-emergence herbicide for control of broadleaf weeds in winter and spring cereals.

Triameen. [Akzo] Tallow dipropylene triamine.

Triamine. [ICI Am.] Dye fixing agent.

Tri-A-Mine. [Scholler] Dye assistants.

Triamphoram. [Ceca SA] Amphoteric; bactericide.

Triangle Brand. [Phelps Dodge Refining] Copper sulfate.

Tri-A-Nol. [Scholler] Anionic scouring agents, some with optical brighteners.

Triax®. [Monsanto; Monsanto Europe] Nylon/ABS or PVC/ABS alloys; engineering thermoplastics.

Tribase. [BASF AG] Intermediate for prod. of dyes.

Tribex EP. [Atochem N. Am.] Plastic additives.

Tribol. [Tribol] High performance lubricants.

Tribunil®. [Bayer] Methabenzthiazuron; herbicide.

Tricap. [Croda Surf. Ltd.] Triethylene glycol dicaprylate/caprate; plasticizer for plastics and rubber.

Trichelock. [Calgon] Chelating agents.

Trichem. [Pacific Anchor] Epoxy resins, diluents, curing agents.

Triclene D. [OxyChem] Trichlorethylene.

Tricocid. [Tricon Colors] Used for acid dyes.

Tricol. [Ivax Industries] Antidusting compd. for cotton processing.

Tricol. [Takemoto Oil & Fat] Anionic/nonionic surfactants with mineral oil; coning oil for polyester yarn.

Tricosol. [Tricon Colors] Textile dyes and pigments, solvent dyes.

Tri-Ethane. [PPG Industries] 1,1,1-Trichloroethane; solvent for cold cleaning, vapor degreasing, photoresist processing, adhesives.

Trifernal. [Firmenich] Fragrance ingredient.

Trifloc. [Giulini Corp.] Iron (III) sulfate sol'n.

Trifluralin 4. [Terra Int'l.] Liq. herbicide.

Triglyme. [Ferro/Grant] Triethylene glycol dimethyl ether; solvent for electrochemistry, polymer/boron chemistry; gas absorp., extraction, stabilization; used in industrial prods. such as fuels, lubricants, textiles, pharmaceuticals, pesticides.

Trigonal®. [Akzo; Akzo Chem. BV] Ketones or benzoin ethers; uv initiator/sensitizer for printing inks, paper coatings, curing polyester and FRP laminates.

Trigonox®. [Akzo; Akzo Chem. BV] Peroxide derivs.; initiator, crosslinking agent.

Triguard. [Dexter] Water treatment chemicals for boilers.

Tri-K. [Tri-K Industries] Cosmetics ingredients.

Trikup. [Malvern Minerals] Surface-modified aluminum trihydrate.

Trilane. [Tri-K Industries] Squalane.

Tri-Lastin 10F. [Tri-K Industries] Hydrolyzed animal elastin.

Trilene®. [Uniroyal] Liq. ethylene-propylene or ethylene-propylene-diene copolymers; reactive processing aids for elastomers, sealants.

Trilin®. [Griffin] Trifluralin; herbicide.

Trilipidina. [Vevy] Glyceryl palmitate/stearate and glyceryl laurate/oleate.

Trillagene®. [Gattefosse; Gattefosse SA] Soluble animal collagen; biological additive.

Trilon®. [BASF; BASF AG] NTA, pentetic acid, EDTA, or HEDTA or their salts; chelating agents.

Trim Bond. [PPG Industries] Interior trim adhesive.

Trimene®. [Uniroyal] Additive for latex

and foam.

Trimet®. [Rhone-Poulenc] Trimethylolethane; raw material for alkyd and polyester resins for paints, syn. lubricants, plasticizers, stabilizers for plastics, coating agents for pigments.

Trim Set. [PPG Industries] One component, water borne, vacuum forming adhesive.

Trinoram. [Ceca SA] Alkyl dipropylene triamine; emulsifier for bitumen; antistripping agent.

Triodan. [Grindsted Prods.; Grindsted Prods. Denmark] Polyglyceryl esters; food emulsifier.

Tri-Ol. [Tri-K Industries] Natural oils.

Triox. [Mines De La Lucette] Antimony trioxide.

Trioxene. [Vevy] Esters or ethoxylates.

Tri-Pak WS. [Apollo] Surfactant for wetting, scouring, leveling for sulfur and indigo dyeings.

Triple A. [Zophar Mills] Asphalt and coal tar coatings.

Triple C. [Metal Lubricants] Metal working prods.

Triple-H. [Hercules] Hydrolyzed vegetable protein with salt; flavoring.

Triple Tool. [Chem-Pak] Penetrant/lubricant/rust preventive.

Triplus. [GE Silicones] Silicone resin.

Triply. [Degussa Metal Group] Braze double clad on copper.

Tripro-5. [Tri-K Industries] TEA-cocohydrolyzed animal protein.

Tri-Pronectin. [Tri-K Industries] Fibronectin and soluble animal collagen.

Triquat-S. [Tri-K Industries] Steartrimonium hydrolyzed animal protein.

Tris Amino®. [Angus] Tris (hydroxymethyl) aminomethane; pigment dispersant, neutralizing amine, corrosion inhibitor, acid-salt catalyst, pH buffer, chemical and pharmaceutical intermediate, solubilizer.

Tri-Scept. [Am. Cyanamid/Ag] Herbicide.

Trisco®. [Scholler] Body builders, hand modifiers, retarder for dyes, leveling agent.

Trisept. [Tri-K Industries] Parabens; preservatives.

Trisiv. [UOP] Adsorbent.

Tris Nitro®. [Angus] Tris (hydroxymethyl) nitromethane; antibacterial agent, preservative for water treatment, metalworking fluids, oil prod., deodorizing; formaldehyde releaser.

Trisolan. [Henkel/Emery] Lanolin-isopropyl ester; emollient, lubricant for hair sprays, makeup.

Tri-Sorb. [United Desiccants-Gates] Molecular sieve based desiccant for adsorption of moisture and/or gasses.

Trisperse. [Leatex] Low foaming desizing and scouring agent.

Trista. [TA Biochemicals] Tris(hydroxymethyl) aminomethane.

Tri-Star. [Tri-Star] White glues and foam control agents.

Tristat. [Tri-K Industries] Preservatives.

Trisulphoil®. [Scholler] Sulfonated castor oils; protective agents, dye leveling agents, penetrants, softeners, dispersants.

Trisyl. [W.R. Grace/Davison] Synthetic adsorbent oils.

Tri Tab. [Rhone-Poulenc Basic] Granular tricalcium phosphate.

Tri-Tein Milk Polypeptide. [Tri-K Industries] Hydrolyzed casein.

Tri-Tein Silk AA. [Tri-K Industries] Silk amino acids.

Tritex. [Ivax Industries] Lubricant for textile processing.

Tritint. [Ivax Industries] Identification tints for synthetic and natural fibers.

Tritisol. [Croda Inc.] Soluble wheat protein; film-former, conditioner, moisturizer for skin and hair care prods.

Triton®. [Union Carbide; Union Carbide Europe] Surfactants; wetting agent, detergent, penetrant, defoamer, emulsifier, hydrotrope, solubilizer, dispersant for household and industrial cleaning, agric., food processing, textiles, metal cleaning, leather, emulsion polymerization.

Triumphnetzer. [Zschimmer & Schwarz] Diisooctyl sulfosuccinate; wetting agent for textile and chemical industry prods.

Trivalon. [Henkel/Emery/Cospha] Walnut aroma chemical.

Trivent. [Trivent] Esters , ethers, or blends.

Tri-XL. [Chattem] Trioxyaluminum triisopropoxide.

Trizma. [Sigma Prodotti Chimici] Tris(hydroxymethyl)aminomethane; biochemical buffer.

Troenan. [Henkel/Emery/Cospha] Privet blossom aroma chemical.

Trogamid. [Hüls Am.] Polyamide transparent thermoplastic.

Trona. [Kerr-McGee] Chemicals.

Tronamang. [Kerr-McGee] Manganese metal and manganese aluminum briquettes.

Tronox. [Kerr-McGee] Titanium dioxide pigments.

Trovicel. [VT Plastics Ltd.] Rigid PVC foam.

Trovidur. [Rias A/S; VT Plastics Ltd.] PVC.

Troycat. [Troy] Catalysts.

Troykyd®. [Troy] Bodying, antifloating, wetting, antisettling, antiskinning, suspending agents, catalyst, antioxidant, stabilizer, surfactant, drier for coatings, caulks, glazing compds., sealants.

Troymax Drier. [Troy] Calcium, cobalt, iron, lead, manganese, zinc, or zirconium driers; for coatings, inks, polyesters.

Troysan®. [Troy] Antimicrobial, fungicide, preservative for food pkg. adhesives, paper coatings, textile, cordage, wood, paints.

Troysan® Polyphase®. [Troy] Broad spectrum fungicide; for interior and exterior coatings, cutting oils, textiles, paper coatings, inks, plastics, adhesives, canvas, cordage, wood preservative.

Troysol®. [Troy] Dispersants, surfactants, wetting agent for paints, aq. and nonaq. systems.

Troythix. [Troy] Polymerized ester; rheology modifier, bodying agent for coatings.

Trübungsmittel 1. [Zschimmer & Schwarz] Pearling component for cosmetics.

True Tone. [Frank D. Davis] Cement and mortar colors.

Truflex. [Teknor Apex] Monomeric, polymeric plasticizers.

Trugreen. [W.A. Cleary] Formulated micronutrients; chelating agent promoting chlorophyll production; for growth and maintenance of turf, trees, shrubs, flowering plants.

Tru-Pro. [Trugman Nash] Milk protein.

Trusurf LP. [Owens/Corning Fiberglas] Low print resins for the marine industry.

Tru-Tan. [AJ & JO Pilar] Synthetic leather tanning agents.

Trutaste. [Crompton & Knowles] Artifical flavors.

Trychem. [Sentry] Soluble lubricants, softeners, specialty sol'ns. for industry.

Trycite. [Dow] Plastic films and sheets.

Trycite Film. [Mec Pac Srl] Biorientated polystyrene.

Trycol®. [Henkel/Emery] Ethoxylated ethers; emulsifier, lubricant, solubilizer, dispersant, wetting agent, detergent, penetrant for textiles, polishes, lubricants, cosmetics, household and industrial cleaners, metal processing.

Trydet®. [Henkel/Emery] Ethoxylated esters; detergent, emulsifier, softener, lubricant, leveling agent, solubilizer, coupling agent for textiles, household and industrial cleaners, lubricants, leather, polishes, agric., paints, paper, cosmetics, pharmaceuticals.

Tryfac®. [Henkel/Emery] Phosphate esters; detergent, wetting agent, antistat, emulsifier, dispersant, hydrotrope, release agent, corrosion inhibitor for heavy duty cleaners, metalworking compds., textile applics., agric., emulsion polymerization.

Trylon®. [Henkel/Emery] Emulsifier, detergent, lubricant, wetting agent, dye assistant.

Trylox®. [Henkel/Emery] Ethoxylated castor oil or esters; emulsifier, dispersant, carrier, lubricant, softener, antistat, humectant, plasticizer, solubilizer.

Trymeen®. [Henkel/Emery] Ethoxylated alkyl amine; antistat, emulsifier, lubricant, dye assistant for textiles, industrial lubricants, metal buffing, latex rubber compding.

Trymer. [Dow] Urethane rigid foam;

thermal insulation.

TS 100. [Degussa] Silica; flatting agent for coatings.

TS-33-F. [Hefti Ltd.] Sorbitan tristearate; emulsifier, stabilizer for cosmetics, pharmaceuticals, textile printing inks, silicone antifoam agents, polishes, wax working.

TSA-70. [Henkel Canada] Toluene sulfonic acid; catalyst in resin prod.; intermediate for salts used as hydrotropes in household and industrial cleaners, metal cleaning compds.

T-San. [Thatcher] 10% Quaternary germicide.

TSE-2000. [TSE] Diphenylmethane diisocyanate; one-coat adhesive to bond millathane to a variety of substrates during vulcanization.

TSE Mold Release®. [TSE] Glycol surfactant; mold release.

T-Soft®. [Harcros] Linear alkylbenzene sulfonic acid; high foaming detergent intermediate.

TTC. [TTC Mouldings BV] Polyethylene.

TTDIA. [Degen] Polymerized tung oil.

T-Tergamide. [Harcros] Fatty acid diethanolamides; foam stabilizer, thickener, visc. modifier for detergents, shampoos.

Tud. [Transene] Ultrasonic detergent.

Tuex®. [Uniroyal] Thiuram accelerator; for tires, air-cured footwear, molded soling, closed-cell sponge, insulated wire.

Tufcote. [E-A-R] Polyether urethane foam; for absorbers, barriers, and composite prods.

Tufel. [GE Silicones] Silicone elastomer system.

Tuffak. [Rohm & Haas] Polycarbonate sheet.

Tufflo. [Lyondell Petrochemical] Process oils for rubber and plastics.

Tuf-Quik. [California Prods.] Oil-modified polyurethane enamels.

Tufset. [Tufnol Ltd.] Rigid polyurethane.

Tuf-Sil. [Sil-Med] Platinum catalyzed silicone rubber formulation.

Tuftane. [Lord Corp. UK] Polyurethane film.

Tufteze®. [Nat'l. Starch & Chem.] Lubricant for carpet yarn tufting.

Tullanox. [Tulco] Hydrophobic precipitated silica; rheology control agent, water repellant, anticaking agent, reinforcing agent, carrier for silicone sealants, coatings, powders, resins, elastomers, elec. and heat insulation, paints, inks, pharmaceuticals, cosmetics, fertilizers, adhesives.

Tupersan. [DuPont/Ag] Herbicide.

Turbo. [Miles/Ag] Soybean and potato herbicide.

Tureen. [Champlain Industries] Autolyzed yeast extracts.

Turfcide. [Uniroyal] Fungicide.

Turgum®. [Whitney & Oettler] Turpene-resin acid blend; plasticizer and conditioner for SBR; retarder for furnace black/rubber stocks.

Turkey Red Oil 100%. [Zschimmer & Schwarz] Sulfated castor oil; solubilizer, refatting agent.

Türkischrotöl. [Zschimmer & Schwarz] Sulfated castor oil; solubilizer for cosmetics.

Turpinal®. [Henkel; Henkel KGaA] Etidronic acid or salts; chelating agent, stabilizer for cosmetics, pharmaceuticals.

Turpol. [Cardolite] Rubber plasticizer.

Tween®. [ICI Spec. Chem.; ICI Surf. Belgium] Ethoxylated sorbitan esters; solubilizer, emulsifier, antistat, fiber lubricant for textile industry.

Twitchell. [Henkel/Emery] Rewetting agent, softener, lubricant, dyestuff deduster, textile auxiliary.

Tybrite. [Dow] Plastic film.

Tycel®. [Lord] Laminating adhesives; for food, health care, personal care, and industrial applics.

Tychem. [Reichhold/Emulsion Polymers] Emulsion polymer thickeners.

Tyfo. [Nat'l. Research & Chem.] Specialty compds.

Tylac®. [Reichhold/Emulsion Polymers] Styrene-butadiene, butadiene acrylonitrile, polystyrene emulsions.

Tylopur. [Hoechst Celanese] Methyl hydroxyethylcellulose.

Tylorol. [Triantaphyllou] Fatty ether sulfates; detergent base for personal care prods.

Tylose®. [Hoechst Celanese/Colorants & Surf.; Hoechst AG] Cellulose derivs.; binder, thickener, plasticizer, dispersant for coatings, mining, batteries, rubber industry, cosmetics, foodstuffs, pharmaceuticals, agric., paper, textile industry.

Tymor. [Morton Int'l.] Extrudable, modified polyolefin adhesive resins.

Tyner. [Bruce Chem.] Nonionic, cationic, and anionic softeners.

Tynex. [DuPont] Nylon filament impregnated with abrasive particulates; for industrial and floor maintenance brushes.

Tynol. [Baker Perf. Chem.] Emulsion preventer.

Typar. [DuPont] Spunbonded polypropylene.

Type-41 Clay, -80 Clay. [Southeastern Clay] Hard rubber clay.

Type 798 Bulked Roving. [PPG Industries/Fiber Glass Prods.] Glass roving; for polyester and vinyl ester resin systems for filament winding and pultrusion applics.

Type BPL. [Calgon Carbon] Activated carbon; for vapor phase applics.

Type CAL, CPG, OL, SGL. [Calgon Carbon] Granular carbon; for fixed or moving beds for purification and decolorization of aq. and organic liqs.

Type GRC-22. [Calgon Carbon] Granular activated carbon; for gold adsorption process recovery operations.

Type HGR. [Calgon Carbon] Granular sulfur-impregnated activated carbon; for mercury removal.

Type IVP, PCB. [Calgon Carbon] Granular activated carbon; for vapor phase applics.

Type PWA. [Calgon Carbon] Pulverized activated carbon; for purification in pharmaceutical and other industries.

Type TOG. [Calgon Carbon] Granular activated carbon; for water treatment applics.

Type WPH, WPL. [Calgon Carbon] Powdered activated carbon; for water treatment applics.

Ty-Ply® BN. [Lord] One-coat adhesive for bonding nitrile, polyacrylate, polyepichlorohydrin, or millable polyurethane rubber to metal or other rigid substrates during vulcanization of the elastomer.

Typopack. [TTC Mouldings BV] Polyethylene.

Typophor®. [BASF; BASF AG] Dye base preparations in olein; for brightening of printing inks.

Tyrez. [Reichhold/Emulsion Polymers] Plastic impact modifiers.

Tyrfil. [Synair; Compounding Ingredients Ltd.] Elastomer urethane; for flatproofing tires.

Tyril. [Dow Plastics] SAN resin; thermoplastic for inj. and blow molding, cosmetic containers, industrial battery cases and caps, high-pressure filter housings, pkg., appliances, blood aspirators, other medical parts, connectors, tumblers, dinnerware, utensils.

Tyrin. [Dow Plastics] Chlorinated polyethylene elastomers; for extruded, calendered, and molded goods, hose, cable jacketing, linings, gasketing, o-rings, for blending and modifying.

Tyrite. [Lord] One-component urethane structural adhesives; moisture-curing systems.

Tyrosilane. [Exsymol] Methylsilanol acetyltyrosine.

Tyrosilane C. [Exsymol] Copper acetyl tyrosinate methylsilanol; tanning activator with anti-aging action for cosmetic and health prods.

Tysul. [DuPont] Peroxygen compd.

Tytanpol. [British Traders & Shippers] Titanium dioxide.

Tyvek. [DuPont] Spunbonded olefin.

Tyzor. [DuPont] Organic titanates; catalysts for esterification and olefin polymerization; crosslinking agent for automotive prods., coatings, elastomers, films/paints, graphic arts, plastics.

U

U-2 CMC. [Perma-Flex Mold] Two-component polyurethane elastomer and acclerator.

U35. [BFGoodrich/Spec. Polymers] Polyester-based aq. aliphatic urethane dispersion; skincoat for transfer coating applics.

U 46® Products. [BASF AG] 2,4-D, MCPA, dichlorprop, mecoprop; for weed control in cereals, maize, sugar cane, rice, perennial crops, etc.

Uantox. [UPI] Antioxidant, preservative.

Ubabond. [Imperial Adhesives] Urethane adhesive.

Ubagrip. [Imperial Adhesives] Neoprene adhesive-solvent.

Ubatol. [HTS High Tech Specialties] Polymer dispersions and wax emulsions.

Ubbink Gutter. [Ubbink Nederland BV] Glass fiber-reinforced PVC.

Ubisoil. [Ubbink Nederland BV] PVC.

U Blue 104. [Presperse] Ultramarine blue, bismuthoxychloride.

UBOB®. [Uniroyal] p-Aminodiphenylamine; intermediate.

Ucar. [UCAR Carbon] Carbon and graphite prods. for industry.

Ucar® Acrylic. [Union Carbide] Modified acrylic emulsion; surfactant for paints.

Ucarcide®. [Union Carbide] Glutaral; Preservative, antimicrobial for cosmetics, toiletries, chemical specialties.

Ucardri. [Union Carbide] Gas dehydration solvent.

Ucare® Polymer. [Amerchol] Polyquaternium-10; hair fixative.

Ucar Filmer IBT. [Union Carbide] Industrial chemical.

Ucarkool. [Union Carbide] Heat transfer fluid for gas compression and line heater applics.

Ucar® Latex. [Union Carbide] Acrylic, PVAc, styrene-acrylic, vinyl/acrylic latexes; for caulks, sealants, mastics, paints, industrial finishes.

Ucarmod. [Union Carbide] Acid-grafted polyethers.

Ucar® Phenoxy Resins. [Union Carbide] Phenoxy resin; for flexible and rigid pkg.

Ucarsan® Sanitizers. [Union Carbide] Glutaraldehyde/detergent; industrial sanitizers.

Ucarsil®. [Union Carbide] Organosilicon; dispersant, stabilizer, processing aid.

Ucar® Silicone [Union Carbide] Amino silicone emulsion; used in car care applics., e.g., auto polishes, vinyl protectants, tire dressings.

Ucartherm. [Union Carbide] Heat transfer fluid.

Ucartritherm. [Union Carbide] Heat transfer fluid.

Ucar® Vehicle. [Union Carbide] Acrylic or styrene-acrylic emulsions; for industrial finishes.

Ucar® V Series. [Union Carbide] Vinyl chloride/VA copolymers; for coatings.

UCC. [Hernon Mfg.] Uv-cured conformal coating systems for printed circuit board and electronic assembly.

Ucefix. [UCB] Thermoplastic polyurethane adhesive.

Uceflex. [UCB] Thermoplastic polyurethane; for inj. molding, extrusion, textile coating.

Ucon®. [Union Carbide] PEG/PPG copolymers or PPG ethers; emollients.

Uconex®. [Union Carbide] Glutaraldehyde; microbicide for metalworking fluids.

Ucon® Hydrolube. [Union Carbide] Water-glycol; hydraulic fluid, high-pressure fluid.

Ucure. [Union Carbide] Reactive modifier for low profile SMC.

U-Dagen DEO. [UPI] Triethyl citrate and BHT.

Udel®. [Amoco; Amoco Europe] Polysulfone, some glass reinforced; for printed circuit boards, chip carriers, connectors, elec. housings, process equip. for severe environments, sterilizable medical devices, food service equip., automotive, plumbing fixtures, microwave cookware.

Udet®. [Witco/Organics] Linear alkylaryl sulfonate; for detergent prods.

UDMH. [Uniroyal] Intermediate.

Ufablend. [Unger Fabrikker AS] Surfactant blends; for hard surface, laundry, shampoo detergents, dishwashing, hand cleaning soaps.

Ufacid. [Unger Fabrikker AS] Alkylbenzene sulfonic acid; base for mfg. of sulfonates used in detergent formulations.

Ufanon. [Unger Fabrikker AS] Coconut DEA; visc. modifier, foam booster/stabilizer for liq. detergents, shampoo, bath preps., leather industry.

Ufapol. [Unger Fabrikker AS] Surfactant blend; for emulsion polymerization.

Ufarol. [Unger Fabrikker AS] Lauryl sulfate salts; for shampoos, bath prods., carpet shampoos, furniture cleaning, textiles.

Ufaryl. [Unger Fabrikker AS] Alkylbenzene sulfonates; detergent emulsifiers.

Ufasan. [Unger Fabrikker AS] Alkylbenzene sulfonates; surfactants for liq. detergents.

Ufasoft. [Unger Fabrikker AS] Quaternary ammonium compds.; cationic fabric softener and conditioner for laundry.

Uformite®. [Reichhold] Urea, melamine, triazine, or urea formaldehyde resins; thermosetting resins.

Ufoxane. [Borregaard LignoTech] Lignosulfonates; dispersant for textile dyestuffs, pesticides.

Ugikral. [Northern Industrial Plastics Ltd.] ABS.

Ukanil. [ICI PLC] POP POE block polymers; defoamer and rinse aid for industrial and household detergents; wool lubricant.

Ukaril. [ICI PLC] Modified alcohol alkoxylate; textile spin-finish component.

Ulasein 15. [Ultra Additives] Casein sol'n.

Ulasperse 994B. [Ultra Additives] Dispersant.

Ultem®. [GE Plastics; GE Plastics Ltd.] Polyetherimide resins, some glass or mineral reinforced; thermoplastic resin for laboratory ware and automotive heat transfer systems, elec./electronic applics., computer circuitry, microwave components, molded parts, electronic housings, structural components, EMI/RFI/ESD applics. in aircraft industry.

Ultimet. [Haynes Int'l.] Cobalt-based alloy; corrosion resistant alloy for nozzles, pump and valve parts, agitators, bolts, fan blades, screw conveyors, filters, and rolls.

Ultra. [Columbian Chem.] Carbon black.

Ultra. [Zipp Industries] Dry fertilizer.

Ultra Anhydrous Lanolin HP-2060. [Henkel/Emery/Cospha] Lanolin; for cosmetics, lip prods; moisturizer.

Ultra Bead. [Michelman] Wax additives.

Ultrablack. [Loctite] RTV silicone.

Ultrablend®. [BASF/Engineering Plastics; BASF AG] PBT/ASA or PBT/PC blends; for inj. molding of tech. parts in automotive and elec./electronics industries.

Ultra Blend. [Witco] Alkylaryl sulfonate blend; detergent.

Ultrabond. [Hernon Mfg.] Uv-curing adhesives for electronics bonding, tacking, potting and encapsulation.

Ultrac. [Allied-Signal; NV Allied Corp. Int'l. SA] Polyethylene/ethylene copolymer mixture.

Ultracarb. [Microfine Minerals Ltd.] Flame retardant.

Ultracast. [Air Prods.] pDI-PTMEG; polyurethane prepolymer.

Ultra-Clear. [Oil Dri Corp. of Am.] Filter aid.

Ultracolor. [Teknor Color] High ratio color conc.

Ultra Cote. [Engelhard] Kaolin; high brightness coating.

Ultra-Cure. [PMC Specialties] Chlorothioxanthone; photoinitiator for uv-cured inks and coatings.

Ultradine. [West Agro] Dairy teat dip conc.

Ultradoss 70. [Mfg's Chem. & Supply] Sodium dioctyl sulfosuccinate; leveling agent, wetting agent, emulsifier for foam dyeing.

Ultra-Dry. [Urban Chem.] Solder fluxes.

Ultradur®. [BASF/Engineering Plastics; BASF AG] PBT resin, some glass and/or mineral reinforced; for engineering parts, paper coatings, sections, sheet, extruded stock, film.

Ultradye 200. [Mfg's Chem. & Supply] Surfactant, retarding and leveling agent for textile dyeing operations.

Ultra Ethylux. [Westlake Plastics] HDPE.

Ultrafast®. [BASF] UV absorbers for textiles.

UltraFine®. [Laurel Industries] Antimony trioxide.

Ultrafine® II. [Laurel Industries] Submicron sized antimony oxide; flame retardant for thermoplastics, thermosets, synthetic fibers; as catalyst for PET mfg.

Ultra Fine Lanolin. [Fanning] Lanolin.

Ultrafinish. [Hall Chem.] Chemicals for surface finishing.

Ultrafix. [Mfg's Chem. & Supply] Fixing agent for dyes.

Ultraflex®. [Petrolite/Polymers] Microcrystalline waxes; plastic wax for hot-melt laminating adhesives for papers, films, and foils; hot-melt coatings.

Ultraflo. [Teknor Color] Liq. color for plastics.

Ultraform®. [BASF/Engineering Plastics; BASF AG] Acetal copolymer; for extrusion and inj. molding.

UltraGard Premier. [Manville] Phenolic foam roof insulation.

Ultra Gel Coats. [Ferro/Plastic Colorants] Gel coats.

Ultragen. [Mfg's Chem. & Supply] Antiprecipitant, leveling agent for dyeings.

Ultraglaze®. [GE Silicones] Silicone sealant.

Ultra Gloss 90. [Engelhard] High-glossing kaolin.

Ultraguard. [Mfg's Chem. & Supply] Stain release prod. for nylon carpet.

Ultrahold® 8. [BASF; BASF AG] Acrylates/acrylamide copolymer; hair spray resin.

Ultrakrome. [Hilton Davis] Pigment dispersions.

Ultra Lantrol® HP-2074. [Henkel/Emery/Cospha] Lanolin oil; for cosmetics, hair and skin prods., bath oils, medicinals; pigment dispersant, emollient for makeup; moisturizer.

Ultralen®. [BASF AG] PET resin; spinning polymer for textile applics.

Ultralev. [Mfg's Chem. & Supply] Leveling agent, dyeing assistant, retarding agent for textiles.

Ultralok. [Mfg's Chem. & Supply] Dyeing auxiliary for beck and continuous dyeing of nylon carpet.

Ultralon®. [Whitford] PTFE release coating; nonstick coatings.

Ultralube. [Mfg's Chem. & Supply] Lubricant for dyebaths, coning, knitting.

Ultramark. [PCR] Mass markers for mass spectroscopists.

Ultramid®. [BASF/Engineering Plastics; BASF AG] Nylon 6, 6/6, 6/9, 6/10, or 6-6/6, some glass or mineral reinforced; for inj. molding and extrusion, machine parts, housings, gears, bearings, pipe, sections, fiber, monofilaments, bristles, stretched tape, coatings.

Ultramite. [ECC Int'l.] Calcium carbonate; for coatings.

Ultramix. [Rayonier] Lignosulfonate specialty chemicals.

Ultramoll. [Bayer] Polyadipate, polyphthalate, or polyurethane; polymeric plasticizers.

Ultra NCS Liq. [Witco] Ammonium cumene sulfonate; hydrotrope, coupling agent, solubilizer.

Ultranox®. [GE Specialty] Antioxidant, stabilizer for plastics, rubber.

Ultra NXS Liq. [Witco] Ammonium xylene sulfonate; hydrotrope, coupling agent, solubilizer.

Ultranyl®. [BASF/Engineering Plastics; BASF AG] Polyamide and polyphenylene ether alloy; for inj. molding of tech. parts.

Ultra Ox. [Liquid Air Corp.] 99.9999% Oxygen.

Ultrapadol. [Mfg's Chem. & Supply] Polyacrylamide copolymer; used in pads for carpet dyeings.

Ultrapas. [Hüls Am.] Melamine formaldehyde; molding compds.

Ultrapek®. [BASF; BASF AG] Poly-aryletherketone, some glass reinforced; for inj. moldings and extrudates.

Ultra-Pflex®. [Pfizer] Surface-treated precipitated calcium carbonate; impact modifier for rigid PVC applics.

Ultraphor®. [BASF; BASF AG] For weed control in cereals, maize, sugar cane, rice, perennial crops, etc.

Ultraphor® CW. [BASF] Optical brightener for textiles.

Ultraphos. [Witco SA] Ethoxylated fatty alcohol phosphate; metalworking additive, detergent.

Ultraprene. [Teknor Apex] PVC elastomers; for architectural glazing seals, weatherstripping, wire and cable, hardware, tool handle grips, automotive interior and exterior parts, hoses, tubing, medical equip., toys, sporting goods.

Ultra Pure. [Olin] Hydrazine propellant.

Ultraquest. [Mfg's Chem. & Supply] Chelating agents.

Ultrar. [Atomergic Chemetals] Acids, solvents.

Ultra-Rez. [Lawter Int'l.] Phenolic modified resin.

Ultrarib. [Wavin Industrial Prods. Ltd.] PVC.

Ultrascour. [Mfg's Chem. & Supply] Detergent, wetting agent, penetrant, scouring agent for textiles.

Ultra SCS Liq. [Witco] Sodium cumene sulfonate; hydrotrope, coupling agent, solubilizer.

Ultrasil®. [Degussa] High purity precipitated silica; catalyst carrier.

Ultrasist. [Mfg's Chem. & Supply] Dispersant for cationic dyes.

Ultrasof®. [Ciba-Geigy/Dyestuffs] Softener and napping aid for textiles.

Ultrasoft. [Mfg's Chem. & Supply] Textile softeners.

Ultrason®. [BASF/Engineering Plastics; BASF AG] Polyethersulfone or poly-sulfone, some glass reinforced; for inj. molding, extrusion, film extrusion, blow molding.

Ultrasperse #1533. [Ultra Additives] Organic flux.

Ultraspin. [Mfg's Chem. & Supply] Lubricant for staple spinning and extrusion.

Ultra Sulfate. [Witco] Sulfates; wetting agent, penetrant, lubricant, emulsifier.

Ultra SXS Liq. [Witco] Sodium xylene sulfonate; hydrotrope, coupling agent, solubilizer.

Ultrasyr. [Montedipe Srl] Modified antishock polymers.

Ultra-Tech. [JacksonLea] Phosphate coating systems.

Ultratex®. [Ciba-Geigy/Dyestuffs] Polysiloxane; textile finish.

Ultrathene®. [Quantum/USI] EVA copolymer; for adhesives, coatings, blown and cast film, inj. molding, blow molding, flexible tubing, profiles, sheet.

Ultravon®. [Ciba-Geigy/Dyestuffs] Proprietary blend; wetting agent, detergent for peroxide bleaching.

Ultrawet. [Arco] Sodium xylenesulfonate.

Ultrawet. [Mfg's Chem. & Supply] Wetting agent, penetrant for textiles.

Ultra White 90. [Engelhard] Kaolin; high brightness paper pigment.

Ultrax®. [BASF AG] Self-reinforcing liq. crystal polymer; for extrusion and inj. molding.

Ultrazine. [Borregaard LignoTech] Calcium or sodium lignosulfonates; dispersant for textile dyestuffs, pesticides, gypsum board, concrete additives.

Ultrazym. [Novo Nordisk] Pectinase; for mash and juice treatment; processing aid in wine industry.

Ultrex. [Lawter Int'l.] High gloss/high transfer ink varnish.

Ultrolon. [Childers Prods.] Tedlar laminated metal.

Ultron. [Monsanto] Filament, staple nylon yarns.

Ultrox. [Atochem N. Am.] Zirconium silicate; glaze opacifier; stabilizes color shades for glazes for sanitary ware, wall tile, glazed brick, structural tile, stoneware, dinnerware, special porcelains, refractory compositions, epoxy formulations, encapsulating resins.

Umulse-E. [UPI] Sorbitan sesquioleate, beeswax, aluminum stearate.

Unads®. [R.T. Vanderbilt] Tetramethylthiuram monosulfide; rubber accelerator.

Unamide®. [Lonza] Fatty acid alkanolamide or ethoxylated amides; foam booster/stabilizer, visc. builder, emulsifier, detergent base for personal care, household and industrial prods.

Unamine®. [Lonza] Fatty imidazolines; surfactant, fungicide, emulsifier, demulsifier, softener, antistat, corrosion inhibitor for drycleaning, leather cleaning.

Undamide. [Vevy] Undecylenamide DEA.

Undebenzofene. [Vevy] Paraben blends.

Undelene. [Vevy] PEG-6 undecylenate.

Unden®. [Bayer] Propoxur; broad spectrum insecticide.

Undenat. [Vevy] Sodium undecylenate.

Undezin. [Vevy] Zinc undecylenate.

Unette. [UPI] Cetearyl alcohol or blends.

Unger. [Unger Fabrikker AS] Softener, antistat, superfatting agent for textiles, skin and hair care prods.

Ungerol. [Unger Fabrikker AS] Lauryl ether sulfates; detergent, shampoos, wallboard mfg., textile industry, drilling auxiliary.

Unibec BDA. [Hickson Danchem] Polymeric beck dyeing auxiliary.

Unibetaine. [UPI] Betaines.

Unibind. [Hart Chem. Ltd.] Binding agent for coal and mineral transportation or storage to prevent wind blown losses.

Unibiovit. [Induchem AG] Farnesol/farnesyl acetate blends.

Unibond. [Unicast Development] Foundry binders.

Uni-Cal 66. [Hüls Am.] Universal industrial colorant.

Unicarrier. [Unitex] Dye carriers.

Unicast. [Emerson & Cuming] One-component epoxy casting resins incl. general purpose, thermally conductive, and fire retardant.

Unicast. [Unicast Development] Foundry binders and chemicals.

Unicast. [UPI] Castor oil.

Unichem. [Colorite Plastics] Flexible and rigid PVC molding and extrusion compds.

Unichem. [UPI] Chemicals.

Unichlor. [Neville] Chlorinated paraffin; used in adhesives, coating, rubber, flame-retardant compds., concrete-curing compds., caulking compds., and cutting oils.

Unichol. [UPI] Cholesterol.

Unichrome. [Atochem N. Am.] Plating process and compds.

Unicide U-13. [Induchem AG; Lipo] Imidazolidinyl urea.

Unicoat. [Emerson & Cuming] Epoxy, acrylic, or silicone conformal coatings.

Unicoat. [Morton Int'l.] Universal automotive coating system.

Unicol. [UPI] Ethoxylates, propoxylates, or blends.

Unicor. [UOP] Corrosion inhibitor.

Unicorn Universal Master Batches. [Begg & Co. Thermoplastics Ltd.] Color masterbatches for plastics.

Unicrepe. [Georgia-Pacific] Creping aid for papermaking.

Unidene. [EniChem Elastomeri Srl] Hydrocarbon rubber.

Uniderm A. [UPI] Allantoin.

Uniderm HOMSAL. [UPI] Homosalate.

Uniderm SSME. [UPI] Sesame oil.

Uniderm WGO. [UPI] Wheat germ oil.

Unidex. [Corn Prods.] Dextrose agglomerated maltodextrin.

Unidri. [Hart Chem. Ltd.] Dewatering agent, low-foaming surfactant for metallic oxides, coal, vacuum filtration of minerals; drainage aid for pulp and paper processing.

Unidye. [C.H. Patrick] Scour and dye in one bath.

Unidyme. [Union Camp] Dimer acids.

Unifat. [UPI] Fatty acids.

Unifilter U-41. [Induchem AG] Butyl methoxydibenzoylmethane, octyl methoxycinnamate, 3-benzylidene camphor.

Uniflex. [Unicast Development] Foundry pattern materials.

Uniflex®. [Union Camp] Polymeric or monomeric plasticizers, processing aid, lubricant esters; for rubbers, thermoplastics, fiber lubricants, metalworking oils.

Uniflex®. [UPI] Dipropylene glycol dibenzoate or blends.

Unifloat. [Union Camp] Ore flotation reagents.

Uniflot C. [Hart Chem. Ltd.] Proprietary blend; coal promoters for flotation operations.

Uniflot SP. [Hart Chem. Ltd.] Sulfhydril; copper or zinc sulfide collectors for copper/zinc flotation operations.

Unifroth. [Hart Chem. Ltd.] Blends; frother for flotation operations.

Unigen. [Unicast Development] Foundry exothermic compds.

Unihib®. [Lonza] Organic phosphonates; dispersant, corrosion and scale inhibitor.

Unihydag WAx. [UPI] Fatty alcohol.

Unihydol. [UPI] Ethoxylated ethers.

Unikon A-22. [Induchem AG] Dichlorobenzyl alcohol.

Unilac. [Unicast Development] Foundry pattern coatings.

Unilene. [Marlin Chem. Ltd.] Hydrocarbon resins.

Unilev. [Unitex] Levelers for dyeing.

Unilex. [UPI] Lecithin.

Unilin®. [Petrolite/Polymers] Polymeric alcohols; functional polymers for modification of plastics.

Unilink®. [UOP] Aromatic diamine; chain extender for polyurethanes.

Unilube. [Unitex] Textile lubricants.

Unilube MB-370. [Nippon Oils & Fats] PPG-40 butyl ether.

Unimate®. [Union Camp] Esters; skin emollient, solubilizer, coupler, spreading agent, lubricant, carrier for lipsticks, shave lotions, hair prods., bath oils, aerosol toilet preparations.

Unimax. [Gabriel-Chemie GmbH] Colored masterbatches.

Unimer. [Induchem AG] PVP copolymers.

Unimin. [Hart Chem. Ltd.] Selective iron sulfide depressant for use in base metal sulfide separation.

Unimoist U-125. [Induchem AG] Glycerin, urea, saccharide hydrolysate, magneisum aspartate, glycine, alanine, creatine.

Unimoll. [Bayer; Miles] Phthalates; plasticizers for PVC, calendering, extrusion, inj. molding, VC copolymers, PVAc, surf. coatings, adhesives, rubber, PS, NC, ethyl cellulose, CAB, acrylic surf. coatings, alkyd resins, polymethacrylate, rubber.

Unimox. [UPI] Amine oxides.

Unimul. [UPI] Esters or ethoxylated ethers.

Unimulgade. [UPI] Cetearyl alcohol blends.

Union Carbide® A. [Union Carbide] Silane derivs.; coupling agent.

Union Carbide® L, LE, R, Y. [Union Carbide] Silicone derivs.; emollient, antifoam, wetting agent, lubricant, penetrant, coemulsifier, softener, release agent; for cosmetics, rubber, plastic mold releases; polishes; textile finishes; paint additive.

Unipabol U-17. [Induchem AG; Lipo] PEG-25 PABA.

Unipart. [Unicast Development] Foundry parting agents.

Unipeg. [UPI] PEGs or ethoxylated esters.

Uniperol®. [BASF; BASF AG] Ethoxylated ether or amine; emulsifier, dyeing auxiliary, leveling agent, dispersant, detergent for textiles.

Unipertain. [Induchem AG] Hydrolyzed animal protein/tyrosine blends.

Unipertan. [Lipo] Hydrolyzed animal collagen/tyrosine blends.

Uniphen P-23. [Induchem AG; Lipo] Phenoxyethanol, methylparaben, ethylparaben, propylparaben, butylparaben.

Unipherol U-14. [Induchem AG] Isopropyl myristate, lecithin, tocopherol.

Uniphyllin SC. [UPI] Chlorophyllin-copper complex.

Uniplex. [Unitex] Plasticizers.

Unipol. [UPI] Sulfates or other surfactants.

Unipon. [UPI] Sodium lauryl sulfate.

Unipoxy. [Fibreglass Evercoat] Epoxy laminating resin.

Unipro. [UPI] Casein.

Uniquart. [UPI] Ethoxylated polyamines or quaternaries.

Uniquat®. [Lonza] Quaternary ammonium compds.; microbicide, corrosion inhibitor.

Unirep U-18. [Induchem AG] Dimethyl phthalate, diethyl toluamide, ethyl hexanediol.

Unirex. [Exxon] Grease.

Uni-Rez®. [Union Camp] Polyamide resin, amidoamine, rosin or rosin esters, maleic resin, or phenolic resin; resins for inks and coatings.

Uniscour. [Unitex] Detergents, emulsifiers, scouring agents.

Unisept. [UPI] Preservatives, fungicides.

Uniset®. [Emerson & Cuming] One-part epoxies and uv-cured systems; circuit assembly materials, dielectric and conductive adhesives.

Unisil. [UPI] Silicone prods.

Unisilkon®. [Kluber Lubrication N. Am.] Silicone-based greases for bearing applics.

Unisist PER. [Unitex] Peroxide bleach bath stabilizer.

Unisize®. [Air Prods.] Tackified polyvinyl alcohol; for specialty paper applics.

Unislip. [Unichema; Unichema France SA] Erucamide or oleamide; lubricant, slip, and antiblock agent for polyolefins; friction modifiers for lubricants; in rubber and inks.

Unislip®. [Goldschmidt] Slip agent for leather industry.

Unisoft. [Unitex] Textile softeners.

Unisoft SAC. [UPI] Stearalkonium chloride.

Unisoil. [Engelhard] Soil conditioner.

Unisol. [ICI Surf. UK] Ethylene oxide condensates; leveling agent, penetration and dyeing assistant for dyestuffs.

Unisol S-22. [Induchem AG] 3-Benzylidene camphor.

Unispar. [Unimin Specialty Minerals] Feldspar; fillers and flatting agents for traffic paint, interior/exterior architectural coatings, protective, maintenance and marine coatings, mastics and adhesives.

Unisperse. [Ciba-Geigy/Pigments] Aq. dispersions, organic pigments.

Unisperse CB. [Unitex] Dispersing agent for disperse dyes on acrylics, nylon and polyester.

Unistab D-33. [Union Carbide] 1,3,5-Trimethyl-2,4,6-tris(3,5-di-t-butyl-4-hydroxybenzyl) benzene (40%) in polymeric organosilicon; antioxidant.

Unistat. [UPI] Sorbic or propionic acids or esters.

Unisuprol S-25. [Induchem AG] Phenoxyethanol, triethylene glycol, dichlorobenzyl alcohol.

Unisweet. [UPI] Sweeteners.

Uni-Tac®. [Union Camp] Rosin or rosin esters; tackifiers for adhesives.

Unitak. [Imperial Adhesives] Water-based pressure-sensitive adhesive.

Unital. [Henkel-Nopco] Depolymerized polypeptide; protective colloid; leveling and retarding agent for vat and naphthol dyes.

Unitane. [Kemira] Titanium dioxide.

Unite. [Aristech] Chemically modified polyolefins with anhydride functionality; compatibilizer for polymer blends and alloys, coupling agent in reinforced and filled polymers, adhesive agent for bonding polyolefins to various substrates.

Uniterge. [UPI] Ethoxylated nonyl phenyl ethers.

Unitex. [UPI] Propionic acid or salts or etidronic acid.

Unitex Creamed. [Guthrie Latex] High ammonia natural rubber creamed latex; for use in extruded thread.

Unithox. [Petrolite/Polymers] Ethoxylated synthetic alcohols ethers.

Uniti. [Kemira] Titanium chemicals.

Unitina. [UPI] Oils or esters.

Unitol. [Union Camp] Tall oil fractions.

Unitolate. [UPI] Esters or ethoxylates.

Unitone. [Manufacturers Chems.] Surfactant, dye assistant, leveling agent.

Unitrenol. [Induchem AG; Lipo] Farnesyl acetate, farnesol blends.

Universal. [Ebonex] Ammonium bicarbonate, carbonate, or carbamate.

Universal MENT. [UPI] Menthol.

Universal ZPS. [UPI] Zinc phenolsulfonate.

Universene. [UPI] EDTA or salts; sequestrants.

Universil. [Universal Scientific] Silica gel; for HPLC columns.

Univis. [Exxon] Hydraulic fluids.

Univit-E Acetate. [UPI] Tocopheryl acetate.

Univolt. [Exxon] Dielectric oil.

Uniwax. [Astor Wax] Synthetic waxes.

Uniwax. [Unichema; Unichema France SA] Fatty acid amides; lubricant, slip and antiblock agent, release agent for plastics; defoamers.

Uniwax 1450. [UPI] PEG-6 and PEG-32.

Uniwet. [Unitex] Detergents, emulsifiers.

Uniwhite AO, KO. [UPI] Titanium dioxide.

Uniwhite Oil. [UPI] Mineral oil.

Unizeen. [UPI] Alkylamidopropyl dimethylamines or ethoxylated amines.

Unizol. [Lubrizol] Hydraulic/transmission fluid additives.

Unocal. [Unocal/Polymers] Styrene acrylic, vinyl acetate, vinyl acrylic, acrylic, SBR, PVDC emulsion polymers.

Unocal Plus. [Unocal/Nitrogen Group] Urea sol'n.; for foliar applic.

Unopol. [Stockhausen] Wire drawing lubricants.

Unsoft 475. [UPI] Quaternium-27.

Upamate. [UPI] Adipates, myristates, or palmitates.

Upamide. [UPI] Fatty acid alkanolamide or ethoxylated fatty amides.

Upamid Resin UPC-1283. [UPI] Nylon.

Upantifoam NS. [Nutex] Nonsilicone defoamer emulsion.

U Pink 113. [Presperse] Ultramarine pink, bismuthoxychloride.

Upiwax. [UPI] PEGs, ethoxylated ethers, or fatty alcohol blends.

Upiwax Synbee. [UPI] Synthetic beeswax.

UR. [Thermoset Plastics] Two-part polyurethane system; for potting applics.

UR-. [Compounding Tech.] Polyurethane, carbon or glass reinforced.

Urac. [Am. Cyanamid] Urea-formaldehyde wood adhesive.

Urac. [Anderson Labs] Reagent for urea determination.

Uracron. [DSM Resins UK Ltd.] Polyacrylic resins.

Urad. [DSM Resins UK Ltd.] Additives.

Uradil. [DSM Resins UK Ltd.] Uralacs for waterborne systems.

Uradur. [DSM Resins UK Ltd.] Moisture-cure polyurethane resins, polyisocyanate resins for one and two-component systems.

Uraflex. [DSM Resins UK Ltd.] Polyurethane resins and elastomers.

Uragum. [DSM Resins UK Ltd.] Rosin maleic, rosin phenolic, ester, gums, hydrocarbon hard resins.

Uralac. [GCA Chem.; DSM Resins UK Ltd.] Alkyds and saturated/unsaturated polyester resins; for powd. coating.

Uralite. [Hexcel] Polyurethanes; for elastomers, casting, surface coatings, adhesives.

Uralloy® Hybrid Polymer. [Olin] Polyurethane polymer in styrene monomer; used in fiber-reinforced unsat. polyester composites.

Uraloid. [Rowe Prods. Distribution] Urethane base coating.

Uramex. [DSM Resins UK Ltd.] Amino resins (plasticized and nonplasticized).

Uramul. [DSM Resins UK Ltd.] Polymer dispersions/emulsions; for architectural paints.

Uranox. [DSM Resins UK Ltd.] Epoxy esters, epoxy resins.

Urathix. [DSM Resins UK Ltd.] Thixotropic Uralacs.

Uraver. [DSM Resins UK Ltd.] Etherified phenolic resins, novolaks, resols, precondensates of phenolics.

Urecoll®. [BASF AG] Urea-formaldehyde or melamine-formaldehyde resins; for prod. of adhesives for paper processing, binders for granular materials, fiber webs, shoe cap materials, for papermaking.

Uredur. [Schering AG] Hardener for epoxy resins.

Ureflex. [Flexible Prods.] Polyurethane elastomer systems; for gaskets, o-rings, potted components, custom molded work.

Ureol®. [Ciba-Geigy Plastics UK] Polyurethane resin/hardener system.

Urepan®. [Bayer] Polyurethane elastomers.

Uresin. [Hoechst Celanese] Melamine and urea resins; for coatings.

Uresoft. [Kao] Anticaking agent for urea fertilizer.

Uresolve. [Dynaloy] Solvents for silicones, urethanes, photo resists, conformal coating removal.

Urestyl. [Resina Chemie BV] MDI.

UrethHALL®. [C.P. Hall] Polyester polyol; for urethanes.

Urex. [Witco] Modified urea; for rubber compounding.

Urez. [Degen] Urethane-modified alkyd.

Uricase S. [Novo Nordisk] Bacterial uricase; enzyme for uric acid determination.

Uroset. [Lawter Int'l.] Ink varnish quick set vehicle.

Urotuf. [Reichhold] Synthetic coating resin.

Ursol®. [BASF AG] Oxidation bases for fur dyeing.

URT. [Parkland Engineering Ltd.] Polypropylene.

Usalco. [U.S. Aluminate] Stabilized sodium aluminate sol'n.

USG 420 Landplaster. [U.S. Gypsum] Agri-gypsum.

USG Perlite. [U.S. Gypsum] Raw perlite ore.

Usol. [Standard Tar Prod.] Liq. wood preservative.

Usolan. [Croda Chem. Ltd.] Ethoxylated lanolin.

USP®. [Witco/Argus] Organic peroxide derivs.; initiator for curing polyester resins.

U.S.P. [Zinc Corp. of Am.] Zinc oxide.

USP Salt. [Akzo Salt] US pharmacopia salt for pharmaceuticals.

Ustilan®. [Bayer] Ethidimuron; herbicide for total weed control on noncrop land.

U-Tanol G. [UPI] Octyldodecanol.

U-Tanol HD. [UPI] Oleyl alcohol.

U-Thane. [Dow] Rigid foam insulation.

UV Absorber. [Fanwood] Benzophenones; uv absorbers.

UV-Absorber. [Bayer] Uv absorbers for protection of plastics and surface coating materials.

Uvasil 299. [Enichem Synthesis SpA] Hindered amine light stabilizer.

Uvasorb. [3-V; Sigma Prodotti Chimici] Benzophenones; uv absorbers for cosmetics, plastic additives.

UV-Chek. [Ferro/Bedford] Uv absorbers, light stabilizers for polymers.

Uvex. [Eastman] Cellulose acetate butyrate plastic sheet.

Uvinul®. [BASF; BASF AG] Benzophenones, acrylic acid derivs., or PABA derivs.; uv absorbers for sunscreen preps., cosmetics.

U Violet 109. [Presperse] Ultramarine violet, bismuthoxychloride.

Uvitex®. [Ciba-Geigy/Dyestuffs; Ciba-Geigy/Additives] Fluorescent whitener for optical brightening of polymers, textiles.

Uvithane. [Morton Int'l.] Urethane oligomers.

V

V-. [Wako Chem. USA] Azonitrile compds.; polymerization initiators.

V-0701 T 1/8″. [M&T Harshaw] Vanadia catalyst.

V-1065, 1066. [Perma-Flex Mold] RTV silicone.

V-9415. [Ferro/Color] NiSbTi; pigment for thermoplastic and thermoset resins.

VA-044, -086. [Wako Chem. GmbH] Azo polymerization initiator.

Vacoil. [Transene] Ultrahigh vacuum diffusion pump fluid.

Vaden-100. [Ronco Labs] Corrosion inhibitor concs.

Valbond®. [Air Prods.] Acrylic emulsions; print binders, stabilizer for textile applics.

Val Calm. [Traco Labs] Valerian root extract.

Valchem. [Air Prods.] Polymers.

Valcoat. [Air Prods.] Release coating.

Valdet. [Air Prods./Valchem] Wetting agent, surfactant, emulsifier, stabilizer.

Valeron Film. [Van Leer Flexibles] Cross laminated HDPE film and cross laminated LLDPE film.

Valfoam Stabilizer 341. [Air Prods.] Bath stabilizer and compatibilizer for textile applics.

Valfor®. [PQ Corp.] Sodium silicoaluminate; ion exchange and selective absorption/adsorption.

Valifast Colors. [Orient Chem.] Solvent soluble dyes.

Valiosol. [Orient Chem.] Liq. form solvent soluble dyes.

Valite. [Valite] Industrial resins and molding compds.

Valloy. [GE] Metal alloys.

Valox®. [GE Plastics; GE Plastics Ltd.] PET or PBT polyester resins, some glass or mineral reinforced; engineering resins for connectors, switches, potentiometers, health care prods., business machine and TV components, lighting components, structural foam applics.

Valscour. [Air Prods./Valchem] Phosphate ester; emulsifier used in textiles.

Valsof. [Air Prods./Valchem] Amine condensate; softener, lubricant.

Valtac®. [Air Prods.] Acrylic emulsion.

Valtec®. [Himont; Enimont Iberica SA] Polypropylene resins; for inj. moldings, blow molding, thermoforming, extrusion coating, sheet, profiles, strapping, multifilament fiber spinning, in prep. of pigment and stabilizer concs., film, appliances and automotive under-the-hood applics.

Value. [Marubishi Oil Chem.] Ethoxylated esters or ethers; emulsifier, dispersant, wetting agent, detergent for textiles, general use.

Valueline. [BASF] Organic pigments.

Valwet. [Air Prods./Valchem] Arylalkyl sulfonate blend; wetting agent, penetrant.

Vamac®. [DuPont; DuPont UK] Ethylene/acrylic elastomer; for hose, tubing, boots, seals, wire and cable, underhood automotive, heavy equipment, and industrial parts, vibration dampers, rolls, seals, gaskets, misc. molded items, and oil well drilling parts, plastic modification, adhesives.

Vanabate. [Stewart Hall] Fuel oil additive.

Vanadaban. [Crowley Tar Prods.] Vanadium remover.

Vanade. [Am. Ingredients] Sorbitan stearate and polysorbate 60; emulsifier blend for food industry.

Vanall. [Am. Ingredients] Sorbitan stearate, mono and diglycerides, polysorbate 60; emulsifier blend for food industry.

Vanamid. [R.T. Vanderbilt] Epoxy resin

curing agents.

Vanax®. [R.T. Vanderbilt] Accelerators, vulcanizing agents for rubber.

Vanchem®. [R.T. Vanderbilt] Chemical intermediate; adhesion promoter in adhesives; corrosion inhibitor and metal deactivator; crosslinking agent for elastomers and plastics.

Vancide®. [R.T. Vanderbilt] Fungicide, antimicrobial, preservative for latex, rubber, cosmetics, cutting fluids, coolants; paper mill slimicide; used in petrol. storage tanks, recirculating cooling towers, paper and paperboard, cotton fabric.

Vancryl. [Air Prods.] Acrylic emulsion; for inks.

Vandar®. [Hoechst Celanese/Engineering Plastics; Hoechst UK] Thermoplastic alloy (elastomer-modified PBT), some glass or mineral filled; used for automotive body components and housings, furniture, ski boots, appliances, clips, fasteners.

Vandex. [R.T. Vanderbilt] Selenium; sec. vulcanizing agent for rubber.

Vanease. [Am. Ingredients] Hydrogenated polysorbate 80, glyceryl lactyl palmitate, sodium carboxymethyl cellulose, sodium propionate, glyceryl tristearate, sodium benzoate, acetic acid; emulsifier for food industry.

Vanesta. [R.T. Vanderbilt] Ethoxylated fatty acid ester; starch additive.

Vanfre®. [R.T. Vanderbilt] Dispersant, processing aid, lubricant for natural and synthetic rubbers, mold lubrication; corrosion inhibitor.

Van Gel® B. [R.T. Vanderbilt] Magnesium aluminum silicate; thickener and visc. stabilizer for dispersions, paints.

Vanguard. [David Michael] Natural flavors.

Vanillin. [Rayonier] Specialty chemicals.

Vanisol BIS sodico-2. [Rhone-Poulenc Geronazzo SpA] Sodium bistridecyl sulfosuccinate; visc. depressant, stabilizer, and emulsifier for emulsion polymerization of PVC; dispersant for resins, pigments in plastics and org. media; base for rust inhibitors.

Vanisperse CB. [Borregaard LignoTech] Fractionated sodium salt of oxylignin; dispersant.

Vanlube. [R.T. Vanderbilt] Antioxidant, antiwear, antiscuff, extreme pressure agent, corrosion inhibitor, metal deactivator for industrial lubricants, petroleum fuels, solvents.

Vannox. [Nippon Nyukazai] Emulsifiers, wetting agents for insecticides.

Vanox®. [R.T. Vanderbilt] Antioxidant, antiozonant, stabilizer for rubbers, plastics.

Vanoxy. [R.T. Vanderbilt] Curing agents for epoxy resins.

Vanplast® R. [R.T. Vanderbilt] Sodium petroleum sulfonate and mixed petroleum process oil; anticorrosion agent.

Vanseal. [R.T. Vanderbilt] Sarcosines or sarcosinates; industrial surfactants, chelating agents for soaps, bath gels, shampoos, shaving creams, dentifrices, textile and leather processing.

Vansil®. [R.T. Vanderbilt] Wollastonite; extender pigment for solvent-thinned and latex paints.

Vanstay. [R.T. Vanderbilt] Organic phosphite; chelator, stabilizer, synergist, processing aid for PVC formulations.

Vantage. [Calgon] Automation and control technology for water treatment.

Vantalc® 6H. [R.T. Vanderbilt] Hydrous magnesium silicate; pigment, filler for coatings.

Vantard®. [R.T. Vanderbilt] N-(Cyclohexylthio)phthalimide.

Vanwax®. [R.T. Vanderbilt] Protective wax, sunchecking inhibitor for elastomers.

Vanzak. [R.T. Vanderbilt] Pitch control for paper mills.

Vanzyme. [R.T. Vanderbilt] Enzymes.

Vapor Screen System. [Surco Prods.] Perimeter odor control system.

Vaposector. [West Chem. Prods.] Odorless insecticide.

Varamide®. [Sherex] Fatty acid alkanolamide; thickener, foam booster/stabilizer, detergent, emulsifier, anticorrosive for cosmetics, household and industrial cleaners, textile scouring, metalworking fluids.

Varcum®. [OxyChem/Durez; Reichhold Chemie AG] Phenolic or furan resins; bonding agent, tackifier, epoxy curative, modifier, impregnating agent.

Varifoam®. [Sherex] Shampoo concs.

Varifos. [Sherex] Phosphate esters; cleaners, emulsifiers.

Varine. [Sherex] Fatty imidazolines; emulsifier, anticorrosive, raw material for shampoos, penetrating oils, antistats, corrosion inhibitors, paints, printing inks, textiles.

Varion®. [Sherex] Betaines, glycinates, sultaines, propionates, or diacetates; amphoteric surfactants for personal care prods., heavy-duty detergents, drilling operations.

Variquat®. [Sherex] Quaternary ammonium salts; germicide, algicide, disinfectant, sanitizer, emulsifier, antistat for swimming pools, water treatment, pesticides, food processing, dairy, restaurant, industrial and household prods., textiles.

Varisoft®. [Sherex] Quaternaries; fabric softeners, base for hair conditioners.

Varonic®. [Sherex] Fatty alcohol blends or ethoxylated glycerides, ethers, esters, or fatty amines; emulsifier, stabilizer, moisturizer, dye leveler, corrosion inhibitor, dispersant, wetting agent for hair conditioners, cosmetics, metalworking, oil field chemicals, textiles, agric., leather, fur, paints.

Varox®. [R.T. Vanderbilt] Peroxide derivs.; crosslinking and vulcanizing agent for elastomers, plastics.

Varox®. [Sherex] Amine oxides; foam booster/stabilizer.

Varsol®. [Exxon] Aliphatic hydrocarbon solvent.

Varstat®. [Sherex] Modified amines and alcohols.

Varsulf®. [Sherex] Sulfosuccinates; detergent, refatting agent for dishwash, fabric wash, rug and upholstery shampoos, personal care prods.

Vaslin. [Purflo DTL SA] Polyester.

Vastolein. [G. Whitfield Richards] Lubricant.

Vaycron. [Hydro Polymers Ltd./Vinyls Div.] Thermoplastic elastomers.

Vazo. [DuPont] Initiators for polymerization; catalyst for vinyl polymerization.

VC-. [Borden] PVC or PVC/PVAc copolymer resins; for blending, calendering, extrusion, pipe, conduit, film, tubing, siding, profiles, wire and cable coating, phonograph records, floor tile, sol'n. coatings, casting, foam applics.

VCR. [Viobin] Cocoa replacer.

VCX 11-548. [Henkel] Water-reducible amidoamine resin; epoxy curing agent.

Vector®. [Dexco] Styrene block copolymers

Vectra®. [Hoechst Celanese/Engineering Plastics; Hoechst UK] Liq. crystal polymers, some mineral, glass, carbon filled; used in electronics, fiber optics, automotive, aircraft/aerospace, chem. processing, industrial, mfg. fields, encapsulation of electronic components.

Vedoc. [Ferro/Powd. Coatings] Thermosetting powder coatings.

Vee Gee. [Vyse Gelatin] Gelatins.

Veegum®. [R.T. Vanderbilt] Magnesium aluminum silicate; thickener, visc. modifier, stabilizer, suspending agent, binder, dispersant for cosmetics, toiletries, toothpaste, pharmaceuticals, paints, textile finishes, chemical specialties, industrial applics.

Veeprex. [Champlain Industries] Autolyzed yeast extracts.

Veepro. [Champlain Industries] Protein hydrolysates.

Vegamino 30-SF. [Brooks Industries] Vegetable amino acids; moisture binding and substantivity agent for hair and skin care cosmetics.

Vegetone. [Kalsec] Yellow food color.

Vehicle A Conc. [Ultra Additives] Wetting agent.

Vein Seal. [DCS Color & Supply] Iron oxide compd.; used to eliminate subsurface porosity and expansion defects in castings.

Veko. [V&E Kohnstamm] Flavors, extracts, and essential oils.

Vekton®. [Norton Performance Plastics] Nylon 6, some filled with graphite or molybdenum disulfide; wear-resistant cast nylon for bearings, bushings, gears, rollers, cable wheels, roll covers, wear plates.

Velate. [Velsicol] Coalescing aid for latex emulsions.

Velex. [Day-Glo Color] Printing ink vehicles.

Velpar. [DuPont/Ag] Herbicide.

Velsan®. [Sandoz] Carboxylates; emollient for cosmetics.

Velustrol KPA. [Hoechst AG] Polyethylene emulsion; softener for oil-repellent finishing on cotton fabrics.

Velvacast. [Dry Branch Kaolin] Kaolin.

Velvacrest. [Reilly-Whiteman] Cationic softener for textiles.

Velvadri®. [Ashland/Foundry Prods.] Core and mold coatings.

Velvalite®. [Ashland/Foundry Prods.] Core and mold coatings.

Velvamine. [Rhone-Poulenc/Textile & Rubber] Alkyl imidazoline deriv.; softener for hair rinses, leather, textiles.

Velvanilla. [Crompton & Knowles] Blend of natural and artificial vanillas.

Velvaplast®. [Ashland/Foundry Prods.] Core and mold coating.

Velvaseal®. [Ashland/Foundry Prods.] Core and mold coatings.

Velvatex. [Vyse Gelatin] 250 Bloom gelatin.

Velvawash®. [Ashland/Foundry Prods.] Core and mold coatings.

Velvetex®. [Henkel/Emery/Cospha] Betaines, propionic acid, acetates; amphoteric surfactants for personal care prods., cleansers, foam drilling and blanketing.

Velvetol®. [Rhone-Poulenc/Textile & Rubber] Softener, lubricant, plasticizer, rewetting agent for textiles.

Velvetouch 1601. [Goldschmidt] Polysiloxane softener for textiles.

Velvet Veil. [Presperse] Mica/silica blends; micronized powders providing velvet-like, lubricious feel to skin prods.

Vendex. [DuPont/Ag] Miticide.

Venmet. [Morton Int'l.] Metal recovery reducing agent.

Venpure. [Morton Int'l.] Organic purification agent, industrial reducing agent.

Vensil. [Morton Int'l.] Silver recovery reducing agent.

Venta. [Ubbink Nederland BV] Polyurethane/polyethylene.

Ventsorb. [Calgon Carbon] Air purification units.

Venvat. [Morton Int'l.] Vat dye reducing agent.

Venyl. [Vecoplas] Polyamide 6, 6/6.

Veoceal Water Repellent. [Wacker Silicones] Solvent-free, all weather sealer.

Verafil®. [Ciba-Geigy GmbH] Long glass fiber reinforced thermoplastics; engineering thermoplastic.

Veragel. [Dr. Madis Labs] Aloe vera derivs.

Verajuice-Cold Processed. [Chemetics Labs] Aloe vera gel.

Veranthrene. [Miles/Organic Prods.] Textile dyes and pigments.

Veratraldehyde. [Rayonier] Specialty chemicals.

Verdoxan. [Henkel/Emery/Cospha] 2,2,5,5-Tetramethyl-4-isopropyl-1,3-dioxane; woody fragrance raw material.

Veriset. [Lawter Int'l.] Ink varnish quick set vehicle.

Veriwet. [Wetronics] Wetting agent.

Verona. [Miles/Organic Prods.] Textile dyes and fixatives.

Verozyme. [Miles/Organic Prods.] Enzymatic desizing agent.

Versabacs. [Dow] Carpet backing.

Versacure. [Henkel/Functional Prods.] Epoxy curing agent.

Versadet. [Oakite Prods.] General cleaning compd.

Versaflex®. [W.R. Grace/Organics] Acrylic or vinylidene chloride emulsion polymers; for factory finishes, paints, ceiling tile, wall board, paper coatings, textile treatments, release coatings.

Versal. [La Roche Chem.] Catalytic grade aluminas.

Versamag®. [Morton Int'l.] Magnesium hydroxide; filler, fire retardant, smoke suppressant for plastics, thermosets, elastomers.

Versamid®. [Henkel/Functional Prods.] Polyamide resin; epoxy curing agent.

Versamine®. [Henkel/Functional Prods.] Aliphatic amines; epoxy curing agent.

Versaprime. [Southern Coatings] Universal metal primer.

Versatint®. [Milliken] Fugitive tint.

Versa TL. [Hart Chem. Ltd.] Dispersants, antistats, crystal modifier for industrial water treatment, electrophotography, films, fibers.

Versa-TL. [Alco; Nat'l. Starch & Chem.] Sulfonated styrene/maleic anhydride; for water treatment, oil fields.

Versatyl-42. [Nat'l. Starch & Chem.] Octyl acrylamide/acrylates copolymer; hairspray polymer for systems containing high proportion of hydrocarbon propellant; aerosol and pump hairsprays, setting lotions, spritzes.

Versene. [Dow; Dow Europe] Chelating agents and micronutrients.

Versenex. [Dow] Chelating agent.

Versenol. [Dow] Chelating agent.

Versicon Condutive Polymer. [Allied-Signal] Inherently conductive polymer.

Versilan. [Harcros; Harcros UK] Surfactants for personal care prods., highly built liquids.

Versilok®. [Lord] Acrylic structural adhesives; for strong, durable bonds between metals, plastics, ceramics, glass, other substrates.

Versilube®. [GE Silicones] Silicone lubricant.

Versol. [Chemetics Labs] Aloe extract.

Vertagreen. [La Roche Industries] Mixed fertilizers.

Verton®. [LNP; ICI PLC] Nylon 6/6 or PPS, long glass fiber reinforced; long fiber thermoplastic composites.

Verv®. [Am. Ingredients] Calcium stearoyl-2-lactylate; starch and protein complexing agent; softener, conditioner for food prods.

Vespel. [DuPont; DuPont UK] Polyimide precision parts.

Vestamelt. [Hüls France SA] Polyamide 12, 6/12, and copolymers.

Vestamid. [Hüls AG] Polyamide 6/12 or 12, some glass reinforced; for inj. moldings, tech. parts, extrusions, cable sheathing, extrusion coatings, tubing, gears, bearings, film, sheets, food pkg.,

sports gear.

Vestenamer®. [Hüls Am.; Hüls AG] Polyoctenamer; blend component for other rubbers.

Vestodur. [Hüls France SA] PBT.

Vestofine. [Astor Wax] Micronized synthetic wax (polyolefin); for printing inks.

Vestolen®. [Hüls Am.; Hüls AG] HDPE or PP; for inj. molding, rotational molding, large containers, tubular film, coatings, extrusion, fibers, pkg., domestic articles, camping equipment, household and disposable articles, pressure pipe, monofilaments, coated fabrics, automotive, elec. goods.

Vestolit®. [Hüls AG] Rigid, emulsion, or microsuspension PVC; for paste processing, molding, extrusion, window profiles, foam.

Vestopal. [Hüls France SA] Unsaturated polyester.

Vestoplast. [Hüls France SA] Copolyolefins.

Vestoran®. [Hüls AG] Polyphenylene ether, some glass reinforced; molding compds. for instrument engineering, elec./electronic engineering, automotive engineering, office equip.

Vesto-Wax. [Astor Wax] Synthetic polyethylene wax; for mold release compounds, solv. and liq. polishes, solid wax compds., hot-melts, automobile rustproofing compds., solv.-dispersed agric. coatings.

Vestowax AS-1550. [Astor Wax] Polyethylene wax, calcium soap; lubricant for films, profiles, inj. molding.

Vestypor. [Hüls AG] Expanded polystyrene; for insulating material in the building and refrigeration industries; moldings used for chair shells, shoe soles, instep supports, net floats, buoys, flower pots, casting molds, and in the pkg. industry.

Vestyron. [Hüls AG] Polystyrene or HIPS; for inj. molding, extrusion, film extrusion, thermoforming, structural foam, pkg., technical parts, disposable cups, household articles, toys, elec. components.

Vetak. [Imperial Adhesives] Packaging

adhesive.

Vetfeedor. [Burlington Bio-Medical] Packaging adhesive.

Vialon®. [BASF; BASF AG] Liq. 1:2 metal complex dyes; for dyeing and printing nylon fibers.

Vibrapair. [Uniroyal] Urethane repair compd.

Vibraspray. [Uniroyal] Polyether-TDI urethane; sprayable liq. urethane for mining, geothermal, and amusement applics.

Vibrathane®. [Uniroyal; Uniroyal Chem. Ltd.] Polyurethane prepolymer; for castings.

Vibrin®. [Owens-Corning Fiberglas] Vinyl ester or polyester resins; corrosion-resistant resins for chemical tanks, pipe, fume handling equip.

VIC®. [Ashland] Polyester and acrylic urethane resins; for gas curing finishes for plastics, metal, heat-sensitive substrates.

Vical. [Pfizer] Ground limestone.

Vicir-E. [Companhia Industrial De Resinas Sinteticas CIRES SA] Emulsion PVC.

Vicir-S. [Companhia Industrial De Resinas Sinteticas CIRES SA] Suspension PVC.

Vicol. [Rhone-Poulenc] Water-soluble resin.

Vicotex. [Ciba-Geigy GmbH] Prepreg.

Vicron. [Pfizer] Ground limestone.

Vicryl. [Childers Prods.] Vinyl acrylic; protective coatings.

Victabrite. [Rhone-Poulenc Basic] Phosphoric acid.

Victastab. [Akzo] Stabilizers.

Victawet®. [Akzo] Wetting agent, penetrant, dispersant, stabilizer; for pkg. dyeing of nylon, acid-type cleaners, emulsion polymerization, starch coatings.

Victor. [Rhone-Poulenc Basic] Dicalcium phosphate dihydrate; for dentifrices.

Victory®. [Petrolite/Polymers] Microcrystalline wax; plastic wax for hot-melt adhesives; hot-melt coatings; in antisunchecking agents in rubber goods; elec. insulating agents, leather treating agents, water repellents for textiles, rustproof coatings, cosmetic ingredients, and as plasticizers.

Victrex®. [ICI Advanced Materials; ICI PLC] PEEK or PEK resins; high temperature resins.

Vicure®. [Akzo] Benzoin ether or methyl phenylglyoxylate; photo sensitizer or photoinitiator for uv-curable systems.

Vidar. [Solvay & Cie] PVDF.

Vifcoll. [Nikko Chem. Co. Ltd.] N-Cocoyl collagen peptide, sodium salt; detergent, emulsifier used in personal care prods., pharmaceuticals, food industry, and household cleaning prods.

Vigilan. [Fanning] Lanolin oil or deriv.; emulsifier, emollient, plasticizer, solubilizer, wetting agent for personal care prods.

Vigilante. [Am. Cyanamid/Ag] Insecticide.

Viking. [Marine Bio Prods.] North Atlantic cod liver oil.

Viking Ship. [Norsk Hydro AS] Fertilizers.

Vikoflex 7170. [Atochem N. Am.] Epoxidized soybean oil.

Vikoflex 7190. [Atochem N. Am.] Epoxidized linseed oil.

Vikol®. [Vikon] Antimicrobial, bacteriostat, algicide, fungistat for textiles.

Vikolev. [Vikon] Proprietary; dye leveler for polyamide, polyacrylamide, and polyester fibers.

Vikolox. [Atochem N. Am.] Olefin oxides.

Vikomul. [Vikon] Emulsifier/detergent blends; for emulsification of waxes and oils, starch and synthetic size removal.

Vikon®. [Vikon] Textile hand builders.

Vikopen VP. [Vikon] Wetting agent for caustic and acid textile processing.

Vikosperse. [Vikon] Sulfonated naphthalene formaldehyde condensate; low dusting, low foaming dispersant for disperse dyes.

Vilex. [Atomergic Chemetals] Denatonium benzoate; bittering aversive agent.

Vilmacupro. [Alcan Rubber & Chem.] Textile dyes and pigments.

Vilmafix. [Alcan Rubber & Chem.] Textile dyes and pigments.

Vilmamin. [Alcan Rubber & Chem.] Textile dyes and pigments.

Vinac®. [Air Prods.] Polyvinyl acetate emulsion, bead, powder; for adhesives, handbuilding of polyester and polyester/cotton blends.

Vinacryl. [Vinamul Ltd.] Acrylic dispersions.

Vinamul. [Vinamul Ltd.] Polymer dispersions.

Vinapol. [Vinamul Ltd.] Polymer powders.

Vinatex PVC. [Hydro Polymers Ltd./ Perf. Prods.] Plastisols.

Vineland MBT. [Vineland] Methylene bis thiocyanate.

Vinex. [Air Prods.] Thermoplastic polyvinyl alcohol copolymer resin; for extrusion, inj. or blow molding, tubular blown film, bottles, fiber for industrial or personal care applics.

Vinidur®. [BASF AG] Vinyl chloride/ acrylate graft copolymer; for weather-resistant parts, profiles for outdoor applics. (window frames), pipes, panels, films.

Viniflex. [Elastoplast SL] PVC granules and compds.

Vinkeel. [Vinings Industries] Chelating agents.

Vinnapas®. [Wacker Chemie GmbH] Vinyl acetate, VAE, or styrene/acrylic copolymers; binder for textiles, glass fiber, paints, paper coatings, thermal insulation systems; adhesive for flooring, walls, foam, tiles.

Vinnol®. [Wacker Chemie GmbH] Vinyl chloride/vinyl acetate, vinyl chloride/ethylene, or PVC/polyacrylate copolymers; binder for fabrics, glass fiber, paper coatings; textile auxiliary; impact modifier for rigid PVC.

Vinofan®. [BASF AG] Vinyl polymer dispersions; for adhesives, pkg. adhesives, binders for coating paper and board, textile coating, and for bonding fiber webs .

Vinoflex®. [BASF; BASF AG] PVC; for extrusion, calendering, and inj. molding (film, pipes, profiles, hollow articles, panels), rigid or flexible parts.

Vinol. [Air Prods.] Polyvinyl alcohol.

Vinsol®. [Hercules] Pinewood resin or derivs.; thermoplastic resin for lacquers, adhesives, elec. insulation, floor coverings, foundry molds; as extender for latexes; water repellent, air entrainment aid for masonry cements; asphalt emulsifier.

Vinuran®. [BASF AG] Polymers for modifying PVC.

Vinyl Acetate Monomer. [Quantum/ USI] Vinyl acetate.

Vinyloid. [Rowe Prods. Distribution] Vinyl base coatings.

Vinyloid. [Universal Chem. & Coatings] Exterior organosol.

Vinylube®. [Lonza] Ester wax; specialty plastic lubricant.

Vinymix. [Maurel Freres] PVC compd.

Vinyzene®. [Morton Int'l.; Morton Int'l. NV SA] 10,10´-Oxybisphenoxarsine sol'n.s; antimicrobial, bacteriostat, fungistat for PVC, other plastics, rubber, polylefins, PU, PS, CPE, hot-melt adhesives, film and sheet, extruded profiles, plastisols, molded goods, organosols, fabric coatings.

Viobin Multi-Germ Oil. [Viobin] Blend of wheat germ oil, corn germ oil, and sunflower oil.

Viobin Oil. [Viobin] Wheat germ oil.

Vipla. [European Vinyls Corp. GmbH] PVC.

Viplex. [Crowley Chem.] Plasticizer/extender for polyurethane, primer coatings, epoxy systems.

Vircol® 82. [Albright & Wilson Am.] Flame retardant for plastics and foams.

Virco Pet®. [Albright & Wilson Am.] Corrosion inhibitor.

Viroc. [Pfizer] Ground limestone.

Viscalex. [Allied Colloids] Thickeners.

Viscarin. [FMC/Marine Colloids] Carrageenan or salts; thickeners.

Viscasil®. [GE Silicones] Silicone fluids; emollient, defoamer, release agent, lubricant for cosmetics, polishes, paint additives, mechanical devices, textile softeners, petroleum refining, rubber and plastic mold release, film modifier in coatings, damping in mechanical/ elec. applics.

Visc-Ayd. [Daniel Prods.] Visc. modifiers for paints.

Viscogel. [Avebe Am.] Potato starches.

Viscol. [Sanyo Chem. Industries] Low m.w. polypropylene; lubricants and pigment dispersants for plastics.

Viscolene. [Vevy] PEG-150 stearate, PEG-150 distearate.

Viscomix. [Vyse Gelatin] 300 Bloom gelatin.

Visconorust. [Viscosity Oil] Rust preventives.

Viscontran. [Henkel] Cellulose derivs.

Visco-Seal R. [United Catalysts] Castor/ organoclay complex; thickener, thixotrope for caulks, sealants, mastics.

Viscosil®. [Boehme Filatex] Silicone emulsions; softeners and antistats for textiles.

Visco-Stab. [Atochem N. Am.] Paint stabilizer, visc. stabilizer for cuprous oxide-containing paints based on bioMeT antifoulant polymers.

Viscotrol-A. [Mooney Chems] Thixotropic thickener.

Viscovoss. [Vosschemie GmbH] Polyester resin systems.

Viscozyme. [Novo Nordisk] Carbohydrase; for cereal and vegetable processing.

Vismul. [Toho Chem. Industry] Nonionic blends; emulsifier, thickener for textile printing.

Visonoil. [Vevy] Hydrogenated mink oil.

Vis-Syn. [Viscosity Oil] Synthetic lubricants.

Vista. [Vista; Vista Chem. Europe] PVC compds.; for extrusions, molding, calendering, profiles, automotive parts, wire and cable jacketing, pipe, flexible and rigid applics.

Vista C-550, -560. [Vista] Sodium dodecylbenzene sulfonate; detergent, foamer.

VistaFlex®. [Advanced Elastomer Systems] Thermoplastic rubber; for appearance parts.

Vistalon®. [Exxon; Exxon Chem. Mediterranea SpA] Ethylene/propylene or EPDM rubbers; for molded goods, calendering, sponge, extrusion, hose, tubing, elec. insulation and jacketing, weatherstrip, conveyor belting, roofing membranes, in blends, for modifying polyolefins.

Vista LPA. [Vista] Mixts. of hydrotreated isoparaffins and naphthenes; solvents for food applics., pesticides, coatings, water, paper, and mining chemicals, textile lubricants, chemical processing, printing inks.

Vistanex®. [Exxon] Polyisobutylene; polymeric additive for rubber processing, in molded and extruded prods. and coated materials.

Vista Purge Dryblend. [Vista] PVC dryblend; extruder purging agent.

Vista SA. [Vista] Linear alkylbenzene sulfonic acid; surfactant intermediate.

Vista STXS. [Vista] Sodium toluene/ xylene sulfonate; hydrotrope, solubilizer.

Vista Xtralife. [Vista] PVC dryblend; for extrusion of weatherable sheet and profiles.

Vistel. [Vista] Inj. moldable rigid PVC.

Vistone. [Exxon] Oiliness and/or film strength improver for lubricants.

Visu-Glow. [La-Co Industries] Fluorescent gas leak detector.

Vitacell LS. [Serobiologiques] Yeast extract.

Vita-Cos®. [CasChem] Wheat germ glycerides.

Vitamarine. [Sederma] Mineral oil, docosahexenoic acid, eicosapentaenoic acid, algae extract.

Vitamin A Palmitate Exsyliposomes. [Exsymol] Vitamin A palmitate/ glycerophospholipids blend; for cosmetics.

Vitamin B Complex CLR. [Henkel/ Cospha] Yeast extract and B vitamins in water-alcohol medium; prods. for treatment of greasy hair, dandruff, and oily skin.

Vitamin F Series. [Henkel/Cospha] Complex of fatty acids; prods. for treatment of dry skin and hair.

Vitaplant CLR Oil-Soluble N. [Henkel/ Cospha] Calendula extract and lipoid extract of pig skins in oil medium; prods. for aging, damaged, and sunburned skin.

Vitaplant CLR Water-Soluble. [Henkel/Cospha] Echinacea extract and aloe juice in water-alcohol medium; prods. for aging, damaged, and sun-burned skin.

Vitavax. [Uniroyal] Fungicide.

Vitazyme®. [Brooks Industries] Complexed vitamins; for cosmetic applics.

Vitel. [Goodyear] Copolyester resins; for sol'n. and hot-melt adhesives and coatings.

Vitexol®. [BASF; BASF AG] Antifoam for textile printing, dyeing, and finishing.

Viton®. [DuPont; DuPont UK] Fluoroelastomers; for inj., compr., and transfer molding, extrusion, o-rings, gaskets, seals, fuel hose, sol'n. coatings, dipped goods, adhesives, blending.

Viton® Curative. [DuPont] Curatives in fluoroelastomer bases; curing systems for Viton®.

Vitrafos. [Rhone-Poulenc Basic] Sodium hexametaphosphate; glassy phosphates.

Vitrex. [Atlas Minerals & Chem.] Silicate mortar.

Vitride®. [Zeeland] Sodium bis(2-methoxyethoxy) aluminum hydride; reducing agent.

Vitrolan. [Sandoz] Textile dyes and pigments.

Vitromix BMC. [DSM Resins Espana SA] Polyester resin/fiberglass preimpregnated composite.

Vitroplast. [Atlas Minerals & Chem.] Polyester cement.

Vitrox C. [W.R. Grace/Davison] Polishing prod.

Vituf. [Goodyear] Pigmented polyester resins.

VLR. [Pfizer] Insecticide diluent.

Vocol®. [Monsanto] Zinc dibutylphosphorodithioate; accelerator for EPDM cures.

Volamine. [Aquatec Chem. Int'l.] Boiler compd.

Volan®. [DuPont] Methacrylato chromic chloride; bonding agent and coupler for glass reinforced laminates.

Volatile Silicone. [Union Carbide] Cyclomethicone; emollient, lubricant for skin creams and lotions, antiperspirants, bath oils, shaving prods., suntan lotions, colognes, hair care prods.

Volaton®. [Bayer] Phoxim; broad-spectrum insecticide.

Volclay. [Am. Colloid] Sodium bentonite; suspending agent, gellant, binder for household, cosmetics, pharmaceutical use, automotive prods., aerosols, paints, enamels.

Volpo. [Croda Inc.; Croda Chem. Ltd.] PEG ethers of oleyl or stearyl alcohols; emollient, lubricant, emulsifier, solubilizer for cosmetics.

Voltalef. [Atochem N. Am.; Atochem UK] Chlorotrifluoroethylene oils; lubricant, EP additive; used for compressor lubricants, hydraulic fluids, pump fluids, damping fluids, heat transfer fluids.

Voracel. [Dow] Carpet backing.

Voracor. [Dow] Polyurethane components.

Voralast. [Dow] Polyurethane components.

Voranate. [Dow; Dow Ahlen] Specialty isocyanates; for polyurethane industry for appliances, specialty foams.

Voranol. [Dow; Dow Ahlen] Polyether polyols; for polyurethane industry for appliances, rigid foams, adhesives, binders, sealants, coatings, flexbile slabstock foams, RIM and structural polymers, dynamic elastomers.

Voraspan. [Dow] Expandable plastic beads.

Voratron. [Dow Ahlen] Encapsulation and casting systems for elec./electronics industries.

Vorflex. [DRG Flexible Pkg.] Film for steam sterilization applics.

Vorite. [CasChem] Polymerized castor oil or urethane prepolymers; plasticizer, lubricant, penetrant, wetting agent, dispersant, coupling solvent, adhesion promoter for cellulose lacquers, inks, adhesives, polish, caulks, leather dressing, hydraulic fluids, rubber compding., gasket cement.

Vos/Hesaglas. [Peerless Prods. Ltd.] Cast acrylic.

Vossenblue. [Degussa] Iron blue pigment.

Vostec. [W.R. Grace/Dearborn] Corrosion inhibitor/passivator.

VPA No. 3 Processing Aid. [DuPont] Additive improving mold release or cure rate for Viton® fluoroelastomers.

VPC®. [Ashland] Vapor permeation curing coating systems.

V-Pyrol. [ISP] N-Vinyl-2-pyrrolidone with stabilizer; reaction rate accelerator for systems incl. adhesives, coatings, cosmetics, textiles, syn. fibers, textile sizes, protective colloids, lube oil additives.

VR-110, -160. [Wako Chem. GmbH] Azo polymerization initiators.

VS-103®. [Air Prods.] Vinyl foam stabilizer.

VSA. [Vinings Industries] Liq. sodium aluminate.

Vueguard. [Panelgraphic] Coating treatment for plastic sheet and molded articles providing resistance to scratching and abrasion.

Vulcabond®. [Akzo] Bonding agent, adhesion promoter for textiles, rubber, metal bonds.

Vulcan®. [Cabot] Oil furnace carbon black; for conductive and antistatic applics., plastics.

Vulcastab®. [Akzo] Stabilizer, thickener, gellant for latex rubber processing.

Vul-Cup. [Hercules] Bisperoxide; vulcanizing agent and polymerization catalyst for elastomers and plastics.

Vulkacit. [Miles/Polysar Rubber] Accelerators for rubbers.

Vulkadur. [Bayer; Miles] Formaldehyde resins; reinforcing and hardening agent, bonding agent for rubber goods, latex dips for textiles, tire cord.

Vulkalent. [Miles] Vulcanization retarders.

Vulkanol. [Bayer; Miles] Plasticizer, tackifier for rubber goods.

Vulkanox. [Miles] Antioxidant, antiozonant, antiflexcracking agents for rubbers.

Vulkaresin. [Hoechst Celanese/Fine Chem.] Phenolic curing resins for rubber.

Vulkasil. [Bayer] Precipitated silica; reinforcing filler for all rubbers and silicone rubber.

Vulkazon. [Miles/Polysar Rubber] Antiozonant for rubber goods.

Vulklor®. [Uniroyal] Additive for tire processing, insulated wire.

Vulkollan®. [Bayer; Miles] Water-crosslinked polyurethane elastomer; for pump diaphragms, coupling elements, solid tires, rollers, pump housings, seals, wipers, bushes and split sections for antifriction bearings and flexible couplings.

Vulnopols. [Alco] Rubber short stops.

Vultac®. [Atochem N. Am.] Alkyl phenol disulfides; vulcanizer, plasticizer, tackifier, accelerator for rubber, adhesives.

Vult-Acet®. [General Latex & Chem.] Polyvinyl acetate emulsion; latex for adhesives, coatings; hand modifiers for textiles.

Vult-Acryl®. [General Latex & Chem.] Acrylic emulsions; used for fabric coatings, adhesives, paper coatings; pigment binders; textile finishes.

Vultafoam. [General Latex & Chem.] Urethane prepolymer rigid foam systems.

Vultamol®. [BASF; BASF AG] Sodium naphthalene sulfonate condensate; dispersant for rubber syntheses and processing.

Vultellan. [Soc Airtec Industrie] Polyurethane.

Vultex®. [General Latex & Chem.] Raw or compounded natural rubber; for dipping compds., adhesives, carpet backing, textile coatings, paper coatings.

VVF. [Pfizer] Very fine pigments.

Vybar®. [Petrolite/Polymers] Synthetic wax; lubricant, anticaking agent, modifier for paraffin, candles, hot melt inks, mold release compds., plastic lubricants, protective coatings, polishes, slip and antimar additives.

Vybex. [Ferro/Engineering Thermoplastics] PBT or PET/PC blends, some glass reinforced; thermoplastics.

Vydate. [DuPont/Ag] Insecticide/nematicide.

Vydax. [DuPont] Fluorotelomer dispersion; mold lubricant and release agent for release coatings, water-based paints and inks; dry lubricant films.

Vydyne®. [Monsanto; Monsanto Europe] Nylon 6/6, 6/9, some glass or mineral reinforced; for inj. molding, extrusion, wire jacketing, elec. and electronic, appliance, and industrial applics.

Vygen®. [Vygen] PVC resins; for extrusion, calendering, inj. molding.

Vykacet. [Croda Food Prods. Ltd.] Acetylated monoglyceride; mold release agent for food industry.

Vykamol. [Croda Food Prods. Ltd.] Sorbitan ester/polysorbate blend; for cake mixes, confectionery coatings, whipped desserts.

Vykasoid. [Croda Chem. Ltd.] Monoglyceride and vegetable oil; emulsifier for dairy spreads.

Vylor. [DuPont] Nylon 6/6 monofilaments.

Vynamon. [ICI Am.] Plastic and textile pigments.

Vynathene®. [Quantum/USI] Vinyl acetate ethylene copolymers; for adhesives, coatings, extruded and molded goods.

Vynel. [Yorkshire Pat-Chem] Hand builders for cotton and blends.

Vyn-Eze®. [Syn. Prods.] Stearamide; antiblocking agent for plastic films.

Vyox. [Vevy] Tocopherol, triethyl citrate, BHA.

Vyram®. [Advanced Elastomer Systems] Thermoplastic elastomer; general-performance rubber replacement.

W13 Stabilizer. [Am. Maize Prods.] Modified amioca starch.

W 180. [Hart Chem. Ltd.] Esters/quaternary amines blend; softener, rewetting agent.

W-200 Alumina. [Universal Scientific] Alumina for chromatography. providing rewetting properties with increased suppleness esp. for felted substrates.

Wachsemulsion 1864. [Zschimmer & Schwarz] Carnauba wax emulsion.

Wacker Belsil. [Wacker Silicones] Silicone derivs.

Wacker HDK. [Wacker Chemie GmbH] Silica.

Wacker Silicone. [Wacker Chemie GmbH] Silicone derivs.

Waco. [Waco Am.] Textile dyes, pigments, antifoams.

Wacofix. [Waco Am.] Fixative for direct and fiber reactive dyes.

Wacor. [Landers-Segal Color] Anticorrosive, nontoxic pigment.

Wacoscour. [Waco Am.] Detergents for textile scouring.

Wacosoft. [Waco Am.] Softener for synthetics and natural fabrics.

Wacosol. [Waco Am.] Scouring and prescouring agent.

Wacowet DOSS-A. [Waco Am.] Sodium dioctylsulfosuccinate; wetting agent for textile applics.

Wadco. [J.C. Whitlam Mfg.] Cutting, threading oils.

Wafex. [Borregaard LignoTech] Unfermented calcium lignosulfonate; industrial binder, dispersant, emulsifier.

Wafolin. [Borregaard LignoTech] Calcium and/or magnesium lignosulfonates; pellet binder, nutritive additive in animal feeds.

Wakal® A. [Grünau] Alginates; gelling agents for desserts and filling creams.

Wakal® J. [Grünau] Locust bean gum; thickener and stabilizer for food industry.

Wakal® K. [Grünau] Carrageenan; gelling agents for desserts and filling creams.

Wallkyd. [Reichhold] Alkyd resins.

Wallpol®. [Reichhold/Emulsion Polymers] Vinyl acetate or vinyl acetate acrylic emulsion copolymers.

Wanin. [Borregaard LignoTech] Unfermented ammonium or sodium lignosulfonates; binder, dispersant, auxiliary tanning agent.

Warbex. [Am. Cyanamid/Ag] Famphur.

Warco® Endust. [Sequa] Minimizes dust and lint formation from polyester/cotton and polyester/rayon fabrics.

Warcoset®. [Sequa] Hand builder for polyester/rayon woven fabrics.

Wareflex®. [Sartomer] Dibutoxyethoxyethyl adipate; plasticizer.

Warel. [Reliance Chem. Prods.] Water repellent.

Wargonin Compact. [Borregaard LignoTech] Desugared sodium/calcium lignosulfonate; water-reducing, strength-increasing, air-excluding concrete additive.

Wargotan. [Borregaard LignoTech] Unfermented calcium lignosulfonate; auxiliary tanning agent, dispersant.

Wash. [Huntington Labs] Lanolized skin cleanser.

Waspaloy. [Haynes Int'l.] Heat-resistant alloys.

Watchung. [Cookson Pigments] Textile dyes and pigments.

Watcon. [Watcon] Chemicals and equip. for water treatment.

Water Lock®. [Grain Processing] Starch/acrylates/acrylamide copolymer; superabsorbent polymer.

Waterproofon. [Apex] Water repellents.

Water-Skipper. [Mateson] Resins.

Waterstop Rx. [Am. Colloid] Waterproofing for construction joints.

Wave®. [Air Prods.] Vinyl-acrylic copolymer emulsion; for paints and coatings.

Wavecore. [Dow] Polyethylene film.

Wavelene. [Flexible Reinforcements Ltd.] Nylon-reinforced polythene.

Wavincel. [Wavin Industrial Prods. Ltd.] PVC.

Wavincoil. [Wavin Industrial Prods. Ltd.] PVC.

Wavingas. [Wavin Industrial Prods. Ltd.] Polyethylene.

Wavin Safe. [Wavin Industrial Prods. Ltd.] PVC.

Wavin Supagas. [Wavin Industrial Prods. Ltd.] Polyethylene.

Wavin Supasure. [Wavin Industrial Prods. Ltd.] Polyethylene.

Wavin Sure. [Wavin Industrial Prods. Ltd.] Polyethylene.

Wavin Surefit. [Wavin Industrial Prods. Ltd.] PVC.

Waxenol®. [CasChem] Esters; binder, emollient, lubricant, emulsifier, solubilizer for cosmetics, toiletries, metalworking.

Waxoline. [ICI Am.] Oil soluble dyes for textiles.

Wayfos. [Olin] Organic phosphate esters; corrosion inhibitor, hydrotrope, detergent, wetting agent, coupling agent, solubilizer, lubricant, antistat, dispersant, emulsifier for pesticides, emulsion polymerization, drycleaning, paper/pulp processing, textiles, plastics, metals.

Wayhib®. [Olin] Phosphate esters; corrosion inhibitor, pipeline scale inhibitor, water circulating systems; for air conditioning, boiler treatment compds.

Wayplex. [Olin] Polyphosphonate or salts; dispersant, scale and corrosion inhibitor, sequestrant for detergents, paper coatings, bottle washing formulations, water treatment.

W&B. [Dryden Oil] Metalworking fluids.

WD. [ECC Int'l.] Kaolin; paper filling clay.

Wear-Dated. [Monsanto] Apparel and home furnishing textile prods.

Weatherpruf. [Kano Labs] Rustproof coating.

Wecobee®. [Stepan/PVO; Stepan Europe] Hydrogenated vegetable oil; synthetic cocoa butter; used in personal care prods., pharmaceuticals, food industry.

Wecotop. [Stepan/PVO] Soya and palm oil vegetable fat; food emulsifier.

Weedar. [Rhone-Poulenc/Ag] Herbicide.

Weed-Free. [Chapman] Herbicides; weed-killing composition.

Weed-Hoe. [Vineland] Monosodium methane arsonate or blends.

Weedone®. [Rhone-Poulenc/Ag; W.A. Cleary] Herbicide for control of annual and perennial broadleaf weeds on golf courses and ornamental turf areas.

Weedtrol 586. [Amax Industrial Prods.] Nonselective and contact herbicide.

Weevilgo. [Adco] Insecticides.

Weighter PM. [Sybron] Weighter for textiles giving soft full hand.

Weldmaster. [Nat'l. Starch & Chem.] Adhesives.

Wellamid. [Wellman; MBS Plastics] Nylon 6 or 6/6 resins, some glass and/ or mineral reinforced; engineering resin for inj. molded parts.

Wellezmid. [CP-Polymer-Technik GmbH] Nylon 6 and 6/6 compds.

Welltex 300 F. [Borregaard LignoTech] Fermented calcium lignosulfonate; paper sizing agent.

Welvic. [European Vinyls Corp. UK Ltd.] PVC compds.

WEP® 662P. [Ashland] Unsaturated polyester resin.

Wescat. [Western Polymer] Cationic potato starch.

Weschem. [Wesco Tech. Ltd.] Powdered lignin sulfonates.

Wescodyne. [West Chem. Prods.] General purpose iodine disinfectant.

Weslig. [Wesco Tech. Ltd.] Liq. lignin sulfonates.

Wessalon. [Degussa] Amorphous precipitated silica; for pesticides industry.

Westchlor. [Westwood] Antiperspirant active ingredients.

Westo-Floc. [Western Water Management] Coagulants/flocculants.

Weston. [GE Specialty] Phosphites; stabilizer, chelating agent, reactive diluent; for adhesives, polymers.

Westo-Pac. [Western Water Management] Polymerized aluminum salt-based; coagulants/flocculants.

Westosan. [West Chem. Prods.] Quaternary disinfectant.

Westsafe. [West Chem. Prods.] Aerosol solvent; for cleaning motors, printed circuits, etc.

Westvaco®. [Westvaco] Tall oil rosins; emulsifier for emulsion polymerization, rubber prod.

Westvaco Diacid®. [Westvaco] Dicarboxylic acid derivs.; intermediates.

Wetaid. [C.H. Patrick] Alkali-stable wetting agent and emulsifier for batch and continuous bleaching and preparation.

Wetfix. [ScanRoad] Asphalt antistripping additive.

Wettable Sulfur. [Cuproquim] 97% Sulfur.

Wettable Sulfur. [FMC/Ag] Fungicide.

Wetting Agent. [Chem-Y BV] Organic sulfonic acid; wetting and washing agent.

Wetting Agent FCGB. [Henkel/Emery] Sodium laureth phosphate; wetting agent for metal surfaces, plating baths.

Wettol®. [BASF AG] Dispersant, emulsifier, wetting agent for agric. formulations.

Wet Zinc. [Morton Int'l.] External release coating for rubber.

WFT-78 Microbiocide. [Buckman Labs] Hexahydro-1,3,5-tris(2-hydroxyethyl)-s-triazine; preservative in soluble cutting fluids and synthetic coolants.

WG Mica. [KMG Minerals] Wet ground mica.

W.G.S. Hydrogenated Fish Glyceride. [Werner G. Smith] Triester of long chain fatty acids and glycerin; for wax compds., textile softeners and sizes, yarn lubricants, grease sticks, polishing compds., crayons, candles, leather stuffings, wire drawing compds., paper coatings, plastics.

Wheat Germ Oil CLR. [Henkel/Cospha] Wheat germ fatty oil, natural vitamin E

carrier; prods. for general skin protection.

Wheat-Pro EN-20. [Brooks Industries] Hydrolyzed wheat protein; for skin and hair care cosmetics.

Wheat-Tein NL. [Maybrook] Hydrolyzed wheat protein; substantivity agent, film-former, anti-irritant, protective, moisturizer for skin and hair care prods.

Whitcon®. [ICI Fluoropolymers] PTFE or FEP-based lubricant; lubricant, thickener, extreme pressure agent for coatings, oils, greases.

White Charcoal. [Ikeda] Silica, magnesium oxide.

White Crystal. [Morton Salt] Variety of rock and solar salt grades.

White Doe. [Virginia Dare Extract] Flavor oils.

White Swan. [Croda Chem. Ltd.] Lanolin BP; conditioner, moisturizer, emulsifier, emollient, superfatting agent for personal care prods., pharmaceuticals.

WHM. [Zimpro Passavant Environmental Systems] Chemical feed systems.

Wibarco. [Wibarco GmbH] Alkylbenzene sulfonic acid; detergent intermediate.

Wicera. [Paramelt Syntac BV] Microcrystalline wax; base wax for hot-melt adhesives, polishes.

Wickenol®. [CasChem] Esters or PPG ethers; lubricant, emollient, solubilizer, vehicle, solvent, plasticizer for cosmetics, toiletries, pharmaceuticals.

Wil-Add. [Akzo Engineering Plastics] Additive concs.

Wilbur-Ellis. [Wilbur-Ellis] Bentonite sulfur.

Wilbur & Williams. [California Prods.] Industrial maintenance coatings.

Wilclor. [George Mann] EPA registered sodium hypochlorite.

Wil-Foam. [Akzo Engineering Plastics] Blowing agent concs.

Wil-Lube. [Akzo Engineering Plastics] Lubricant concs.

Wilson. [Akzo Engineering Plastics Sweden AB] Color concs. and masterbatches.

Wiltrol. [Olin] Surface-treated phthalic

anhydride; scorch inhibitor for rubber stocks.

Wingdale White. [Georgia Marble] Calcium carbonate; general use filler.

Wingstay®. [Goodyear; Goodyear Europe; R.T. Vanderbilt] Rubber antioxidants and antiozonants.

Wingtack®. [Goodyear; Goodyear Europe] C5 hydrocarbon resins; tackifying and modifying resins.

Winnofil. [ICI Resins] Precipitated stearate-coated calcium carbonate; filler, impact modifier, processing aid for thermoplastics and paints.

Winnofos Mark II Bind. [ICI Am.] Aluminum chlorophosphate hydrate.

Winsor. [Witco] Cutting oils and rust preventives.

Winterex. [Avatar] Winterized soybean oil.

Winterking. [E&F King] Salt wetting sol'ns.

Winterpoxy. [Porter Int'l.] Low temp. curing polyamide epoxy primer and topcoat.

Winterzinc. [Porter Int'l.] Low temp. curing polyamide epoxy organic zinc-rich primer.

Wipe. [Merix] Antistatic agent.

Wisprofloc. [Avebe Am.] Potato starches; for water treatment.

Witafrol®. [Hüls Am.] Antifoams for dairy industry, sugar industry, jams, food flavors, fruit juices, seasonings, other food prods.

Witarix®. [Hüls Am.] Hydrogenated vegetable oils or glycerides' special fats for chocolate and confectionery industry.

Witbreak. [Witco/Organics; Witco SA] Glycol esters, oxalkylated phenolics, or polymeric amine salts; demulsifier for petroleum industry and waste disposal of processing oils in metals.

Witcamide®. [Witco/Organics; Witco SA] Fatty acid alkanolamide; detergent, emulsifier, lubricant, wetting agent, penetrant, dye dispersant, conditioner, foam booster/stabilizer, visc. modifier, gellant for cosmetics, textiles, detergents, drilling fluids, metal processing, coatings.

Witcamine®. [Witco/Organics; Witco SA] Amines; emulsifier, lubricant, wetting agent, penetrant, corrosion inhibitor, antistat, dispersant for water treatment, petroleum industry, metal processing, textiles, ore flotation, leather.

Witcast. [Witco SA] Castable polyurethane.

Witcat. [Witco] Specialty catalysts.

Witco®. [Witco; Witco SA] Industrial detergents, dispersants, emulsifiers, wetting agents, hydrotropes.

Witcobond®. [Witco; Witco SA] Polyurethane aq. dispersions; for high-performance adhesives and coatings.

Witcodet. [Witco/Organics] Formulated detergent conc. base; for dishwash, carwash, shampoo, carpet and upholstery cleaners.

Witcolate. [Witco/Organics; Witco SA] Sulfated surfactants; detergent, emulsifier, foamer, coupler, solubilizer, penetrant, lubricant for personal care and industrial applics.

Witcomul. [Witco/Organics] Esters; emulsifier, thickener, emulsion stabilizer for industrial applics., drilling fluid additive for petroleum industry.

Witconate. [Witco/Organics; Witco SA] Sulfonic acid or salts; detergent base, emulsifier, foamer, wetting agent, coupler, stabilizer, dispersant for industrial and household detergents, emulsion polymerization, textiles, cosmetics, petroleum industry, metal processing, lube oils, paints.

Witconol. [Witco/Organics; Witco SA] Nonionic surfactants; wetting agent, emulsifier, lubricant, antifoamer, solubilizer for industrial uses, food processing, cosmetics.

Witcopaque. [Witco] Modified polystyrene latexes; opacifier for dishwash, shampoo, personal care prods.

Witcopearl 15. [Witco SA] Anionic surfactant/pearlizing agents; pearlescent additive for cosmetics.

Witcor. [Witco] Corrosion and scale inhibitor, surfactant for petroleum industry, water treatment.

Witcosperse. [Witco] Wetting agent for

pesticide formulations.

Witepsol®. [Hüls Am.] Hydrogenated coco-glycerides; suppository bases.

Witflow. [Witco] Surfactants for coatings.

Witocan®. [Hüls Am.] Vegetable triglyceride; cocoa butter replacement.

Witresin. [Witco] Hard hydrocarbon extender and plasticizer for rubber compds.

Witsol. [Witco/Sonneborn] Aliphatic ink oil solvent.

Wolfaid. [Union Camp] Dispersant, processing aid.

Wolfamid. [Union Camp] Nonreactive polyamides.

Wolfkur. [Union Camp] Reactive polyamides.

Wolflex. [Union Camp] Polymeric plasticizers.

Wollastokup®. [Nyco Minerals; Malvern Minerals] Surface-modified wollastonite.

Wolverine. [Crompton & Knowles] Essential oils.

Wonder Shortening. [Van Den Bergh Foods] Partially hydrogenated soybean oil, partially hydrogenated cottonseed oil, water, vegetable mono and diglycerides, salt, nonfat dry milk, TBHQ.

Wondrop. [Lockrey] Photographic wetting agent.

Woodepox. [Abatron] Wood filler and patching compd.

Woodguard. [Abatron] Wood coating.

Wood-Lok. [Nat'l. Starch & Chem.] Adhesives.

Wood Nu'N'Lite. [Chapman] Wood bleach.

Wood-Stik. [Nat'l. Casein] Adhesive.

Woodthane. [Flexible Prods.] Rigid polyurethane systems; for furniture, mirror and picture frames, decorative molding, carvings.

Woodtreat C8. [Kop-Coat] Copper-8-quinolinolate sol'ns.; mold, stain, and mildew control for wood.

Woodtreat MB/NB. [Kop-Coat] Formulated water repellent preservative for millwork and similar treatments.

Woodtreat WB. [Kop-Coat] Water-based repellent preservative for millwork treatment.

Wood Tuff. [Chapman] Wood coating.

Worblex-CA. [Albrow Prods. Ltd.] Cellulose acetate.

Worblex-CAB. [Albrow Prods. Ltd.] Cellulose acetobutyrate.

Worblex-CP. [Albrow Prods. Ltd.] Cellulose propionate.

Worblex PE. [Albrow Prods. Ltd.] Polyethylene.

Worblex-PS. [Albrow Prods. Ltd.] Polystyrene.

Wrens Filler. [ECC Int'l.] Kaolin; paper filling clay.

Wrico®. [Drew Ind. Div.] Antifoulant.

Wubalen. [Finke-Farbstoffe] Dry pigments.

Wyk. [Upright] Absorbents for containment and clean-up of acids, caustics, solvents, toxic, hazardous and nontoxic materials.

Wytox®. [Uniroyal] Phosphites or hindered phenols; antioxidant, stabilizer for plastics and rubber.

XY

X-12. [Spartan Flame Retardants] Flame retardant compd.

X50-S. [Degussa] Si69 and N330 carbon black; reinforcing filler for rubber industry.

X78-2. [Reilly-Whiteman] Sulfated oil; leather lubricant; synthetic sulfated sperm oil replacement.

X-743. [Neville] Aromatic plasticizer; for adhesives (mastic, pressure sensitive), rubber (cements, mechanical and molded goods, tires), and caulking compds.

Xalidrene. [Vevy] PEG-20 myristate and PEG-20 palmitate.

Xalifin 15. [Vevy] C12-10 acid PEG-8 ester.

Xanco-Frac®. [Kelco/Oil Field Prods.] Heteropolysaccharide prod.; gum used as viscosifier in oil field hydraulic fracturing fluids; suspending agent.

Xanflood®. [Kelco] Industrial grade xanthan gum; foam stabilizer, flocculant, suspending agent, gellant, rheology modifier, lubricant for industrial applics. esp. for sec. and tert. oil recovery.

Xantar. [Xantar Polycarbonates VoF] Polycarbonate, some glass-filled; for household appliances, computer systems, precision instruments, toys, medical prods., food pkg.

Xanter. [DSM Verkoopkantoor Polymeren Nederland] PC.

Xanthates. [Hoechst AG] Sodium xanthogenates; flotation collector for sulfide and sulfidized minerals.

XC 9. [Releasomers] Semipermanent release dispersion.

XCE-89. [Air Prods.] Chain extender.

X-Coat. [E/M Corp.] EMI/RFI shielding coating.

XEA 9361. [Hysol Aerospace Prods.] Two-component adhesive.

Xenoy®. [GE Plastics; GE Plastics Ltd.] Thermoplastic alloys, some glass reinforced; for blow molding, structural foam applics., automotive exterior body parts, fluid handling equip., medical prods., lightweight lawnmower casings.

X-E-Tex Jet 200. [Leatex] Low foaming fixative for acid dyes on nylon and wool.

XK 22. [Releasomers] Nonsilicone semipermanent mold release agent.

X-Link. [Nat'l. Starch & Chem.] Cross-linking vinyl acetate copolymer.

XO White. [Georgia Marble] Ground calcium carbonate; filler for neutralization of acids, syn. marble, aggregate for cement finishes, vinyl asbestos tile, welding rods, polyester and epoxy floor tiles.

XOX. [Thoro Prods.] Bleach.

X-Pal. [W.R. Grace/Davison] Polishing prod.

X-Pand'R. [A.E. Staley Mfg.] Modified corn starch.

XR 7. [Releasomers] Semipermanent mold release agent.

XSA. [Hart Chem. Ltd.] Xylene sulfonic acid; catalyst for resins.

XT 66. [Releasomers] Mold sealer for tool sealing in the composite industry.

X-Tan® Special C. [Sybron] Suppresses chlorine release in chlorite bleaching; prevents metal corrosion.

XTherm. [Transene] Heat sink compd. for electronic applics.

Xtol. [Georgia-Pacific] Tall oil fatty acid.

XT® Polymer. [Cyro Industries] Acrylic multipolymer; sheet for thermoformed pkg. applics.

XTR-155A. [Himont] Thermoplastic elastomer; for seals and gaskets.

Xtru-Sets. [Westlake Plastics] Extruded thermosets.

Xtrusorb. [Calgon Carbon] Activated

carbon; for vapor phase applics.

XU. [Dow Plastics] Resins.

XXX-1. [CasChem] Lubricant for trolley lubrication in meat packing plants, conveyor and equip. for bakeries, canneries, food/beverage operations.

Xycon Hybrid Resin. [Amoco] Two-component polymer; thermoset for truck-body panels, after-market automotive parts, recreational vehicles, outdoor equip., tubs and showers, elec. parts.

Xydar®. [Amoco; Amoco Europe] Liq. crystal polymer, some glass and/or mineral filled; for appliances, elec./electronics, aerospace, automotive, ordnance, lighting applics.

Xylene®. [Sandoz] Specialty dyes for coloring aq. media.

Xyligen®. [BASF AG] Active ingredients for wood preservative formulations; additives to glues for wood applic.

Xytrex®. [EGC] Polyetherimide, PES, polyketone, PEEK, or PPS, some carbon or glass filled; for inj. molding, compr. molding, extrusion, coating, bearing applics.

Xzene. [Colgate-Palmolive] Room deodorant ingredient.

Y-25 Paint Additives. [United Catalysts] Suspension agent, antisettling agent.

Y-40. [United Catalysts] Organic paste; antisettling agents.

Yeastal. [Champlain Industries] Brewers yeast.

Yelkin. [Ross & Rowe] Soy lecithins.

Yellow 201. [Presperse] Iron oxides, bismuthoxychloride.

Yellowstone. [Montana Sulphur & Chem.] Sulfur flakes.

Yeoman. [Croda Chem. Ltd.] Anhydrous lanolin BP; emollient for personal care prods., pharmaceuticals.

Yogurtab. [Pharmachem Labs] Directly compressable granulation of lactodacillus, acidophillus, and bulgaricus.

Yoracryl. [Yorkshire Pat-Chem] Cationic basic dyes for acrylic, modified acrylic, and other fibers.

York. [United Catalysts] Castor oil; dye solvent, gloss agent, emollient for lipsticks, cosmetics.

York White. [R.E. Carroll] Ground limestone; filler.

Youmex. [Sanyo Chem. Industries] 2,5-Furandion, polymer with 1-propene or ethene; dispersants for fillers and pigments.

YSE-Cure. [Ajinomoto] Amines; Epoxy curing agents.

Z

Zaclon®. [C.P. Hall] Galvanizing fluxes.

Zap. [East West Minerals] All-purpose absorbent.

Zapon®. [BASF; BASF AG] Metal complex dyes; for lacquers, baking finishes, polyurethane lacquers, wood stains, textiles.

Zappit. [PPG Industries] Shock treatment and super chlorinator.

Zeeantimigrant. [DeeZee] Stops bleeding of dyes prior to aftertreatment.

Zeebleach C. [DeeZee] Mild peroxygen bleaching and oxidizing agent for cellulosics, synthetics, and blends.

Zeeclean. [DeeZee] Alkyl sulfonates/ nonionic blends; all-purpose cleaner, grease emulsifier, machine cleaning, textile washing.

Zeedefoam C. [DeeZee] Silicone defoamer.

Zee-emul. [DeeZee] Alkyl sulfonates/ nonionic blend; emulsifier for carriers in textile dyeing, pesticides.

Zeefix. [DeeZee] Direct dye fixative.

Zeelev. [DeeZee] Leveling agent.

Zeelube. [DeeZee] Antistat, lubricant for dyeing and finishing of natural and synthetic yarn and piece goods.

Zeenap. [DeeZee] Napping agent for polyester/cotton and acrylic/cotton blends.

Zeeospheres®. [Zeelan] Silica-alumina ceramic; inert hollow spheres for use as filler for a variety of plastic resins in inj. molding, extrusion, SMC, BMC, RTM, compression molding, potting/ encapsulating, adhesives, tooling, casting, flooring, grouting, sealants, mastics, coatings, films.

Zeequest. [DeeZee] Chelating agent for bleaching, scouring, dyeing, and other wet processing of textiles.

Zeescour. [DeeZee] Alkyl sulfonates/ nonionic blend; detergent, emulsifier.

Zeesoft. [DeeZee] Softener for textiles.

Zeesperse. [DeeZee] Dispersant and leveling agent for disperse dyes on nylon, polyester, and acrylic fibers.

Zeestat. [DeeZee] Antistats for dyeing, finishing, spinning and carding systems.

Zeestrip. [DeeZee] Desizing agents.

Zeeterge. [DeeZee] Carboxylated alcohol; specialty cleaner.

Zeewhite. [DeeZee] Stilbene deriv.; optical brightener for cellulosics and nylon.

Zelcon®. [DuPont] Fabric conditioner, soil release, finish for polyester.

Zelec®. [DuPont] Fatty alcohol phosphates; antistat for textiles, plastics, films.

Zellamid. [Senova Kunststoffe GmbH] Polyamide.

Zelux W®. [Westlake Plastics] Polycarbonate; for windows in combat helicopters, visors on space helmets, protective enclosures for bank tellers, portholes in pressure chambers.

Zemid®. [DuPont] Toughened HDPE resin, reinforced; for inj. and blow molding, sheet, structural foam, thermoforming, lawn and garden equip., industrial machinery, cold weather applics.

Zeniplex 2. [Atochem N. Am.] Extreme pressure grease.

Zeo® 49. [J.M. Huber] Hydrated silica; polishing agent, catalyst support.

Zeoclean. [Western Water Management] Zeolite softener resin cleaner.

Zeodent®. [J.M. Huber] Hydrated silica; polishing agent, antiskid agent.

Zeofree®. [J.M. Huber] Hydrated silica; carrier, filler, defoamer, anticaking agent, free-flow aid.

Zeolex®. [J.M. Huber] Sodium silicoaluminate; reinforcing filler, conditioner, anticaking agent, absorbent for rubber, paper, printing inks, powdered detergents.

Zeolum. [Tosoh] Crystalline hydrous alumino-silicate; adsorbent for natural gas, petrochemical, refrigeration, paint, insulated gas, gas chromatography.

Zeomatt 155. [J.M. Huber] Silica; flatting agegnt.

Zeonet. [Zeon] Curing agent, accelerator, retarder for rubbers.

Zeoquest. [W.R. Grace/Dearborn] Resin cleaner.

Zeospan. [Zeon] Polyether elastomer; for automotive components where cold, heat, ozone, and oil resistance are desired.

Zeosyl®. [J.M. Huber] Hydrated silica; carrier, reinforcing agent, rheology agent, thickener, adsorbent for liq. detergents.

Zeothix®. [J.M. Huber] Hydrated silica; thickener, thixotrope, defoamer, anticaking and free-flow agent, flatting agent.

Zepel. [DuPont] Fluoropolymer; fabric fluoridizer imparting oil and water repellency.

Zerogen®. [Solem Industries] Halogen-free proprietary compds.; flame retardant and smoke suppressant for thermoplastics and elastomers.

Zerol. [Shrieve Chem. Prods.] Refrigeration compressor lubricant.

Zetabon. [Dow] Plastic clad metals.

Zetax®. [R.T. Vanderbilt] Zinc 2-mercaptobenzothiazole; accelerator for latex foam curing systems.

Zetesap. [Zschimmer & Schwarz] Disodium lauryl sulfosuccinate blends; basic material for synthetic toilet soap bars.

Zetesol. [Zschimmer & Schwarz] Sulfates or blends; detergent, emulsifier for personal care and household prods.

Zetpol®. [Zeon] Nitrile rubbers; for fuel hose, fuel diaphragms, o-rings, packings, gaskets, oil seals, belt, rolls.

Zewakol. [Borregaard LignoTech] Modified lignosulfonates; extenders of urea-formaldehyde resins in particle board mfg.

Zewalon FN. [Borregaard LignoTech]

Sodium lignosulfonate; auxiliary tanning agent.

Zewa SL 2. [Borregaard LignoTech] Desugared sodium lignosulfonate; water-reducing concrete additive; dispersant for pesticides.

Zimag Bar. [Akrochem] Dispersed zinc oxide and magensium oxide in bar form.

Zimate®. [R.T. Vanderbilt] Zinc dithiocarbamates; accelerator for natural and polyisoprene rubbers.

Zimpro. [Zimpro Passavant Environmental Systems] Wet oxidation unit, high pressure pump.

Zinar. [Arizona] Zinc resinate.

Zincidone®. [UCIB] Zinc PCA; cicatrizing and tissue hardening agent; for dermatological soap, shampoo, shower gel, deodorants, nutritive creams.

Zinc-Lock. [Porter Int'l.] Zinc-rich coatings.

Zinc Omadine®. [Olin] Zinc pyrithione; antimicrobial, antidandruff agent for cosmetics, metal coolant and cutting fluids, PVC plastics, fabrics.

Zinc Oxide 35. [Akrochem] Precipitated zinc oxide; accelerator/activator for rubber.

Zinc Oxide No. 185, 318. [Eagle Zinc] Zinc oxide; accelerator/activator, pigment, reinforcing agent for rubber.

Zinc Oxide Transparent. [Miles/Polysar Rubber] Precipitated zinc oxide; vulcanization accelerator/activator for transparent rubber goods.

Zinc Oxide USP 66. [Whittaker, Clark & Daniels] Zinc oxide.

Zinc Pyrion®. [Pyrion-Chemie GmbH] Zinc pyrithione; antidandruff agent, preservative, antibacterial, antimicrobial.

Zinc Stearate. [Witco] Zinc stearate; lubricant, antitack agent, mold release agent for plastics, rubber.

Zincum. [Bärlocher GmbH] Metallic soaps.

Zink-Gro. [Am. MicroTrace] Zinc sulfate monohydrate + 35.5% zinc; high water soluble; for use in dry fertilizers.

Zink-Gro AS. [Am. MicroTrace] Zinc sulfate monohydrate + 35.5% zinc; for ammoniation applics. and for direct applic.

Zinkoxyd Activ. [Miles] Precipitated zinc oxide; vulcanization accelerator/activator for rubber.

Zink Pyrion 48%. [Ruetgers-Nease; Pyrion-Chemie GmbH] Zinc pyridinethione; antibacterial, topical antifungal, antiseborrheic.

Zinox. [Am. Chemet] Zinc oxide.

Zinplex 15. [Ultra Additives] Crosslinker for ink, floor polish, emulsion polymers.

Zinpol. [Zinchem] Various polymers, sol'ns., emulsions.

Zinstabe. [Zinc Corp. of Am.] Zinc oxide-based; activator, stabilizer for PVC foams.

Zipp. [Stewart Hall] Elec. equipment degreaser.

Zipp. [Zipp Industries] Dry fertilizer.

Zip Stik. [Ashland/Foundry Prods.] Core paste.

Zip-Slip®. [Ashland/Foundry Prods.] Release agent.

Ziram. [Atochem N. Am.] Fungicides.

Ziram. [FMC/Ag] Fungicide.

Zircadyne. [Teledyne Wah Chang Albany] Zirconium alloys.

Zircar. [Zircar Prods.] Zirconium oxide; fabricated prods., fibrous ceramic thermal insulation.

Zircat. [Ultra Additives] Ink, paint and varnish driers.

Zircopax. [TAM Ceramics] Zircon opacifier.

Zircore. [DuPont] Zirconium silicate/aluminum silicate sand.

Zircotan. [Rohm & Haas] Zirconium; synthetic tanning agent.

Zirex. [Arizona] Zinc resinate.

Zirgel K. [MEI (Magnesium Elektron Inc.)] Paint gellant.

Zirox. [TAM Ceramics] Zirconium dioxide.

Zirtung. [GTE Prods.] Zirconium dioxide doped tungsten electrodes.

Zisnet F-PT. [Zeon] 2,4,6-Trimercapto-s-triazine; curing agent for epichlorohydrin rubber.

Zitrilon® 10%. [BASF AG] Zinc chelate with 10% Zin; for foliar applic.

ZMBT. [Akrochem] Zinc 2-mercaptobenzothiazole; accelerator.

Zn-. [M&T Harshaw] Zinc chromite or zinc oxide; catalysts.

ZN-7. [Advanced Refractory Tech.] Zirconium diboride; for oxidation-resistant composites, burnable absorber of neutrons, elec. contacts, molten metal crucibles, refractory toughener, cutting tool composites, structural ceramics, wear components, metal matrix composites.

ZO-9. [Disco] Fatty acid salts and petroleum derivs.; plasticizer and lubricant for polymers.

Zohar. [Zohar Detergent Factory] Esters; stabilizer, opacifier, pearlescent, emollient, conditioner, superfatting agent for cosmetics, pharmaceuticals.

Zohar. [Zohar Detergent Factory] Sodium alkylbenzene sulfonate and builders; detergent powders.

Zoharconc. [Zohar Detergent Factory] Shampoo, bubble bath, soap, dishwashing, laundry, floor cleaner, rinse aid, textile softener concs.

Zoharex. [Zohar Detergent Factory] Fatty acid polyglycol ester; detergent for laundry prods.

Zoharfoam. [Zohar Detergent Factory] Foaming agent for drilling operations.

Zoharlab. [Zohar Detergent Factory] Alkylbenzene sulfonate; raw material for mfg. of detergents.

Zoharphos A-3. [Zohar Detergent Factory] Potassium alkyl ether phosphate; antistat for syn. fibers.

Zoharpon. [Zohar Detergent Factory] Sulfates or sulfosuccinates; raw material for personal care prods., household cleaners, pharmaceuticals.

Zoharquat. [Zohar Detergent Factory] Benzalkonium chloride; bacteriocide, algicide.

Zoharsoft. [Zohar Detergent Factory] Fabric softener base.

Zoharsyl. [Zohar Detergent Factory] Sodium lauroyl sarcosinate; raw material for mfg. of shampoos, conditioners, toothpastes, carpet and upholstery shampoos.

Zohartain. [Zohar Detergent Factory]

Alkyl betaine; foam booster, mild surfactant for shampoos and detergents.

Zohartaine TM. [Zohar Detergent Factory] Dihydroxyethyl tallow glycinate; thickener, anticorrosive agent for tech. acid formulations, mild shampoos.

Zoharteric. [Zohar Detergent Factory] Amphoteric surfactants for personal care prods., cleaners.

Zoldine®. [Angus] Oxazolidines; crosslinking agent, catalyst, resin reactant, formaldehyde substitute, corrosion inhibitor, tanning agent, raw material for polymer synthesis.

Zonarez®. [Arizona] Polyterpene resins; thermoplastic polymers for adhesives, rubber cements, emulsion adhesives, hot melt adhesives/coatings, can sealants, caulking and general sealants., ink, paints, concrete waterproofing agents, varnishes, chewing and bubble gum bases.

Zonatac®. [Arizona] Terpene hydrocarbon resins; thermoplastic tackifying resins for adhesives and coatings.

Zone Defense. [In-Cide Tech.] Insecticide.

Zonester®. [Arizona] Rosin esters; thermoplastic resin, tackifier for rubber, adhesives, contact cements, coatings.

Zonolite. [W.R. Grace/Construction Prods.] Vermiculite, insulation materials.

Zontes. [Matsumoto Yushi-Seiyaku] Alkyl polyamide deriv.; softener for textiles.

Zonyl®. [DuPont] Fluorochemical surfactant; fluoridizer, emulsifier, lubricant, wetting agent, dispersant, corrosion inhibitor, foamer, flotation agent for textiles, adhesives, agric., cleaners, paper coatings, polishes, polymerization, paints, fire fighting, ink, oil, plastics.

Zoramide CM. [Zohar Detergent Factory] Cocamide MEA; foam booster, thickener, superfatting agent.

Zoramox. [Zohar Detergent Factory] Coconut amido alkyl amine oxide; wetting agent, foam booster/stabilizer, visc. builder for shampoos, bubble baths.

Zorapol. [Zohar Detergent Factory] Sodium lauryl sulfate; foaming agent, emulsifier for synthetic latexes, emulsion polymerization, household detergents.

ZS-7. [Advanced Refractory Tech.] Zirconium diboride; for oxidation-resistant composites, burnable absorber of neutrons, elec. contacts, molten metal crucibles, refractory toughener, cutting tool composites, structural ceramics, wear components, metal matrix composites.

Z-Thane. [UTI] Polyurethane systems; compds., elastomers, foams, rigid and flexible grades.

Zusolat. [Zschimmer & Schwarz] Alkyl polyglycol ether; dispersant, emulsifier, wetting agent.

Zusomin. [Zschimmer & Schwarz] Fatty amine ethoxylate; basic material for textile and dyeing auxiliaries.

Zylac. [Petrolite/Polymers] Plastics additives.

Zylar®. [Novacor] Methyl methacrylate butadiene styrene terpolymer; for displays, medical devices, office accessories, small appliances, toys.

Zymo-Best. [PMP Fermentation Prods.] Feed enzyme.

Zytel®. [DuPont; DuPont UK] Nylon 6, 6/6, 6/12, some glass reinforced; engineering thermoplastic for gears, bushings, bearings, elec. connectors, automotive parts, wire jacketing, film, rod, tubing, fiber optics.

Zytel®. [Sandoz] UV absorber for exhaust dyeing.

# #

#1, 15, 30, 40 Oil. [CasChem] Castor oil and polymerized castor oils; emollient for industrial, cosmetic, pharmaceutical applics.; plasticizer, lubricant, penetrant, wetting agent, dispersant, coupling solvent, adhesion promoter for cellulose lacquers, inks, adhesives, polish, caulks, leather dressing, hydraulic fluids

1,1,1-Trichloroethane. [Ethyl] Trichlorethane; chlorinated solv. for cold cleaning, vapor degreasing, in formulations.

1-K, 10-KS, 100-K Mica. [KMG Minerals] Dry ground mica.

2-1-5 Acid. [Molecular Rearrangement] 2-Diazo-1-naphthol-5-sulfonic acid, sodium salt.

2-1-5 Chloride. [Molecular Rearrangement] 2-Diazo-1-naphthol-5-sulfonyl chloride.

2 Plus 2. [ISK Biotech] MCPP + 2,4-D amine; postemergence turf care herbicide.

3 CC. [Sigma Prodotti Chimici] 3,4,4′-Trichlorocarbanilide; bacteriostat.

4-K Mica. [KMG Minerals] Dry ground mica.

9-11. [CasChem] Acid.

10X Breaker. [Rhone-Poulenc/Perf. Resins & Coatings] Enzyme prod. for hydrolyzing soluble polysaccharides (derivatized guar, cellulose ethers).

12-HSA. [CasChem] Hydroxystearic acid; chemical intermediate.

20-K Mica Flakes. [KMG Minerals] Dry ground mica.

76 RES. [Unocal/Polymers] Emulsion polymers.

90 Ram HS. [Premier Refractories] Chemically bonded high alumina erosion-resistant plastic.

112 Concentrate. [West Agro] Iodine hand cleaner conc.

150 Bloom. [Vyse Gelatin] 150 Bloom gelatin.

200® Fluid. [Dow Corning] Silicone fluid; liq. dielectric, coolant, antifoam, surfactant, release agent, water repellent, for personal care prods., food pkg. and processing, release of acrylics, phenolics and urethanes.

220 Solvent. [Total Petroleum] Paraffin-based oil.

225 Bloom. [Vyse Gelatin] 225 Bloom gelatin.

282. [ECC Int'l.] Kaolin; paper coating clay.

288. [Tiodize] Rust preventer, penetrant.

311 Hand Cream. [West Agro] Protective skin cream.

400 Stabilizer. [Am. Maize Prods.] Modified amioca starch.

404 Ochre. [New Riverside Ochre] Natural yellow iron oxide pigments.

1900 UHMW Polymers. [Hercules] Ultrahigh m.w. HDPE resins; high-performance materials for fabricating equip. and mech. parts subjected to extraordinary wear; used in chemical, food, beverage, mining, metals processing, paper, and textile industries for processing equipment and materials handling systems.

Manufacturers' Directory

Aakash Chemicals & Dye-Stuffs Inc.
447 Vista Ave., Addison, IL 60101 (Tel.: 708-543-0864; FAX 615-472-6158)

A-Aroma Tech, Inc.
197 Meister Ave., Somerville, NJ 08876 (Tel.: 908-707-0707; 800-542-7662; FAX 908-707-1704)

Abatron, Inc.
33 Center Dr., Gilberts, IL 60136 (Tel.: 708-426-2200; 800-445-1754; FAX 708-426-5966)

Abco Industries, Ltd.
200 Railroad St., PO Box 335, Roebuck, SC 29376 (Tel.: 803-576-6821; 800-476-4476; FAX 803-576-9378; Telex: 628 17731)

Abso-Clean Industries Inc.
17325-T Lamont St., Detroit, MI 48212 (Tel.: 800-837-5000 x 901; FAX 313-366-4334)

ABS Technology Ltd.
Birmingham, B46 3BB, UK (Tel.: 0675-462-233; FAX 0675-467-485)

Accurate Chemical & Scientific Corp.
300 Shames Dr., Westbury, NY 11590 (Tel.: 516-333-2221; 800-645-6264; FAX 516-997-4948; Telex: 4972582)

Aceto Chemical Co., Inc.
1 Hollow Lane, Suite 201, Lake Success, NY 11042-1215 (Tel.: 516-627-6000; FAX 516-627-6093; Telex: 62662)

Acheson
Acheson Colloids Co., 1600 Washington Ave., PO Box 611747, Port Huron, MI 48061-1747 (Tel.: 313-984-5581; 800-255-1908)
Acheson Colloids (Canada) Ltd., PO Box 665, Brantford, Ontario, N3T 5P9, Canada (Tel.: 519-752-5461)
Acheson Industries (Europe) Ltd., Sun Life House, 85 Queens Rd., Reading, Berkshire, RG1 4PT, UK (Tel.: 0734-588844)
Acheson (Japan Ltd., PO Box 538, Kobe Port, 651-01, Japan (Tel.: 078-332-3601)

ACL Inc.
1960 E. Devon Ave., Elk Grove Village, IL 60007 (Tel.: 708-981-9212; 800-782-8420; FAX 708-981-9278; Telex: 4330251)

Acla-Werke GmbH
Frankfurter Str. 142-190, 5000 Köln 80, Germany (Tel.: 0221-699 98-0; FAX 0221-697121; Telex: 8873423)

Acme Div./Allied Products Corp.
166 Chappel, PO Box 1404, New Haven, CT 06505 (Tel.: 203-562-2171; Telex: 295503 ACME UR)

Acme Resin Corp./Subsid. of Borden Inc.
10330 W. Roosevelt Rd., Westchester, IL 60153 (Tel.: 312-343-1900)

Acrol Ltd.
Everite Rd., Ditton, Widnes, Cheshire, WA8 8PT, UK (Tel.: 051-424-1341; FAX 051-495-1853; Telex: 628440 acrol g)

Activated Metals & Chemicals, Inc.
Reagan Ind. Park, PO Box 4130, Sevierville, TN 37864-4130 (Tel.: 615-453-7177; FAX 615-428-3446; Telex: 533077)

Active Organics, Inc.
6849 Hayvenhurst Ave., Van Nuys, CA 91406 (Tel.: 818-786-3310; FAX 818-786-3313)

Acton Technologies, Inc.
101 Thompson St., Pittston, PA 18640 (Tel.: 717-654-0612, x 311; FAX 717-654-2810)

Adco, Inc.
900-T W. Main St., PO Box 999, Sedalia, MO 65301 (Tel.: 816-826-3300; FAX 816-826-1361)

Adell Plastics, Inc.
4530 Annapolis Rd., Baltimore, MD 21227 (Tel.: 301-789-7780)

Adhesive Products Corp.
1660 Boone Ave., Bronx, NY 10460 (Tel.: 212-542-4600)

Adhesives & Chemicals Inc.
131 Brown St., Yatesville, PA 18640 (Tel.: 717-654-6735)

ADM Corn Processing/Div. Archer Daniels Midland Co.
7627 West Lake St., River Forest, IL 60305 (Tel.: 708-771-2990; 800-323-0735)

Adshead Ratcliffe & Co. Ltd.
Derby Rd., Belper, DE5 1WJ, UK (Tel.: 0773-826661; FAX 0773-821215; Telex: 377184 arbo g)

Advance Chem Co Inc.
3841 West Wisconsin Ave., Milwaukee, WI 53213 (Tel.: 414-344-0880; 800-222-5326 (WI))

Advance Coatings, Inc.
Depot Rd., Westminster, MA 01473 (Tel.: 508-874-5921)

Advanced Elastomer
Advanced Elastomer Systems, L.P., 260 Springside Dr., PO Box 5584, Akron, OH 44334-0584 (Tel.: 216-668-3600, 216-668-8242)
Advanced Elastomer Systems Canada Inc., Box 787, Streetsville Postal Station, Mississauga, Ontario, L5M 2G4, Canada (Tel.: 416-826-9575)
Advanced Elastomer Systems NV/SA, Ave. de Tervuren 270-272, PO Box 1, B-1150 Brussels,, Belgium (Tel.: 32-2 761-41-11)
Monsanto Japan Ltd., Room 520, Kokusai Bldg., 1-Marunouchi 3-Chome, Chiyoda-Ku, Tokyo, 100, Japan (Tel.: 81-3 287-1251, 81-3 287-1250; Telex: J 22614)

Advanced Polymer Systems, Inc.
3696 Haven Ave., Redwood City, CA 94063 (Tel.: 415-366-2626; FAX 415-365-6490; Telex: 361-290 aps incud)

Advanced Refractory Technologies, Inc.
699 Hertel Ave., Buffalo, NY 14207 (Tel.: 716-875-4091; FAX 716-875-0106)

Advanced Web Products, Inc.
PO Box 117, Ingomar, PA 15127 (Tel.: 412-367-8820; FAX 412-367-8848)

AEI Compounds
AEI Cables Ltd., Power Cable Div., Gravesend, Kent, DA11 9AF, UK (Tel.: 44-474-564466; FAX 44-474-564386)

Aerochem Corp.
PO Box 2330, High Point, NC 27261-2330 (Tel.: 919-841-4000; FAX 919-841-7116)

Aero Consultants UK Ltd.
Stonehill, Stukeley Meadows Industrial Estate, Huntingdon, Cambs, PE18 6ED, UK (Tel.: 0480-432111; FAX 0480-412910)

Aerojet Propulsion Div.

PO Box 134222, Bldg. 2015B/D5172, Sacramento, CA 95813-6000 (Tel.: 916-355-1000; FAX 916-355-2572; Telex: 377-400 aspc sac)

Aetna Chemical Corp.

Wallace St. Ext., Elmwood Park, NJ 07407 (Tel.: 201-796-0230; 800-345-2518; FAX 201-796-4845)

Agrashell Inc./Industrial Flour

5934 Keystone Dr., Bath, PA 18014 (Tel.: 215-837-6705; FAX 215-837-8802)

Agrico Chemical Co.

1615 Poydras St., New Orleans, LA 70112 (Tel.: 504-585-3335; 800-535-7094; Telex: 497523 Agrico NO)

Air Products

Air Products and Chemicals, Inc., 7201 Hamilton Blvd., Allentown, PA 18195 (Tel.: 215-481-4911; FAX 215-481-5900; Telex: 275425)

Air Products and Chemicals, Inc., Polymer Chemicals Div., Allentown, PA 18195 (Tel.: 215-481-6799; 800-345-3148; FAX 215-481-5900)

Air Products and Chemicals, Inc., Polyurethane Chemicals Div, 7201 Hamilton Blvd., Allentown, PA 18195-1501 (Tel.: 800-345-3148; FAX 215-481-5900)

Air Products, Valchem Div., 403 Carline Rd., Langley, SC 29834 (Tel.: 803-593-4466)

Air Products Nederland B.V., Herculesplein 359, PO Box 85075, NL 3508 AB Utrecht, Netherlands (Tel.: 31-30-511828)

Air Products Pacific, Inc., Sakurabashi Yachiyo Bldg., 5-6, Umeda 2-Chome, Kita-Ku, Osaka, 530, Japan

Pacific Anchor Chemical Corp., 5701 S. Eastern Ave., Suite 530, Los Angeles, CA 90040 (Tel.: 213-725-1800; 800-423-4391; FAX 213-725-1915)

St. Lawrence Chemical Inc., 5405 Pare St., Montreal, Quebec, H4P 1P7, Canada (Tel.: 514-731-3628)

Air-Scent Int'l.

PO Box 1000, 215 Eighth Ave., Braddock, PA 15104 (Tel.: 412-351-2100; 800-247-0770; FAX 412-351-7701)

Aiscondel SA

Aragon 182, 08011 Barcelona, Spain (Tel.: 93-323-1020; FAX 93-323-7921; Telex: 97887 etin e)

Ajay Chemicals, Inc.

1400 Industry Rd., Powder Springs, GA 30073 (Tel.: 404-943-6202; FAX 404-439-0369; Telex: 804-468 ATL)

Ajinomoto

Ajinomoto Co., Inc., Mina Mi 5-32, Koenji Suginami, Tokyo, Japan (Tel.: 33-314-3211; FAX 33-312-7207)

Ajinomoto USA, Inc., Glenpointe Centre West, 500 Frank W. Burr Blvd., Teaneck, NJ 07666-6894 (Tel.: 201-488-1212; FAX 201-488-6472; Telex: 275425 (AJNJ)

Akrochem Chemical Co.

255 Fountain St., Akron, OH 44304 (Tel.: 216-535-2108)

Akzo

Akzo Chemicals Inc., 300 S. Riverside Plaza, Chicago, IL 60606 (Tel.: 312-906-7500; 800-257-8292; FAX 312-906-7680; Telex: 25-3233)

Akzo Coatings Inc., US Hwy. 341 E., Baxley, GA 31513 (Tel.: 912-367-3616; FAX 912-367-5754; Telex: 810-788-5450)

Akzo Chemicals Ltd., 1-5 Queens Rd., Hersham, Walton-on-Thames, Surrey, KT12 5NL, UK (Tel.: 0932-247891; FAX 0932-231204))

Akzo Chemie S.A., 13, Ave. Marnix, 1050 Bruxelles, Belgium (Tel.: 02-5180411)

Akzo Chemicals GmbH, Phillippsstrasse 27, PO Box 100132, 5160 Dueren, Germany

Akzo Chemicals Ltd., PO Box 80, Parramatta N.S.W., 2150, Australia
Akzo Chemicals Ltd., 100 University Ave., Suite 906, Toronto, Ontario, MSJ IV6, Canada
Akzo Chemicals BV, PO Box 975, 3800 AZ Amersfoort, Netherlands (Tel.: 033-643911; FAX 033-637448; Telex: 79322)
Akzo Chemie Italia SpA, Via Vismara, 20020 Arese, Milano, Italy
Akzo Japan Ltd., Godo Kaikan Bldg., 3-27 Kioi-cho, Chiyoda-Ku, Tokyo, 102, Japan
Akzo Engineering Plastics, Inc., PO Box 3333, 2267 West Mill Rd., Evansville, IN 47732 (Tel.: 812-424-3831; 800-457-3764; FAX 812-424-0892)
Akzo Engineering Plastics BV, Velperweg 76, PO Box 9300, 6800 SB Arnhem, Netherlands (Tel.: 085-664422; FAX 085-665140; Telex: 45204 ENKA NL)
Akzo Engineering Plastics Sweden AB, Box 10, S-424 21 Angered, Sweden (Tel.: 031-94 36 90; FAX 031 94 37 84; Telex: 27756)
Akzo Salt Co., Abington Executive Park, Clarks Summit, PA 18411 (Tel.: 717-587-5131; FAX 717-586-6278; Telex: 756470)

Alba International Inc.

508 Clearwater Dr., N. Aurora, IL 60542 (Tel.: 708-897-4200; 800-669-9333; FAX 708-377-5330)

Alban Muller Int'l.

212, rue de Rosny, 93100 Montreuil, France (Tel.: 33-1-48-58-30-25; FAX 33-1-48-58-03-71; Telex: 236030 F)

Albion Kaolin/Div. of United Catalysts Inc.

PO Box 32370, Louisville, KY 40232 (Tel.: 502-634-7502; FAX 502-634-7727)

Albis

Albis Corp., PO Box 711, Rosenberg, TX 77471 (Tel.: 713-342-3311; 800-231-5911)
Albis UK Ltd., Parkgate Lane Ind. Est., Knutsford, Cheshire, WA16 8DX, UK (Tel.: 0565-755777; FAX 0565-755196; Telex: 668516)
Albis Plastics GmbH Hamburg, PO Box 261162, Muhlenhagen 35 D-2000, Hamburg, 28, Germany

Albright & Wilson

Albright & Wilson Americas, PO Box 26229, Richmond, VA 23260-6229 (Tel.: 804-550-4300; 800-446-3700; FAX 804-550-4385)
Albright & Wilson Ltd., European Hdqtrs., PO Box 3, 210-222 Hagley Rd. West, Oldbury, Warley, West Midlands, B68 0NN, UK (Tel.: 44-21-429-4942; FAX 44-21-420-5151; Telex: 336291)
Albright & Wilson (Australia) Ltd., PO Box 20, Yarraville, Victoria, Australia (Tel.: 3-688-7777; FAX 3-688-7788)
Albright & Wilson Am. (Canada), 2 Gibbs Rd., Islington, Ontario, M9B 1R1, Canada (Tel.: 416-234-7000; 800-268-2520; FAX 416-237-1064)
Marchon Espanola SA, Carretera Montblack Km 2, 4, Alcover (Tarragona), Spain
Marchon France SA, BP 19, F-55300, St. Mihiel, France
Marchon Italiana SpA, Casella Postale No. 30, 1-46043, Castiglione delle Stivier, Italy

Albrow Products Ltd.

Unit 11E Middlewich Rd., Byley, N. Middlewich, Cheshire, CW10 9NX, UK (Tel.: 060684-2701-3184; FAX 060684-5836)

Alcan

Alcan Chemicals, Box 6977, Cleveland, OH 44101-1977 (Tel.: 216-523-2813; 800-321-3864)
Alcan Chemicals, 3690 Orange Place, Suite 400, Cleveland, OH 44122-4438 (Tel.: 216-765-2550; 800-321-3864; FAX 216-765-2570)
Alcan Rubber & Chemical Inc., 3570 Executive Dr., Suite 102, Uniontown, OH 44203 (Tel.: 216-896-0355; FAX 216-896-9816; Telex: 986477)
British Alcan Aluminum PLC, Chalfont Park, Gerrards Cross, Buckinghamshire, SL9 0QB, UK (Tel.: 0753-887373)

BA Chemicals Ltd., Chalfont Park Gerrards Cross, Buckinghamshire, SL9 0QB, UK (Tel.: 0753-887373; FAX 0753-889602; Telex: 847343)

Alchemie Ltd.

Brookhampton Lane, Kineton, Warwickshire, CV35 0JA, UK (Tel.: 0926-641600; FAX 0926-641698; Telex: 311890)

Alco Chemical/Div. of National Starch & Chem.

909 Mueller Dr., PO Box 5401, Chattanooga, TN 37406 (Tel.: 615-629-1405; 800-251-1080; FAX 615-698-8723; Telex: 755002)

Alcoa Industrial Chemicals Div.

PO Box 67, Bauxite, AR 72011 (Tel.: 501-776-4981; 800-643-8771; FAX 501-776-4685; Telex: 536447)

Alconox Inc.

215 Park Ave. So., New York, NY 10003 (Tel.: 212-473-1300; FAX 212-353-1342)

Aldoa Co., Inc.

12727 Westwood, Detroit, MI 48223-1214 (Tel.: 313-273-5705; FAX 313-273-0310)

Alfelder Kunststoffwerke Herm. Meyer GmbH

Hildesheimer St. 78, Postfach 1155, 3220 Alfeld, Germany (Tel.: 05181 80 18-0; FAX 05181 1877; Telex: 92908)

Alfordshire Ltd.

Unit 16 Station Lane Ind. Est. Witney, Oxon, OX8 6XZ, UK (Tel.: 0993 771813; FAX 0993 703104)

Aframine Corp.

72 Putnam St., Paterson, NJ 07524 (Tel.: 201-279-5334; FAX 201-279-5335)

Alken-Murray Corp.

417 Canal St., New York, NY 10013 (Tel.: 212-431-4020; FAX 212-431-4944)

Alkor Plastics UK Ltd.

Odhams Trading Estate, St. Albans Rd., Watford, Herts, WD2 5DG, UK (Tel.: 0923-49511; FAX 0923-227427; Telex: 923993)

Allied Colloids

Allied Colloids Inc., 2301 Wilroy Rd., PO Box 820, Suffolk, VA 23434 (Tel.: 804-934-3700; FAX 804-934-3989)

Allied Colloids Ltd., PO Box 38, Low Moor, Bradford, Yorkshire, BD12 0JZ, UK

Allied-Signal

Allied-Signal, Inc., PO Box 2332R, Columbia Rd. & Park Ave., Morristown, NJ 07960 (Tel.: 201-455-2000; 800-446-1800; FAX 201-445-4807)

Allied-Signal, Inc./Fibers Div., 1411 Broadway, New York, NY 10018 (Tel.: 212-391-5000)

Allied Corp. Int'l. NV-SA, Haasrode Research Park, B-3030 Heverlee, Belgium

Allied Corp. Int'l. NV-SA, International House, Bickenhill Lane, Birmingham, B37 7HQ, UK

Allied Chem. Int'l. Corp., PO Box 99067, Tsimshatsui Post Office, Hong Kong

NV Allied Corp. Int'l. SA, Haasrode Research Park, 3001 Heverlee, Belgium (Tel.: 016-211-211; FAX 016-203-269; Telex: 26283)

Allrim SA

Z I des Arbletiers, 25400 Audincourt, France (Tel.: 81 30 47 40; Telex: 361561)

Alma SA

CH-1783 Pensier, Fribourg, Switzerland (Tel.: 037-265363; FAX 037-265171; Telex: 942246 ALMA CH)

Aloe Laboratories Inc.

PO Box 831, Harlingen, TX 78551 (Tel.: 512-428-8416; FAX 512-428-8482)

Alox Corp.
3943 Buffalo Ave., PO Box 517, Niagara Falls, NY 14302 (Tel.: 716-282-1295)

Alpe SRL
Via Delle Orchidee 12, 230020 Vanzaghello, Milano, Italy (Tel.: 0331-657115; FAX 0331-657501; Telex: 352245 VANALP I)

Alpha Calcit Fullstoff GmbH & Co. KG
Otto-Hahn-Strasse, Postfach 11 06, D-5000 Köln 50, Germany (Tel.: 02236-8914-0; FAX 02236-40644; Telex: 8886960)

Alpha Chemical & Plastics Corp.
9635 Industrial Dr., PO Box 490, Pineville, NC 28134 (Tel.: 704-554-8675)

Alphaflex Industries, Inc.
3339 W. 16th St., Indianapolis, IN 46222, USA)

Alpha Metals, Inc.
600 Route 440, Jersey City, NJ 07438 (Tel.: 201-434-6778)

Alpine Aromatics Int'l. Inc.
51 Ethel Rd. West, CN 1348, Piscataway, NJ 08854-1348 (Tel.: 201-572-5600; 800-631-5389; Telex: 844-528 ALPAROINT)

AluChem Inc.
One Landy Lane, Reading, OH 45215 (Tel.: 513-733-8519; FAX 513-733-0608; Telex: 298252 ALUC UR)

Alveo AG
Ward Royal Parade, Alma Road, Windsor, Berkshire, SL4 3HR, UK (Tel.: 0753-856641; FAX 0753-859941; Telex: 348922)

Alvin Products Inc.
PO Box 942, 20-22 Houghton St., Worcester, MA 01613 (Tel.: 508-754-8604; FAX 508-753-7092)

Alzo Inc.
6 Gulfstream Blvd., Matawan, NJ 07747 (Tel.: 908-446-3270; FAX 908-446-5225)

Amano Int'l. Enzymes
PO Box 1000, Troy, VA 22974 (Tel.: 804-589-8278; 800-446-7652; Telex: 822-438)

Amari Plastics
Humphreys Rd., Woodside Estate, Dunstable, Bedfordshire, LU5 4TP, UK (Tel.: 0582-666773; FAX 0582-699375)

Amax Industrial Products Div.
960 S. Third St., Louisville, KY 40203 (Tel.: 800-662-0023)

Ambersil Ltd.
Wylds Road, Castlefield Industrial Estate, Bridgewater, Somerset, TA6 4DD, UK (Tel.: 0278-424200; FAX 0278-425644; Telex: 46796)

Amerchol
Amerchol Corp., PO Box 4051, 136 Talmadge Rd., Edison, NJ 08818 (Tel.: 908-248-6000; FAX 908-287-4186; Telex: 833472)
Amerchol Europe, Havenstraat 86, B-1800 Vilvoorde, Belgium (Tel.: 32-2-252-4012; FAX 32-2-252-4909; Telex: 846-69105)
Amerchol, D.F. Anstead Ltd., Victoria House, Radford Way, Bellericay, Essex, CM12 0DE, UK
Amerchol, Ikeda Corp., New Tokyo Bldg., No. 3-1, Marunouchi 3-Chome, Chiyoda-Ku, Tokyo, 100, Japan

Ameribrom. See Dead Sea Bromine

American Bio-Synthetics Corp.
710 W. National Ave., Milwaukee, WI 53204 (Tel.: 414-384-7017)

American Chemet Corp.
PO Box 437, 400 County Line Rd., Deerfield, IL 60015 (Tel.: 708-948-0800; FAX 708-948-0811; Telex: 72-4301)

American Chemical Services
PO Box 190, Griffith, IN 46319 (Tel.: 219-924-4370)

American Chrome & Chemicals Inc.
PO Box 9912, Buddy Lawrence Dr., Corpus Christi, TX 78469 (Tel.: 512-883-3202; FAX 512-883-5145; Telex: 778428)

American Colloid Co.
1500 W. Shure Dr., Arlington Hts., IL 60004-1434 (Tel.: 708-392-4600; FAX 708-506-6199; Telex: 4330321)

American Cyanamid
American Cyanamid/Corporate Headquarters, One Cyanamid Plaza, Wayne, NJ 07470 (Tel.: 201-831-4111; 800-922-0187; FAX 201-839-8847)

American Cyanamid/Engineered Materials Dept., 1300 Revolution St., Havre de Grace, MD 21078 (Tel.: 301-939-1910; FAX 301-939-0930)

American Cyanamid/Polymer Additives Dept., One Cyanamid Plaza, Wayne, NJ 07470 (Tel.: 201-831-2000)

American Cyanamid/Process Chems., Fine Chems., Textile Chem, One Cyanamid Plaza, Wayne, NJ 07470 (Tel.: 800-438-5615)

American Cyanamid/Textiles

American Cyanamid/Agricultural Div., One Cyanamid Plaza, Wayne, NJ 07470 (Tel.: 201-831-2000; 800-443-0443)

Cyanamid Canada Inc./Carbide Products Div., 88 McNabb St., Markham, Ontario, L3R 6E6, Canada (Tel.: 416-470-3600; FAX 416-470-3852; Telex: 06-966602)

Cyanamid Aerospace Products Ltd., Abenbury Way, Wrexham Industrial Estate, Wrexham, CLWYD LL 139UF, Wales, UK

Cyanamid BV, Postbus 1523, 3000 BM, Rotterdam, The Netherlands (Tel.: 010-4116340; FAX 010-4136788; Telex: 23554)

Cyanamid India Ltd., Nyloc House, 254-D2 Dr. Annie Besant Rd., Bombay, 400 025, India

Cyanamid Quimica do Brasil Ltda., Av. Imperatriz Leopoldina, 86, Sao Paulo, Brazil

Cyanamid Taiwan Corp., 8/F Union Commercial Bldg., 137, Nanking E. Rd., Sec. 2, Taipei, Taiwan, R.O.C.

American Emulsions Co.
PO Box 3787, Dalton, GA 30721 (Tel.: 404-226-7028; FAX 404-278-5183)

American Fillers & Abrasives
14 Industrial Park Dr., Bangor, MI 49013 (Tel.: 616-427-7955)

American Grease Stick Co.
2651 Hoyt St., PO Box 729, Muskegon, MI 49443 (Tel.: 616-733-2101; 800-253-0403; FAX 616-733-1784)

American Ingredients
American Ingredients Co., 14622 S. Lakeside Ave., Dolton, IL 60419 (Tel.: 708-849-8590)

Patco Specialty Chemicals Div., 3947 Broadway, Kansas City, MO 64111 (Tel.: 816-561-9050; 800-821-2250; FAX 816-561-9909)

American Labs Inc.
4410 S 102 St., Omaha, NE 68127 (Tel.: 402-339-2494; 800-445-5989; FAX 402-339-0801; Telex: 3735593 CACOMA)

American Lecithin. See Rhone-Poulenc

American Lignite Products Co.
4655 Coal Mine Rd., Ione, CA 95640 (Tel.: 209-274-2407; FAX 209-274-2846)

American Maize Products Co.
1100 Indianapolis Blvd., Hammond, IN 46320-1094 (Tel.: 219-659-2000; 800-348-9896)

American MicroTrace Corp.
569 Central Dr., Suite 201, Virginia Beach, VA 23454 (Tel.: 804-463-5013; 800-544-3155; FAX 804-463-8156)

American Minerals
Hickory Hill Plaza, Ste. 328, 151 S. Warner Rd., Wayne, PA 19087 (Tel.: 215-971-1500; FAX 215-971-1506; Telex: 881075)

American Norit Co., Inc.
1050 Crown Pointe Pkwy., Suite 1500, Atlanta, GA 30338 (Tel.: 404-512-4610; 800-641-9245; FAX 404-512-4622)

American Oil & Supply Co.
238 Wilson Ave., Newark, NJ 07105 (Tel.: 201-589-0250; 800-582-3267; FAX 201-465-5083)

American Resin Corp.
6250 Southwest Pkwy., PO Box 4505, Wichita Falls, TX 76308 (Tel.: 817-692-8011; FAX 817-692-8014)

American Synthetic Rubber Corp.
PO Box 32960, 4500 Camp Ground Rd., Louisville, KY 40232 (Tel.: 502-449-8300; 800-262-9253; FAX 502-449-8468)

Americhem, Inc.
225 Broadway East, PO Box 375, Cuyahoga Falls, OH 44222-0375 (Tel.: 216-929-4213; FAX 216-929-4144)

Ameripol Synpol Co./Div. of Uniroyal Goodrich Tire Co.
146 S. High St., 7th floor, Akron, OH 44308-1493 (Tel.: 800-962-TECH)

Ameron Protective Coatings Div.
201 N. Berry St., Brea, CA 92621 (Tel.: 714-529-1951; 800-854-3118)

Amico, Inc.
PO Box 4056, Atlanta, GA 30302 (Tel.: 404-522-7060; FAX 404-581-0116)

Amoco
Amoco Chemicals Corp., 200 East Randolph Dr., Chicago, IL 60601 (Tel.: 312-856-3806; 800-621-8888)
Amoco Lubricants Business Unit, MC 1102, 200 East Randolph Dr., Chicago, IL 60601 (Tel.: 312-856-4599)
Amoco Chemical Europe S.A., 15, Rue Rothschild, CH-1211, Geneva, 21, Switzerland (Tel.: 022-31-02-81)
Amoco Performance Products, Japan Ltd., 10th Floor, Tonichi Bldg., 2-31 Roppongi 6-Chome, Mianto Ku, Tokyo, 106, Japan

Ampacet Corp.
660 White Plains Rd., Tarrytown, NY 10591 (Tel.: 914-631-6600; FAX 914-631-7197)

Amphoterics International Ltd.
Unit 3, Berrington Rd., Syndenham Industrial Estate, Leamington Spa, UK (Tel.: 0926-883-551; FAX 0926-882-578; Telex: 312306 AMPHO G)

Ampion Corp.
4-88 47th Ave., Long Island City, NY 11101 (Tel.: 718-784-3374)

AMRESCO
30175 Solon Industrial Pkwy., Solon, OH 44139 (Tel.: 216-349-2802; 800-829-2802; FAX 216-349-1182; Telex: 985582)

Amspec Chemical Corp.
Foot of Water St., Gloucester City, NJ 08030 (Tel.: 609-456-3930; 800-5AMSPEC; FAX 609-456-6704; Telex: 136714 GLCY)

Amstat Industries, Inc.
3012 N. Lake Terrace, Glenview, IL 60025-5794 (Tel.: 708-998-6210; FAX 708-998-6218)

Analytichem International, Inc.
24201 Frampton Ave., Harbor City, CA 90710 (Tel.: 213-539-6490; 800-421-2825; FAX 213-539-4270; Telex: 664832 ANACHEM HRBO)

Anar Chemical Co.
1765F Cortland Court, Addison, IL 60101 (Tel.: 708-953-1660; 800-344-1660; FAX 708-953-1698)

Anderson Development Co.
1415 E. Michigan St., Adrian, MI 49221 (Tel.: 517-263-2121; FAX 517-263-1000)

Anderson Laboratories, Inc.
5901 Fitzhugh St., Fort Worth, TX 76119 (Tel.: 817-457-4474)

The Andersons
1200 Dussel Dr., Maumee, OH 43537 (Tel.: 419-893-5050; 800-537-3370; FAX 419-891-6539)

Andertech Plastteknik A/S
Bakkegaardsvej 406B, 3050 Humlebaek, Denmark (Tel.: 42-19-41-61, 42-19-41-66; Telex: 41152 Atech Dk)

Andrea Aromatics
PO Box 3091, Princeton, NJ 08543-3091 (Tel.: 609-695-7710; FAX 609-392-8914)

Angus
Angus Chemical Co., 2211 Sanders Rd., Northbrook, IL 60062 (Tel.: 708-498-6700; 800-323-6209; FAX 708-498-6706; Telex: 275422 ANGUSUR)
Angus Chemie GmbH, Huyssenallee 5, 4300 Essen 1, Germany (Tel.: 0201-233531)
Angus Chemie GmbH, Le Bonaparte, Centre d'Affaires Paris-Nord, F-93153 Le Blanc Mesnil, Paris, France
Angus Chemie GmbH, 19, Moorgate St., Rotherham, S60 2DA, UK
Angus Chemical Co., 101 Cecil St., #14-07 Tong Eng Bldg., 0106, Singapore

Anilac NV/SA
724 Schaerbeeklei, 1800 Vilvoorde, Belgium (Tel.: 02 251 25 25, 02 251 59 86; Telex: 22696 Anilac b)

Apex Chemical Co., Inc.
200 South First St., Elizabethport, NJ 07206 (Tel.: 908-354-5420; FAX 908-354-2640; Telex: 178326 APEX UT)

API Applicazion Plastiche Industriali
Via Dante Alighieri 27, Mussolente, 36065, Italy (Tel.: 0424-878444 RA; FAX 039-424-87677; Telex: 480640 API I)

Apollo Chemical Corp.
PO Box 2176, Burlington, NC 27216 (Tel.: 919-226-1161; FAX 919-228-6963)

Applied Textile Technologies Ltd.
PO Box 5022, Winston-Salem, NC 27113-5022 (Tel.: 919-764-2500; FAX 919-764-1544)

Aquafil SpA
Via Linfano, 38062 Arco/TN Italy (Tel.: 4 64 51 73 91; FAX 4 64 53 22 67; Telex: 4 00 203)

Aqualon
Aqualon Co., PO Box 15417, 2711 Centreville Rd., Wilmington, DE, 19850-5417 (Tel.: 302-996-2000; 800-345-8104; FAX 302-996-2049; Telex: 4761123)

Aqualon Canada Inc., 5407 Eglinton Ave. West, Suite 103, Etobicoke, Ontario, M9C 5K6, Canada (Tel.: 416-620-5400)

Aqualon UK Ltd., Genesis Centre, Garrett Field, Birchwood, Warrington, Cheshire, WA3 7BH, UK (Tel.: 44-925-830077)

Aqualon France, 44, Ave. de Chatou, F-92508 Rueil Malmaison Cedex, France (Tel.: 1-4751-2919)

Aqualon GmbH, PO Box 130125, Paul Thomas Strasse 58, D-4000 Düsseldorf 13, Germany (Tel.: 49-211-7491-0)

Aquaness Chemicals

3920 Essex Lane, PO Box 27714, Houston, TX 77227 (Tel.: 713-599-7400; FAX 713-599-7460; Telex: 4620058)

Aqua Process Inc.

10801 Galveston Rd., Houston, TX 77034-4896 (Tel.: 713-481-9720; FAX 713-481-3333)

Aquatec Chem. Int'l.

408 Auburn Ave., Pontiac, MI 48058 (Tel.: 313-334-4747; FAX 313-334-1461)

Aquatec Quimica SA

Rua Sampaio Viana, 425, CEP 0400 Postal 4885, Sao Paulo, Brazil (Tel.: 011-884-4466 ext.329; FAX 11-884-0747; Telex: 1121312)

Arakawa Chemical (USA) Inc.

625 N. Michigan Ave., Suite 1700, Chicago, IL 60611 (Tel.: 312-642-1750; FAX 312-642-0089; Telex: 26-5514)

Arcadian Corp.

6750 Poplar Ave., Suite 600, Memphis, TN 38138-7419 (Tel.: 901-758-5200; 800-654-4514; FAX 901-758-5203; Telex: 54-5404)

Arco

Arco Chemical/Headquarters, Research & Engineering Center, 3801 West Chester Pike, Newtown Sq., PA 19073 (Tel.: 215-359-2000; 800-321-7000)

Arco Chemical Canada Inc., 100 Consilium Pl., Suite 306, Scarborough, Ontario, M1H 3E3, Canada)

Arco Chemical Europe, Inc., Bridge Ave., Maidenhead, Berkshire, SL6 1YP, UK (Tel.: 44-628-775000)

Arco Chemical Pan American, Inc., Paseo de la Reforma, 390 Decimo Piso, 06600 Mexico City, Mexico)

Arco Chemical Asia/Pacific Ltd., Toranomon 37 Mori Bldg., 5th Floor, 5-1 Toranomon 3-Chome, Minato-Ku, Tokyo, 105, Japan

Argeville SA

rue Ferrier 9, CH-1202 Geneva, Switzerland (Tel.: 022-32 09 80; FAX 022-732 04 42; Telex: 289-770 arpa CH)

M. Argueso & Co., Inc.

441 Waverly Ave., PO Box 3, Mamaroneck, NY 10543 (Tel.: 914-698-8500; FAX 914-698-0325)

Arista Industries, Inc.

1082 Post Rd., Darien, CT 06820 (Tel.: 203-655-0881; 800-255-6457; FAX 203-656-0328; Telex: 996493)

Aristech Chemical Corp.

600 Grant St., Room 1028, Pittsburgh, PA 15219-2704 (Tel.: 412-433-2747; 800-526-4032; FAX 412-433-1816)

Arizona Chemical Co./Div. of International Paper

1001 E. Business Hwy. 98, Panama City, FL, 32401 (Tel.: 904-785-6700; 800-526-5294; FAX 904-785-2203; Telex: 514411)

John L. Armitage & Co.
1259 Route 46, Parsippany, NJ 07054 (Tel.: 201-402-9000; 800-446-1600; FAX 201-402-9811)

Armite Labs
1845 Randolph St., Los Angeles, CA 90001 (Tel.: 213-587-7768; FAX 213-587-5075; Telex: 67-3476)

Arnco
One Centerpointe Dr., #260, La Palma, CA 90623-1094 (Tel.: 714-739-7900; FAX 714-739-1764)

Arol Chemical Products Co.
649 Ferry St., Newark, NJ 07105 (Tel.: 201-344-1510; FAX 201-344-7127)

Arsynco Co. Inc.
PO Box 8, Carlstadt, NJ 07072 (Tel.: 516-627-6000; FAX 516-627-6093; Telex: 62662)

Artilabo SA
Paardenstraat 12, B-9070 Destelbergen, Belgium (Tel.: 091-55 27 42; FAX 091-55 27 53)

Arto Chemicals Ltd.
Arto House, London Road, Binfield, Berkshire, RG12 5BU, UK (Tel.: 0344 860737; FAX 0344 860820; Telex: 847713 ARTO G)

Asahi Denka Kogyo K.K.
4-1, Higashi-Ogu 8-Chome, Arakawa-Ku, Tokyo, 116, Japan (Tel.: 03-270-7511)

Asahi Glass America, Inc.
1185 Ave. of the Americas, 30th floor, New York, NY 10036 (Tel.: 212-764-3155; FAX 212-764-3384; Telex: 275830 ASAGL UR)

Asarco Inc.
180 Maiden Lane, New York, NY 10038 (Tel.: 212-510-2000; Telex: ITT 420585)

Asarco Tech.
3422 S. 700 West, Salt Lake City, UT, 84119 (Tel.: 801-262-2459)

Asbury Graphite Mills, Inc.
41 Main St., Asbury, NJ 08802 (Tel.: 201-537-2155; FAX 201-537-3908; Telex: 834457)

Ash Grove Cement Co.
8900 Indian Creek Pkwy., Overland Park, KS, 66225 (Tel.: 913-451-8900)

Ashland
Ashland Chemical, Inc./Subsid. of Ashland Oil, Inc., 5200 Paul G. Blazer Memorial Pkwy., Dublin, OH 43017 (Tel.: 614-889-3333)

Ashland Chemical Inc./Industrial Chemicals & Solvents, PO Box 2219, Columbus, OH 43216 (Tel.: 614-889-3333; FAX 614-889-3465)

Ashland Chemical Inc./Specialty Polymers & Adhesives, PO Box 2219, Columbus, OH 43216 (Tel.: 614-889-3527; FAX 614-889-3206)

Ashland Chemical Inc./General Polymers Div., PO Box 2219, Columbus, OH 43216 (Tel.: 614-889-3855; 800-828-7659; FAX 614-889-3465)

Ashland Chemical Inc./Composite Polymers Div., PO Box 2219, Columbus, OH 43216 (Tel.: 614-889-3571; 800-327-8720; FAX 614-889-3735)

Ashland Chemical Inc./Foundry Products Div., PO Box 2219, Columbus, OH 43216 (Tel.: 614-889-3514; 800-848-7485; FAX 614-889-3285; Telex: 245385 ASHCHEM)

Drew Industrial Div., One Drew Plaza, Boonton, NJ 07005 (Tel.: 201-263-7800; 800-526-1015 x7800; FAX 201-263-4483; Telex: DREWCHEMS BOON)

Ashland-Südchemie-Kernfest GmbH, Reisholzstrasse 16, 4010 Hilden, Germany (Tel.: 021-711030; FAX 0211-7110335; Telex: 8582889)

Ashley Polymers, Inc.
5114 Fort Hamilton Pkwy., Brooklyn, NY 11219 (Tel.: 718-851-8111; FAX 718-972-3256)

Assessa-Industria Comercio and Exportacao, Ltd.

Rua Cardoso Quintao, 110-CEP 21 381, Rio de Janeiro, Brazil (Tel.: 021-591-4345; Telex: 21-39602 SESX-BR)

H.A. Astlett & Co.

8 King St. East, Toronto, Ontario, M5C 1B5, Canada (Tel.: 416-366-2747; 800-387-1152 (U.S.); FAX 416-366-6389; Telex: 06-218801)

Astor Wax Corp.

200 Piedmont Ct., Doraville, GA 30340 (Tel.: 404-448-8083; FAX 404-840-0954)

Astro Industries/Div. of Borden Inc.

PO Box 2559, Morganton, NC 28655 (Tel.: 704-584-3800; FAX 704-584-3885)

Atsaun Chemical Corp.

23, Janki Niwas, N.C. Kelkar Rd., Dadar, Bombay, 400 028, India (Tel.: 430-1454-422-3145)

Atlas Minerals & Chemicals, Inc.

Farmington Rd., Mertztown, PA 19539 (Tel.: 215-682-7171; 800-523-8269; FAX 215-682-9200)

Atlas Refinery, Inc.

142 Lockwood St., Newark, NJ 07105 (Tel.: 201-589-2002; FAX 201-589-7377; Telex: 138-425 ATLASOIL)

Atochem

Atochem, Groupe elf aquitaine, 4, cours Michelet, La Défense 10-Cedex 42, 92091 Paris La Défense, France (Tel.: 011-33-1-4900-8080; FAX 01133-14900-7447)

Atochem North America Inc., Three Parkway, Philadelphia, PA 19102 (Tel.: 215-587-7000; 800-225-7788; FAX 215-587-7591)

Atochem North America Inc./Organic Peroxides Div., 1740 Military Rd., PO Box 1048, Buffalo, NY 14240 (Tel.: 716-877-1740; 800-558-5575; FAX 716-877-1541)

Atochem North America Inc./Plastics Dept., Three Parkway, Philadelphia, PA 19102 (Tel.: 215-587-7000; 800-328-2811; FAX 215-587-7497)

Atochem/Plastics Additives Div., 4 King James South, 24500 Center Ridge Rd., Room 180, Cleveland, OH 44145 (Tel.: 216-835-5030; FAX 216-835-9185)

Atochem Inc./Wire Mill Products Dept., 43 James St., Homer, NY 13077 (Tel.: 607-749-2652)

Atochem North America Inc./Textile Chemicals Dept., Three Parkway, Room 820, Philadelphia, PA 19102 (Tel.: 215-587-7380; FAX 215-587-7911)

Atochem UK Ltd., Colthrop Lane, Thatcham, Newbury, Berkshire, RG13 4LW, UK (Tel.: 0635-70000; FAX 0635-61212; Telex: 847689 ATOKEM G)

Pennwalt Plastics Ltd., Cherwell House, St. Clements, Oxford, OX4 1BD, UK (Tel.: 0865-726961)

Atochem Deutschland GmbH, Niederlassung Düsseldorf, Uerdiger Strasse 5, Postfach 30 01 52, D-4000 Düsseldorf 30, Germany (Tel.: 0211-4552-0; FAX 0211-14552-112; Telex: 8584682)

Atochem Yoshitomo, Ltd., 6-9 Hiranomachi, 2-Chome, Chuo-Ku, Osaka, 541, Japan (Tel.: 201-1161)

Pennwalt Japan Ltd., Sumitomo Jimbo-Cho Bldg., 3-25, Kanda Jimbo-Cho, Chiyoda-Ku, Tokyo, 101, Japan (Tel.: 81-3-262-8481)

Ceca SA, 22 Place de l'Irise, Cedex 54 92062, Paris-La Defense 2, France (Tel.: 33-147-96-9090; FAX 33-147-96-9234; Telex: 611444 ckd)

M&T Harshaw, Two Riverview Dr., PO Box 6768, Somerset, NJ 08875-6768 (Tel.: 908-302-3500; FAX 908-271-8960)

Atomergic Chemetals Corp.

222 Sherwood Ave., Farmingdale, NY 11735-1718 (Tel.: 516-694-9000; FAX 516-694-9177; Telex: 6852289)

Atramax Inc.

PO Box 278, Hawthorne, NJ 07507 (Tel.: 212-882-2263; FAX 212-798-2546)

Auralux Corp.,
PO Box 113, Yantic, CT 06389 (Tel.: 203-886-2616; FAX 203-823-1252)

Auschem SpA
Via Baertsch, 1, 24100 Bergamo, Italy (Tel.: 35-346282; FAX 35-340386; Telex: 300275)

Ausimont USA Inc.
44 Whippany Rd., Morristown, NJ 07962-1838 (Tel.: 201-292-6250; 800-323-AUSI; FAX 201-292-0886)

Avatar Corp.
7728 W. 99 St., Hickory Hills, IL 60457 (Tel.: 708-430-4200; 800-255-3181)

Avebe America Inc.
4 Independence Way, Princeton, NJ 08540 (Tel.: 609-520-1400; FAX 609-520-1473; Telex: 0820713)

Axel Plastics Research Labs
PO Box 855, 58-20 Broadway, Woodside, NY 11377 (Tel.: 718-672-8300; FAX 718-565-7447; Telex: 429033 AXELLAB)

Axial SA
7 Rue D'Alsace, BP 21, 57192 Florange Cedex, France (Tel.: 82585109; FAX 82 58 8200; Telex: 860678)

Ayer's Cliff Chemical Products Ltd.
PO Box 322, Ayer's Cliff, Quebec, J0B 1C0, Canada (Tel.: 514-649-0503; FAX 514-922-5166)

Azdel Inc./GE-PPG Joint Venture
925 Washburn Switch Rd., Shelby, NC 28150 (Tel.: 704-434-2271)

BA Chem. Ltd. See Alcan

Bacon Industries Inc.
192 Pleasant St., Watertown, MA 02172 (Tel.: 617-926-2550)

H.J. Baker & Bro. Inc.
100 E. 42 St., New York, NY 10017 (Tel.: 212-867-0200; FAX 212-370-1639; Telex: 420944)

Baker Performance Chemical, Inc./A Baker Hughes Co.
3920 Essex Lane, Houston, TX 77027 (Tel.: 713-599-7400; 800-231-3606; FAX 713-599-7460; Telex: ITT-4620058)

Bamberger Polymers
Bamberger Polymers, Inc., 690 Jersey Ave., Bldg. #4, New Brunswick, NJ 08901 (Tel.: 908-214-8867; 800-888-8959; FAX 908-214-8868)
Bamberger Polymers (Canada) Inc., 62 Selby Rd., Brampton, Ontario, L6W 3L4, Canada (Tel.: 416-456-0851)
Bamberger Polymers (Europe) BV, Steurweg 2, 4941 VR Raamsdonksveer (Tel.: 31-1621-20240)
Bamberger Polymers De Mexico S.A., Rio Lerma 143-103, Col. Cuauhtemoc, Mexico, DF)

Barbe America, Inc.
West White Road, PO Box 69, Flowery Branch, GA 30542 (Tel.: 404-967-4305; FAX 404-967-4445)

Barco Chemical Products Inc.
327 S. LaSalle St., Chicago, IL 60604 (Tel.: 312-427-2916)

Barium & Chemicals Inc.
County Road 44, PO Box 218, Steubenville, OH 43952 (Tel.: 614-282-9776)

Barkley Plastics Ltd.
120-121 Highgate St., Balsall Heath, Birmingham, B12 0XR, UK (Tel.: 021-440-1303; FAX 021-440-4902)

Otto Bärlocher GmbH,
    Postfach 500108, 8000 München 50, Germany (Tel.: 089-1488-0; FAX 089-1488-312; Telex: 5215701)

Barnebey & Sutcliffe Corp.
    835 N. Cassady Ave., PO Box 2526, Columbus, OH 43216 (Tel.: 614-258-9501; FAX 7614-258-3464)

BASF
    BASF AG, ESA/WA-H 201, D-6700 Ludwigshafen, Germany (Tel.: 0621-60-99603; FAX 0621-60-41787; Telex: 469499-0 BAS D)
    BASF Corp., 100 Cherry Hill Rd., Parsippany, NJ 07054 (Tel.: 201-316-3000; 800-669-BASF; FAX 201-402-1832)
    BASF Corp./Chemicals Div., Thermoplastic PU, Elastomers, 1609 Biddle Ave., Wyandotte, MI 48192 (Tel.: 313-246-6323)
    BASF Corp./Engineering Plastics, 100 Cherry Hill Rd., Parsippany, NJ 07054 (Tel.: 609-467-2456; 800-227-3746)
    BASF Corp./Fibers Div., Textile Colors & Chemicals, 9401 Arrowpoint Blvd., Suite 200, Charlotte, NC 28273 (Tel.: 800-247-0557; FAX 704-527-3503)
    BASF Canada Ltd., PO Box 430, Montreal, Quebec, H4L 4V8, Canada)
    BASF PLC, PO Box 4, Earl Road, Cheadle Hulme, Cheshire, SK8 6QG, UK (Tel.: 061-485-6222; FAX 061-486-0891; Telex: 669211)
    BASF Belgium S.A., Ave. Hamoir-Iaan 14, B-1180 Brussels, Belgium
    BASF Espanola S.A., Apartado 762, Barcelona, 8, Spain
    BASF S.A. Compagnie Francaise, MC-NT, 140, Rue Jules Guesde, 92303 Levallois-Perret, France
    BASF India, Ltd., Maybaker House, S.K. Ahire Marg., PO Box 19108, Bombay, 400 025, India
    BASF Japan Ltd., C.P.O. Box 1757, Tokyo, 100-91, Japan
    Elastogran Kunststoff-Technik GmbH, Worms Plant, Flosshafenstasse 40, D-6520 Worms, Germany (Tel.: 06241-844-0; FAX 06241-844-113; Telex: 467748 ektwo d)

Baxenden Chemical Co. Ltd.
    Union Lane, Droitwich, Worcestershire, WR9 9BB, UK (Tel.: 0905-772 454; FAX 0905-794-002; Telex: 338310)

Bay Resins, Inc.
    PO Box 630, Route 313, Millington, MD 21651 (Tel.: 410-928-3083; FAX 410-928-5412)

Bayer
    Bayer AG/Plastics Business Group, D-5090 Leverkusen, Germany (FAX 0214-3031482; Telex: 85103-249 by d)
    Bayer UK Ltd., Bayer House, Strawberry Hill, Newbury, Berkshire, RG13 1JA, UK
    Miles Inc./Polymers Div., Mobay Rd., Pittsburgh, PA 15205-9741 (Tel.: 412-777-2000; 800-526-4550)
    Miles Inc./Polysar Rubber Div., 2603 W. Market St., Akron, OH 44313 (Tel.: 216-836-0451; FAX 216-836-4614)

BDJ (England) Ltd.
    Maple Cross Ind. Est., Denham Way, Rickmansworth, Hertfordshire, WD3 2RZ, UK (Tel.: 0923-779959; FAX 0923-896270)

Beacon Chemical Co.
    PO Box 2500, 125 Macquiesten Pkwy., Mt. Vernon, NY 10550 (Tel.: 914-699-3400; FAX 914-699-2783)

Begg & Co. Thermoplastics Ltd.
    71 Hailey Road, Erith, Kent, DA18 4AW, UK (Tel.: 081 310 1236; FAX 081 310-4371; Telex: 8966166)

Belding Chemical Industries
    1430 Broadway, New York, NY 10018 (Tel.: 212-944-6040)

Bell Flavors & Fragrances, Inc.
500 Academy Dr., Northbrook, IL USA (Tel.: 312-291-8300; 800-323-4387; Telex: 910-686-0653)

Bell Laboratories, Inc.
3699 Kinsman Blvd., Madison, WI 53704 (Tel.: 608-241-0202; FAX 608-241-9631; Telex: 910 286 2775)

Belzak Corp.
850 Bloomfield Ave., Clifton, NJ 07012 (Tel.: 201-773-0602)

Bercen, Inc.
1381 Cranston St., Cranston, RI, 02920 (Tel.: 401-943-7400; 800-525-0595; Telex: 927637)

Berncolors-Poughkeepsie Inc.
PO Box 29, Poughkeepsie, NY 12602 (Tel.: 914-454-6700; FAX 914-454-4742; Telex: 926011)

Bernel Chemical Co., Inc.
174 Grand Ave., Englewood, NJ 07631 (Tel.: 201-569-8934; FAX 201-569-1741)

Berol Nobel
Berol Nobel, Rue Gachard 88, Bte 9, B-1050 Bruxelles, Belgium (Tel.: 32-02-640-5065; FAX 32-2-640-6997; Telex: 62812)
Berol Nobel AB, S-444, 85 Stenungsund, Sweden (Tel.: 46-303-850-00; FAX 46-303-843-71; Telex: 27038)
Berol Nobel Inc., Meritt 8 Corporate Park, 99 Hawley Lane, Stratford, CT 06497

Betz Industrial
Somerton Rd., Trevose, PA 19047 (Tel.: 215-355-3300; FAX 215-953-2473; Telex: 173148)

Biddle Sawyer Corp.
2 Penn Plaza, New York, NY 10121 (Tel.: 212-736-1580)

Bio-Botanica, Inc.
75 Commerce Dr., Hauppauge, NY 11788 (Tel.: 516-231-5522; 800-645-5720; FAX 516-231-7332)

Bioplast SRL
Via Longare 9, Torri di Quartesolo VI, Italy (Tel.: 0444-580082/581547; FAX 0444-581851; Telex: 434396 BIOPTST I)

Bio-Rad Laboratories
3300 Regatta Blvd., Richmond, CA 94804 (Tel.: 415-232-7000; 800-227-5589; FAX 415-232-4257; Telex: 71-3720184)

Bioscience, Inc.
1530 Valley Center Pkwy., Suite 120, Bethlehem, PA 18017 (Tel.: 215-974-9693; 800-627-3069; FAX 215-691-2170)

BIP Chemicals Ltd.
Popes Lane, Oldbury, Warley, West Midlands, B69 4PD, UK (Tel.: 021-552-1551; FAX 021-552-4267; Telex: 337261)

Blackman Uhler Chemical Div./Synalloy Corp.
PO Box 5627, Spartanburg, SC 29304 (Tel.: 803-585-3661; FAX 803-596-1501; Telex: 810-282-2582)

Blew Chemical Co.
PO Box 501, Palos Heights, IL 60463 (Tel.: 708-448-5780; FAX 708-448-5781)

Blomocan Ltd.
Pitfield Kiln Farm, Milton Keynes, MK11 3LE, UK (Tel.: 0908-566666; FAX 0908-566244; Telex: 825755)

BMC, Inc./Bulk Molding Compounds Inc.
3N497 N. 17th St., St. Charles, IL 60174 (Tel.: 708-377-1065; FAX 312-377-7395)

Boehme Filatex Inc.
209 Watlington Industrial Dr., Reidsville, NC 27320 (Tel.: 919-342-6631; FAX 919-342-1208)

Boehringer Ingelheim
Boehringer Ingelheim KG, Binger Strasse 172, Postfach 200, D-6507 Ingelheim, Germany (Tel.: 061-32-77-6748; FAX 061-32-77-3755; Telex: 418 791-22 bi d)

Henley Chemicals, Inc., 50 Chestnut Ridge Rd., Montvale, NJ 07645 (Tel.: 201-307-0422)

Boliden Compound Trelleborg AB
231 81, Trelleborg, Sweden (Tel.: 0410-51781; FAX 0410-16981; Telex: 32739)

Boliden-Intertrade Inc.
3400 Peachtree Rd. NE, Suite 401, Atlanta, GA 30326 (Tel.: 404-239-6700; 800-241-1912; FAX 404-239-6701; Telex: 981036)

Bonar Polymers Ltd.
Horndale Ave., Newton Aycliffe, Co Durham, DL5 6YE, UK (Tel.: 0325-300990; FAX 0325-314925)

Bond Chemicals Inc.
1500 Brookpar Rd., Cleveland, OH 44109 (Tel.: 216-741-6935; 800-837-6500; FAX 216-741-8374)

Bondabelt
Unit 506, Ketley Business Park, Ketley, Telford, Shropshire, TF1 4JD, UK (Tel.: 0952-641001; FAX 0952-641474; Telex: 35810 B Belt)

Bondaglass-Voss Ltd.
158-160 Ravenscroft Rd., Beckenham, Kent, BR3 4TW, UK (Tel.: 081-778-0071; FAX 081-659-5297)

Borax Consolidated Ltd.
Borax House, Carlisle Place, London, SW1 1HT, UK (Tel.: 071-834-9070; FAX 071-973-9075; Telex: 920453 BORAXL G)

Borden Chemical Div.
180 E. Broad St., Columbus, OH 43215 (Tel.: 614-225-4000; 800-225-8044; FAX 614-225-3476)

Borregaard LignoTech
Borregaard LignoTech, PO Box 162, N-1701 Sarpsborg, Norway (Tel.: (09)11 80 00; FAX (09) 11 87 70)

LignoTech USA, Inc., 100 Highway 51 South, Rothschild, WI 54474-1198 (Tel.: 715-359-6544; FAX 715-355-3648)

LignoTech Canada Inc., 1950, Rue Léon Harmel, Quebec, P.Q., GIn HK3, Canada (Tel.: 418-684-3000; FAX 418-684-3005)

LignoTech (U.K.) Ltd., Clayton Rd., Birchwood, Warrington, Cheshire, WA3 6QQ, UK (Tel.: (0925)824511; FAX (0925) 812186)

Bostik Div./Emhart Corp.
Boston St., Middleton, MA 01949 (Tel.: 508-777-0100; 800-726-7845; FAX 508-774-7376; Telex: 200282)

Bowdene-Redcar Ltd.
Longbeck Works, Marske, Redcar, Cleveland, TS11 6HW, UK (Tel.: 0642-490190; FAX 0642 482376)

BP Chemicals
BP Chemicals Ltd., Belgrave House, 76 Buckingham Palace Rd., London, SW1W 0SU, UK (Tel.: 071-581-1388; FAX 071-581-6411)

BP Chemicals Inc., PO Box 411, Keasbey, NJ 08832 (Tel.: 908-417-3099)

BP Chemicals Inc., 4440 Warrensville Center Rd., Cleveland, OH 44128-2837 (Tel.: 800-447-2724; FAX 216-581-5211)

BP Performance Polymers, Newburg Rd., PO Box 400, Hackettstown, NJ 07840 (Tel.: 908-852-1100; 800-526-4636, 908-850-7282; Telex: 510-235-2736)

R.F. Bright Enterprises Ltd.

London Road, West Kingston, Sevenoaks, Kent, TN15 6AP, UK (Tel.: 0474 852 852; FAX 0474 853 944)

British Cork Mills Ltd.

Vulcan St., Bootle, Merseyside, L20 4HL, UK (Tel.: 051-922-1917; FAX 051-922-5843)

British Traders & Shippers Ltd.

6/7 Merrilands Crescent, Dagenham, Essex, RM9 6SL, UK (Tel.: 081-595-4211; FAX 081-593-0933; Telex: 897438)

Brooks Industries Inc.

70 Tyler Place, South Plainfield, NJ 07080 (Tel.: 908-561-5200; FAX 908-561-9174)

Bruce Chemical Corp.

PO Box 7343, Macon, GA 31209 (Tel.: 912-743-1111; FAX 912-742-3547)

The Brulin Corp.

2920 Dr. Andrew J. Brown Ave., PO Box 270, Indianapolis, IN 46206-0270 (Tel.: 317-923-3211; 800-776-7149)

Bruna Coatings Ltd.

Oakenclough Mill, Oakenclough, Garstang, Lancashire, PR3 1TB, UK (Tel.: 0995 604981; Telex: 677426)

Bubble & Foam Industries NV

Gijzelbrechtegemstraat 8, 8570 Anzegem, Belgium (Tel.: 056 68 09 41; FAX 056 68 93 91; Telex: 86356)

Buckman Labs Int'l., Inc.

1256 North McLean Blvd., Memphis, TN 38108 (Tel.: 901-278-0330; 800-727-2772; Telex: 68-28020)

Buffalo Color Corp.

959 Rte. 46 East, Suite 403, Parsippany, NJ 07054 (Tel.: 201-316-5600; 800-631-0171; FAX 201-316-5828; Telex: 7109885924)

Bullen Chemical Co.

PO Box 37, Folcroft, PA 19032 (Tel.: 215-724-8100; FAX 215-534-8912)

Buller Plastics Ltd.

St. James Mill Road, Northampton, NN5 5JP, UK (Tel.: 0604-755 616; FAX 6064 755 616; Telex: 317196 BULLER G)

Burgess Pigment Co.

PO Box 349, Sandersville, GA 31082 (Tel.: 912-552-2544; 800-841-8999; FAX 912-746-4882; Telex: 804523)

Burlington Bio-Medical Corp.

222 Sherwood Ave., Farmingdale, NY 11735-1718 (Tel.: 516-694-9000; FAX 516-694-9177; Telex: 6852289)

Burlington Chemical Co., Inc.

PO Box 111, Burlington, NC 27216 (Tel.: 919-584-0111; 800-334-8550; FAX 919-584-3548; Telex: 9102502503)

Burn Tubes Ltd.

Burcol Works, Radway Road, Shirley, Solihull, West Midlands, B90 4NS, UK (Tel.: 021 704 2211; FAX 021 704 2217; Telex: 339758)

BXL Plastics Ltd.
Huddersfield Rd., Darton, Barnsley, South Yorkshire, S75 5NA, UK (Tel.: 0226-390039; FAX 0226-390125; Telex: 54157)

BYK-Chemie USA
524 S. Cherry St., Wallingford, CT 06492 (Tel.: 203-265-2086; Telex: 643378)

Cabot
Cabot Corp./Cab-O-Sil Div., PO Box 188, Tuscola, IL 61953 (Tel.: 217-253-3370; 800-222-6745; FAX 217-253-4334; Telex: 910-663-2542)
Cabot GmbH/Cab-O-Sil Div., PO Box 1766, D-7888 Rheinfelden, Germany (Tel.: 7623-9090; FAX 7623-90932; Telex: 773451)
Cabot Plastics Ltd., Gate St., Dunkinfield, SK16 4RU, UK (Tel.: 061-330-5051; FAX 061-308-2641)
Cabot Plastics Belgium SA, Rue E Vandervelde 131, B4431 Ans (Loncin), Belgium
Cabot Plastics Italiana SpA, Z.I. 38055 Grigno (TN), Italy
Cabot Plastics Hong Kong Ltd., 914 Sun Plaza, 28 Canton Rd., Tsim Sha Tsui, Kowloon, Hong Kong

Cadauta sas di R Fornasero & C
Via Torino snc. 10020 S Sebastiano Dapo, Torino, Italy (Tel.: 011 9191284; Telex: 21114 CSINDI)

Cadillac Plastic Ltd.
Rivermead Drive, Westlea, Swindon, Wiltshire, SN5 7YT, UK (Tel.: 0793 514 949; FAX 0793 511 762; Telex: 449537)

Cairn Chemicals Ltd.
Cairn House, Elgiva Lane, Chesham, Buckinghamshire, HP5 2JD, UK (Tel.: 0494 786066; FAX 0494 791816; Telex: 837075)

Calbar Inc.
2626 N. Marta St., Philadelpha, PA 19125 (Tel.: 215-739-9141; 800-347-8910; FAX 215-739-3258; Telex: 215-739-3258)

Calgon
Calgon Corp., PO Box 1346, Pittsburgh, PA 15230 (Tel.: 412-777-8000; FAX 412-777-8927)
Calgon Carbon Corp., PO Box 717, Pittsburgh, PA 15230-0717 (Tel.: 412-787-6700; 800-422-7266; FAX 412-787-6825)

California Products Corp.
PO Box 569, 169 Waverly St., Cambridge, MA 02139 (Tel.: 617-547-5300; 800-225-1141; FAX 617-547-6934; Telex: 951587 CALPRO CAM)

Callaway Chemical Co.
6003 Hamilton Rd., Columbus, GA 31909 (Tel.: 706-576-2000; FAX 706-576-6455)

Cal Polymers, Inc.
2115 Gaylord St., Long Beach, CA 90813 (Tel.: 213-436-7372)

Calumet Lubricants Co.
2780 Waterfront Pkwy. E. Drive, Suite 200, Indianapolis, IN 46214 (Tel.: 317-328-5660; FAX 317-328-5668)

Cambridge Industries
Cambridge Industries Co., Inc., 440 Arsenal St., Watertown, MA 02172, USA)
Cambridge Industries Co. of America, 7-33 Amsterdam St., Newark, NJ 07105 (Tel.: 201-465-4565; FAX 201-465-7713)

Camie-Campbell, Inc.
9225 Watson Industrial Park, St. Louis, MO 63126-1581 (Tel.: 314-968-3222; 800-325-9572; FAX 314-968-0741)

John Campbell & Co.
PO Box 96, Perkasie, -PA 18944 (Tel.: 215-257-2708; FAX 215-257-2709)

Campine SA (Compagnie Chimique et Métallurgique)
Ave. Louise 522, bte 16, B-1050 Brussels, Belgium (Tel.: 2 647 98 70; FAX 2 640 35 68; Telex: 24 045)

Canada Packers
Canada Packers Inc., Edible Oils Div., 30 Weston Rd., Toronto, Ontario, M6N 3P4, Canada (Tel.: 416-766-4311; FAX 416-761-4452; Telex: 069-69539)
Canada Packers Inc., Food Ingredients Group, 30 Weston Rd., Toronto, Ontario, M6N 3P4, Canada (Tel.: 416-761-4231; FAX 416-761-4452)

Cancarb Ltd.
1702 Brier Park Cresc. N.W., PO Box 310, Medicine Hat, Alberta, T1A 7G1, Canada (Tel.: 403-527-1121; FAX 403-529-6093; Telex: 03-824866)

Carboline Co.
350 Hanley Indl. Crt., St. Louis, MO 63144 (Tel.: 314-644-1000; FAX 314-644-4617)

Carbonic Industries Corp.
3340 Rosebud Rd., Loganville, GA 30249 (Tel.: 404-979-0250; 800-241-5882)

Carborundum Co.
PO Box 156, Niagara Falls, NY 14302 (Tel.: 716-278-2000; FAX 716-278-2900)

Carbose Corp.
100 Maple St., Somerset, PA 15501 (Tel.: 814-443-1611; FAX 814-445-8951)

Cardolite Corp.
500 Doremus Ave., Newark, NJ 07105-4805 (Tel.: 201-344-5015; FAX 201-344-1197)

Cardox/Div. of Liquid Air Corp.
2121 N. California Blvd., Walnut Creek, CA 94596 (Tel.: 415-977-6500; Telex: 433-0467 LAC-UL)

Carey Industries Inc.
PO Box 620, 190 Rear White St., Danbury, CT 06813-0620 (Tel.: 203-744-7280; FAX 203-791-2762; Telex: 969642)

Cargill
Cargill, Inc., Box 5630, Minneapolis, MN 55440 (Tel.: 612-475-6478; Telex: CGL MPS 290625)
Cargill Inc./Salt Div., PO Box 5621, Minneapolis, MN 55440 (Tel.: 800-544-2498; FAX 612-475-5144; Telex: 290625)

Carib International
PO Box 111, Elmwood Park, NJ 07407 (Tel.: 201-791-6700; FAX 201-791-0038)

R.H. Carlson Co./Northern Labs
41 Chestnut St., Greenwich, CT 06830 (Tel.: 203-531-5500; 800-243-5404)

Carolina Color & Chemical Co.
PO Box 5642, Charlotte, NC 28225 (Tel.: 704-333-5101; FAX 704-342-3023; Telex: 575207)

Carpenter Technology Corp.
PO Box 14662, 101 W. Bern St., Reading, PA 19612-4662 (Tel.: 215-371-2000)

Carroll Co.
2900 W. Kingsley, Garland, TX 75041 (Tel.: 214-278-1304; 800-527-5722; FAX 214-840-0678)

R.E. Carroll, Inc.
1570 N. Olden Ave., Trenton, NJ 08638 (Tel.: 609-695-6211; 800-257-9365; FAX 609-695-0102)

Carus Chemical Co.
> PO Box 1500, 1001 Boyce Memorial Dr., Ottawa, IL 61350 (Tel.: 815-433-9070; 800-435-6856; FAX 815-433-9075; Telex: 797551)

CasChem Inc.
> 40 Ave. A, Bayonne, NJ 07002 (Tel.: 201-858-7900; 800-526-1467; FAX 201-437-2728; Telex: 710-729-4466)

Castrol Industrial East Inc.
> 775 Louis Dr., Warminster, PA 18974-0357 (Tel.: 215-443-5220; Telex: 201507)

Catalyst Resources, Inc.
> 2190 North Loop West, Suite 400, Houston, TX 77018 (Tel.: 713-682-5300; FAX 713-957-6839; Telex: 910-881-7119)

Catalytic Products Int'l., Inc.
> 980 Ensell Rd., Lake Zurich, IL 60047 (Tel.: 708-438-0334; FAX 708-438-0944)

Cataphote, Inc.
> PO Box 2369, Jackson, MS, 39225-2369 (Tel.: 601-939-4612; 800-221-2574; FAX 601-932-5339; Telex: 58-5467)

Catawba-Charlab, Inc.
> 5046 Pineville Rd., PO Box 240497, Charlotte, NC 240497 (Tel.: 704-523-4242; FAX 704-522-8142)

CC Pollen Co.
> 3627 E. Indian School Rd., Suite 209, Phoenix, AZ, 85018-5126 (Tel.: 602-957-0096; 800-950-0096; FAX 602-381-3130; Telex: 559834)

CDC International Inc.
> 22 Portsmouth Rd., Amesbury, MA 01913 (Tel.: 508-388-2221)

Ceca SA. See Atochem

Celite
> Celite Corp., PO Box 519, Lompoc, CA 93438-0519 (Tel.: 805-735-7791; FAX 805-735-5699; Telex: 62776493 ESL UD)
> Celite France., 9 rue du Colonel-de-Rochebrune B.P. 240, 92504 Rueil-Malmaison Cedex, France (Tel.: (14)749-0560; FAX (14) 749-5824; Telex: 203089 MANVL F)
> Manville Great Britain Ltd., Regal House, 1st Floor, London Rd., Twickenham, Middlesex, TW13QE, UK (Tel.: 4481-891-0813; FAX (4481) 892-9325; Telex: 928635 MANVIL G)
> Manville Japan Ltd., Toranomon Asahi Bldg., 10F, 1-11-3, Nishi-shimbashi, Minato-ku, Tokyo, 105, Japan (Tel.: (03)580-3791; FAX (03) 580-3793; Telex: 2222487 MVL J)

CE Minerals, Div. Combustion Engrg.
> 901 E. Eighth Ave., King of Prussia, PA 19406 (Tel.: 215-265-6880)

Centrachem AG
> Im Feld, 4626 Härkingen, Switzerland (Tel.: 062-61 21 61; FAX 062-61 34 04; Telex: 982 862)

Central Chemicals Co. Ltd.
> Unit 35 Planetary Rd., Willenhall, Wolverhamtpon, W. Midlands, WV13 3XB, UK (Tel.: 0902 727544; FAX 0902 864260; Telex: 337226)

Central Soya
> Central Soya Co. of America, 1300 Ft. Wayne Nat'l. Bank Bldg., PO Box 1400, Fort Wayne, IN 46801 (Tel.: 219-425-5100; 800-348-0960; FAX 219-425-5485; Telex: 276170)
> Central Soya, PO Box 5063, 3008 AB Rotterdam, Netherlands (Tel.: 31-10-42-39-600; FAX 31-10-42-30-897; Telex: 20041 CNSOY NL)

Century Chemical
> PO Box 1442, 28790 CR20-W, Elkhart, IN 46515 (Tel.: 219-293-9521; 800-348-3505; FAX 219-522-5723)

Cerac, Inc.
PO Box 1178, 407 N. 13th St., Milwaukee, WI 53201 (Tel.: 414-289-9800; FAX 414-289-9805; Telex: RCA 286122)

Ceralox
Rue De La Piquette No. 8, 78890 Garancieres, France (Tel.: 34 86 40 90; FAX 1 34 86 56 74; Telex: 695567)

Certified Processing Corp.
US Hwy. 22, Hillside, NJ 07205 (Tel.: 201-923-5200)

Cestidur Industries
Cestidur Industries, Zone Industrielle Rue Rene Desgrand, BP 1270, 69608 Villeurbanne Cedex, France (Tel.: 72 04 7878; FAX 72 04 6985; Telex: 370678)
Erta Cestidur Industries, Zone Industrielle Rue Rene Desgrand, BP 1270, 69608 Villeurbanne Cedex, France (Tel.: 72 04 7878; FAX 72 04 6985; Telex: 370678)

Chamberlain Plastics Div. Ltd.
North End, Higham Ferrers, Wellingborough, Northants, NN9 8JD, UK (Tel.: 0933 53875; FAX 0933 410206; Telex: 311238)

Champlain Industries Inc.
25 Styertowne Rd., Clifton, NJ 07012 (Tel.: 201-778-4900; 800-222-4904; Telex: 3725769)

Chapman Chemical Corp.
416 E. Brooks Rd., PO Box 9158, Memphis, TN 38109-1058 (Tel.: 901-396-5151; 800-238-2523; FAX 901-396-5159; Telex: 53-3201)

Chattem Chemicals
1715 W 38 St., Chattanooga, TN 37409 (Tel.: 615-821-4571; FAX 615-821-0395; Telex: 55-8463)

Chemach NV
Sobieskilaan 13, 1020 Brussels, Belgium (Tel.: 02 479 99 57; Telex: 65075 CHEMA)

Chemag AG
Postfach 970167, D-6000 Frankfurt am Main, Germany (Tel.: 069-74-34-0)

Chemax, Inc.
PO Box 6067, Highway 25 South, Greenville, SC 29606 (Tel.: 803-277-7000; 800-334-6234; FAX 803-277-7807; Telex: 570412 IPM15SC)

Chemclean Corp.
130-45 180 St., Springfield Gardnes, NY 11434 (Tel.: 718-525-4500; FAX 718-481-6470)

Chemetall
Chemetall GmbH, Reuterweg 14, Postfach 14, D-6000 Frankfurt, Germany (Tel.: 069-159-2833)
Ore & Chemical Corp., 520 Madison Ave., New York, NY 10022 (Tel.: 212-715-5236; FAX 212-715-5291; Telex: ITT 422681)
Ore & Chemical Corp./Latex Div., 520 Madison Ave., New York, NY 10022 (Tel.: 212-715-5200; 800-632-6009; FAX 212-826-5516)

Chemetals Inc.
711 Pittman Rd., Baltimore, MD 21226 (Tel.: 410-636-7100; 800-876-3464; FAX 410-636-7113)

Chemetics Laboratories, Inc.
2954 Congressman Lane, Dallas, TX 75220 (Tel.: 214-351-2434; FAX 214-358-0426; Telex: 734037 CHEMETICS)

Chemfax Inc.
3 Rivers Rd., PO Box 1390, Gulfport, MS, 39502 (Tel.: 601-863-6511; FAX 601-863-6547)

Chemical Corp. of America
2 Carlton Ave., E. Rutherford, NJ 07073 (Tel.: 201-438-5800; FAX 201-438-0041)

Chemical & Pigment Co.
600 Nichols Rd., Pittsburg, CA 94565 (Tel.: 415-689-2030; FAX 415-458-2410)

Chemical Processing Co./Div. of RJI
2316 S. Blvd., Charlotte, NC 28203 (Tel.: 704-333-8930)

Chemical Products Corp.
PO Box 449, Cartersville, GA 30120 (Tel.: 404-382-2144; FAX 706-386-6053; Telex: 542268)

Chemical Specialties, Inc.
One Woodlawn Green, Suite 250, Charlotte, NC 28217 (Tel.: 704-522-0825)

Chemithon Corp.
5430 W. Marginal Way SW, Seattle, WA 98106 (Tel.: 206-937-9954; FAX 206-932-3786; Telex: 328910)

Chemlease Inc.
PO Box 386, Groveport, OH 43125 (Tel.: 614-836-2331; FAX 614-836-2868)

Chem-Lig International Industries, Inc.
30 Main St., Danbury, CT 06810 (Tel.: 203-744-6622; Telex: 969648)

Chemloid Inc.
8501 Pelham Rd., Greenville, SC 29615 (Tel.: 803-288-8331; FAX 803-288-8339)

Chemol Co.
2410 Randolph Ave., PO Drawer 20687, Greensboro, NC 27406 (Tel.: 919-333-3060; FAX 919-273-4645)

Chemonic Industries Inc.
PO Box 9601, Greensboro, NC 27408 (Tel.: 919-373-0145; FAX 919-373-0145)

Chem-Pak, Inc.
11 Oates Ave., PO Box 1685, Winchester, VA 22601 (Tel.: 703-667-1341; 800-336-9828; FAX 703-722-3993; Telex: 3791728)

Chemron Corp.
PO Box 2299, Paso Robles, CA 93447 (Tel.: 805-239-1550; FAX 805-239-8551)

Chemsal Chemicals & Co. KG/Chem-Y GmbH/Salim joint venture
Kupferstrasse 1, PO Box 100262, D-4240 Emmerich 1, Germany (Tel.: 02822 711-0; FAX 02822 18294; Telex: 8125124)

Chem/Serv Inc.
715 SE 8 St., Minneapolis, MN 55414 (Tel.: 612-379-4411; 800-966-2436; FAX 612-379-8244)

Chemson
Chemson GmbH, Post Box 12, A-9601 Arnoldstein, Austria (Tel.: 04255 2226; FAX 04255 2435; Telex: 45596)
Chemson Ltd., Hayhole Works, Willington Quay, Wallsend, Tyne and Wear, NE28 0PB, UK (Tel.: 091 258 5892; FAX 091 258 1549; Telex: 537726)

Chemtech Industries, Inc.
1655 Des Peres Rd., St. Louis, MO 63131 (Tel.: 314-966-9900; 800-325-3332)

Chem-Tex Laboratories Inc.
1016 N. 185 Service Rd., Charlotte, NC 28297-6264 (Tel.: 704-392-9322; 800-532-5361; FAX 704-392-9325)

Chemtrade
Chemtrade, Inc., 551 Lancaster Ave., Haverford, PA 19041 (Tel.: 215-527-0808; FAX 215-527-6422; Telex: 272746)

Chemtrade International BV, Steynlaan 2, 3743 CH, Baarn, Netherlands (Tel.: 02154 12325; FAX 02159 32127; Telex: 43776 INCO NL)

Chem-Trend

Chem-Trend Inc., 1445 W. McPherson Park Dr., PO Box 860, Howell, MI 48844-0860 (Tel.: 517-546-4520; 800-248-4056)

Chem-Trend A/S, Smedeland 14, PO Box 13 84, DK-2600 Glostrup, Denmark (Tel.: 42 45 6711; FAX 43 63 03 50; Telex: 33187 CMTREND DK)

Chemurgy Prods Inc.

Box 3977, 450 Furman Hall Rd., Greenville, SC 29608-3977 (Tel.: 803-232-7697)

Chem-Y

Chem-Y Fabriek Van Chemische Producten B.V., PO Box 50, 2410 AB, Bodegraven, Netherlands (Tel.: 028-22/711-0; FAX 49282218294; Telex: 8125 124)

Chem-Y GmbH, Kupferstrasse 1, D4240 Emmerich, Germany (Tel.: 49-2822/7110; FAX 49-2822/18294; Telex: 8125124)

Chevron

Chevron Chemical Co., PO Box 3766, Houston, TX 77253 (Tel.: 713-754-4290; Telex: 762799)

Chevron Chemical Co./Olefin & Derivs., PO Box 3766, Houston, TX 77253 (Tel.: 800-231-3826)

Chevron Chemical Co./Agricultural Chemicals Div., 6750 Poplar, Suite 300, Memphis, TN 38138 (Tel.: 901-754-3400)

Chiba Flour Milling Co. Ltd.

17, Sinminato, Chiba-shi, Chiba-ken, Japan (Tel.: 472-41-0111; FAX 472-47-8282)

Childers Prods Co. Inc.

35555 Curtis Blvd., Eastlake, OH 44095 (Tel.: 216-953-5200; 800-321-7994)

Chilean Nitrate Corp.

150 Boush St., Suite 701, Norfolk, VA 23510 (Tel.: 804-640-7270; 800-648-6827; FAX 804-640-7271; Telex: 823427)

Chimex

16 rue Maurice-Berteaux, Le Thillay, 95500, Gonesse, France (Tel.: 842-695784)

Chinghall Ltd.

Ward Road, Bletchley, Milton Keyes, MK1 1JA, UK (Tel.: 0908 376227; FAX 0908 271735; Telex: 826196)

Chiorino SpA

Via Sant'agata 9, PO Box 460, 13051 Biella, Italy (Tel.: 015 8489 1; FAX 015 401159; Telex: 214095 CHIOR I)

Chlorinators Inc.

4125 SW Martin Hwy., Suite 2, Palm City, FL, 34990-5524 (Tel.: 407-288-4854; 800-327-9761; FAX 407-287-3238; Telex: 350341 CHLORINC)

Chomarat et Cie

Les Fils D'Auguste, South View, Bamford House, Bamford, Rochdale, OL11 5HU, UK (Tel.: 0706 350994; FAX 0706 42543)

CHR Industries. See Furon

Chrostiki SA

PO Box 22, GR-194 00, Koropi Attkis, Greece (Tel.: 01 6624692; FAX 01 662 3873; Telex: 21-8000)

Charles B. Chrystal Co. Inc.

25 Ann St., New York, NY 10038 (Tel.: 212-227-2151; FAX 212-233-7916; Telex: 420803 CBCC)

Church & Dwight Co. Inc./Specialty Prods. Div.
Box CN5297, 469 N. Harrison St., Princeton, NJ 08543-5297 (Tel.: 609-497-7116; 800-221-0453; FAX 609-497-7176; Telex: 752226)

Ciba-Geigy
Ciba-Geigy Corp., 444 Saw Mill River Rd., Ardsley, NY 10502 (Tel.: 914-478-3131; 800-431-1874)

Ciba-Geigy Corp./Additives Div., Seven Skyline Dr., Hawthorne, NY 10532-2188 (Tel.: 914-785-4461; 800-431-1900)

Ciba-Geigy Corp./Dyestuffs & Chemicals Div., PO Box 18300, Greensboro, NC 27419 (Tel.: 800-334-9481; FAX 919-632-7098)

Ciba-Geigy Corp/ Pigments Div., 7 Skyline Dr., Hawthorne, NY 10532 (Tel.: 914-785-2000; 800-431-1900; FAX 914-347-2202)

Ciba-Geigy Corp./Plastics Div., Seven Skyline Dr., Hawthorne, NY 10532 (Tel.: 914-347-6600; 800-222-1906)

Ciba-Geigy PLC, 30 Buckingham Gate, London, SW1E 6LH, UK

Ciba-Geigy Plastics UK, Duxford, Cambridge, CB2 4QA, UK (Tel.: 0223-832121; FAX 0223-838404; Telex: 0223-81101)

Ciba-Geigy Corp./Dyestuffs & Chemicals UK, Ashton New Road, Clayton, Manchester, M11 4AR UK

Ciba-Geigy Pigments, Ashton New Road, Clayton, Manchester, M11 4AR UK (Tel.: 061 223 1341; FAX 061 231 7422; Telex: 668083)

Ciba-Geigy AG, CH-4002, Basel, Switzerland (Tel.: 061 223 1341; FAX 061 231 7422; Telex: 668083)

Ciba-Geigy Marienberg GmbH, Postfach 1253, D-6140 Bensheim 1, Germany (Tel.: 06254 79 0; FAX 06254 79493; Telex: 625491)

Clark Chemical Inc.
103 Walnut Grove Rd., Cartersville, GA 30120 (Tel.: 404-386-6397; FAX 404-386-6393)

Clean-Chem Corp./Subsid. Chemical Consulting Co. Inc.
21640 Wyoming Crt., Oakpark, MI 48237 (Tel.: 313-399-4600)

Clean Room Products, Inc.
1800 Ocean Ave., Ronkonkoma, NY 11779 (Tel.: 516-588-7000; FAX 516-588-7863)

W.A. Cleary Chemical Corp.
Southview Industrial Park, 178 Route #522 Suite A, Dayton, NJ 08810 (Tel.: 908-329-8399; 800-524-1662; FAX 908-274-0894)

Climax Molybdenum
Climax Molybdenum Co., 1370 Washington Pike, Bridgeville, PA 15017 (Tel.: 412-257-1560; FAX 412-257-0540)

Climax Molybdenum UK Ltd., 50-52 Great Eastern St., London, EC2A 3EP, UK (Tel.: 071 739 6422; FAX 071 729 1964; Telex: 27316)

Climax Molybdenum Asia, Ltd., Akasaka Twin Tower, Main Bldg., 1-22 Akasaka, 2-chome, Minato-ku, Tokyo, 107, Japan

Climax Performance Materials Corp.
7666 W. 63rd St., Summit, IL 60501 (Tel.: 708-458-8450; 800-323-3231; FAX 708-458-0286)

Climax Specialty Chem.
101 Merritt 7 Corporate Park, PO Box 5113, Norwalk, CT 06856-5113 (Tel.: 203-845-2951; FAX 203-845-2953)

Clough Chemical Co., Ltd.
178 St. Pierre, PO Box 1017, St-Jean-sur-Richelieu, Quebec, J3B 7B5, Canada (Tel.: 514-346-6848; 800-363-9284; FAX 514-346-7263)

CNC International, Limited Partnership
PO Box 3000, Woonsocket, RI, 02895 (Tel.: 401-769-6100; FAX 401-769-4509)

Coal Fillers Inc.
Box 1063, Bluefield, VA 24605 (Tel.: 703-322-4675)

Coated Specialities Ltd.
Chester Hall Lane, Basildon, Essex, SS14 3BG, UK (Tel.: 0268 530331; FAX 0268 527211; Telex: 995514)

Coating Systems Inc.
55 Crown St., Nashua, NH 03060 (Tel.: 603-883-0554)

Colgate-Palmolive Co.
300 Park Ave., New York, NY 10022 (Tel.: 212-310-2000)

Colonial Chemical, Inc.
9431 Mountain Shadows Dr., Chattanooga, TN 37421 (Tel.: 615-267-8947; FAX 615-266-0770)

Colorant GmbH
Justus Staudt Str 1, D-6250 Limburg, Germany (Tel.: 06431 53391; FAX 06431 53199; Telex: 4821044 COLO D)

Color-Chem International Corp.
7 Plymouth Rd., Glen Rock, NJ 07452-1216 (Tel.: 201-444-8563; FAX 201-670-4106)

Colorco Inc.
1261 W. Elizabeth Ave., Linden, NJ 07036 (Tel.: 201-862-3011; Telex: 710-996-5929)

Colorcon
Moyer Blvd., West Point, PA 19486 (Tel.: 215-699-7733)

Colores Hispania SA
Josep Pla 149, 08019 Barcelona, Spain (Tel.: 3 307 13 50; FAX 3 303 2505; Telex: 54116 COHI)

Colorite Plastics Co.
101 Railraod Ave., Ridgefield, NJ 07657 (Tel.: 201-941-2900; 800-631-1577; Telex: 134442)

Columbia Filter Co.
PO Box 1597, Kingston, NY 12401 (Tel.: 914-331-0113; 800-336-0056; Telex: 145-338)

Columbian Chemicals Co.
1600 Parkwood Circle,Suite 400, Atlanta, GA 30339 (Tel.: 404-951-5700; 800-235-4003)

Comco Plastics Ltd.
Unit 2, Balmoral Industrial Estate, Balmoral Road, Belfast, BT12 6HR, N. Ireland, UK (Tel.: 668358)

Cominco American Inc.
PO Box 3087, W. 601 Riverside Ave., Spokane, WA 99220 (Tel.: 509-747-6111; FAX 509-459-4440)

Commercial Quimica Insular SA
Pol. Indl. Las Rubiesas Nave-3, 35200 Cruce De Lemenara-Telde, Las Palmas de Gran Canaria, Spain (Tel.: 928 680960; FAX 928-694347)

Companhia Industrial De Resinas Sinteticas CIRES SA
Apartado 20, 3861 Estarreja Codex, Portugal (Tel.: 0343 41432; FAX 034 41077; Telex: 37037 Cires P)

Composition Materials Co., Inc.
1375 Kings Hwy. East, Fairfield, CT 06430 (Tel.: 203-384-6111; 800-262-7763; FAX 203-335-9728; Telex: 131454)

Compounding Ingredients Ltd.
Unit 217, Walton Summit, Bamber Bridge, Preston, Lancs, PR5 8AL, UK (Tel.: 0772 322888; FAX 0722 315853; Telex: 677621 CIL G)

Compounding Technology Inc.
13435 Estelle St., Corona, CA 91720 (Tel.: 714-371-7701; 800-325-1564; FAX 714-371-7724)

Comprifalt
Rue De La Piquette No. 8, 78890 Garancieres, France (Tel.: 34 86 40 90; FAX 1 34 86 56 74; Telex: 695567)

Conap, Inc.
1405 Buffalo St., Olean, NY 14760 (Tel.: 716-372-9650; FAX 716-372-1594)

Concept Polymer Technologies Inc.
12755 60th St. N., Clearwater, FL, 34620 (Tel.: 813-535-6500; 800-541-6880; FAX 813-535-3887)

Condea Chemie GmbH
Überseering 40, 2000 Hamburg 60, Germany (Tel.: 40 6375-0; FAX 40 6375 3595)

Consolidated Polymers Ltd.
Dawson House, 1A Lytherton Ave., Cadishead, Manchester, M30 5BU, UK (Tel.: 061 776 1028; Telex: 665947 MAMET G)

Consos, Inc.
PO Box 34186, Charlotte, NC 28234 (Tel.: 704-596-2813; FAX 704-596-4861)

Continental Chemical Co.
270 Clifton Blvd., Clifton, NJ 07015 (Tel.: 201-472-5000; FAX 201-472-5221; Telex: 133572)

Continental Polymers Inc.
2225 East Del Amo Blvd., Compton, CA 90220 (Tel.: 310-637-2103; 800-441-3943, 310-637-2415; Telex: 4720686 CPIINC)

Continental Sulfur Co.
7500 San Felipe, Suite 410, Houston, TX 77063 (Tel.: 713-782-5513; FAX 713-782-3071)

Convert SA
B.P. 1007, Chemin du Grand Moulin, F 01101 Oyonnax, France (Tel.: 74 77 46 22; FAX 74 77 83 08; Telex: 340 948 convert f)

Cook Composites & Polymers
Cook Composites & Polymers, PO Box 419389, Kansas City, MO 64141-6389 (Tel.: 816-391-6000)
Cook Composites & Polymers, 217 Freeman Dr., PO Box 996, Pt. Washington, WI 53074-0996 (Tel.: 414-284-5541; 800-745-5541; FAX 414-284-7519; Telex: 26737)

Cookson Minerals Ltd.
Cookson House, Willington Quay, Wallsend, Tyne & Wear, NE28 6UQ, UK (Tel.: 091 262 2211; FAX 091 263 4491; Telex: 537357)

Cookson Pigments Inc.
256 Vanderpool St., Newark, NJ 07114 (Tel.: 201-242-1800; FAX 201-242-7274)

The Cooper Co. Inc.
PO Box 726, Gulf Breeze, FL, 32562-0726 (Tel.: 904-932-5005; FAX 904-932-1923)

Copolymer Rubber & Chemical Corp./A DSM Co.
Scenic Highway, PO Box 2591, Baton Rouge, LA 70821 (Tel.: 504-355-5655; 800-535-9960; FAX 504-355-8056; Telex: 586419)

Corn Products/Unit of CPC Int'l.
6500 Archer Rd., Summit-Argo, IL 60501 (Tel.: 708-563-2400; FAX 708-563-6852; Telex: 708-563-6763)

Cornelius Chemical Group Ltd.
St. James's House, 27-43 Eastern Road, Romford, Essex, RM1 3NN, UK (Tel.: 0708 722300; FAX 0708 768204; Telex: 885589 CORNEL G)

Corradini Gustavo & C SpA
Ria Circondaria 5, 2919 Correggio RE, Italy (Tel.: 059 565022; Telex: 510571 CORSOL I)

Corstyrene
Route d'Antisanti, 20270 Aleria, France (Tel.: 95570387; FAX 95 57 0705; Telex: 460132)

Cosmetochem USA, Inc.
Industrial West, Clifton, NJ 07012 (Tel.: 201-471-8301; FAX 201-471-3783; Telex: 642643)

Cosmic Plastics, Inc.
12314 Gladstone Ave., San Fernando, CA 91342-5381 (Tel.: 818-365-3249; 800-423-5613)

Costa Floros O.E.
157A Patission St., Athens, 112 52, Greece (Tel.: 8656419; FAX 1 8651726; Telex: 216407 FLOR GR)

Costec, Inc.
PO Box 693, Palatine, IL 60013 (Tel.: 708-359-5713)

Cote Color Corp.
PO Box 3584, Spartanburg, SC 29304 (Tel.: 803-579-0535; FAX 803-579-1839)

Covan Ltd.
901 Conshohocken Rd., Conshohocken, PA 19428 (Tel.: 215-834-5744)

CP-Polymer-Technik GmbH
Berliner Strasse 3-5, Postfach 11 58, D-2863 Ritterhude, Germany (Tel.: 0 42 92 10 34; FAX 042 92 10 39; Telex: 249926)

CPS Chem. Co., Inc.
PO Box 162, Old Bridge, NJ 08857 (Tel.: 908-727-3100; FAX 908-727-2260; Telex: 844532-CPSOLDB)

CPS Kemi Aps
Hejreskovv, 22, 3490 Kvistagaard, Denmark (Tel.: 2 890533; FAX 42 23 80 77)

Crain Chemical Co. Inc.
2630 Andjon Dr., Dallas, TX 75354 (Tel.: 214-358-3301; FAX 214-358-3304)

Craynor. See Sartomer

Cray Valley Products Inc.
Box 247A, Stuyvesant, NY 12173 (Tel.: 518-828-4383; FAX 518-828-4382)

Cri-Tech, Inc.
85 Winter St., Hanover, MA 02339 (Tel.: 617-826-5600; FAX 617-826-5770)

CR Minerals Corp.
14142 Denver W. Pkwy., Suite 250, Golden, CO, 80401 (Tel.: 303-278-1706; 800-527-7315)

Croda
Croda Inc., 7 Century Dr., Parsippany, NJ 07054-4698 (Tel.: 201-644-4900; FAX 201-644-9222)
Croda Canada Ltd., 78 Tisdale Ave., Toronto, Ontario, M4A 1Y7, Canada (Tel.: 416-751-3571; FAX 416-751-9611)
Croda Chemicals Ltd., Cowick Hall, Snaith Goole, North Humberside, DN14 9AA, UK (Tel.: 0405-8605551; FAX 0405-860205; Telex: 57601)
Croda Surfactants Ltd., Cowick Hall, Snaith, Goole, North Humberside, DN14 9AA, UK (Tel.: 0405 860551; FAX 0405 860205; Telex: 57601)
Croda Universal Ltd., Cowick Hall, Snaith, Goole, North Humberside, DN14 9AA, UK (Tel.: 0405 860551; FAX 0405 860205; Telex: 57601)
Croda Food Products Ltd., Cowick Hall, Snaith, Goole, North Humberside, DN14 9AA, UK (Tel.: 0405 860551; FAX 0405 860205; Telex: 57601)
Croda Italiana Srl, Via Grocco, N917 27036, Mortara (PV), Italy
Croda Chemicals Group Pty. Ltd., PO Box 1012, Richmond, North Victoria, 3121, Australia

Croda Japan KK, Aceman Bldg., 5F 3 7, Tokuicho 1-Chome Highashi-ku, Osaka, 540, Japan (Tel.: 6-942-1791)

Croda do Brazil Ltda., Rua Croda 230 Distrito Industrial, CEP 13.053, Campinas/SP-C.P. 1098, Brazil

Crompton & Knowles

Crompton & Knowles Corp./Dyes & Chems. Div., PO Box 33188, Charlotte, NC 28233 (Tel.: 704-372-5890; FAX 704-372-1522)

Crompton & Knowles Corp./Ingredient Tech. Div., 1595 MacArthur Blvd., Mahwah, NJ 07430 (Tel.: 201-818-1200; 800-631-8305)

Crosfield Chemicals, Inc.

101 Ingalls Ave., Joliet, IL 60435 (Tel.: 815-727-3651; 800-727-3651; FAX 815-727-5312)

Crowley Chemical Co.

261 Madison Ave., New York, NY 10016 (Tel.: 212-682-1200; 800-424-9300)

Crowley Tar Products Co., Inc.

261 Madison Ave., New York, NY 10016 (Tel.: 212-682-1200)

Crown Technology, Inc./Chemical Div.

7513 E. 96 St., PO Box 50426, Indianapolis, IN 46250 (Tel.: 317-845-0045; FAX 317-845-9086)

Croxton & Garry Ltd.

Curtis Rd. Industrial Estate, Dorking, Surrey, RH4 1XA, UK (Tel.: 0306-886688; FAX 0306-887780; Telex: 859567/8 cand g)

Crucible Chemical Co.

PO Box 6786, Donaldson Center, Greenville, SC 29606 (Tel.: 803-277-1284; 800-845-8873)

CSI. See Chemical Specialties Inc.

Cuproquim Corp.

9601 Katy Freeway, Suite 350, Houston, TX 77024-1333 (Tel.: 713-464-1103; 800-488-2224; FAX 713-464-1421; Telex: 910 240 6712 PDC)

Custom Fibers

Custom Fibers International, 28130 Ave. Crocker, #311, Valencia, CA 91355 (Tel.: 805-295-0990; 800-321-5324; FAX 805-295-5148)

Custom Fibers Europe, 13 Rassau Indust. Esatate, Ebbw Vale, Gwent, Wales, UK (Tel.: 495-350-655)

Cuyahoga Plastics

1265 Babbitt Rd., Cleveland, OH 44132-2798 (Tel.: 216-261-2744; FAX 216-261-3537)

CVC Specialty Chemicals, Inc.

600 Deer Rd., Cherry Hill, NJ 08034 (Tel.: 609-354-0040; FAX 609-354-6226)

Cyanamid. See American Cyanamid

Cyprus Industrial Minerals

9100 East Mineral Circle, PO Box 3299, Englewood, CO, 80155 (Tel.: 303-643-5484; FAX 303-643-5168)

Cyro

Cyro Industries, 100 Valley Rd., PO Box 950, Mt. Arlington, NJ 07856 (Tel.: 201-770-3000; 800-631-5384; FAX 201-770-6117)

Cyro Canada Inc., 360 Carlingview Dr., Etobicoke, Ontario, M9W 5X9, Canada (Tel.: 416-675-9433)

Dacar Chem Co.

1007 McCartney St., Pittsburgh, PA 15220 (Tel.: 412-921-3620; 800-223-8875; FAX 412-921-4478)

Dai-ichi Kogyo Seiyaku Co., Ltd.
Miki Bldg., 3-12-1, Nihombashi, Chuo-ku, Tokyo, 103, Japan (Tel.: 03-3274-6731; FAX 03-3274-4128; Telex: 222 6258)

Dainippon Ink & Chemicals, Inc.
7-20, Nihonbashi 3-Chome, Chuo-ku, Tokyo, 103, Japan (Tel.: 03 272-4511; FAX 0434-98-2229; Telex: 222-2977)

Dalau Ltd.
Ford Road, Clacton on Sea, Essex, UK (Tel.: 0255 220220; FAX 0255 221122; Telex: 98593 DALAU G)

Richard Daleman Ltd.
Old Wolverton Raod, Milton Keynes, Bucks, MK12 5PS, UK (Tel.: 0908 312108; Telex: 825487)

Daniel Products Co.
400 Claremont Ave., Jersey City, NJ 07304 (Tel.: 201-432-0800; FAX 201-432-0266; Telex: 126-304)

Davathane Ltd.
Isandula Road, Basford, Nottingham, UK (Tel.: 0602 785116; FAX 0602 422253)

Davidson Metals
1615 Wilson Ave., Youngstown, OH 44506 (Tel.: 216-743-3001; 800-343-3506; FAX 216-743-2817)

Frank D. Davis Co.
7011 Muirkirk Rd., Beltsville, MD 20705 (Tel.: 301-776-2400; 800-638-4444; FAX 301-776-4967; Telex: 910-997-5179)

Day-Glo Color Corp./Subsid. of Nalco Chemical Co.
4515 St. Clair Ave., Cleveland, OH 44103 (Tel.: 216-391-7070; FAX 216-391-7751; Telex: 960687)

DCS Color & Supply Co., Inc.
2011 S. Allis St., Milwaukee, WI 53207 (Tel.: 414-769-2580; FAX 414-769-2598)

Dead Sea Bromine
Dead Sea Bromine Co. Ltd., Bromine Compds. Ltd., Makleff House, PO Box 180, Beer-Sheva, 84101, Israel (Tel.: 972-57-667222; FAX 972-57-32444; Telex: 5335, 5343)
Ameribrom, Inc., 52 Vanderbilt Ave., New York, NY 10017 (Tel.: 212-286-4000; FAX 212-286-4475; Telex: RCA 220531)
Bromine & Chemicals Ltd., England (Tel.: 44-71-493-9711; FAX 44-71-493-9714; Telex: 23845)

DeeZee Chemical Inc.
14010 Orange Ave., Paramount, CA 90723 (Tel.: 310-529-2556; FAX 310-529-8037)

The Degen Co.
200 Kellogg St., PO Box 5240, Jersey City, NJ 07305 (Tel.: 201-432-1192; FAX 201-432-8483; Telex: 325999 DEGEN CO)

Degussa
Degussa Corp., 65 Challenger Rd., Ridgefield Park, NJ 07660 (Tel.: 201-641-6100; FAX 201-807-3183)
Degussa Corp./Pigment Group, 425 Metro Place N., Suite 550, Dublin, OH 43017 (Tel.: 614-761-0658; FAX 614-761-9441)
Degussa AG, GB Industrie-und Feinchemikallen Anwendungstechnik, Postfach 1345, D-6450 Hanau 1, W. Germany (Tel.: 6181-59-2119; FAX 6181-59-4099)
Degussa Metal Group, 3900 S. Clinton Ave., S. Plainfield, NJ 07080 (Tel.: 201-561-1100; FAX 201-769-9456)

J.W.S. Delavau Co. Inc.
2140 Germantown Ave., Philadelphia, PA 19122 (Tel.: 215-235-1100)

Dennis Chem Co.
2700 Papin St., St. Louis, MO 63103 (Tel.: 314-771-1800; FAX 314-771-8399)

Dependable Extrusions Ltd.
Oxnam Road, Jedburgh, Roxburghshire, TD8 6NN, UK (Tel.: 0835 62591; FAX 0835 63879; Telex: 727338)

Detrex Corp./Chemicals Div.
PO Box 1398, Ashtabula, OH 44004 (Tel.: 216-997-6131; 800-362-6915)

Development Associates, Inc.
300 Old Baptist Rd., No. Kingstown, RI, 02852 (Tel.: 401-884-1350; 800-648-9966; FAX 401-885-7888)

Deventer Benelux BV
Spinveld 34, 4834 JP Breda, Netherlands (Tel.: 076 226070; FAX 076 226443; Telex: 54428)

Devoe Coatings Co.
PO Box 7600, Louisville, KY 40257 (Tel.: 502-897-9861; FAX 502-893-1475)

Dexco Polymers/ A Dow/Exxon Partnership
Exxon Chem., Polymers Group, 13501 Katy Freeway, Houston, TX 77079 (Tel.: 713-870-6055)

Dexter
Dexter Chemical Corp., 845 Edgewater Rd., Bronx, NY 10474 (Tel.: 212-542-7700; FAX 212-991-7684; Telex: 127061)
Dexter Corp./Frekote Products, One Dexter Dr., Seabrook, NH, 03874 (Tel.: 603-474-5541; FAX 603-474-5545; Telex: 6817306 HYSEA)
Dexter Corp./Hysol Engineering Adhesives, One Dexter Dr., Seabrook, NH, 03874 (Tel.: 603-474-5545; FAX 603-474-5545)
Hysol Aerospace Prods., 2850 Willow Pass Rd., PO Box 312, Pittsburg, CA 94565 (Tel.: 415-687-4201; FAX 415-687-4205; Telex: TWX: 910-387-0363)
Dexter Specialty Coatings, One East Water St., Waukegan, IL 60085 (Tel.: 312-623-4200)

DIC Trading (USA)
222 Bridge Plaza So., Fort Lee, NJ 07024 (Tel.: 201-592-5100; FAX 201-592-8232)

Disco, Inc./Assoc. Polymer Prods. Div.
1010 Greenwood Lake Tpk., Ringwood, NJ 07456 (Tel.: 201-728-7731)

Ditta Salt SpA
Via Canton Santo 5, 21050 Borsano VA Italy (Tel.: 033 1342111; FAX 033 1340108; Telex: 330598)

Diversified Compounders
5701 E. Union Pacific Ave., Los Angeles, CA 90022 (Tel.: 213-728-3000; FAX 213-725-7806)

Dixo Co., Inc.
158 Central Ave., PO Box 6, Rochelle Park, NJ 07662 (Tel.: 201-845-6000; FAX 201-845-6004)

D.J. Enterprises
PO Box 31366, Cleveland, OH 44131 (Tel.: 216-524-3879)

D M & C Watering BV
Herengracht 400, 1017 BX Amsterdam, Netherlands (Tel.: 020 266 555; Telex: 11015 DMCWA NL)

Dock Resins Corp.
1512 W. Elizabeth Ave., Linden, NJ 07036 (Tel.: 908-862-2351; FAX 908-862-4015)

Ronald T. Dodge Co.
55 Westpark Rd., Dayton, OH 45459 (Tel.: 513-439-4497; FAX 513-439-1704; Telex: 910 250 1073)

M. Dohmen USA Inc./D&G Dyes
9 Distribution Ct., Greer, SC 29650-9116 (Tel.: 803-676-1669; FAX 803-676-0114)

Dolphin Paint & Chem Co.
922 Locust St., Toledo, OH 43604 (Tel.: 419-241-8267)

John C. Dolph Co.
PO Box 267 W. New Rd., Monmouth Junction, NJ 08852 (Tel.: 201-329-2333; Telex: 362260)

Domingo Pascual Carbo SA
Camprodón 55, E17240 Llagostera, Spain (Tel.: 72 830025; FAX 72 830759; Telex: 57275 Taps)

Dominion Products Inc.
882 Third Ave., Brooklyn, NY 11232 (Tel.: 718-499-3050)

Dong Jin (USA)
555 Canal St., Manchester, NH, 03101 (Tel.: 800-666-5583; FAX 313-429-5572)

Dooley Chemical Co.
PO Box 71951, Chattanooga, TN 37407 (Tel.: 615-624-0086; FAX 615-622-4848)

S. Dory Ltd.
490 Honeypot Lane, Stanmore, Middlesex, HA7 1JX, UK (Tel.: 081 951 3232; FAX 081 952 8631; Telex: 299357)

Dover Chemical Corp.
Davis at West 15th St., PO Box 40, Dover, OH 44622 (Tel.: 216-343-7711; 800-321-8805/6; FAX 216-364-1579; Telex: 983466)

Dow
Dow Chemical U.S.A., 2020 W.H. Dow Center, Midland, MI 48674 (Tel.: 517-636-1000; 800-441-4DOW)
Dow Plastics, 2040 W.H. Dow Center, Midland, MI 48674 (Tel.: 800-441-4DOW; FAX 517-638-9942)
Dow Chemical Canada Inc., 1086 Modeland Rd., PO Box 1012, Sarnia, Ontario, N7T 7K7, Canada (Tel.: 519-339-3131; 800-363-6250)
Dow Chemical Co. Ltd., Stana Place, Fairfield Ave., Staines, Middlesex, TW18 4SX, UK
Dow Chemical Europe S.A., Bachtobelstrasse 3, CH-8810 Horgen, Switzerland (Tel.: 41-1-728-2111; FAX 41-1-728-2935; Telex: 826940)
Dow Chemical Pacific ltd., 39th Floor, Sun Hung Kai Centre, 30 Harbour Rd., Wanchai, PO Box 711, Hong Kong
Dow Quimica S.A., Sao Paulo, Brazil
Dow Ahlen, Theodor-Schwarte-Str. 39, D-4730 Ahlen 1, Germany (Tel.: 02382-891-0; FAX 49-2382-5151; Telex: 820768)

Dow Corning
Dow Corning Corp., PO Box 0994, Midland, MI 48686-0994 (Tel.: 517-496-4000; Telex: 227450)
Dow Corning Ltd., Reading Bridge House, Reading, Berkshire, RG1 8PW, UK
Dow Corning STI, 45550 Helm St., Plymouth, MI 48170
Dow Corning STI, Eastern Region, 50 Commerce Dr., Trumbull, CT 06611 (Tel.: 203-377-3800)
Dow Corning France SA, Le Britannia A10, 20 Bid E Deruelle, 69432 Lyon Cedex 03, France (Tel.: 78 60 51 48; FAX 78 62 78 98; Telex: 300537)
Dow Corning/Wickhen, East Coast Service Center, Bracken Rd., PO Box 384, Montgomery, NY 12549 (Tel.: 914-456-9631)

Dragoco Inc.
Gordon Dr., Totowa, NJ 07512 (Tel.: 201-256-3850; FAX 201-256-6420)

Drew Industrial Div. See Ashland

DRG Flexible Packaging
Filwood, Fishponds, Bristol, BS16 3RY, UK (Tel.: 0272 656232; FAX 0272 651617; Telex: 44811 DRG FP G)

Dry Branch Kaolin
Rt. 1, Box 468D, Dry Branch, GA 31020 (Tel.: 201-851-2820)

Drycolor AB
Box 9053, 200 39 Malmö, Sweden (Tel.: 40 210150; FAX 40 223346; Telex: 32608 DRYCOL S)

Dryden Oil Co.
7131 Westfield, PO Box 04067, Detroit, MI 48204 (Tel.: 313-834-3600; FAX 313-834-2846)

DSM
DSM Engineering Plastics, North Amer. Inc., 501 Crescent Ave., PO Box 15051, Reading, PA 19612-5051 (Tel.: 215-320-6918; 800-366-6923; FAX 215-320-6930)
Sheffield Plastics, Inc., Salisbury Rd., Sheffield, MA 01257 (Tel.: 800-628-5084)
DSM Resins UK Ltd., PO Box 8, Ellesmere Port, South Wirral, L65 0HB, UK (Tel.: 051 355 6170; FAX 051 357 1282; Telex: 628213)
DSM United Kingdom Ltd., Kingfisher House, Kingfisher Walk, Redditch, Worcestershire, BG7 4EZ, UK (Tel.: 0527 68254; FAX 0527 62949; Telex: 3398618)
DSM France SA, Immeuble Périsud, 5, Rue Legeune, 92128 Montrouge (Cédex), France
DSM Polymers International, Postbus 43, 6130 AA Sittard, The Netherlands
DSM Verkoopkantoor Polymeren Nederland, PO Box 3204, 3502 GE Utrecht, Churchill Laan 11, Utrecht, Netherlands (Tel.: 030 921911; FAX 030 943990; Telex: 47918)
DSM Deutschland GmbH & Co., Tersteegenstrasse 77, 4 Düsseldorf 30, Germany (Tel.: 0211 454940; FAX 0211 4370917; Telex: 2114365)
DSM Resins Espana SA, C'an Jané-Coll de la Manya, 08400 Granollers, Barcelona, Spain (Tel.: 93 849 55 22; FAX 93 849 85)

Ducey Chemical, Inc.
PO Box 776, Princeton Jct., NJ 08550 (Tel.: 609-799-3191; FAX 609-799-2085)

Duflex Ltd.
Bridge Court, Bridge Street, Long Eaton, Nottingham, Notts., NG10 4QQ, UK (Tel.: 0602 462828; FAX 0602 462860; Telex: 377366)

Dumo Plastics NV
Wijnendalestraat 171, 8800 Roeselare, Belgium (Tel.: 051 23 02 10; FAX 051 23 02 38; Telex: 81946)

Dunlopillo UK
Station Road, Pannal, Harrogate, North Yorkshire, HG3 1JL, UK (Tel.: 0423 872411; FAX 0423 879232; Telex: 57857)

DuPont
E.I. DuPont de Nemours & Co., Inc., 1007 Market St., Wilmington, DE, 19898 (Tel.: 302-774-7573; 800-441-9442; FAX 302-774-7573; Telex: 6717325)
DuPont/Petrochemicals Dept., Wilmington, DE, 19898 (Tel.: 302-999-5053; 800-231-0998)
DuPont/Polymer Products Div., Wilmington, DE, 19880-0011 (Tel.: 800-441-7111)
DuPont Canada Inc., Box 2200, Streetsville, Mississauga, Ontario, L5M 2H3, Canada (Tel.: 416-821-5612)
DuPont UK Ltd., Wedgewood Way, Stevenage, Herts, SG1 4QN, UK
DuPont UK Ltd./Polymer Products Dept., Maylands Ave., Hemel Hempstead, Herts, HP2 7DP, UK
DuPont de Nemours International S.A., Polymer Products Dept., 2, Chemin du Pavillon, PO Box 50, CH-1218 Le Grand Saconnex, Geneva, Switzerland (Tel.: 022-717-51-11)

DuPont de Nemours (France) S.A., 9, Rue de Vienne, 75008-Paris, France
DuPont Far East Inc., Kowa Bldg. No. 2, 11-39 Akasaka 1-Chome, Minato-Ku, Tokyo, 107, Japan (Tel.: 585-5511)
DuPont S.A. de C.V., Homero 206, Col. Polanco, Mexico, 5, D.F. Mexico
DuPont Automotive Products Dept., 950 Stephenson Hwy., Troy, MI 48006-7013 (Tel.: 313-583-8000)
DuPont Co./Agricultural Products, 1007 N. Market St., Wilmington, DE, 19898 (Tel.: 302-992-6400; 800-441-7515; Telex: 6717325)
DuPont Medical Products, Scintillation Chemicals, 549-4 Albany St., Boston, MA 02118 (Tel.: 617-482-9595; 800-323-8903; FAX 617-542-8463)

Dwight Products, Inc.
10 Stuyvesant Ave., Lyndhurst, NJ 07071 (Tel.: 201-438-3377; FAX 201-438-0594)

The Dycho Co.
PO Box 513, Niota, TN 37826 (Tel.: 615-568-2112; FAX 615-568-2116)

Dyetech Inc.
PO Box 2761, Dalton, GA 30722-2761 (Tel.: 404-278-0536; FAX 404-226-2191)

Dylon Industries, Inc.
7700 Clinton Rd., Cleveland, OH 44144 (Tel.: 216-651-1300; 800-237-8246; FAX 216-651-1777)

Dymax Corp.
51 Greenwoods Rd., Torrington, CT 06790 (Tel.: 203-482-1010)

DynaGel Inc.
Wentworth Ave. & Plummer St., Calumet City, IL 60409 (Tel.: 708-891-8400; FAX 708-891-8432; Telex: 211666)

Dynaloy, Inc.
7 Great Meadow Lane, Hanover, NJ 07936 (Tel.: 201-887-9270; FAX 201-887-3678; Telex: 642033)

Eaglebrook, Inc.
1150 Junction, Schererville, IN 46375 (Tel.: 219-322-2560; 800-428-3311; FAX 219-322-8533)

Eagle Chem Co., Inc.
PO Box 107, Mobile, AL 36601 (Tel.: 205-452-9624; FAX 205-452-3383)

Eagle-Picher
Eagle-Picher Industries, Inc., C & Porter Sts., Joplin, MO 64802 (Tel.: 417-623-8000)
Eagle-Picher Research Laboratory, 200 9th Ave. N.W., Miami, OK 74354-3305 (Tel.: 918-542-1801; 800-331-3144)

Eagle Zinc Co.
30 Rockefeller Plaza, New York, NY 10112 (Tel.: 212-582-0420; FAX 212-582-3412)

E-A-R Specialty Composites
7911 Zionsville Rd., Indianapolis, IN 46268 (Tel.: 317-872-1111; FAX 317-872-0618)

Eastern Color & Chemical Co.
35 Livingston St., PO Box 6161, Providence, RI 02904 (Tel.: 401-331-9000; FAX 401-331-2155)

Eastman
Eastman Chemical Co., PO Box 431, Kingsport, TN 37662 (Tel.: 615-229-2000; 800-EASTMAN; FAX 615-229-1064; Telex: 6715569)
Eastman Chemical UK Ltd., Hemel Hempstead, PO Box 66, Kodak House, Station Road, Herts, HP1 1JU, UK (Tel.: 44-442-41171; FAX 044 24177; Telex: 826502)
Eastman Japan Ltd., Nishi-Shinbashi Mitsui Bldg., 1-24-14 Nishi-Shinbashi, Minato-Ku, Tokyo, 105, Japan

Eastmanchem, Inc., 155 Gordon Baker Rd., Suite 213, Willowdale, Ontario, M2H 3N7, Canada (Tel.: 416-497-7222)

Eastman Chemical Brasileira Ltda., Rua George Eastman, 213, Caixa Postal 225, Sao Paulo, Brazil (Tel.: 55-11-543-5122)

Eastman Chemical International AG, Hertizentrum 6, CH-6300 Zug 6, Switzerland (Tel.: 042 23 25 25; FAX 042 21 12 52; Telex: 868 824)

Kodak Mexicana, Calzada de Tlalpan No. 2980, Mexico D.F., 04870, Mexico

East West Minerals Inc.

100 Shoreline Hwy. S., Mill Valley, CA 94941 (Tel.: 415-331-8880; 800-237-6089; FAX 415-331-5937)

Ebonex Corp.

2380 S. Wabash St., Melvindale, MI 48122 (Tel.: 313-388-0060)

ECC

ECC International, 5775 Peachtree-Dunwoody Rd. NE Suite 200G, Atlanta, GA 30342 (Tel.: 404-843-1551; 800-334-0122; FAX 404-843-8872)

ECC America Inc./Calcium Prods., PO Box 330-9981, Sylacauga, AL 35150 (Tel.: 205-249-4901; 800-251-6327)

ECC International, John Keay House, St. Austell, Cornwall, PL25 4DJ, UK (Tel.: 0726 74482; FAX 0726 623019; Telex: 45526)

Eclipse Colours Ltd.

Hillam Road, Bradford, West Yorks, BD2 1QN, UK (Tel.: 0274 731552; FAX 0274 731552)

Edlon Products, Inc.

117 State Rd., Avondale, PA 19311 (Tel.: 215-268-3101; 800-75-EDLON; FAX 215-268-8898)

Edulan A/S UK

250 Wellington Road South, Stockport, Cheshire, SK2 6NW, UK (Tel.: 061 429 7809; Telex: 667609 EDULAN G)

EGC Corp.

Box 16080, Houston, TX 77222 (Tel.: 713-447-8611; 800-342-7677; FAX 713-931-2201)

Egil Sterud Agentur

Kapellveien 15, 1410 Kolbotn, Norway (Tel.: 02 805088; FAX 02 800603; Telex: 79440 ESSA N)

Eki Eeknoff Kunstof Verwerkende Industrie

De Biezen 1, 6541 BN Nejmegen, Netherlands (Tel.: 080 773378; Telex: 48337)

Elastochem, Inc.

145 Parker Court, Chardon, OH 44024 (Tel.: 216-285-3547; FAX 216-285-2464)

Elastogran. See BASF

Elastoplast SL

Blas Valero 34, 03201 Eiche, Alicante, Spain (Tel.: 96 5443203)

The Elco Corp.

PO Box 609168, 1000 Belt Line, Cleveland, OH 44109 (Tel.: 216-749-2605; 800-321-0467; FAX 216-749-7462)

Elders Exsud Ltd.

247 Tottenham Court Road, London, W1P 0BU, UK (Tel.: 071 782 0011; FAX 071 631 0320; Telex: 264751)

Electro Abrasives Corp.

701 Willet Rd., Buffalo, NY 14218 (Tel.: 716-822-2500; 800-284-GRIT)

Electronic Materials Inc.

PO Box 1014, New Milford, CT 06776 (Tel.: 203-355-3749)

Electro Science Labs Inc.
416 E. Church Rd., King of Prussia, PA 19406 (Tel.: 215-272-8000; 800-257-8340, 215-272-6759; Telex: 834788)

Elkem A/S Nefelin
PO Box 4283, Torshov 0401 0510 4, Norway (Tel.: 2 450100; FAX 2 450205; Telex: 77756)

Eltech International Corp.
12850 Bournewood Dr., Sugar Land, TX 77478 (Tel.: 713-240-6770; FAX 713-240-6762; Telex: 795459)

E/M Corporation
PO Box 2400, 2801 Kent Ave., W. Lafayette, IN 47906 (Tel.: 317-497-6346; 800-428-7802; FAX 317-497-6348)

Emco Services Inc.
PO Box 2191, Taunton, MA 02780 (Tel.: 508-823-8852; FAX 508-822-1931)

EM Industries, Inc./Pigment Div.
5 Skyline Drive, Hawthorne, NY 10532 (Tel.: 914-592-4350; FAX 914-592-9469)

Emkay Chemical Co.
319-325 Second St., PO Box 42, Elizabeth, NJ 07206 (Tel.: 908-352-7053)

Emmerich (Berlon) Ltd.
Wotton Road, Ashford, Kent, UK (Tel.: 0233 622684; FAX 0233 645801)

EMS
EMS-American Grilon, Inc., PO Box 1717 Industrial Park & Corporate Way, Sumter, SC 29151-1717 (Tel.: 803-481-3172; 800-845-8501; FAX 803-481-3820)

EMS-Grilon UK Ltd., Astonfields Industrial Estate, Drummond Rd., GB-Stafford, ST16 3EL, UK (Tel.: 0785-59121; FAX 0785 213068; Telex: 36254)

EMS-Chemie AG, Selnaustrasse 16, CH-8039 Zurich, Switzerland (Tel.: 01 2 015411; FAX 01 201 0155; Telex: 815312)

EMS France SA, 204 Ave. Maréchal Juin, B P 52, 92105 Boulogne Cedex, France (Tel.: 1 46047065; FAX 1 48255607; Telex: 633430)

EMS-Japan Corp., Kikuchi Bldg., 4-6 Nihonbashi Hongoku-cho, 3-Chome, Chuo-ku, Tokyo, 103, Japan

Emulan Inc.
3726 Roosevelt Rd., PO Box 582, Kenosha, WI 53141 (Tel.: 414-654-0734; FAX 414-654-3410)

Emulsion Systems Inc.
70 East Sunrise Hwy., Valley Stream, NY 11581-1233 (Tel.: 516-825-3232; 800-ESI-CRYL; FAX 516-825-3233)

Engelhard Corp./Performance Minerals Group
101 Wood Ave., Iselin, NJ 08830-0770 (Tel.: 908-205-6244; FAX 908-205-6711; Telex: 219984 ENGL UR)

Enichem
Enichem Elastomeri Srl, Milanofiori Strada 3, Palazzo B1, 20090 Assago, Milan, Italy (Tel.: 02 5201; FAX 0039 2 52026077; Telex: 310246)

Enichem Synthesis SpA, Via Medici del Vascello 40, 20138 Milano, Italy (Tel.: 02 5201 39218; FAX 02 5203 9385; Telex: 310246)

Enichem America, Inc., 1211 Ave. of the Americas, New York, NY 10036 (Tel.: 212-382-6531; FAX 212-382-6520; Telex: 6801159 ENICHEM)

Enimont
Enimont Iberica SA, Avda. Diagonal 652-656, Edificio B 31a, 08034 Barcelona, Spain (Tel.: 3 2051911; FAX 3 2051310; Telex: 97964)

Enimont UK Ltd., Enimont House, 111 Upper Richmond, Putney, London, SW15 2TG, UK (Tel.: 081 780 2000; Telex: 918743)

Eni-Trade Grafiske AS
Badevej 2, PO Box 194, 3000 Helsingør, Denmark (Tel.: 0 42102033; FAX 0 42103023; Telex: 41177 eni dk)

Ennar Latex, Inc.
PO Box 247, Middlebury, CT 06762 (Tel.: 203-597-9275; FAX 203-758-8654)

Enodim SA
Chignat, 639100 Vertaizon, France (Tel.: 73 68 10 06; FAX 73 62 93 98; Telex: 990509)

Enterprise Chemical Corp. Ltd.
137 E. Iron Ave., Dover, OH 44622 (Tel.: 216-343-8861; 800-875-8861; FAX 216-343-8853)

Enzyme Development Corp.
2 Penn Plaza, New York, NY 10121 (Tel.: 212-736-1580; FAX 212-239-1089; Telex: 427471 BSCE)

Epoleon
Epoleon Corp. of Am., 2858 Carson St., Suite 121, Torrance, CA 90503 (Tel.: 213-316-4242; 800-448-6367; FAX 213-316-4463)
Epoleon Corp./Div. of Aikoh Co., Ltd., D.S. Bldg. 1-39 Ikenohata 2-Chome, Taito-Ku, Tokyo, 110, Japan (Tel.: 03-823-1111)

Epolin, Inc.
358-364 Adams St., Newark, NJ 07105 (Tel.: 201-465-9495; FAX 201-465-5353)

Erie Foods International Inc.
PO Box 30, Rochelle, IL 61068 (Tel.: 815-562-3441; FAX 309-659-7270; Telex: 671-1864)

Erta Cestidur Industries. See Cestidur Industries

Erta NV
Industriepark Noord, Galgeveldstraat 10, 8700 Tielt, Belgium (Tel.: 51-423211; FAX 51 423360; Telex: 81800)

Esperis SpA
Via Ambrogio Binda, 29, 20143 Milan, Italy (Tel.: 2-891-22219-27-36; FAX 02-891-22257; Telex: 310-485)

Essential Industries Inc.
28391 Essential Rd., PO Box 12, Merton, WI 53056-0012 (Tel.: 414-538-1122; FAX 414-538-1354)

Essex Specialty Products Inc./Subsid. of Dow Chemical Co.
1135 Broad St., Clifton, NJ 07015 (Tel.: 201-773-6300; FAX 201-778-3280; Telex: 62953879)

Estra-Kunststoff GmbH
Postfach 825, D-7410 Reutlingen, Germany (Tel.: 0751-48513; Telex: 732312)

Estron Chemical, Inc.
P.O. Box 127, Highway 95, Calvert City, KY 42029 (Tel.: 502-395-4195; FAX 502-395-5070)

Ethox Chemicals, Inc.
PO Box 5094, Greenville, SC 29606 (Tel.: 803-277-1620; FAX 803-277-8981)

Ethyl
Ethyl Corp., 451 Florida Blvd., Baton Rouge, LA 70801 (Tel.: 504-388-7040; 800-535-3030; FAX 504-388-7686; Telex: 586441, 586431)
Ethyl Petroleum Additives, Inc., 20 South 4th St., St. Louis, MO 63102-1886 (Tel.: 314-421-3930; FAX 314-421-5321; Telex: 447137)
Ethyl Canada, Inc., 350 Burnhamthorpe Rd. West, Suite 600, Mississauga, Ontario, L5B 3JI, Canada (Tel.: 416-566-9222; FAX 416-566-99962)

Ethyl Corp. UK Goldlay House, 114 Parkway, Chelmsford, Essex, CM2 7PP, UK (Tel.: 245-287-577)

Ethyl S.A., 523 Ave. Louise, Box 19, B-1050 Brussels, Belgium (Tel.: 32-2-642-4411; FAX 32-2648-0560; Telex: 22549)

Ethyl Asia Pacific Co., #13-06 PUB Bldg., Devonshire Wing, 111 Somerset Rd., Singapore 0923, Rep. of Singapore

Ethyl Japan, Christy Bldg. 2/F, 1-22 Moto Azabu 3-Chome, Minato-Ku, Tokyo, 106, Japan

Etna Products Inc.
PO Box 630, 16824 Park Circle Dr., Chagrin Falls, OH 44022 (Tel.: 216-543-9845; FAX 216-543-1789; Telex: 980131 WDMR)

Eurobrom BV
PO Box 158, NL-2280 AD Rijswijk, Netherlands (Tel.: 070 340 84 08; FAX 070 399 90 35; Telex: 32137)

Europa Plastics BV
Gladsaxe 31, 7327 JZ Apeldoorn, Netherlands (Tel.: 055 422616; FAX 055 422588)

European Master Batch
17-17 Diepemeers, B-8970 Poperinge, Belgium (Tel.: 0 57/33 66 73; FAX 0 57/33 35 23)

European Vinyls
European Vinyls Corp. (Deutschland) GmbH, Emil-von-Behring-Strasse 2, D-6000 Frankfurt M.50, Germany (Tel.: 0 69 58 01 01; FAX 069 5801 640; Telex: 4189387)

European Vinyls Corp. UK Ltd., Norton House, Crown Gate, Runcorn, Cheshire, WA7 2UX, UK (Tel.: 0928 714482; FAX 0928 715101; Telex: 269415 EVC UK

Eval Co. of America
1001 Warrenville Rd., Suite 201, Lisle, IL 60532 (Tel.: 708-719-4610; 800-423-9762; FAX 708-719-4622)

Evans Chemetics. See W.R. Grace

Evans Clay Co.
PO Box 6, Summit, NJ 07902-0006 (Tel.: 908-273-2500; FAX 908-273-8718; Telex: 5106007803)

Everlight Chemical Industrial Corp.
6th Floor, 695 Tun Hua South Rd., Taipei, Taiwan, ROC (Tel.: 886-2-706-6006; FAX 886-2-708-1254; Telex: 20964 EVERCHEM)

Evode
Evode Plastics Ltd., Wanlip Rd., Syston, Leicester, LE7 8PD, UK (Tel.: 0533-696752/7; FAX 0533-692960)

Evode Speciality Adhesives Ltd., Wanlip Road, Syston, Leicester, LE7 8PD, UK (Tel.: 0533 606001; FAX 0533 692411; Telex: 34485)

Evode-Tanner Industries Inc.
PO Box 1967, Greenville, SC 29602 (Tel.: 803-232-3893; FAX 803-232-3094)

Exolon-Esk Co.
1000 E. Niagara St., Tonawanda, NY 14150 (Tel.: 716-693-4550; 800-962-1100; FAX 716-693-0151; Telex: 91217)

Expancel. See Nobel Industries

Exsymol
4 Ave. Prince Hereditaire Albert, Zone F-Bloc C, MC 98000, Monaco (Tel.: 93-30-13-08; FAX 93-50-43-47)

Exxon
Exxon Chemical Co., PO Box 3272, Houston, TX 77253-3272 (Tel.: 713-870-6000; 800-526-0749; FAX 713-870-6661; Telex: 794588)

Exxon Chemical Co./Tomah Products, 1012 Terra Dr., PO Box 388, Milton, WI 53563 (Tel.:

608-868-6811; 800-441-0708; FAX 608-868-6810; Telex: 910-280-1401)

Exxon Chemical Ltd., Arundel Towers, Portland Terrace, Southampton, SO9 2GW, UK (Tel.: 0703-634191; Telex: 47437)

Exxon Chemical Int'l. Marketing Inc., Mechelsesteenweg 363, B-1950 Kraainem, Belgium (Tel.: 02-769-3111; Telex: 24733)

Deutsche Exxon Chemical GmbH, Dompropst Ketzer-Str. 1-9, 5000 Köln 1, Germany (Tel.: 221 16150; FAX 221 160 5320)

Exxon Chemical Mediterranea SpA, Via Paleocapa 7, 20121 Milano, Italy (Tel.: 2 88031; FAX 2 8803231; Telex: 311561 ESSOCH I)

Exxon Chemical Japan Ltd., TBS Kaikan Bldg., 3-3, Akasaka 5-Chome, Minato-Ku, Tokyo, 107, Japan (Tel.: 03-582-9243; Telex: 22846)

Fabricolor Inc.

PO Box 2398, Paterson, NJ 07509 (Tel.: 800-873-7030; FAX 201-742-5038; Telex: 4933272 FABRIUI)

Fabriquimica S.R.L.

Calle 32 No. 3313, San Martin 1650, Argentina (Tel.: 54-1-755-7290)

Faesy & Besthoff, Inc.

143 River Rd., Edgewater, NJ 07020 (Tel.: 201-945-6200; FAX 201-945-6145)

Fairmount Chemical Co., Inc.

117 Blanchard St., Newark, NJ 07105 (Tel.: 201-344-5790; 800-872-9999; FAX 201-690-5298; Telex: 138905)

The Fanning Corp.

1775 W. Diversity Pkwy., Chicago, IL 60614-1009 (Tel.: 312-248-5700; FAX 312-248-6810; Telex: 910-221-1335)

Fanwood

219 Martine Ave. North, PO Box 159, Fanwood, NJ 07023 (Tel.: 201-322-8440)

The Feldspar Corp.

One West Pack Square, Suite 700, Asheville, NC 28801 (Tel.: 704-254-7400; FAX 704-255-4909)

Fel-Pro Inc.

7450 N. McCormick Blvd., Box 1103, Skokie, IL 60076-8103 (Tel.: 708-674-7700; Telex: 4330156)

Fenocast SA

Gran Vial, No. 4, 08170 Montornés Del Valles, Spain (Tel.: 5 68 23 04; FAX 5 68 48 54)

Alex C. Fergusson, Inc.

Spring Mill Dr., Frazer, PA 19355 (Tel.: 215-647-3300; 800-345-1329; FAX 215-644-8240)

Ferro

Ferro Corp./World Headquarters, 1000 Lakeside Ave., Cleveland, OH 44114 (Tel.: 216-641-8580)

Ferro Corp./Bedford Chemical Div., 7050 Krick Rd., Bedford, OH 44146 (Tel.: 216-641-8580; 800-321-9946; Telex: 98-165)

Ferro Corp./Color Div., 4150 E. 56th St., PO Box 6550, Cleveland, OH 44101 (Tel.: 216-641-8580; FAX 216-641-8831; Telex: 98-0165)

Ferro Corp./Grant Chemical Div., 7050 Krick Rd., Bedford, OH 44146, USA)

Ferro Corp./Engineering Thermoplastics Div., 7500 East Pleasant Valley Rd., Independence, OH 44131 (Tel.: 216-641-8580; FAX 216-524-0493)

Ferro Corp./Keil Chemical Div., 3000 Sheffield Ave., Hammond, IN 46320 (Tel.: 219-931-2630; FAX 219-931-6318; Telex: 725484)

Ferro Corp./Plastic Colorants & Dispersions Div., 3 Railroad Ave., Stryker, OH 43557 (Tel.: 419-682-3311; FAX 419-682-4924)

Ferro Corp./Refractories Div., 661 Willet Rd., Buffalo, NY 14218 (Tel.: 716-825-7900; FAX 716-825-0421; Telex: 980165)

Ferro Corp./Transelco Div., Box 217, Penn Yan, NY 14527 (Tel.: 315-536-3357; FAX 315-536-8091; Telex: 97 8373)

Ferro Enamel Espanola SA, A.C. 232, Castellon, 12080, Spain (Tel.: 522211; FAX 534051; Telex: 65523)

Ferro-Plast Srl, Via Modigliani 2, Milano, Italy (Tel.: 02 2137295; FAX 02 2139576; Telex: 340397 FERPLA)

Fers SA

Arquimedes No. 1, 08930 Sant Adria De Besos, Barcelona, Spain (Tel.: 3 381 20 22; FAX 3 381 78 66; Telex: 50196)

Fiberite. See ICI

Fibertint & Chemicals Inc.

Box 352, Fairforest, SC 29336 (Tel.: 803-574-0334; FAX 803-574-0674)

Fibreglass Evercoat

6600 Cornell Rd., Cincinnati, OH 45242 (Tel.: 513-489-7600; 800-826-8987; FAX 513-489-9229; Telex: 710 1107200)

Fibro Chem Inc.

PO Box 3004, Dalton, GA 30721 (Tel.: 404-278-3514; FAX 404-275-0846)

Filmolux

327 Rue de Charenton, 75012 Paris, France (Tel.: 1 43450400; FAX 1 43411284; Telex: 230564 F)

Filtec Ltd.

Suite A, Constance House, Constance Ind. Est., Waterloo Road, Widnes, Cheshire, WA8 0QR, UK (Tel.: 051 495 1988; Telex: 628151 Filtec G)

Filterite/Memtec America Corp.

2033 Greenspring Dr., Timonium, MD 21093 (Tel.: 301-252-0800; 800-FILTERS; FAX 301-252-6027)

Filter-Media Inc.

3603 Westcenter Dr., Houston, TX 77224-9156 (Tel.: 713-780-9000; FAX 713-781-4320; Telex: 775144 FEMCO HOU)

Fina

Fina Oil and Chemical Co., 8350 North Central Expressway, PO Box 2159, Dallas, TX 75221 (Tel.: 214-750-2400; 800-344-FINA)

Fina Chemicals, Nijverheldsstraat, 52, Rue de l'Industrie, B-1040 Brussels, Belgium (Tel.: 32-2-288-9111; FAX 32-2-288-3388; Telex: 21 556 PFINA B)

Oleofina UK Petrofina House, 1 Ashley Ave., Epsom, Surrey, KT18 5AD, UK (Tel.: 44-03727-26226; FAX 44-03727-45821)

Oleofina Far East, 138 Cecil St., #17-01, Cecil Court, RS Singapore 0106

Finetex Inc.

418 Falmouth Ave., PO Box 216, Elmwood Park, NJ 07407 (Tel.: 201-797-4686; FAX 201-797-6558; Telex: 710-988-2239)

Represented by: Pennine Chemical Ltd., Kent Works, Thomas St., Conglton, Cheshire, CW12 1QZ, UK

Finke-Farbstoffe

Karl Finke, Hatzfelder Strasse 174-176, D-5600 Wuppertal 2, Germany (Tel.: 02 02 70 90 6 0; FAX 02 02 70 39 29; Telex: 8 592 403)

Finnsugar Bioproducts, Inc.

1400 N. Meacham Rd., Schaumburg, IL 60173-4808 (Tel.: 312-843-3200; 800-626-5363; FAX 312-843-3368)

Firestone

Firestone Synthetic Rubber & Latex Co., PO Box 26611, 381 W. Wilbeth Rd., Akron, OH 44319 (Tel.: 216-379-7759; 800-282-0222; FAX 216-379-7483)

Firestone Textiles Co., PO Box 486, Woodstock, Ontario, N4S 7Y9, Canada (Tel.: 800-265-2237)

Firmenich Inc., Chem-Fleur Div.
928-964 Doremus Ave., Port Newark, NJ 07114 (Tel.: 201-589-3443; FAX 201-344-2999)

First Preference Prod. Corp.
PO Box 630C, Pocket Knife Sq., Lakeville, CT 06039 (Tel.: 203-435-0881; 800-553-2701; FAX 203-435-9289)

Flame Control Coatings Inc.
PO Box 786, Niagara Falls, NY 14302 (Tel.: 716-282-1399; FAX 716-285-6303; Telex: 466 840)

Fleming Labs, Inc.
PO Box 34384, Charlotte, NC 28234 (Tel.: 704-372-5613; FAX 704-343-9357; Telex: 572 479)

Flexello Castors (Sales) Ltd.
Blackworth Industrial Estate, Highworth, Swindon, Wiltshire, SN6 7NA, UK (Tel.: 0793 763819; FAX 0793 766460; Telex: 848261)

Flexible Products
Flexible Products Co., 1007 Industrial Dr., PO Box 3190, Marietta, GA 30061 (Tel.: 404-428-2684; FAX 404-590-3658)

Flexible Products Co./Urethane Div., 1007 Industrial Park Dr., Box 3190, Marietta, GA 30061 (Tel.: 404-428-2684; FAX 404-421-6495)

Flexible Reinforcements Ltd.
Queensway House, Queensway, Clitheroe, Lancashire, BB7 1AU, UK (Tel.: 0200 25241; FAX 0200 28960; Telex: 635528 RECTEL G)

Florasynth, Inc.
410 E. 62 St., New York, NY 10021 (Tel.: 212-371-7700; FAX 212-752-4012; Telex: FLORA 620426)

Florida Food Prods., Inc./Aloe Div.
PO Box 1300, Eustis, FL, 32727 (Tel.: 904-357-4141)

Floridin Co.
PO Box 510, Quincy, FL, 32351-0510 (Tel.: 904-62707688; 800-228-1131; FAX 904-875-4408; Telex: 4931835 FLOQYUI)

FMC
FMC Corp./Chemical Products Group, 1735 Market St., Philadelphia, PA 19103 (Tel.: 215-299-6000; FAX 215-299-5999; Telex: 685-1326)

FMC Corp./Ag Chem Group, 1735 Market St., Philadelpha, PA 19103 (Tel.: 215-299-6000; FAX 215-299-5993)

FMC Corp./Lithium Div., 449 N. Cox Road, Gastonia, NC 28054 (Tel.: 704-868-5300; Telex: 6843132 LITHCONC)

FMC Corp./Marine Colloids Div., 1735 Market St., Philadelphia, PA 19103 (Tel.: 215-299-6242; 800-526-3649; FAX 215-299-6291; Telex: 6851326)

FMC Corp. Canada, 11475 Cote de Liesse Rd., Dorval, Quebec, H9P 1B3, Canada

FMC Corp. Belgium, Ave. Louise 523, Box 1, B-1050 Brussels, Belgium

Foamink Technologies/Div. Chubb National Foam
150 Gordon Dr., Box 270, Exton, PA 19341-1350 (Tel.: 215-363-6403; FAX 215-692-7947)

Foamtek, Inc.
1151 Atlantic Dr., Unit 5, West Chicago, IL 60185 (Tel.: 708-293-9299; FAX 708-293-9363)

Franklin Industrial Minerals
Franklin Industrial Minerals, 612 Tenth Ave. North, Nashville, TN 37203 (Tel.: 615-259-4222; FAX 615-726-2693)

Franklin Industrial Minerals, 821 Tilton Bridge Rd., S.E., Dalton, GA 30721 (Tel.: 404-277-3740; FAX 404-277-9827)

Franklin Mineral Products/Div. of Mearl Corp.
Drawer 390, Hartwell, GA 30673 (Tel.: 404-376-3174)

Freeman. See Cook Composites & Polymers

Frekote. See Dexter

Fries & Fries
1199 Edison Dr., Cincinnati, OH 45216 (Tel.: 513-948-8000; 800-543-4643; Telex: 21-4348)

FSW Coatings Ltd.
Virginia, Co. Caven, Rep. of Ireland (Tel.: 049 47209; FAX 049 47470)

H.B. Fuller Co.
3530 Lexington Ave. North, St. Paul, MN 55126 (Tel.: 612-481-1588; 800-468-6358; FAX 612-481-1863)

Furane Products/Isochem Operation
99 Cook St., Lincoln, RI 02865 (Tel.: 401-723-2100)

Furon
Furon Co., 386 Metacom Ave., Bristol, RI 02809 (Tel.: 401-253-2000; 800-336-3534; FAX 401-253-8211)

CHR Industries/Furon, 407 East St., New Haven, CT 06509-9988 (Tel.: 203-777-3631; 800-525-2523)

Futura Coatings, Inc.
9200 Latty Ave., Hazelwood, MO 63042 (Tel.: 314-521-4100; FAX 314-521-7255)

Gabriel-Chemie
Gabriel-Chemie GmbH, Stipcakeg 6, A1234 Vienna, PO Box 15, Austria (Tel.: 222 693610; FAX 222 694374; Telex: 131376 GC A)

Gabriel-Chemie UK Ltd., Transfesa Rd., Paddock Wood, Kent, TN12 6UT, UK (Tel.: 089 283 6566; FAX 089 283 6979; Telex: 957503)

GAF Corp. See ISP Technologies
1361 Alps Rd., Wayne, NJ 07470 (Tel.: 201-628-3000; 800-622-4423)

J. Gaillon SA
Route de Nuites, 69830 St. Georges de Reneins, France (Tel.: 74677733; FAX 74676675; Telex: 370364)

Gard Corp.
2727 Roe Lane, Kansas City, KS 66103 (Tel.: 913-236-5000; FAX 913-432-8309; Telex: 510-100-6629)

Gaston Chemicals Inc.
PO Box 12653, Gastonia, NC 28053 (Tel.: 704-867-0614; FAX 704-853-3850)

Gattefosse
Gattefosse SA, 36 Chemin de Genas, BP 603, 69800 Saint Priest, France (Tel.: 78-90-63-11; FAX 78-90-4567; Telex: 340 240)

Gattefosse Corp., 189 Kinderkamack Rd., Westwood, NJ 07675 (Tel.: 201-573-1700; FAX 201-573-9671)

Represented by: Alfa Chemicals Ltd., Broadway House, 7-9 Shute End, Workingham, Berkshire, RG11 1BH, UK

Gaylord Chemical Co.
106 Galeria Blvd., PO Box 1209, Slidell, LA 70459-1209 (Tel.: 504-649-5464; 800-426-6620; Telex: 629-20-353)

GCA Chemical Corp.
Research Park, 9 Viaduct Rd., Stamford, CT 06907 (Tel.: 203-322-5880; FAX 203-968-1279; Telex: 4750170)

Geltech, Inc.

Two Innovation Dr., Alachua, FL 32615 (Tel.: 904-462-2358; FAX 904-462-2993)

GenCorp Polymer

GenCorp Polymer Products, 165 So. Cleveland Ave., Mogadore, OH 44260 (Tel.: 216-628-6542; FAX 216-628-6501; Telex: 1561448)

Gencorp Polymer Products, 85 The Broadway, London, W13 9BP, UK (Tel.: 081 840 6730; FAX 081 840 6919)

Genencor Int'l. Inc.

4010 Winnetka Ave., Rolling Meadows, IL 60008 (Tel.: 708-870-1030; 800-847-5311; FAX 708-870-9977)

General Chemical

General Chemical Corp., 90 East Halsey Rd., PO Box 393, Parsippany, NJ 07054-0393 (Tel.: 201-515-0900; 800-631-8050; FAX 201-515-2468; Telex: 362262)

General Chem. Canada Ltd., 201 City Center Dr., Mississauga, Ontario, L5B 3A3, Canada (Tel.: 416-896-9595; 800-668-0433; FAX 416-276-6594)

General Electric

General Electric Company, 3135 Easton Tpke., Fairfield, CT 06431 (Tel.: 800-626-2004)

General Electric Co./Plastics Div., One Plastics Ave., Pittsfield, MA 01201 (Tel.: 413-448-7110; 800-845-0600; Telex: 926430)

General Electric Co./Silicone Products Div., 260 Hudson River Rd., Waterford, NY 12188 (Tel.: 518-237-3330; 800-255-8886)

GE Specialty Chemicals, 5th and Avery Sts., Parkersburg, WV 26102 (Tel.: 304-424-5411)

GE Plastics-Canada, 2300 Meadowvale Blvd., Mississauga, Ontario, L5N 5P9, Canada (Tel.: 416-858-5774)

GE Plastics Ltd., Birchwood Park, Risley, Warrington, Cheshire, WA3 6DA, UK (Tel.: 44-925-811522)

GE Europe, Cyprusweg 2, 1044 AA Amsterdam, The Netherlands (Tel.: 31-20-5806911)

General Electric Plastics BV, Plasticslaan 1, PO Box 117, 4600 AC Bergen op Zoom, The Netherlands

General Electric (USA) Plastics Japan Ltd., No.3 5 Kowa Bldg. 5F, 14-14 Akasaka 1-Chome, Minato-ku, Tokyo, 107, Japan

GE Hong Kong, 15/F Convention Center, No. 1 Harbor Rd., Wanchai, Hong Kong (Tel.: 852-5-8105616)

General Latex & Chemical Corp.

67 High St., N. Bellerica, MA 01862 (Tel.: 508-663-3485)

General Plastics/Div. PMC Inc.

55 La France Ave., Bloomfield, NJ 07003 (Tel.: 201-748-5500; FAX 201-748-3988)

Generichem Corp.

85 Main St., PO Box 369, Little Falls, NJ 07424 (Tel.: 201-256-9266; FAX 201-256-0069; Telex: 510-601-6431)

Genesis Polymers. See Novacor

Genstar Stone Products Co.

Executive Plaza IV, 11350 McCormick Rd., Hunt Valley, MD 21031 (Tel.: 301-628-4000; FAX 301-527-4535)

Geoliquids, Inc.

1618 Barclay Blvd., Buffalo Grove, IL 60089 (Tel.: 708-215-0938; FAX 708-215-9821)

Georgia Gulf Corp./PVC Div.

PO Box 629, Plaquemine, LA 70765-0629 (Tel.: 504-685-1200; 800-PVC-VYCM)

Georgia Marble Co.

1201 Roberts Blvd., Bldg. 100, Kennesaw, GA 30144-3619 (Tel.: 404-521-4711)

Georgia-Pacific

Georgia-Pacific Corp., 133 Peachtree St. N.E., PO Box 105605, Atlanta, GA 30348 (Tel.: 404-521-4711)

Georgia-Pacific Chemical Div., 1754 Thorne Rd., Tacoma, WA 98421 (Tel.: 206-572-8181)

Getzner Chemie GmbH

Herrenau 5, Postfach 159, A-6700 Bludenz/Burs, Austria (Tel.: 05552 63310 0; FAX 05552 66864; Telex: 35 52 300)

Gist-Brocades Food Ingredients, Inc.

2200 Renaissance Blvd., Suite 150, King of Prussia, PA 19406 (Tel.: 215-272-4040; 800-662-4478; Telex: 216902)

Giulini Adolfomer Industrias Quimicas SA

Rua Ferreira Viana, 656, Sao Paulo, 04761, Brazil (Tel.: 55-11-523-4877; FAX 55-11 247-0648; Telex: 11 57698 GAIQ BR)

Giulini Corp.

105 E. Union Ave., Boundbrook, NJ 08805 (Tel.: 201-469-6504; FAX 201-469-8418; Telex: 700179)

Givaudan

Givaudan-Roure Corp., 100 Delawanna Ave., Clifton, NJ 07014 (Tel.: 201-365-8277; FAX 201-777-9304)

Givaudan & Cie SA, L., CH-1214 Vernier/Geneva, Switzerland

Glastic Corp./Subsid. of Kobe Steel

4321 Glenridge Rd., Cleveland, OH 44121-2891 (Tel.: 216-486-0100; FAX 216-486-1091)

Global United Industries, Inc.

13609 Industrial Rd., Suite 117, Houston, TX 77015 (Tel.: 713-453-2400; FAX 713-451-5005; Telex: 798465)

Glo-Mold, Inc.

3261 Copley Rd., Copley, OH 44321 (Tel.: 216-668-6513)

Glo-Tex Chemicals Inc.

PO Box 1019, Roebuck, SC 29376 (Tel.: 803-576-6771; FAX 803-576-0794)

Golden Cat Corp.

348 S. Columbia St., South Bend, IN 46601 (Tel.: 219-234-8191; 800-233-5693; FAX 219-232-2934)

Goldschmidt

Goldschmidt AG, Th., Goldschmidtstrasse 100, Postfach 101461, D-4300 Essen 1, Germany (Tel.: 0201-173-2947; FAX 201-173-2160; Telex: 857170)

Goldschmidt Chemical Corp., 914 E. Randolph Rd., PO Box 1299, Hopewell, VA 23860 (Tel.: 804-541-8658; 800-446-1809; FAX 804-541-2783; Telex: 710-958-1350)

Goldschmidt Ltd., Th., Tego House, Victoria Rd., Ruislip, Middlesex, HA4 0YL, UK (Tel.: 01-4227788; FAX 01-8648159)

Goldschmidt Japan KK, Th., Rm. 1113, Shuwa Kioi-cho TBR Bldg. No. 7, 5-Chome, Koji-machi, Chiyoda-ku, Tokyo, 102, Japan

Tego Chemie Service USA, PO Box 1299, 914 E. Randolph Rd., Hopewell, VA 23860 (Tel.: 804-541-8658; 800-446-1809; FAX 804-541-2783)

Tego Chemie Service GmbH, Goldschmidstr. 100, Postfach 101461, D-4300 Essen 1, Germany (Tel.: 0201-1732571; FAX 0201-1732639; Telex: 85717-20tgd)

Goldsmith & Eggleton

Goldsmith & Eggleton, Inc., 2550 Gilchrist Rd., PO Box 1784, Akron, OH 44309 (Tel.: 216-733-7565; FAX 216-733-7560)

Goldsmith & Eggleton Spain, Plaza Urquinaona, 6, 10D, EDIF Torre Urquinaona, 08010 Barcelona, Spain

Goo Chemical Industries Co., Ltd.

Ijiri-58, Iseda-Cho, Uji-shi, Kyoto, Japan (Tel.: 0774-41-6143; FAX 0774-43-3552; Telex: 5453-643 GOOJ)

Goodpasture, Inc.

PO Box 912, Brownfield, TX 79316 (Tel.: 806-637-2541; 800-692-4450)

BFGoodrich

BFGoodrich Co./Specialty Polymers & Chem. Div., 9921 Brecksville Rd., Brecksville, OH 44141 (Tel.: 800-331-1144; Telex: 423313)

BFGoodrich/Geon Vinyl Div., 6100 Oak Tree Blvd., Cleveland, OH 44131 (Tel.: 216-447-6000; 800-438-4366)

BFGoodrich Canada, 195 Columbia St. West, Waterloo, Ontario, N2J 4N9, Canada

BFGoodrich Chemical UK Ltd., The Lawn, 100 Lampton Road, Hounslow, Middlesex, TW3 4EB, UK

BFGoodrich Chemical (Deutschland) GmbH, Goerlitzer Str. 1, 4040 Neuss 1, Germany

Goodyear

Goodyear Tire & Rubber Co., 1485 E. Archwood Ave., Akron, OH 44316 (Tel.: 216-796-8755; FAX 216-796-2617; Telex: 640550 Gdyr)

Goodyear Tire & Rubber Co./Fills Div. & Polyester Div., 1144 East Market St., Akron, OH 44316 (Tel.: 216-796-3845; 800-321-2385)

Goodyear Canada Inc., 45 Raynes Ave., Bowmanville, Ontario, Canada

Goodyear Chemicals Europe, Ave. des Tropiques, Z.A. de Courtaboeuf, 91952 Les Ullis Cedex, France (Tel.: 33-1 64463660; FAX 33-1 64465551; Telex: 602895F)

Goodyear Int'l. Corp., Sankaido Bldg., 1-9-13 Akasaka, Minato-Ku, Tokyo, 107, Japan (Tel.: 81-3 582-0926; FAX 81-3 582-1877)

Gotham Ink & Color Co., Inc.

5-19 47th Ave., Long Island City, NY 11101 (Tel.: 718-729-2347; 800-221-4441; FAX 718-784-4531)

George A. Goulston

PO Box 5025, Monroe, NC 28111-5025 (Tel.: 704-289-6464; FAX 704-283-9110; Telex: 810-649-1355)

W.R. Grace

Grace & Co., W.R./Organic Chemicals Div., 55 Hayden Ave., Lexington, MA 02173 (Tel.: 617-861-6600; 800-232-6100; FAX 617-862-3869; Telex: 200076)

W.R. Grace/Construction Product Div., 62 Whittemore Ave., Cambridge, MA 02140 (Tel.: 617-876-1400; Telex: 174167 GRACE UT)

W.R. Grace/Davison Chemical Div., PO Box 2117, Baltimore, MD 21203 (Tel.: 301-659-9000)

W.R. Grace/Dearborn Div., 300 Genesse St., Lake Zurich, IL 60047 (Tel.: 708-438-1800; FAX 708-540-1588)

W.R. Grace/Emerson & Cuming, Inc., 25 Hartwell Ave., Lexington, MA 02173 (Tel.: 800-DIE BOND)

W.R. Grace/Evans Chemetics Div., 55 Hayden Ave., Lexington, MA 02173 (Tel.: 617-861-6600 ext. 2331; Telex 200076 GRLX UR)

W.R. Grace/Hampshire & Polymer Prods., 55 Hayden Ave., Lexington, MA 02173 (Tel.: 617-861-6600 ext. 2314)

Cryovac, Duncan, SC 29334

Cryovac Canada, 2365 Dixie Rd., Mississauga, Ontario, L4Y 2A2, Canada

Grace & Co. of Canada Ltd., W.R., 3455 Harvester Rd., Unit #7, Burlington, Ontario, L7N 3P2, Canada (Tel.: 416-681-0285)

W.R. Grace Ltd., Northdale House, North Circular Rd., London, NW10 7UH, UK

Grace NV, Nijverheidsstraat 7, 2260 Westerlo, Belgium (Tel.: 014 57 56 11; FAX 014 58 55 30; Telex: 31500)

Grace Rexolin Chem. AB, Box 622, S-25106 Helsingborg, Sweden (Tel.: 42261460; FAX 42260051; Telex: 72353)

Graden Chemical Co., Inc.
426 Bryan St., Havertown, PA 19083 (Tel.: 215-449-3808)

Grain Processing Corp.
1600 Oregon St., Muscatine, IA, 52761 (Tel.: 319-264-4265; FAX 319-264-4289; Telex: 46-8497)

Granby Plastics Ltd.
New Star Road, Leicester, LE4 7JD, UK (Tel.: 0533 760958, 0533 460348; Telex: 34694)

Grant Industries Inc.
PO Box 360, 125 Main St., Elmwood Park, NJ 07407 (Tel.: 201-791-6700; FAX 201-791-0038)

Graphite Products Corp.
5756 Warren-Sharon Rd., Brookfield, OH 44403 (Tel.: 216-394-1617; 800-321-7521; FAX 216-394-2389; Telex: 271 512 GPI UR)

Graver Chemical/Div. of The Graver Co.
2720 US Highway 22 East, Union, NJ 07083 (Tel.: 908-964-0768; FAX 908-964-2481; Telex: 6853093)

Great Lakes Chemical Corp.
PO Box 2200, W. Lafayette, IN 47906 (Tel.: 317-497-6100; 800-428-7947)

R.W. Greeff & Co., Inc.
1445 East Putnam Ave., Old Greenwich, CT 06870 (Tel.: 203-637-4371)

Grefco Inc.
3435 W. Lomita Blvd., Torrance, CA 90509 (Tel.: 213-517-0700; Telex: 664266 GREFC LSA)

Gresco Mfg., Inc.
216 E. Hollyhill Rd., Thomasville, NC 27360 (Tel.: 919-475-8101; FAX 919-475-0100)

Griffin Corp.
PO Box 1847, Rock Ford Rd., Valdosta, GA 31601 (Tel.: 912-242-8635; 800-237-1854)

Grindsted
Grindsted Products Inc., 201 Industrial Pkwy., PO Box 26, Industrial Airport, KS 66031 (Tel.: 913-764-8100; 800-255-6837; FAX 913-764-5407; Telex: 4-37295)

Grindsted Products A/S, Edwin Rahrs Vej 38, DK-8220 Brabrand, Denmark (Tel.: 45-06-25-3366; FAX 45-06-25-1077; Telex: 64177)

Grindsted Products Ltd., Northern Way, Bury St. Edmunds, Suffolk, IP32 6NP, UK (Tel.: 44284769631)

Grinsted France S.A.R.L., Parc D'Activités de Tissaloup, Ave. Jean D'Alembert, F-78190 Trappes, France

Grindstedvaerket GmbH, Roberts-Bosch Strabe, D-2085 Quickborn, Deutschland, Germany

Grindsted do Brazil, Ind. Ecom Ltda., Rodovia Regisé Bitten Court, KM 275, 5 Cx. Postal 172, 06800 Embú S.P., Brazil

Grünau
Chemische Fabrick Grünau GmbH/A Henkel Group Co., Robert-Hansen Str. 1, Postfach 1063, W-7918 Illertissen, Bavaria, W. Germany (Tel.: (07303)13-0 (07303)13206; Telex: 719114 gruea-d)

GTE Products Corp./Chemical & Metallurgical Div.
Hawes St., Towanda, PA 18848 (Tel.: 717-265-2121; 800-828-7280; Telex: 834610)

Guardian Laboratories/Div. of United-Guardian, Inc.
PO Box 2500, Smithtown, NY 11787 (Tel.: 516-273-0900; 800-645-5566; FAX 516-273-0858)

Guelph Soap Co. Inc.
34 York St., Elora, Ontario, N0B 1S0, Canada (Tel.: 519-846-0934; FAX 519-846-9552)

Gulbrandsen Co. Inc.
PO Box 508, Milford, NJ 08848 (Tel.: 908-995-7759; 800-255-7759; Telex: 201-995-9482)

Guthrie Latex, Inc.
7400 N. Oracle, Suite 330, Tucson, AZ, 85704 (Tel.: 602-742-3087; FAX 602-575-0511; Telex: 187150 guth ut)

Haarmann & Reimer Corp.
PO Box 175, 70 Diamond Road, Springfield, NJ 07081 (Tel.: 201-467-5600; 800-422-1559; FAX 201-912-0499; Telex: 219134)

The Hall Chemical Co.
PO Box 200, 28960 Lakeland Blvd., Wickliffe, OH 44092 (Tel.: 216-944-8500; 800-322-6666; FAX 216-944-1298; Telex: 980707)

C.P. Hall Co.
7300 South Central Ave., Chicago, IL 60638-0428 (Tel.: 312-767-4600; FAX 708-458-0428)

Howard Hall
Howard Hall Int'l., 223 E. Putnam Ave., PO Box 199, Cos Cob, CT 06807 (Tel.: 203-869-4504; FAX 203-869-1439; Telex: 681 9012)

Halocarbon Products Corp.
887 Kinderkamack Rd., Riveredge, NJ 07661 (Tel.: 201-262-8899; FAX 201-262-0019; Telex: 134378)

W.A. Hammond Drierite Co.
138 Dayton Ave., Xenia, OH 45385 (Tel.: 513-376-2927)

Halstab
3100 Michigan St., Hammond, IN 46323 (Tel.: 219-844-3980; FAX 219-844-7287)

Hammond Lead Products Inc.
PO Box 6408, 5231 Hohman Ave., Hammond, IN 46325-6408 (Tel.: 219-931-9360; FAX 219-931-2140)

Handy & Harman
850 Third Ave., New York, NY 10022 (Tel.: 212-752-3400; FAX 212-207-2614; Telex: 126288)

Hanno-Werk Werner Kemper KG Nachf GmbH & Co.
Julius-Fengler-Str. 53, D-3014 Laatzen 4, Germany (Tel.: 05102 871; FAX 05102 1711; Telex: 922286)

Hansen & Rosenthal KG
Heilholtkamp 11, D-2000 Hamburg 60, Germany (Tel.: 040-5130-930; FAX 040-51309340; Telex: 211902 hur d)

Harborchem
186 North Ave. E., PO Box 630, Cranford, NJ 07016 (Tel.: 908-272-7070; FAX 908-272-8966)

Harcros
Harcros Chemicals UK Ltd., Lankro House, PO Box 1, Eccles, Manchester, M30 0BH, UK (Tel.: 44-61-789-7300; FAX 44-61-788-7886; Telex: 667725)
Harcros Chemicals Inc., 5200 Speaker Rd., PO Box 2930, Kansas City, KS, 66106-1095 (Tel.: 913-321-3131; FAX 913-621-7718; Telex: 477266)
Harcros Chemicals BV, Haagen House, PO Box 44, 6040 AA Roermond, The Netherlands
Harcros Chemicals France S.a.r.l., BP 40, 441220 St. Laurent, Nouan, France
Harcros Chemicals Scandia ApS, Vesterbrogade 14A, 1620 Copenhagen V, Denmark (Tel.: 31 21 42 00; FAX 31 21 42 27; Telex: 16 152 LANKRO DK)
Lankro Chemicals, 140 Chia Hsin Bldg., 96 Chung Shan N. Rd., Sec. 2, Taipei 10449, Taiwan, R.O.C.

Hardman Inc./A Harcros Chemical Group Company

600 Cortlandt St., Belleville, NJ 07109 (Tel.: 201-751-3000; FAX 201-751-8407; Telex: TWX: 710-995-4940)

Harlow Chemical Co. Ltd.

Central Road, Templefields, Harlow, Essex, CM20 2BH, UK (Tel.: 0279 436211; FAX 0279 444025; Telex: 817528 HARCO G)

A. Harrison & Co., Inc.

PO Box 494, Pawtucket, RI, 02862 (Tel.: 401-725-7450; 800-544-2942; FAX 401-725-3570)

Hart Chemicals Ltd.

256 Victoria Rd. South, Guelph, Ontario, N1H 6K8, Canada (Tel.: 519-824-3280; FAX 519-824-0755; Telex: 06956537)

Hart Products Corp.

173 Sussex St., Jersey City, NJ 07302 (Tel.: 201-433-6665)

Harwick Chemical Corp.

60 S. Seiberling St., PO Box 9360, Akron, OH 44305-0360 (Tel.: 216-798-9300; FAX 216-798-0214; Telex: TWX: 810-431-2126)

Hastings Plastics Co.

1704 Colorado Ave., Santa Monica, CA 90404 (Tel.: 213-829-3449; FAX 213-328-6820)

Hatco Corp.

King George Post Rd., Fords, NJ 08863-0601 (Tel.: 908-738-1000; FAX 908-738-9385; Telex: 84-4545)

W. Hawley & Son Ltd.

Colour Works, Duffield, Derby, DE6 4FG, UK (Tel.: 0332 840294; FAX 0332 842570; Telex: 37286 HAWLEY G)

Haynes International

Haynes Int'l. Inc., 1020 West Park Ave., PO Box 9013, Kokomo, IN 46904-9013 (Tel.: 800-342-9637; 800-354-0806; FAX 317-456-6905; Telex: 272280)

Haynes Int'l. Ltd., PO Box 10, Parkhouse St., Openshaw, Manchester, M11 2ER, UK (Tel.: (061)223-5054; FAX (061-223-2412; Telex: 667611)

Haynes Int'l. S.A.R.L., 43 Rue de Bellevue, Boite Postale No. 47, 92101 Boulogne Cedex, France (Tel.: (01)48-25-22-76; FAX (01)48250251)

Hefti Ltd

Chemical Products, PO Box 1623, CH-8048 Zurich, Switzerland (Tel.: 41-01-432-1340; FAX 41-01-432-2940; Telex: 822225)

Heico Chemicals, Inc../A Cambrex Company

Route 611, PO Box 160, Delaware Water Gap, PA 18327-0160 (Tel.: 717-420-3900; 800-34-HEICO; FAX 717-421-9012)

Henkel

Henkel Corp./Organic Products Div., 300 Brookside Ave., Ambler, PA 19022 (Tel.: 215-627-1000; 800-531-0815; FAX 215-628-1200)

Henkel Corp./Coating Chemicals, 300 Brookside Ave., Ambler, PA 19002 (Tel.: 800-445-2207)

Henkel Corp./Cospha, 300 Brookside Ave., Ambler, PA 19002 (Tel.: 215-628-1476; 800-531-0815 (sales); FAX 215-628-1450)

Henkel Corp./Functional Products, 300 Brookside Ave., Ambler, PA 19002 (Tel.: 215-628-1466; 800-654-7588; FAX 215-628-1155)

Henkel Corp./Textile Chemicals, 11709 Fruehauf Dr., Charlotte, NC 28273-6507 (Tel.: 800-634-2436; FAX 704-587-3804)

Henkel Polymers Div., Fine Chemicals Div., 5325 South 9th Ave., La Grange, IL 60525-3602 (Tel.: 708-579-6150; 800-237-4037; FAX 312-579-6152)

Henkel Canada Ltd., 2290 Argentia Rd., Mississauga, Ontario, L5N 6H9, Canada (Tel.: 416-542-7550; FAX 416-542-7588)

Henkel Chemicals Ltd., Henkel House, 292-308 Southbury Rd., Enfield, EN1 1TS, UK (Tel.: 081 804 3343; FAX 081 443 2777; Telex: 922708)

Henkel Chemicals Ltd./Organic Products Div., Merit House, The Hyde, Edgeware Rd., London, NW9 5AB, UK

Henkel KGaA/Cospha, Postfach 1100, D-4000, Dusseldorf 1, Germany (Tel.: 49-211-797-2289; FAX 49-211-798-7696; Telex: 085817-0)

Henkel KGaA/Dehydag, Postfach 1100, D-4000 Düsseldorf 1, Germany (Tel.: 49-211 797-4221; FAX 49-211 798-8558; Telex: 085817-122)

Henkel-Nopco SA/Process Chem. Div., 185 Ave. de Fontainebleau, 77310 St. Fargeau, Ponthierry, France (Tel.: 33-60-65-9090; FAX 33-60-65-7880; Telex: 692027)

Henkel Argentina S.A., Avda. E. Madero Piso 14, 1106 Capital Federal, Argentina

Parker & Amchem/Henkel Corp. Distribution & Resale Div., 300 Welsh Rd., Bldg. 2, Ste. 1, Horsham, PA 19044 (Tel.: 215-830-1200; 800-521-6895; Telex: 211872)

Henkel Corp./Emery Group, 11501 Northlake Dr., Cincinnati, OH 45249 (Tel.: 513-530-7300; 800-543-7370; FAX 513-530-7581)

Henkel Corp./Emery Group, 1301 Jefferson St., Hoboken, NJ 07030 (Tel.: 800-234-4365)

Henkel Corp./Emery Group/OPG, 3300 Westinghouse Blvd., Charlotte, NC 28217 (Tel.: 800-634-2436)

Henkel Corp./Emery Chemicals Ltd., 365 Evans Ave., Toronto, Ontario, M8Z 1K2, Canada (Tel.: 416-259-3751)

Henkel Corp./Emery Group Japan, PO Box 191, World Trade Center, 2-4-1 Hamamatsu-cho, Minato-Ku, Tokyo, 105, Japan (Tel.: 4355611-2)

Henley. See Boehringer Ingelheim

Hepar Industries, Inc.
160 Industrial Dr., Franklin, OH 45005-0338 (Tel.: 513-746-3603; FAX 513-746-9855; Telex: 9102407902)

Hercules
Hercules Inc., Hercules Plaza-6205SW, Wilmington, DE, 19894 (Tel.: 302-594-6500; 800-247-4372; FAX 302-594-5400; Telex: 835-479)

Hercules Inc./Carbon Fiber Materials, PO Box 98, Magna, UT, 84044

Hercules Inc./Fragrance & Food Ingred. Group, 33 Sprague Ave., Middletown, NY 10940 (Tel.: 914-343-1900; FAX 914-343-8794)

Hercules BV, 8 Veraartlaan, PO Box 5822, 2280 HV Rijswijk, The Netherlands (Tel.: 31-070-150-000; Telex: 31172)

Hercules Ltd./European Hdqts., 20 Red Lion St., London, WC1R 4PB, UK

Heresite Protective Coatings Inc.
PO Box 249, 822 S. 14th St., Manitowoc, WI 54220 (Tel.: 414-684-6646)

Hernon Mfg. Inc.
121 Tech Dr., Sanford, FL, 32771 (Tel.: 407-322-4000; 800-527-0004; FAX 407-321-9700)

Heterene Chemical Co., Inc.
PO Box 247, 792 21st Ave., Paterson, NJ 07543 (Tel.: 201-278-2000; FAX 201-278-7512; Telex: 883358)

Heterochem Corp.
111 E. Hawthorne Ave., Valley Stream, NY 11580 (Tel.: 516-561-8225)

Heveatex Corp.
106 Ferry St., PO Box 2573, Fall River, MA 02722 (Tel.: 508-675-0181)

Hewchem
PO Box 188, Gulfport, MS, 39502 (Tel.: 601-863-6600; Telex: 955439)

Hexagon Enterprises Inc./Chemical Components Div.
20 De Forest Ave., Hanover, NJ 07936 (Tel.: 201-887-7750)

Hexcel
Hexcel Corp., 11711 Dublin Blvd., Dublin, CA 94566 (Tel.: 415-828-4200)

Hexcel Corp./Resins Chemical Div., 20701 Nordhoff St., Chatsworth, CA 91311 (Tel.: 213-322-8050; 800-423-5451; FAX 818-709-0399)

Hickson Danchem Corp.
PO Box 4000, Danville, VA 24543 (Tel.: 804-797-8100; FAX 804-799-2814; Telex: 940103 WU PUBTLXBSN)

High Point Chemical Corp.
PO Box 2316, 243 Woodbine St., High Point, NC 27261 (Tel.: 919-884-2214; 800-727-2214; FAX 919-884-5039)

Hill Brothers Chem. Co.
1675 N. Main St., Orange, CA 92667 (Tel.: 714-998-8800; 800-821-7234; FAX 714-998-6310)

Hillcrest Labs, Inc.
17 Ohio Ave., PO Box 116, Spring Valley, NY 10977 (Tel.: 914-356-0433)

Hilton Davis
Hilton Davis Chemical Co., 2235 Langdon Farm Rd., Cincinnati, OH 45237 (Tel.: 513-841-4000; 800-477-1022; FAX 800-477-4565)
Represented by: Censtead Ltd., D.F., Victoria House, Radford Way, Billericay, Essex, CM12 0DE, UK

Himac, Inc.
457 Main St., Suite 4B, Danbury, CT 06811 (Tel.: 203-743-6069; FAX 203-743-0428)

Himont
Himont U.S.A., Inc., Three Little Falls Centre, 2801 Centerville Rd., PO Box 15439, Wilmington, DE, 19850-5439 (Tel.: 302-996-6000; 800-545-7719)
Himont Advanced Materials, 2727 Alliance Dr., PO Box 26248, Lansing, MI 48909 (Tel.: 517-336-9600; 800-874-4668)
Himont Canada Inc., 3 Robert Speck Pkwy., Mississauga, Ontario, L4Z 2G5, Canada (Tel.: 416-848-8800)

HIP Petrohemija DP
Spoljnostarcevacka 82, 26000 Pancevo, Yugoslavia (Tel.: 13 44 122; FAX 13 44 187; Telex: 13 181)

HiTech Polymers, Inc.
PO Box 30041, Cincinnati, OH 45230 (Tel.: 513-231-4043)

Hitemco.
6421 Lozano, Houston, TX 77041 (Tel.: 713-466-9655; FAX 713-466-1762)

Hitox Corp. of America
PO Box 2544, Corpus Christi, TX 78403 (Tel.: 512-882-5175; FAX 512-882-6948)

Hodag Corp.
7247 N. Central Park Ave., Skokie, IL 60076 (Tel.: 708-675-3950; FAX 708-675-3013)

Hoechst Celanese
Hoechst Celanese/Int'l. Headqtrs., 26 Main St., Chatham, NJ 07928 (Tel.: 201-635-2600; 800-235-2637; FAX 201-635-4330; Telex: 136346)
Hoechst Celanese/Colorants & Surfactants Div., 5200 77 Center Dr., Charlotte, NC 28217 (Tel.: 704-527-6000; 800-255-6189; FAX 704-559-6323)
Hoechst Celanese/Engineering Plastics Div., 26 Main St., Chatham, NJ 07928 (Tel.: 201-635-2600; FAX 201-635-4300)
Hoechst Celanese/Fine Chemicals Div., Portsmouth Tech. Ctr., 3340 West Norfolk Rd., Portsmouth, VA 23703 (Tel.: 804-483-7320; FAX 804-483-7460)
Hoechst Celanese/Specialty Chem. /Polymer Additives Group, 5200 77 Center Dr., PO Box 1026, Charlotte, NC 28217 (Tel.: 704-559-6027; FAX 704-559-6305)
Hoechst Celanese/Spec. Polymers/Waxes Div., PO Box 58160, Houston, TX 77258 (Tel.: 713-474-6026; 800-242-UHMW; FAX 713-474-6025)

Polymer Composites, Inc./Subsid. of Hoechst Celanese Corp., PO Box 30010, 4610 Theurer Blvd., Winona, MN 55987 (Tel.: 507-454-4150; 800-526-4960; FAX 507-457-4040)

Hoechst Canada Inc./Plastic Prods. Dept., 4045 Cote Vertu Blvd., Montreal, Quebec, H4R 1R6, Canada (Tel.: 514-333-3500; FAX 514-331-1526)

Hoechst UK Ltd./Polymers Div., Walton Manor, Wlaton, Milton Keynes, Bucks, MK7 7A3, UK (Tel.: 0908 665050; FAX 0908 680516; Telex: 826300)

Hoechst UK Ltd./Films Div., Hoechst House, Salisbury Rd., Hounslow, Middlesex, TW4 6JH, UK (Tel.: 081 570 7712; FAX 081 572 4854; Telex: 23284)

Hoechst AG, Postfach 80 03 20, D-6230 Frankfurt am Main 90, Germany (Tel.: 49-69-305-03113/7043; FAX 49-69-303665/66; Telex: 6990936)

Celanese France S.a.r.l., Z.I. Le Broteau 69540, Irigny, France

Hoechst Celanese Plastics Ltd., 78-80 St. Albans Rd., Watford, Herts, WD2 4AP, UK (Tel.: 923-33616)

Celanese GmbH, Hordenbachstrasse #40, D-5600 Wuppertal 21, Germany

Celanese Mexicana S.A., Av. Revolucion No. 1425, Mexico 20, D.F., Mexico

Hoechst Japan, 10-33, 4-Chome-Akasaka, Minato-Ku, Tokyo, Japan

Hoechst Celanese Far East Ltd., 801 Hong Kong Club Bldg., 3A Chater Rd., Central, Hong Kong

Hoffmann-La Roche Inc.

340 Kingsland St., Nutley, NJ 07110 (Tel.: 201-235-8080; 800-526-0189)

Hoffmann Mineral

Franz Hoffmann & Söhne KG, Postfach 1460, W-8858 Neuburg/Donau, Germany (Tel.: 08431/53-0; FAX 08431/53-330; Telex: 55223 hond-d)

Holland Colours Apeldoorn BV

Halvemaanweg 1-7323 RW, PO Box 720, 7300 AS Apeldoorn, Netherlands (Tel.: 0 55 66 31 43; FAX 0 55 66 29 81; Telex: 49791 holco ni)

Holland Colours UK

PO Box 96, Berkeley House, Rochester, Kent, ME1 1EQ, UK (Tel.: 0634 845691; FAX 0634 848943)

Hope Chemical Corp.

PO Box 908, Pawtucket, RI, 02862 (Tel.: 401-724-8000; FAX 401-724-8076)

Hormel & Co., Geo. A.

501 16th Ave. NE, Box 800, Austin, MN 55912 (Tel.: 507-437-5676; FAX 507-437-5120)

Houghton Chemical Corp.

PO Box 307, Allston, MA 02134 (Tel.: 617-254-1010; 800-777-2466; FAX 617-254-2713)

E.F. Houghton & Co.

Valley Forge Tech. Ctr., PO Box 930, Valley Forge, PA 19482 (Tel.: 215-666-4000; FAX 215-666-7354; Telex: 510-6604518)

Housmex Inc.

411 Sam Houston Pkwy. East, Suite 360, Houston, TX 77060 (Tel.: 713-999-5862; FAX 713-999-5865)

H&S Chem./Div. Huntington Labs, Inc.

970 E. Tipton St., Huntington, IN 46750 (Tel.: 219-356-7073; 800-448-6522; FAX 219-356-6485)

HTS High Tech Specialties

PO Box 24, 3830 AA Leusden-C, Netherlands (Tel.: 033 941921; FAX 033 944033)

J.M. Huber

J.M. Huber Corp./Chemicals Div., PO Box 310, Havre de Grace, MD 21078 (Tel.: 301-939-3500; FAX 301-939-0394)

J.M. Huber Corp./Carbon Black Div., PO Box 2831, Borger, TX 79008-2831 (Tel.: 806-274-6331; 800-631-6331)

J.M. Huber Corp./Clay Div., One Huber Rd., Macon, GA 31298 (Tel.: 912-745-4751; 800-TRY-HUBER; FAX 912-745-1116; Telex: 544438)

Solem Industries Div./J.M. Huber Corp., 4940 Peachtree Industrial Blvd., Norcross, GA 30071 (Tel.: 404-441-1301; FAX 404-368-9908)

Hubron Ltd.

Albion St. Works, Failsworth, Manchester, M35 0FP, UK (Tel.: 061 681 2691; FAX 061 683 4658; Telex: 666852)

The Huge Company, Inc.

7625 Page Blvd., St. Louis, MO 63133 (Tel.: 314-725-2555)

Hukill Chemical Corp.

7013 Krick Rd., Bedford, OH 44146 (Tel.: 216-232-9400; FAX 216-232-9477)

Hules Mexicanos S.A.

Ave. Constituventes No. 117, Col. San Miguel Chapultepec, Del. M. Hudalgo C.P. 11850, Mexico D.F. (Tel.: 905-514-2050)

Hüls

Hüls AG, Postfach 1320, D-4370 Marl 1, Germany (Tel.: 49-02365-49-1; FAX 49-02365-49-2000; Telex: 829211-0)

Hüls America Inc., PO Box 456, 80 Centennial Ave., Piscataway, NJ 08855-0456 (Tel.: 908-980-6800; 800-631-5275; FAX 908-980-6970)

Hüls Canada, Inc., 235 Orenda Rd., Brampton, Ontario, L6T 1E6, Canada (Tel.: 416-451-3810)

Hüls UK Ltd., Edinburgh House, 43-51 Windsor Rd., Slough, Berks, SL1 2HL, UK (Tel.: 0753-71851)

Hüls France SA, 49-51 Quai de Dion Bouton, 92815 Puteaux Cedex, France (Tel.: 1 49 06 55 00; FAX 1 47 73 97 65; Telex: 611868)

Humphrey Chemical Co., Inc., A Cambrex Co.

Devine St., North Haven, CT 06473-0325 (Tel.: 800-652-3456; FAX 203-287-9197; Telex: 994487)

Humus Products of America, Inc.

7319 North Park Dr., Richmond, TX 77469 (Tel.: 713-341-5045; 800-228-4322; FAX 713-232-9987; Telex: 170-524 HUMU-UT)

Hunt Chemicals Inc.

530 Permalume Pl. NW, Atlanta, GA 30318 (Tel.: 404-352-1418; FAX 404-352-0395)

Huntington Laboratories, Inc.

970 East Tipton St., Huntington, IN 46750 (Tel.: 219-356-8100; 800-537-5724; FAX 219-356-6485)

Huntsman

Huntsman Chemical Corp., 2000 Eagle Gate Tower, Salt Lake City, UT, 84111 (Tel.: 801-532-5200; FAX 801-355-6629)

Huntsman Chemical Co. of Canada, Inc., Bellevue St., C.P./PO Box 231, Mansonville, Quebec, J0E 1X0, Canada

Huntsman Chemical Co. Ltd., Carrington, Urmston, Manchester, M31 4AJ, UK (Tel.: 44-61-775-5321; FAX 44-61-775-0105)

Huntsman Pacific Chem. Corp., 42-1 Hsu Chang St., Taipei, 100, Taiwan, R.O.C.

General Electric-Huntsman Corp. (GEH), One Noryl Ave., Selkirk, NY 12158 (Tel.: 518-475-5734; FAX 518-475-5583)

Huntsman Polypropylene Corp., 15 Salt Creek Lane, Suite 410, Hinsdale, IL 60521 (Tel.: 708-920-9155; 800-221-1789; FAX 708-920-0138)

Huntsman Expandable Polystyrene, 5100 Bainbridge Blvd., Chesapeake, VA 23320 (Tel.: 804-494-2500; FAX 804-494-2770; Telex: 9103334117)

HVC, Inc.

4600 Dues Dr., Cincinnati, OH 45246 (Tel.: 513-874-9261; 800-347-7300; FAX 513-874-0350)

Hybri-Chem Inc.
212 West Taft Ave., Orange, CA USA (Tel.: 714-921-2300; FAX 714-921-9643)

Hydrolabs, Inc.
27 E. 33 St., Paterson, NJ 07514 (Tel.: 201-345-5100; FAX 201-345-7628)

Hydro Polymers Ltd.
Hydro Polymers Ltd./Vinyls Div., Aycliffe Ind. Est., Newton Aycliffe, Co. Durham, DL5 6EA, UK (Tel.: 0325 300555; FAX 0325 300215; Telex: 58322)

Hydro Polymers Ltd./Performance Prods. Div., New Lane, Havant, Hampshire, PO9 2NQ, UK (Tel.: 0705 486350; FAX 0705 472388; Telex: 86720)

Hysan Corp.
4309 S. Morgan St., Chicago, IL 60609 (Tel.: 312-376-8981)

Hysol. See Dexter

ICC Industries Inc.
720 Fifth Ave., New York, NY 10019 (Tel.: 212-903-1730; Telex: 212-903-1794)

ICI
ICI Americas, Inc., New Murphy Rd. & Concord Pike, Wilmington, DE, 19897 (Tel.: 302-886-3000; 800-456-3669; FAX 302-886-2972; Telex: 4945649)

ICI Acrylics-KSH Inc., 10091 Manchester Rd., St. Louis, MO 63010 (Tel.: 314-996-3111; 800-325-9577; FAX 800-44-0803)

ICI Advanced Materials, Rt. 202 & Murphy Rd., Wilmington, DE, 19897 (Tel.: 302-886-3290; FAX 302-886-5333)

ICI Advanced Materials, 475 Creamery Way, Exton, PA 19341 (Tel.: 800-854-8774)

ICI Biocides (Tel.: 302-886-3070)

ICI Colors, Charlotte, NC)

ICI Fiberite Molding Materials, 501 East 3rd St., Winona, MN 55987 (Tel.: 507-454-3611)

ICI Films/Polyester Polymer Group, Wilmington, DE, 19897, USA)

ICI Fluoropolymers, 475 Creamery Way, Exton, PA 19341 (Tel.: 215-363-4746; 800-842-8739)

ICI Polymer Additives, PO Box 751, Wilmington, DE, 19897 (Tel.: 302-886-3564; 800-822-8215; FAX 302-886-5267)

ICI Polyurethanes Group/Formulated Products Div., 6555 Fifteen Mile Rd., Sterling Heights, MI 48077 (Tel.: 313-826-7660; 800-553-8624)

ICI Polyurethanes Group/West Deptford Div., Mantua Grove Rd., West Deptford, NJ 08066 (Tel.: 609-423-8300; 800-257-5547)

ICI Resins US, 730 Main St., Wilmington, MA 01887-0677 (Tel.: 508-658-6600; 800-225-0947; FAX 508-657-7978)

ICI Specialty Chemicals, Cherry Lane, PO Box 231, New Castle, DE, 19720 (Tel.: 302-575-4700; 800-456-3669; FAX 302-652-7035)

ICI PLC/Petrochemicals & Plastics Div., PO Box 90, Wilton, Middlesborough, Cleveland, TS6 8JE, UK (Tel.: 0642 454144; FAX 0642 432444; Telex: 587 461)

ICI Chemicals & Polymers Ltd./Acrylics Business, PO Box 14, The Heath, Runcorn, Cheshire, WA7 4QF, UK (Tel.: 0928 514444)

ICI Surfactants Ltd. (UK, Smith's Rd., Bolton, Lancashire, BL3 2QJ, UK (Tel.: 44-204-21971/4; FAX 44-204-363676; Telex: 667841)

ICI Advanced Materials UK PO Box 6, Shire Park, Bessemer Rd., Welwyn Garden City, Herts, AL7 1HD, UK (Tel.: 0707-323400; FAX 0642-432444)

ICI Surfactants (Belgium), Everslaan 45, B-3078 Everberg, Belgium (Tel.: 32-02-758-9361; FAX 32-02-758-9686)

ICI Polyurethanes Group/Canada, 2795 Slough St., Mississauga, Ontario, L4T 1G2, Canada (Tel.: 416-678-9150)

Atkemix Inc., 70 Market St., PO Box 1085, Brantford, Ontario, Canada

ICI Japan Ltd., Osaka Green Bldg., 1,3-Chome Kitahama, Higashi-Ku, Osaka, 541, Japan

ICI Australia Operations Pty. Ltd., ICI House, 1 Nicholson St., Melbourne, 300, Australia (Tel.: 61-03-665-7396; FAX 61-03-665-7009; Telex: 30192)

Fiberite Europe, Industriestrasse 1, D-7524 Oestringen, Germany (Tel.: 49-7253-81870)

Kasei Fiberite Ltd., Mitsubishi Bldg., 11th Floor, 5-2, Marunouchi 2-Chome, Chiyoda-Ku, Tokyo, Japan

LNP Engineering Plastics/An ICI Company, 475 Creamery Way, Exton, PA 19341 (Tel.: 215-363-4500; 800-854-8774)

LNP Plastics Nederland B.V., Ottergeerde 24, Raamsdonksveer, The Netherlands

LNP Plastics UK Unit 25 Monkspath Business Park, Solihull, West Midlands, B90 4NX, UK (Tel.: 021-744-9922)

Ideas Inc.

PO Box 262, Wood Dale, IL 60191 (Tel.: 708-766-2326; FAX 708-350-3121)

IGI Baychem Inc.

5625 FM 1960 W., Suite 400, Houston, TX 77069 (Tel.: 713-537-7434; FAX 713-537-7835)

IGI Boler, Inc.

85 Old Eagle School Rd., Wayne, PA 19087 (Tel.: 215-687-9030; 800-852-6537)

IGI Petroleum Specialities

164 Sheridan St., Perth Amboy, NJ 08861 (Tel.: 201-826-0140; FAX 201-826-0641)

Igoplast-Faigle AG

Werkstrasse, CH-9434 Au/SG, Switzerland (Tel.: 071 716666; FAX 071 716669; Telex: 881876)

Ikeda Corp.

New Tokyo Bldg., 3-1-Marunouchi, 3-Chome, Chiyoda-ku, Tokyo, 100, Japan (Tel.: 81/3-212-8791; FAX 81/3-215-5069; Telex: J26370)

IMC Fertilizer, Inc.

501 E. Lange St., Mundelein, IL 60060 (Tel.: 708-949-3700; 800-323-5523)

Impco Inc.

335 Valley St., Providence, RI, 02908 (Tel.: 800-243-1220)

Imperial Adhesives

6315 Wiehe Rd., Cincinnati, OH 45237 (Tel.: 513-351-1300; 800-365-1301; FAX 513-351-1994)

Ina-Oki

Zitnjak b.b., PO Box 402, 41000 Zagreb, Yugoslavia (Tel.: 041 231666; FAX 231975/218124; Telex: 21226)

In-Cide Technologies Inc.

50 N. 41 Ave., Phoenix, AZ 85009 (Tel.: 602-233-0756; 800-777-4569; FAX 602-272-5864)

Inco Alloys Int'l. Inc.

3200 Riverside Dr., Huntington, WV 25720 (Tel.: 304-526-5100; FAX 304-526-5441; Telex: 886413)

Indium Corp. of America

1676 Lincoln Ave., Utica, NY 13502 (Tel.: 315-768-6400; 800-4-INDIUM; FAX 315-768-6362; Telex: 937363)

Indol Chem. Co.

Lefferts St., Carteret, NJ 07008 (Tel.: 201-242-1300; 800-631-4461; FAX 201-242-4087; Telex: 467259)

Indspec

Indspec Chemical Corp., 411 Seventh Ave., Suite 300, Pittsburgh, PA 15219 (Tel.: 412-765-1200; FAX 412-765-0439; Telex: 199187 Indspec)

Indspec Chemical Corp./European Sales, Gebouw de Goudsesingel, Kipstraat 8-10, 3011 RT Rotterdam, The Netherlands (Tel.: 011-31-10-4-120-122; Telex: 21253 Indsp NL)

Induchem AG

Lagerstrasse 14, CH-8600 Duebendorf 1, Switzerland (Tel.: 1/820-11 61; FAX 1/820 21 13; Telex: 828-455)

Indulor-Chemie GmbH
Schulstr. 3, 4554 Ankum, Germany

Industeel Co.
37 St. & Smallman, Pittsburgh, PA 15201 (Tel.: 412-682-0866)

Industria Engineering Products Ltd.
Eskdale Road, Uxbridge, Middlesex, UB8 2SL, UK (Tel.: 0895 37971; FAX 0895 72191; Telex: 23946)

Industrial Adhesives Co./Bond-Plus Adhesives & Coatings Div.
130 N. Campbell Ave., Chicago, IL 60612 (Tel.: 312-666-2686; 800-336-2686 (IL); FAX 312-666-1824)

Industrial Dielectrics, Inc.
PO Box 357, Noblesville, IN 46060 (Tel.: 317-773-1766)

Industrial Fibers, Inc.
2889 North Nagel Court, Lake Bluff, IL 60044 (Tel.: 708-295-0046; FAX 708-295-0520)

Industrial Polymers UK/Div. of Ashland Plastics (UK) Ltd.
Charter House, 177 Angel Rd., Edmonton, London, N18 3BW, UK (Tel.: 081 884 4969; FAX 081 884 0680; Telex: 896228)

Industrial Polymer Services Ltd.
Andre House, Salisbury Sq., Hatfield, Herts, AL9 5BH, UK (Tel.: 07072 62200; FAX 07072 65595; Telex: 25102 CHACOM IPS)

Industrial Raw Materials Corp.
575 Madison Ave., New York, NY 10022 (Tel.: 212-688-8080; 800-3-INDRAW; FAX 212-759-3696; Telex: 232636)

Industrias Nioco SA
Calle Esmeralda Na 9, 08950 Esplugues De Liobregat, Barcelona, Spain (Tel.: 371 21 94; FAX 372 49 54; Telex: 54378 Ioco-E)

Industriplas Ltd.
Weston Coyney Rd., Longton, Stoke on Trent, UK (Tel.: 0782 311311; FAX 0782 343400; Telex: 367184)

Innovative Engineering of Michigan, Inc.
941 West Round Lake Rd., DeWitt, MI 48220 (Tel.: 517-669-1591; FAX 516-669-9120)

Inolex
Inolex Chemical Co., Jackson & Swanson Sts., Philadelphia, PA 19148-3497 (Tel.: 215-271-0800; 800-521-9891; FAX 215-271-2621; Telex: 834617)
Represented by: Stanley Black, Ltd., 30/31 Islington Green, London, UK

Insl-X Products Corp.
38 Wells Ave., Yonkers, NY 10701 (Tel.: 914-969-8000; 800-225-5554)

Insta-Foam Products Inc.
1500 Cedarwood Dr., Joliet, IL 60435 (Tel.: 815-741-6800; 800-800-FOAM; FAX 815-741-6822; Telex: 72-3415)

Interensco NV
Stoopstraat 1, B-2000 Antwerpen, Belgium (Tel.: 03 2325807; FAX 03 2250737; Telex: 31379 INTECO B)

Interface Research Corp.
100 Chastain Center Blvd., Suite 165, Kennesaw, GA 30144 (Tel.: 404-421-9555; FAX 404-424-1888)

Interfibe Corp.
30000 Aurora Rd. #A-202, Solon, OH 44139 (Tel.: 216-248-2266; 800-262-3771; FAX 216-248-2132)

Intergen Company
Two Manhattanville Rd., Purchase, NY 10577 (Tel.: 914-694-1700; 800-431-4505; FAX 914-694-1429; Telex: 426973)

International Bio-Synthetics Inc.
International Bio-Synthetics Inc., PO Box 241068, Charlotte, NC 28224 (Tel.: 704-527-9000; 800-438-1361; FAX 704-527-8184)
Represented by: Ward Blenkensop & Co., Ltd., Halebank Widnes, Cheshire, WA8 8NS, UK

International Dioxcide, Inc.
136 Central Ave., Clark, NJ 07066 (Tel.: 908-499-9660; FAX 908-388-3648; Telex: 642582 PROLINE)

International Dyestuffs Corp.
PO Box 2169, Clifton, NJ 07015 (Tel.: 201-778-0122; FAX 201-778-1124; Telex: 642634 LAROPCO)

International Specialty Products, Inc. See ISP

International Sulphur, Inc.
PO Box 611, Interstate 30 W., Mt. Pleasant, TX 75455 (Tel.: 903-577-5500; 800-828-7857; FAX 903-577-5540)

International Wax Refining Co., Inc.
181 E. Jamaica Ave., PO Box 221, Valley Stream, NY 11582 (Tel.: 516-561-2500)

Interox
Interox America, 3333 Richmond Ave., Houston, TX 77098 (Tel.: 713-522-4155; 800-468-3769; FAX 713-524-9032; Telex: 166307)
Interox Chemicals Ltd., PO Box 7, Warrington, Cheshire, WA4 6HB, UK (Tel.: 0925 51277; FAX 0925 232207; Telex: 627834 LAPORT G)

Interpolymer Corp.
330 Pine St., Canton, MA 02021 (Tel.: 617-828-7120; 800-262-1281; FAX 617-821-2485; Telex: 6974364)

Intertex World Resources Ltd.
94 Shiawassee Ave., Akron, OH 44333 (Tel.: 216-867-8770; FAX 216-867-8619)

Intex Chemical, Inc., Div. of EZE Prods., Inc.
8 N. Kings Rd., Greenville, SC 29606 (Tel.: 803-242-6152; 800-233-8606; FAX 803-299-3475)

IPEL Ltd.
Speke Hall Ind. Est., Speke Hall Ave., Merseyside, L24 1UU, UK (Tel.: 051 486 9666; FAX 051 448 1152; Telex: 627452)

IPE Overseas BV
Charles House, 108-110 Finchley Rd., London, NW3 5JJ, UK (Tel.: 071 794 5743; FAX 071 794 7928; Telex: 8951038 PEBV)

Irpen SA
Gran Via de las Cortes Catalanas 814, 08013 Barcelona, Spain (Tel.: 3 2452705; FAX 3 232 0917; Telex: 98356 IRPN E)

ISA Industria Solventi E Adesivi Srl
Via Salisburgo No. 3, 37135 Verona, Italy (Tel.: 045 5051 74)

ISK Biotech Corp.
5966 Heisley Rd., PO Box 8000, Mentor, OH 44061-8000 (Tel.: 216-357-4100; FAX 216-354-9506; Telex: 196191 ISKB UT)

Isochem Inc.
635-F Atando Ave., Charlotte, NC 28206 (Tel.: 704-334-4633; 800-786-4633; FAX 704-334-4634)

ISP
ISP Technologies Inc., PO Box 1006, Bound Brook, NJ 08805 (Tel.: 908-271-0111; 800-622-4423)
GAF Europe, 40 Alan Turing Rd., Surrey Research Park, Buildford, Surrey, UK
GAF Great Britain Ltd., Tilson Rd., Roundthrone, Wythenshawe, Manchester, M23 9PH, UK

Ivar Laboratories Inc.
625 Marina Vista, Martinez, CA 94553-1132 (Tel.: 510-372-6966)

Ivax Industries Inc./Textile Products Div.
PO Box 10027, Rock Hill, SC 29731 (Tel.: 803-366-9411; FAX 803-366-7256)

JacksonLea, A Unit of Jason Inc.
Hwy. 70 East, PO Box 699, Conover, NC 28613 (Tel.: 704-464-1376; FAX 704-464-7094; Telex: 962426)

Jahres Fabrikker A/S
PO Box 2051, N-3201 Sandefjord, Norway (Tel.: 3474700; FAX 3474720; Telex: 8320351)

James River Corp./Berlin-Gorham Group
650 Main St., Berlin, NH, 03570 (Tel.: 603-752-4600)

Jarchem Industries
414 Wilson Ave., Newark, NJ 07105 (Tel.: 201-344-0600; FAX 201-344-5743; Telex: 362-660)

Jersey Ind. Chem. Prod. Inc.
PO Box 430, Elmwood Park, NJ 07407 (Tel.: 201-796-4092; FAX 201-796-4845)

Jetco Chemicals, Inc.
PO Box 1898, Corsicana, TX 75110 (Tel.: 214-872-3011; FAX 214-872-4216; Telex: 75110)

Jet Inc.
750 Alpha Dr., Cleveland, OH 44143 (Tel.: 216-461-2000; 800-321-6960; FAX 216-442-9008; Telex: 251616 JET UR)

S.C. Johnson Wax
1525 Howe St., Racine, WI 53403 (Tel.: 414-631-4875; 800-231-7868)

Jojoba Growers & Processors Inc.
2267 S. Coconino Dr., Apache Junction, AZ 85220 (Tel.: 602-982-1125; FAX 602-982-4183; Telex: 754858)

Jonas Chemical Corp./Specialty Chemical Div.
1682 59th St., Brooklyn, NY 11204 (Tel.: 718-236-1666; FAX 718-236-2248; Telex: 423616)

Jotun (Deutschland) GmbH
Winsbergring 25, D-2000 Hamburg 54, Germany (Tel.: 040 853136 0; FAX 040 85 62 34)

Joy Chemical Co.
35 Livingston St., Providence, RI 02904 (Tel.: 401-331-9000; FAX 401-331-2155)

J-Von Inc.
25 Litchfield St., Leominster, MA 01453 (Tel.: 508-537-4721; FAX 508-537-4724)

Kahl & Co./Wax Refiners & Chemical Manufacturers
Otto-Hahn-Strasse 2, D-2077 Trittau, Germany (Tel.: 014154-3011; FAX 04154-81508)

Kalama Chemical, Inc.
1110 Bank of California Center, Seattle, WA 98164 (Tel.: 2006-682-7890; 800-233-7799)

Kalker
5 Rue des Bruyeres, 23260 Les Lilas, France (Tel.: 1 43617045; FAX 1 43632020; Telex: 220089)

Kalsec Inc.
PO Box 511, 3713 W. Main St., Kalamazoo, MI 49005 (Tel.: 616-349-9711; Telex: 295181)

Kamco Ltd.
27 Bullens Green Lane, Colney Heath, St. Albans, Herts, AL4 0QR, UK (Tel.: 0727 27490; FAX 0727 25981)

Kane International Corp.
411 Theo Fremd Ave., Rye, NY 10580 (Tel.: 914-921-3100; FAX 914-921-3180; Telex: 6818109)

Kano Laboratories, Inc.
1079-K Thompson Lane, Nashville, TN 37211 (Tel.: 615-833-4101; FAX 615-833-5790; Telex: 554-322)

Kao
Kao Corp./Int'l. Chemicals Dept., 14-10 Nihonbashi, Kayabacho 1-Chome, Chuo-ku, Tokyo, 103, Japan

Kao Corp./Edible Fat & Oil Div., 14-10 Nihonbashi, Kayabacho 1-Chome, Chuo-ku, Tokyo, 103, Japan (Tel.: 81-3-660-7862; FAX 81-3-660-7964; Telex: J24816)

Kao Corp. S.A., Puig dels Tudons, 10, 08210 Barbera Del Valles, Barcelona, Spain (Tel.: 34-3-729-0000; FAX 34-3-718-9829; Telex: 59749)

Kaopolite, Inc.
2444 Morris Ave., Union, NJ 07083 (Tel.: 908-789-0609; FAX 908-851-2974)

Karlshamns
Karlshamns Lipids for Care, S-374 82, Karlshamn, Sweden (Tel.: 46-454-823-00; FAX 46-454-129-11; Telex: 4500 fopart s)

Karlshamns Lipids for Care, PO Box 569, Columbus, OH 43216 (Tel.: 614-299-3131; 800-526-4547; FAX 614-299-8279; Telex: 245494 capctyprdcol)

Katalistiks International/A Unit of UOP
1501 Sulgrave Ave., Baltimore, MD 21209 (Tel.: 301-367-1000; FAX 301-367-5254; Telex: 6711629 KATAIUW)

Kawasaki Steel Corp.
Hibiya Kokusai Bldg., 2-3 Uchisaiwaicho, 2-Chome, Chiyoda-ku, Tokyo, 100, Japan (Tel.: 03/597-4622; FAX 03/597-3633; Telex: 0222-3673 KWST TJ)

KD Thermoplastics Ltd.
119 Guildford St., Chertsey, Surrey, KT16 9AL, UK (Tel.: 0932 566033; FAX 0932 560363; Telex: 928547)

Kelco
Kelco/Div. of Merck & Co., Inc., 8355 Aero Dr., San Diego, CA 92123 (Tel.: 619-292-4900; 800-535-2656; FAX 619-467-6520; Telex: WUD 695228)

Kelco, Westminster Tower, 3 Albert Embankment, London, SE1 7RZ, UK (Tel.: 01-735-0333)

Kelco/Oil Field Prods. Div.

E.H. Kellogg & Co., Inc.
63 E. First St., Mt. Vernon, NY 10550 (Tel.: 914-664-3045; FAX 914-664-0350)

Kemira, Inc.
Box 368, Savannah, GA 31402 (Tel.: 912-236-6171; 800-4-KEMIRA; FAX 912-234-3584)

Elektrochemische Fabrik Kempen GmbH
Postfach 100 260, D-4152 Kempen 1, W. Germany

Kenrich Petrochemicals, Inc.
140 E. 22nd St., PO Box 32, Bayonne, NJ 07002-0032 (Tel.: 201-823-9000; 800-LICA KPI; FAX 201-823-0691; Telex: 12-5023)

Kerr-McGee Chemical Corp.
Kerr-McGee Ctr., PO Box 25861, Oklahoma City, OK, 73125 (Tel.: 405-270-1313; 800-654-3911; Telex: 747-128)

Kerrypak
Longbrook House, Ashton Vale Road, Bristol, BS3 2HA, UK (Tel.: 0272 669684; 800-654-3911; FAX 0272 231251; Telex: 444277 Kerry G)

Ketro A/S
Postboks 323, 1324 Lysaker, Norway (Tel.: 2 530507; FAX 2 58 02 65; Telex: 74198 KETRO N)

Key Polymer Corp.
Jacob's Way, Lawrence Ind. Park, Lawrence, MA 01842 (Tel.: 508-683-9411; FAX 508-686-7729; Telex: 940 103)

Keystone Aniline Corp.
2501 W. Fulton St., Chicago, IL 60612 (Tel.: 312-666-2015; FAX 312-666-8530)

Kincaid Enterprises, Inc.
PO Box 549, Plant Rd., Nitro, WV 25143 (Tel.: 304-755-3377; FAX 304-755-4547)

E&F King & Co. Inc.
640 Pleasant St., Norwood, MA 02062 (Tel.: 617-762-3113; FAX 617-762-7743)

King Industries, Inc.
Science Rd., Norwalk, CT 06852 (Tel.: 203-866-5551; 800-431-7900; FAX 203-866-1268; Telex: 710-468-0247)

Kingston Technologies, Inc.
2235-B Route 130, Dayton, NJ 08810 (Tel.: 908-274-2288)

Klinger Engineering Plastics Div.
Sidcup, Kent, DA14 5AG, UK (Tel.: 081-300-7777)

Kluber Lubrication North America Inc.
54 Wentworth Ave., Londonderry, NH 03053 (Tel.: 603-434-7704; FAX 603-434-8046)

KMG Minerals, Inc.
PO Box 729, 1433 Grover Rd., Kings Mountain, NC 28086 (Tel.: 704-739-3616; Telex: 703063)

KMZ Chemicals Ltd.
48 Station Road, Stoke D'Abermon, Cobham, Surrey, KT11 3BN, UK (Tel.: 0932 866426; FAX 0932 867099; Telex: 929624)

Kobo Prods., Inc.
607 Montrose Ave., South Plainfield, NJ 07080 (Tel.: 201-757-0033; FAX 201-757-0905; Telex: 405609 KOBO UD)

Koch Chemical Co./Muskegon Specialites Div.
1725 Warner St., Whitehall, MI 49461 (Tel.: 616-894-4018; FAX 616-893-4141; Telex: 4938994)

V&E Kohnstamm
Bldg. #10-Bush Terminal, 3 Ave. & 33 St., Brooklyn, NY 11232 (Tel.: 718-788-6320; Telex: 425707)

Kolmar Laboratories, Inc.
123 Pike St., Port Jervis, NY 12771 (Tel.: 914-856-5311; FAX 914-856-5507; Telex: 230-646260)

Kop-Coat, Inc./Protection Products Div.
5137 Southwest Ave., St. Louis, MO 63110 (Tel.: 314-772-2200)

Kromachem Ltd.
Unit 10, Moor Park Industrial Centre, Tolpits Lane, Watford, WD1 8SP, UK (Tel.: 0923 223368; FAX 0923 39308; Telex: 933219)

Kronos
Kronos, Inc., PO Box 60087, 3000 N. Sam Houston Pkwy. East, Houston, TX 77205 (Tel.: 713-987-6300; 800-866-5600; FAX 713-987-6358)

Kronos Canada, Inc., 4 Place Ville-Marie, Suite 500, Montreal, Quebec, H3B 4M5, Canada (Tel.: 514-397-3501; FAX 514-393-1186; Telex: 5268667)

Kunstoplast-Chemie GmbH
An den Drei Hasen 37, Postfach 14 30, D-6370 Oberursel bei Frankfurt, Germany (Tel.: 061 71 636 0; FAX 061 71 40 21; Telex: 410 900 akc d)

Kureha Chem. Industry Co., Ltd./Plastics Dept.
1-9-11 Nihonbashi Horidome-cho, Chuo-ku, Tokyo, 103, Japan (Tel.: 03-3249-4625; Telex: 03-3661-8380)

Kyowa
Kyowa Chemical Industry Co., Ltd., 305, Yashimanishi-machi, Takamatsu, Kagawa, 761-01, Japan (Tel.: (0878)41-9179; FAX (0878)41-6147; Telex: 5822220)

Kyowa America Corp., 2500-T, S.E. Main St., Irvine, CA 92714 (Tel.: 714-641-0411)

Kyros Corp.
5428 Lake Mendota Dr., Madison, WI 53705 (Tel.: 608-238-3587)

Kyung-In Synthetic Co.
990 Ave. of Americas, Suite 15R, New York, NY 10018 (Tel.: 212-594-0839; FAX 212-594-6607)

Labbco Inc.
2903 Dupree Circle, Houston, TX 77054 (Tel.: 713-747-8710; 800-726-9133; FAX 713-747-8925)

La-Co Industries Inc.
250 N. Washtenaw Ave., Chicago, IL 60612 (Tel.: 312-826-1700; FAX 312-826-7130)

Lanaetex Products, Inc.
151-157 Third Ave., Elizabeth, NJ 07206 (Tel.: 201-351-9700; FAX 201-351-8753; Telex: 3792268 TLAP1)

Landers-Segal Color Co. Inc.
84 Dayton Ave., Passaic, NJ 07055 (Tel.: 201-779-5001; FAX 201-779-8948; Telex: 6971185)

C & R Landsman
29 Endersleigh Gardens, Hendon, London, NW4 4SD, UK (Tel.: 081 202 7057)

Lankro Polychem AB
Box 6, 44221 Kungälv, Sweden (Tel.: 30316600; FAX 303 10807; Telex: 27451)

Laporte
Laporte Inc., 3701 Algonquin Rd., Suite 390, Rolling Meadows, IL 60008 (Tel.: 708-670-8870; FAX 708-670—8874)

Laporte Inc., Southern Clay Prods. Inc., 1212 Church St., PO Box 44, Gonzales, TX 78629 (Tel.: 512-672-2891; 800-324-2891; FAX 512-672-3930)

La Roche Chemicals Inc.
PO Box 1031 Airline Hwy., Baton Rouge, LA 70821 (Tel.: 504-356-8463; 800-524-2586; FAX 504-356-8551; Telex: 196137)

La Roche Industries Inc.
1100 Johnson Ferry Rd. NE Atlanta, GA 30342 (Tel.: 404-851-0300; FAX 404-851-0476)

Laserson & Sabetay
B.P. 57 Zone Industrielle, 91150 Etampes, Cedex, France (Tel.: 1-64-94-31-24; FAX 1-64-94-98-97; Telex: 601-532 LASAROM)

Lati SpA
Via F Baracca, no. 7, 1-21040 Vedano Olona, Italy (Tel.: 03 32 4091 11; FAX 03 32 4092 35; Telex: 380114)

Laurel Industries
30000 Chagrin Blvd., Cleveland, OH 44124-5794 (Tel.: 216-831-5747; 800-221-1304; FAX 216-831-8479)

Lauricidin, Inc.
4841 Southbridge Rd., Toledo, OH 43623 (Tel.: 419-841-1821)

Laur Silicone Rubber Compounding Inc.
4930 S. M-18, Box 509, Beaverton, MI 48612 (Tel.: 517-435-7706; FAX 517-535-7707)

Lawrence Industries PLC
PO Box 3, Mitcham, Surrey, CR4 3ZD, UK (Tel.: 081 648 2272; FAX 081 640 9932; Telex: 946846)

Lawter International, Inc.
990 Skokie Blvd., Northbrook, IL 60062 (Tel.: 708-498-4700; Telex: RCA 210013)

Leadertech Colors Inc.
11 Louisberg Square, Verona, NJ 07044 (Tel.: 201-857-8585; FAX 201-857-9881)

The Lea Mfg. Co.
237 E. Aurora St., PO Box 71, Waterbury, CT 06720 (Tel.: 203-753-5116; FAX 203-754-3770; Telex: 962 426 LEACO)

Leatex Chemical Co.
2722 N. Hancock St., Philadelphia, PA 19133 (Tel.: 215-739-6324; FAX 215-739-5910)

LeChem, Inc.
12537 Scenic Hwy., Baton Rouge, LA 70807 (Tel.: 504-775-1801; FAX 504-775-5431; Telex: 62903974)

Leepoxy Plastics, Inc.
3324 Ferguson Rd., Fort Wayne, IN 46809, USA)

Lenmar Chemical Corp.
PO Box 571, Dalton, GA 30722-0571 (Tel.: 404-226-8603; FAX 404-226-3432)

Lenox Chemical Co.
PO Box 6422, Providence, RI 02904 (Tel.: 401-521-7474; FAX 401-331-2155)

Lensfield Prods. Ltd.
Maulden Rd.-Flitwick, Bedford, MK 45 5, UK

Lenzing AG
Abt. P84, A-4860 Lenzing, Austria (Tel.: 07672-701-3602; FAX 07672-74823; Telex: 026-606 lenfa a)

Lermont Plastics SA
Can Buscarons De Baix, Nave 4, 08170 Montornes Del Valles, Barcelona, Spain (Tel.: 93 5683961; FAX 93 5682755)

Lever Bros. Co./Industrial Div.
390 Park Ave., New York, NY 10022 (Tel.: 212-688-6000)

C. Lever Co., Inc.
736 Dunks Ferry Rd., Bensalem, PA 19020 (Tel.: 215-639-8640)

Lewcott Corp.
86 Providence Rd., Millbury, MA 0001527 (Tel.: 508-865-1791; 800-225-7725; FAX 508-865-0302)

Lewis Industrial Products
Ottery Moor Lane, Honiton, Devon, EX14 8AR UK (Tel.: 0404 44194; FAX 0404 45102; Telex: 42583 Lewis G)

Leyland & Birmingham Rubber Co. Ltd.
Golden Hill Lane, Leyland, Lancs PR5 1UB, UK (Tel.: 0772 421434; FAX 0772 436401; Telex: 67126)

Licharz GmbH Nylontechnik
In der Holle 2-6, D-5205 Sankt Augustin 1, Germany (Tel.: 02241 332021; FAX 02241 336581; Telex: 889 459)

LignoTech. See Borregaard LignoTech

Lilly Chemicals
210 E. Alondra Blvd., Gardena, CA 90248 (Tel.: 213-321-0710; 800-472-6243)

Limburgse Vinyl Maatschappij LVM
H. Hartlaan, Industriepak Schoonhees West, 3980 Tessendeerlo, Belgium (Tel.: 013 666112; FAX 013 668406; Telex: 39780 LVM B)

Lindau Chems, Inc.
PO Box 13565, Columbia, SC 29201 (Tel.: 803-799-6863; FAX 803-256-3639)

Lindley Laboratories Inc.
PO Box 341, Burlington, NC 27216-0341 (Tel.: 919-449-7521; FAX 919-449-0624)

Lion Corp.
2-22,1-Chome Yokoami, Sumida-ku, Tokyo, 130, Japan (Tel.: 81-3-3621-6675; FAX 81-3-3621-6738; Telex: 262-2114)

LiphaTech, Inc.
3101 W. Custer Ave., Milwaukee, WI 53209 (Tel.: 414-462-7600; 800-558-1003)

Lipo
Lipo Chemicals, Inc., 207 19th Ave., Paterson, NJ 07504 (Tel.: 201-345-8600; FAX 201-345-8365; Telex: 130117)
Represented by: Blagden Campbell & Campbell Ltd., A.M.P. House, Dingwall Rd., Croydon, CR9 3QU, UK

Liquid Air Corp.
2121 North California Blvd., Suite 350, Walnut Creek, CA 94596 (Tel.: 415-977-6500; Telex: ITT 470020)

Liquid Carbonic
135 S. LaSalle St., Chicago, IL 60603 (Tel.: 312-855-2500; Telex: 254165)

LNP. See ICI

The Lockrey Co. Inc.
PO Box 1269, Merchantville, NJ 08109 (Tel.: 609-665-4794)

Loctite Corp.
705 N. Mountain Rd., Newington, CT 06111 (Tel.: 203-278-1280; 800-562-0560; Telex: 275207)

Mario Lombardini Srl
Via Simone da Corbetta 66, 20011 Corbetta, Milano, Italy (Tel.: 02 9779597; FAX 02 9771 793)

London Chemical Co., Inc.
240 Foster Ave., Bensenville, IL 60106 (Tel.: 312-766-5902; 800-323-9625; Telex: 726360)

Lonza Inc.
17-17 Route 208, Fair Lawn, NJ 07410 (Tel.: 201-794-2400; 800-777-1875 (tech.); FAX 201-703-2028)

Lord

Lord Corp./Elastomer Prods. Div., Industrial Adhesives, 2000 West Grandview Blvd., PO Box 10038, Erie, PA 16514-0038 (Tel.: 814-868-3611; FAX 814-864-3452; Telex: 291935)

Lord Corporation, Stretford Motorway Estate, Barton Dock Rd., Stretford, Manchester, UK (Tel.: 061 865 8048; FAX 061 865 0096; Telex: 665701 HUGMAN)

Represented by: Qyadra Chemical Ltd., 2121 Argentia Rd., Suite 303, Mississauga, Ontario, L5N 2X4, Canada)

Lowenstein

Lowenstein Dyes & Cosmetics Inc., 420 Morgan Ave., Brooklyn, NY 11222 (Tel.: 718-388-5410; FAX 718-387-3806)

Represented by: Siber Hegner UK Ltd., 221-241 Beckenham Rd., Makenzie House, Beckenham, Kent, BR3 4UF, UK

Lowi

Lowi Chem. Corp., 7 West Bowery St., Suite 811, Akron, OH 44308 (Tel.: 216-762-8614; FAX 216-762-1135)

Chemische Werke Lowi GmbH & Co., Postfach 1660, Teplitzerstr., D-8264 Waldkraiburg, Germany (Tel.: 08638-6080; FAX 08638-608 200; Telex: 77863884)

Lubrication Engineers, Inc.

3851 Airport Freeway, Fort Worth, TX 76111 (Tel.: 817-834-6321)

Lubrizol

Lubrizol Corp., 29400 Lakeland Blvd., Wickliffe, OH 44092 (Tel.: 216-943-4200; FAX 216-943-5337)

Lubrizol Ltd., Waldron House, 57-63 Old Church St., London, SW3 5BS, UK (Tel.: 351-3311-20; FAX 351-3310)

Lubrizol GmbH, Bogenallee 10, 2000 Hamburg 13, Germany

Lubrizol France, Tour Europe, 92400 Courbevoie, France

Lubrizol Japan Ltd., No. 23 Mori Bldg., 5th Floor, 23-7 Toranomon, 1-Chome, Minato-ku, Tokyo, 105, Japan

Lucky Ltd.

20 Yoido-dong, Yongdungpo-gu, Seoul, 150-721, Korea (Tel.: 82-2-787-7317; FAX 82-2-782-6549; Telex: 24235 LUCKYSL K)

Lunds of Bingley

PO Box 4, Bingley, West Yorkshire, UK (Tel.: 0274 562301; FAX 0274 569002; Telex: 51519)

Luzenac, Inc.

1075 North Service Rd. W., Suite 14, Oakville, Ontario, L6M 2G2, Canada (Tel.: 416-825-3930; FAX 416-825-3932)

Lydall, Inc.

12 Davis St., Hoosick Falls, NY 12090 (Tel.: 518-686-7313)

Lyondell Petrochemical Co.

12000 Lawndale, Houston, TX 77017 (Tel.: 713-475-4111; 800-447-4LPC; FAX 713-475-4125)

3M

3M Co./Engineered Materials, Industrial Specialty Div., 3M Center Bldg., St. Paul, MN 55144-1000 (Tel.: 612-736-9700)

3M Co./Performance Polymers and Additives, 3M Center Bldg., St. Paul, MN 55144-1000 (Tel.: 612-736-9700)

3M Canada Inc., PO Box 5757 Terminal A, 1840 Oxford St. East, London, Ontario, N6A 4T1, Canada

3M Belgium N.V./S.A., Canadastraat 11, 2730 Zwijndrecht, Belgium

3M Deutschland GmbH, PO Box 100422, D-4040 Neuss 1, Germany

3M United Kingdom PLC, 3M House, PO Box 1, Bracknell, Berkshire, RG12 1JU, UK

3M Australia Pty. Ltd., 950 Pacific Hwy., PO Box 99, Pymble N.S.W. 2073, Australia
Sumitomo/3M Ltd., Central PO Box 490, 33-1 Tamagawadai 2-Chome, Setagaya-ku, Tokyo, 158, Japan
3M Co./Industrial Chem. Prods. Div., 3M Center, St. Paul, MN 55144 (Tel.: 612-736-1394)

Maag Agrochemicals Inc./Subsid. Ciba-Geigy Ltd.
PO Box 6430, Vero Beach, FL, 32961-6430 (Tel.: 407-567-7506; FAX 407-567-7626)

MacAndrews & Forbes Co.
3rd St. & Jefferson Ave., Camden, NJ 08104 (Tel.: 609-964-8840; Telex: 84-5337)

Mace Adhesives & Coatings Co., Inc.
Roberts Rd., PO Box 37, West Main Branch, Dudley, MA 01571 (Tel.: 508-943-9052; FAX 508-943-6527)

Mach-1 Compounding
775 E. Highland Rd., Macedonia, OH 44056 (Tel.: 216-467-8108)

Mackenzie Chem Works, Inc.
55 G Brook Ave., Deer Park, NY 11729 (Tel.: 516-243-3737; FAX 516-243-0123)

MacPherson Polymers
Station Road, Birch Vale, Nr Stockport, Cheshire, SK12 5BR, UK (Tel.: 0663 746518; FAX 0663 746605; Telex: 669258)

Dr. Madis Labs Inc.
375 Huyler St., South Hackensack, NJ 07606 (Tel.: 201-440-5000; FAX 201-342-8000; Telex: 134-200)

Magie Bros. Oil/Div. Pennzoil Products Co.
9101 Fullerton Ave., Franklin Park, IL 60131 (Tel.: 708-455-4500; 800-MAGIE 47; FAX 708-455-0383)

Magruder Color Co.
1029 Newark Ave., Elizabeth, NJ 07201 (Tel.: 201-242-1300; 800-631-4461; FAX 201-242-4087; Telex: 467259)

Magyar Viscosagyár
H-2537 Nyergesujfalu, Hungary (Tel.: 33 55088; FAX 33 55541; Telex: 22 5626)

Malco Products, Inc.
361 Fairview Ave., Barberton, OH 44203 (Tel.: 216-753-0361; FAX 216-753-2025)

Mallinckrodt
Mallinckrodt Specialty Chemicals, 16305 Swingley Ridge Dr., Chesterfield, MO 63017 (Tel.: 314-895-2000; 800-325-7155; FAX 314-530-2562)
Mallinckrodt Canada Inc., 7500 Trans Canada Hwy., Pointe Claire, Quebec, H9R 5H8, Canada (Tel.: 514-695-1220)
Mallinckrodt GmbH, Josef-Dietzgen-Strasse 1, PO Box 1462, D-5202 Hennef/sieg 1, Germany (Tel.: 02242-887-0; FAX 02242-887-195)

Malvern Minerals Co.
PO Box 1238, Hot Springs, AR 71901-1238 (Tel.: 501-623-8893; FAX 501-623-5113; Telex: 536-120)

Mammoth Int'l. Chem.
2825 Wilcrest, Suite 309, Houston, TX 77042-3358 (Tel.: 713-266-4936; FAX 713-266-2031; Telex: 790361 MAMMOTH HOU)

Manchem
Manchem Inc., 77 Maple Dr., Hudson, OH 44238
Manchem Ltd., Aston New Rd., Manchester, M11 4AT, UK

Man-Gill Chemical Co.
23002-T St. Clair Ave., Cleveland, OH 44117 (Tel.: 800-627-6422)

George Mann & Co., Inc.
123 Georgia Ave., PO Box 9066, Providence, RI 02940 (Tel.: 401-781-5600; 800-556-2426; FAX 401-941-0830)

Manro Products Ltd.
Bridge St., Stalybridge, Cheshire, SK15 1PH, UK (Tel.: 44-61-338-5511; FAX 44-61-303-2991; Telex: 668442)

Mantrose-Haeuser Co.
500 Post Road East, Westport, CT 06880 (Tel.: 203-454-1800; 800-344-4229; FAX 203-227-0558)

Manufacturers Chemicals Corp.
PO Box 2788, Cleveland, TN 37320 (Tel.: 615-476-6518; FAX 615-479-7284)

Manville. See also Celite
Manville, PO Box 5108, Denver, CO 80217-5108 (Tel.: 303-978-4900; 800-654-3103)
Manville Filtration & Minerals, PO Box 519, Lompoc, CA 93438-0519 (Tel.: 805-735-7791; FAX 805-736-7856)

Marbo France
155 rue Pierre Corneille, 69.003 Lyon, France (Tel.: 78 62 24 35; FAX 78 95 33 70; Telex: 305167 MARBO F)

MarChem Corp.
2500 Adie Rd., Maryland Heights, MO 63043 (Tel.: 314-872-8700; FAX 314-872-8750)

Marchon. See Albright & Wilson

A. Margolis & Sons Corp.
1504 Atlantic Ave., Brooklyn, NY 11216 (Tel.: 718-773-6270)

Marine Bio Products Inc.
333 W 1 St., Boston, MA 02127 (Tel.: 617-268-0758)

Marine Magnesium Co.
995 Beaver Grade Rd., Coraopolis, PA 15108 (Tel.: 412-264-0200; FAX 412-264-9020)

Markal Co.
250 N. Washtenaw Ave., Chicago, IL 60612 (Tel.: 312-826-1700; FAX 312-826-7130)

Marlin Chemicals Ltd.
Tubs Hill House, London Road, Sevenoaks, Kent, TN13 1BL, UK (Tel.: 0732 450066; Telex: 957054)

Marlowe-Van Loan Corp.
PO Box 1851, High Point, NC 27261 (Tel.: 919-886-7126; 800-422-4MVL; FAX 919-889-6663; Telex: TWX: 510-926-1589)

Marnic PLC
Marnic House, 37 Shooters Hill Rd., London, SE3 7HS, UK (Tel.: 081 858 8100; FAX 081 305 1305; Telex: 8956061 MARNIC)

Martin Marietta Magnesia Specialties
Executive Plaza II, Hunt Valley, MD 21030 (Tel.: 301-527-3700; 800-648-7400; FAX 301-527-3861; Telex: 710-862-2630)

Marubishi Oil Chemical Co., Ltd.
3-7-12 Tomobuchi-cho, Miyakojima-ku, Osaka, 534, Japan (Tel.: 81-6-928-0331; FAX 81-6-928-0310)

Marval Industries Inc.
315 Hoyt Ave., Mamaroneck, NY 10543 (Tel.: 914-381-2400)

Mason
Mason Chemical Co., 5253 West Belmont Ave., Chicago, IL 60641 (Tel.: 312-282-0200; 800-362-1855; FAX 312-282-0821)

Mason Chemical Co. Ltd., Carolyn House, Dingwall Rd., Croydon, CR0 9XF, UK (Tel.: 081 686 5625; Telex: 929278 MASTON G)

Master Bond Inc.
154 Hobart St., Hackensack, NJ 07601 (Tel.: 201-343-8983)

Master Builders Inc./Ceilcote Prods. Group
23700 Charin Blvd., Cleveland, OH 44122 (Tel.: 216-831-5500; 800-227-3350; FAX 216-831-6460)

Masterplast Ltd.
Maudlins Industrial Estate, Dublin Road, Naas, Co. Kildare, Rep. of Ireland (Tel.: 045 66565; FAX 045 75765; Telex: 747149 PITMAN G (UK

Mateson Chemical Corp.
1025 E. Montgomery Ave., Philadelphia, PA 19125 (Tel.: 215-423-3200)

Matheson Gas Products
30 Seaview Dr., Secaucus, NJ 07096 (Tel.: 201-867-4100)

Matsumoto Yushi-Seiyaku Co., Ltd.
1-3, 2-Chome, Shibukawa-cho, Yao City, Osaka, Japan

Maurel Freres
59 Rue De LAncienne Mairie, BP 106, 92106 Boulougne Billancourt, France (Tel.: 1 46 05 93 01; FAX 1 46 05 87 38; Telex: 270447 F)

Maybrook Inc.
570 Broadway, PO Box 68, Lawrence, MA 01842 (Tel.: 508-682-1853; FAX 508-682-2544)

Mayco Oil & Chemical Co.
775 Louis Dr., Warminster, PA 18974 (Tel.: 215-672-6600; FAX 215-443-7094)

Mayo Chemical Co., Inc.
5544 Oakdale Rd., SE, Smyrna, GA 30082 (Tel.: 404-696-6711; 800-962-6296; FAX 404-696-7463)

Mazer. See PPG

MBS Plastics (MB Sweda AB)
Box 48, 401 20 Göteborg, Sweden (Tel.: 031 83 80 00; FAX 010 46 3184 2572; Telex: 27408)

M-CAP Technologies Int'l.
PO Box 7136, Wilmington, DE, 19803-0136 (Tel.: 302-695-5616; FAX 302-695-5681)

McGean-Rohco, Inc./McGean Div.
1252 Terminal Tower, Cleveland, OH 44113 (Tel.: 216-621-6425; 800-932-7006; FAX 216-621-9697; Telex: 98-0403)

McLube Div. of McGee Industries Inc.
9 Crozerville Rd., Aston, PA 19014 (Tel.: 215-459-1890; 800-2-MCLUBE; FAX 215-459-9538; Telex: 910- 3807571 MCLUBE)

McIntyre Chemical Co., Ltd.
1000 Governors Hwy., University Park, IL 60466 (Tel.: 708-534-6200; FAX 708-534-6216)

McLaughlin Gormley King Co.
8810 10th Ave. N., Minneapolis, MN 55427 (Tel.: 612-544-0341; Telex: 290 544 MACK GOVY)

Mearl
Mearl Corp., 41 E. 42nd St., New York, NY 10017 (Tel.: 212-573-8500; FAX 212-557-0742; Telex: 421841)

Mearl Corp., 217 North Highland Ave., PO Box 960, Ossining, NY 10562 (Tel.: 914-941-7450; FAX 914-941-7858)

Represented by: Cornelius Chemical Co. Ltd., St. James House, 27-43 Eastern Rd., Romford, Essex, London, RM1 3NN, UK

Mec Pac Srl
>Via Lattanzio 23, 20137 Milano, Italy (Tel.: 02 551 0357; Telex: 02 55189000)

Medical Chemical Corp.
>1909 Centinela Ave., Santa Monica, CA 90404 (Tel.: 213-829-4304; FAX 213-453-1212)

Meer Corp.
>PO Box 9006, 9500 Railroad Ave., N. Bergen, NJ 07047 (Tel.: 201-861-9500; FAX 201-861-9267; Telex: 219130)

MEI (Magnesium Elektron Inc.)
>500 Point Breeze Rd., Flemington, NJ 08822 (Tel.: 908-782-5800; 800-366-9596; FAX 908-782-7768)

Melamine Chems, Inc.
>PO Box 748, Donaldsonville, LA 70346 (Tel.: 504-473-3121; FAX 504-473-0550; Telex: 8109526516)

Mendell Co., Inc., Edward/A Penwest Company
>2981 Rt. 22, Patterson, NY 12563-9970 (Tel.: 914-878-3414; 800-431-2457; FAX 914-878-3484; Telex: 4971034)

E. Merck
>Postfach 4119, Frankfurter Strasse 250, D 6100 Darmstadt 1, Germany (Tel.: 06151-72-0; FAX 06151-72-3368; Telex: 421-468)

Mereco Products Div./Metachem Resins Corp.
>1505 Main St., W. Warwick, RI 02893 (Tel.: 401-822-9300; 800-556-7164; FAX 401-822-9311)

Meridian Petroleum Co.
>330 S. Wells St., Chicago, IL 60606 (Tel.: 312-939-2693; FAX 312-939-7853)

Meridionale des Plastiques SA
>ZI les Paluds, 636 av de la Fleuride, 13685 Aubagne Cedex, France (Tel.: 42 82 90 82; FAX 42 82 20 36; Telex: 401 839 F)

Merix Chemical Co.
>2234 E. 75th St., Chicago, IL 60649 (Tel.: 312-221-8242)

Merrand Int'l. Corp.
>1236 Weathervane Lane, Suite #203, Akron, OH 44313 (Tel.: 216-869-5800; FAX 216-869-9629)

Mersey Plastics Ltd.
>Window Lane, Garston, Liverpool, Lancashire, L19 8EN, UK (Tel.: 051 494 9000)

Metal Lubricants Co.
>17050 Lathrop Ave., Harvey, IL 60426 (Tel.: 708-333-8900)

Metalspecialties Inc.
>515 Commerce Dr., Fairfield, CT 06430 (Tel.: 203-384-0332; Telex: 964289)

Met-Pro Corp.
>160 Cassell Rd., Box 144, Harleysville, PA 19438 (Tel.: 215-723-6751; Telex: 244 903)

Lucas Meyer GmbH & Co.
>PO Box 261665, D-2000 Hamburg 26, Germany (Tel.: 49-40-789-550; FAX 49-40-789-8329; Telex: 2163220 myer d)

Mfg's Chemical & Supply Inc.
>PO Box 4359, Dalton, GA 30721 (Tel.: 404-226-4114; FAX 404-275-6044)

David Michael & Co., Inc.
>10801 Decatur Rd., Philadelphia, PA 19154 (Tel.: 215-632-3100)

M. Michel & Co., Inc
90 Broad St., New York, NY 10004 (Tel.: 212-344-3878; FAX 212-344-3880; Telex: 421468)

Michelman, Inc.
9089 Shell Rd., Cincinnati, OH 45236-1299 (Tel.: 513-793-7766; FAX 513-793-2504)

Michigan Chrome & Chem. Co.
8615 Grinnell Ave., Detroit, MI 48213 (Tel.: 313-267-5215)

Michlin Diazo Products Corp.
10501 Haggerty St., Dearborn, MI 48126 (Tel.: 313-846-5700; 800-521-3240; FAX 313-846-0741; Telex: 798-658)

Microban Germicide Co.
PO Box 777, Eighth & Pine Aves., Braddock, PA 15104 (Tel.: 412-351-7702; FAX 412-351-7701)

Microfine Minerals Ltd.
Raynesway, Derby, DE2 7BE, UK (Tel.: 0332 673131; FAX 0332 677590; Telex: 37261 MICMIN)

Microfluidics Corp./Subs. of Biotechnology Development Corp.
90 Oak St., PO Box 9101, Newton, MA 02164-9101 (Tel.: 617-969-5452; 800-370-5452; FAX 617-965-1213)

Micropol Ltd.
Stamford Works, Bayley St., Stalybridge, Cheshire, SK15 1QQ, UK (Tel.: 061 330 5570; FAX 061 330 5576; Telex: 667549)

Micro Powders, Inc.
580 White Plains Rd., Tarrytown, NY 10591 (Tel.: 914-793-4058; FAX 914-472-7098; Telex: 996584)

Midwest Elastomers Inc.
700 Industrial Dr., PO Box 412, Wapakoneta, OH 45895 (Tel.: 419-738-9634; FAX 419-738-4504)

Midwest Rubber Reclaiming Div.
PO Box 2349, E. St. Louis, IL 62202-2349 (Tel.: 618-337-6400)

Miles. See also Bayer
Miles Inc./Organic Products Div., Bldg. 14, Mobay Rd., Pittsburgh, PA 15205-9741 (Tel.: 412-777-2000; FAX 412-777-7840; Telex: 1561261)
Miles/Agricultural Chem. Div., PO Box 4913, Kansas City, MO 64120-0013 (Tel.: 816-242-2000; FAX 816-242-2592)
Miles/Polyurethane Div., Mobay Rd., Pittsburgh, PA 15205-9741 (Tel.: 412-777-2000; 800-662-2927)

Miller Chem. & Fertilizer Corp.
Radio Rd., PO Box 333, Hanover, PA 17331 (Tel.: 717-632-8921)

Harry Miller Corp.
4 & Bristol Sts., Philadephlia, PA 19140 (Tel.: 215-324-4000)

Miller-Stephenson Chem. Co., Inc.
George Washington Hwy., Danbury, CT 06810 (Tel.: 203-743-4447; 800-992-2424)

Milliken
Milliken Chemicals, PO Box 1927, Spartanburg, SC 29304 (Tel.: 803-573-2200; FAX 803-573-2430; Telex: 810-282-2580)
Milliken Chemicals, PO Box 817, Inman, SC 29349 (Tel.: 803-472-7208; FAX 803-472-4129)

Milport Chemical Co.
2829 S. 5th Court, Milwaukee, WI 53207 (Tel.: 414-769-7350; 800-236-7350; FAX 414-769-0617)

Mineral Research & Development Corp.
One Woodlawn Green, Charlotte, NC 28217 (Tel.: 704-525-2771; 800-334-0417; FAX 704-527-8232)

Mines De La Lucette
4 Rue De Rome, 75008 Paris, France (Tel.: 1 43 87 33 50; FAX 1 42 93 44 05; Telex: 640 437)

Mississippi Chemical Corp.
PO Box 388, Yazoo City, MS 39194 (Tel.: 601-746-4131; FAX 601-746-9158; Telex: 585 449)

Mitsubishi Gas Chemical Co., Inc.
5-2, 2-Chome, Marunouchi, Chiyoda-ku, Tokyo, Japan (Tel.: 03-283-4800; FAX 03-214-0938; Telex: 222-2624 MGCHO J)
Mitsubishi Gas Chemical Co., Inc., 520 Madison Ave., 9th Floor, New York, NY 10022 (Tel.: 212-752-4620; FAX 212-758-4012; Telex: 649545 MGC UR NYK)
Represented by: Franklin Polymers, Inc., 2 Bala Plaza, Suite 300, Bala Cynwyd, PA 19004 (Tel.: 215-668-4353; FAX 215-667-8174)

Mitsubishi International Corp./Chemicals Dept.
277 Park Ave., New York, NY 10172 (Tel.: 212-922-5054)

Mitsubishi Kasei
Mitsubishi Kasei Corp., 5-2, Marunouchi 2-Chome, Chiyoda-ku, Tokyo, 100, Japan
Mitsubishi Kasei America, Inc., 81 Main St., Suite 401, White Plains, NY 10601 (Tel.: 914-761-9450; FAX 914-681-0760)
Mitsubishi Kasei, Europe Office, Am Seestern, Niederkasseler LohWeg 8, 4000 Duesseldorf 11, Germany

Mitsui Petrochemicals
Mitsui Petrochemical Industries, Ltd., Kasumigaseki Bldg., 2-5 Kasumigaseki 3-Chome, Chiyoda-ku, Tokyo, 100, Japan (Tel.: 03-593-1630; FAX 03-593-0979)
Mitsui Petrochemicals (America) Ltd., 250 Park Ave., Suite 950, New York, NY 10017 (Tel.: 212-682-2366; FAX 212-490-6694)
Mitsui Petrochemical Industries, Ltd. UK, Temple Court, 11 Queen Victoria St., London, EC4N 4SB, UK (Tel.: 236-2463; FAX 236-2220)

Mobil
Mobil Chemical Co./Films Div., 1150 Pittsford-Victor Rd., Pittsford, NY 14534 (Tel.: 800-654-3436 (NY); 800-828-6381)
Mobil Chemical Co./Polystyrene Business Group, Rt. 27 & Vinyard Rd., Box 3029, Edison, NJ 08818-3029 (Tel.: 908-321-3500; 800-922-0380; FAX 908-321-3501)

Molecular Rearrangement Inc./MRI Int'l. Corp.
44-50 Clinton St., Newton, NJ 07860 (Tel.: 201-383-3645; FAX 201-383-6672; Telex: 178102 MRI)

Mona Industries Inc.
PO Box 425, 76 E. 24th St., Paterson, NJ 07544 (Tel.: 201-345-8220; FAX 201-345-3527; Telex: 130308)

Monmouth Plastics, Inc.
Box 921, 814 Asbury Ave., Asbury Park, NJ 07712 (Tel.: 201-775-5100; 800-526-2820; FAX 201-775-9068)

Monofil Technology Ltd.
Craven St., Leicester, LE1 4BX, UK (Tel.: 0533 517 640; FAX 0533 516 191; Telex: 341975)

Monomer-Polymer & Dajac Laboratories, Inc./Div. MTM Research
PO Box 28, Trevose, PA 19053 (Tel.: 215-938-1750; FAX 215-938-7410)

Monoplas Industries Ltd.
Allen Works, Badger Lane, Hipperholme, Halifax, West Yorkshire, HX3 8PW, UK (Tel.: 0422 201161; FAX 0422 205676; Telex: 517344 Monin DG)

Monsanto
Monsanto Chemical Co., 800 N. Lindbergh Blvd., St. Louis, MO 63167 (Tel.: 314-694-1000; 800-325-4330; FAX 314-694-7625; Telex: 44-7282)

Monsanto Canada Inc., 2330 Argentia Rd., Box 787, Streetsville Postal Station, Mississauga, Ontario, L5M 2G4, Canada (Tel.: 416-826-9222)

Monsanto PLC, Monsanto House, Chineham Court, Chineham, Basingstoke, Hants, RG24 0UL, UK (Tel.: 0256-572-88; FAX 0256-54995; Telex: 858837)

Monsanto Europe SA, Ave. de Tervuren 270-272, B-1150 Brussels, Belgium (Tel.: 2-761-41-11; FAX 2-761-40-40; Telex: 62927 Mesab)

Monsanto Japan Ltd., Room 520, Kokusai Bldg., 1-Marunouchi 3-Chome, Chiyoda-Ku, Tokyo, 100, Japan

Monson Chemicals Inc.
154 Pioneer Dr., Leominster, MA 01453 (Tel.: 508-534-1425; FAX 508-840-1060)

Montana Sulphur & Chem. Co.
PO Box 31118, Billings, MT 59107-1118 (Tel.: 406-252-9324; FAX 406-252-8250)

Montana Talc Co.
28769 Sappington Rd., Three Forks, MT 59752 (Tel.: 406-285-3286; FAX 406-285-3530)

Montedipe Srl
Via Rosellini, 15-17, 20124 Milan, Italy (Tel.: 02-63331; FAX 2 6270 8255; Telex: 310679 MONTED I)

Montefluos
Via Principe Eugenio, 1/5, 20155 Milano, Italy (Tel.: 02-6270-3427; FAX 02-6270-3948; Telex: 310679 Monted I)

Mooney Chems, Inc.
2301 Scranton Rd., Cleveland, OH 44113 (Tel.: 216-781-8383; 800-321-9696)

Mooney Plastics Ltd.
Braintree Road, Ruislip, Middx, HA4 0XX, UK (Tel.: 081 841 4211; Telex: 935058)

Moretex Chemical Products Inc.
PO Box 1799, Spartanburg, SC 29304 (Tel.: 803-583-8441; FAX 803-591-1909)

Morflex, Inc.
2110 High Point Rd., Greensboro, NC 27403 (Tel.: 919-292-1781)

Morgan Matroc Ltd.
Park Royal Div., Cambridge Road, Sandy, Bedfordshire, SG19 1QQ, UK (Tel.: 0767 680305; FAX 0767 682668)

Morlot Color & Chemical Co.
111 Ethel Ave., Hawthorne, NJ 07506 (Tel.: 201-423-0600; FAX 201-423-4096)

Morton International
Morton International, Inc., 150 Andover St., Danvers, MA 01923 (Tel.: 508-774-3100; 800-621-2847; FAX 508-777-5672)

Morton International, Inc./Dayton, 10 S. Electric St., West Alexandria, OH 45381 (Tel.: 513-839-4612; 800-348-8846; FAX 513-839-5615)

Morton International, Inc./Specialty Chemicals Group, 2000 West St., Cincinnati, OH 45215-3431 (Tel.: 513-733-2100)

Morton International, Inc./Automotive & Industrial Finishes, 2700 E. 170 St., Lansing, IL 60438 (Tel.: 708-474-7000; 800-323-3224; FAX 708-868-7490)

Morton International Ltd./Specialty Chem, Ind. Chem & Addit, 7900-A Taschereau Blvd., Suite 106, Brossard, Quebec, J4X 1C2, Canada (Tel.: 514-466-7764; FAX 514-466-7771)

Morton International UK Ltd./Automotive & Industrial Finishes, Greville House, Hibernia Road, Hounslow, Middlesex, TW3 3RX, UK (Tel.: 081 570 7766; FAX 081 570 6943; Telex: 262002)

Morton International NV SA, Chaussee de la Hulpe 130, Boite 5, B-1050 Brussels, Belgium (Tel.: 32-2-6602909; FAX 32-2-6604702; Telex: 23708)

Morton International Inc., Room 2501 Dominion Centre, 37-59A Queen's Road East, Wanchai, Hong Kong

Morton Salt, 100 North Riverside Plaza, Chicago, IL 60606 (Tel.: 312-807-2562; FAX 312-807-2228; Telex: 25-4433)

MRC Polymers Inc.
1716 West Webster Ave., Chicago, IL 60614 (Tel.: 312-276-6345)

M-R-S Chemicals, Inc.
2494 Adle Rd., Maryland Heights, MO 63043 (Tel.: 314-872-8750)

M&T Harshaw. See Atochem

H. Muehlstein & Co., Inc./Rubber Div.
3296 West Market St., Akron, OH 44313 (Tel.: 216-836-9135; FAX 216-836-1412)

Multi-Kem Corp.
PO Box 538, Ridgefield, NJ 07657-0538 (Tel.: 201-941-4520; FAX 201-941-5239)

Multiplastic SpA
Via Privata Multiplastic, 22070 Portichetto di Luisago, pv Como, Italy (Tel.: 031 928076; Telex: 380090)

Multitherm Corp.
125 S. Front St., Colwyn, PA 19023 (Tel.: 215-461-6442; 800-225-7440; FAX 215-532-1289; Telex: 510-669-0032)

Murphy-Phoenix Co.
PO Box 22930, Beachwood, OH 44122

Mutchler Chemical Co., Inc.
99-C Kinderkermack Rd., Westwood, NJ 07675 (Tel.: 201-666-7002; FAX 201-666-3652; Telex: 219320)

Mykroy/Mycalex Ceramics
125 Clifton Blvd., Clifton, NJ 07011 (Tel.: 201-779-8866; FAX 201-779-2013)

Nalco Chemical Co.
One Nalco Center, Naperville, IL 60563-1198 (Tel.: 708-305-1000; 800-527-7753)

Napex Ltd.
27 Beethoven St., London, W10 4LL, UK (Tel.: 081 969 5353; FAX 081 960 4960; Telex: 922824)

National Ammonia Co.
Tacony & Vankirk St., Philadelphia, PA 19135 (Tel.: 215-535-7530; 800-NH3-NACO)

National Casein Co.
601 W. 80 St., Chicago, IL 60620 (Tel.: 312-846-7300; FAX 312-487-5709)

National Chemical & Plastics Co.
3610 Milford Mill Rd., Baltimore, MD 21207 (Tel.: 301-655-0400)

National Magnesia Chemicals/Div. Nat'l. Refractories & Min.
Highway One, Moss Landing, CA 95039 (Tel.: 408-633-2499; 800-622-2499; FAX 408-633-2904; Telex: 650-2730561 (MCI)

National Refractories & Minerals Corp.
PO Box 1938, 11771 Old Stage Rd., Salinas, CA 93902 (Tel.: 408-449-7206; FAX 408-443-5599)

National Research & Chem. Co.
15600 New Century Dr., Gardena, CA 90248 (Tel.: 213-515-1700; 800-338-7160; FAX 213-515-4937)

National Starch & Chemical
National Starch & Chemical Corp., Box 6500, 10 Finderne Ave., Bridgewater, NJ 08807 (Tel.: 908-685-5000; 800-726-0450; FAX 908-685-5005)
Represented by: National Adhesives & Resins Ltd., Braunston Daventry, Northante, NN11 7JL, UK

National Wax Co.
3650 Touhy Ave., PO Box 549, Skokie, IL 60076 (Tel.: 708-679-6300; 800-628-9299; FAX 708-679-6312; Telex: 724434)

Natrochem, Inc.
PO Box 1205, Exley Ave., Savannah, GA 31498 (Tel.: 912-236-4464; FAX 912-236-1919)

Natural Gas Odorizing, Inc.
PO Box 1429, Baytown, TX 77522-1429 (Tel.: 713-424-5568; FAX 713-424-3681)

Nepera
Nepera, Inc., Route 17, Harriman, NY 10926 (Tel.: 914-782-1200; FAX 914-783-9713; Telex: 510-249-4847)
Nepera, Ltd., 4 Heath Sq., Boltro Rd., Haywards Heath, W. Sussex, RH16 1BL, UK (Tel.: 0444-441270; FAX (0444) 441127)

Neste
Neste Chemicals International NV-SA, Bazellaan, 1-1, Ave. de Bale, B-1140 Brussels, Belgium (Tel.: 32-2-2444211)
Neste Polyeten AB, S-44486 Stenungsund, Sweden
Neste Composite Materials, Läkkisepänkuja 5, 00620 Helsinki, Finland (Tel.: 358-0-450-5293; FAX (358)0-450-5260)

Neville
Neville Chemical Co., 2800 Neville Rd., Pittsburgh, PA 15225-1496 (Tel.: 412-331-4200; FAX 412-777-4234)
Newark Chemical Co., 44 Genessee Ave., Oceanport, NJ 07757 (Tel.: 908-222-4460; FAX 908-222-4330; Telex: 134611)

Newgate Simms Ltd.
PO Box 32, Chester, CH4 0EJ, UK (Tel.: 0244 660771; FAX 0244 661220; Telex: 61597 SIMMS G)

New Riverside Ochre Co., Inc.
75 River Road SE, PO Box 387, Cartersville, GA 30120 (Tel.: 404-382-4568; FAX 404-387-1658)

New Zealand Milk Prods., Inc.
PO Box 808016, Petaluma, CA 94975-8016 (Tel.: 707-664-1000)

Ney Products, Inc.
269 Freeman St., Brooklyn, NY 11222 (Tel.: 718-389-4900; FAX 718-349-2313; Telex: 424174 NEYUI (ITT)

Niacet Corp.
PO Box 258, 400 47th St., Niagara Falls, NY 14304 (Tel.: 716-285-1474; 800-828-1207; FAX 716-285-1497; Telex: 6730170)

Nicca USA Inc.
PO Box 1600, Fountain Inn, SC 29644 (Tel.: 803-862-1426; FAX 803-862-1427)

Nichimen Italia SpA
Corsa Europa 7, 20122 Milano, Italy (Tel.: 02 783251; FAX 02 782258; Telex: 310489 MICHMI I)

Nihon Emulsion Co., Ltd.
Minami 5-32-7, Koenji, Suginami-ku, Tokyo, Japan (Tel.: 03-314-3211; FAX 03-312-7207; Telex: 2322358 EMALEX J)

Nihon Surfactants Kogyo KK
3-24-3 Hasune, Itabashi-Ku, Tokyo, Japan (Tel.: 03-966-7331)

Nikko Chemical Co., Ltd.
1-4-8 Nihonbashi-Bakurocho, Chuoku, Tokyo, 103, Japan (Tel.: 81-3-662-0371; FAX 81-3-664-8620; Telex: 2522744 NIKKOL J)

Nipa Laboratories, Inc.
3411 Silverside Rd., 104 Hagley Bldg., Wilmington, DE, 19810 (Tel.: 302-478-1522; FAX 302-478-4097; Telex: 905030)

Nippon Chemical Co., Ltd.
3-1 Iwamoto-Cho, 2-Chome, Chiyoda-Ku, Tokyo, Japan (Tel.: 03-861-2291)

Nippon Nyukazai Co., Ltd.
9-19 Ginza 3-Chome, Chuo-ku, Tokyo, 104, Japan (Tel.: 81-3-543-8571)

Nippon Oils & Fats Co., Ltd.
10-1, Yaraku-Cho, 1-Chome, Chiyoda-Ku, Tokyo, 100, Japan (Tel.: 81-3-283-7140; FAX 81-3-283-7134; Telex: 222-2041)

Nippon Senka Chemical Industries, Ltd.
17-34 Hanaten-Higashi, 1-Chome, Tsurami-ku, Osaka, Japan

Nisshin Oil Mills, Ltd.
23-1 1-Chome, Shinkawa, Chuo-Ku, Tokyo, 104, Japan (Tel.: 03-555-6843)

Nobel Industries
Nobel Industries Sweden, Box 13000, S-850 13 Sundsvall, Sweden (Tel.: 46-60-13-40-00, 46-60-56-95-18; Telex: 71399 expancl s)
Eka Nobel Inc./Int'l. Resins Div., 7 Beeches Lane, Woodstock, CT 06281 (Tel.: 203-963-2061)
Eka Nobel Ltd., Unit 304 Worle Pkwy., Summer Lane, Worle, Weston-Super-Mare, BS22 OWA UK (Tel.: 934-522244)
Eka Nobel (Australia) Pty Ltd., 22 Commercial Dr., Dandenong, Victoria, 3175, Australia (Tel.: (03)706-4488)
Expancel., 1519 Johnson Ferry Rd., Suite 200, Marietta, GA 30062 (Tel.: 404-971-8005; FAX 404-578-1359)
ScanRoad, Nobel Industries Sweden, PO Box 7677, Waco, TX 76714-7677 (Tel.: 817-772-7677; FAX 817-772-7246; Telex: 73-0195 SS WAC)

The Norac Co. Inc.
405 S. Motor Ave., Azusa, CA 91702 (Tel.: 818-334-2908; Telex: 882552 NORAC CO AZSA)

Noramco Inc.
410 George St., New Brunswick, NJ 08901 (Tel.: 908-524-3900; FAX 908-524-1947)

Nordchem SpA
Via Spilimbergo 160, 33035 Martignacco Udine, Italy (Tel.: 0432 677361; FAX 0432 678648; Telex: 450192 NCHEM I)

Norland Products Inc.
695 Joyce Kilmer Ave., New Brunswick, NJ 08902 (Tel.: 908-545-7828; FAX 908-545-9542)

Norman, Fox & Co.
5511 S. Boyle Ave., PO Box 58727, Vernon, CA 90058 (Tel.: 213-583-0016; 800-632-1777; FAX 213-583-9769)

Norold Composites Inc.
4255 Sherwoodtown Blvd., 3rd Floor, Mississauga, Ontario, L4Z 1Y5, Canada (Tel.: 416-279-2740; FAX 416-848-0455)

Norplast Ltd.
Unit 2, Adamzes Industrial Estate, Scotswood Road, Newcastle Upon Tyne, UK (Tel.: 0912 749777; Telex: 53587)

Norsk Hydro
Norsk Hydro AS, Bygdoyalle 2, N 0240 Oslo 2, Norway (Tel.: 243 2100; Telex: 78350 HYDRO N)
Norsk Hydro Plast Oy, Sabiansgatan 12 G49, PL 299, 00131 Helsinki, Finland (Tel.: 0660446)

Norsohaas SA
10 Ave. de Bergoide, 60550 Verneuil en Halatte, France (Tel.: 33 44617878; FAX 33 44268044; Telex: 155628)

North American Carbon, Inc.
432 McCormick Blvd., Columbus, OH 43213 (Tel.: 614-864-8100; FAX 614-864-9914; Telex: 246515)

North American Refractories Co.
500 Halle Bldg., Cleveland, OH 44115 (Tel.: 216-621-5200)

North American Salt Co.
650 Westlake Ctr., 4555 Lake Forest Dr., Cincinnati, OH 45242 (Tel.: 513-683-9135; 800-962-9819; FAX 513-683-9139; Telex: 801-731-4881)

North Coast Compounders, Inc.
4935 Mills Industrial Pkwy., North Ridgeville, OH 44039 (Tel.: 216-327-0485; FAX 216-327-2770)

Northern Industrial Plastics Ltd.
Stock Lane, Off Peel Street, Chadderton, Oldham, Lancs, OL9 9EY, UK (Tel.: 061 624 9479; FAX 061 678 8877)

Northern Products, Inc.
PO Box 1175, 153 Hamlet Ave., Woonsocket, RI 02895 (Tel.: 401-766-2240; FAX 401-766-2287)

Norton Chemical Process Products
PO Box 350, Akron, OH 44309 (Tel.: 216-673-5860; FAX 216-673-5868; Telex: 433-8012)

Norton Materiaux Avances
Rue de L'Ambassadeur, BP 148, 78702 Conflans ste Honorine, France (Tel.: 1 34 90 40 03; FAX 1 34 90 07 88; Telex: 695115)

Norton Pampus GmbH
Postfach 80, Am Nordkanal 37, 4156 Willich 3, Germany (Tel.: 02154 600; FAX 02154 60310; Telex: 8 531 924 ffe d)

Norton Performance Plastics
Norton Performance Plastics, PO Box 3660, Akron, OH 44309 (Tel.: 216-798-9240)
Norton Performance Plastics UK Chesterton Works, Loomer Rd., Newcastle, Staffordshire, ST5 7HR, UK (Tel.: 0782 563726; FAX 0782 562455; Telex: 36389)

Norwegian Talc UK Ltd.
205 Cotton Exchange Bldg., Old Hall St., Liverpool, Merseyside, L3 9LA UK (Tel.: 051 236 6435; FAX 051 227 5903; Telex: 627012)

Novachem Corp.
PO Box 6379, High Point, NC 27262 (Tel.: 919-885-0041; FAX 919-885-4964)

Novacor
Novacor Chemicals Ltd., PO Box 2535, Station M, Calgary, Alberta, T2P 2N6, Canada (Tel.:

403-290-8977; 800-661-1548; FAX 403-264-6012; Telex: 038-25775)
Novacor Chemicals Inc., Suite 840 South, 1515 Woodfield Rd., Schaumburg, IL 60173-5437 (Tel.: 708-605-1836; 800-222-7213; FAX 708-517-7781)
Novacor Chemicals Inc., 690 Mechanic St., Leominster, MA 01453-4451 (Tel.: 508-537-1111; 800-243-4750; FAX 508-537-2272)
Genesis Polymers, 2550 Busha Hwy., Marysville, MI 48040 (Tel.: 313-364-5555; 800-627-1221; FAX 313-364-4670)

Novamont North America, Inc.
1114 Ave. of the Americas, Suite 3300, New York, NY 10036 (Tel.: 212-997-7035)

Nova Polymers, Inc.
4004 East Morgan Ave., Evansville, IN 47715 (Tel.: 812-476-0339)

Novatec Plastics & Chemicals Co., Inc.
PO Box 597, 275 Industrial Way West, Eatontown, NJ 07724 (Tel.: 908-542-6600; 800-PVC-NOVA; FAX 908-389-0431)

Novkabel
21000 Novi Sad, Put Nov.par.odreda 4, Yugoslavia (Tel.: 021 338 199; FAX 338025; Telex: 14157 yu)

Novon/Div. of Warner-Lambert Co.
182 Tabor Rd., Morris Plains, NJ 07950 (Tel.: 201-540-4332; FAX 201-540-4487)

Novo Nordisk
Novo Nordisk A/S, Bioindustrial Group, Novo Allé, DK-2880 Bagsvaerd, Denmark (Tel.: 45-4444-8888; FAX 45-4444-6088; Telex: 37173)
Novo Nordisk Bioindustrials Inc., 33 Turner Rd., Danbury, CT 06813-1907 (Tel.: 800-251-6686; FAX 203-790-2748)
Novo Nordisk Bioindustries UK Ltd., 4 St. Georges Yard, Castle St., Farnham, Surrey, GU9 7LW, UK (Tel.: (0252)711212 (0252)711187)
Novo Nordisk Bioindustry Ltd., Makuhari Techno Garden CB-6, 3, Nakase 1-chome, Chiba-shi, 261-01, Japan (Tel.: (0472)966767; FAX (0472)966760)

Novosystems Farben und Additive GmbH
Grenzkehre 7, 2100 Hamburg 90, Germany (Tel.: 040 768 7025; FAX 040 768 7306; Telex: 2165433)

NRC Inc.
45 Industrial Pl., Newton, MA 02164 (Tel.: 617-969-7690; FAX 617-332-3947)

Nutex Inc.
110 Catalina Dr., Greenville, SC 29609 (Tel.: 803-244-5555; FAX 803-244-2735)

Nyco Minerals, Inc.
Mountain View Dr., Willsboro, NY 12996 (Tel.: 518-963-4262; FAX 518-963-4187; Telex: 957014)

Nylon Corp. of America
333 Sundial Ave., Manchester, NH 03103 (Tel.: 800-851-2001; FAX 603-627-5154)

Oakite Products, Inc.
50 Valley Rd., Berkeley Hts., NJ 07922 (Tel.: 908-464-6900; 800-526-4473; FAX 908-464-6031)

O'Brien Industries, Inc./A Zinkan Enterprises Co.
10574 Ravenna Rd., Twinsburg, OH 44087 (Tel.: 216-487-1500; FAX 216-425-8202)

Obron Atlantic Corp.
830 E. Erie St., Painesville, OH 44077 (Tel.: 216-354-0600; 800-556-1111; FAX 216-354-6224)

O&C Corp.
PO Box 681380, Indianapolis, IN 46251 (Tel.: 317-290-5000; FAX 317-290-5011)

Occidental

Occidental Chemical Corp., 360 Rainbow Blvd. South, PO Box 728, Niagara Falls, NY 14302 (Tel.: 716-286-3000; FAX 716-286-3441)

Occidental Chemical Corp., 5005 LBJ Freeway, Dallas, TX 75244 (Tel.: 214-404-3925; 800-752-5151)

Occidental Chem. Europe, Holidaystraat 5, 1920 Diegem, Belgium (Tel.: 2 721 24 20; FAX 2 721 45 66; Telex: 23046)

Occidental Chemical Corp. (Australia Pty. Ltd.), Suite 15, Gateway Court, 81-91 Military Rd., Neury Bay, NSW 2089, Australia)

Occidental Chemical Asia Ltd., Toranomon 34 Mori Bldg., 9th Floor, 25-5 Toranomon 1-Chome, Minato-ku, Tokyo, 105, Japan

Octagon Process, Inc.

725 River Rd., The Marketplace, Edgewater, NJ 07020 (Tel.: 201-945-9400; FAX 201-945-1203; Telex: 4754324)

Oil Dri Corp. of America

520 N. Michigan Ave., Chicago, IL 60611 (Tel.: 312-321-1515; 800-621-7191)

Old Bridge Chemicals Inc.

Old Waterworks Rd., Old Bridge, NJ 08857 (Tel.: 201-727-2225)

Oleofina. See Fina

Olin

Olin Chemicals, 120 Long Ridge Rd., PO Box 1355, Stamford, CT 06904 (Tel.: 203-356-2000; 800-243-9171)

Olin Corp./Agricultural Products Dept., PO Box 991, Little Rock, AR 72203

Olin UK Ltd., Suite 7, Kidderminster Rd., Cutnall Green, Worchestershire, WR9 0NS, UK

Olin Europe S.A., 108-110 Blvd. Haussmann, 75008 Paris, France (Tel.: 33-1-293-3210)

Olin Australia Ltd., 1-3 Atchison St., PO Box 141, St. Leonards 2065, N.S.W., Australia

Olin Brasil Limitada, Rua Galeno de Castro, 165, Jurubatuba, Santo Amaro, 04696 Sao Paulo, SP, Brazil

Olin Japan Inc., Shiozaki Bldg., 7-1 Hirakawa-Cho 2-Chome, Chiyoda-ku, Tokyo, 102, Japan (Tel.: 81-3-263-4615)

Omega

Omega Chemicals Inc., PO Box 1723, Spartanburg, SC 29304 (Tel.: 803-582-5346; FAX 803-463-4330; Telex: 805133 Omega SC)

Omega Chemical/SprayCore, 900 Industrial Dr., Wildwood, FL, 34785 (Tel.: 904-748-5200; 800-346-3734; FAX 904-748-5001)

OmniChem NV

Industrial Research Park, B-1348, Louvain-la-Neuve, Belgium (Tel.: 32-10-450031; FAX 32-10-450693)

Omni-Ajax

Drawer P Tomchicken Rd., Sugarloaf, PA 18249 (Tel.: 717-788-6000)

Omya

Omya Inc., 61 Main St., Proctor, UT, 05765 (Tel.: 802-459-3311; 800-451-4468; Telex: 945-658)

Omya GmbH, Brohlerstr. 11, 5000 Köln 51, Germany (Tel.: 0221 3775 0; FAX 0221 3775347; Telex: 8882676)

Omya SA, 35 Quai Andre Citroen, 75015 Paris, France (Tel.: 1 40 584400; FAX 1 45 797352; Telex: 202385 F)

Opssa Resinas Sinteticas

Avenida Bertrán y Güell 59, E-08850 Gavá (Barcelona), Spain (Tel.: 662 12 00; FAX 662 93 71; Telex: 54694 OQ)

Opticolor

Mess-und Regelanlagen, Vor dem Dorf 10, 3202 Bad Salzdetfurth Hockeln, Germany (Tel.: 5064 664; FAX 5064 8187; Telex: 927 305 D)

Ore & Chem. Corp. See Chemetall

Oregon Metallurgical Corp.
530 W. 34 St., Albany, OR 97321 (Tel.: 503-926-4281; 800-547-8090; FAX 503-967-8669; Telex: 510-595-0974)

Organic Dyestuffs Corp.
PO Box 14258, East Providence, RI 02914 (Tel.: 401-434-3300; FAX 401-438-8136; Telex: 952024)

Organika Zachem Chem. Works
Aleje LWP 65, 85-825 Bydgoszcz, Poland (Tel.: 52 617011; FAX 52 610282; Telex: 0562383 ZCHB PL)

Orient Chemical Corp.
1201 Corbin St., Port Elizabeth, NJ 07201 (Tel.: 908-355-4010; FAX 908-355-5931; Telex: 289793 QUEST-UR)

Original Bradford Soap Works Inc.
200 Providence St., West Warwick, RI 02893-0907 (Tel.: 401-821-2141; FAX 401-821-5960; Telex: 952 240)

C.J. Osborn, Div. Suvar Corp.
820 Sherman Ave., Pennsauken, NJ 08110 (Tel.: 609-662-0128; 800-526-4476; FAX 609-488-4668)

Otsuka Chemical Co., Ltd.
747 Third Ave., 26th Floor, New York, NY 10017 (Tel.: 212-826-4374; FAX 212-826-5094)

Ouest Isol
Zone Industrielle, BP 15, 27460 Alizay, France (Tel.: 35 23 01 10; FAX 35 23 04 85; Telex: 180263)

Owens-Corning Fiberglas Corp.
Fiberglas Tower, Toledo, OH 43659 (Tel.: 419-248-8000; FAX 419-248-6712)

Oxid Inc.
3813 Buffalo Speedway, Houston, TX 77098 (Tel.: 713-526-8291; FAX 713-522-9392)

OxyChem
OxyChem®, 5005 LBJ Freeway, Dallas, TX 75244 (Tel.: 214-404-3800; 800-752-5151; FAX 214-404-3669; Telex: 229835)
OxyChem/Agricultural Prods., Tampa, FL (Tel.: 813-286-3800)
OxyChem/Alathon Polymers Div., PO Box 27702, Houston, TX 77227-7702 (Tel.: 800-521-2905)
OxyChem/Durez Div., 673 Walck Rd., North Tonawanda, NY 14120 (Tel.: 716-696-6000; 800-733-3339)
OxyChem/Polymers & Plastics, PO Box 1772, Berwyn, PA USA (Tel.: 215-251-1000)
OxyChem/Vinyls Div., 300 Berwyn Park, Suite 300, Berwyn, PA 19312 (Tel.: 215-251-1000; FAX 215-251-5873)

Pacific Anchor. See Air Products

Pacific Dyes & Chemicals Inc.
PO Box 52-6743, Miami, FL, 33152 (Tel.: 305-591-2131; FAX 305-591-9570; Telex: 16812062 USADYE)

Padanaplast SpA
Strada Paganina 3, Roccabianca 43010 (PR), Italy (Tel.: 0521 870 421; FAX 0521 870 427; Telex: 530216)

M.S. Paisner Inc.
53 Beaumont St., PO Box 358, Canton, MA 02021 (Tel.: 617-828-2040; FAX 617-828-2202)

Pall Corp.
2200 Northern Blvd., East Hills, NY 11548-1289 (Tel.: 516-484-5400; 800-645-6532; FAX 516-484-6164; Telex: 968855)

Pamplast ApS
Foldgaardsvej 15, Vester Nebel, 6715 Esbjerg N., Denmark (Tel.: 75 169 470; FAX 75 169 471)

Panef Mfg. Co.
5700 W. Douglas Ave., Milwaukee, WI 53218 (Tel.: 414-464-7200)

Panelgraphic Corp.
10 Henderson Dr., West Caldwell, NJ 07006 (Tel.: 201-227-1500; 800-222-0618; FAX 201-227-7750)

Paniplus
100 Paniplus Rdwy., Olathe, KS, 66061

Pan Polymers Ltd.
Unit K Penfold Works, Imperial Way, Watford, Herts, WD2 4YY, UK (Tel.: 0923 211244; FAX 0923 53530; Telex: 915107)

Pantasote Polymers Inc.
26 Jefferson St., Passaic, NJ 07055 (Tel.: 201-777-8500; FAX 201-777-3370)

Paradigm Labs Inc.
PO Box 448, Bernville, PA 19506-0448 (Tel.: 215-432-4328; FAX 215-432-4327)

Paramelt Syntac BV
PO Box 86, NL-1700, AB Heerhugowaard, The Netherlands (Tel.: 31 2207 50600; FAX 31 2207 50699)

Parish Chemical Co.
145 N. Geneva Rd., Vineyard, UT 84058 (Tel.: 801-226-2018; FAX 801-226-8496)

Parker & Amchem. See Henkel

Parkland Engineering Ltd.
Hoults Estate, Walker Rd., Newcastle, NE6 2HL, UK (Tel.: 091 265 2991; FAX 091 276 2910)

Particle Dynamics Inc.
2601 S. Hanley Rd., St. Louis, MO 63144 (Tel.: 314-968-2376; FAX 314-968-5208; Telex: 434182)

Passaic Color & Chemical Co.
28-36 Paterson St., Paterson, NJ 07501 (Tel.: 201-279-0400; Telex: 820907 PASDYE UD)

Patco. See American Ingredients

C.H. Patrick & Co.
PO Box 2526, Greenville, SC 29602 (Tel.: 803-244-4831; FAX 803-292-0652)

Paxon Polymer Co.
12875 Scenic Hwy., PO Box 53006, Baton Rouge, LA 70892-3006 (Tel.: 504-775-4330)

PCD Polymere GmbH
St. Peter-Strasse 25, Postfach 675, A-4021 Linz, Austria (Tel.: 0732 59 81 0; FAX 0732 5981 5244)

PCR, Inc.
8570 Phillips Hwy., Suite 101, Jacksonville, FL, 32256-8208 (Tel.: 904-376-8246; 800-331-6313; FAX 904-371-6246; Telex: 810-825-6342)

Pea Ridge Iron Ore Co., Inc.
HC-65, Sullivan, MO 63080 (Tel.: 314-468-7211; FAX 314-468-7202)

Peerless Prods. Ltd.
Unit 8, Commercial Rd., March, Cambs, PE15 8QP, UK (Tel.: 0354 56600; FAX 0354 55442)

Pelron Corp.
7847 W. 47 St., Lyons, IL 60534 (Tel.: 708-442-9100)

Penco of Lyndhurst Inc.
540 New York Ave., Lyndhurst, NJ 07071 (Tel.: 201-935-6600; FAX 201-896-0539; Telex: 132641)

Penetone Corp./Subsid. of West Chem. Prods. Inc.
74 Hudson Ave., Tenafly, NJ 07670 (Tel.: 201-567-3000; 800-631-1652; FAX 201-569-5340)

Penford Products Co.
1001 First St. SW, Cedar Rapids, IA 52406 (Tel.: 319-398-1232; 800-553-7294)

Peninsula Copper Industries Inc.
1700 Duncan Ave., PO Box 509, Hubbell, MI 49934 (Tel.: 906-296-9918; FAX 906-296-9484)

Penn Color, Inc.
400 Old Dublin Pike, Doylestown, PA 18901 (Tel.: 215-345-6550)

Pennwalt. See Atochem

Penreco
Penreco/Div. of Pennzoil Products Co., RD 2, Box 1, Karns City, PA 16041 (Tel.: 412-756-0110)
Penreco, 4401 Park Ave., Dickinson, TX 77539 (Tel.: 800-458-5845; FAX 713-337-2341)

Pentapharm AG
Engelgasse 109, CH-4002 Basel, Switzerland (Tel.: 061-312-9680; FAX 061-311-2049; Telex: 910-845-2193)

Pentek, Inc.
1026 Fourth Ave., Coraopolis, PA 15108 (Tel.: 412-262-0725; 800-262-0725 (PA); FAX 412-262-0731)

Percy Harms Corp.
PO Box 156, 430 S. Wheeling Rd., Wheeling, IL 60090 (Tel.: 708-541-7220; 800-323-6433; FAX 708-541-7986)

Peridot Chemicals (NJ) Inc.
1680 Rt. 23 North, Wayne, NJ 07470 (Tel.: 201-696-9000; 800-222-0121; FAX 201-696-2501)

Perkin-Elmer Corp.
761 Main Ave., Norwalk, CT 06859-0012 (Tel.: 203-762-1000; FAX 203-762-6000; Telex: 965-954)

Perma-Flex Mold Co.
1919 East Livingston Ave., Columbus, OH 43209 (Tel.: 614-242-8034; 800-736-6653 ; FAX 614-252-8572)

Permali Gloucester Ltd.
125 Bristol Rd., Gloucestershire, GL1 5SU, UK (Tel.: 0452 28282; FAX 0452 507409; Telex: 43293)

Perrite Plastic Compds.
Taylor Industrial Estate, Risley, Warrington, WA3 6BL, UK (Tel.: 0925 764142; FAX 0925 766597; Telex: 627529 Perwar)

Perstorp Chemitec
Sebastian Kohl Gasse 3-9, A-1211 Wien, Austria (Tel.: 0222 383 6360; FAX 0222 383 6361; Telex: 134990)

Peterson Chemical Corp.
PO Box 102, 710 Forest Ave., Sheboygan Falls, WI 53085 (Tel.: 414-467-2471)

Petrokem Corp. Inc.
101 Oliver St., PO Box 1888, Paterson, NJ 07509 (Tel.: 201-773-7770; 800-822-2436)

Petrolite
Petrolite Corp./Polymers Div., 6910 E. 14th St., Tulsa, OK, 74112 (Tel.: 918-836-1601; 800-331-5516; FAX 918-834-9718)
Petrolite Ltd., 137 Finchley Rd., London, NW3 6JE, UK

Petroplas Ltd.
Wellington House, 2 Kentwood Hill, Reading, Berkshire, RG3 6DE, UK (Tel.: 0734 420440; FAX 0734 420 631; Telex: 848197)

Pettibone Laboratories Inc.
136 Central Ave., Clark, NJ 07066 (Tel.: 908-499-9660; FAX 908-388-3648)

Pfaltz & Bauer Inc.
172 E. Aurora St., Waterbury, CT 06708 (Tel.: 203-574-0075; 800-225-5172; FAX 203-574-3181; Telex: 996471)

Pfanstiehl Laboratories, Inc.
1219 Glen Rock Ave., PO Box 439, Waukegan, IL 60085 (Tel.: 708-623-0370; 800-383-0126; FAX 708-623-9173; Telex: 25 3672 PFANLAB)

Geo. Pfau's Sons., Co., Inc.
PO Box 7, Jeffersonville, IN 47131 (Tel.: 800-PFAU-OIL; FAX 812-283-0765; Telex: 20-4135)

Pfizer
Pfizer Inc./Chem. Div., Minerals, Pigments, Metals, 235 E. 42nd St., New York, NY 10017 (Tel.: 212-573-7217; 800-336-9008; FAX 212-573-1876; Telex: 420440)
Pfizer Chemicals Europe & Africa, 10 Dover Rd., Sandwich, Kent, CT13 0BN, UK (Tel.: 0304 615518; FAX 0304 615529; Telex: 966555)
Pfizer/Brewery & Dairy Prods., 4215 N. Port Washington Rd., Milwaukee, WI 53212 (Tel.: 414-332-3545; 800-231-1590; Telex: ITT 420440)

Pharmachem Labs, Inc.
130 Wesley St., South Hackensack, NJ 07606 (Tel.: 201-343-5000; 800-526-0609; FAX 201-343-5807; Telex: 710-990-5026)

Phelps Dodge Refining Corp.
7001 N. Loop Rd., El Paso, TX 79998 (Tel.: 915-775-8825; 800-223-8567; FAX 915-772-6284)

Phillips
Phillips 66 Co., 376 Phillips Bldg. Annex, Bartlesville, OK 74004 (Tel.: 806-274-5236; 800-858-4327; FAX 806-274-5230)
Phillips Petrol. Chem. SA/NV, Brusselsesteenweg 355, 3090 Overijse, Belgium (Tel.: 02 689 1211; FAX 02 689 1472; Telex: 23866/22197)

Phoenix Chemical, Inc.
322 Courtyard Dr., Somerville, NJ 08876 (Tel.: 201-707-0232; FAX 201-707-0186)

Piedmont Chem. Industries, Inc.
PO Box 2728, High Point, NC 27261 (Tel.: 919-885-5131; FAX 919-887-5563)

Piemme Srl
Via la Casella 13/15, 52010 Capolona (AR), Italy (Tel.: 0575 48152; Telex: 570436 PIEMMEI)

Pierce Chem. Co.
3747 Meridian Rd., Rockford, IL 61105 (Tel.: 815-968-0747; 800-874-3723; Telex: 910-631-3419)

Pierce & Stevens Corp./A Pratt & Lambert Co.
PO Box 1092, Buffalo, NY 14240-9990 (Tel.: 716-856-4910; FAX 716-856-7530)

Pigment & Chem. Inc.
604 Main St. E., Milton, Ontario, L9T 1R2, Canada (Tel.: 416-878-8858; Telex: 055-66339)

Pigment Dispersions, Inc.
54 Kellog Court, Edison, NJ 08817 (Tel.: 908-985-7300)

AJ & JO Pilar, Inc.
145 Chapel St., Newark, NJ 07105-4198 (Tel.: 201-589-3808; FAX 201-589-0836)

Pillo Pak, Setinel Group
PO Box 8, 6960 AA Eerbeek, Netherlands (Tel.: 08338 76911; FAX 08338 53549; Telex: 35304)

Pilot Chemical Co.
11756 Burke St., Santa Fe Springs, CA 90670 (Tel.: 213-723-0036; FAX 213-945-1877)

Pilzecker's Industrie BV
Pb3, Willem Alexander str. 11, 6690 AA Gendt, Netherlands (Tel.: 0 8812 1944; FAX 0 8812 4985)

Plaskon Electronic Materials
Independence Mall W., Philadelphia, PA 19105 (Tel.: 215-592-2099; 800-537-3350; FAX 215-592-2295; Telex: 845-247)

Plastec Iberica SA
Hostal del PI S/N, Abrera, 08630 Barcelona, Spain (Tel.: 7700726; FAX 7702058; Telex: 94853)

Plastenterprise Pemu
Pest Megyei Muanyagipari Vallalat, Terstyanszky ut 89, H-2083 Solymar, Hungary (Tel.: 06 26 39033; Telex: 22 6370)

Plastic Coatings Ltd.
Woodbridge Industrial Estate, Guildford, Surrey, GU1 1BG, UK (Tel.: 0483 31155; FAX 0483 33534; Telex: 859237)

Plasticolors, Inc.
2600 Michigan Ave., Ashtabula, OH 44004 (Tel.: 216-997-5137; FAX 216-992-3613)

Plastics & Chemicals, Inc.
PO Box 306, Cedar Grove, NJ 07009 (Tel.: 908-221-0002; FAX 908-221-1095; Telex: 219744)

Plastics Europe
Rue Jeanne D'Arc, BP23, 52101 Saint-Dizier, France (Tel.: 25 05 91 00; Telex: 840637)

Plastiques Obra SA
Rue Paradis 48, 4000 Liege, Belgium (Tel.: 041 53 22 13; Telex: 41923 OBRA)

Plastribution Ltd.
Plastribution House, 81 Market St., Ashby-de-la-Zouch, LE6 5AH, UK (Tel.: 0530 560 560; FAX 0530 560 303; Telex: 341323 PLSTRB G)

Plenco Plastics Engineering Co.
3518 Lakeshore Rd., PO Box 758, Sheboygan, WI 53081 (Tel.: 414-458-1923; FAX 414-458-1923)

Plysolene Ltd.
Southwater Business Park, Worthing Rd., Southwater nr. Horsham, West Sussex, RH13 7HE, UK (Tel.: 0403 730 032; FAX 0403 731 783; Telex: 877283 PLYSOL)

PMC, Inc./Specialties Group
501 Murray Rd., Cincinnati, OH 45217 (Tel.: 513-242-3300; 800-543-2466)

PMP Fermentation Products, Inc.
9525 W. Bryn Mawr Ave., Suite 725, Rosemont, IL 60018 (Tel.: 708-928-0050; 800-558-1031; FAX 708-928-0065)

Pokonobe Industries Inc.
163 Amsterdam Ave., Suite 311, New York, NY 10023 (Tel.: 212-496-5585; FAX 212-721-1627)

Polarome Mfg. Co. Inc.
200 Theodore Conrad Dr., Jersey City, NJ 07305 (Tel.: 201-309-4500; FAX 201-433-0638; Telex: RCA 233 176)

Polialden Petroquimica SA
Rua Geraldo Flausino Gomes, 78-8 andar-Brooklin Novo, Sao Paulo, Brazil

Polimex SpA
Via Frigimelica 2, Padova, Italy (Tel.: 49 664 855; FAX 49 664 460; Telex: 432021 frigim i)

Polyad Co.
113 Rose Terrace, Barrington, IL 60010-1320 (Tel.: 708-526-3322; FAX 708-526-3326)

Polybac Corp.
3894 Courtney St., Bethlehem, PA 18107-8999 (Tel.: 215-867-7338; 800-523-9385; Telex: 350247)

Polychem Systems Srl
Via Roma 52, Casalromano (MN), Mantova 46040, Italy (Tel.: 0376 76183; FAX 0376 76374; Telex: 303138 POLCHE I)

Polychim
Nijverheidsstraat 2, 2340 Beerse, Belgium (Tel.: 014 60 16 11; FAX 014 61 75 97; Telex: 31 955)

PolyColors, Inc.
PO Box 291, Bull Run Rd., Greene, ME, 04236 (Tel.: 207-946-7339)

Polycom Huntsman
90 W. Chestnut St., Washington, PA 15301 (Tel.: 412-225-2220; 800-538-3149)

Polyemballages SA
198a Rue Henry Chene, 5350-Ohey, Belgium (Tel.: 085 612 152; FAX 02 5368601; Telex: 61344 ext 406)

Polyfol Klepsch & Co. GmbH
Scheydgasse 30, A-1210 Wien, Austria (Tel.: 0222 30 70 70 0; FAX 0222 38 76 24 40; Telex: 115770 poly a)

Poly-Hi Menasha Corp.
2710 American Way, PO Box 9086, Fort Wayne, IN 46899 (Tel.: 219-747-1611)

Polykemi AB
Box 14, S 271 21, Ystad, Sweden (Tel.: 411 17030; FAX 41117630; Telex: 32278)

Polymed Ltd.
Unit 1 Clos Menter, Excelsior Industrial Estate, Western Ave., Cardiff, CF4 3AT, UK (Tel.: 0222 521 234; FAX 0222 521 221)

Polymer Applications
3445 River Rd., Tonawanda, NY 14150 (Tel.: 716-875-0775; FAX 716-875-7317)

Polymer Composites. See Hoechst Celanese

The Polymer Corp.,
PO Box 14235, Reading, PA 19603 (Tel.: 215-320-6600; 800-366-0300; FAX 215-320-6866)

Polymer Extruded Products Inc.
297 Ferry St., Newark, NJ 07105 (Tel.: 201-344-2700)

Polymerics, Inc.
2828 Second St., Cuyahoga Falls, OH 44221 (Tel.: 216-928-2210; FAX 216-929-8819)

Polymer Industries
Viaduct Rd., Stamford, CT 06907

Polymer Research Corp. of America
2186 Mill Ave., Brooklyn, NY 11234 (Tel.: 718-444-4300)

Polymer Systems Corp.
10223 General Dr., Orlando, FL 32824 (Tel.: 407-855-5590)

Polymer Valley Chemicals, Inc.
806 W. Market St., Akron, OH 44303 (Tel.: 216-535-8499; FAX 216-535-6057)

Poly Organix, Inc.
9 Opportunity Way, Newburyport, MA 01950 (Tel.: 508-462-5555; 800-992-8506; FAX 508-465-2057)

Polyproducts Corp.
PO Box 42, Roseville, MI 48066 (Tel.: 313-774-2500; 800-521-1005; FAX 313-778-7775; Telex: 287082 SAPUR)

Polypure, Inc.
One Gatehall Dr., Parsippany, NJ 07054 (Tel.: 201-292-2900; 800-848-POLY; FAX 201-292-5295)

Poly Research Corp.
125 Corporate Dr., Holtsville, NY 11742 (Tel.: 516-758-0460; FAX 516-758-0471)

Polysar
Polysar Inc., 17 Woodland Rd., Madison, CT 06643-2399 (Tel.: 203-245-0441; FAX 203-245-4004)
Polysar Inc., 690 Mechanical St., Leominster, MA 01453 (Tel.: 508-537-1111; FAX 508-537-2272)

Polysat Inc.
7240 State Rd., Philadelphia, PA 19135 (Tel.: 215-332-7700; 800-858-2828; FAX 215-332-9997; Telex: 831869)

Polysciences Inc.
400 Valley Rd., Warrington, PA 18976 (Tel.: 215-343-6484; 800-523-2575; FAX 215-343-0214; Telex: 510-665-8542)

Polyurethane Corp. of America
PO Box 8, Everett, MA 02149 (Tel.: 617-389-7889)

Polyurethane Specialties Co. Inc.
624 Schuyler Ave., Lyndhurst, NJ 07071 (Tel.: 201-438-2325)

Polyvel, Inc.
120 No. White Horse Pike, Hammonton, NJ 08037 (Tel.: 609-567-0080)

Porter Int'l.
400 S. 13 St., Louisville, KY 40203 (Tel.: 502-588-9200; 800-828-2994; FAX 502-588-9338; Telex: 20 4279)

Potters Industries Inc./Affiliate of The PQ Corp.
Waterview Corp. Center, 20 Waterview Blvd., Parsippany, NJ 07054 (Tel.: 201-299-2900; FAX 201-335-9350; Telex: 219054)

PPG
PPG Industries Inc., One PPG Place, Pittsburgh, PA 15272 (Tel.: 412-434-3131; 800-CHEM-PPG; Telex: 86 6570)

PPG Industries/Specialty Chemicals, 3938 Porett Dr., Gurnee, IL 60031 (Tel.: 708-244-3410; 800-323-0856; FAX 708-244-9633; Telex: 25-3310)

PPG Industries/Fiber Glass Products, One PPG Place, Pittsburgh, PA 15272 (Tel.: 412-434-3250; FAX 412-434-2197)

PPG Industries/Adhesives & Sealants, 1400 E. Avis Dr., Madison Hts., MI 48071 (Tel.: 313-588-1500; FAX 313-585-3999)

PPG Canada Inc., 2 Robert Speck Pkwy., Suite 750, Mississauga, Ontario, L4Z 1H8, Canada (Tel.: 416-848-2500; FAX 416-848-2501)

PPG Industries/Specialty Chem. UK Carrington Business Park, Carrington, Urmston, Manchester, M31 4DD, UK (Tel.: 44-61-777-9203; FAX 44-61-777-9064)

PPG Industrial do Brazil Ltda., Edificio Grande Avenida, Paulista Ave. 1754, Suite 153, Sao Paulo, Brazil 01310

PPG Industries Int'l. Inc./Taiwan, Suite 601, Worldwide House, No. 131, Min Sheng East Rd., Sec. 3, Taipei, 105, Taiwan R.O.C. (Tel.: 886-2-514-8052)

Mazer Chemicals UK Ltd., Carrington Business Park, Carrington, Urmston, Manchester, M31 4DD, UK

Mazer de Mexico S.A. de C.V., Londres 226-4 Piso, Colonia Juarez, Mexico D.F. 06600

Mazer Chemicals (Canada), Mississauga, Ontario, L4Z 1H8, Canada

PQ Corp.

PO Box 840, Valley Forge, PA 19482 (Tel.: 215-293-7200; FAX 215-293-7456)

Pre-Formed Components Ltd.

Garth Rd., Morden, Surrey, UK (Tel.: 081 330 6522; FAX 081 391 4325; Telex: 888711 CORFIL G)

Premier Malt Products, Inc.

1137 North 8th St., Milwaukee, WI 53201 (Tel.: 414-271-4272)

Premier Refractories & Chemicals, Inc.

7887 Hub Pkwy., Valley View, OH 44125 (Tel.: 216-328-0200)

Premier Services Corp.

7887 Hub Parkway, Cleveland, OH 44125 (Tel.: 216-328-0200; 800-227-4287)

Premix OY

Box 12, 05201 Rajamäki, Finland (Tel.: 90 2901066; FAX 90 2903135; Telex: 15179)

Prentiss Drug & Chem. Co. Inc.

21 Vernon St., CB 2000, Floral Park, NY 11001 (Tel.: 516-326-1919; Telex: 236854)

Presperse Inc.

PO Box 735, South Plainfield, NJ 07080 (Tel.: 908-756-2033; FAX 908-756-8754; Telex: 2409388)

Price-Driscoll Corp.

75 Milbar Blvd., Farmingdale, NY 11735 (Tel.: 516-249-4200)

Prince Mfg. Co.

One Prince Plaza, Box 1009, Quincy, IL 62301 (Tel.: 217-222-8854; FAX 217-222-5098)

Proalan SA

Apartado de Correos 301, 08400, Barcelona, Spain (Tel.: 93-849-5399; FAX 93-849-1018; Telex: 94129GNSA)

Procedyne Corp.

11 Industrial Dr., New Brunswick, NJ 08901 (Tel.: 908-249-8347; FAX 908-249-7220)

The Process Control Co.

Griffin Lane, Aylesbury, Bucks, HP19 3BP, UK (Tel.: 0296 84877; FAX 0296 393122)

Prochem Inc.

890 Fern Hill Rd., West Chester, PA 19380 (Tel.: 215-436-4812)

Pro Chem Chemicals Inc.
1670 English Rd., High Point, NC 27262 (Tel.: 919-882-3308; FAX 919-889-6047)

Prochimie Int'l.
488 Madison Ave., New York, NY 10022 (Tel.: 212-688-9240; FAX 212-755-7951)

Procter & Gamble
Procter & Gamble Co./Chem. Div., 120 W. Fifth St., Suite 502, Cincinnati, OH 45202 (Tel.: 513-562-2655; 800-543-1580; FAX 513-579-9582)
Procter & Gamble Inc. Canada, 4711 Yonge St., PO Box 355, Station A, Toronto, Ontario, M5W 1C5, Canada (Tel.: 416-730-4059; FAX 416-730-4122)

Products Research & Chemical Corp.
Products Research & Chemical Corp., 5430 San Fernando Rd., PO Box 1800, Glendale, CA 91209 (Tel.: 818-240-2060)
Products Research & Chemical Corp., 21800 Burbank Blvd., Woodland Hills, CA 91365-4226 (Tel.: 818-720-8900; 800-423-2411; FAX 818-702-7499; Telex: 67 4208)

Protameen Chemicals, Inc.
375 Minnisink Rd., PO Box 166, Totowa, NJ 07511 (Tel.: 201-256-4374; FAX 201-256-6764; Telex: 130125)

Protan
Protan, Inc., Suite 201, 135 Commerce Way, Portsmouth, NH 03801 (Tel.: 603-433-1231; 800-223-9030; FAX 603-433-1348)
Protan, Postboks 420, N-3002 Drammen, Norway (Tel.: 47-3837660)

Protein Tech. Int'l./Polymer Prods. Group
14T, One Checkerboard Square, St. Louis, MO 63164 (Tel.: 314-982-1220; 800-325-7137; FAX 314-982-1190; Telex: 44 7240 KAL PRO STL)

Protex
B.P. 177, 6, rue Barbès, 92305 Levallois-Paris, France (Tel.: 47-57-74-00; FAX 47-57-69-28; Telex: 620987)

Provital
Centro Industrial Santiaga, Talleres 6, no. 15, Apartado Correos 78, Barcelona, Spain (Tel.: 93-718-80-12; FAX 93-718-38-30; Telex: 98476 DITT E)

PSG Group Ltd.
49-53 Glengall Rd., Peckham, London, SE15 6NF, UK (Tel.: 071 639 2075; FAX 071 635 2075; Telex: 884268)

Pulcra SA
Sector E C/42, Barcelona, 08040, Spain (Tel.: 34-3-323-5914; FAX 34-3-323-6760; Telex: 98301)

Purac America, Inc.
111 Barclay Blvd., Suite 280, Lincolnshire, IL 60069 (Tel.: 708-634-6330; FAX 708-634-1992; Telex: 280231 PURACINC ARHT)

Purflo DTL SA
Route Saint Laurent de la Plaine, France (Tel.: 41 78 06 30; FAX 41 78 06 40; Telex: 720964F)

E.L. Puskas Co.
448 E. South St., PO Box 229, Akron, OH 44309 (Tel.: 216-376-5161; FAX 216-376-2124; Telex: 751040)

PVS Chemicals, Inc.
11001 Harper Ave., Detroit, MI 48213 (Tel.: 313-921-1200; FAX 313-921-1378)

Pylam Products Co.
1001 Stewart Ave., Garden City, NY 11530 (Tel.: 516-222-1750; FAX 516-222-1988; Telex: 425900 UIPPC)

Pyrion-Chemie GmbH
Sandhoferstrasse 96, D-6800 Mannheim 31, Germany (Tel.: 0621-7501-419; FAX 0621-7501-447; Telex: 462334 CFAM D)

Pyron Corp.
Box E La Salle Sta., Niagara Falls, NY 14304 (Tel.: 716-285-3451)

QO
QO Chemicals, Inc./Subsid. of Great Lakes Chem. Corp., 2801 Kent Ave., Box 2500, West Lafayette, IN 47906 (Tel.: 317-497-6300; 800-621-9521; FAX 317-497-6287; Telex: 446968)
QO Chemicals Inc.-UK Spring House, 231 Glossop Rd., Sheffield, S10 2GW, UK (Tel.: 0742-767842)
QO Chemicals Inc. Belgium, Industriedok 391, B-2030 Antwerp, Belgium
QO Chemicals GmbH, Hermannstrasse 54, D-6078 Neu Isenburg, Germany
QO Chemicals (Australia) Inc., Suite #4, 21 Drummond Place, Carlton, Victoria, 3053, Australia)
Kao-Quaker Co., Ltd., 14-10 Nihonbashi Kayabacho, 1-Chome, Chuo-ku, Tokyo, 103, Japan

Quantum Composites, Inc./Subsidiary of Premix, Inc.
4702 James Savage Rd., Midland, MI 48642-8642 (Tel.: 517-496-2884; FAX 517-496-2333)

Quantum
Quantum Chemical Corp./USI Div., 11500 Northlake Dr., PO Box 429550, Cincinnati, OH 45249 (Tel.: 513-530-6500; 800-543-7900)
Quantum Chemical Europe BV, Lange Bunder 7, 4854 MB Bavel, The Netherlands (Tel.: 001613-6600)

Ques Industries
PO Box 02590, Cleveland, OH 44102 (Tel.: 216-961-7445)

Quest International
Quest International, Lindtsedijk 8, 3336 le Zwijndrecht, The Netherlands (Tel.: 31-78-128511; FAX 31-78-195279; Telex: 29477)
Quest International, 400 International Dr., Mt. Olive, NJ 07828 (Tel.: 201-691-7100; FAX 201-691-7479)

Quigley Co. Inc./Subsid. Pfizer Inc.
235 E. 42 St., New York, NY 10017 (Tel.: 212-573-3444; Telex: 961247)

Quimigal-Quimica de Portugal E.P.
Av. Infante Santo No. 2, 1300 Lisboa, Portugal (Tel.: 351-1-604040; Telex: 12301)

Rad-Cure Corp.
112 Naylon Ave., Livingston, NJ 07039 (Tel.: 201-994-4334; FAX 201-994-3341)

Radici Novacips SpA
Via Provinciale 11, 1-24020 Villa D'Ogna (BG), Italy (Tel.: 0346 224 53; FAX 0346 23730; Telex: 303344)

Radilon Inc.
PO Box 18367, Spartanburg, SC 29318 (Tel.: 803-579-2729)

Raffi & Swanson Inc.
100 Eames St., Wilmington, MA 01887-3389 (Tel.: 617-933-4200; FAX 508-658-3366)

Rampf Giessharzsysteme GmbH
Albstr. 37, 7441 Grafenberg, Germany (Tel.: 07123 34389; FAX 07123 34177)

Ranbar Technology Inc.
1114 William Flinn Hwy., Glenshaw, PA 15116 (Tel.: 412-486-1111; FAX 412-487-3313; Telex: 9102500271)

Randolph Products Co.
PO Box 830, Carlstadt, NJ 07072 (Tel.: 201-438-3700; FAX 201-438-4231)

RapidPurge Corp.
2285 Reservoir Ave., Trumbull, CT 06611 (Tel.: 203-372-5677; 800-243-4203)

Raschig
Raschig AG, Mundenheimer Str. 100, D-6700 Ludwigshafen, Germany (Tel.: 0621 56180; FAX 0621 582885; Telex: 464877 ralu d)

Raschig Corp., PO Box 7656, Richmond, VA 23231 (Tel.: 804-222-9516; FAX 804-226-1569)

Rayonier Inc.
18000 Pacific Hwy. South, Suite 900, Seattle, WA 98188 (Tel.: 206-246-3400; FAX 206-248-4162; Telex: 4949619)

RBH Dispersions
L5 Factory Lane, Boundbrook, NJ 08805 (Tel.: 201-356-1800; 800-631-5278; FAX 201-356-8369)

RCMS Polysales
Weybridge Trading Estate, Hamm Moor Lane, Addlestone, Surrey, KT15 2SA, UK (Tel.: 0932 856221; FAX 0932 854373)

Reade Advanced Materials
PO Box 15039, Riverside, RI, 02915-0039 (Tel.: 401-433-7000; FAX 401-433-7001)

Reade Manufacturing Co./Div. of Magnesium Elektron, Inc.
100 Ridgeway Blvd., Lakehurst, NJ 08733 (Tel.: 908-657-6451; FAX 908-657-6628; Telex: 642676 Remco LKHS)

Reed Plastics/Div. of Sandoz Chemicals Corp.
Holden Industrial Park, Holden, MA 01520 (Tel.: 508-829-6321; FAX 508-829-2118)

Reedy International Corp.
42 First St., Keyport, NJ 07735 (Tel.: 908-264-1777; FAX 908-264-1189)

Regent Chem. Ltd.
Suite 4, Inchs Yard, Newbury, Berkshire, RG14 4DP, UK (Tel.: 0635 41844; FAX 0635 49829)

Regis Chemical Co.
8210 Austin Ave., Morton Grove, IL 60053 (Tel.: 708-967-6000; 800-323-8144; FAX 708-967-5876; Telex: 910 223 0808)

Reheis
Reheis Inc., 235 Snyder Ave., Berkeley Heights, NJ 07922 (Tel.: 201-464-1500; FAX 201-464-7726; Telex: 219463 RCCA UR)

Reheis (Ireland), Kilbarrack Rd., Dublin, 5, Ireland

Reichhold
Reichhold Chemicals, Inc., PO Box 13582, Research Triangle Park, NC 27709 (Tel.: 919-544-9225; 800-448-3482)

Reichhold Chemicals/Emulsion Polymers Div., PO Drawer K, Dover, DE, 19903 (Tel.: 302-736-9100; 800-441-6461)

Reichhold Chemicals/Reactive Polymers Div., 8540 Baycenter Rd., PO Box 19129, Jacksonville, FL, 32245 (Tel.: 904-739-2170)

Reichhold Chemicals/Coating Polymers & Resins Div., PO Box 1433, Pensacola, FL, 32596-1433 (Tel.: 904-433-7621; 800-874-0868; FAX 904-433-3655)

Reichhold Chemie AG, CH 5212 Hausen b. Brugg, Switzerland (Tel.: 056 482222; FAX 056 482412; Telex: 825109 RCAG CH)

Reilly Industries Inc.
1510 Market Sq. Ct., 151 N. Delaware St., Indianapolis, IN 46204 (Tel.: 317-248-6411; FAX 317-248-6413; Telex: 27 404)

Reilly-Whiteman Inc.
 801 Washington St., Conshohocken, PA 19428 (Tel.: 215-828-3800; 800-533-4514; FAX 215-739-6136)

Releasomers, Inc.
 PO Box 82, Bradfordwoods, PA 15015 (Tel.: 412-452-4474; FAX 412-452-1965)

Reliance Chemical Products Co.
 PO Box 336, 64 Ave. A, Bayonne, NJ 07002 (Tel.: 201-437-4144; FAX 201-437-0003)

Renite Co., Lubrication Engineers
 PO Box 30830, Columbus, OH 43230 (Tel.: 614-253-5509; FAX 614-253-1333; Telex: 6874646)

Renosol Corp.
 PO Box 1424, Ann Arbor, MI 48106 (Tel.: 313-429-5418)

Resina Chemie BV
 Korte Groningerweg 1A, 9607 PS Foxhol, Netherlands (Tel.: 05980 17911; FAX 05980 90437; Telex: 77246 Resch nl)

Resinas y Complementos Castro
 Urzaiz 77, 36204 Vigo, Pontevedra, Spain (Tel.: 986 439282; FAX 986 439940)

Resinoid Engineering Corp./Materials Div.
 7557 St. Louis Ave., Skokie, IL 60076 (Tel.: 708-673-1050; FAX 708-673-2160)

Reter Srl
 Via Oliviero Forzetta 32, 31100 Teviso, Italy (Tel.: 0422 306771; FAX 0422 420470; Telex: 420383 RETER I)

Reusche & Co. of TWS, Inc.
 2 Lister Ave., Newark, NJ 07105 (Tel.: 201-589-2040)

Revertex Ltd.
 Temple Fields, Harlow, Essex, CM20 2AH, UK (Tel.: Harlow 29555)

Rewo
 Rewo Chemische Werke GmbH, Postfach 1160, Industriegebiet West, D-6497 Steinau, Germany (Tel.: 49-06663-540; FAX 49-06663-54-129; Telex: 493589)
 Rewo Chemical Ltd., Gorsey lane, Widnes, Cheshire, WA8 0HE, UK (Tel.: 051-495-1989; FAX 051-495-2003; Telex: 627434 SIPWID G)

Rexene Products Co.
 5005 LBJ Freeway, Occidental Tower, Dallas, TX 75244 (Tel.: 214-450-9000; 800-233-1159; FAX 214-450-9028)

Rhein
 Rhein Chemie Corp., 1008 Whitehead Road Ext., Trenton, NJ 08638 (Tel.: 609-771-9100; FAX 609-771-0232)
 Rheinau GmbH, Muelheimerstrasse, 21-28, D-6800 Mannheim 81, Germany

Rheox
 Rheox, Inc., PO Box 700, Hightstown, NJ 08520 (Tel.: 609-443-2500; FAX 609-443-2446)
 Rheox, Inc., Rue de l'Hôpital 31, B-1000 Brussels, Belgium (Tel.: 02-512-0048)

Rho-Chem Corp.
 425 Isis Ave., PO Box 6021, Inglewood, CA 90301 (Tel.: 213-776-6233; FAX 213-645-6379)

Rhone-Poulenc
 Rhone-Poulenc Chimie (France), Cedex 29, F-92097 Paris La Defense, France (Tel.: 33-47-68-1234; FAX 33-47-68-0900)
 Rhone-Poulenc S.A., Dept. Biochimie, 18, ave. d'Alsace-F, 92400 Courbevoie, Paris, France
 Rhone-Poulenc, Inc./Chemicals Div., Box 125, Monmouth Junction, NJ 08852 (Tel.: 201-297-0100; FAX 201-297-1597)

Rhone-Poulenc, Inc./Film Div., 2754 West Park Dr., Holcomb, NY 14469 (Tel.: 716-657-5800; FAX 716-657-5838)

Rhone-Poulenc, Inc./Performance Resins & Coatings Div., 9808 Bluegrass Pkwy., Louisville, KY 40299 (Tel.: 502-499-4011; 800-922-2189; FAX 502-499-4578)

Rhone-Poulenc, Inc./Performance Resins & Coatings, 1525 Church St. Ext., Marietta, GA 30060 (Tel.: 404-422-1250; FAX 404-427-0874; Telex: 542-112)

Rhone-Poulenc, Inc./Surfactants & Specialties, CN 7500, Prospect Plains Rd., Cranberry, NJ 08512-7500 (Tel.: 609-395-8300; 800-922-2189; FAX 609-395-7626)

Rhone-Poulenc, Inc./Textile & Rubber Div., PO Box 1740, Dalton, GA 30720 (Tel.: 404-259-4831; FAX 404-259-5979)

Rhone-Poulenc Ag Co., PO Box 12014, 2 TW Alexander Dr., Research Triangle Park, NC 27709 (Tel.: 919-549-2101; 800-334-8577; FAX 919-549-9639; Telex: 4999378 APC RTP)

Rhone-Poulenc Basic Chemical Co., One Corporate Dr., Box 881, Shelton, CT 06484 (Tel.: 203-925-3300; 800-642-4200; FAX 203-925-3627)

Rhone-Poulenc Geronazzo SpA, Via Milano 78, 20021 Ospiate Di Bollate, Milano, Italy (Tel.: 39-2-350-3212; FAX 39-2-350-1770; Telex: 331547 GERO I)

Rhone-Poulenc Chem. Ltd., Perf. Prods. Group, Woodley, Stockport, Cheshire, SK6 1PQ, UK (Tel.: 44-61-430-4391; FAX 44-61-430-4364; Telex: 667835)

American Lecithin Co., Inc., PO Box 1908, 33 Turner Rd., Danbury, CT 06813-1908 (Tel.: 203-790-2700; FAX 203-790-2705)

Nattermann Phospholipid GmbH, Nattermannallee 1, D-5000 Cologne 30, Germany (Tel.: 02-21-509-2239; FAX 02 21-509-2816)

Rias A/S

Industrivej 9-17, 4000 Roskilde, Denmark (Tel.: 2757200; FAX 2757210; Telex: 43130)

G. Whitfield Richards Co.

4202-10 Main St., Philadelphia, PA 19127 (Tel.: 215-487-1202; FAX 215-487-3090; Telex: 845-322 GWRCO)

Richmond Oil Soap & Chem. Co. Inc.

190 W. Glenwood Ave., Philadelphia, PA 19140 (Tel.: 215-GA6-4305)

Chemische Laboratorium Dr. Kurt Richter GmbH

PO Box 410480, Bennigsenstrasse 25, Berlin 41 (West) D-1000, Germany (Tel.: 30-852-70-75; FAX 30-851-18-22; Telex: 184626 clr d)

Ricon Resins, Inc.

569 24 1/4 Road, Grand Junction, CO, 81505 (Tel.: 303-245-8148; FAX 303-245-4348)

Ricsan Industries Inc.

PO Box 18082, Memphis, TN 38118 (Tel.: 901-332-0781)

Riddle & Hobb (Specialist Plastics) Ltd.

Hobb House, Unit 8 Zobel Close, Sweetbriar Industrial Estate, Norwich, Norfolk, NR3 2BY, UK (Tel.: 0603 400920; FAX 0603 400945; Telex: CHA COMM)

Riechem/Mount Vernon Mills Inc.

PO Box 329, Ware Shoals, SC 29692 (Tel.: 803-456-7464; FAX 803-456-7467)

Riken Vitamin Oil Co., Ltd.

2-9-18 Misaki-Cho, Chiyoda-Ku, Tokyo, Japan (Tel.: 03-5275-5111; FAX 03-261-2628; Telex: 02322783)

RITA

RITA Corp., 1725 Kilkenny Court, PO Box 585, Woodstock, IL 60098 (Tel.: 815-337-2500; 800-426-7759; FAX 815-337-2522; Telex: 72-2438)

Represented by: Maprecos, 4, Rue des Passe-Loups, 7770 Fontaine, Le Port, France)

Rit-Chem Co. Inc.

109 Wheeler Ave., PO Box 435, Pleasantville, NY 10570 (Tel.: 914-769-9110; FAX 914-769-1408; Telex: 229 639 RTCH)

Rite Industries Inc.
PO Box 1747, High Point, NC 27261 (Tel.: 919-886-5173; FAX 919-884-8785)

Riverside Products Corp.
PO Box 729, Cartersville, GA 30120 (Tel.: 404-386-3115; FAX 404-386-7985; Telex: 54 2268)

RMc Minerals, Inc.
111 E. Drake Rd., Suite 7104, Fort Collins, CO 80525 (Tel.: 303-223-7790; FAX 303-226-5617)

Robeco Chemicals Inc.
99 Park Ave., New York, NY 10016 (Tel.: 212-986-6410; FAX 212-986-6419; Telex: 23-3053)

Robnorganic Systems Ltd.
Highworth Road, Swindon, Wiltshire, SN3 4TE, UK (Tel.: 0793 823 741; FAX 0793 827 033; Telex: 449 905)

Roche Vitamins & Fine Chemicals/Div. of Hoffman-Laroche Inc.
340 Kingsland St., Nutley, NJ 07110-1199 (Tel.: 201-235-5000; FAX 201-535-7606)

Röchling Sustaplast KG
Fritz-Erler-Strasse, Postfach 12 50, D-5420 Lahnstein, Germany (Tel.: 02621 6930; FAX 02621 69358; Telex: 869828)

Rockwell Lime Co.
4110 Rockwood Rd., Manitowoc, WI 54220 (Tel.: 414-682-7771; 800-558-7711; FAX 414-682-7972)

Rockwood Systems Corp.
640 E. Main St., Lancaster, TX 75146 (Tel.: 214-227-3100; Telex: 910-860-5634)

Rogers Anti-Static Chemicals
120 W. Madison St., Room 1118, Chicago, IL 60602 (Tel.: 312-276-0665; FAX 312-276-4371)

Rogers Corp.
Rogers Corp./Molding Materials Div., Box 550, Manchester, CT 06040 (Tel.: 203-646-5500; 800-227-6437; FAX 203-646-5503)
Rogers Corp./Poron Materials Div., Box 158, E. Woodstock, CT 06244 (Tel.: 203-774-9605; FAX 203-928-7843)

Rohm & Haas
Rohm & Haas Co., Independence Mall West, Philadelphia, PA 19105 (Tel.: 215-592-3000; 800-323-4165; FAX 215-592-2285)
Rohm & Haas Canada Inc., 2 Manse Rd., West Hill, Ontario, M1E 3T9, Canada
Rohm & Haas Co. European Operations, Chesterfield House, Bloomsbury Way 15-19, London, WC1A 2TP, UK (Tel.: 071 242 4455; FAX 071 404 4126; Telex: 24139)
Rohm & Haas UK Ltd., Bloomsbury Way, Chesterfield House, London, UK (Tel.: 71-242-4455; FAX 71-404-4126)
Röhm GmbH Chemische Fabrik, Kirschenallee, Postfach 4242, D-6100 Darmstadt, Germany (Tel.: 6151 18 01; FAX 6151 184007)
Rohm & Haas (Australia) Pty. Ltd., 969 Burke Rd., PO Box 11, Camberwell, Victoria, 3124, Australia
Rohm & Haas Asia Ltd., Kaisei Bldg., 8-10 Azabudai 1-Chome, Minato-ku, Tokyo, 106, Japan

Roma Color Inc.
749 Quequechan St., PO Box 5360, Fall River, MA 02723 (Tel.: 800-284-8829; FAX 508-676-9011)

Rona
Rona, 5 Skyline Dr., Hawthorne, NY 10532 (Tel.: 914-592-4660; FAX 914-592-9469)

Represented by: S. Black (Import & Export) Ltd., The Colonnade, High St., Chesunt Hents, EN8 0DJ, UK

Ronco Labs Inc.
1039 Lilac St., Pittsburgh, PA 15217 (Tel.: 412-422 0119; FAX 412-422 7616)

Ronsheim & Moore Ltd.
Ings Lane, Castleford, Yorkshire, WF10 2JT, UK (Tel.: 44-977-556565; FAX 44-977-518058; Telex: 55378)

Roquette Corp.
1550 Northwestern Ave., Gurnee, IL 60031 (Tel.: 708-249-5950; 800-223-5305; FAX 708-578-1027; Telex: 687 1679 ROQ ILL)

Roseville Charcoal & Mfg. Co.
PO Box1148, Zanesville, OH 43701 (Tel.: 614-452-5473)

Ross Chemical, Inc.
303 Dale Dr., Fountain Inn, SC 29644 (Tel.: 803-862-4474; 800-521-8246; FAX 803-862-2912)

Frank B. Ross
Frank B. Ross Co., Inc., 22 Halladay St., PO Box 4085, Jersey City, NJ 07304-0085 (Tel.: 201-433-4512; FAX 201-332-3555)
Represented by: Harrisons & Crosfield (Canada) Ltd., St. Laurent, Quebec, H4M 2N6, Canada (Tel.: 514-748-7911)

Ross & Rowe Inc./Subsid. of Archer Daniels Midland
PO Box 1409, Decatur, IL 62525 (Tel.: 217-424-5803; 800-637-5843; Telex: 250121)

Rostone Corp.
PO Box 7497, 2450 Sagamore Pkwy. South, Lafayette, IN 47903 (Tel.: 317-474-2421; 800-637-4851; FAX 317-474-8785)

Rotec Chemicals Ltd.
3 Francis Court, Wellingborough Rd., Rushden, Northants, NN10 9AY, UK (Tel.: 0933 315 500; FAX 0933 58721; Telex: 311366 ROTEC G)

Rotuba
Rotuba Plastics, 1401 Park Ave. South, Linden, NJ 07036-1698 (Tel.: 201-486-1000; FAX 201-486-0874; Telex: 910-240-3809)
Rotuba Extruders, Inc., 1401 Park Ave. S., Linden, NJ 07036 (Tel.: 201-486-1000)

Rowa GmbH
Siemensstr. 1-3, Postfach 1929, 2080 Pinneberg, Germany (Tel.: 041 01 70 60; FAX 041 01 706 200; Telex: 2189 088)

Rowe Products Distribution, Inc.
3857 Hyde Park Blvd., Niagara Falls, NY 14305 (Tel.: 716-285-9348; FAX 716-285-9431)

Royalite Plastics Ltd.
Cliftonhall Rd., Newbridge, Midlothian, EH28 8TW, UK (Tel.: 031 333 2819; FAX 031 333 4603; Telex: 727805)

RTD Chemical Corp.
164 Main St., Hackettstown, NJ 07840 (Tel.: 908-852-6128; FAX 908-852-1335)

RTP Co.
580 E. Front St., PO Box 439, Winona, MN 55987-5439 (Tel.: 507-454-6900; 800-433-4787; FAX 507-454-8130)

Ruco Polymer Corp.
New South Rd., Hicksville, NY 11802 (Tel.: 516-931-8100; FAX 516-931-8179; Telex: 5102220856)

Rudd Co. Inc.
1630 15 Ave. W., Seattle, WA 98119 (Tel.: 206-284-5400; 800-444-7833; FAX 206-286-8179)

Ruetgers-Nease Chemical Co., Inc.
201 Struble Rd., State College, PA 16801 (Tel.: 814-238-2424; FAX 814-238-1567)

Ruger Chemical Co. Inc.
85 Cordier St., Irvington, NJ 07111 (Tel.: 201-926-0331; 800-631-7844; FAX 201-926-4921)

W.J. Ruscoe Co.
483 Kenmore Blvd., Akron, OH 44301 (Tel.: 216-253-8148)

Ruscon Plastics
Ruscon Works, Rotherham Rd., Park Gate, Rotherham, S62 6EZ, UK (Tel.: 0709 527 751; FAX 0709 523 298; Telex: 547791)

Sachtleben Chemie GmbH
Pestalozzistrasse 4, D-4100 Duisburg-Homberg, Germany (Tel.: 02136-22-0; FAX 021-36-22-660)

L.A. Salomon Inc.
150 River Rd., Suite L-3B, Montville, NJ 07045 (Tel.: 201-335-8300; FAX 201-335-1236)

Sanchem Inc./Subsid. of Santell Chem. Co.
1600 S. Canal St., Chicago, IL 60616 (Tel.: 312-733-6100; 800-733-7432; Telex: 25 6229)

Sandoz
Sandoz Chemicals Corp., 4000 Monroe Rd., Charlotte, NC 28205 (Tel.: 704-331-7000; 800-444-4225; FAX 704-372-0210; Telex: 216-922)
Sandoz Chemicals Corp. Canada, Dorva, Quebec, H9R 4PR, Canada
Sandoz Prods. Ltd./Chemicals Div., Lichtstrasse 35, CH-4002 Basel, Switzerland (Tel.: 41-61-324-1111; FAX 41-61-324-6080; Telex: 96505049)
Sandoz Ltd., Calverley Lane, Horsforth, Leeds, LS18 4RP, UK (Tel.: 0532 584646; FAX 0532 390063; Telex: 557114)

Sanitized
Sanitized AG, Lyssachstrasse 95, 3401 Burgdorf, Switzerland (Tel.: 034 222 055; FAX 034 222 058)
Sanitized Inc., 57 Litchfield Rd., PO Box 2211, New Preston, CT 06777 (Tel.: 203-868-9491; FAX 203-868-9494)

Sanncor Industries, Inc.
300 Whitney St., PO Box 703, Leominster, MA 01453 (Tel.: 508-537-4748; FAX 508-537-8245; Telex: 701662)

Sanraku, Inc.
5-8, 1-Chome, Kyobashi, Chuo-Ku, Tokyo, Japan (Tel.: 03-231-3917; FAX 03-276-0151; Telex: 03-252-2761 SROKJ)

Santech Inc.
35 Jutland Rd., Toronto, Ontario, M8Z 2G6, Canada (Tel.: 416-252-5997; FAX 416-251-9315)

Sanyo Chemical Industries, Ltd.
11-1 Ikkyo Nomoto-cho Higashiyama-ku, Kyoto, 605, Japan (Tel.: 81-75-541-4311; FAX 81-75-551-2557; Telex: 05422110)

Sapici SpA (Soc Azion per Ind Chimica Italia)
Via Bergamo 2 Zai, 20063 Cernusco Sul Naviglio, Milano, Italy (Tel.: 02 9655121; FAX 02 9659459; Telex: 325343)

Saramco, Inc.
8915 Sorensen Ave., PO Box 599, Santa Fe, CA 90670 (Tel.: 213-685-4388; FAX 213-698-7571)

Sarma SpA
Via Lainate 26, 20010 Pogliano MI Italy (Tel.: 02 93256813; Telex: 331066 SARMA I)

Sartomer
Sartomer Co., Oaklands Corp. Center, 468 Thomas Jones Way, Exton, PA 19341 (Tel.: 215-363-4100; 800-345-8247; FAX 215-363-4140)

Sartomer International, Inc., Kingswick House, Sunninghill, Berkshire, SL5 7BH, UK (Tel.: 44-99023491; FAX 44-99026176)

Sartomer International, Inc., Bankastraat 131d, 2585 El Den Haag, The Netherlands

Sartomer International Deutschland GmbH, Konigsallee 60F, 4000 Dusseldorf 1, Germany

Sartomer International, Inc., 7500A Beach Rd., Unit No. 14-313, The Plaza, Singapore

Craynor, 74-80 rue Roque de Fillol, 92800 Puteaux, France (Tel.: 1-47-78-94-35)

Saudi Basic Industries Corp. (SABIC)
PO Box 5101, Riyadh, 11422, Saudi Arabia (Tel.: 966 1 401 2033; FAX 966 1 401 3831; Telex: 401177 SABIC SJ)

Sauereisen Cements Co.
160 Gamma Dr., Pittsburgh, PA 15238-2989 (Tel.: 412-963-0303; FAX 412-963-7620)

Schaefer Salt & Chemical
1255 Magie Ave., PO Box 236, Elizabeth, NJ 07207 (Tel.: 908-352-7010; 800-631-7300; FAX 908-352-7329; Telex: 139423)

Schaefer Technologies Inc.
3000 Carrollton Rd., Saginaw, MI 48604 (Tel.: 517-753-1877; 800-444-9034)

Alexander Scheiner AG
PO Box 163, CH-8057 Zurich, Switzerland (Tel.: 1-362-5209; FAX 1-362-5075; Telex: 816 999)

Schenectady
Schenectady Chemicals, Inc., PO Box 1046, Schenectady, NY 12301 (Tel.: 518-370-4200; FAX 518-382-8129; Telex: 145457)

Schenectady Chemicals Canada Ltd., 319 Comstock Rd., Scarborough, Ontario, M1L 2H3, Canada)

Schenectady-Midland Ltd., Four Ashes, Wolverhampton, WV10 7BT, UK (Telex: 339075)

Nisshoku Schenectady Chemicals, Inc., Kogin Bldg., 5-1 Koraibashi, Higashi-Ku, Osaka, 541, Japan (Telex: 5225243)

Scher
Scher Chemicals, Inc., Industrial West & Styertowne Rd., PO Box 4317, Clifton, NJ 07012 (Tel.: 201-471-1300; FAX 201-471-3783; Telex: 642643)

Represented by: Chesham Chemicals Ltd., Cunningham House, Bessborough Rd., Harrow, HA1 3DU, UK

Schering AG
Waldstrasse 14, Postfach 15 40, D-4709 Bergkamen, Germany (Tel.: 0 2307 65 1)

Schering Berlin Polymers Inc.
4868 Blazer Memorial Pkwy., PO Box 1227, Dublin, OH 43017 (Tel.: 614-793-7700; FAX 614-793-7711)

Schering Industrial Chemicals
Gorsey Lane, Widnes, Cheshire, WA8 0HE, UK (Tel.: 051 495 1989; FAX 051 495 2003; Telex: 627434 SIPWID G)

Scholler Inc.
3320 Collins St., Philadelphia, PA 19134 (Tel.: 215-739-0900; FAX 215-739-0905)

Schuller Mats & Reinforcements/Div. of Schuller Int'l.
PO Box 517, Toledo, OH 43697-0517 (Tel.: 419-878-8111; Telex: 5225243)

A. Schulman

A. Schulman Inc., 3550 West Market St., PO Box 1710, Akron, OH 44309-1710 (Tel.: 216-666-3751; FAX 216-668-7204)

A. Schulman Canada Ltd., 170 Attwell Dr., Suite 503, Ontario, M9W 5Z5, Canada (Tel.: 416-675-7878)

A. Schulman Inc. Ltd., Croespenmaen Industrial Estate, Crumlin, Newport, Gwent NP 14AG, South Wales, UK

Schumacher

1969 Palomar Oaks Way, Carlsbad, CA 92009 (Tel.: 619-931-9555; 800-545-9241; FAX 619-931-7819; Telex: 910 322 1382)

Schwartz Chemical Co.

50-01 Second St., Long Island City, NY 11101 (Tel.: 718-784-7592)

SCM

SCM Chemicals, 7 St. Paul St., Suite 1010, Baltimore, MD 21202 (Tel.: 301-783-1120; 800-638-3234)

SCM Chemicals Ltd., PO Box 26, Grimsby, South Humberside, DN37 8DP, UK (Tel.: 44-469-571000)

SCM Chemicals Ltd., Hop Hing Centre, 22nd Floor, 8-12 Hennessy Rd., Wanchai, Hong Kong

SCM Chemicals Ltd., PO Box 465, Auburn, N.S.W., 2144, Australia (Tel.: 61-2-647-2566)

SCM Glidco Organics

PO Box 389, Jacksonville, FL, 32201 (Tel.: 904-768-5800; 800-231-6728; FAX 904-768-2200; Telex: 441763)

SCM Metal Products, Inc.

11000 Cedar Ave., Cleveland, OH 44106 (Tel.: 216-795-5000; FAX 216-795-4916; Telex: 196 072)

Scor SAS

Via Montello 15, 2100040 Gormate Oloma (VA), Italy (Tel.: 0331 82 00 38)

Scott Bader Co. Ltd.

Wollaston, Wellingborough, Northamptonshire, NN9 7RL, UK (Tel.: 0933-663100; FAX 0933-663474)

Sederma

7110 Fort Hamilton Pkwy., Brooklyn, NY 11228 (Tel.: 718-833-1046; FAX 718-833-7028)

Semiplastic Ltd.

Unit 1 Stafford Park 12, Telford, TF3 3BJ, UK (Tel.: 0952 292026; FAX 0952 292121; Telex: 35812 SENOUR G)

Senova Kunststoffe GmbH

Postfach 8, A-5721 Piesendorf, Austria (Tel.: 06949 7840; FAX 06549 7941 788; Telex: 66669)

Sentry/Custom Services Corp., Specialty Prods. Div.

Box 193, Allamuchy, NJ 07820 (Tel.: 717-421-2574; 800-235-2100 x 497; FAX 717-421-2511; Telex: 475-4556 KWKTLUI)

Seppic

75 Quai d'Orsay, 75321 Paris Cedex 07, France (Tel.: 33-40-62-59-01; Telex: 290665)

Sequa Chemicals Inc.

One Sequa Dr., Chester, SC 29706 (Tel.: 803-385-5181; FAX 803-377-3542)

Laboratories Serobiologiques, Inc.

161 Chambers Brook Rd., Branchburg Township, Somerville, NJ 08876 (Tel.: 201-218-0330; FAX 201-218-0333; Telex: 709485 LABSEBIO)

Serviplast
Montée des Pins les Gabelles ZI, 13340 Rognac, France (Tel.: 428 701 18; Telex: 400895 Servip)

Servo Chemische Fabriek B.V.
Postbus 1, 7490 AA Delden, The Netherlands (Tel.: 31-5407-63535; FAX 31-5407-64125; Telex: 44347)

Seydel-Woolley & Co.
PO Box 169, Pendergrass, GA 30567 (Tel.: 404-693-2266; FAX 404-693-4163; Telex: 668-7010)

Shamokin Filler Co., Inc.
Venn Access Rd., Shamokin, PA 17872 (Tel.: 717-644-0437; FAX 717-648-4094)

Shamrock Specialties Inc.
PO Box 1369, 3503 W. Oak, Palestine, TX 75801 (Tel.: 214-729-7831)

Shanti GmbH
Postfach 1301, 53 Bonn 1, Germany (Tel.: 0228 663 813; FAX 0228 667 160; Telex: 8869995 Shtt D)

Sharp Chem. Co.
6700 Dixie Dr., Houston, TX 77087 (Tel.: 713-641-1444; FAX 713-641-5121)

Sheffield Plastics. See DSM

Shell
Shell Chemical Co., PO Box 2463, 1 Shell Plaza, Houston, TX 77002 (Tel.: 713-241-0981; FAX 713-241-6916; Telex: 762248)
For Int'l. Sales Represented by: Pecten Chemicals, Inc., One Shell Plaza, Houston, TX 77252-9932 (Tel.: 713-241-6161)
Shell Chemicals UK Ltd., 1 Northumberland Ave., London, WC2N 5LA UK (Tel.: 71-934-1234)
Shell Chimie France, 27 Rue de Berri, 7539 Paris Cedex 08, France
Shell Nederland Chemie B.V., PO Box 187, 2501 CD The Hague, The Netherlands

Sherex Chemical Co., Inc.
PO Box 646, 5777 Frantz Rd., Dublin, OH 43017 (Tel.: 614-764-6500; 800-848-7370; FAX 614-764-6544; Telex: 245356)

Sherex Polymers, Inc.
2525 S. Combee Rd., Lakeland, FL, 33801 (Tel.: 813-665-6226)

Sherwin-Williams Co./Chemical Coatings Div.
11541 S. Champlain St., Chicago, IL 60628 (Tel.: 312-821-3182)

Shieldalloy Metallurgical Corp.
West Blvd., Newfield, NJ 08344 (Tel.: 609-692-4200; 800-SMC-2020; FAX 609-692-4017; Telex: 510-687-8918)

Shin Etsu Silicones of America, Inc.
431 Amapola Ave., Torrance, CA 90501 (Tel.: 213-533-6961; FAX 213-533-8936)

Shrieve Chemical Products Co.
1717 Woodstead Ct., Suite 205, Woodlands, TX 77380 (Tel.: 713-367-4226; 800-367-4226; FAX 713-292-2014; Telex: 508818)

Shuman Plastics
35 Neoga St., Depew, NY 14043 (Tel.: 716-685-2121; FAX 716-685-3236; Telex: 91 9121)

Shyamac International Inc.
485 Pheasant Ridge Rd., Lake Zurich, IL 60047 (Tel.: 708-540-7241; FAX 708-540-7246)

Sigma Prodotti Chimici
PO Box 1, 7490 AA Delden, Holland (Tel.: 035-212274-233010; FAX 035-239569; Telex: 300108)

Silberline Ltd.
> Banbeath Rd., Leven, Fife, KY8 5HD, UK (Tel.: 0333 24734; FAX 0333 21369; Telex: 727373 SILBER G)

Silbrico Corp.
> 6300 River Rd., Hodgkins, IL 60525-4257 (Tel.: 708-354-3350; FAX 708-354-6698)

Sil-Med Corp.
> 700 Warner Blvd., Taunton, MA 02780 (Tel.: 508-823-7701; FAX 508-823-1438; Telex: 951962)

Siltech Inc.
> 4437 Park Dr., Suite E, Norcross, GA 30093 (Tel.: 404-279-8601; FAX 404-279-8535)

Simco (Nederland) BV
> Postubs 11, NL-7240 AA Lochem, Netherlands (Tel.: 05730 88333; FAX 05730 57319)

Simlak Ltd.
> Radnor Park Trading Estate, Back Lane, Congleton, Cheshire, CW12 4XJ, UK (Tel.: 0260 299164; FAX 0260 278263; Telex: 666207 SIMLAK G)

Sino-Japan Chemical Co.
> 3 fl. 237 Sec. 1, Chien Kuo South Rd., Taipei, Hsien, Taiwan (Tel.: 886-2-700-1422; FAX 886-2-707-3921)

Sintimid Hochleistungkunststoffe GmbH
> Postfach, A-4860 Lenzing (Oberösterr), Austria (Tel.: 07672 72511 3299; FAX 07672 74826; Telex: 026 606 lenfa a)

Siral-Kunststoffwerk Siebauer
> Niedermauker Str. 8-10, 8541 Roettenbach-Bayern, Germany (Tel.: 09172 454; Telex: 624756)

Sir Industriale-Gruppo Montedison
> Via Grazioli 33, 20161 Milano, Italy (Tel.: 02 63331; Telex: 310679)

Smith Chemical & Color Co., Inc.
> 104-20 Dunkirk St., Jamaica, NY 11412 (Tel.: 718-454-9400; FAX 718-454-7101)

Smith Lime Flour Co.
> 60-70 Central Ave., S. Kearny, NJ 07032 (Tel.: 201-344-1700; FAX 201-690-5936)

Werner G. Smith, Inc.
> 1730 Train Ave., Cleveland, OH 44113 (Tel.: 216-861-3676; 800-535-8343; FAX 216-861-3680)

Wilfrid Smith Ltd.
> Gemini House, High St., Edgware, UK (Tel.: 081 952 6655; FAX 081 952 6694; Telex: 261259)

Smits Neuchatel
> PO Box 30, 3430 AA Nieuwegein, Netherlands (Tel.: 3402 32004; FAX 3402 42854)

Smooth-On, Inc.
> 1000 Valley Rd., Gillette, NJ 07933 (Tel.: 201-647-5800; FAX 201-604-2224; Telex: 882833)

Sniatechnopolimeri
> Via Stabilimenti 11, 20020 Ceriano Laghetto, Milano, Italy (Tel.: 02 96161; FAX 02 9616547; Telex: 314594 Tecpol)

Soc Airtec Industrie
> 88 Rue Saint Leger, 78100 Saitn Germain en Laye, France (Tel.: 16 39 73 55 52; Telex: 69 9 98 9)

Societa Ital. Emulsionanti
Via R. Cozzi 34, 20125 Milano, Italy (Tel.: 39-2-642-4041; FAX 39-2-643-0820; Telex: 315006)

Softer SpA
Via Cardano 8, 47100 Forli, Italy (Tel.: 0543 725702; FAX 0543 722748; Telex: 551013)

Solem. See J.M. Huber

Solico-Southwest Vermiculite Co., Inc.
5119 Edith, NE Box 6287, Albuquerque, NM, 87197 (Tel.: 505-345-1633)

Solidur
Solidur Plastics Co., Solidur Bldg., PO Box 407, 200 Industrial Dr., Delmont, PA 15626 (Tel.: 412-468-6868; 800-322-8469; FAX 412-468-4044)
Solidur Canada, 281 Ambassador Dr., Mississauga, Ontario, L5T 2J3, Canada (Tel.: 416-564-6870; FAX 416-564-6901)
Solidur Kunststoffewerk Pennekamp + Huesker KG, Weberstrasse 2, Postfach 12 64, D-4426 Vreden, Germany (Tel.: 02564 4008; FAX 02564 4008; Telex: 89739 solid d)

Soluol Chemical Co.
Green Hill & Market Sts., Box 112, W. Warwick, RI 02893 (Tel.: 401-821-8100; FAX 401-823-6673)

Solvay
Solvay & Cie, Rue du Prince Albert 33, 1050 Brussels, Belgium (Tel.: 2/509-6111; FAX 2/509-6617; Telex: 21337)
Solvay Chemicals Ltd., Unit 1, Grovelands Business Centre, Boundary Way, Hemel Hempstead, Herts, HP2 7TE, UK (Tel.: 0442-236555)
Solvay Duphar BV, Postbus 900, 1380 DA Weesp, The Netherlands (Tel.: 31-2940-77711; FAX 31-2940-80253; Telex: 14232)
Solvay Polymers Inc., 3333 Richmond Ave., PO Box 27328, Houston, TX 77227-7328 (Tel.: 713-522-1781; 800-231-6313; FAX 713-522-2435; Telex: 023-166307)
Solvay Enzymes, Inc., PO Box 4859, Elkhart, IN 46514-0859 (Tel.: 219-523-3700; 800-342-2097)

Solvit Inc.
7001 Raywood Rd., Madison, WI 53713 (Tel.: 608-222-8624)

Sondex AB
Limhamnsvägen 108-110, S-2613 Malmö, Sweden (Tel.: 40155020; FAX 40163428; Telex: 33130)

Sorbothane, Inc.
2144 State Route 59, PO Box 178, Kent, OH 44240 (Tel.: 216-678-9444; FAX 216-678-1303)

Southeastern Adhesives Co.
815 D Virginia St., PO Box 2070, Lenoir, NC 28645 (Tel.: 704-754-3493; FAX 704-754-0052)

Southeastern Chemical Corp.
PO Box 110, Elon College, NC 27244 (Tel.: 919-584-8862; FAX 919-584-8872)

Southeastern Clay Co.
PO Box 1055, Aiken, SC 29802 (Tel.: 803-648-3248; FAX 803-649-5701)

Southeastern Minerals, Inc.
PO Box 1866, 1100 Dothan Rd., Bainbridge, GA 31717 (Tel.: 912-246-3396; FAX 212-246-7309; Telex: 804620 SEM BBRG (WU)

Southern Clay Prods. See Laporte

Southern Coatings, Inc./Subsid. of Pratt & Lambert Inc.
PO Box 160, Sumter, SC 29151 (Tel.: 803-775-6351; 800-845-0487)

Southern Dye & Chemical Co.
PO Box 26914, Greenville, SC 29616 (Tel.: 404-782-7233; FAX 404-782-7502)

Sovereign Chemicals Co.
2115 Lindbergh Ave., Cuyahoga Falls, OH 44223 (Tel.: 216-928-6642; FAX 216-928-0036)

Sovitec SA/Subsid. of Glaverbel
Zoning Industriel, 6220 Fleurus, Belgium (Tel.: 071 81 50 11; FAX 071 817673; Telex: 51447 Svitec B)

Spartan Flame Retardants, Inc.
345 E. Terra Cotta Ave., PO Box 395, Crystal Lake, IL 60014 (Tel.: 815-459-8500; 800-435-5700; FAX 815-459-8560; Telex: RCA 297135)

Spaulding Composites Co.
310 Wheeler St., Tonawanda, NY 14150 (Tel.: 716-692-2000; FAX 716-692-4410)

S P & C Ltd.
Colthrop Way, Thatcham, Newbury, Berkshire, RG13 4LW, UK (Tel.: 0635 70000; FAX 0635 61212; Telex: 847689 ATOKEM G)

Specialty Chemical Co.
PO Box 2606, Cleveland, TN 37311 (Tel.: 615-479-9664; FAX 615-472-6158)

Specialty Products Co.
75 Montgomery St., PO Box 306, Jersey City, NJ 07303-0306 (Tel.: 201-434-4700; 800-321-8506; FAX 201-434-6052)

Specmat Ltd.
Street Court, Kingsland, Leominster, HR6 9QA, UK (Tel.: 056881 744; FAX 056881 713; Telex: 837264)

Spectra Colors Corp./Bachmeier Colors Div.
25 Rizzolo Rd., Kearny, NJ 07032 (Tel.: 201-997-0606; FAX 201-997-0504)

Spectra Dyestuffs Inc.
9837 Belmont St., Bellflower, CA 90706 (Tel.: 310-804-7644; FAX 310-804-7645)

Spectra Polymer Inc.
Maple Ave., PO Box 308, Ashburnham, MA 01430 (Tel.: 508-827-6732)

Spectro Color & Chemical Inc.
PO Box 35468, Charlotte, NC 28235-5468 (Tel.: 704-335-1603; 800-622-9540; FAX 704-335-1607)

Spectrum Chemical Mfg. Corp.
14422 S. San Pedro St., Gardena, CA 90248 (Tel.: 213-516-8000; 800-772-8786; FAX 213-516-9843; Telex: 182395)

Spice King Corp.
6009 Washington Blvd., Culver City, CA 90232-7488 (Tel.: 213-836-7770; FAX 213-836-6454; Telex: 664350)

H.C. Spinks Clay Co.
PO Box 820, Paris, TN 38242 (Tel.: 901-642-5414; FAX 901-642-5493)

Spray Products Corp.
PO Box 737, Norristown, PA 19404 (Tel.: 215-277-1010)

Sidney Springer Co.
PO Box 21368, Los Angeles, CA 90021 (Tel.: 213-626-1233; FAX 213-626-6948)

SP Systems (Structural Polymer Systems Ltd.)
Love Lane, Cowes, Isle of Wight, PO31 7EU, UK (Tel.: 0983 298 451; FAX 0983 298 453)

Stag Polymers & Sealants Ltd.
Tavistock Road, West Drayton, Middlesex, UV7 7RA, UK (Tel.: 0895 445511; FAX 0895 449199; Telex: 28559)

A.E. Staley Manufacturing Co.
2200 E. Eldorado St., Decatur, IL 62525 (Tel.: 217-423-4411)

Standard Chemical Co.
Mill Lane, Cheadle, Cheshire, SK8 2NX, UK (Tel.: 44-61-428-5225; FAX 44-61-428-0890; Telex: 666413)

Standard Polymers Corp.
2233 Nesconset Hwy., Lake Grove, NY 11755 (Tel.: 516-467-5656)

Standard Tar Prod. Co. Inc.
2456 W. Cornell St., Milwaukee, WI 53209-6294 (Tel.: 414-873-7650; 800-825-7650; FAX 414-873-7737)

Stanley Plastics Ltd.
Holmbush Industrial Estate, Midhurst, Sussex, GU29 9HX, UK (Tel.: 0730 816221; FAX 0730 812877)

Statoil
Statoil Petrokemi AB/Plastics Div., S-44481 Stenungsund, Sweden (Tel.: 0303 87450; FAX 0303 87460; Telex: 27346 Staplas)

Statoil UK Ltd., Taplow House, Clivemont Rd., Maidenhead, Berks, SL6 7BU, UK (Tel.: 0628 37543; FAX 0628 73047; Telex: 849236)

Steelcote Mfg. Co.
One Steelcote Square, St. Louis, MO 63103 (Tel.: 314-771-8053; 800-444-0282; FAX 314-771-7581)

Steetley Magnesia Products Ltd.
PO Box 8, Hartlepool, Cleveland, TS24 0BY, UK (Tel.: 0429 267071; FAX 0429 266600; Telex: 58649 Pclase G)

Steetley Minerals Ltd.
PO Box 2, Retford Rd., Worksop, Notts, S81 8AF, UK (Tel.: 0909 475511; FAX 0909 486532; Telex: 547901)

Steetley Quarry Products US Inc./Ohio Lime Co.
PO Box 128, Woodville, OH 43469 (Tel.: 419-849-2321; 800-445-3930; FAX 419-849-3589; Telex: 241043)

Stepan
Stepan Co., 22 West Frontage Rd., Northfield, IL 60093 (Tel.: 708-446-7500; 800-228-8312; FAX 708-501-2443)

Stepan Co./PVO Dept., 100 West Hunter Ave., Maywood, NJ 07607 (Tel.: 201-845-3030; FAX 201-845-6754; Telex: 710-990-5170)

Stepan Canada, 90 Matheson Blvd. W., Suite 201, Mississauga, Ontario, L5R 3P3, Canada (Tel.: 416-507-1631; FAX 416-507-1633)

Stepan Europe, BP127, 38340 Voreppe, France (Tel.: 33-7650-8133; FAX 33-7656-7165; Telex: 320511 F)

Stevens Elastomerics
395 Pleasant St., Northampton, MA 01060 (Tel.: 413-586-8750; FAX 413-584-6348)

Stevenson Brothers & Co.
PO Box 38349, 1039 West Venango St., Philadelphia, PA 19140 (Tel.: 215-223-2600; FAX 215-223-3597)

Stewart Hall Chem. Corp.
222 Washington St., Mt. Vernon, NY 10553 (Tel.: 914-668-6300)

Stiles-Kem/Div. of Met Pro Corp.
3301 Sheridan Rd., Zion, IL 60099 (Tel.: 708-746-8334)

Stockhausen, Inc.
2408 Doyle St., Greensboro, NC 27406 (Tel.: 919-333-3500; FAX 919-333-3545; Telex: 574405)

Stokvis Plastics BV
Borchwerf 10, 4704 RG Roosendaal, PO Box 1575, 4700 BN Roosendaal, Netherlands (Tel.: 01650 62700; FAX 01650 62900)

Stoner Inc.
1070 Robert Fulton Hwy., PO Box 65, Quarryville, PA 17566 (Tel.: 717-786-7355; 800-227-5538; FAX 717-786-9088)

Strahl & Pitsch, Inc.
230 Great E. Neck Rd., W. Babylon, NY 11704 (Tel.: 516-587-9000; FAX 516-587-9120; Telex: 221636 STRALUR)

Streett Industries Inc.
PO Box 6509, St. Louis, MO 63125 (Tel.: 314-892-2958)

Stroblite Co., Inc.
430 W. 14 St., Room 507, New York, NY 10014 (Tel.: 212-929-3778)

Struktol
Struktol Co., 201 E. Steels Corner Rd., PO Box 1649, Stow, OH 44224-0649 (Tel.: 216-928-5188; 800-327-2709; FAX 216-928-8726)
Struktol Co., Ltd., 60 Venture Dr., Unit 23, Scarborough, Ontario, M1B 3S4, Canada (Tel.: 416-286-4040; FAX 416-286-4043)

SA Sturge
65 Rue Sully, 80000 Amiens, France (Tel.: 22 43 04 22; FAX 22 43 31 33; Telex: 140928)

Sullivan Chem Coatings
410 N. Hart St., Chicago, IL 60622 (Tel.: 312-666-8080)

Sunbelt Corp.
PO Box 2589 CRS, Rock Hill, SC 29732 (Tel.: 803-329-9787; FAX 803-329-3350)

Sun Chemical Corp.
411 Sun Ave., Cincinnati, OH 45232 (Tel.: 513-681-5950; FAX 513-641-3269; Telex: 214502)

Sun Chemical Enterprises
1 Rue Larmoyer, 1301 Bierges, Belgium (Tel.: 10 411303)

Superior Graphite Co.
120 S. Riverside Plaza, Chicago, IL 60606 (Tel.: 312-559-2999; FAX 312-559-9064; Telex: 201456)

Surco Products, Inc.
PO Box 777-Eighth & Pine Aves., Braddock, PA 15104 (Tel.: 412-351-7700; 800-556-0111; FAX 412-351-7701)

Surpass Chemical Co.
PO Box 4165, Albany, NY 12204 (Tel.: 518-434-8101; FAX 518-434-2798)

Sutton Laboratories, Inc./Member of the ISP Inc. Group
Sutton Laboratories, Inc., 116 Summit Ave., PO Box 837, Chatham, NJ 07928-0837 (Tel.: 201-635-1551; FAX 201-635-4964; Telex: 710-999-5607)
Represented by: Blagden Campbell Chemicals Ltd., A.M.P. House, Dengwall Rd., Croydon, Surrey, CR9 3QU, UK

SVO Enterprises
35585-B Curtis Blvd., Eastlake, OH 44095 (Tel.: 800-292-4786; FAX 216-942-1045; Telex: 4938879 AGCEAKE)

Swastik Household & Industrial Products Ltd.
Shahibag House, 13 Walchand Hirachand Marg, Ballard Estate, Bombay, 400 038, India)

Sybron
Sybron Chemicals Inc., PO Box 125, Wellford, SC 29385 (Tel.: 803-439-6333; 800-677-3500; FAX 803-439-1612)

Sybron Chemicals Canada Ltd., 120 Norfinch Dr., Unit 1, Downsview, Ontario, M3N 1X2, Canada (Tel.: 416-663-7166)

Sybron/Biochemical
Birmingham Rd., Birmingham, NJ 08011 (Tel.: 609-893-1100; 800-678-0020; FAX 609-894-8641)

Sydplast AB
Box 22035, 250 22 Helsingborg, Sweden (Tel.: 42 25 02 00; FAX 42 15 40 08; Telex: 72448)

Symalit AG/A Company of the Royal Dutch/Shell Group
CH-5600 Lenzburg, Switzerland (Tel.: (0)64 50 81 50; FAX (0)64 50 83 83; Telex: 981 352 syma ch)

Synair Corp.
2003 Amnicola Hwy., PO Box 5269, Chattanooga, TN 37406 (Tel.: 615-698-8801; 800-251-7642; FAX 615-624-0321)

Synergy Production Group
1315 Marsten Rd., Burlingame, CA 94010 (Tel.: 317-885-7671; 800-274-7711; FAX 301-876-4651)

Synfleur/Subsid. of Bell Flavors & Fragrances Inc.
33 Lakewood Ave., Monticello, NY 12701 (Tel.: 914-794-4444; 800-323-4387)

Synthetic Products Co.
1000 Wayside Rd., Cleveland, OH 44110 (Tel.: 216-531-6010; FAX 216-486-6638)

Synthetic Rubber Technologies
3898 Shawnee St. N.W., PO Box 639, Uniontown, OH 44685 (Tel.: 216-699-1256; FAX 216-699-1404)

Synthetic Surfaces Inc.
PO Box 241, Scotch Plains, NJ 07076 (Tel.: 908-233-6803; FAX 908-233-6844; Telex: 833231 Att: 663)

Synthron Inc.
PO Box 1111, Morganton, NC 28655 (Tel.: 704-437-8611; FAX 704-437-4126)

TA Biochemicals Inc.
PO Box 1146, Depew, NY 14043 (Tel.: 716-685-4390; FAX 716-683-0471)

TACC International Corp.
Air Station Industrial Park, PO Box 535, Rockland, MA 02370 (Tel.: 617-878-7015; FAX 617-871-6727)

Taiwan Surfactant Corp.
No. 106, 8-1 Floor, Sec. 2, Chung An E. Rd., Taipei, Taiwan, R.O.C. (Tel.: 886-2-507-9155; FAX 886-2-507-7011; Telex: 27568 surfact)

Taiyo Kagaku Co., Ltd.
9-5 Akahori-Shinmachi, Yokkaichi, Mie-Pref., Japan

Takeda USA, Inc.
8 Corporate Dr., Orangeburg, NY 10962-2614 (Tel.: 914-365-2080; 800-825-3328; FAX 914-365-2786; Telex: 421149)

Takemoto Oil & Fat Co., Ltd.
No. 5, Sec. 2, Minato-Machi, Gamagori, Aichi, 443, Japan (Tel.: 81-533-68-2117; FAX 81-533-67-3496; Telex: 4324604)

TAM Ceramics Inc.
4511 Hyde Park Blvd., Niagara Falls, NY 14305 (Tel.: 716-278-9400; FAX 716-285-3026; Telex: 710-524-1659)

H.B. Taylor Co.
4830 S. Christiana Ave., Chicago, IL 60632 (Tel.: 312-254-4805; FAX 312-254-4563)

TDF Tiofine BV
Patentlaan 5, 2288 EE Rijswijk, Netherlands (Tel.: 070 3956666; FAX 070 3956699; Telex: 33196)

Technic, Inc.
1 Spectacle St., Cranston, RI, 02910 (Tel.: 401-781-6100; FAX 401-781-2890; Telex: ESYLINK 927 514)

Technical Chemicals, Inc.
RR-4 Box 165A, Sheidy Rd., Bernville, PA 19506 (Tel.: 215-488-7525; FAX 215-488-6243; Telex: 493 1039)

Technopol GmbH
Kirchwerder Elbdeich 154, 2050 Hamburg 80, Germany (Tel.: 040 7239049; FAX 040 7238658)

Tecindes-Technicas Industriales Espanolas
Torrente De La Bomba 14, PO Box 156, E-08190 San Cugat del Vallés, Spain (Tel.: 93 6746048; FAX 93 674 60 48)

Tego. See Goldschmidt

Teknor
Teknor Apex Co., 505 Central Ave., Pawtucket, RI, 02861 (Tel.: 401-725-8000; 800-556-3864; FAX 401-725-8095; Telex: 927530)
Teknor Color Co., 505 Central Ave., Pawtucket, RI, 02861 (Tel.: 401-725-8000)

Telechemische Inc.
222 Dupont Ave., #14, Newburgh, NY 12550 (Tel.: 914-561-3237; FAX 914-561-3622; Telex: 62954502 WU)

Teledyne Wah Chang Albany
PO Box 460, 1600 NE Old Salem Rd., Albany, OR 97321 (Tel.: 503-926-4211; FAX 503-967-6994; Telex: 360 741)

Temfibre Inc.
C.P. 3000, Temiscaming, Quebec, J0Z 3R0, Canada (Tel.: 819-627-9505; FAX 819-627-3622; Telex: 067-76281)

Tenmat Ltd.
Bowdon House, Ashburton Rd. West, Trafford Park, Manchester, M17 1RU, UK (Tel.: 061 872 2181; FAX 061 872 7596; Telex: 667638 TACMANG)

Terbrack Kunststoff GmbH & Co. KG
Ölbachstr. 50, Postfach 13 53, D 4426 Vreden, Germany (Tel.: 0 25 64 393 30; FAX 0 25 64 393 60; Telex: 17 2 56 414)

Terra International Inc.
Terra Centre, 600 4, St. Sioux City, IA 51101 (Tel.: 712-277-1340; 800-831-1002; Telex: 9109681700)

Terry Laboratories
3270 Pineda Ave., Melbourne, FL, 32940 (Tel.: 407-259-1630; 800-367-2563; FAX 407-242-0625)

Tetra Chemicals
25231 Grogans Mill Rd., The Woodlands, TX 77380 (Tel.: 713-364-2233; 800-327-7817; FAX 713-367-6471; Telex: 214 874 TETRA UR)

Tetradur Kunststoff-Produktion GmbH
Brookdamm 3, Postfach 2242, 2105 Seevetal 2, Germany (Tel.: 040 7687005; FAX 040 7681840; Telex: 2166081)

Texaco
Texaco Chemical Co., PO Box 15730, Austin, TX 78761 (Tel.: 512-483-0053; 800-231-3107; FAX 512-483-0925; Telex: 776-408)
S.A. Texaco Belgium N.V., Int'l. Congress Center, Citadel Park, B-900 Ghent, Belgium (Tel.: 011-32-91-41-5920)
Texaco Chemical Deutschland GmbH, Baumwall 5, 2000 Hamburg 11, Germany (Tel.: 011-49-40-36-3737)
Texaco Ltd., 195 Knightsbridge, London, SW7 1RU, UK (Tel.: 011-411-584-5000)
Texaco France S.A., 5, rue Bellini, Tour Arago, F-92806 Puteaux Cedex, France (Tel.: 011-33-1-47-78-1655)
Sanseki-Texaco Chemicals Co., Ltd., PO Box 90, 2,5, 3-Chome Kasumigaseki, Chiyoda-Ku, Tokyo, 100, Japan (Tel.: 03-580-3611)

Texapol Corp.
177 Mikron Rd., Bethlehem, PA 18017 (Tel.: 215-759-8222; 800-523-9242; FAX 215-759-9433)

Texasgulf Inc.
3101 Glenwood Ave., Raleigh, NC 27622 (Tel.: 919-881-2700; Telex: 6844904)

Texo Corp.
2801 Highland Ave., Cincinnati, OH 45212 (Tel.: 513-731-3400)

Textile Rubber & Chem. Co.. Tiarco Chem. Div. See Tiarco

Thatcher Co.
1900 Fortune Rd., PO Box 27407, Salt Lake City, UT 84127 (Tel.: 801-972-4587)

Theodor Leonhard Wax Co., Inc.
136 Church St., Haledon, NJ 07508 (Tel.: 201-956-1444)

Thermal Ceramics
2102 Old Savannah Rd., Augusta, GA 30906 (Tel.: 404-796-4200; FAX 404-796-4398; Telex: 545423 THERMAL)

Thermofil, Inc.
PO Box 489, 6150 Whitmore Lake Rd., Brighton, MI 48116-0489 (Tel.: 800-444-4408; FAX 313-227-3824)

Thermoset Plastics Inc.
5101 East 65th St., PO Box 20902, Indianapolis, IN 46220 (Tel.: 317-259-4161; FAX 317-252-8402)

Thertec SA
Zl des Ebizoires, 4 rue des Frères Lumière, 78370 Plaisir, France (Tel.: 1 30 55 42 70; FAX 1 30 55 90 10; Telex: 696106)

Thibaut & Walker Co.
PO Box 296, 49 Rutehrford St., Newark, NJ 07101 (Tel.: 201-589-3331)

Thiele Kaolin Co.
Box 1056, Sandersville, GA 31082 (Tel.: 912-552-3951; FAX 912-552-4131)

Thor Chemicals, Inc.
Brook House, 37 North Ave., Norwalk, CT 06851 (Tel.: 203-846-8613; FAX 203-846-4810; Telex: 888630)

Thoro Products Co.
6611 W. 58 Pl., Arvada, CO 80002 (Tel.: 303-422-0335; FAX 303-420-0999)

Tiarco Chemical Div./Textile Rubber & Chemical Co.
1300 Tiarco Dr., Dalton, GA 30720 (Tel.: 404-277-1300; FAX 404-277-3738)

TIC Gums, Inc.
4609 Richlynn Dr., Belcamp, MD 21017 (Tel.: 301-273-7300; FAX 301-273-6469; Telex: 221049)

Tifa (CI) Ltd.
> 50 Division Ave., Millington, NJ 07946 (Tel.: 908-647-4570; FAX 908-647-2517; Telex: 178098)

Tiodize Co., Inc.
> 15701 Industry Lane, Huntington Beach, CA 92649 (Tel.: 714-898-4377; FAX 714-891-7467)

Tioga Coatings Corp.
> 208 Quaker Rd., Rockford, IL 61104-7088 (Tel.: 815-962-4200; FAX 815-962-1712)

Tioxide Group PLC
> Tioxide House, 137-143 Hammersmith Rd., London, W14 0QL, UK (Tel.: 071 602 7121; FAX 081 784 0019; Telex: 920900)

Tiszai Vegyi Kombinát
> PO Box 20, H-3581 Leninváros, Hungary (Tel.: 49 22 222; FAX 49 21 322; Telex: 22 5330)

Titan Chemical Products, Inc.
> PO Box 20, Short Hills, NJ 07078

Titanium Metals Corp.
> 1999 Broadway, Suite 4300, Denver, CO, 80202 (Tel.: 303-296-5600; FAX 614-537-5753; Telex: 825750)

A.M. Todd Co.
> 1717 Douglas Ave., Kalamazoo, MI 49007 (Tel.: 616-343-2603; FAX 616-343-4913; Telex: 224416 amtodd)

Toho Chemical Industry Co., Ltd.
> No. 2-5, 1-chome, Ningyo-cho, Nihonbashi, Chuo-ku, Tokyo, 103, Japan (Tel.: 81-3-3668-2271; FAX 81-3-3668-2278; Telex: 252-2332 TOHO K J)

Tokai Seiyu Ind. Co. Ltd.
> 67, 2-Chome, Yamadahigashimachi, Higashi-ku, Nagoya, Japan (Tel.: 81-52-721-2611; FAX 81-52-721-8775)

Tolson Holland BV
> Overgoo 1, PO Box 400, 2260 AK Leidschendam, Netherlands (Tel.: 70 3571671; FAX 70 3274915; Telex: 34050)

Tomah Products. See Exxon

Tosoh
> Tosoh Corp., 1-7-7 Akasaka, Minato-ku, Tokyo, 107, Japan (Tel.: 03-585-4289; FAX 03-582-7846; Telex: J24475tosoh)
>
> Tosoh USA Inc., 1700 Water Place, Suite 204, Atlanta, GA 30339 (Tel.: 404-956-1100; FAX 404-956-7368; Telex: 542272 tosoh atl)
>
> Tosoh Canada Ltd., 1200 Sheppard Ave. East, Suite 511, Willowdale, Ontario, M2K 2S5, Canada (Tel.: 416-756-2226; FAX 416-756-2750)
>
> Tosoh Europe BV, World Trade Centre Amsterdam, Tower C, Floor 13, Strawinskylaan 1351, 1077 XX Amsterdam, The Netherlands (Tel.: 020-644026, 020-623412; Telex: 18573tosoh nl)

Total Petroleum Inc.
> East Superior St., Alma, MI 48801 (Tel.: 517-463-1161; FAX 517-463-9623)

Tower Chemical Corp.
> 2703 Freemansburg Ave., PO Box 3070, Palmer, PA 18043 (Tel.: 215-253-6206; FAX 215-258-9695)

Toyomenka (America) Inc.
> Suite 1806, 11 Greenway Plaza, Houston, TX 77046 (Tel.: 713-626-3610; FAX 713-871-1129)

Traco Labs, Inc.
604 Country Fair Dr., Champaign, IL 61821 (Tel.: 217-359-3703; 800-79-TRACO; FAX 217-359-7113; Telex: 217-359-7113)

Tra-Con, Inc.
55 North St., Medford, MA 02155 (Tel.: 617-391-5550; 800-872-2661; FAX 617-391-7380)

Trade Search Associates
525 East 82nd St., New York, NY 10028 (Tel.: 212-744-9582; FAX 212-628-8244; Telex: 291470 ELMA URTSA)

Tradex Colori SAS
Via Goldoni 1, 20129 Milano, Italy (Tel.: 76002134; FAX 783144; Telex: 313123)

Tramico SA
52 Quai de Dion Bouton, 92800 Puteaux, France (Tel.: 1 47 76 43 14; FAX 1 47 78 46 21; Telex: 630905 F)

Transene Co. Inc.
Route 1, Rowley, MA 01969 (Tel.: 508-948-2501; FAX 508-948-2206)

Transformaciones Quimico-Industriales SA
Circunvalacion 27, 08210 Barbera del Valles, Spain (Tel.: 729 04 14; FAX 217 9241; Telex: 57411 Traqe)

Transmare BV
PO Box 21118, 3001 AC Rotterdam, Netherlands (Tel.: 10 4137037; FAX 10 4136267; Telex: 21479)

T&R Chemicals, Inc.
700 Celum Rd., Box 330, Clint, TX 79836 (Tel.: 915-851-2761)

Thomas Triantaphyllou SA
405 Tatoiou Ave., TK 136 71, Acharnes, Athens, Greece (Tel.: 30-1-807-6413; Telex: 216370)

Tribol
21031 Ventura Blvd., Woodland Hills, CA 91364-2297 (Tel.: 818-888-0808; Telex: 4720069)

Tricon Colors Inc.
16 Leliarts Ln., Elmwood Park, NJ 07407 (Tel.: 201-794-3800; FAX 201-797-4660; Telex: 4991537 TRICN)

Tri-K Industries, Inc.
466 Old Hook Rd., PO Box 312, Emerson, NJ 07630 (Tel.: 201-261-2800; 800-526-0372; FAX 201-261-1432)

Tri-Star Chem. Co.
PO Box 38627, Dallas, TX 75238 (Tel.: 214-341-0054)

Trivent Chemical Co., Inc.
45 Ridge Rd., PO Box 597, South River, NJ 08882 (Tel.: 201-251-1116; FAX 201-251-0967)

Tropag GmbH
Bundesstr. 4, 2000 Hamburg 13, Germany (Tel.: 040 414 013-0; FAX 040 414 013-20; Telex: 2 161 945 TOP D)

Troy Chemical Corp.
One Avenue L, Newark, NJ 07105 (Tel.: 201-589-2500)

Trugman Nash Inc.
90 West St., New York, NY 10006 (Tel.: 212-964-9350; FAX 212-791-1863; Telex: 232098-RCA)

TSE Industries, Inc./Millathane Div.
5260 113th Ave. North, PO Box 17225, Clearwater, FL, 33520-7225 (Tel.: 813-576-5643; 800-621-4141; FAX 813-572-0487)

TTC Mouldings BV
Post Box 268, 7570 AG Oldenzaal, Netherlands (Tel.: 5410 19025; FAX 5410 22425)

T&T Industries Corp.
6th Floor, No. 124 Naking East Rd., Sec. 2, Taipei, Taiwan, ROC (Tel.: 02-5064107; FAX 02-5060618; Telex: 23105 TAILONCO)

Tufnol Ltd.
PO Box 376, Perry Barr, Birmingham, West Midlands, B42 2TB, UK (Tel.: 021 356 9351; FAX 021 331 4235; Telex: 339730)

Tulco, Inc.
9 Bishop Rd., Ayer, MA 01432 (Tel.: 508-772-4412; FAX 508-772-1751)

Ubbink Nederland BV
Verheuellweg 9, Postbox 26, 6984 AA Doesburg, Netherlands (Tel.: 08334 71080; Telex: 35388)

UCAR Carbon Co., Inc.
39 Old Ridgebury Rd.-J4, Danbury, CT 06817 (Tel.: 203-794-3684; 800-342-3698)

UCB
UCB, Anderlechstraat 33, 1620 Drogenbos, Belgium (Tel.: 02 371 45 11; FAX 02 378 39 44; Telex: 22342 UCBOS B)

UCB NV Filmsektor, Ottergemsesteenweg 801, PO Box 369, 9000 Gent, Belgium (Tel.: 091 40 32 11; FAX 091 40 88 00; Telex: 11280 SIDAC B)

UCIB/Distrib. by SST Corp.
635 Brighton Rd., PO Box 1649, Clifton, NJ 07015 (Tel.: 201-473-4300; FAX 201-473-4326; Telex: RCA 219149)

Ultra Additives, Inc.
PO Box 98, Park Station, Paterson, NJ 07543 (Tel.: 201-279-1306; 800-524-0055; FAX 201-279-0602)

Uncle Sam Chem. Co. Inc.
573-577 W. 131 St., New York, NY 10027 (Tel.: 212-281-6100; FAX 212-368-7055)

Unger Fabrikker AS
PO Boks 254, N-1601 Fredrikstad, Norway (Tel.: 47-9-32-0020; FAX 47-9-32-3775; Telex: 76382 unger n)

Unicast Development Corp.
345 Tompkins Ave., PO Box 242, Pleasantville, NY 10570 (Tel.: 914-769-9100; FAX 914-769-9373; Telex: 13 7344)

Unichema
Unichema North America, 4650 S. Racine Ave., Chicago, IL 60609 (Tel.: 312-376-9000; 800-833-2864; FAX 312-376-0095)

Unichema Chemie GmbH, Postfach 1280, D-4240 Emmerich, Germany (Tel.: 49-0-2822-720; FAX 49-0-2822-72276; Telex: 8125113)

Unichema Chemicals Ltd., Begington, Wirral, Merseycide, L62 4UF, UK (Tel.: 44-0-51-645-2020; FAX 44-0-51-645-9197; Telex: 629408)

Unichema France SA, 148 Boulevard Haussemann, 75008 Paris, France (Tel.: 1 45630863; FAX 1 42563188; Telex: 643217)

Unichema Japan, Sankei Bldg. 7F 708, 4-9, Umeda 2-chome, Kita-ku, Osaka, 530, Japan (Tel.: 81-6341-7221; FAX 81-6341-7725)

Unimin Specialty Minerals
Unimin Specialty Minerals Inc., 258 Elm St., New Canaan, CT 06840 (Tel.: 203-966-8880; 800-243-9004)

Unimin Specialty Minerals Inc., PO Box 33, Rt. 127, Elco, IL 62929 (Tel.: 618-747-2311; FAX 618-747-9318)

Union Camp
Union Camp Corp., 1600 Valley Rd., Wayne, NJ 07470 (Tel.: 201-628-2680; 800-628-9220;
Telex: 130735)
Union Camp Corp./Chem. Prods. Div., PO Box 60369, Jacksonville, FL 32236

Union Carbide
Union Carbide Chem. & Plastics Co. Inc./Specialty Chem. Div., 39 Old Ridgebury Rd.,
Danbury, CT 06817-0001 (Tel.: 203-794-2000; 800-621-1972)
Union Carbide/Ucar Emulsion Systems, 410 Gregson Dr., Cary, NC 27511
Union Carbide Corp./Agricultural Products Co., Inc., PO Box 12014, Research Triangle Park,
NC 27709
Union Carbide Canada Ltd., 10455 Metropolitan E., Montreal East, Quebec, H1B 1A1,
Canada (Tel.: 514-493-2610)
Union Carbide UK Ltd., Rickmansworth, Herts, WD3 1RB, UK
Union Carbide Europe S.A., 15 Chemin Louis-Dunant, CH-1211 Geneve 20, Switzerland
(Tel.: 41-22-739-6111; FAX 41-22-739-6545; Telex: 419207)
Union Carbide Brazil, Rua Dr. Eduardo De Souza Aranha, 153, Sao Paulo, 04530, Brazil
Union Carbide Japan KK, Toranomon 45 Mori Bldg., 1-5 Toranomon, 5-Chome Minato-Ku,
Tokyo, 105, Japan (Tel.: 813431-7281)

Union Derivan SA
Av. Meridiana 133, Barcelona 08026, Spain (Tel.: 2322113; FAX 2323951; Telex: 98204)

Uniroyal
Uniroyal Chemical Co., Inc., World Headquarters, Middlebury, CT 06749 (Tel.: 203-573-
3880; 800-243-3024; FAX 203-573-3393; Telex: 6710383 uniroyal)
Uniroyal/Plastics Div. (Tel.: 219-255-2181)
Uniroyal Chemical Ltd., Kennet House, 4 Langley Quay, Waterside Drive, Slough, Berkshire,
SL3 6EH, UK (Tel.: 0753 580888; FAX 0753 591352; Telex: 84 9934 UCHEM G)
Uniroyal Chimica, SpA, Corso Vinzaglio 35, 10121 Torino, Italy
Uniroyal Chemical (Singapore), Cathay Bldg., 11, Dhoby Ghaut, Suite #14-05, Singapore
0922
Uniroyal Quimica S.A., Avenida Morumbi, 7.029 (CEP 05650) Caixa Postal 30.380, 0100 Sao
Paulo, SP, Brazil
Uniroyal Adhesives & Sealants Co. Inc., 312 N. Hill St., PO Box 2000, Mishawaka, IN 46544
(Tel.: 219-256-8598; 800-336-1973; Telex: 810-297-7127)

United Catalysts Inc.
PO Box 32370, Louisville, KY 40232 (Tel.: 502-634-7500; 800-468-7210; FAX 502-634-
7727; Telex: 204190)

United Coconut Chemicals, Inc./Cocochem
UCPB Bldg., 17th Fl., Makat Ave., Makati, Metro Manila, Philippines (Tel.: 818-8361; FAX
(00632) 817-2251; Telex: 66928 COCOCHEM PN)

United Composites, Inc.
Lyn Creek Facility, 2203 Webb Lynn Rd., Arlington, TX 76018 (Tel.: 817-468-2929; FAX
817-468-3122)

United Desiccants-Gates
1227 S. 12 St., Louisville, KY 40210 (Tel.: 502-634-6800; FAX 502-634-7727; Telex:
204190)
United States Aluminate Co., Inc., 701 Chesapeake Ave., Baltimore, MD 21225 (Tel.: 301-
354-3020; FAX 301-354-4784)

United States Aluminum Inc.
Route 202, PO Box 2190, Flemington, NJ 08822 (Tel.: 908-782-5454; FAX 908-782-3489;
Telex: 833488)

United States Biochemical Corp.
PO Box 22400, Cleveland, OH 44122 (Tel.: 216-765-5000; 800-321-9322; FAX 216-464-
5075; Telex: 980718)

United States Borax & Chemical Corp.
3075 Wilshire Blvd., Los Angeles, CA 90010 (Tel.: 213-251-5400; 800-USB-ORAX; FAX 213-251-5455)

United States Cosmetics
313 Lake Rd., PO Box 859, Dayville, CT 06241 (Tel.: 203-779-3990; 800-752-0490; FAX 203-779-3994)

United States Gypsum Co./DAP Inc., Subsid. of USG Corp.
101 S. Wacker Dr., Chicago, IL 60606-4385 (Tel.: 312-606-4000; FAX 312-606-4093)

United States Polymeric
700 E. Dyer Rd., Santa Ana, CA 92705 (Tel.: 714-549-1101)

United States Polymers Inc.
30 E. Primm St., St. Louis, MO 63111 (Tel.: 314-638-1632)

United States Silica Co.
PO Box 187, Berkeley Springs, WV 25411 (Tel.: 304-258-2500; 800-243-7500; FAX 304-258-3500; Telex: 4942414)

Unitex Chemical Corp.
PO Box 16344, 520 Broome Rd., Greensboro, NC 27406 (Tel.: 919-378-0965)

Universal Chemicals & Coatings Inc.
1975 Fox Lane, Elgin, IL 60123 (Tel.: 708-931-1700; FAX 708-931-1799)

Universal Laboratories Inc.
3 Terminal Rd., New Brunswick, NJ 08901 (Tel.: 201-545-3130; 800-872-0101; FAX 201-214-1210)

Universal Scientific Inc.
PO Box 80402, Atlanta, GA 30366-0402 (Tel.: 404-441-1134; FAX 404-441-3402)

Unocal
Unocal Chemicals/Chemicals Distribution Div., 1700 East Golf Rd., Schaumburg, IL 60173 (Tel.: 708-619-2539; 800-CHE-MS76; FAX 708-619-2515)
Unocal Chemicals/Unocal Polymers, 1700 East Golf Rd., Schaumburg, IL 60173-5862 (Tel.: 800-548-0162)
Unocal Chemicals/Nitrogen Group, 1201 W. 5th St., Los Angeles, CA 90017 (Tel.: 213-977-7600)

UOP, Inc./Universal Oil Prods.
25 East Algonquin Rd., Box 5017, Des Plaines, IL 60017-5017 (Tel.: 312-391-3300; 800-348-0832; FAX 312-391-2758)

UPI/Universal Preserv-A-Chem, Inc.
297 North 7th St., Brooklyn, NY 11211 (Tel.: 718-782-7429)

Upright, Inc.
2640 Creve Coeur Dr., St. Louis, MO 63144 (Tel.: 314-961-3711; 800-248-7007; FAX 314-961-3264)

Urban Chem. Co.
151 S. Pfingsten Rd., Deerfield, IL 60015 (Tel.: 708-498-3080; FAX 708-498-0491)

UTI Chemicals, Inc.
#7 Whatney, Irvine, CA 92718 (Tel.: 714-837-0800; FAX 714-837-0871)

UVP, Inc.
5100 Walnut Grove Ave., PO Box 1501, San Gabriel, CA 91778 (Tel.: 818-285-3123; 800-452-6788; Telex: 371 6381)

3-V
1500 Harbor Blvd., Weehawken, NJ 07087 (Tel.: 201-865-3600; FAX 201-865-1892)

Valite Div./Valentine Sugars Inc.
Rt. 2 Box 625, Lockport, LA 70374 (Tel.: 504-532-2541; 800-678-8662; FAX 504-532-6806)

VAMP Srl
Viale Teodorico 19/2, 20149 Milano, Italy (Tel.: 2 3493231; FAX 2 3492281; Telex: 322546 vamp i)

Van Den Bergh Foods Co.
2200 Cabot Dr., Lisle, IL 60532 (Tel.: 708-955-5276; FAX 708-955-5497)

R.T. Vanderbilt Co., Inc.
30 Winfield St., PO Box 5150, Norwalk, CT 06855 (Tel.: 203-853-1400; FAX 203-853-1452; Telex: 6813581 RTVAN)

Van Dyk & Co., Inc./An ISP Co.
11 William St., Belleville, NJ 07109 (Tel.: 201-450-3264; FAX 201-759-5279; Telex: 710-995-4928)

Van Leer Flexibles
9505 Bamboo Rd., Houston, TX 77041 (Tel.: 713-462-6111; 800-925-3766; FAX 713-690-2746)

Van Raaijen Kunststoffen
Heinsiuslaan 21, 3818 JE Amersfoort, Netherlands (Tel.: 33 61 83 27; FAX 33 65 90 45)

Van Schuppen Chemie
Nieuweweg, Veenendaal, The Netherlands

Vanson Chemical Co.
8840 152 Ave NE Redmond, WA 98052 (Tel.: 206-881-6464; 800-876-POLY; FAX 206-882-2476; Telex: 329 473 Burgess Sea)

Van Waters & Rogers Inc.
1600 Norton Bldg., 801 Second Ave., Seattle, WA 98104 (Tel.: 206-447-5911)

Vapor Blast Mfg. Co.
3019 W. Atkinson Ave., Milwaukee, WI 53209 (Tel.: 414-871-6500 x 20; FAX 414-871-7683; Telex: 26-639)

Varobel BVBA
Villalaan 31, 1630 Linkebeek, Belgium (Tel.: 02 380 88 82; FAX 02 380 9230; Telex: 64920)

Vecoplas
21 Rue St. Jean Cublize, 69550 Amplepuis, France (Tel.: 74 89 50 02; FAX 74 89 55 81; Telex: 900535)

Vega Biotechnologies Inc.
1250 E. Aero Park Blvd., Tucson, AZ 85706 (Tel.: 602-746-1401; 800-528-4882; FAX 602-889-4139)

Velsicol Chemical Corp.
5600 N. River Rd., Rosemont, IL 60018-5119 (Tel.: 800-843-7759; FAX 708-698-9714)

Vevy Europe SpA
Via P. Semeria 18, 16131 Genova, Italy (Tel.: 010-314193-4; FAX 010-316343; Telex: 281257 Vevy-1)

Vigar SA
Poligono Industrial La Bastida, Carretera de Molins a Caldas KM 13.500, Apartado 123, Rubi 08191 Barcelona, Spain (Tel.: 93 699 8611; FAX 92 699 9251; Telex: 94584 vigar e)

Vikon Chemical Co., Inc.
PO Box 1520, Burlington, NC 27215 (Tel.: 919-226-6331; FAX 919-222-9568)

Vinamul Ltd.
Mill Lane, Carshalton, Surrey, SM5 2JU, UK (Tel.: 081 669 4422; FAX 081 669 4422; Telex: 266264)

Vineland Chemical Co., Inc.
1611 West Wheat Rd., Vineland, NJ 08360 (Tel.: 609-691-3535; FAX 609-691-3615)

Vinings Industries
3950 Cumberland Pkwy., Atlanta, GA 30339 (Tel.: 404-436-1542; 800-347-1542; FAX 404-436-3432; Telex: 54-2481)

Viobin Corp.
226 West Livingston St., Monticello, IL 61856 (Tel.: 217-762-2561; FAX 217-762-2489; Telex: 26 5479)

Virginia Dare Extract Co., Inc.
882 Third Ave., Brooklyn, NY 11232 (Tel.: 718-788-1776; 800-847-4500 (ex.NY); FAX 718-768-3978; Telex: 425707 DARE U1)

Viscosity Oil/A Tenneco Co.
3200 S. Western Ave., Chicago, IL 60608 (Tel.: 312-847-0224)

Vista
Vista Chemical Co., PO Box 19029, 900 Threadneedle, Houston, TX 77224 (Tel.: 713-588-3000; 800-231-8216; FAX 713-588-3236; Telex: 794557)
Vista Chemical Europe, Hilton Tower, Blvd. de Waterloo #39, 81000 Brussels, Belgium (Tel.: 32-2-513-7490)
Vista Chemical Far East Inc., Kasumigaseki Bldg., 25th Floor, PO Box 110, Tokyo, 100, Japan

Vita Cortex
Vita Cortex (NI) Ltd., Dunmurry Industrial Estate, Belfast, BT17 9HU, UK (Tel.: 0232 618625; FAX 0232 619479)
Vita Cortex Ltd., Kinsale Road, Cork, Rep. of Ireland (Tel.: 021 964377; FAX 021 313943)

Vitamins, Inc.
200 E. Randolph Dr., Chicago, IL 60601 (Tel.: 312-861-0700; FAX 312-861-0708; Telex: 25 4717)

Vosschemie GmbH
Esinger Steinweg 50, Postfach 1355, D-2082 Uetersen, Germany (Tel.: 04122 7170; FAX 04122 717158; Telex: 218526)

VT Plastics Ltd.
Snaithing Grange, Snaithing Lane, Sheffield, S10 3LF, UK (Tel.: 0742 306500; FAX 0742 630570)

Vulcaflex SpA
Direziona Generale Milano, Via Boncompagni 3, Stabilimenti (Cotignola), Ravenna, Italy (Tel.: 025 34721; Telex: 313516)

Vulcan Materials Co./Chemicals Div.
PO Box 530390, Birmingham, AL, 35253-0390 (Tel.: 205-877-3000; FAX 205-877-3448)

Vygen Corp.
Ashtabula, OH

Vyse Gelatin Co.
5010 N. Rose St., Schiller Park, IL 60176 (Tel.: 708-678-4780)

Wacker
Wacker Chemie GmbH, Div. L, Prinzregentenstrasse 22, D-8000, Munchen 22, Germany (Tel.: 089-2109-0; FAX 089-2109-1772; Telex: 529-121-56)
Wacker Chemicals Ltd., The Clock Tower, Mount Felix, Bridge St., Walton-on-Thames, Surrey, KT12 1AS, UK (Tel.: 932-246111; FAX 932-240141)
Wacker Quimica Ibérica, S.A., Córcega, 303-2 3a, E-08008 Barcelona, Spain

Wacker Chemie Danmark A/S, Park Alle 380 A, Postboks 170, DK-2625 Vallensbaek, Denmark)

Wacker Chemicals (USA) Inc., 50 Locust Ave., New Canaan, CT 06840 (Tel.: 203-966-9999; Telex: 643 444)

Wacker Silicones Corp.

3301 Sutton Rd., Adrian, MI 49221-9397 (Tel.: 517-264-8500; 800-248-0063; FAX 517-264-8246; Telex: 510-450-2700)

Waco America Inc.

PO Box 1001, Chattanooga, TN 37401 (Tel.: 615-622-5855; FAX 615-622-5856)

Wako

Wako Chemicals USA, Inc., 1600 Bellwood Rd., Richmond, VA 23237 (Tel.: 804-271-7677; FAX 804-271-7791; Telex: 293208 wako ur(rca)

Wako Chemicals GmbH, Nissanstr. 2, 4040 Neuss 1, Germany (Tel.: (02101) 35011; FAX (02101) 39879; Telex: 8517001 wako d)

Wako Pure Chemical Industries, Ltd., 1,2-Doshomachi 3-Chome, Chuo-ku, Osaka, 541, Japan (Tel.: (06)203-3741; FAX (06)222-1203; Telex: 65188 wakoos j)

Warner-Graham Ltd. Partnership, The Warner-Graham Co.

160 Curch Lane, PO Box 249, Cockeysville, MD 21030 (Tel.: 301-667-6200; 800-872-2300; FAX 301-628-0617)

Warner-Jenkenson Co.

2526 Baldwin St., St. Louis, MO 63106 (Tel.: 314-658-7469; 800-325-8110; FAX 314-658-7431; Telex: 44 7184)

Warren Laboratories, Inc.

12603 Executive Dr., Suite 806, Stafford, TX 77477 (Tel.: 713-240-2563)

Washington Penn Plastic Co., Inc.

2080 North Main St., PO Box 236, Washington, PA 15301 (Tel.: 412-228-1260; 800-245-1520)

Watcon, Inc.

2215 S. Main St., Box 2829, South Bend, IN 46613 (Tel.: 219-287-3397; FAX 219-287-2427)

Wavin Industrial Prods. Ltd.

Meadowfield Industrial Estate, Meadowfield, Durham, DH7 8RJ, UK (Tel.: 091 378 0841; FAX 091 378 0841; Telex: 537117)

WCC Industries Inc.

439 S. Bolmar St., PO Box 39, W. Chester, PA 19381 (Tel.: 215-696-9220; FAX 215-344-7519)

J & W Wegman BV

Computerweg 5, PO Box 1648, 3600 BP Maarssen, Netherlands (Tel.: 03465 72864; FAX 03465 72842; Telex: 40346)

Wellman Inc./Plastics Div.

Hwy. 41, Johnsonville, SC 29555 (Tel.: 803-386-2011; 800-845-6709)

Wesco Technologies Ltd.

PO Box 3880, San Clemente, CA 92674-3880 (Tel.: 714-661-1142; 800-223-3878 (CA); FAX 714-492-6025; Telex: GRT 3718658)

West Agro, Inc.

11100 N. Congress, Kansas City, MO 64153 (Tel.: 312-298-5505; 800-421-1905; Telex: 437012)

Westbrook Lanolin Co.

Argonaut Works, Laisterdyke, Bradford, BD4 8AU, UK (Tel.: 44-274-663331; FAX 44-274-667665; Telex: 51502)

West Chemical Products, Inc.
1000-T Herrontown Rd., Princeton, NJ 08540 (Tel.: 609-921-0501; FAX 201-569-5340; Telex: 843-462)

Western Nutrients Corp.
245 Industrial St., Bakersfield, CA 93307 (Tel.: 805-327-9604; 800-542-6664 (CA), 805-327-1740; Telex: 689 411 IPM 02CA)

Western Polymer Corp.
PO Drawer N, Exit 184, I-90 Raugust Rd., Moses Lake, WA 98837 (Tel.: 509-765-1803)

Western Water Management, Inc.
1345 Taney, N. Kansas City, MO 64116 (Tel.: 816-842-0560; 800-821-7784; FAX 816-842-6388)

Westlake Plastics Co.
PO Box 127, W. Lenni Rd., Lenni, PA 19052 (Tel.: 215-459-1000; FAX 215-459-1084; Telex: 83-5406)

Westvaco Chemical Div.
PO Box 70848, Charleston Hts., SC 29415-0848 (Tel.: 803-740-2300; 800-336-2211; FAX 803-747-2270)

Westwood Chemical Corp.
46 Tower Dr., Middletown, NY 10940 (Tel.: 914-692-6721)

Wetronics, Inc.
708 Cadmus Rd., Pottstown, PA 19464 (Tel.: 215-469-0715)

Whitford Corp.
PO Box 507, West Chester, PA 19381 (Tel.: 215-296-3200)

J.C. Whitlam Mfg. Co.
Box 71, Wadsworth, OH 44281 (Tel.: 216-334-2524; 800-321-8358; FAX 216-334-3005)

Whitney & Oettler
PO Box 8024, Savannah, GA 31402 (Tel.: 912-232-7166; Telex: 546 443)

Whittaker, Clark & Daniels
1000 Coolidge St., South Plainfield, NJ 07080 (Tel.: 201-561-6100; 800-732-0562; FAX 800-833-8139)

Chemische Fabrik Wibarco GmbH
Postfach 1662, D-4530 Ibbenburen 1, Germany (Tel.: 0-54-59-590; FAX 0-54-59-59104; Telex: 94505)

Wilbur-Ellis Co.
PO Box 1286, Fresno, CA 93715 (Tel.: 209-442-1220; FAX 209-442-4089)

Wilford Plastics Ltd.
Cosgrove Way, Luton, Bedfordshire, LU1 1XL, UK (Tel.: 0582 36961; FAX 0582 451488)

Wills Engineered Polymers Ltd.
Dunball Park, Dunball, Bridgwater, Somerset, TA6 4TP, UK (Tel.: 0278 684888; FAX 0278 685051; Telex: 46207)

Witco
Witco Corp., 520 Madison Ave., New York, NY 10022 (Tel.: 212-605-3680; FAX 212-486-4198)

Witco Corp./Argus Chem. Div., Bussey Rd., PO Box 1439, Marshall, TX 75671-1439 (Tel.: 903-938-5141; 800-431-1413; FAX 903-938-2647)

Witco Corp./Argus Div./Int'l. Sales, 520 Madison Ave., New York, NY 10022-4236 (Tel.: 212-605-3999; FAX 212-759-5739; Telex: 422186)

Witco Corp./Concarb, 10500 Richmond, PO Box 42817, Houston, TX 77042 (Tel.: 713-978-5700; FAX 713-266-9984)

Witco Corp./Golden Bear Div., 10100 Santa Monica Blvd., Los Angeles, CA 90067-4183 (Tel.: 213-277-4511; FAX 213-201-0383)

Witco Corp./Humko Chem. Div., PO Box 125, Memphis, TN 38101-0125 (Tel.: 901-684-7000; FAX 901-682-6531; Telex: 53-928)

Witco Corp./Inorganic Specialties Div., 520 Madison Ave., New York, NY 10022 (Tel.: 212-605-3645; 800-634-4010)

Witco Corp./Organics Div., 1000 Convery Blvd., Perth Amboy, NJ 08862-1932 (Tel.: 201-826-7777; 800-231-1542)

Witco Corp./Sonneborn Div., 520 Madison Ave., New York, NY 10022-4236 (Tel.: 212-605-3981; FAX 212-754-5676; Telex: 62470)

Witco Canada Ltd., 2 Lansing Sq., Suite 1200, Willowdale, Ontario, M2J 4Z4, Canada (Tel.: 416-497-9991)

Witco B.V., PO Box 5, Koogaan de Zaan, The Netherlands

Witco Chemical Ltd. UK, Union Lane, Droitwich, Worcester, WR9 9BB, UK

Witco SA, 10 Rue Cambaceres, 75008 Paris, France (Tel.: 42-65-99-03; FAX 42-65-67-61; Telex: 290233)

Witco Ltd., PO Box 10245, 26112 Haifa Bay, Israel (Tel.: 972-4-469-111; FAX 972-4-469-137; Telex: 45198)

WJP Engineering Plastics Ltd.

Albert Works, Albert Ave., Bobbers Mill, Nottingham, Notts, NG8 5BE, UK (Tel.: 0602 299 555; FAX 0602 290 422; Telex: 377 494 WJP)

Wyo-Ben, Inc.

PO Box 1979, Billings, MT, 59103 (Tel.: 406-652-6351; 800-548-7055; Telex: WYOBEN)

Xantar Polycarbonates VoF

501 Crescent Ave., PO Box 15051, Reading, PA 15051 (Tel.: 215-320-6918; 800-366-6923; FAX 215-320-6930)

Xenox, Inc.

10810 Katy Rd., PO Box 79773, Houston, TX 77279 (Tel.: 713-467-9723)

XymaX Inc.

PO Box 825, Boca Raton, FL, 33429 (Tel.: 407-395-4405; 800-332-3136; FAX 407-395-5262; Telex: 522133)

Yorkshire Nachem, Inc.

1099 Hingham St., Rockland, MA 02370 (Tel.: 617-871-4440; 800-622-4361; FAX 617-871-4777; Telex: 951346)

Yorkshire Pat-Chem Inc.

11 Worley Rd., PO Box 1926, Greenville, SC 29602 (Tel.: 803-233-3941; 800-443-9358; FAX 803-232-3542)

Yoshimura Oil Chemical Co., Ltd.

Minami 5-Chome-1-1, Honan-cho, Toyonaki-shi, Osaki, 561, Japan (Tel.: 81-6-334-3331-7; FAX 81-6-331-4078)

Zeelan Industries Inc.

141 East Fourth St., #220, St. Paul, MN 55101-1620 (Tel.: 612-292-9271; FAX 612-297-6138)

Zeeland Chemicals, Inc./A Cambrex Company

215 N. Centennial St., Zeeland, MI 49464 (Tel.: 616-772-2193; 800-223-0453; FAX 616-772-7344; Telex: 226375)

Zeon Chemicals, Inc.

Three Continental Towers, Suite 1211, 1701 Golf Rd., Rolling Meadows, IL 60008 (Tel.: 312-437-9770; 800-735-3388; FAX 312-437-9773)

Zeus Industrial Products Inc.

PO Box 2167, Orangeburg, SC 29116 (Tel.: 803-531-2174; 800-526-3842; FAX 803-533-5694)

Zimpro Passavant Environmental Systems, Inc.
301 W. Military Rd., Rothchild, WI 54474 (Tel.: 715-359-7211; 800-826-1476; FAX 715-355-3219; Telex: 29 0495)

Zinc Corp. of America
Zinc Corp. of America, Fourth & Delaware, Palmerton, PA 18071 (Tel.: 412-773-2295; 800-962-7500; FAX 412-773-2217)
Zinc Corp. of America, 300 Frankfort Rd., Monaca, PA 15061 (Tel.: 412-773-2295; FAX 412-773-2248)

Zinchem, Inc.
39 Belmont Dr., Somerset, NJ 08875 (Tel.: 908-469-8100; FAX 908-469-4539)

Zinkan Enterprises, Inc.
10574 Ravenna Rd., Twinsbury, OH 44087 (Tel.: 216-487-1500; FAX 216-425-8202)

Zipperling Kessler & Co.
Postfach 1464, D-2070 Ahrensburg, Germany (Tel.: 04102-5151-0; FAX 04102-5151-69; Telex: 21 89 842)

Zipp Industries Inc.
PO Box 7248, Amarillo, TX 79114-7248 (Tel.: 806-353-5581; FAX 806-358-0353)

Zircar Products Inc.
110 N. Main St., Florida, NY 10921 (Tel.: 914-651-4481; FAX 914-651-0441; Telex: 996608)

Zohar Detergent Factory
PO Box 11 300, Tel-Aviv, 61 112, Israel (Tel.: 03-528-7236; FAX 03-5287239; Telex: 33557 zohar il)

M.T. Zouros & Co.
Maleme 15, 15237 Filothei, Greece (Tel.: 6817192/9323828/30; FAX 1 9353718; Telex: 21-9363 MT2 GR)

Zophar Mills, Inc.
112 26 St., Brooklyn, NY 11232 (Tel.: 718-768-0907; FAX 718-768-0910; Telex: PHARMILLS, NY)

Zschimmer & Schwarz
Zschimmer & Schwarz GmbH & Co., 4-5 Max-Schwarz-Str., Postfach 2179, D-5420 Lahnstein, Germany (Tel.: 49-02621 12 0; Telex: 86 9816 750)
Zschimmer & Schwarz France S.a.r.l., 10 rue Saint-Marc, F-75002 Paris, France
Zschimmer & Schwarz Italiana SpA, Casella Postale N.1, I-13038 Tricerro (Vc), Italy
Zschimmer & Schwarz Argentina S.A., Bdo. de Irigoyen 556-5 B, Buenos Aires, Argentina

Zymet Inc.
7 Great Meadow Lane, East Hanover, NJ 07936 (Tel.: 201-428-5245)